Time Series Analysis

WILEY SERIES IN PROBABILITY AND STATISTICS

Established by WALTER A. SHEWHART and SAMUEL S. WILKS

Editors: *Vic Barnett, Ralph A. Bradley, Nicholas I. Fisher, J. Stuart Hunter, J. B. Kadane, David G. Kendall, David W. Scott, Adrian F. M. Smith, Jozef L. Teugels, Geoffrey S. Watson*

A complete list of the titles in this series appears at the end of this volume

Time Series Analysis

Nonstationary and Noninvertible Distribution Theory

KATSUTO TANAKA
Hitotsubashi University

A Wiley-Interscience Publication
JOHN WILEY & SONS, INC.
New York • Chichester • Brisbane • Toronto • Singapore

This text is printed on acid-free paper.

Copyright © 1996 by John Wiley & Sons, Inc.

All rights reserved. Published simultaneously in Canada.

Reproduction or translation of any part of this work beyond that permitted by Section 107 or 108 of the 1976 United States Copyright Act without the permission of the copyright owner is unlawful. Requests for permission or further information should be addressed to the Permissions Department, John Wiley & Sons, Inc., 605 Third Avenue, New York, NY 10158-0012.

Library of Congress Cataloging-in-Publication Data:
Tanaka, Katsuto, 1950-
    Time series analysis : nonstationary and noninvertible distribution theory / Katsuto Tanaka.
      p.   cm. — (Wiley series in probability and statistics)
    "A Wiley-Interscience publication."
    Includes bibliographical references and indexes.
    ISBN 0-471-14191-7 (cloth : alk. paper)
    1. Time-series analysis. I. Title. II. Series.
QA280.T35   1996
519.5'5—dc20   95-46270
                                                                 CIP

Printed in the United States of America

10 9 8 7 6 5 4 3 2

Contents

Preface   ix

1 Motivating Examples   1

    1.1 The Test Statistic for the Parameter Constancy, 1
    1.2 The Test Statistic for a Moving Average Unit Root, 5
    1.3 Statistics from the One-Dimensional Random Walk, 10
    1.4 Statistics from the Two-Dimensional Random Walk, 19
    1.5 Statistics from the Cointegrated Process, 28

2 Stochastic Calculus in Mean Square   35

    2.1 The Space L_2 of Random Variables, 35
    2.2 The Standard Brownian Motion and the Brownian Bridge, 39
    2.3 Mean Square Integration, 41
    2.4 The Integrated Brownian Motion, 48
    2.5 The Mean Square Ito Integral: The Scalar Case, 53
    2.6 The Mean Square Ito Integral: The Vector Case, 55
    2.7 The Ito Calculus, 57

3 Functional Central Limit Theorems   65

    3.1 Function Space C, 65
    3.2 Weak Convergence of Stochastic Processes in C, 66
    3.3 The Functional Central Limit Theorem, 68
    3.4 Continuous Mappings and Related Theorems, 70
    3.5 FCLT for Linear Processes: Case 1, 77
    3.6 FCLT for Martingale Differences, 79
    3.7 FCLT for Linear Processes: Case 2, 83
    3.8 Weak Convergence to the Integrated Brownian Motion, 86
    3.9 Weak Convergence to the Ornstein–Uhlenbeck Process, 90
    3.10 Weak Convergence of Vector-Valued Stochastic Processes, 95
    3.11 Weak Convergence to the Ito Integral, 102

4 The Stochastic Process Approach — 109

- 4.1 Girsanov's Theorem: Case 1, 109
- 4.2 Girsanov's Theorem: Case 2, 117
- 4.3 Girsanov's Theorem: Case 3, 121
- 4.4 The Cameron–Martin Formula, 124
- 4.5 Advantages and Disadvantages of the Present Approach, 125

5 The Fredholm Approach — 129

- 5.1 Motivating Examples, 129
- 5.2 The Fredholm Theory: The Homogeneous Case, 132
- 5.3 The c.f. of the Quadratic Brownian Functional, 136
- 5.4 Various Fredholm Determinants, 144
- 5.5 The Fredholm Theory: The Nonhomogeneous Case, 157
- 5.6 Weak Convergence of Quadratic Forms, 170

6 Numerical Integration — 181

- 6.1 Introduction, 181
- 6.2 Numerical Integration: The Nonnegative Case, 182
- 6.3 Numerical Integration: The Oscillating Case, 186
- 6.4 Numerical Integration: The General Case, 196
- 6.5 Computation of Percent Points, 203
- 6.6 The Saddlepoint Approximation, 207

7 Estimation Problems in Nonstationary Autoregressive Models — 213

- 7.1 Nonstationary Autoregressive Models, 213
- 7.2 Convergence in Distribution of LSEs, 218
- 7.3 The Negative Unit Root Case, 230
- 7.4 The c.f.s for the Limiting Distributions of LSEs, 232
- 7.5 Tables and Figures of Limiting Distributions, 239
- 7.6 Approximations to the Distributions of the LSEs, 249
- 7.7 Nearly Nonstationary Seasonal AR Models, 253
- 7.8 Complex Roots on the Unit Circle, 264
- 7.9 Autoregressive Models with Multiple Unit Roots, 267

8 Estimation Problems in Noninvertible Moving Average Models — 279

- 8.1 Noninvertible Moving Average Models, 279
- 8.2 The Local MLE in the Stationary Case, 282
- 8.3 The Local MLE in the Conditional Case, 294
- 8.4 Noninvertible Seasonal Models, 300
- 8.5 The Pseudolocal MLE, 307
- 8.6 Probability of the Local MLE at Unity, 311
- 8.7 The Relationship with the State Space Model, 314

9 Unit Root Tests in Autoregressive Models 321

9.1 Introduction, 321
9.2 Optimal Tests, 323
9.3 Equivalence of the LM Test with the LBI or LBIU Test, 328
9.4 Various Unit Root Tests, 333
9.5 Integral Expressions for the Limiting Powers, 335
9.6 Limiting Power Envelopes and Point Optimal Tests, 342
9.7 Computation of the Limiting Powers, 346
9.8 Seasonal Unit Root Tests, 355
9.9 Unit Root Tests in the Dependent Case, 362
9.10 The Unit Root Testing Problem Revisited, 367

10 Unit Root Tests in Moving Average Models 373

10.1 Introduction, 373
10.2 The LBI and LBIU Tests, 374
10.3 The Relationship with the Test Statistics in Differenced Form, 383
10.4 Performance of the LBI and LBIU Tests, 385
10.5 Seasonal Unit Root Tests, 392
10.6 Unit Root Tests in the Dependent Case, 402
10.7 The Relationship with Testing in the State Space Model, 405

11 Statistical Analysis of Cointegration 417

11.1 Introduction, 417
11.2 Case of No Cointegration, 419
11.3 Cointegration Distributions: The Independent Case, 424
11.4 Cointegration Distributions: The Dependent Case, 433
11.5 The Sampling Behavior of Cointegration Distributions, 438
11.6 Testing for Cointegration, 445
11.7 Determination of the Cointegration Rank, 453
11.8 Higher Order Cointegration, 458

12 Solutions to Problems 469

References 609

Author Index 617

Subject Index 619

List of Series Titles 625

Preface

This book attempts to describe nonstandard theory for linear time series models that are nonstationary and/or noninvertible. Nonstandard aspects of the departure from stationarity or invertibility have attracted much attention in the field of time series econometrics during the last ten years. Since there seem few books concerned with the theory for such nonstandard aspects, I have been at liberty to choose my way. Throughout this book, attention is oriented toward the most interesting theoretical issue; that is, the asymptotic distributional aspect of nonstandard statistics. The subtitle of the book reflects this.

Chapter 1 is a prelude to the main theme. By using simple examples, various asymptotic distributions of nonstandard statistics are derived by a classical approach, which I call the *eigenvalue approach*. It turns out that, if more complicated problems are to be dealt with, the eigenvalue approach breaks down, and the introduction of notions such as the Brownian motion, the Ito integral, the functional central limit theorem, and so on is inevitable. These notions are now developed very deeply in probability theory. In this book, however, a knowledge of such probability notions is required only at a moderate level, which I explain in Chapters 2 and 3 in an easily accessible way.

Probability theory, in particular the functional central limit theorem, enables us to establish weak convergence of nonstandard statistics and to realize that limiting forms can be expressed by functionals of the Brownian motion. However, more important from a statistical point of view is how to compute limiting distributions of those statistics. For this purpose I do not simply resort to simulations, but employ numerical integration. To make the computation possible, we first need to derive limiting characteristic functions of nonstandard statistics. To this end, two approaches are presented. Chapter 4 discusses one approach, which I call the *stochastic process approach*, while Chapter 5 discusses the other, which I call the *Fredholm approach*. The two approaches originate from quite different mathematical theories, which I explain fully, indicating the advantage and disadvantage of each approach for judicious use.

Chapter 6 discusses and illustrates numerical integration for computing distribution functions via inversion of characteristic functions. This chapter is necessary because a direct application of any computer package for integration cannot do a proper job. We overcome the difficulty by employing Simpson's rule, which can be executed on a desktop computer. The necessity for accurate computation based on numerical integration is recognized, for instance, when close comparison has to be made between limiting local powers of competing nonstandard tests.

Chapters 7 through 11 deal with statistical and econometric problems to which the nonstandard theory discussed in previous chapters applies. Chapter 7 considers the estimation problems associated with nonstationary autoregressive models, while Chapter 8 considers those with noninvertible moving average models. The corresponding testing problems, called the *unit root tests*, are discussed in Chapters 9 and 10, respectively. Chapter 11 is concerned with cointegration, which is a stochastic collinearity relationship among multiple nonstationary time series. The problems discussed in these chapters originate in time series econometrics. I describe in detail how to derive and compute limiting nonstandard distributions of various estimators and test statistics.

Chapter 12, the last chapter, gives a complete set of solutions to problems posed at the end of most sections of each chapter. Most of the problems are concerned with corroborating the results described in the text, so that one can gain a better understanding of details of the discussions.

There are about 90 figures and 50 tables. Most of these are of limiting distributions of nonstandard statistics. They are all produced by the methods described in this book and include many distributions that have never appeared in the literature. Among these are limiting powers and power envelopes of various nonstandard tests under a sequence of local alternatives.

This book may be used as a textbook for graduate students majoring in econometrics or time series analysis. A general knowledge of mathematical statistics, including the theory of stationary processes, is presupposed, although the necessary material is offered in the text and problems of this book. Some knowledge of a programming language like FORTRAN and computerized algebra like REDUCE is also useful.

The late Professor E. J. Hannan gave me valuable comments on the early version of my manuscript. I would like to thank him for his kindness and for pleasant memories extending over the years since my student days. This book grew out of joint work with Professor S. Nabeya, another respected teacher of mine. He read substantial parts of the manuscript and corrected a number of errors in its preliminary stages, for which I am most grateful. I am also grateful to Professors C. W. Helstrom, S. Kusuoka, and P. Saikkonen for helpful discussions and to Professor G. S. Watson for help of various kinds. Most of the manuscript was keyboarded, many times over, by Ms. M. Yuasa, and some parts were done by Ms. Y. Fukushima, to both of whom I am greatly indebted. Finally, I thank my wife, Yoshiko, who has always been a source of encouragement.

Tokyo, Japan   KATSUTO TANAKA

CHAPTER 1

Motivating Examples

We deal with linear time series models on which stationarity or invertibility is not imposed. Using simple examples arising from estimation and testing problems, we indicate nonstandard aspects of the departure from stationarity or invertibility. In particular, asymptotic distributions of various statistics are derived by the eigenvalue approach under the normality assumption on the underlying processes. As a prelude to discussions in later chapters, we also present equivalent expressions based on other approaches for limiting random variables.

1.1 THE TEST STATISTIC FOR THE PARAMETER CONSTANCY

Let us consider the following model:

(1.1)
$$y_t = \beta_t + \varepsilon_t,$$
$$\beta_t = \beta_{t-1} + u_t, \quad \beta_0 = 0, \quad (t = 1, \ldots, T),$$

where

i) $\{y_t\}$ is an observable sequence, whereas $\{\beta_t\}$ is an unobservable sequence starting from $\beta_0 = 0$;
ii) $\{\varepsilon_t\}$ and $\{u_t\}$ are error sequences assumed to be independent of each other;
iii) $\{\varepsilon_t\}$ is normally independently distributed (NID) with common mean 0 and variance $\sigma_\varepsilon^2(>0)$, which will be abbreviated as $\{\varepsilon_t\} \sim \text{NID}(0, \sigma_\varepsilon^2)$; it is also assumed that $\{u_t\} \sim \text{NID}(0, \sigma_u^2)(\sigma_u^2 \geq 0)$.

The model (1.1) is the so-called *state space model* or the random walk plus noise model, and our concern is to test if β_t is constant, that is, $\beta_t = 0$ for all t. This is equivalent to testing

$$H_0 : \rho = \frac{\sigma_u^2}{\sigma_\varepsilon^2} = 0 \quad \text{versus} \quad H_1 : \rho > 0.$$

Since $y_t = u_1 + \cdots + u_t + \varepsilon_t$, the observation vector $y = (y_1, \ldots, y_T)'$ has the distribution

(1.2)
$$y = Cu + \varepsilon \sim \text{N}(0, \sigma_\varepsilon^2(I_T + \rho CC')),$$

1

where $u = (u_1, \ldots, u_T)'$, $\varepsilon = (\varepsilon_1, \ldots, \varepsilon_T)'$, and I_T is the $T \times T$ identity matrix, while

$$(1.3) \qquad C = \begin{pmatrix} 1 & & & 0 \\ \cdot & \cdot & & \\ \cdot & & \cdot & \\ \cdot & & & \cdot \\ 1 & \cdot & \cdot & 1 \end{pmatrix}, \quad C^{-1} = \begin{pmatrix} 1 & & & 0 \\ -1 & \cdot & & \\ & \cdot & \cdot & \\ & & \cdot & \cdot \\ 0 & & -1 & 1 \end{pmatrix}.$$

The matrix C necessarily appears from the random walk process $\beta_t = \beta_{t-1} + u_t$ and may be called the *random walk generating matrix*. Note that the (s,t)th element of CC' is $\min(s,t)$ and the tth largest eigenvalue λ_t of CC' is given by Rutherford (1946) (see also Problem 1.1 in this chapter) as

$$(1.4) \qquad \lambda_t = \frac{1}{4}\left(\sin\frac{t - \frac{1}{2}}{2T + 1}\pi\right)^{-2}.$$

For the present problem, we consider the *Lagrange multiplier* (LM) or *score* test based on the derivative of the log-likelihood evaluated under H_0. The optimality of the LM test will be discussed in Chapter 9. Suppose, for simplicity, that σ_ε^2 is known and is assumed to be unity. Then the log-likelihood $L(\rho)$ is given by

$$(1.5) \qquad L(\rho) = -\frac{T}{2}\log 2\pi - \frac{1}{2}\log |I_T + \rho CC'| - \frac{1}{2}y'(I_T + \rho CC')^{-1}y$$

so that

$$(1.6) \qquad \left.\frac{dL(\rho)}{d\rho}\right|_{\rho=0} = -\frac{1}{2}\mathrm{tr}(CC') + \frac{1}{2}y'CC'y$$

$$= \frac{1}{2}\left(\varepsilon'CC'\varepsilon - \frac{T(T+1)}{2}\right).$$

The resulting statistic is a quadratic form in NID(0, 1) random variables plus a constant. Thus its exact distribution can be computed by Imhof's (1961) formula.

An asymptotic expansion for this distribution can also be obtained. In fact we have (Problem 1.2) that the characteristic function (c.f.) $\phi_T(\theta)$ of $\varepsilon'CC'\varepsilon/(T + \frac{1}{2})^2$ can be expanded, up to $O(T^{-2})$, as

$$(1.7) \qquad \phi_T(\theta) = \prod_{t=1}^{T}(1 - 2i\theta\lambda_t)^{-1/2}$$

$$\sim (\cos\sqrt{2i\theta})^{-1/2}\left[1 - \frac{i\theta}{8T^2}\left(1 - \frac{\sqrt{2i\theta}}{3}\tan\sqrt{2i\theta}\right)\right].$$

Note that the term of the order T^{-1} vanishes, while it can be verified that the c.f. of $\varepsilon'CC'\varepsilon/T^2$ contains the term of the order T^{-1}.

As for the limiting distribution, it is known that, if regularity conditions hold, the first derivative of the log-likelihood divided by $\sqrt{T}$ tends to normality, but this is not the case with the present situation. One might argue that this is because the parameter ρ to be tested is on the boundary of the parameter space $\rho \geq 0$, so one of the regularity conditions does not hold. The LM statistic, however, tends to normality without this condition, in general, unlike the likelihood ratio and Wald statistics. In fact, if the testing problem is such that $H_0 : \rho = \rho_0 > 0$ versus $H_1 : \rho > \rho_0$ for which the parameter space is $\rho_0 \leq \rho < \infty$, then $dL(\rho)/d\rho|_{\rho=\rho_0}$ tends to normality [Tanaka (1983a, b) and Problem 1.3].

Here we can make use of the knowledge of eigenvalues given in (1.4) to derive the asymptotic distribution of (1.6), whose approach may be referred to as the *eigenvalue approach*. Two other general approaches applicable to cases where eigenvalues are unknown and $\{\varepsilon_t\}$ is not necessarily normal and is dependent will be presented in later chapters. From (1.6) we have, by diagonalization,

$$\frac{2}{T^2} \left. \frac{dL(\rho)}{d\rho} \right|_{\rho=0} = \frac{1}{T^2} \sum_{t=1}^{T} \lambda_t \xi_t^2 - \frac{T+1}{2T},$$

where $\{\xi_t\} \sim \text{NID}(0, 1)$. Denoting as V_T the first term on the right side, it can be shown (Problem 1.4) that

(1.8) $$\text{plim}_{T \to \infty} \left(V_T - \sum_{t=1}^{T} \frac{\xi_t^2}{\left(t - \frac{1}{2}\right)^2 \pi^2} \right) = 0.$$

Thus it holds that, as $T \to \infty$,

(1.9) $$\mathcal{L}(V_T) \to \mathcal{L}(V) = \mathcal{L} \left(\sum_{n=1}^{\infty} \frac{\xi_n^2}{\left(n - \frac{1}{2}\right)^2 \pi^2} \right),$$

where $\mathcal{L}(X)$ denotes the probability law of X so that $\mathcal{L}(2dL(\rho)/(T^2 d\rho)|_{\rho=0}) \to \mathcal{L}\left(V - \frac{1}{2}\right)$ as $T \to \infty$. Note that V is an infinite, weighted sum of independent $\chi^2(1)$ random variables with each weight being nonnegligible to the sum of weights $\sum_{n=1}^{\infty} (1/(n - \frac{1}{2})^2 \pi^2) = 0.5$, although the first weight $1/(\pi/2)^2 = 0.4053$ is dominant.

The limiting distribution of $\varepsilon' CC'\varepsilon/T^2$ or $\varepsilon' C'C\varepsilon/T^2 = \sum_{t=1}^{T} S_t^2/T^2$ with $S_t = \sum_{j=1}^{t} \varepsilon_j$ was first dealt with by Erdös and Kac (1946) without assuming normality on $\{\varepsilon_t\}$. Their assumption is that $\{\varepsilon_t\}$ is independent and identically distributed (i.i.d.) with common mean 0 and variance 1, which will be abbreviated as $\{\varepsilon_t\} \sim$ i.i.d. (0,1). Their work opened the way to the so-called *functional central limit theorem* or *invariance principle* to be discussed in Chapter 3.

The present testing problem was first discussed in Nyblom and Mäkeläinen (1983) and Tanaka (1983b), which was generalized by Nabeya and Tanaka (1988) and Nabeya (1989) to cases where eigenvalues cannot be obtained explicitly. A different testing problem leading to the same asymptotic result as in (1.9) was earlier discussed in Sen and Srivastava (1973) [see also Gardner (1969) and MacNeill (1974)]. We shall return to this problem later, where general cases are treated together with the limiting distributions under local alternatives.

In passing we note that

$$\phi(\theta) = E(e^{i\theta V}) = \prod_{n=1}^{\infty} \left(1 - \frac{2i\theta}{\left(n - \frac{1}{2}\right)^2 \pi^2}\right)^{-1/2} = (\cos\sqrt{2i\theta})^{-1/2}$$

and it can be shown [Sen and Srivastava (1973) and Problem 1.5] that

$$(1.10) \qquad P(V \leq x) = 2\sqrt{2} \sum_{n=0}^{\infty} \binom{-\frac{1}{2}}{n} \Phi\left(-\frac{2n + \frac{1}{2}}{\sqrt{x}}\right),$$

where Φ is the distribution function of $N(0, 1)$.

As a prelude to subsequent discussions, we give four equivalent expressions for V in the sense of distribution, that is,

$$(1.11) \qquad \mathcal{L}(V) = \mathcal{L}\left(\sum_{n=1}^{\infty} \frac{\xi_n^2}{\left(n - \frac{1}{2}\right)^2 \pi^2}\right)$$

$$= \mathcal{L}\left(\int_0^1 w^2(t)\, dt\right)$$

$$= \mathcal{L}\left(\int_0^1 \int_0^1 (1 - \max(s, t))\, dw(s)\, dw(t)\right)$$

$$= \mathcal{L}\left(\int_0^1 \int_0^1 \min(s, t)\, dw(s)\, dw(t)\right),$$

where $\{w(t)\}$ is the *standard Brownian motion* defined on $[0,1]$, the definition and properties of which will be described in Chapter 2 together with the integrals involved in these expressions. The first of these expressions refers to the eigenvalue approach we have taken above and will take in this chapter, the second to the *stochastic process approach* to be discussed in Chapter 4, and the third and fourth to the *Fredholm approach* to be discussed in Chapter 5. The proof for the equivalence of the four expressions in (1.11) will be deferred until Chapter 5.

We have assumed the initial value β_0 to be zero. If β_0 is an unknown constant, the LM test statistic becomes different; so is the limiting distribution. In any case we have realized that the present model does produce nonstandard results. An intuitive reasoning may be that $\{y_t\}$ becomes an i.i.d. sequence under H_0, while it is nonstationary with an autoregressive (AR) unit root under H_1. Weakening the i.i.d. assumption on $\{\varepsilon_t\}$ to the extent of stationarity, the present test may be interpreted as testing the null hypothesis of stationarity against nonstationarity. Moreover, by incorporating a deterministic trend into the model, the null hypothesis of stationarity around the trend can be tested against the AR unit root hypothesis. This idea is explored in Kwiatkowski, Phillips, Schmidt, and Shin (1992). More details will be discussed in Chapters 9 and 10.

PROBLEMS

1.1 Show that the eigenvalues of $(CC')^{-1}$ are $4\sin^2((t-\frac{1}{2})\pi/(2T+1))$ so that (1.4) results.

1.2 Derive the expansion in (1.7).

1.3 For the model (1.1) with $\sigma_\varepsilon^2 = 1$, derive the LM test for testing $H_0 : \rho = \rho_0$ (> 0) versus $H_1 : \rho > \rho_0$ and show that the statistic tends to normality under H_0. More specifically, show that

$$\frac{1}{\sqrt{T}} \frac{dL(\rho)}{d\rho}\bigg|_{\rho=\rho_0} \to N(0, \sigma^2),$$

where $\sigma^2 = (\rho_0 + 2)/(4(\rho_0(\rho_0 + 4))^{3/2})$.

1.4 Prove that

$$\text{plim}_{T\to\infty} \left(\frac{4}{(2T+1)^2} \sum_{t=1}^T \lambda_t \xi_t^2 - \sum_{t=1}^T \frac{\xi_t^2}{\left(t-\frac{1}{2}\right)^2 \pi^2} \right) = 0$$

so that (1.8) holds.

1.5 Establish the formula (1.10), using the fact that the inverse Laplace transform of $e^{-c\sqrt{\theta}}$ is $c \exp(-c^2/(4x))/(2\sqrt{\pi x^3})$ for $c > 0$.

1.2 THE TEST STATISTIC FOR A MOVING AVERAGE UNIT ROOT

Let us next consider the first-order moving average [MA(1)] model

(1.12) $$y_t = \varepsilon_t - \alpha \varepsilon_{t-1} \quad (t = 1, \ldots, T),$$

where $\varepsilon_0, \varepsilon_1, \ldots,$ are NID$(0, \sigma^2)$ random variables. The parameter α is restricted to be $|\alpha| \leq 1$ because of the identifiability condition. The MA(1) model (1.12) is said to be *noninvertible* when $|\alpha| = 1$. The nonstandard nature of the noninvertible MA(1) model was first recognized by Kang (1975), which was followed by the theoretical work of Cryer and Ledolter (1981), Sargan and Bhargava (1983), Anderson and Takemura (1986), Tanaka and Satchell (1989), and Davis and Dunsmuir (1996).

Our purpose here is to test if the MA(1) model is noninvertible, that is, to test

$$H_0 : \alpha = 1 \quad \text{versus} \quad H_1 : \alpha < 1 .$$

Following Tanaka (1990b), we consider an LM-type test, the optimality of which will be demonstrated in Chapter 10. The log-likelihood $L(\alpha, \sigma^2)$ for $y = (y_1, \ldots, y_T)'$ is given by

$$L(\alpha, \sigma^2) = -\frac{T}{2} \log(2\pi\sigma^2) - \frac{1}{2} \log |\Omega(\alpha)| - \frac{1}{2\sigma^2} y' \Omega^{-1}(\alpha) y ,$$

where

$$\Omega(\alpha) = \begin{pmatrix} 1+\alpha^2 & -\alpha & & & 0 \\ -\alpha & 1+\alpha^2 & \cdot & & \\ & \cdot & \cdot & \cdot & \\ & & \cdot & \cdot & -\alpha \\ 0 & & & -\alpha & 1+\alpha^2 \end{pmatrix}.$$

The maximum likelihood estimators (MLEs) of α and σ^2 under H_0 are $\hat{\alpha} = 1$ and $\hat{\sigma}^2 = y'\Omega^{-1}y/T$ with $\Omega^{-1} = \Omega^{-1}(1)$. It can be checked easily that

$$\left.\frac{d\Omega(\alpha)}{d\alpha}\right|_{\alpha=1} = \Omega, \quad \left.\frac{d^2\Omega(\alpha)}{d\alpha^2}\right|_{\alpha=1} = 2I_T,$$

$$\frac{\partial L(\alpha, \sigma^2)}{\partial \alpha} = -\frac{1}{2}\text{tr}\left(\Omega^{-1}(\alpha)\frac{d\Omega(\alpha)}{d\alpha}\right) - \frac{1}{2\sigma^2} y' \frac{d\Omega^{-1}(\alpha)}{d\alpha} y,$$

$$\frac{\partial^2 L(\alpha, \sigma^2)}{\partial \alpha^2} = -\frac{1}{2}\text{tr}\left(\frac{d\Omega^{-1}(\alpha)}{d\alpha}\frac{d\Omega(\alpha)}{d\alpha} + \Omega^{-1}(\alpha)\frac{d^2\Omega(\alpha)}{d\alpha^2}\right) - \frac{1}{2\sigma^2} y' \frac{d^2\Omega^{-1}(\alpha)}{d\alpha^2} y.$$

These yield $\partial L(\alpha, \sigma^2)/\partial \alpha |_{\alpha=1, \sigma^2=\hat{\sigma}^2} = 0$ and

$$\left.\frac{\partial^2 L(\alpha, \sigma^2)}{\partial \alpha^2}\right|_{\alpha=1, \sigma^2=\hat{\sigma}^2} = -\frac{1}{2}\text{tr}(-I_T + 2\Omega^{-1}) - T\frac{y'(\Omega^{-1} - \Omega^{-2})y}{y'\Omega^{-1}y}$$

$$= -\frac{T(T+5)}{6} + T\frac{y'\Omega^{-2}y}{y'\Omega^{-1}y} ,$$

where we have used the fact that

$$\Omega^{-1} = [(CC')^{-1} + e_T e_T']^{-1} = CC' - \frac{1}{T+1} Cee'C'$$

$$= [(C'C)^{-1} + e_1 e_1']^{-1} = C'C - \frac{1}{T+1} C'ee'C$$

with $e = (1, \ldots, 1)'$, $e_1 = (1, 0, \ldots, 0)' : T \times 1$, and $e_T = (0, \ldots, 0, 1)' : T \times 1$.
The LM test considered here rejects H_0 if

(1.13) $$S_T = \frac{1}{T^2} \left(\frac{\partial^2 L(\alpha, \sigma^2)}{\partial \alpha^2} \bigg|_{\alpha=1, \sigma^2=\hat{\sigma}^2} + \frac{T(T+5)}{6} \right)$$

$$= \frac{1}{T} \frac{y' \Omega^{-2} y}{y' \Omega^{-1} y}$$

takes large values. The limiting distribution under H_0 can be derived by the eigenvalue approach as follows. Put $\xi = \Omega^{-1/2} y / \sigma$ so that $\xi \sim N(0, I_T)$ and

$$\frac{1}{T\sigma^2} y' \Omega^{-1} y = \frac{1}{T} \xi' \xi \to 1 \quad \text{in probability,}$$

$$\frac{1}{T^2 \sigma^2} y' \Omega^{-2} y = \frac{1}{T^2} \xi' \Omega^{-1} \xi = \frac{1}{T^2} \sum_{t=1}^{T} \delta_t \xi_t^2,$$

where δ_t is the tth largest eigenvalue of Ω^{-1} given by Anderson (1971) (see also Problem 2.1) as

(1.14) $$\delta_t = \frac{1}{4} \left(\sin \frac{t\pi}{2(T+1)} \right)^{-2}.$$

Here the c.f. $\phi_T(\theta)$ of $\xi' \Omega^{-1} \xi / (T+1)^2$ can be expanded (Problem 2.2), up to $O(T^{-2})$, as

(1.15) $$\phi_T(\theta) = \prod_{t=1}^{T} (1 - 2i\theta \delta_t)^{-1/2}$$

$$\sim \left(\frac{\sin \sqrt{2i\theta}}{\sqrt{2i\theta}} \right)^{-1/2} \left[1 - \frac{i\theta}{8T^2} \left(1 + \frac{\sqrt{2i\theta}}{3} \cot \sqrt{2i\theta} \right) \right].$$

Note that, as in (1.7), the term of the order T^{-1} vanishes, while it can be verified that the c.f. of $\xi' \Omega^{-1} \xi / T^2$ contains the term of the order T^{-1}.

For the limiting distribution we first have [Nyblom and Mäkeläinen (1983) and Problem 2.3]

$$(1.16) \quad \text{plim}_{T \to \infty} \left(\frac{1}{T^2} \sum_{t=1}^{T} \delta_t \xi_t^2 - \sum_{t=1}^{T} \frac{\xi_t^2}{t^2 \pi^2} \right) = 0$$

so that, as $T \to \infty$ under H_0,

$$(1.17) \quad \mathcal{L}(S_T) \to \mathcal{L}(W) = \mathcal{L} \left(\sum_{n=1}^{\infty} \frac{\xi_n^2}{n^2 \pi^2} \right).$$

Thus we have

$$\phi(\theta) = E(e^{i\theta W}) = \prod_{n=1}^{\infty} \left(1 - \frac{2i\theta}{n^2 \pi^2} \right)^{-1/2} = \left(\frac{\sin \sqrt{2i\theta}}{\sqrt{2i\theta}} \right)^{-1/2}.$$

The limiting distribution in (1.17) was first dealt with by Anderson and Darling (1952) in connection with goodness-of-fit tests. They showed that

$$P(W \leq x) = \frac{1}{\pi \sqrt{x}} \sum_{n=0}^{\infty} (-1)^n \binom{-\frac{1}{2}}{n} \sqrt{4n+1} \, e^{-b_n/x} K_{1/4}\left(\frac{b_n}{x} \right),$$

where $b_n = (4n+1)^2/16$ and $K_\nu(z)$ is the modified Bessel function defined by

$$K_\nu(z) = \frac{\sqrt{\pi}(z/2)^\nu}{\Gamma\left(\nu + \frac{1}{2}\right)} \int_0^\infty e^{-z \cosh x} \sinh^{2\nu} x \, dx,$$

and tabulated percent points. It will be recognized in Chapter 5 that

$$(1.18) \quad \mathcal{L}(W) = \mathcal{L} \left(\sum_{n=1}^{\infty} \frac{\xi_n^2}{n^2 \pi^2} \right)$$

$$= \mathcal{L} \left(\int_0^1 \left(w(t) - tw(1) \right)^2 dt \right)$$

$$= \mathcal{L} \left(\int_0^1 \int_0^1 \left(\min(s,t) - st \right) dw(s) \, dw(t) \right)$$

$$= \mathcal{L} \left(\int_0^1 \int_0^1 \left(\frac{1}{3} - \max(s,t) + \frac{s^2 + t^2}{2} \right) dw(s) \, dw(t) \right),$$

where $\{w(t) - tw(1)\}$ is the *Brownian bridge process* to be introduced in Chapter 2. The four expressions in (1.18) are comparable with those in (1.11).

It is important to note that the assumption on the initial value ε_0 is very crucial. If we assume $\varepsilon_0 = 0$, which may be referred to as the conditional case, so that $\{y_t\}$ is not stationary, the LM test becomes different; so is the limiting distribution of the LM statistic (Problem 2.5).

An MA unit root is often caused by overdifferencing of the original time series. From this point of view, Saikkonen and Luukkonen (1993a) suggested the following model:

$$
\begin{aligned}
y_1 &= \mu + \varepsilon_1, \\
\Delta y_t &= y_t - y_{t-1} = \varepsilon_t - \alpha \varepsilon_{t-1}, \quad (t = 2, \ldots, T),
\end{aligned}
\tag{1.19}
$$

where μ is a constant and $\varepsilon_1, \ldots, \varepsilon_T \sim \text{NID}(0, \sigma^2)$. Then the null hypothesis $H_0 : \alpha = 1$ implies overdifferencing. Note that, if μ is known and is assumed to be zero, $(y_1, y_2 - y_1, \ldots, y_T - y_{T-1})'$ follows the conditional MA(1) model. The LM test in the present case rejects H_0 when $\left(\sum_{t=1}^T y_t\right)^2 / \sum_{t=1}^T y_t^2$ takes large values (Problem 2.6).

Suppose that the constant μ in (1.19) is unknown. Then the LM test rejects H_0 for large values of

$$
\text{SL}_T = \frac{1}{T-1} \frac{y'MCC'My}{y'My},
\tag{1.20}
$$

where $M = I_T - ee'/T$ with $e = (1, \ldots, 1)'$ and C is defined in (1.3) [Saikkonen and Luukkonen (1993a) and Problem 2.7]. It can be shown (Problem 2.8) that SL_T in (1.20) is rewritten as

$$
\text{SL}_T = \frac{1}{T-1} \frac{(\Delta y)' \Omega_*^{-2} (\Delta y)}{(\Delta y)' \Omega_*^{-1} (\Delta y)},
\tag{1.21}
$$

where $\Delta y = (y_2 - y_1, \ldots, y_T - y_{T-1})' : (T-1) \times 1$ and Ω_* is the first $(T-1) \times (T-1)$ submatrix of Ω. Comparing SL_T with S_T in (1.13) we can conclude that the LM statistic for the model (1.19) is derived completely in the same way as in (1.13) just by disregarding the first equation in (1.19) and replacing y_t by Δy_t ($t = 2, \ldots, T$). Nonetheless, the formulation (1.19) is meaningful in connection with the determination of the order of integration of $\{y_t\}$, that is, the order of the AR unit root. If $\{\Delta^{d+1} y_t\}$ is found to have an MA unit root, while $\{\Delta^d y_t\}$ does not, then the order of integration of $\{y_t\}$ is supposed to be d. The MA unit root test may be useful for that purpose.

PROBLEMS

2.1 Derive (1.14) by computing first the eigenvalues of Ω.

2.2 Derive the expansion in (1.15).

2.3 Establish (1.16).

2.4 Obtain the limiting c.f. of $\varepsilon'\Omega^{-2}\varepsilon/T^4$ for $\varepsilon \sim N(0, I_T)$.

2.5 Derive the LM test for testing $H_0 : \alpha = 1$ in the model (1.12) with $\varepsilon_0 = 0$. Obtain also the asymptotic distribution of the LM statistic under H_0.

2.6 Show that the LM test for testing $H_0 : \alpha = 1$ in the model (1.19) with $\mu = 0$ rejects H_0 for large values of $\left(\sum_{t=1}^{T} y_t\right)^2 / \sum_{t=1}^{T} y_t^2$.

2.7 Derive the LM statistic in (1.20) for testing $\alpha = 1$ in the model (1.19) with μ being unknown.

2.8 Show that the statistic SL_T in (1.20) is rewritten as in (1.21).

1.3 STATISTICS FROM THE ONE-DIMENSIONAL RANDOM WALK

So far we have discussed the limiting distributions of $\varepsilon'CC'\varepsilon/T^2$ in (1.9) and $\varepsilon'\Omega^{-1}\varepsilon/T^2$ in (1.17) assuming $\varepsilon \sim N(0, I_T)$. Here we deal with these statistics from a different point of view. For this purpose, consider the one-dimensional random walk

(1.22) $$y_t = y_{t-1} + \varepsilon_t, \quad y_0 = 0, \quad (t = 1, \ldots, T),$$

where $\varepsilon_1, \varepsilon_2, \ldots$ are NID(0, 1) random variables and define

$$S_{1T} = \frac{1}{T^2} \sum_{t=1}^{T} y_t^2 = \frac{1}{T^2} \varepsilon'C'C\varepsilon,$$

$$S_{2T} = \frac{1}{T^2} \sum_{t=1}^{T} (y_t - \bar{y})^2 = \frac{1}{T^2} \varepsilon'C'MC\varepsilon,$$

where $M = I_T - ee'/T$ with $e = (1, \ldots, 1)'$. The statistic S_{1T} is the one discussed in Erdös and Kac (1946), as was described in Section 1.1, and it has the same distribution as that of $\varepsilon'CC'\varepsilon/T^2$. For S_{2T}, it can be shown (Problem 3.1) that

(1.23) $$\mathcal{L}(S_{2T}) = \mathcal{L}\left(\frac{1}{T^2} \varepsilon'MCC'M\varepsilon\right)$$

$$= \mathcal{L}\left(\frac{1}{T^2} \sum_{t=1}^{T-1} \gamma_t \xi_t^2\right),$$

where $\{\xi_t\} \sim$ NID(0, 1) and

(1.24) $$\gamma_t = \frac{1}{4}\left(\sin \frac{t\pi}{2T}\right)^{-2}, \quad (t = 1, \ldots, T-1).$$

Therefore S_{2T} has the same limiting distribution as that of $\varepsilon'\Omega^{-1}\varepsilon/T^2$.

To summarize we have

(1.25) $$\mathcal{L}\left(\frac{1}{T^2}\sum_{t=1}^{T} y_t^2\right) \to \mathcal{L}\left(\sum_{n=1}^{\infty} \frac{\xi_n^2}{(n-\frac{1}{2})^2 \pi^2}\right),$$

(1.26) $$\mathcal{L}\left(\frac{1}{T^2}\sum_{t=1}^{T} (y_t - \bar{y})^2\right) \to \mathcal{L}\left(\sum_{n=1}^{\infty} \frac{\xi_n^2}{n^2 \pi^2}\right).$$

These expressions tell us clearly that mean correction does affect the asymptotic distribution, unlike in the stationary case.

Figure 1.1 draws the densities of distributions in (1.25) for $T = 10, 20, 50$, and ∞. The densities $f_T(x) = dP(V_T \leq x)/dx$, where $V_T = \sum_{t=1}^{T} y_t^2/T^2$, were computed numerically following the inversion formula (Problem 3.2)

(1.27) $$f_T(x) = \frac{1}{2\pi}\int_{-\infty}^{\infty} e^{-i\theta x} \phi_T(\theta)\, d\theta = \frac{1}{\pi}\int_0^{\infty} \operatorname{Re}\left(e^{-i\theta x}\phi_T(\theta)\right) d\theta,$$

where Re(z) is the real part of z; $\phi_T(\theta) = (\cos\sqrt{2i\theta})^{-1/2}$ for $T = \infty$, while, for T finite,

$$\phi_T(\theta) = \prod_{t=1}^{T} (1 - 2i\theta\lambda_t)^{-1/2}$$

with λ_t defined in (1.4). The numerical computation involves the square root of complex variables, and how to compute this together with numerical integration as in

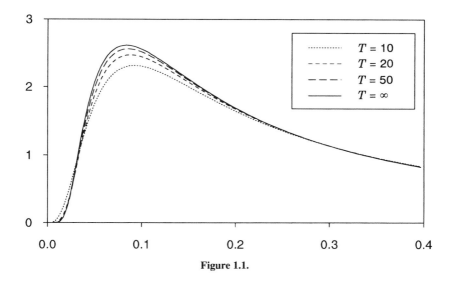

Figure 1.1.

(1.27) will be discussed in Chapter 6. It is seen from Figure 1.1 that the finite sample densities converge rapidly to the limiting density, although the former have a heavier right-hand tail.

Figure 1.2 draws the densities of distributions in (1.26) for $T = 10, 20$, and ∞. The densities $g_T(x) = dP(W_T \leq x)/dx$ with $W_T = \sum_{t=1}^{T}(y_t - \bar{y})^2/T^2$ were computed following (1.27) with $\phi_T(\theta) = (\sin\sqrt{2i\theta}/\sqrt{2i\theta})^{-1/2}$ for $T = \infty$ and, for T finite,

$$\phi_T(\theta) = \prod_{t=1}^{T-1}(1 - 2i\theta\gamma_t)^{-1/2},$$

where γ_t is defined in (1.24). Note that Figure 1.2 does not contain the density for $T = 50$ because it was found to be very close to that for $T = \infty$, while it is not as close in Figure 1.1. This is related to theoretical findings about asymptotic expansions given in (1.7) and (1.15). The normalizer $\left(T + \frac{1}{2}\right)^2$, instead of T^2 in (1.25), could make finite sample densities closer to the limiting density, which tables for percent points will exemplify.

Table 1.1 reports percent points and means for distributions of $\sum_{t=1}^{T} y_t^2 / \left(T + \frac{1}{2}\right)^2$ for $T = 10, 20, 50$, and ∞, where "E" stands for exact distributions, and "A" for distributions based on the asymptotic expansion given in (1.7). Table 1.2 contains percent points for distributions of $\sum_{t=1}^{T}(y_t - \bar{y})^2/T^2$, where the asymptotic expansion "A" is based on (1.15). It can be seen from these tables that the finite sample distributions are really close to the limiting distribution. Especially, percent points for $T = 50$ are identical with those for $T = \infty$ within the deviation of $3/10{,}000$. Asymptotic expansions also give a fairly good approximation to finite sample distributions. In most cases, they give a correct value up to the fourth decimal point.

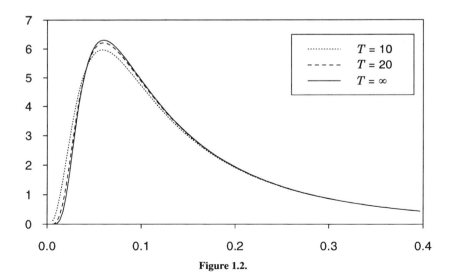

Figure 1.2.

STATISTICS FROM THE ONE-DIMENSIONAL RANDOM WALK 13

Table 1.1. Percent Points for Distributions of $\sum_{t=1}^{T} y_t^2 / (T + \frac{1}{2})^2$

Case		0.01	0.05	0.1	0.5	0.9	0.95	0.99	Mean
$T = 10$	E	0.0292	0.0527	0.0736	0.2894	1.1962	1.6570	2.7908	0.4989
	A	0.0285	0.0522	0.0733	0.2893	1.1963	1.6571	2.7911	0.4988
$T = 20$	E	0.0330	0.0555	0.0758	0.2902	1.1959	1.6561	2.7883	0.4997
	A	0.0329	0.0554	0.0757	0.2902	1.1959	1.6561	2.7884	0.4997
$T = 50$	E	0.0342	0.0563	0.0764	0.2904	1.1958	1.6558	2.7876	0.49995
	A	0.0342	0.0563	0.0764	0.2904	1.1958	1.6558	2.7876	0.49995
$T = \infty$		0.0345	0.0565	0.0765	0.2905	1.1958	1.6557	2.7875	0.5000

It is an easy matter to compute moments of these distributions. Let $\kappa_{1T}^{(j)}$ be the jth order cumulant for the distribution of $\sum_{t=1}^{T} y_t^2 / (T + \frac{1}{2})^2$ based on the asymptotic expansion in (1.7). Define $\kappa_{2T}^{(j)}$ similarly for the distribution of $\sum_{t=1}^{T} (y_t - \bar{y})^2 / T^2$ based on the asymptotic expansion in (1.15). Then we have (Problem 3.3), up to $O(T^{-2})$,

$$
\begin{aligned}
\kappa_{1T}^{(1)} &\sim \frac{1}{2} - \frac{1}{8T^2}, & \kappa_{2T}^{(1)} &\sim \frac{1}{6} - \frac{1}{6T^2}, \\
\kappa_{1T}^{(2)} &\sim \frac{1}{3} + \frac{1}{6T^2}, & \kappa_{2T}^{(2)} &\sim \frac{1}{45} + \frac{1}{18T^2}, \\
\kappa_{1T}^{(3)} &\sim \frac{8}{15} + \frac{1}{3T^2}, & \kappa_{2T}^{(3)} &\sim \frac{8}{945} + \frac{1}{45T^2}, \\
\kappa_{1T}^{(4)} &\sim \frac{136}{105} + \frac{16}{15T^2}, & \kappa_{2T}^{(4)} &\sim \frac{8}{1575} + \frac{16}{945T^2}.
\end{aligned}
\tag{1.28}
$$

Table 1.2. Percent Points for Distributions of $\sum_{t=1}^{T}(y_t - \bar{y})^2 / T^2$

Case		0.01	0.05	0.1	0.5	0.9	0.95	0.99	Mean
$T = 10$	E	0.0181	0.0313	0.0416	0.1174	0.3480	0.4629	0.7472	0.1650
	A	0.0181	0.0296	0.0401	0.1171	0.3481	0.4632	0.7480	0.1650
$T = 20$	E	0.0228	0.0352	0.0449	0.1185	0.3475	0.4617	0.7444	0.1663
	A	0.0226	0.0350	0.0448	0.1185	0.3475	0.4618	0.7445	0.1663
$T = 50$	E	0.0245	0.0363	0.0458	0.1188	0.3473	0.4614	0.7436	0.1666
	A	0.0245	0.0363	0.0458	0.1188	0.3473	0.4614	0.7436	0.1666
$T = \infty$		0.0248	0.0366	0.0460	0.1189	0.3473	0.4614	0.7435	0.1667

For the limiting distributions, we have (Problem 3.4)

(1.29) $$\kappa_1^{(j)} = \lim_{T\to\infty} \kappa_{1T}^{(j)} = \frac{(j-1)!\, 2^{3j-2}(2^{2j}-1)}{(2j)!} B_j,$$

(1.30) $$\kappa_2^{(j)} = \lim_{T\to\infty} \kappa_{2T}^{(j)} = \frac{(j-1)!\, 2^{3j-2}}{(2j)!} B_j,$$

where the B_js are the Bernoulli numbers $B_1 = \frac{1}{6}, B_2 = \frac{1}{30}, B_3 = \frac{1}{42}, B_4 = \frac{1}{30}$, and so on. The skewness $\kappa_1^{(3)}/\left(\kappa_1^{(2)}\right)^{3/2}$ and kurtosis $\kappa_1^{(4)}/\left(\kappa_1^{(2)}\right)^2 - 3$ are 2.771 and 8.657, respectively, while $\kappa_2^{(3)}/\left(\kappa_2^{(2)}\right)^{3/2} = 2.556$ and $\kappa_2^{(4)}/\left(\kappa_2^{(2)}\right)^2 - 3 = 7.286$.

As another example, let us consider the following statistic:

$$\hat{\rho}_\delta = \sum_{t=2}^{T} y_{t-1} y_t \bigg/ \left(\sum_{t=2}^{T} y_{t-1}^2 + \delta y_T^2 \right),$$

where y_t follows (1.22) and δ is a given constant. The statistic $\hat{\rho}_\delta$ may be regarded as an estimator for the model

(1.31) $$y_t = \rho y_{t-1} + \varepsilon_t, \quad y_0 = 0, \quad (t = 1, \ldots, T),$$

where $\{\varepsilon_t\} \sim \text{NID}(0, 1)$. In particular, $\hat{\rho}_0$ becomes both the least squares estimator (LSE) and the MLE of ρ for the model (1.31), while $\hat{\rho}_1$ is called the Yule–Walker estimator. We shall show that the asymptotic distribution of $\hat{\rho}_\delta$ does depend on the value of δ when $\rho = 1$, unlike in the stationary case, by deriving the limiting distribution of a suitably normalized quantity of $\hat{\rho}_\delta$. White (1958) first obtained the limiting c.f. associated with $T(\hat{\rho}_0 - 1)$ as $T \to \infty$ under $|\rho| \geq 1$. Here we continue to assume that $\rho = 1$ and follow his approach.

Consider now $T(\hat{\rho}_\delta - 1) = U_T/V_T$, where

$$U_T = \frac{1}{T} \sum_{t=2}^{T} y_{t-1} \varepsilon_t - \frac{\delta}{T} y_T^2,$$

$$V_T = \frac{1}{T^2} \sum_{t=2}^{T} y_{t-1}^2 + \frac{\delta}{T^2} y_T^2.$$

Then we have (Problem 3.5), for any real x,

(1.32) $$X_T = xV_T - U_T$$

$$= \varepsilon' \left[\frac{x}{T^2} C' \begin{pmatrix} 1 & & 0 \\ & \ddots & \\ & & 1 \\ 0 & & \delta \end{pmatrix} C - \frac{1-2\delta}{2T} ee' + \frac{1}{2T} I_T \right] \varepsilon,$$

where $\varepsilon = (\varepsilon_1, \ldots, \varepsilon_T)'$, $e = (1, \ldots, 1)' : T \times 1$, and C is the random walk generating matrix defined in (1.3). Note that $P(T(\hat{\rho}_\delta - 1) \leq x) = P(xV_T - U_T \geq 0) = P(X_T \geq 0)$. We can show [White (1958) and Problem 3.6] that the moment generating function (m.g.f.) $m_T(\theta)$ of X_T is given by

$$(1.33) \quad m_T(\theta) = \left[r^T \left\{ \cos T\omega - \frac{r \cos \omega - d}{r \sin \omega} \sin T\omega \right\} \right]^{-1/2},$$

where

$$r = 1 - \frac{\theta}{T}, \quad d = 1 - \frac{2\delta\theta}{T} - \frac{2\delta\theta x}{T^2},$$

$$\cos \omega = 1 - \frac{\theta x}{rT^2}, \quad \sin \omega = \frac{1}{rT} \sqrt{2r\theta x - \frac{\theta^2 x^2}{T^2}}.$$

The m.g.f. $m_T(\theta)$ may be expanded [Knight and Satchell (1993) and Problem 3.7], up to $O(T^{-1})$, as

$$m_T(\theta) \sim e^{\theta/2} \left[\cos A + \theta(1 - 2\delta) \frac{\sin A}{A} \right]^{-1/2}$$

$$\times \left[1 + \frac{2\delta\theta^2 \cos A + \theta\{(\theta - 1)(\theta(1 - 2\delta) + 2x) + 4\delta x\}(\sin A/A)}{4T \{\cos A + \theta(1 - 2\delta)(\sin A/A)\}} \right],$$

(1.34)

where $A = \sqrt{2\theta x}$, and thus the limiting c.f. $\phi_\delta(\theta; x)$ of X_T is given by

$$(1.35) \quad \phi_\delta(\theta; x) = e^{i\theta/2} \left[\cos \sqrt{2i\theta x} + i\theta(1 - 2\delta) \frac{\sin \sqrt{2i\theta x}}{\sqrt{2i\theta x}} \right]^{-1/2}.$$

Unlike the asymptotic expansions obtained before, it seems impossible to find a normalizer with which the term of the order T^{-1} is eliminated.

In any case we have

$$(1.36) \quad F(x; \delta) = \lim_{T \to \infty} P(T(\hat{\rho}_\delta - 1) \leq x)$$

$$= \lim_{T \to \infty} P(X_T \geq 0)$$

$$= \frac{1}{2} + \frac{1}{\pi} \int_0^\infty \frac{1}{\theta} \operatorname{Im}(\phi_\delta(\theta; x)) \, d\theta,$$

where $\operatorname{Im}(z)$ is the imaginary part of z. The limiting probability density $f(x; \delta)$ of $T(\hat{\rho}_\delta - 1)$ is computed as $f(x; \delta) = \partial F(x; \delta)/\partial x$. The following equivalent expressions will be obtained in later chapters for the weak convergence of X_T in (1.32):

(1.37)
$$\mathcal{L}(X_T) \to \mathcal{L}\left(\sum_{n=1}^{\infty} \frac{\xi_n^2}{\lambda_n} + \frac{1}{2}\right)$$
$$= \mathcal{L}\left(x \int_0^1 w^2(t)\,dt - (1-2\delta)\int_0^1 w(t)\,dw(t) + \delta\right)$$
$$= \mathcal{L}\left(x \int_0^1 \int_0^1 \bigl(1 - \max(s,t)\bigr)\,dw(s)\,dw(t)\right.$$
$$\left. - \frac{1-\delta}{2}\int_0^1 \int_0^1 dw(s)\,dw(t) + \frac{1}{2}\right),$$

where $\{\lambda_n\}$ is a sequence of solutions to

$$\cos\sqrt{\lambda x} + \frac{\lambda(1-2\delta)}{2}\frac{\sin\sqrt{\lambda x}}{\sqrt{\lambda x}} = 0,$$

while the integral $\int_0^1 w(t)\,dw(t)$ is called the *Ito integral*, whose definition will be given in Chapter 2. Alternatively we shall have

(1.38)
$$\mathcal{L}\bigl(T(\hat{\rho}_\delta - 1)\bigr) = \mathcal{L}(U_T/V_T)$$
$$\to \mathcal{L}\left(\frac{(1-2\delta)\int_0^1 w(t)\,dw(t) - \delta}{\int_0^1 w^2(t)\,dt}\right)$$
$$= \mathcal{L}\left(\frac{\frac{1-2\delta}{2}\int_0^1 \int_0^1 dw(s)\,dw(t) - \frac{1}{2}}{\int_0^1 \int_0^1 \bigl(1-\max(s,t)\bigr)\,dw(s)\,dw(t)}\right).$$

We note in passing that the first expression in (1.37) cannot be converted explicitly into the ratio form as in (1.38).

Moments of the limiting distribution of $T(\hat{\rho}_\delta - 1)$ can be derived following Evans and Savin (1981b). Suppose that $\mathcal{L}(T(\hat{\rho}_\delta - 1)) \to \mathcal{L}(U/V)$ with $P(V > 0) = 1$ and put $\psi(\theta_1, \theta_2) = E[\exp(\theta_1 U + \theta_2 V)]$. Then the kth order raw moment $\mu_\delta(k)$ of $F(x;\delta)$ in (1.36) is given by

(1.39)
$$\mu_\delta(k) = \frac{1}{(k-1)!}\int_0^\infty \theta_2^{k-1} E(U^k e^{-\theta_2 V})\,d\theta_2$$
$$= \frac{1}{(k-1)!}\int_0^\infty \theta_2^{k-1} \left.\frac{\partial^k \psi(\theta_1, -\theta_2)}{\partial \theta_1^k}\right|_{\theta_1=0} d\theta_2,$$

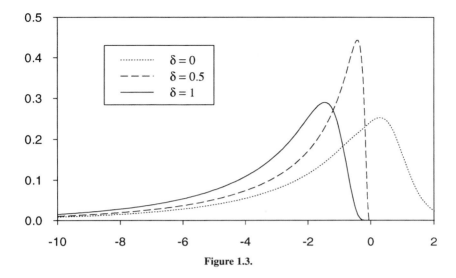

Figure 1.3.

where

$$\psi(\theta_1, -\theta_2) = E\left[\exp\left\{(-\theta_1)\left(\frac{\theta_2}{\theta_1}V - U\right)\right\}\right]$$

$$= \phi_\delta(i\theta_1; \theta_2/\theta_1)$$

$$= e^{-\theta_1/2}\left[\cosh\sqrt{2\theta_2} - \theta_1(1-2\delta)\frac{\sinh\sqrt{2\theta_2}}{\sqrt{2\theta_2}}\right]^{-1/2}.$$

Figure 1.3 draws the limiting probability densities $f(x; \delta)$ of $T(\hat{\rho}_\delta - 1)$ for $\delta = 0, 0.5$, and 1. It is seen that $f(x; 1)$ is located to the left of $f(x; 0)$, as is expected from the definition of $T(\hat{\rho}_\delta - 1)$, though these are not congruent. Table 1.3 reports percent points, means, and standard deviations (SDs) of the limiting distributions of $T(\hat{\rho}_\delta - 1)$ for various values of δ. The limiting distribution of $T(\hat{\rho}_0 - 1)$ was earlier tabulated in Fuller (1976) by simulations, while tables based on numerical integration were provided by Evans and Savin (1981a), Bobkoski (1983), Perron

Table 1.3. Percent Points for Limiting Distributions of $T(\hat{\rho}_\delta - 1)$

	Probability of a Smaller Value								
δ	0.01	0.05	0.1	0.5	0.9	0.95	0.99	Mean	SD
−1	−12.687	−7.036	−4.714	0.578	4.717	6.095	9.337	0.219	4.152
−0.5	−13.124	−7.469	−5.144	−0.141	2.732	3.567	5.495	−0.781	3.576
0	−13.695	−8.039	−5.714	−0.853	0.928	1.285	2.033	−1.781	3.180
0.5	−14.510	−8.856	−6.533	−1.721	−0.418	−0.302	−0.179	−2.781	3.037
1	−15.803	−10.107	−7.752	−2.757	−1.142	−0.925	−0.636	−3.781	3.180
1.5	−17.860	−11.852	−9.325	−3.743	−1.576	−1.236	−0.791	−4.781	3.576
2	−20.666	−13.942	−11.092	−4.662	−1.895	−1.446	−0.883	−5.781	4.152

(1989), and Nabeya and Tanaka (1990a). A closer look at the values of means and SDs in Table 1.3 leads us to conjecture the following: Let the limit in distribution of $T(\hat{\rho}_\delta - 1)$ be U_δ/V, where

$$U_\delta = (1 - 2\delta) \int_0^1 w(t)\,dw(t) - \delta = U_{1-\delta} + 2(1 - 2\delta)\left(U_0 + \frac{1}{2}\right),$$

$$V = \int_0^1 w^2(t)\,dt.$$

Then we have

(1.40) $\quad E\left(\dfrac{U_0 + \frac{1}{2}}{V}\right) = E\left(\dfrac{\int_0^1 w(t)\,dw(t) + \frac{1}{2}}{\int_0^1 w^2(t)\,dt}\right)$

$\quad\quad\quad\quad\quad\quad\quad = E\left(\dfrac{1}{2(1 - 2\delta)}\dfrac{U_\delta - U_{1-\delta}}{V}\right) = 1,$

(1.41) $\quad\quad \operatorname{Var}\left(\dfrac{U_\delta}{V}\right) = \operatorname{Var}\left(\dfrac{U_{1-\delta}}{V}\right).$

The relation in (1.40) can be proved (Problem 3.8) by showing that the mean $E((U_0 + \frac{1}{2})/V)$ of the limiting distribution of

(1.42) $\quad T(\hat{\rho}_0 - 1) + \dfrac{\frac{1}{2}}{\dfrac{1}{T^2}\sum_{t=2}^{T} y_{t-1}^2} = \dfrac{\dfrac{1}{T}\sum_{t=2}^{T} y_{t-1}\varepsilon_t + \dfrac{1}{2}}{\dfrac{1}{T^2}\sum_{t=2}^{T} y_{t-1}^2}$

is equal to 1. On the other hand, the relation in (1.41) can be proved (Problem 3.9) by showing that

(1.43) $\quad\quad \mu_\delta(2) - \big(\mu_\delta(1)\big)^2 = \mu_{1-\delta}(2) - \big(\mu_{1-\delta}(1)\big)^2.$

The estimator $\hat{\rho}_\delta$ may be used to test the unit root hypothesis $H_0 : \rho = 1$ versus $H_1 : \rho < 1$ for the model (1.31) with $|\rho| \leq 1$. The limiting local powers will be computed in Chapter 9 by numerical integration for various values of δ, and it will be found that the test based on $\delta = 0$ is the best of all the tests based on $\hat{\rho}_\delta$.

PROBLEMS

3.1 Prove the distributional equivalence in (1.23).
3.2 Derive the last equality in (1.27) from the second.

3.3 Compute cumulants in (1.28).

3.4 Obtain the expressions for cumulants in (1.29) and (1.30).

3.5 Obtain the last expression in (1.32).

3.6 Show that the m.g.f. $m_T(\theta)$ of X_T in (1.32) is given by (1.33).

3.7 Derive the asymptotic expansion of $m_T(\theta)$ given in (1.34).

3.8 Prove $E\left(\left(U_0 + \frac{1}{2}\right)/V\right) = 1$ in (1.40) by showing that the mean of the limiting distribution of (1.42) is equal to 1.

3.9 Prove (1.41) by showing that (1.43) holds.

1.4 STATISTICS FROM THE TWO-DIMENSIONAL RANDOM WALK

As a sequel to the previous section we consider the two-dimensional random walk

$$(1.44) \qquad y_t = \begin{pmatrix} y_{1t} \\ y_{2t} \end{pmatrix} = \begin{pmatrix} y_{1,t-1} \\ y_{2,t-1} \end{pmatrix} + \begin{pmatrix} \varepsilon_{1t} \\ \varepsilon_{2t} \end{pmatrix}, \quad (t = 1, \ldots, T),$$

where $y_0 = 0$ and $\varepsilon_t = (\varepsilon_{1t}, \varepsilon_{2t})' \sim \text{NID}(0, I_2)$. Under this last assumption $\{y_{1t}\}$ and $\{y_{2t}\}$ are independent of each other so that $\text{Cov}(y_{1s}, y_{2t}) = 0$ for any s and t.

The nonstandard nature of statistics arising from the model in (1.44) can be best seen from the following example. Consider

$$(1.45) \qquad S_T = \frac{1}{T^2} \sum_{t=1}^{T} y_{1t} y_{2t} = \frac{1}{T^2} \varepsilon_1' C' C \varepsilon_2$$

$$= \frac{1}{2T^2} \varepsilon' \begin{pmatrix} 0 & C'C \\ C'C & 0 \end{pmatrix} \varepsilon,$$

where $\varepsilon_j = (\varepsilon_{j1}, \ldots, \varepsilon_{jT})'$ $(j = 1, 2)$, $\varepsilon = (\varepsilon_1', \varepsilon_2')'$, and C is defined in (1.3). Then the c.f. $\phi_T(\theta)$ of S_T is given (Problem 4.1) by

$$\phi_T(\theta) = \prod_{t=1}^{T} \left(1 + \frac{\theta^2}{T^4} \lambda_t^2\right)^{-1/2} = \prod_{t=1}^{T} \left[\left(1 - \frac{2i\theta}{2T^2} \lambda_t\right)\left(1 + \frac{2i\theta}{2T^2} \lambda_t\right)\right]^{-1/2},$$

(1.46)

where λ_t is the tth largest eigenvalue of $C'C$ or CC' given in (1.4). It is noted that the distribution of S_T is symmetric about the origin since $\phi_T(\theta)$ is real. From the expression in (1.46) we have

$$\mathcal{L}(S_T) = \mathcal{L}\left(\frac{1}{2T^2} \sum_{t=1}^{T} \lambda_t(\xi_{1t}^2 - \xi_{2t}^2)\right),$$

where $(\xi_{1t}, \xi_{2t})' \sim \text{NID}(0, I_2)$ and thus

$$\mathcal{L}(S_T) \to \mathcal{L}(S) = \mathcal{L}\left(\frac{1}{2}\sum_{n=1}^{\infty} \frac{\xi_{1n}^2 - \xi_{2n}^2}{\left(n - \frac{1}{2}\right)^2 \pi^2}\right),$$

$$\phi(\theta) = E(e^{i\theta S}) = \prod_{n=1}^{\infty}\left(1 + \frac{\theta^2}{\left(n - \frac{1}{2}\right)^4 \pi^4}\right)^{-1/2}$$

$$= (\cos\sqrt{i\theta})^{-1/2}(\cosh\sqrt{i\theta})^{-1/2}.$$

Therefore $\sum_{t=1}^{T} y_{1t}y_{2t}/T^2$ has a nondegenerate limiting distribution even if $\{y_{1t}\}$ and $\{y_{2t}\}$ are independent of each other with $E(y_{1t}) = E(y_{2t}) = 0$. Note that the limiting distribution is also symmetric about the origin. Three equivalent expressions for the limiting random variable S are

(1.47) $$\mathcal{L}(S) = \mathcal{L}\left(\frac{1}{2}\sum_{n=1}^{\infty}\frac{\xi_{1n}^2 - \xi_{2n}^2}{\left(n - \frac{1}{2}\right)^2 \pi^2}\right)$$

$$= \mathcal{L}\left(\int_0^1 w'(t)Hw(t)\,dt\right)$$

$$= \mathcal{L}\left(\int_0^1\int_0^1 (1 - \max(s,t))\,dw'(s)H\,dw(t)\right),$$

where $w(t) = (w_1(t), w_2(t))'$ is the two-dimensional standard Brownian motion to be introduced in Chapter 2, while

$$H = \begin{pmatrix} 0 & \frac{1}{2} \\ \frac{1}{2} & 0 \end{pmatrix}.$$

As the next example consider

(1.48) $$U_T = \frac{1}{T}\sum_{t=1}^{T} y_{1t}\varepsilon_{2t} = \frac{1}{2T}\varepsilon'\begin{pmatrix} 0 & C' \\ C & 0 \end{pmatrix}\varepsilon,$$

which has the c.f.

$$\phi_T(\theta) = \left|I_T + \frac{\theta^2}{T^2}C'C\right|^{-1/2} = \prod_{t=1}^{T}\left(1 + \frac{\theta^2}{T^2}\lambda_t\right)^{-1/2}.$$

STATISTICS FROM THE TWO-DIMENSIONAL RANDOM WALK

As $T \to \infty$ it holds that

(1.49)
$$\mathcal{L}(U_T) \to \mathcal{L}(U) = \mathcal{L}\left(\frac{1}{2}\sum_{n=1}^{\infty}\frac{\xi_{1n}^2 - \xi_{2n}^2}{(n-\frac{1}{2})\pi}\right)$$
$$= \mathcal{L}\left(\int_0^1 w_1(t)\,dw_2(t)\right),$$

where the integral is the Ito integral to be introduced in Chapter 2. In the present case, we cannot express the limit in distribution in (1.49) using a double integral with a continuous integrand as in the last expression in (1.47). The c.f. $\phi(\theta)$ of U in (1.49) is $\phi(\theta) = (\cosh\theta)^{-1/2}$. If we consider, instead of U_T in (1.48),

(1.50)
$$V_T = \frac{1}{T}\sum_{t=1}^{T} y_{1t}\varepsilon_{1t},$$

it holds (Problem 4.2) that

(1.51)
$$\mathcal{L}(V_T) \to \mathcal{L}\left(\frac{1}{2}(\xi^2 + 1)\right)$$
$$= \mathcal{L}\left(\int_0^1 w_1(t)\,dw_1(t) + 1\right)$$
$$= \mathcal{L}\left(\frac{1}{2}\int_0^1\int_0^1 dw_1(s)\,dw_1(t) + \frac{1}{2}\right),$$

where $\xi \sim N(0,1)$ and the single integral is again the Ito integral. The double integral expression is possible in the present case, although an additive constant term emerges.

As the third example let us consider

(1.52) $\quad W_{1T} = \dfrac{1}{2T}\sum_{t=1}^{T}(y_{1t}\varepsilon_{2t} + y_{2t}\varepsilon_{1t}) = \dfrac{1}{4T}\,\varepsilon'\begin{pmatrix} 0 & C'+C \\ C'+C & 0 \end{pmatrix}\varepsilon,$

(1.53) $\quad W_{2T} = \dfrac{1}{2T}\sum_{t=1}^{T}(y_{1t}\varepsilon_{2t} - y_{2t}\varepsilon_{1t}) = \dfrac{1}{4T}\,\varepsilon'\begin{pmatrix} 0 & C'-C \\ C-C' & 0 \end{pmatrix}\varepsilon,$

which are mixed versions of U_T in (1.48). The c.f.'s ϕ_{jT}s of W_{jT}s ($j = 1, 2$) are given (Problem 4.3) by

(1.54)
$$\phi_{1T}(\theta) = \left[\left(1 + \frac{\theta^2}{4}\left(1 + \frac{1}{T}\right)^2\right)\left(1 + \frac{\theta^2}{4T^2}\right)^{T-1}\right]^{-1/2},$$

(1.55) $$\phi_{2T}(\theta) = \left[\frac{1}{2}\left\{\left(1+\frac{\theta}{2T}\right)^T + \left(1-\frac{\theta}{2T}\right)^T\right\}\right]^{-1}.$$

Therefore the distributions of W_{1T} and W_{2T} are symmetric about the origin. It evidently holds that

(1.56) $$\lim_{T\to\infty} \phi_{1T}(\theta) = \phi_1(\theta) = \left(1+\frac{\theta^2}{4}\right)^{-1/2},$$

(1.57) $$\lim_{T\to\infty} \phi_{2T}(\theta) = \phi_2(\theta) = \left(\cosh\frac{\theta}{2}\right)^{-1} = \prod_{n=1}^{\infty}\left(1+\frac{\theta^2}{((2n-1)\pi)^2}\right)^{-1}.$$

We shall have the following equivalent expressions:

(1.58) $$\mathcal{L}(W_{1T}) \to \mathcal{L}\left(\frac{1}{4}(\xi_1^2 - \xi_2^2)\right)$$
$$= \mathcal{L}\left(\frac{1}{2}\int_0^1 w'(t)\begin{pmatrix}0 & 1\\ 1 & 0\end{pmatrix}dw(t)\right)$$
$$= \mathcal{L}\left(\frac{1}{4}\int_0^1\int_0^1 dw'(s)\begin{pmatrix}1 & 0\\ 0 & -1\end{pmatrix}dw(t)\right),$$

(1.59) $$\mathcal{L}(W_{2T}) \to \mathcal{L}\left(\frac{1}{4}\sum_{n=1}^{\infty}\frac{\xi_{1n}^2 + \xi_{2n}^2 - \xi_{3n}^2 - \xi_{4n}^2}{\left(n-\frac{1}{2}\right)\pi}\right)$$
$$= \mathcal{L}\left(\frac{1}{2}\int_0^1 w'(t)\begin{pmatrix}0 & 1\\ -1 & 0\end{pmatrix}dw(t)\right),$$

where $(\xi_1, \xi_2)' \sim N(0, I_2)$, $(\xi_{1n}, \xi_{2n}, \xi_{3n}, \xi_{4n})' \sim \text{NID}(0, I_4)$, and $w(t)$ is the two-dimensional standard Brownian motion.

The limiting distributions of W_{1T} and W_{2T} are also symmetric about the origin. The former can be interpreted from the first expression in (1.58) as the distribution of the difference of two independent $\chi^2(1)/4$ random variables, while the latter is known as the distribution of Lévy's stochastic area [Hida (1980)]. In the latter case, the double integral expression is not possible, unlike in (1.58). This is closely related with the fact that the matrix appearing in (1.59) is not symmetric. Detailed discussions will be given in Chapter 3. Comparing (1.59) with (1.49), the following relation is seen to hold:

$$\mathcal{L}\left(\int_0^1 (w_1(t)\,dw_2(t) - w_2(t)\,dw_1(t))\right) = \mathcal{L}\left(\int_0^1 (w_1(t)\,dw_2(t) + w_3(t)\,dw_4(t))\right),$$

where $(w_1(t), w_2(t), w_3(t), w_4(t))'$ is the four-dimensional standard Brownian motion.

It is easy to obtain cumulants $\kappa_1^{(j)}$ and $\kappa_2^{(j)}$ for the limiting distributions in (1.58) and (1.59), respectively. We have (Problem 4.4)

$$(1.60) \qquad \kappa_1^{(j)} = \begin{cases} 0 & j : \text{odd} \\ \dfrac{(2l)!}{l\, 2^{2l+1}} & j = 2l, \end{cases}$$

$$(1.61) \qquad \kappa_2^{(j)} = \begin{cases} 0 & j : \text{odd} \\ \dfrac{2^{2l} - 1}{2l} B_l & j = 2l, \end{cases}$$

where B_l is the Bernoulli number.

Figure 1.4 draws the limiting probability densities $f_1(x)$ and $f_2(x)$ of $2 \times W_{1T}$ in (1.58) and $2 \times W_{2T}$ in (1.59), respectively, together with the density of $N(0, 1)$. The three distributions have means 0 and variances 1. We computed $f_1(x)$ and $f_2(x)$ following

$$(1.62) \qquad f_1(x) = \frac{1}{\pi} \int_0^\infty \frac{\cos \theta x}{\sqrt{1 + \theta^2}} d\theta,$$

$$(1.63) \qquad f_2(x) = \frac{1}{\pi} \int_0^\infty \frac{\cos \theta x}{\cosh \theta} d\theta = \frac{1}{2} \frac{1}{\cosh(\pi x/2)}.$$

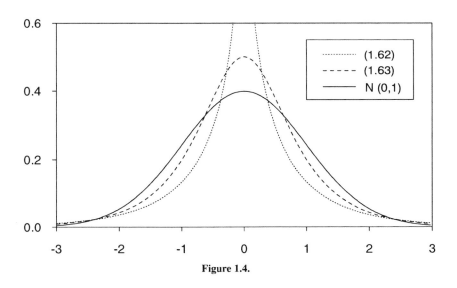

Figure 1.4.

From the computational point of view it is quite easy to deal with (1.63), but we have difficulty in computing (1.62) since the integrand is oscillating and $1/\sqrt{1+\theta^2}$ approaches 0 rather slowly. Chapter 6, from which Figure 1.4 has been produced, will suggest a method for overcoming this difficulty. Percent points for the three distributions are tabulated in Table 1.4.

Finally let us consider the statistics:

$$(1.64) \qquad \hat{\beta}_1 = \sum_{t=1}^{T} y_{1t} y_{2t} \Big/ \sum_{t=1}^{T} y_{1t}^2 ,$$

$$(1.65) \qquad \hat{\beta}_2 = \sum_{t=1}^{T} (y_{1t} - \bar{y}_1)(y_{2t} - \bar{y}_2) \Big/ \sum_{t=1}^{T} (y_{1t} - \bar{y}_1)^2 ,$$

which may be interpreted as the LSEs derived from the regression relations $y_{2t} = \hat{\beta}_1 y_{1t} + \hat{v}_{1t}$ and $y_{2t} = \hat{\alpha} + \hat{\beta}_2 y_{1t} + \hat{v}_{2t}$, respectively. These are called *spurious regressions* following Granger and Newbold (1974) because the regressor $\{y_{1t}\}$ is independent of the regressand $\{y_{2t}\}$. Nonetheless the LSEs $\hat{\beta}_1$ and $\hat{\beta}_2$ have nondegenerate limiting distributions, which we show below.

As for $\hat{\beta}_1$ put $P(\hat{\beta}_1 \leq x) = P(X_{1T} \geq 0)$, where

$$(1.66) \qquad X_{1T} = \frac{x}{T^2} \sum_{t=1}^{T} y_{1t}^2 - \frac{1}{T^2} \sum_{t=1}^{T} y_{1t} y_{2t}$$

$$= \varepsilon' \begin{pmatrix} \dfrac{x}{T^2} C'C & -\dfrac{1}{2T^2} C'C \\ -\dfrac{1}{2T^2} C'C & 0 \end{pmatrix} \varepsilon .$$

The c.f. $\phi_{1T}(\theta; x)$ of X_{1T} in (1.66) is given (Problem 4.5) by

$$(1.67) \qquad \phi_{1T}(\theta; x) = \prod_{t=1}^{T} \left[\left(1 - \frac{2i\theta a(x)}{T^2} \lambda_t\right) \left(1 - \frac{2i\theta b(x)}{T^2} \lambda_t\right) \right]^{-1/2} ,$$

where $a(x) = (x + \sqrt{x^2+1})/2$, $b(x) = (x - \sqrt{x^2+1})/2$, and λ_t is the tth largest eigenvalue of CC' or $C'C$ given in (1.4). It is noted from the expression for $\phi_{1T}(\theta; x)$

Table 1.4. Percent Points for (1.62), (1.63), and N(0, 1)

Case	Probability of a Smaller Value							
	0.5	0.6	0.7	0.8	0.9	0.95	0.975	0.99
(1.62)	0	0.0887	0.2494	0.5169	1.0344	1.5951	2.1819	2.9838
(1.63)	0	0.2034	0.4293	0.7157	1.1731	1.6183	2.0606	2.6442
N(0, 1)	0	0.2533	0.5244	0.8416	1.2816	1.6449	1.9600	2.3263

that

$$\mathcal{L}(X_{1T}) = \mathcal{L}\left(\frac{a(x)}{T^2}\boldsymbol{\xi}_1'C'C\boldsymbol{\xi}_1 + \frac{b(x)}{T^2}\boldsymbol{\xi}_2'C'C\boldsymbol{\xi}_2\right),$$

where $\boldsymbol{\xi}_1$ and $\boldsymbol{\xi}_2$ are independent of each other and both follow N$(0, I_T)$. Arguing as before, it is an easy matter to derive

(1.68) $\phi_1(\theta; x) = \lim_{T \to \infty} \phi_{1T}(\theta; x)$

$$= \prod_{n=1}^{\infty}\left[\left(1 - \frac{2i\theta a(x)}{\left(n - \frac{1}{2}\right)^2 \pi^2}\right)\left(1 - \frac{2i\theta b(x)}{\left(n - \frac{1}{2}\right)^2 \pi^2}\right)\right]^{-1/2}$$

$$= \left[D_1(2i\theta a(x)) D_1(2i\theta b(x))\right]^{-1/2},$$

where $D_1(\lambda) = \cos\sqrt{\lambda}$.

We can deal with $\hat{\beta}_2$ in (1.65) similarly. Let us put $P(\hat{\beta}_2 \leq x) = P(X_{2T} \geq 0)$, where

(1.69) $X_{2T} = \frac{x}{T^2}\sum_{t=1}^{T}(y_{1t} - \bar{y}_1)^2 - \frac{1}{T^2}\sum_{t=1}^{T}(y_{1t} - \bar{y}_1)(y_{2t} - \bar{y}_2).$

The c.f. $\phi_{2T}(\theta; x)$ of X_{2T} in (1.69) is given (Problem 4.6) by

(1.70) $\phi_{2T}(\theta; x) = \prod_{t=1}^{T-1}\left[\left(1 - \frac{2i\theta a(x)}{T^2}\gamma_t\right)\left(1 - \frac{2i\theta b(x)}{T^2}\gamma_t\right)\right]^{-1/2},$

where γ_t $(t = 1, \ldots, T - 1)$ is the tth largest eigenvalue of CMC' or $C'MC$ defined in (1.24). Then we have

(1.71) $\phi_2(\theta; x) = \lim_{T \to \infty} \phi_{2T}(\theta; x)$

$$= \prod_{n=1}^{\infty}\left[\left(1 - \frac{2i\theta a(x)}{n^2 \pi^2}\right)\left(1 - \frac{2i\theta b(x)}{n^2 \pi^2}\right)\right]^{-1/2}$$

$$= \left[D_2(2i\theta a(x)) D_2(2i\theta b(x))\right]^{-1/2},$$

where $D_2(\lambda) = (\sin\sqrt{\lambda})/\sqrt{\lambda}$.

Figure 1.5 draws the limiting probability densities $f_j(x)$ of $\hat{\beta}_j$, which were numerically computed from $f_j(x) = dF_j(x)/dx$, where

(1.72) $F_j(x) = \lim_{T \to \infty} P(\hat{\beta}_j \leq x) = \lim_{T \to \infty} P(X_{jT} \geq 0)$

$$= \frac{1}{2} + \frac{1}{\pi}\int_0^{\infty} \frac{1}{\theta}\operatorname{Im}(\phi_j(\theta; x))\, d\theta.$$

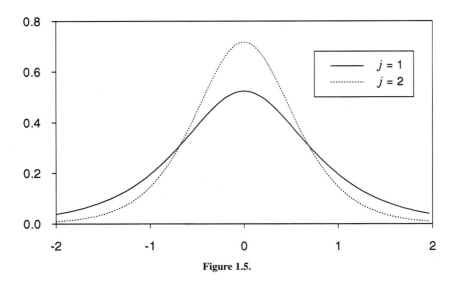

Figure 1.5.

Moments of $F_j(x)$ can also be computed following formula (1.39). In particular we have (Problem 4.7)

$$\mu_1(1) = \mu_2(1) = \mu_1(3) = \mu_2(3) = 0,$$

$$\mu_1(2) = \frac{1}{4} \int_0^\infty \frac{u}{\sqrt{\cosh u}} du - \frac{1}{2} = 0.8907,$$

$$\mu_2(2) = \frac{1}{12} \int_0^\infty \frac{u^{3/2}}{\sqrt{\sinh u}} du - \frac{1}{2} = 0.3965,$$

$$\mu_1(4) = \frac{7}{192} \int_0^\infty \frac{u^3}{\sqrt{\cosh u}} du - \mu_1(2) - \frac{1}{8} = 3.9304,$$

$$\mu_2(4) = \frac{1}{320} \int_0^\infty \frac{u^{7/2}}{\sqrt{\sinh u}} du - \mu_2(2) - \frac{1}{8} = 0.6421,$$

(1.73)

where $\mu_j(k)$ is the kth order raw moment of $F_j(x)$. It can also be shown (Problem 4.8) that $F_1(x)$ and $F_2(x)$ are both symmetric about the origin. Table 1.5 reports percent points for $F_j(x)$ ($x \geq 0$).

Table 1.5. Percent Points of $F_j(x)$ in (1.72)

Case	Probability of a Smaller Value							
	0.5	0.6	0.7	0.8	0.9	0.95	0.975	0.99
$j = 1$	0	0.1940	0.4089	0.6800	1.1100	1.5262	1.9392	2.4852
$j = 2$	0	0.1414	0.2955	0.4836	0.7662	1.0251	1.2723	1.5890

The following equivalent expressions will emerge for the weak convergence of X_{1T} in (1.66):

$$(1.74) \quad \mathcal{L}(X_{1T}) \to \mathcal{L}\left(a(x)\sum_{n=1}^{\infty}\frac{\xi_{1n}^2}{(n-\frac{1}{2})^2\pi^2} + b(x)\sum_{n=1}^{\infty}\frac{\xi_{2n}^2}{(n-\frac{1}{2})^2\pi^2}\right)$$

$$= \mathcal{L}\left(\int_0^1 w'(t)\begin{pmatrix} x & -\frac{1}{2} \\ -\frac{1}{2} & 0 \end{pmatrix} w(t)\,dt\right)$$

$$= \mathcal{L}\left(\int_0^1\int_0^1 (1-\max(s,t))\,dw'(s)\begin{pmatrix} x & -\frac{1}{2} \\ -\frac{1}{2} & 0 \end{pmatrix} dw(t)\right),$$

where $(\xi_{1n},\xi_{2n})' \sim \text{NID}(0,I_2)$ and $w(t) = (w_1(t),w_2(t))'$ is the two-dimensional standard Brownian motion. We shall also have

$$(1.75) \quad \mathcal{L}(X_{2T}) \to \mathcal{L}\left(a(x)\sum_{n=1}^{\infty}\frac{\xi_{1n}^2}{n^2\pi^2} + b(x)\sum_{n=1}^{\infty}\frac{\xi_{2n}^2}{n^2\pi^2}\right)$$

$$= \mathcal{L}\left(\int_0^1 \tilde{w}'(t)\begin{pmatrix} x & -\frac{1}{2} \\ -\frac{1}{2} & 0 \end{pmatrix}\tilde{w}(t)\,dt\right)$$

$$= \mathcal{L}\left(\int_0^1\int_0^1 (\min(s,t)-st)\,dw'(s)\begin{pmatrix} x & -\frac{1}{2} \\ -\frac{1}{2} & 0 \end{pmatrix} dw(t)\right),$$

where $\tilde{w}(t) = w(t) - \int_0^1 w(t)\,dt$ is the two-dimensional *demeaned Brownian motion*. In terms of the weak convergence of $\hat{\beta}_j$, we shall have the following expressions:

$$(1.76) \quad \mathcal{L}(\hat{\beta}_1) \to \mathcal{L}\left(\frac{\int_0^1 w_1(t)w_2(t)\,dt}{\int_0^1 w_1^2(t)\,dt}\right)$$

$$= \mathcal{L}\left(\frac{\int_0^1\int_0^1 (1-\max(s,t))\,dw_1(s)\,dw_2(t)}{\int_0^1\int_0^1 (1-\max(s,t))\,dw_1(s)\,dw_1(t)}\right),$$

$$(1.77) \quad \mathcal{L}(\hat{\beta}_2) \to \mathcal{L}\left(\frac{\int_0^1 \tilde{w}_1(t)\tilde{w}_2(t)\,dt}{\int_0^1 \tilde{w}_1^2(t)\,dt}\right)$$

$$= \mathcal{L}\left(\frac{\int_0^1\int_0^1 (\min(s,t)-st)\,dw_1(s)\,dw_2(t)}{\int_0^1\int_0^1 (\min(s,t)-st)\,dw_1(s)\,dw_1(t)}\right).$$

The first expressions in (1.74) and (1.75) cannot be converted explicitly into the ratio form as above.

PROBLEMS

4.1 Show that the c.f. $\phi_T(\theta)$ of S_T in (1.45) is given by (1.46).

4.2 Prove that V_T in (1.50) converges in distribution to $(\xi^2 + 1)/2$ given in (1.51).

4.3 Show that the c.f. $\phi_{1T}(\theta)$ of W_{1T} in (1.52) is given by (1.54) and the c.f. $\phi_{2T}(\theta)$ of W_{2T} in (1.53) by (1.55).

4.4 Derive the expressions for cumulants in (1.60) and (1.61).

4.5 Show that the c.f. $\phi_{1T}(\theta; x)$ of X_{1T} in (1.66) is given by (1.67).

4.6 Show that the c.f. $\phi_{2T}(\theta; x)$ of X_{2T} in (1.69) is given by (1.70).

4.7 Obtain moments of $F_1(x)$ and $F_2(x)$ given in (1.73).

4.8 Show that $F_j(x)$ ($j = 1, 2$) in (1.72) are symmetric about the origin. (In fact the finite sample distributions are also symmetric about the origin.)

1.5 STATISTICS FROM THE COINTEGRATED PROCESS

Let us consider the model

(1.78)
$$y_{2t} = \beta y_{1t} + \varepsilon_{2t},$$
$$y_{1t} = y_{1,t-1} + \varepsilon_{1t}, \quad y_{10} = 0, \quad (t = 1, \ldots, T),$$

where $\beta \neq 0$, $\{y_{1t}\}$ and $\{y_{2t}\}$ are observable, and $(\varepsilon_{1t}, \varepsilon_{2t})'$ follows NID$(0, I_2)$. This model might be thought to be equivalent to the state space model or the random walk plus noise model dealt with in Section 1.1, but this is not the case because the random walk $\{y_{1t}\}$ is observable. The model is also different from the two-dimensional random walk dealt with in Section 1.4 because the latter assumes that $\{y_{1t}\}$ and $\{y_{2t}\}$ are independent of each other.

The present model is a simplified version of the *cointegrated system* to be discussed in later chapters. Note that y_{2t} is not a random walk because $\Delta y_{2t} = y_{2t} - y_{2,t-1} = \beta \varepsilon_{1t} + \Delta \varepsilon_{2t}$ is not independent, though stationary. The process $\{y_{2t}\}$ is called an *integrated process* of order 1, which is denoted as an I(1) process. A random walk process is also a special case of I(1) processes. The implication of (1.78) is that $y_{2t} - \beta y_{1t}$, a linear combination of two I(1) processes $\{y_{1t}\}$ and $\{y_{2t}\}$, follows NID(0, 1). In general, following Engle and Granger (1987), a vector-valued process $\{y_t\}$ is said to be integrated of order d if $\{\Delta^d y_t\}$ is stationary, and is called a *cointegrated process of order* (d, b) if there exists a linear combination $\alpha' y_t$ ($\alpha \neq 0$), say, which is I($d - b$).

Here we consider the estimators $\hat{\beta}_1$ and $\hat{\beta}_2$ of the cointegration parameter β defined by

(1.79) $$\hat{\beta}_1 = \sum_{t=1}^{T} y_{1t}\, y_{2t} \Big/ \sum_{t=1}^{T} y_{1t}^2 ,$$

(1.80) $$\hat{\beta}_2 = \sum_{t=1}^{T} (y_{1t} - \bar{y}_1)(y_{2t} - \bar{y}_2) \Big/ \sum_{t=1}^{T} (y_{1t} - \bar{y}_1)^2 .$$

Unlike the spurious regressions discussed in Section 1.4, the estimators $\hat{\beta}_1$ and $\hat{\beta}_2$ are consistent, and $T(\hat{\beta}_j - \beta)$ ($j = 1, 2$) have nondegenerate limiting distributions, which we now show.

Put $P(T(\hat{\beta}_1 - \beta) \leq x) = P(X_{1T} \geq 0)$, where

(1.81) $$X_{1T} = \frac{x}{T^2} \sum_{t=1}^{T} y_{1t}^2 - \frac{1}{T} \sum_{t=1}^{T} y_{1t}\, \varepsilon_{2t}$$

$$= \boldsymbol{\varepsilon}' \begin{pmatrix} \dfrac{x}{T^2} C'C & -\dfrac{1}{2T} C' \\ -\dfrac{1}{2T} C & 0 \end{pmatrix} \boldsymbol{\varepsilon}$$

with $\boldsymbol{\varepsilon} = (\varepsilon_{11}, \ldots, \varepsilon_{1T}, \varepsilon_{21}, \ldots, \varepsilon_{2T})'$. The c.f. $\phi_{1T}(\theta; x)$ of X_{1T} in (1.81) is given (Problem 5.1) by

(1.82) $$\phi_{1T}(\theta; x) = \prod_{t=1}^{T} \left(1 - (2i\theta x - \theta^2)\frac{\lambda_t}{T^2}\right)^{-1/2}$$

$$= \prod_{t=1}^{T} \left[(1 - 2i\theta a_t)(1 - 2i\theta b_t)\right]^{-1/2} ,$$

where λ_t is defined in (1.4), while

$$a_t, b_t = \frac{1}{2}\left[\frac{x\lambda_t}{T^2} \pm \sqrt{\frac{\lambda_t}{T^2}\left(\frac{x^2 \lambda_t}{T^2} + 1\right)}\right] .$$

From the last expression for $\phi_{1T}(\theta; x)$ in (1.82) we have

$$\mathcal{L}(X_{1T}) = \mathcal{L}\left(\sum_{t=1}^{T}(a_t \xi_{1t}^2 + b_t \xi_{2t}^2)\right) ,$$

where $(\xi_{1t}, \xi_{2t})' \sim \text{NID}(0, I_2)$. It now holds that

(1.83)
$$\phi_1(\theta; x) = \lim_{T \to \infty} \phi_{1T}(\theta; x)$$
$$= \prod_{n=1}^{\infty} \left(1 - \frac{2i\theta x - \theta^2}{\left(n - \frac{1}{2}\right)^2 \pi^2}\right)^{-1/2}$$
$$= \left(D_1(2i\theta x - \theta^2)\right)^{-1/2},$$

where $D_1(\lambda) = \cos \sqrt{\lambda}$.

We also put $P(T(\hat{\beta}_2 - \beta) \leq x) = P(X_{2T} \geq 0)$, where

(1.84)
$$X_{2T} = \frac{x}{T^2} \sum_{t=1}^{T} (y_{1t} - \bar{y}_1)^2 - \frac{1}{T} \sum_{t=1}^{T} (y_{1t} - \bar{y}_1)\varepsilon_{2t}.$$

The c.f. $\phi_{2T}(\theta; x)$ of X_{2T} in (1.84) is given (Problem 5.2) by

(1.85)
$$\phi_{2T}(\theta; x) = \prod_{t=1}^{T-1} \left(1 - (2i\theta x - \theta^2)\frac{\gamma_t}{T^2}\right)^{-1/2}$$
$$= \prod_{t=1}^{T-1} \left[(1 - 2i\theta c_t)(1 - 2i\theta d_t)\right]^{-1/2},$$

where γ_t is defined in (1.24), while

$$c_t, d_t = \frac{1}{2}\left[\frac{x\gamma_t}{T^2} \pm \sqrt{\frac{\gamma_t}{T^2}\left(\frac{x^2\gamma_t}{T^2} + 1\right)}\right].$$

Thus we have

$$\mathcal{L}(X_{2T}) = \mathcal{L}\left(\sum_{t=1}^{T-1}(c_t \xi_{1t}^2 + d_t \xi_{2t}^2)\right).$$

It now holds that

(1.86)
$$\phi_2(\theta; x) = \lim_{T \to \infty} \phi_{2T}(\theta; x)$$
$$= \prod_{n=1}^{\infty} \left(1 - \frac{2i\theta x - \theta^2}{n^2 \pi^2}\right)^{-1/2}$$
$$= \left(D_2(2i\theta x - \theta^2)\right)^{-1/2},$$

where $D_2(\lambda) = (\sin \sqrt{\lambda})/\sqrt{\lambda}$.

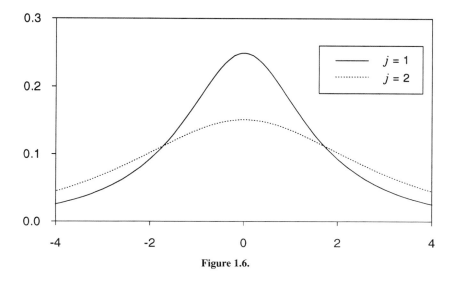

Figure 1.6.

Figure 1.6 draws the limiting probability densities $f_j(x) = dF_j(x)/dx$ of $T(\hat{\beta}_j - \beta)$, where $F_j(x)$ was computed following the last line of (1.72). Moments of $F_j(x)$ can also be computed following formula (1.39). In particular we have (Problem 5.3)

(1.87) $\mu_1(1) = \mu_2(1) = \mu_1(3) = \mu_2(3) = 0$,

$$\mu_1(2) = \int_0^\infty \frac{u}{\sqrt{\cosh u}}\, du = 5.5629,$$

$$\mu_2(2) = \int_0^\infty \frac{u^{3/2}}{\sqrt{\sinh u}}\, du = 10.7583,$$

$$\mu_1(4) = \frac{3}{64}\int_0^\infty \frac{u^4}{\sqrt{\cosh u}}\left(u\tanh^2 u + \frac{2}{3}\tanh u - \frac{2}{3}u\right) du = 203.4937,$$

$$\mu_2(4) = \frac{3}{64}\int_0^\infty \frac{u^{7/2}}{\sqrt{\sinh u}}\left(u^2\coth^2 u - \coth u - \frac{2}{3}u^2\right) du = 558.5358,$$

where $\mu_j(k)$ is the kth order raw moment of $F_j(x)$. It can also be shown (Problem 5.4) that $F_1(x)$ and $F_2(x)$ are both symmetric about the origin. Table 1.6 reports percent points for $F_j(x)$ ($x \geq 0$).

Table 1.6. Percent Points for Limiting Distributions of $T(\hat{\beta}_j - \beta)$

Case	Probability of a Smaller Value							
	0.5	0.6	0.7	0.8	0.9	0.95	0.975	0.99
$j = 1$	0	0.4113	0.8823	1.5163	2.6191	3.7716	4.9646	6.5855
$j = 2$	0	0.6747	1.4242	2.3740	3.8846	5.3489	6.8141	8.8364

The following equivalent expressions will emerge for the weak convergence of X_{1T} in (1.81):

$$\mathcal{L}(X_{1T}) \to \mathcal{L}\left(\sum_{n=1}^{\infty}(A_n\xi_{1n}^2 + B_n\xi_{2n}^2)\right) \tag{1.88}$$

$$= \mathcal{L}\left(x\int_0^1 w_1^2(t)\,dt - \int_0^1 w_1(t)\,dw_2(t)\right),$$

where $(\xi_{1n}, \xi_{2n})' \sim \text{NID}(0, I_2)$ and $(w_1(t), w_2(t))'$ is the two-dimensional standard Brownian motion, while

$$A_n, B_n = \frac{1}{2}\left[\frac{x}{\left(n-\frac{1}{2}\right)^2 \pi^2} \pm \sqrt{\frac{1}{\left(n-\frac{1}{2}\right)^2 \pi^2}\left(\frac{x^2}{\left(n-\frac{1}{2}\right)^2 \pi^2} + 1\right)}\right].$$

The double integral expression is not possible in the present case, unlike the case of spurious regressions. We shall also have

$$\mathcal{L}(X_{2T}) \to \mathcal{L}\left(\sum_{n=1}^{\infty}(C_n\xi_{1n}^2 + D_n\xi_{2n}^2)\right) \tag{1.89}$$

$$= \mathcal{L}\left(x\int_0^1 \tilde{w}_1^2(t)\,dt - \int_0^1 \tilde{w}_1(t)\,d\tilde{w}_2(t)\right),$$

where $(\tilde{w}_1(t), \tilde{w}_2(t))'$ is the two-dimensional demeaned Brownian motion, while

$$C_n, D_n = \frac{1}{2}\left[\frac{x}{n^2\pi^2} \pm \sqrt{\frac{1}{n^2\pi^2}\left(\frac{x^2}{n^2\pi^2} + 1\right)}\right].$$

For the weak convergence of $T(\hat{\beta}_1 - \beta)$ and $T(\hat{\beta}_2 - \beta)$, we shall have the following expressions:

$$\mathcal{L}(T(\hat{\beta}_1 - \beta)) \to \mathcal{L}\left(\frac{\int_0^1 w_1(t)\,dw_2(t)}{\int_0^1 w_1^2(t)\,dt}\right), \tag{1.90}$$

$$\mathcal{L}(T(\hat{\beta}_2 - \beta)) \to \mathcal{L}\left(\frac{\int_0^1 \tilde{w}_1(t)\,d\tilde{w}_2(t)}{\int_0^1 \tilde{w}_1^2(t)\,dt}\right). \tag{1.91}$$

The cointegrated system (1.78) is quite restricted. More generally, components of $\varepsilon_t = (\varepsilon_{1t}, \varepsilon_{2t})'$ are correlated and $\{\varepsilon_t\}$ may be dependent. Then the LSEs of β will have a different distribution. We also need to test if cointegration exists among the components of multiple time series. We will discuss those topics in Chapter 11.

PROBLEMS

5.1 Show that the c.f. $\phi_{1T}(\theta; x)$ of X_{1T} in (1.81) is given by (1.82).

5.2 Show that the c.f. $\phi_{2T}(\theta; x)$ of X_{2T} in (1.84) is given by (1.85).

5.3 Obtain moments of $F_1(x)$ and $F_2(x)$ given in (1.87).

5.4 Show that the limiting distributions of $T(\hat{\beta}_1 - \beta)$ and $T(\hat{\beta}_2 - \beta)$ are both symmetric about the origin. (In fact, the finite sample distributions are also symmetric.)

CHAPTER 2

Stochastic Calculus in Mean Square

As partly presented in Chapter 1, stochastic integrals often appear as limits in the sense of convergence in distribution, even if we deal with discrete-time processes. This chapter discusses three types of stochastic integrals defined in the mean square (m.s.) sense: the m.s. Riemann integral, the m.s. Riemann–Stieltjes integral, and the Ito integral. We naturally require that stochastic processes belong to the space L_2—the space of random variables with finite second moment.

Among important stochastic processes are the (integrated) Brownian motion, Brownian bridge, and Ornstein–Uhlenbeck processes; we examine various properties of their integrals. The so-called Ito calculus is also introduced to deal with stochastic differential equations.

The space L_2, however, is not always suitable when we discuss weak convergence of statistics. For that purpose, we shall usually need to define another space and give another definition of stochastic integrals, which will be discussed in Chapter 3.

2.1 THE SPACE L_2 OF RANDOM VARIABLES

Let us denote as L_2 the space of scalar random variables defined on a common probability space with finite second moment, where we do not distinguish between any two random variables X and Y for which $P(X = Y) = 1$. Let $\{X_n\}$ be a sequence in L_2 and suppose that $E[(X_n - X)^2] \to 0$ as $n \to \infty$. Since the convergence to 0 is possible only if $E[(X_n - X)^2]$ is finite from some value of n on, X also belongs to L_2.

Definition of m.s. Convergence. A sequence $\{X_n\}$ in L_2 is said to converge in mean square (m.s.) to X in L_2 if $E[(X_n - X)^2] \to 0$ as $n \to \infty$, which we denote as l.i.m.$_{n\to\infty} X_n = X$. □

Note that the limit X is unique with probability 1 (Problem 1.1). The following is an essential property of the L_2 space.

L_2-Completeness Theorem [Loève (1977, p. 163)]. *The space L_2 is complete in the sense that* $\text{l.i.m.}_{n\to\infty} X_n = X$ *for some $X \in L_2$ if and only if $E[(X_m - X_n)^2] \to 0$ as $m, n \to \infty$ in any manner.*

As an application of the L_2-completeness theorem it is easy to see (Problem 1.2) that the independent sequence $\{X_n\}$ defined by $P(X_n = \sqrt{n}) = 1/n$ and $P(X_n = 0) = 1 - 1/n$ does not converge in the m.s. sense, though it converges in probability to 0.

We now consider $\{Y_n(t)\}$, $t \in [a, b]$, which is a sequence of q-dimensional stochastic processes, where every element of $Y_n(t)$ belongs to L_2. If $\lim_{n\to\infty} E[(Y_n(t) - Y(t))'(Y_n(t) - Y(t))] = 0$ for each $t \in [a, b]$, then $\{Y_n(t)\}$ is said to converge in m.s. to $\{Y(t)\}$, which we denote as $\text{l.i.m.}_{n\to\infty}\{Y_n(t)\} = \{Y(t)\}$. If a particular point t is our concern, the m.s. convergence at t is denoted as $\text{l.i.m.}_{n\to\infty} Y_n(t) = Y(t)$. It holds as before that every element of $E(Y(t)Y'(t))$ is finite and the limit $Y(t)$ is unique with probability 1. Occasionally we shall need to deal with a matrix-valued stochastic process, $\{Z_n(t)\}$, say. In that case, we assume that $E[\text{tr}(Z_n'(t)Z_n(t))] < \infty$, and $\text{l.i.m.}_{n\to\infty}\{Z_n(t)\} = \{Z(t)\}$ means that $\lim_{n\to\infty} E[\text{tr}\{(Z_n(t) - Z(t))'(Z_n(t) - Z(t))\}] = 0$ for each t.

The following theorem describes an operational property of the m.s. convergence, whose proof is left as Problem 1.3 in this chapter.

Theorem 2.1. *Suppose that $\{X_n(t)\}$ and $\{Y_n(t)\}$, $t \in [a, b]$ are sequences of q-dimensional stochastic processes for which $E(X_n'(t)X_n(t)) < \infty$, $E(Y_n'(t)Y_n(t)) < \infty$, $\text{l.i.m.}_{n\to\infty} X_n(t) = X(t)$, and $\text{l.i.m.}_{n\to\infty} Y_n(t) = Y(t)$. Then it holds that*

(2.1)
$$\lim_{n\to\infty} E(aX_n(t) + bY_n(t)) = aE(X(t)) + bE(Y(t)),$$

$$\lim_{n\to\infty} E(X_n'(t)Y_n(t)) = E(X'(t)Y(t)),$$

where a and b are scalar constants.

The above theorem tells us that l.i.m. and E commute. We further point out that l.i.m. and Gaussianity commute. This is because m.s. convergence implies convergence in distribution, but the following is a proof for the present statement: Suppose that $\{X_n(t)\}$ is a q-dimensional Gaussian sequence; that is, the finite-dimensional distributions of $X_n(t_1), \ldots, X_n(t_k)$ for each finite k and each collection $t_1 < \cdots < t_k$ on $[a, b]$ are normal for all n. Then, if $\text{l.i.m.}_{n\to\infty}\{X_n(t)\} = \{X(t)\}$, $\{X(t)\}$ is also Gaussian. In fact, putting $X_n = (X_n'(t_1), \ldots, X_n'(t_k))'$ and $X = (X'(t_1), \ldots, X'(t_k))'$, we have

$$\phi_n(\theta) = E\{\exp(i\theta'X_n)\}$$

$$= \exp\left\{i\theta'E(X_n) - \frac{1}{2}\theta'V(X_n)\theta\right\},$$

where $E(X_n) \to E(X)$ and $V(X_n) \to V(X)$ by Theorem 2.1. Then the c.f. of X must be $\phi(\theta) = \exp\{i\theta' E(X) - \frac{1}{2}\theta' V(X)\theta\}$ since

$$\begin{aligned}|\phi_n(\theta) - \phi(\theta)|^2 &\leq \left[E\left\{\left|\exp(i\theta' X_n) - \exp(i\theta' X)\right|\right\}\right]^2 \\ &\leq 2E\{(1 - \cos\theta'(X_n - X))\} \\ &\leq \theta' E\{(X_n - X)(X_n - X)'\}\theta \to 0.\end{aligned}$$

The interchangeability of l.i.m. and Gaussianity will be carried over to derivatives and integrals defined subsequently.

The next theorem relates the existence of a limit in m.s. with an operational moment condition, the proof of which can be found in Loève (1978, p. 135). See also Problem 1.4.

Theorem 2.2. *Let $\{Y_n(t)\}$, $t \in [a, b]$, be a sequence of q-dimensional stochastic processes for which $E(Y_n'(t)Y_n(t)) < \infty$ for all t. Then $\text{l.i.m.}_{n\to\infty}\{Y_n(t)\}$ exists if and only if $E(Y_m'(t)Y_n(t))$ converges to a finite function on $[a, b]$ as $m, n \to \infty$ in any manner.*

We now introduce some notions associated with the m.s. convergence described in Loève (1978) extending to vector - and matrix-valued stochastic processes.

Definition of m.s. Continuity. $\{Y(t)\}$ is m.s.continuous at $t \in [a, b]$ if

$$\underset{h\to 0}{\text{l.i.m.}}\, Y(t + h) = Y(t) \qquad \text{for} \qquad t + h \in [a, b]. \qquad \square$$

Here the left side is interpreted as follows. Let $\{h_n\}$ be any sequence which converges to 0 as $n \to \infty$ maintaining $t + h_n \in [a, b]$ and put $Y_n(t) = Y(t + h_n)$. Then the above definition means that $\text{l.i.m.}_{n\to\infty} Y_n(t) = Y(t)$. The process $\{Y(t)\}$ is m.s. continuous at t if and only if $E(Y'(s)Y(t))$ is continuous at (t, t) [Loève (1978, p. 136) and Problem 1.5]. Moreover, $\{Y(t)\}$ is m.s. continuous at every $t \in [a, b]$ if and only if $E(Y'(s)Y(t))$ is continuous at every $(t, t) \in [a, b] \times [a, b]$. In this case $E(Y'(s)Y(t))$ is necessarily continuous on $[a, b] \times [a, b]$ (Problem 1.6).

The m.s. continuity does not necessarily imply the sample path continuity or the m.s. differentiability defined below. An example is the sample paths of Poisson processes, though such processes are not dealt with in this book (Problem 1.7).

Definition of m.s. Differentiability. $\{Y(t)\}$ is m.s. differentiable at $t \in [a, b]$ if

$$\underset{h\to 0}{\text{l.i.m.}}\, \frac{Y(t + h) - Y(t)}{h} = \dot{Y}(t) \qquad \text{for} \qquad t + h \in [a, b]. \qquad \square$$

The left side is interpreted in the same way as above, putting $Y_n(t) = (Y(t + h_n) - Y(t))/h_n$. Because of Theorem 2.2, a necessary and sufficient condition for the m.s. differentiability of $\{Y(t)\}$ at t is that $E[(Y(t + h_1) - Y(t))'(Y(t + h_2) - Y(t))]/(h_1 h_2)$

converges to a finite number as $h_1, h_2 \to 0$ in any manner. If $\{Y(t)\}$ is m.s. differentiable and Gaussian, the derivative process $\{\dot{Y}(t)\}$ is also Gaussian since $Y_n(t) = (Y(t + h_n) - Y(t))/h_n$ is Gaussian and $\text{l.i.m.}_{n\to\infty}\{Y_n(t)\} = \{\dot{Y}(t)\}$. If $\{Y(t)\}$ is m.s. differentiable at s and t, then it can be shown (Problem 1.8) that

$$E(\dot{Y}(t)) = \frac{d}{dt} E(Y(t)), \tag{2.2}$$

$$E(\dot{Y}'(s)\dot{Y}(t)) = \frac{\partial^2}{\partial s \partial t} E(Y'(s)Y(t)). \tag{2.3}$$

Of course, the m.s. differentiability implies the m.s. continuity.

We also need to define the m.s. integral, but it is more involved and thus is treated separately. Since we need to consider the so-called Ito integral as well, which requires knowledge of the Brownian motion, we discuss the Brownian motion prior to integration.

PROBLEMS

1.1 Prove that if a scalar sequence $\{X_n\}$ converges in m.s. to X then X is unique with probability 1.

1.2 Show that the independent sequence $\{X_n\}$ defined by $P(X_n = \sqrt{n}) = 1/n$ and $P(X_n = 0) = 1 - 1/n$ does not converge in the m.s. sense, though it converges in probability to 0.

1.3 Prove Theorem 2.1.

1.4 Using Theorem 2.2, show that the one-dimensional independent stochastic process $\{Y_n(t)\}$ defined by

$$Y_n(t) = \begin{cases} n & P(Y_n(t) = n) = \dfrac{1}{n^2} \\ 1 & P(Y_n(t) = 1) = 1 - \dfrac{1}{n^2} \end{cases}$$

does not converge in the m.s. sense.

1.5 Prove that a q-dimensional stochastic process $\{Y(t)\}$ for which $E(Y'(t)Y(t)) < \infty$ is m.s. continuous at t if and only if $E(Y'(s)Y(t))$ is continuous at (t, t).

1.6 Prove that if $\{Y(t)\}$ is m.s. continuous at every $t \in [a, b]$ then $E(Y'(s)Y(t))$ is continuous on $[a, b] \times [a, b]$.

1.7 The scalar Poisson process $\{X(t)\}$ defined on $[0, \infty)$ has independent increments and is characterized by

$$P(X(t) = k) = \frac{e^{-\lambda t}(\lambda t)^k}{k!}, \quad (k = 0, 1, 2, \ldots).$$

Show that $\{X(t)\}$ is m.s. continuous at all t, but is nowhere m.s. differentiable.

1.8 Show that (2.2) and (2.3) hold if $\{Y(t)\}$ is m.s. differentiable.

1.9 Show that the one-dimensional stochastic process $\{Y(t)\}$ defined by $Y(t) = \cos(\omega t + U)$ is m.s. differentiable, where ω is a constant and U is uniformly distributed on $[0, \pi]$.

2.2 THE STANDARD BROWNIAN MOTION AND THE BROWNIAN BRIDGE

In this section we become more specific about the stochastic process $\{Y(t)\}$ in L_2 and introduce two important processes frequently used throughout this book.

Definition of the Standard Brownian Motion. We call a q-dimensional stochastic process $\{w(t)\}$ defined on $[0, 1]$ the q-dimensional standard Brownian motion if

i) $P(w(0) = 0) = 1$;
ii) $w(t_1) - w(t_0), w(t_2) - w(t_1), \cdots, w(t_n) - w(t_{n-1})$ are independent for any positive integer n and time points $0 \leq t_0 < t_1 < \cdots < t_n \leq 1$;
iii) $w(t) - w(s) \sim N(0, (t-s)I_q)$ for $0 \leq s < t \leq 1$. □

It follows that $\{w(t)\}$ is a zero-mean nonstationary Gaussian process with $\text{Cov}(w(s), w(t)) = \min(s, t)I_q$ and has stationary independent increments. It is known that the sample path of $\{w(t)\}$ is continuous with probability 1, while it is nowhere differentiable [see, for example, Billingsley (1986, Section 37)] and of unbounded variation on any interval subset of $[0, 1]$ (Hida (1980, p. 57)). In terms of the m.s. calculus, it can be shown (Problem 2.1) that

(a) $\{w(t)\}$ is m.s. continuous;
(b) $\{w(t)\}$ is nowhere m.s. differentiable;
(c) $\text{l.i.m.}_{\substack{n \to \infty \\ \Delta_n \to 0}} \sum_{i=1}^{n}(w(t_i) - w(t_{i-1}))'(w(t_i) - w(t_{i-1})) = (b-a)q$; where $0 \leq a = t_0 < t_1 < \cdots < t_n = b \leq 1$ and $\Delta_n = \max_i(t_i - t_{i-1})$.

The property (c) implies that $w(t)$ is of unbounded variation in the m.s. sense.

An example of the q-dimensional standard Brownian motion is [see Chan and Wei (1988) and Problem 2.2]

$$(2.4) \qquad w(t) = \sum_{n=1}^{\infty} \frac{\sqrt{2} \sin\left(n - \frac{1}{2}\right)\pi t}{\left(n - \frac{1}{2}\right)\pi} Z_n,$$

where $\{Z_n\} \sim \text{NID}(0, I_q)$. Note that the right hand variable exists in the m.s. sense because of the L_2-completeness theorem and $\sum_{n=1}^{\infty} 1/((n - \frac{1}{2})\pi)^2 = \frac{1}{2} < \infty$.

Definition of the Brownian Bridge. We call a q-dimensional stochastic process $\{\bar{w}(t)\}$ defined on $[0, 1]$ the q-dimensional Brownian bridge if

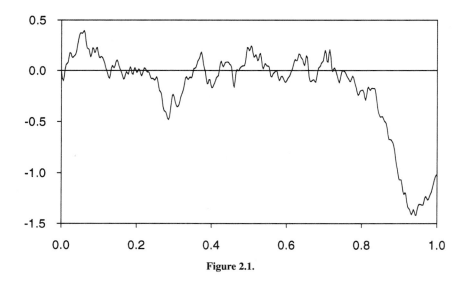

Figure 2.1.

i) $\{\bar{w}(t)\}$ is Gaussian;
ii) $E(\bar{w}(t)) = 0$ and $\text{Cov}(\bar{w}(s), \bar{w}(t)) = (\min(s,t) - st)I_q$ for $0 \le s, t \le 1$. □

It follows that $\{\bar{w}(t)\}$ is the q-dimensional standard Brownian motion $\{w(t)\}$ conditioned on $P(w(1) = 0) = 1$ (Problem 2.3). The processes $\{w(t) - tw(1)\}$ and

$$\bar{w}(t) = \sum_{n=1}^{\infty} \frac{\sqrt{2} \sin n\pi t}{n\pi} Z_n, \tag{2.5}$$

where $\{Z_n\} \sim \text{NID}(0, I_q)$, are examples of the q-dimensional Brownian bridge (Problem 2.4).

Figure 2.1 shows a sample path of the one-dimensional standard Brownian motion simulated from (2.4), while Figure 2.2 that of the one-dimensional Brownian bridge simulated from (2.5). We observe an erratic nature of both processes because of nondifferentiability.

PROBLEMS

2.1 When $\{w(t)\}$ is the q-dimensional standard Brownian motion, show that (a) $\{w(t)\}$ is m.s. continuous; (b) $\{w(t)\}$ is nowhere m.s. differentiable; and

(c) $\underset{\substack{n \to \infty \\ \Delta_n \to 0}}{\text{l.i.m.}} \sum_{i=1}^{n} (w(t_i) - w(t_{i-1}))'(w(t_i) - w(t_{i-1})) = (b-a)q$;

where $0 \le a = t_0 < t_1 < \cdots < t_n = b \le 1$ and $\Delta_n = \max_i(t_i - t_{i-1})$.

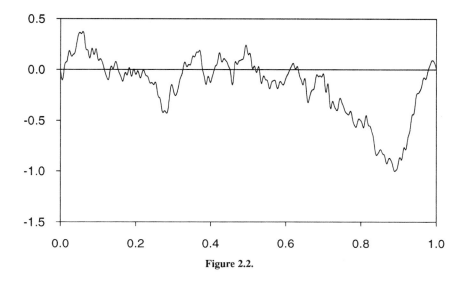

Figure 2.2.

2.2 Show that $\{w(t)\}$ defined in (2.4) is the q-dimensional standard Brownian motion, using the formula

$$\sum_{n=1}^{\infty} \frac{\cos\left(n - \frac{1}{2}\right)\pi x}{\left(\left(n - \frac{1}{2}\right)\pi\right)^2} = \frac{1}{2}(1 - |x|) \quad \text{for } |x| \leq 2.$$

2.3 Show that $\text{Cov}(w(s), w(t)|w(1) = 0) = (\min(s,t) - st)I_q$, where $\{w(t)\}$ is the q-dimensional standard Brownian motion.

2.4 Show that $\{\bar{w}(t)\}$ defined in (2.5) is the q-dimensional Brownian bridge, using the formula

$$\sum_{n=1}^{\infty} \frac{\cos n\pi x}{n^2 \pi^2} = \frac{1}{4}(x-1)^2 - \frac{1}{12} \quad \text{for } 0 \leq x \leq 2.$$

2.3 MEAN SQUARE INTEGRATION

Here we assume that $\{Y(t)\}$ is a q-dimensional stochastic process defined on $[a, b]$ for which $E(Y'(t)Y(t)) < \infty$. We can think of $Y(t)$ as the q-dimensional standard Brownian motion $w(t)$, but, of course, we can think of any $Y(t)$ with $E(Y'(t)Y(t)) < \infty$.

Soong (1973) gives a good exposition for m.s. integrals, which we follow, extending to the vector case. Consider a collection of all finite partitions $\{p_m\}$ of an interval $[a, b]$. The partition p_m is defined by

(2.6) $$p_m : \quad a = s_0 < s_1 < \cdots < s_m = b.$$

Put $\Delta_m = \max_i(s_i - s_{i-1})$ and let s_i' be an arbitrary point in the interval $[s_{i-1}, s_i)$. Suppose that $f(s, t)$ is an ordinary real-valued function defined on $[a, b] \times [a, b]$ and Riemann integrable with respect to s for every $t \in [a, b]$.

2.3.1 The Mean Square Riemann Integral

We define

$$(2.7) \qquad V(t) = \int_a^b f(s, t) Y(s)\, ds$$

if $\underset{\substack{m \to \infty \\ \Delta_m \to 0}}{\text{l.i.m.}} V_m(t)$ exists for any sequence of subdivisions p_m and for any $s_i' \in [s_{i-1}, s_i)$, $i = 1, \ldots, m$, where

$$V_m(t) = \sum_{i=1}^m f(s_i', t) Y(s_i')(s_i - s_{i-1}).$$

The m.s. integral (2.7) is independent of the sequence of subdivisions p_m as well as the positions of $s_i' \in [s_{i-1}, s_i)$. The existence condition for (2.7) can be related with the usual Riemann integral and is given by the following theorem [Soong (1973, Theorem 4.5.1) and Problem 3.1].

Theorem 2.3. *The* m.s. *integral (2.7) exists if and only if the ordinary double Riemann integral*

$$(2.8) \qquad \int_a^b \int_a^b f(r, t) f(s, t) E\big(Y'(r) Y(s)\big)\, dr\, ds$$

exists and is finite.

It is clear that, if the m.s. integral is well defined, "expectation" up to the second order and "integration" commute, i.e.,

$$E\left(\int_a^b f(s, t) Y(s)\, ds\right) = \int_a^b f(s, t) E\big(Y(s)\big)\, ds,$$

$$E\left(\int_a^b \int_a^b f(r, t) f(s, t) Y'(r) Y(s)\, dr\, ds\right) = \int_a^b \int_a^b f(r, t) f(s, t) E\big(Y'(r) Y(s)\big)\, dr\, ds.$$

If $\{Y(t)\}$ is Gaussian and $f(s, t) \equiv 1$, then the integral $V(t)$ in (2.7) is evidently Gaussian. Moreover we have [Soong (1973, Theorem 4.6.4)]

Theorem 2.4. *If the* m.s. *integral*

$$V(t) = \int_a^b f(s, t) Y(s)\, ds$$

of a normal process $\{Y(s)\}$ exists, then $\{V(t)\}$ is also normal with

$$E(V(t)) = \int_a^b f(s,t) E(Y(s)) \, ds,$$

$$E(V'(s)V(t)) = \int_a^b \int_a^b f(r,s) f(u,t) E(Y'(r)Y(u)) \, dr \, du.$$

Some other properties of the m.s. Riemann integral follow [Soong (1973) and Problem 3.2].

(a) If $Y(t)$ is m.s. continuous on $[a,b]$, $Y(t)$ is m.s. integrable on $[a,b]$;
(b) The m.s. integral of $Y(t)$ on $[a,b]$, if it exists, is unique;
(c) If $Y(t)$ is m.s. continuous on $[a,b]$, then $X(t) = \int_a^t Y(s) \, ds$ ($t \in [a,b]$) is m.s. differentiable on $[a,b]$ with $\dot{X}(t) = Y(t)$;
(d) If $\dot{Y}(t)$ is m.s. integrable on $[a,b]$, then

$$Y(t) - Y(a) = \int_a^t \dot{Y}(s) \, ds, \qquad [a,t] \subset [a,b].$$

We have already presented one-dimensional m.s. integrals in (1.11) and (1.18):

(2.9) $$V = \int_0^1 w^2(t) \, dt, \quad W = \int_0^1 \left(w(t) - tw(1)\right)^2 dt.$$

It is evident that V and W are well defined, and it holds (Problem 3.3) that $E(V) = \frac{1}{2}$, $E(V^2) = \frac{7}{12}$, $E(W) = \frac{1}{6}$, and $E(W^2) = \frac{1}{20}$. Moments of all orders are finite in these cases since the c.f.s of V and W were found in Chapter 1 to be $(\cos \sqrt{2i\theta})^{-1/2}$ and $(\sin \sqrt{2i\theta}/\sqrt{2i\theta})^{-1/2}$, respectively. As another example, consider the q-dimensional integrated Brownian motion

(2.10) $$V(t) = \int_0^t w(s) \, ds = \int_0^1 I_{[0,t]}(s) w(s) \, ds,$$

where $I_A(s)$ is the indicator function of the set A. It can be checked (Problem 3.4) that the integral $V(t)$ is well defined and $V(t) \sim N(0, t^3 I_q/3)$.

2.3.2 The Mean Square Riemann–Stieltjes Integral

We first define integrals of the following types:

(2.11) $$V_1 = \int_a^b f(s) \, dY(s),$$

(2.12) $$V_2 = \int_a^b Y(s)\,df(s),$$

when $\text{l.i.m.}_{\substack{m\to\infty \\ \Delta_m\to 0}} V_{1m}$ and $\text{l.i.m.}_{\substack{m\to\infty \\ \Delta_m\to 0}} V_{2m}$ exist, respectively, for any sequence of subdivisions p_m, where $f(t)$ is an ordinary real-valued function and

$$V_{1m} = \sum_{i=1}^m f(s_i')[Y(s_i) - Y(s_{i-1})],$$

$$V_{2m} = \sum_{i=1}^m Y(s_i')[f(s_i) - f(s_{i-1})].$$

The integrals in (2.11) and (2.12) are called the *m.s. Riemann–Stieltjes integrals* and are independent of the sequence of subdivisions as well as the positions of $s_i' \in [s_{i-1}, s_i)$.

Normality is retained in the integrals (2.11) and (2.12). Namely, V_1 and V_2 are both normal if $\{Y(t)\}$ is normal. The existence conditions for (2.11) and (2.12) are derived from Theorem 2.2 and stated as follows.

Theorem 2.5. *The m.s. Riemann–Stieltjes integrals in (2.11) and (2.12) exist if and only if the ordinary double Riemann–Stieltjes integrals*

(2.13) $$\int_a^b \int_a^b f(s)f(t) E(dY'(s)\,dY(t)) = E(V_1'V_1),$$

(2.14) $$\int_a^b \int_a^b E(Y'(s)Y(t))\,df(s)\,df(t) = E(V_2'V_2),$$

exist and are finite, respectively, where the integrals are defined in a self-evident manner as limits of approximating sums.

A sufficient condition for the existence of (2.11) is that $E(Y'(s)Y(t))$ is of bounded variation and $f(t)$ is continuous, while (2.12) exists if $E(Y'(s)Y(t))$ is continuous and $f(t)$ is of bounded variation. If these are the cases, (2.13) and (2.14) can be computed as

(2.15) $$E(V_1'V_1) = \lim_{\substack{m\to\infty \\ \Delta_m\to 0}} \sum_{i=1}^m \sum_{j=1}^m f(s_i)f(s_j) E\big[(Y(s_i) - Y(s_{i-1}))'(Y(s_j) - Y(s_{j-1}))\big],$$

(2.16) $$E(V_2'V_2) = \lim_{\substack{m\to\infty \\ \Delta_m\to 0}} \sum_{i=1}^m \sum_{j=1}^m E(Y'(s_i)Y(s_j))[f(s_i) - f(s_{i-1})][f(s_j) - f(s_{j-1})].$$

As an example, take $[a, b] = [0, 1]$ and consider

(2.17) $$V_1 = \int_0^1 (1 - s)\,dw(s), \quad (f(s) = 1 - s, \; Y(s) = w(s)).$$

Then V_1 is well defined (Problem 3.5) and is normal with $E(V_1) = 0$ and (2.15) leads us to

$$E(V_1' V_1) = \lim_{\substack{m \to \infty \\ \Delta_m \to 0}} \sum_{i=1}^m (1 - s_i)^2 (s_i - s_{j-1}) q$$

$$= q \int_0^1 (1 - s)^2\,ds = \frac{q}{3}.$$

Note that we also have $E(V_1 V_1') = I_q/3$ and that to compute $E(V_1' V_1)$ and $E(V_1 V_1')$ we may use the following relation:

$$E\big(dw'(s)\,dw(t)\big) = \begin{cases} q\,ds & (s = t) \\ 0 & (s \neq t). \end{cases}$$

The two integrals (2.11) and (2.12) can be combined together in the following theorem for *integration by parts* [Soong (1973, Theorem 4.5.3)].

Theorem 2.6. *If either V_1 in (2.11) or V_2 in (2.12) exists, then both integrals exist, and*

(2.18) $$\int_a^b f(s)\,dY(s) = \big[f(s)Y(s)\big]_a^b - \int_a^b Y(s)\,df(s).$$

For example we have

$$\int_0^1 dw(s) = w(1), \quad \int_0^1 (1 - s)\,dw(s) = \int_0^1 w(s)\,ds.$$

For later discussions we also consider the following m.s. double Riemann–Stieltjes integral defined on $[0, 1] \times [0, 1]$:

(2.19) $$X = \int_0^1 \int_0^1 K(s, t)\,dw'(s) H\,dw(t),$$

where $K(s, t)$ is symmetric on $[0, 1] \times [0, 1]$, while H is a $q \times q$ symmetric, constant matrix. Define the partition $p_{m,n}$ of $[0, 1] \times [0, 1]$ by

(2.20) $\quad p_{m,n} : 0 = s_0 < s_1 < \cdots < s_m = 1; \; 0 = t_0 < t_1 < \cdots < t_n = 1$

and put

$$\Delta_{m,n} = \max(s_1 - s_0, \ldots, s_m - s_{m-1}, t_1 - t_0, \ldots, t_n - t_{n-1}).$$

Then the m.s. double integral X in (2.19) is well defined if

$$(2.21) \quad X_{m,n} = \sum_{i=1}^{m} \sum_{j=1}^{n} K(s_i', t_j')[w(s_i) - w(s_{i-1})]'H[w(t_j) - w(t_{j-1})]$$

converges in m.s. as $m, n \to \infty$ and $\Delta_{m,n} \to 0$ for any sequence of subdivisions $p_{m,n}$. The integral X is independent of the choice of $p_{m,n}$ as well as the choice of $s_i' \in [s_{i-1}, s_i)$ and $t_j' \in [t_{j-1}, t_j)$.

The existence condition for the integral (2.19) is similarly given as in Theorem 2.5 using the quadruplex integral. A sufficient condition is that $K(s,t)$ is continuous on $[0,1] \times [0,1]$ (Problem 3.6). If this is the case, it holds (Problem 3.7) that

$$(2.22) \quad E(X) = \int_0^1 K(s,s)\,ds \times \text{tr}(H),$$

$$(2.23) \quad E(X^2) = 2\int_0^1 \int_0^1 K^2(s,t)\,ds\,dt \times \text{tr}(H^2) + \left(\int_0^1 K(s,s)\,dt \times \text{tr}(H)\right)^2.$$

Throughout this book, it will be assumed that $K(s,t)$ is symmetric and continuous on $[0,1] \times [0,1]$, although this assumption can be relaxed [see Anderson and Darling (1952)] for the existence of the integral (2.19).

Some examples of (2.19) follow. Suppose first that $K(s,t) = g(s)g(t)$, where $g(t)$ is a continuous function on $[0,1]$. Then the double integral can be reduced to the product of single integrals

$$(2.24) \quad \int_0^1 \int_0^1 K(s,t)\,dw'(s)H\,dw(t) = \int_0^1 g(s)\,dw'(s)H \int_0^1 g(t)\,dw(t),$$

whose distribution is equal to that of $\int_0^1 g^2(t)\,dt \sum_{i=1}^{q} \lambda_i Z_i^2$, where λ_is are the eigenvalues of H and $\{Z_i\} \sim \text{NID}(0,1)$ (Problem 3.8). Thus the distribution of (2.24) is that of a finite sum of weighted $\chi^2(1)$ random variables. In general the symmetric and continuous function $K(s,t)$ is said to be *degenerate* if it can be expressed as the sum of a finite number of terms, each of which is the product of a function of s only and a function of t only, that is,

$$K(s,t) = \sum_{i=1}^{n} g_i(s)g_i(t).$$

If this is not the case, $K(s,t)$ is said to be *nondegenerate*.

Let us present a few nondegenerate cases of $K(s,t)$ relating to the m.s. Riemann integral. For this purpose we first take up $K(s,t) = 1 - \max(s,t)$. Then, using integration by parts, we obtain

(2.25)
$$\int_0^1 \int_0^1 \left[1 - \max(s,t)\right] dw'(s)\, dw(t)$$
$$= \int_0^1 \left[(1-t) \int_0^t dw'(s) + \int_t^1 (1-s)\, dw'(s) \right] dw(t)$$
$$= \int_0^1 \left[\int_t^1 w'(s)\, ds \right] dw(t) = \int_0^1 w'(t) w(t)\, dt.$$

A more general relation may be derived in a reversed direction as follows. Let $g(t)$ be continuous on $[0,1]$. Then, putting $g_j = g(j/n)$ and $\Delta w_j = w(j/n) - w((j-1)/n)$, we have

$$\int_0^1 g(t) w'(t) w(t)\, dt = \operatorname*{l.i.m.}_{n \to \infty} \frac{1}{n} \sum_{l=1}^n \left[g_l \sum_{j=1}^l \Delta w'_j \sum_{k=1}^l \Delta w_k \right]$$
$$= \operatorname*{l.i.m.}_{n \to \infty} \sum_{j=1}^n \sum_{k=1}^n \left(\frac{1}{n} \sum_{l=\max(j,k)}^n g_l \right) \Delta w'_j \Delta w_k$$
$$= \operatorname*{l.i.m.}_{n \to \infty} \sum_{j=1}^n \sum_{k=1}^n \left(\int_{\max(j,k)/n}^1 g(r)\, dr \right) \Delta w'_j \Delta w_k$$
$$= \int_0^1 \int_0^1 \left(\int_{\max(s,t)}^1 g(r)\, dr \right) dw'(s)\, dw(t).$$

This implies that we can change the order of integration to get

$$\int_0^1 g(t) w'(t) w(t)\, dt = \int_0^1 g(t) \left[\int_0^t \int_0^t dw'(u)\, dw(v) \right] dt$$
$$= \int_0^1 \int_0^1 \left[\int_{\max(u,v)}^1 g(t)\, dt \right] dw'(u)\, dw(v).$$

Note that we have encountered in (1.11) the distributional equivalence of the first and last expressions in (2.25) for the scalar case. We have just seen that these are, in fact, the same in the m.s. sense. The first expression in (1.11) indicates that $K(s,t) = 1 - \max(s,t)$ is nondegenerate, which will be further studied in Chapter 5.

The following relations can also be obtained similarly (Problem 3.9).

$$\int_0^1 \int_0^1 [\min(s,t) - st] \, dw'(s) H \, dw(t) = \int_0^1 \tilde{w}'(t) H \tilde{w}(t) \, dt, \tag{2.26}$$

$$\int_0^1 \int_0^1 \left[\frac{1}{3} - \max(s,t) + \frac{s^2 + t^2}{2}\right] dw'(s) H \, dw(t) = \int_0^1 \bar{w}'(t) H \bar{w}(t) \, dt, \tag{2.27}$$

where $\tilde{w}(t) = w(t) - \int_0^1 w(t) \, dt$ is the demeaned Brownian motion and $\bar{w}(t) = w(t) - tw(1)$ is the Brownian bridge. We note in passing that (1.18) has given the distributional equivalence of the left side in (2.26) and the right side in (2.27). This, however, cannot be carried over to the equivalence in the m.s. sense. The functions $\min(s,t) - st$ and $\frac{1}{3} - \max(s,t) + (s^2 + t^2)/2$ are also nondegenerate. The proofs of these will also be deferred until Chapter 5.

PROBLEMS

3.1 Prove Theorem 2.3.

3.2 Prove that, if $\{Y(t)\}$ is m.s. continuous on $[a, b]$, then $\{Y(t)\}$ is m.s. integrable on $[a, b]$.

3.3 Show that $E(V) = \frac{1}{2}$, $E(V^2) = \frac{7}{12}$, $E(W) = \frac{1}{6}$, and $E(W^2) = \frac{1}{20}$ in (2.9).

3.4 Prove that the integral $V(t)$ in (2.10) is well defined and that $V(t) \sim N(0, t^3 I_q/3)$.

3.5 Prove that the integral V_1 in (2.17) is well defined.

3.6 Show that the integral in (2.19) exists and is finite if $K(s,t)$ is symmetric and continuous on $[0, 1] \times [0, 1]$.

3.7 Derive (2.22) and (2.23).

3.8 Show that

$$\mathcal{L}\left(\int_0^1 g(s) \, dw'(s) H \int_0^1 g(t) \, dw(t)\right) = \mathcal{L}\left(\int_0^1 g^2(t) \, dt \sum_{i=1}^q \lambda_i Z_i^2\right),$$

where λ_is are the eigenvalues of H and $\{Z_i\} \sim \text{NID}(0, 1)$.

3.9 Prove the relations in (2.26) and (2.27).

2.4 THE INTEGRATED BROWNIAN MOTION

As noted in the previous section, the Brownian motion itself is quite erratic in the sense that the sample path is nowhere differentiable, though it is continuous. The integral of the Brownian motion, however, becomes smooth. Here we consider a special case of such integrals defined by

(2.28) $$F_g(t) = \int_0^t F_{g-1}(s)\,ds, \quad (g = 1, 2, \ldots), \quad F_0(t) = w(t),$$

where $\{w(t)\}$ is the q-dimensional standard Brownian motion. The process $\{F_g(t)\}$ defined recursively in this way may be called the *g-fold integrated Brownian motion* [Chan and Wei (1988)]. It is clear that the m.s. integral in (2.28) is well defined for any positive integer g. Thus $F_g(t)$ with $g \geq 1$ is g-times m.s. continuously differentiable, though $F_0(t) = w(t)$ is not differentiable.

Figure 2.3 shows a sample path of $\{F_1(t)\}$ for $q = 1$, which was simulated, on the basis of (2.4), from

$$F_1(t) = \int_0^t w(s)\,ds = \sum_{n=1}^{\infty} \frac{\sqrt{2}}{(n-\frac{1}{2})^2 \pi^2} \left\{1 - \cos\left(n - \frac{1}{2}\right)\pi t\right\} Z_n.$$

Figure 2.4 shows a sample path of $\{F_2(t)\}$ for $q = 1$, simulated similarly from

$$F_2(t) = \int_0^t F_1(s)\,ds = \sum_{n=1}^{\infty} \frac{\sqrt{2}}{(n-\frac{1}{2})^3 \pi^3} \left\{\left(n - \frac{1}{2}\right)\pi t - \sin\left(n - \frac{1}{2}\right)\pi t\right\} Z_n.$$

In comparison with the sample path of $\{F_0(t)\} = \{w(t)\}$ shown in Figure 2.1, we recognize that the sample paths of $\{F_g(t)\}$ become smoother as g gets large because of g-times differentiability. We also observe a decrease in the variation of $\{F_g(t)\}$ with g. In fact, it holds that when $q = 1$, $V(F_0(t)) = t$, $V(F_1(t)) = t^3/3$, and $V(F_2(t)) = $

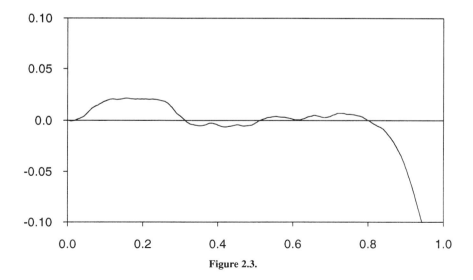

Figure 2.3.

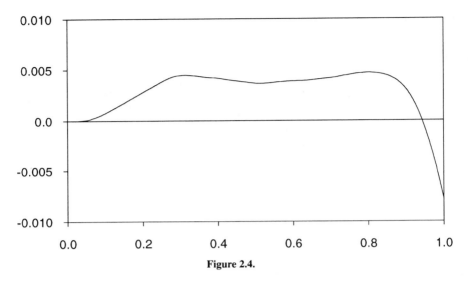

Figure 2.4.

$t^5/20$. In general we have $V(F_g(t)) = t^{2g+1}/((2g+1)(g!)^2)$, which can be proved easily by using Theorem 2.7 below.

We also define

(2.29) $\quad \bar{F}_g(t) = F_g(t) - tF_g(1) \quad\quad$ g-fold integrated Brownian bridge,

(2.30) $\quad \tilde{F}_g(t) = F_g(t) - \int_0^1 F_g(s)\,ds \quad$ g-fold integrated demeaned Brownian motion.

For $\{F_g(t)\}$ we have the following equivalent expression, the proof of which is left as Problem 4.1.

Theorem 2.7. *The stochastic process $\{F_g(t)\}$ in (2.28) can be expressed by the Riemann–Stieltjes integral as*

(2.31) $$F_g(t) = \int_0^t \frac{(t-s)^g}{g!}\,dw(s), \quad (g = 0, 1, \ldots).$$

The superiority of the expression (2.31) over (2.28) is that the former may be defined for real g replacing $g!$ by $\Gamma(g+1)$. In fact, the expression (2.31) with $g!$ replaced by $\Gamma(g+1)$ was earlier introduced as the Holmgren–Riemann–Liouville fractional integral [Mandelbrot and Van Ness (1968)], where g may be any value greater than $-\frac{1}{2}$. The fractional case is certainly interesting, but the analysis needs a separate treatment. We maintain in this book that g is a nonnegative integer.

Using (2.31), we obtain the following integral relations concerning $\{F_g(t)\}$, $\{\bar{F}_g(t)\}$, and $\{\tilde{F}_g(t)\}$ (Problem 4.2):

$$\int_0^1 F_g(t)\,dt = \int_0^1 \frac{(1-s)^{g+1}}{(g+1)!}\,dw(s),$$

$$\int_0^1 \bar{F}_g(t)\,dt = \int_0^1 \left[\frac{(1-s)^{g+1}}{(g+1)!} - \frac{(1-s)^g}{2(g!)}\right]dw(s),$$

$$\int_0^1 \tilde{F}_g(t)\,dt = 0,$$

$$\int_0^1 F'_g(t) F_g(t)\,dt = \int_0^1 \int_0^1 K_g(s,t)\,dw'(s)\,dw(t),$$

$$\int_0^1 \bar{F}'_g(t) \bar{F}_g(t)\,dt = \int_0^1 \int_0^1 [K_g(s,t) + L_g(s,t)]\,dw'(s)\,dw(t),$$

$$\int_0^1 \tilde{F}'_g(t) \tilde{F}_g(t)\,dt = \int_0^1 \int_0^1 \left[K_g(s,t) - \frac{((1-s)(1-t))^{g+1}}{((g+1)!)^2}\right]dw'(s)\,dw(t),$$

where

(2.32) $$K_g(s,t) = \int_{\max(s,t)}^1 \frac{((u-s)(u-t))^g}{(g!)^2}\,du,$$

$$L_g(s,t) = \frac{(1-s)^{g+2} + (1-t)^{g+2}}{g!(g+2)!} + \frac{((1-s)(1-t))^g}{g!}\left(\frac{1}{3(g!)} - \frac{2-s-t}{(g+1)!}\right).$$

It can be checked easily that, when $g = 0$, the above formulas reduce to those obtained in the previous section. The expressions on the right sides above may be more useful for computing moments and deriving their distributions, which will be discussed in Chapter 5.

We note in passing that the integrated Brownian motion naturally appears from a simple I(d) process:

$$(1-L)^d y_t = \varepsilon_t, \quad (t = 1,\ldots,T),$$

where L is the lag operator, d is a positive integer, $\{\varepsilon_t\} \sim$ i.i.d.$(0, I_q)$, and the initial values y_t ($t = -(d-1), -(d-2), \ldots, 0$) are all set at 0. We shall show in Chapter 3 that

$$\mathcal{L}\left(\frac{1}{T^{d-1/2}} y_T\right) \to \mathcal{L}(F_{d-1}(1)),$$

$$\mathcal{L}\left(\frac{1}{T^{d+1/2}} \sum_{t=1}^{T} y_t\right) \to \mathcal{L}\left(\int_0^1 F_{d-1}(t)\, dt\right),$$

$$\mathcal{L}\left(\frac{1}{T^{2d}} \sum_{t=1}^{T} y'_t y_t\right) \to \mathcal{L}\left(\int_0^1 F'_{d-1}(t) F_{d-1}(t)\, dt\right),$$

and so on. Thus a simple I(d) process $\{y_t\}$ divided by $T^{d-1/2}$ is essentially the $(d-1)$-fold integrated Brownian motion.

We remark finally that, for $g \geq 1$, $F_g(t)$ is of bounded variation and

$$\operatorname*{l.i.m.}_{\substack{m \to \infty \\ \Delta_m \to 0}} \sum_{i=1}^{m} (F_g(t_i) - F_g(t_{i-1}))'(F_g(t_i) - F_g(t_{i-1})) = 0.$$

Thus we can define the m.s. integral $\int_0^t F_g(s)\, dF'_g(s)$. Then we can use integration by parts to obtain (Problem 4.3)

Theorem 2.8. *For any positive integer g, it holds that*

$$(2.33) \quad \int_0^t F_g(s)\, dF'_g(s) + \left(\int_0^t F_g(s)\, dF'_g(s)\right)' = F_g(t) F'_g(t), \quad (g = 1, 2, \ldots).$$

Note that the formula (2.33) does not apply to the case $g = 0$, that is, the case where $F_g(t) = w(t)$. We need a separate analysis for this case, which is discussed in the next section.

PROBLEMS

4.1 Derive the formula (2.31) from (2.28).

4.2 Prove that

$$\int_0^1 \bar{F}_g(t)\, dt = \int_0^1 \left[\frac{(1-s)^{g+1}}{(g+1)!} - \frac{(1-s)^g}{2(g!)}\right] dw(s),$$

$$\int_0^1 \tilde{F}'_g(t)\tilde{F}_g(t)\, dt = \int_0^1 \int_0^1 \left[K_g(s,t) - \frac{((1-s)(1-t))^{g+1}}{((g+1)!)^2}\right] dw'(s)\, dw(t),$$

where $\bar{F}_g(t)$ and $\tilde{F}_g(t)$ are defined in (2.29) and (2.30), respectively, while $K_g(s,t)$ is defined in (2.32).

4.3 Prove that the relation (2.33) holds.

2.5 THE MEAN SQUARE ITO INTEGRAL: THE SCALAR CASE

In this section we deal with the integral of the form

(2.34) $$U(t) = \int_0^t X(s)\, dw(s),$$

where $0 \le t \le 1$, $\{w(t)\}$ is the one-dimensional standard Brownian motion, and $\{X(t)\}$ is a scalar stochastic process in L_2.

We define the *Ito integral* for (2.34) as follows: Let the partition p_m of $[0,t]$ be $0 = s_0 < s_1 < \cdots < s_m = t$ and put $\Delta_m = \max_i(s_i - s_{i-1})$. Then, form the random variable

(2.35) $$U_m(t) = \sum_{i=1}^m X(s_{i-1})\bigl(w(s_i) - w(s_{i-1})\bigr).$$

The Ito integral for (2.34) is said to be well defined if $\underset{\substack{m\to\infty\\ \Delta_m\to 0}}{\text{l.i.m.}}\, U_m(t)$ exists. The following theorem is the existence and uniqueness theorem for the Ito integral. For the proof see Jazwinski (1970, Theorem 4.2) and Soong (1973, Theorem 5.2.1).

Existence and Uniqueness Theorem for the Ito Integral. *Suppose that*

i) $X(t)$ is m.s. continuous on $[0,1]$;
ii) $X(t)$ is independent of $\{w(t_j) - w(t_i);\ 0 \le t \le t_i \le t_j \le 1\}$ for all $t \in [0,1]$.

Then the Ito integral defined above exists and is unique.

We notice in (2.35) that the value of $X(s)$ taken in the interval $[s_{i-1}, s_i)$ is $X(s_{i-1})$, unlike the m.s. Riemann or Riemann–Stieltjes integral. If we take $s_i'\, (\ne s_{i-1})$ in $[s_{i-1}, s_i)$, then the m.s. limit will be different, as exemplified later. It is clear that $E(U(t)) = 0$ and, for $h > 0$,

$$E\bigl[(U(t+h) - U(t))^2\bigr] = \int_t^{t+h} E(X^2(s))\, ds$$

$$\le h \max_{t \le s \le t+h} E(X^2(s))$$

so that $U(t)$ is m.s. continuous. The m.s. differentiability of $U(t)$, however, is not ensured in the usual sense. Nonetheless we formally write (2.34) as $dU(t) = X(t)\, dw(t)$.

As an example of the Ito integral let us consider

(2.36) $$U(t) = \int_0^t w(s)\, dw(s).$$

Since $X(t) = w(t)$ satisfies conditions i) and ii) in the above theorem, the sum

$$U_m(t) = \sum_{i=1}^{m} w(s_{i-1})(w(s_i) - w(s_{i-1}))$$

$$= -\frac{1}{2}\left[\sum_{i=1}^{m}(w(s_i) - w(s_{i-1}))^2 - \sum_{i=1}^{m} w^2(s_i) + \sum_{i=1}^{n} w^2(s_{i-1})\right]$$

$$= \frac{1}{2}w^2(t) - \frac{1}{2}\sum_{i=1}^{m}(w(s_i) - w(s_{i-1}))^2$$

must converge in the m.s. sense. We have already shown (see Problem 2.1) that

(2.37) $$\operatorname*{l.i.m.}_{\substack{m\to\infty \\ \Delta_m\to 0}} \sum_{i=1}^{m}(w(s_i) - w(s_{i-1}))^2 = t,$$

where $0 = s_0 < s_1 < \cdots < s_m = t$ and $\Delta_m = \max_i (s_i - s_{i-1})$. Therefore

(2.38) $$\int_0^t w(s)\,dw(s) = \frac{1}{2}(w^2(t) - t).$$

We now formally have $d(\frac{1}{2}(w^2(t) - t)) = w(t)\,dw(t)$ or

(2.39) $$d(w^2(t)) = 2w(t)\,dw(t) + dt.$$

This is a simplified version of the Ito calculus discussed in the next section.

We now show that the m.s. limit of a sum such as (2.35) crucially depends on the choice of values of $X(s)$ in the intervals $[s_{i-1}, s_i)$. In fact, we have

(2.40) $$\operatorname*{l.i.m.}_{\substack{m\to\infty \\ \Delta_m\to 0}} \sum_{i=1}^{m} w((1-a)s_{i-1} + as_i)(w(s_i) - w(s_{i-1})) = \frac{1}{2}(w^2(t) - t) + at,$$

where $0 \le a \le 1$ (Problem 5.1). The case $a = 0$ corresponds to the Ito integral, while the integral with $a = \frac{1}{2}$ is called the *Stratonovich integral*. Some simple and convenient properties of the Ito integral are that $U(t)$ in (2.34) has a zero mean and is a *martingale*, that is, for any $s \le t$, $E(U(t)|U(s)) = U(s)$ with probability 1 (Problem 5.2).

According to the definition of the Ito integral, the following relations can also be established (Problem 5.3).

(2.41) $$\int_0^t X(s)(dw(s))^2 = \int_0^t X(s)\,ds,$$

(2.42) $$\int_0^t X(s)(dw(s))^3 = 0,$$

where $X(t)$ satisfies conditions i) and ii) in the existence and uniqueness theorem, while

$$\int_0^t X(s)(dw(s))^j = \underset{\substack{m\to\infty \\ \Delta_m \to 0}}{\text{l.i.m.}} \sum_{i=1}^m X(s_{i-1})(w(s_i) - w(s_{i-1}))^j.$$

In particular, putting $X(s) \equiv 1$, we formally have $(dw(s))^2 = ds$ and $(dw(s))^3 = 0$.

In connection with the Ito Integral in (2.34), we also wish to define integrals like

$$\int_0^t w(s)\,dX(s), \qquad \int_0^t X(s)\,dX(s)$$

for some stochastic process $\{X(t)\}$. Of course these integrals are not always well defined. In Section 2.7 we define the above integrals using the Ito calculus.

We note in passing that the Ito integral in its simplest form naturally appears in

(2.43) $\quad \mathcal{L}\left(\dfrac{1}{T}\sum_{t=1}^T y_{t-1}\varepsilon_t\right) \to \mathcal{L}\left(\int_0^1 w(t)\,dw(t)\right) = \mathcal{L}\left(\dfrac{1}{2}(w^2(1)-1)\right),$

where $y_t = y_{t-1} + \varepsilon_t$, $y_0 = 0$, and $\{\varepsilon_t\} \sim$ i.i.d.(0,1). We have already encountered a similar expression in (1.51). Although we discuss weak convergence as in (2.43) more generally in Chapter 3, it is easy to prove (2.43) at this stage (Problem 5.4).

PROBLEMS

5.1 Show that the relation (2.40) holds.

5.2 Prove that $U(t)$ in (2.34) is an m.s. continuous martingale.

5.3 Derive the formulas in (2.41) and (2.42).

5.4 Prove the weak convergence in (2.43).

2.6 THE MEAN SQUARE ITO INTEGRAL: THE VECTOR CASE

In this section we extend the scalar m.s. Ito integral to the vector case. Although various cases may be possible, we concentrate here on the integral of the form

(2.44) $\qquad\qquad V(t) = \int_0^t Y(s)\,dw'(s),$

where $0 \le t \le 1$, $\{w(t)\}$ is the q-dimensional standard Brownian motion, and $\{Y(t)\}$ is a p-dimensional stochastic process, each component of which belongs to L_2. Note that $V(t)$ in (2.44) is a $p \times q$ matrix. As in the scalar case, we assume that

i) $\{Y(t)\}$ is m.s. continuous on $[0,1]$;
ii) $Y(t)$ is independent of $\{w(t_j) - w(t_i); 0 \leq t \leq t_i \leq t_j \leq 1\}$ for all $t \in [0, 1]$.

Then the m.s. Ito integral for (2.44) exists and is unique.

As the simplest example, let us put $Y(t) = w(t)$ so that

$$(2.45) \qquad V(t) = \int_0^t w(s)\, dw'(s).$$

The (i, j)th element $V_{ij}(t)$ of $V(t)$ with $i \neq j$ and $t = 1$ has the following distributional relation (Problem 6.1):

$$(2.46) \qquad \mathcal{L}(V_{ij}(1)) = \mathcal{L}\left(\frac{1}{2} \sum_{n=1}^{\infty} \frac{\xi_{1n}^2 - \xi_{2n}^2}{(n - \frac{1}{2})\pi}\right), \qquad (i \neq j),$$

where $(\xi_{1n}, \xi_{2n})' \sim \text{NID}(0, I_2)$. This relation was earlier introduced in (1.49). Note that, unlike the scalar case, we have

$$\int_0^t w(s)\, dw'(s) \neq \frac{1}{2}\left[w(t)w'(t) - tI_q\right],$$

since the right side is symmetric while the left side is not. We have (Problem 6.2)

$$(2.47) \qquad \int_0^t w(s)\, dw'(s) + \left(\int_0^t w(s)\, dw'(s)\right)' = w(t)w'(t) - tI_q.$$

The m.s. Ito integral for (2.45) with $t = 1$ appears in

$$(2.48) \qquad \mathcal{L}\left(\frac{1}{T} \sum_{t=1}^{T} y_{t-1}\varepsilon_t'\right) \to \mathcal{L}\left(\int_0^1 w(s)\, dw'(s)\right),$$

where $y_t = y_{t-1} + \varepsilon_t$, $y_0 = 0$ and $\{\varepsilon_t\} \sim$ i.i.d.$(0, I_q)$. The proof for (2.48) is very involved, unlike the scalar case. This will be discussed in Chapter 3.

As another example, let us put $Y(t) = Aw(t)$ with A being any $q \times q$ constant matrix and consider $\text{tr}(V(t))$ in (2.44), which is

$$(2.49) \qquad \text{tr}\left(\int_0^t A w(s)\, dw'(s)\right) = \int_0^t w'(s)A'\, dw(s).$$

We have already presented this integral in (1.58) and (1.59), where the former corresponds to A's being symmetric, while the latter corresponds to A's being skew symmetric ($A' = -A$). It can be shown (Problem 6.3) that, when A is symmetric with $\text{tr}(A) = 0$, (2.49) with $t = 1$ follows the distribution of a finite sum of weighted, independent $\chi^2(1)$ random variables. When A is skew symmetric, the distribution of

THE ITO CALCULUS

(2.49) is equal, in general, to that of an infinite sum of independent random variables, as was indicated in (1.59).

The m.s. Ito integral for (2.49) with $t = 1$ appears in

$$(2.50) \quad \mathcal{L}\left(\frac{1}{T}\sum_{t=1}^{T} y'_{t-1} A' \varepsilon_t\right) \to \mathcal{L}\left(\int_0^1 w'(s) A' \, dw(s)\right),$$

where $y_t = y_{t-1} + \varepsilon_t$, $y_0 = 0$, and $\{\varepsilon_t\} \sim$ i.i.d.$(0, I_q)$. The weak convergence in (2.50) will also be discussed in Chapter 3.

PROBLEMS

6.1 Prove the distributional equivalence in (2.46).

6.2 Prove that the relation in (2.47) holds.

6.3 Prove that the distribution of (2.49) is equal to that of a finite sum of weighted independent $\chi^2(1)$ random variables when A is symmetric and $\text{tr}(A) = 0$.

2.7 THE ITO CALCULUS

This section applies the m.s. integrals defined in previous sections to the scalar integral equation of the form

$$(2.51) \quad X(t) = X(0) + \int_0^t \mu(X(s), s) \, ds + \int_0^t \sigma(X(s), s) \, dw(s), \quad 0 \le t \le 1,$$

where we notice that there are two types of integrals. One is the Riemann integral and the other the Ito integral. Whether these integrals can be defined in the m.s. sense, and, more importantly, whether this integral equation has a unique m.s. solution $X(t)$ can be answered in the affirmative by the following theorem, whose proof is given in Jazwinski (1970, p. 105).

Theorem 2.9. *Suppose that*

i) $X(0)$ *is any random variable with* $E(X^2(0)) < \infty$ *and is independent of* $\{w(t_j) - w(t_i); \ 0 \le t_i \le t_j \le 1\}$;

ii) *there is a positive constant K such that*

$$|\mu(x, t) - \mu(y, t)| \le K|x - y|, \quad |\sigma(x, t) - \sigma(y, t)| \le K|x - y|,$$
$$|\mu(x, s) - \mu(x, t)| \le K|s - t|, \quad |\sigma(x, s) - \sigma(x, t)| \le K|s - t|,$$
$$|\mu(x, t)| \le K(1 + x^2)^{1/2}, \quad |\sigma(x, t)| \le K(1 + x^2)^{1/2}.$$

Then (2.51) has a unique m.s. continuous solution $X(t)$ on $[0, 1]$ such that $X(t) - X(0)$ is independent of $\{w(t_j) - w(t_i);\ 0 \leq t \leq t_i \leq t_j\}$ for every $t \in [0, 1]$.

This theorem ensures that the two integrals appearing in (2.51) are well defined in the m.s. sense. Note also that the solution process $\{X(t)\}$ is not m.s. differentiable in general. Nonetheless we write (2.51) as

$$(2.52) \quad dX(t) = \mu(X(t), t)\, dt + \sigma(X(t), t)\, dw(t), \quad 0 \leq t \leq 1.$$

Here $dX(t)$ is called the *stochastic differential* of $X(t)$ and we call this equation the Ito *stochastic differential equation* (SDE), which is always understood in terms of the integral equation (2.51).

The idea of stochastic differentials can be further developed for functions of $X(t)$ and t. Namely, we can consider another SDE that $f(X(t), t)$ satisfies on the basis of the following theorem [Jazwinski (1970, p.112)].

Ito's Theorem. *Suppose that $X(t)$ has the stochastic differential (2.52), where conditions* i) *and* ii) *in Theorem 2.9 are satisfied. Let $f(x, t)$ denote a continuous function on $(-\infty, \infty) \times [0, 1]$ with continuous partial derivatives $f_x = \partial f(x, t)/\partial x$, $f_{xx} = \partial^2 f(x, t)/\partial x^2$, and $f_t = \partial f(x, t)/\partial t$. Assume further that $f_t(X(t), t)$ and $f_{xx}(X(t), t)\sigma^2(X(t), t)$ are m.s. Riemann integrable. Then $f(X(t), t)$ has the stochastic differential*

$$df(X(t), t) = f_x(X(t), t)\, dX(t) + \left(f_t(X(t), t) + \frac{1}{2} f_{xx}(X(t), t)\sigma^2(X(t), t) \right) dt.$$
(2.53)

Ito's theorem tells us that if $X(t)$ satisfies the SDE in (2.52) then $f(X(t), t)$ satisfies the SDE in (2.53). The SDE in (2.53) should be interpreted in terms of an integral equation like (2.51).

Some implications of Ito's theorem follow. If we expand $df(X(t), t)$ formally as

$$df \sim f_x dX(t) + f_t\, dt + \frac{1}{2}\left(f_{xx} d^2 X(t) + 2 f_{xt}\, dX(t)\, dt + f_{tt} d^2 t \right),$$

then (2.53) implies that we may put $d^2 X(t) = \sigma^2(X(t), t)\, dt$, $dX(t)\, dt = 0$ and $d^2 t = 0$. The corresponding integral equation to (2.53) may be written as

$$(2.54) \quad \int_a^b f_x(X(t), t)\, dX(t) = f(X(b), b) - f(X(a), a)$$

$$- \int_a^b \left(f_t(X(t), t) + \frac{1}{2} f_{xx}(X(t), t)\sigma^2(X(t), t) \right) dt.$$

The integral on the left side has never appeared before, but it can also be defined by using the Riemann integral as on the right side. This is an important message that Ito's theorem conveys to us. Ito's theorem is also useful for solving SDEs, as will be exemplified shortly.

Three examples of stochastic differentials follow (Problem 7.1).

$$(2.55) \quad d\left(X^n(t)\right) = nX^{n-1}(t)\,dX(t) + \frac{n(n-1)}{2}X^{n-2}(t)\sigma^2\left(X(t),t\right)dt,$$

$$(2.56) \quad d\left(w^n(t)\right) = nw^{n-1}(t)\,dw(t) + \frac{n(n-1)}{2}w^{n-2}(t)\,dt,$$

$$(2.57) \quad d(e^{w(t)}) = e^{w(t)}dw(t) + \frac{1}{2}e^{w(t)}dt.$$

Note that, when $n = 2$, (2.55) implies

$$(2.58) \quad \int_0^t X(s)\,dX(s) = \frac{1}{2}\left(X^2(t) - X^2(0)\right) - \frac{1}{2}\int_0^t \sigma^2\left(X(s),s\right)ds,$$

which reduces to (2.38) if $X(t) = w(t)$. The relation in (2.58) is quite important and will be frequently used in later chapters. We shall also use the following relation derived from (2.56):

$$(2.59) \quad \int_0^t w^n(s)\,dw(s) = \frac{1}{n+1}w^{n+1}(t) - \frac{n}{2}\int_0^t w^{n-1}(s)\,ds,$$

which also reduces to (2.38) if $n = 1$. The formulas (2.58) and (2.59) enable us to convert (extended) Ito integrals into the usual Riemann integrals.

Ito's theorem can be used to obtain a solution to the SDE (2.52). Two examples follow. The first is

$$(2.60) \quad dX(t) = X(t)\,dw(t),$$

which has a solution $X(t) = X(0)\exp\{w(t) - t/2\}$ (Problem 7.2). The process $\{X(t)\}$ in this case is called the *geometric Brownian motion*. The second is

$$(2.61) \quad dX(t) = \left(\alpha X(t) + \beta\right)dt + \gamma\,dw(t),$$

where α, β and γ are constants. We obtain (Problem 7.3)

$$(2.62) \quad X(t) = e^{\alpha t}X(0) + \frac{\beta(e^{\alpha t}-1)}{\alpha} + \gamma e^{\alpha t}\int_0^t e^{-\alpha s}\,dw(s).$$

When $\beta = 0$ and $\gamma = 1$, we have

Definition of the Ornstein–Uhlenbeck Process. The process $\{X(t)\}$ defined by

$$(2.63) \quad dX(t) = \alpha X(t)\,dt + dw(t) \iff X(t) = e^{\alpha t}X(0) + e^{\alpha t}\int_0^t e^{-\alpha s}\,dw(s)$$

is called the Ornstein–Uhlenbeck (O–U) process, where $X(0)$ is independent of increments of $\{w(t)\}$. □

Note that $\{X(t)\}$ reduces to $\{w(t)\}$ when $\alpha = 0$ and $X(0) = 0$. Note also that, because of (2.63), the following integral can be well defined:

$$\int_0^t w(s)\,dX(s) = \alpha \int_0^t w(s) X(s)\,ds + \int_0^t w(s)\,dw(s),$$

where the integral on the left side has never appeared before. It can also be shown (Problem 7.4) that, for the O–U process (2.63), $E(X(t)) = e^{\alpha t} E(X(0))$ and

(2.64) $\operatorname{Cov}(X(s), X(t)) = e^{\alpha(s+t)} \left[V(X(0)) + \dfrac{1 - e^{-2\alpha \min(s,t)}}{2\alpha} \right].$

It turns out that, if α is positive, $V(X(t))$ is greater than $V(w(t))$ and increasing with t. On the other hand, if $\alpha < 0$ and $X(0) \sim N(0, -1/(2\alpha))$, then $\{X(t)\}$ is a Gaussian, stationary process on $[0,1]$ with $E(X(t)) = 0$ and $\operatorname{Cov}(X(s), X(t)) = -e^{\alpha|s-t|}/(2\alpha)$ so that $V(X(t)) = -1/(2\alpha)$.

An example of the O–U process can be constructed, by substituting (2.4) into (2.63), as

(2.65) $X(t) = e^{\alpha t} X(0)$

$$+ \sqrt{2} \sum_{n=1}^{\infty} \frac{\alpha e^{\alpha t} - \alpha \cos\left(n - \tfrac{1}{2}\right)\pi t + \left(n - \tfrac{1}{2}\right)\pi \sin\left(n - \tfrac{1}{2}\right)\pi t}{\left(n - \tfrac{1}{2}\right)^2 \pi^2 + \alpha^2} Z_n.$$

Figure 2.5 shows a sample path of the nonstationary O–U process simulated from (2.65) with $X(0) = 0$ and $\alpha = 1$, while Figure 2.6 shows a sample path of the

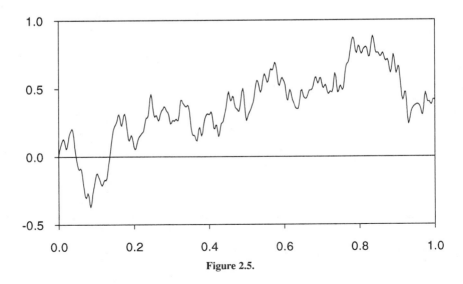

Figure 2.5.

THE ITO CALCULUS

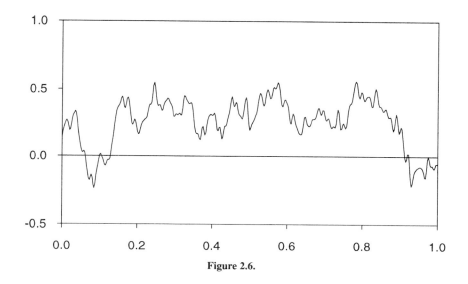

Figure 2.6.

stationary O–U process with $X(0) \sim N(0, 1/(-2\alpha))$ and $\alpha = -1$. The O–U process covers a wider class of stochastic processes than the Brownian motion.

The O–U process $\{X(t)\}$ in (2.63) will play an important role in subsequent chapters. It will be shown in Chapter 3 that $\{X(t)\}$ naturally appears in

$$\mathcal{L}\left(\frac{1}{\sqrt{T}} y_{[tT]}\right) \to \mathcal{L}(X(t)),$$

where $[x]$ denotes the greatest integer not exceeding x, and

(2.66) $$y_t = \left(1 + \frac{\alpha}{T}\right) y_{t-1} + \varepsilon_t, \quad (t = 1, \ldots, T),$$

with $\{\varepsilon_t\} \sim$ i.i.d. $(0, 1)$, while the initial value y_0 is of the form $y_0 = \sqrt{T} X(0)$. The discrete-time process $\{y_t\}$ in (2.66) may be called the *near random walk* or the *near integrated process*. We shall also have

(2.67) $$\mathcal{L}\left(\frac{1}{T^2} \sum_{t=1}^{T} y_t^2\right) \to \mathcal{L}\left(\int_0^1 X^2(t)\, dt\right).$$

It can be shown (Problem 7.5) that the m.s. Riemann integral in (2.67) can be expressed by using the m.s. Riemann-Stieltjes integral as

(2.68) $$\int_0^1 X^2(t)\, dt = \frac{e^{2\alpha} - 1}{2\alpha} X^2(0) + X(0) \int_0^1 \frac{e^{\alpha(2-t)} - e^{\alpha t}}{\alpha} dw(t)$$
$$+ \int_0^1 \int_0^1 \frac{e^{\alpha(2-s-t)} - e^{\alpha|s-t|}}{2\alpha} dw(s)\, dw(t).$$

Two approaches to deriving the c.f. of (2.68) will be presented in Chapters 4 and 5.

The discussions so far have been restricted to scalar cases but can be extended to vector cases. We briefly mention an extended version of Ito's theorem [Jazwinski (1970, p. 112)].

An Extended Version of Ito's Theorem. *Let the p-dimensional stochastic process $\{Y(t)\}$ be the unique solution of the Ito SDE:*

$$(2.69) \qquad dY(t) = \mu(Y(t), t)\, dt + G(Y(t), t)\, dw(t),$$

where μ is $p \times 1$ and G is $p \times q$, while $\{w(t)\}$ is the q-dimensional standard Brownian motion. Let $g(y, t)$ be a real-valued function of $y : p \times 1$ and $t \in [0, 1]$ with continuous partial derivatives $g_y = \partial g/\partial y$, $g_{yy} = \partial^2 g/(\partial y \partial y')$, and $g_t = \partial g/\partial t$. Assume further that $g_t(Y(t), t)$ and $G(Y(t), t)G'(Y(t), t)g_{yy}(Y(t), t)$ are m.s. Riemann integrable. Then $g(Y(t), t)$ has the stochastic differential

$$(2.70) \qquad dg = g_y' dY(t) + \left(g_t + \frac{1}{2} \operatorname{tr}(g_{yy} G G')\right) dt.$$

As an application of this theorem let us consider the differential of $Y_1(t)Y_2(t)$, where $Y(t) = (Y_1(t), Y_2(t))'$ satisfies (2.69). Since $g(y, t) = y_1 y_2$, we have

$$g_y = \begin{pmatrix} y_2 \\ y_1 \end{pmatrix}, \quad g_{yy} = \begin{pmatrix} 0 & 1 \\ 1 & 0 \end{pmatrix}, \quad g_t = 0$$

so that (2.70) yields

$$(2.71) \quad d(Y_1(t)Y_2(t)) = Y_1(t)\, dY_2(t) + Y_2(t)\, dY_1(t) + G_1'(Y(t), t) G_2(Y(t), t)\, dt,$$

where $G(Y(t), t) = (G_1(Y(t), t), G_2(Y(t), t))'$.

Similarly, we can show (Problem 7.6) that

$$(2.72) \qquad d\left(e^{\alpha t} w(t) \int_0^t e^{-\alpha s} dw(s)\right) = w(t)\, dw(t) + dt$$

$$+ e^{\alpha t}(\alpha w(t)\, dt + dw(t)) \int_0^t e^{-\alpha s}\, dw(s),$$

$$(2.73) \quad d\left(\exp\left(w(t) \int_0^t e^{-\alpha s} dw(s)\right)\right) = \left[\left(e^{-\alpha t} w(t) + \int_0^t e^{-\alpha s} dw(s)\right) dw(t)\right.$$

$$\left. + e^{-\alpha t} dt + \frac{1}{2}\left(e^{-\alpha t} w(t) + \int_0^t e^{-\alpha s} dw(s)\right)^2 dt\right] \exp\left(w(t) \int_0^t e^{-\alpha s} dw(s)\right),$$

where $\{w(t)\}$ is the one-dimensional standard Brownian motion.

PROBLEMS

7.1 Obtain stochastic differentials in (2.55), (2.56), and (2.57).
7.2 Show that the SDE in (2.60) has a solution $X(t) = X(0) \exp\{w(t) - t/2\}$.
7.3 Show that the solution to the SDE in (2.61) is given by (2.62).
7.4 Derive the covariance in (2.64).
7.5 Prove that the relation in (2.68) holds.
7.6 Derive the relations in (2.72) and (2.73), where $\{w(t)\}$ is the one-dimensional standard Brownian motion.

CHAPTER 3

Functional Central Limit Theorems

Weak convergence of a stochastic process defined on a function space is discussed. In doing so, we explore various weak convergence results generically called *functional central limit theorems* or *invariance principles*. Emphasis is placed on how to apply those theorems to deal with statistics arising from nonstationary linear time series models. It turns out that, in most cases, the continuous mapping theorem is quite powerful for obtaining limiting random variables of statistics in the sense of weak convergence. In some cases, however, the continuous mapping theorem does not apply. In those cases, limiting forms involve the Ito integral.

3.1 FUNCTION SPACE C

As a sequel to the last chapter, we continue to assume that the stochastic process $\{X(t)\}$ belongs to L_2. This assumption, however, needs to be strengthened for subsequent discussions. Let $C = C[0, 1]$ be the space of all real-valued continuous functions defined on $[0, 1]$ and $(C, \mathcal{B}(C))$ a *measurable space*, where $\mathcal{B}(C)$ is the σ-*field* generated by the subsets of C that are open with respect to the *uniform metric* ρ defined by

$$(3.1) \qquad \rho(x, y) = \sup_{0 \le t \le 1} |x(t) - y(t)|$$

for any $x, y \in C$. The uniform metric $\rho(x, y)$ is a continuous function of x and y (Problem 1.1 in this chapter), that is, $|\rho(x, y) - \rho(\tilde{x}, \tilde{y})| \to 0$ as $\rho(x, \tilde{x}) \to 0$ and $\rho(y, \tilde{y}) \to 0$.

Then we assume that the stochastic process $X = \{X(t)\}$ to be treated below is a measurable mapping from an arbitrary *probability space* $(\Omega, \mathcal{F}, P)$ into C, that is, $X^{-1}(A) \in \mathcal{F}$ for every $A \in \mathcal{B}(C)$. Thus we exclusively consider those stochastic processes which belong to L_2 and have continuous sample paths.

We do not extend the space C in this book to $D = D[0, 1]$, which is the space of all real-valued functions on $[0,1]$ that are right continuous and have finite left limits. This is just because we avoid paying the cost of greater topological complexity associated with metrics with which we equip D.

The space C is known to be *complete* and *separable* under ρ, where completeness means that each fundamental sequence, which is a sequence $\{x_n(t)\}$ that satisfies $\rho(x_m, x_n) \to 0$ as $m, n \to \infty$, converges to some point of the space, while separability means that the space contains a countable, dense set [see, for more details, Billingsley (1968, p. 220) and Problem 1.2]. In this sense the space C is much like the real line, but each element of C is a function so that C is a function space and the distance of two elements of C is defined by the uniform metric ρ. Completeness and separability facilitate subsequent discussions concerning weak convergence of stochastic processes. In particular, it is because of separability that $\rho(X, Y) = \sup_{0 \le t \le 1} |X(t) - Y(t)|$ becomes a random variable when $\{X(t)\}$ and $\{Y(t)\}$ are stochastic processes in C [Billingsley (1968, p. 25)].

The space C, however, is not *compact*, where compactness means that any sequence in the space contains a convergent subsequence. In fact, if we think of $\{x_n(t)\} = \{n\}$, where $x_n(t) = n$ ($n = 1, 2, \ldots$) is a constant-valued element of C, it is clear that $\{n\}$ does not contain any convergent subsequence. The situation is again the same as for sequences on the real line.

When we discuss convergence in distribution, the lack of compactness of the space of distribution functions becomes serious. As an example, consider a sequence $\{F_n\}$ of distribution functions defined by

$$F_n(x) = \begin{cases} 1 & x \ge n, \\ 0 & x < n. \end{cases}$$

It is evident that $F_n(x)$ converges to $G(x) \equiv 0$ for each x, but the limiting function $G(x)$ is not a distribution function. This example shows that the space of distribution functions is not compact. Thus we need a condition like compactness that will prevent mass from escaping to infinity [Shiryayev (1984, p.315)].

A further difficulty arises if we deal with weak convergence of stochastic processes in C. Even if a sequence $\{F_{n,t}\}$ of distribution functions corresponding to $\{X_n(t)\}$ converges properly for each t, it does not necessarily imply weak convergence of $\{X_n(t)\}$ as a whole. In the next section, we shall give a sufficient condition for the proper convergence.

PROBLEMS

1.1 Prove that $\rho(x, y)$ is a continuous function of x and y.
1.2 Show that the space C is complete and separable under the uniform metric ρ.

3.2 WEAK CONVERGENCE OF STOCHASTIC PROCESSES IN C

The stochastic process $X = \{X(t) : 0 \le t \le 1\}$ induces a *probability measure* $Q = PX^{-1}$ on $(C, \mathcal{B}(C))$ by the relation

$$Q(A) = P(X \in A) = P(X^{-1}(A)), \quad A \in \mathcal{B}(C).$$

Similarly any sequence $\{X_n\}$ of stochastic processes, where each stochastic process X_n is given by $\{X_n(t) : 0 \leq t \leq 1\}$, induces a sequence $\{Q_n\}$ of probability measures determined by

$$Q_n(A) = P(X_n \in A) = P(X_n^{-1}(A)), \quad A \in \mathcal{B}(C).$$

We say that $\{X_n\}$ *converges in distribution* to X, and we write $\mathcal{L}(X_n) \to \mathcal{L}(X)$ if

$$\lim_{n \to \infty} E\{f(X_n)\} = \lim_{n \to \infty} \int_C f(x) Q_n(dx)$$

$$= \int_C f(x) Q(dx)$$

$$= E\{f(X)\}$$

for each f in the class of bounded, continuous real functions defined on C. There are some other equivalent conditions for $\mathcal{L}(X_n) \to \mathcal{L}(X)$ [Billingsley (1968, p. 24)].

The difficulty with $\mathcal{L}(X_n) \to \mathcal{L}(X)$ is that, unlike random variables or vectors, the finite-dimensional distributions $\mathcal{L}(X_n(t_1), \ldots, X_n(t_k))$ for each finite k and each collection $0 \leq t_1 < t_2 < \cdots < t_k \leq 1$ by no means determine $\mathcal{L}(X)$. Namely the finite-dimensional sets do not form a *convergence determining class* in C, although they do form a *determining class* [Billingsley (1968, p. 19)]. An example is also found in Billingsley (1968, p. 20). We require a condition referred to as *relative compactness* of $\{X_n\}$, which means that the sequence $\{Q_n\}$ of induced probability measures on $(C, \mathcal{B}(C))$ contains a subsequence which converges weakly to a probability measure on $(C, \mathcal{B}(C))$. The limiting measure need not be a member of $\{Q_n\}$. This is the reason why the adjective "relative" comes in. The relative compactness condition is difficult to verify in general, but, by Prohorov's theorem [Billingsley (1968, p. 35)], that condition is equivalent, under completeness and separability of C, to the more operational condition *tightness*. This condition says that, for each positive ε, there exists a compact set K such that $Q_n(K) > 1 - \varepsilon$ for all n. Tightness prohibits probability mass from escaping to infinity and stipulates that the X_n do not oscillate too violently.

The following is a fundamental theorem concerning weak convergence of $\{X_n\}$ in C [Billingsley (1968, p. 54)].

Theorem 3.1. *If the finite-dimensional distributions $\mathcal{L}(X_n(t_1), \ldots, X_n(t_k))$ converge weakly to $\mathcal{L}(X(t_1), \ldots, X(t_k))$, and if $\{X_n\}$ is tight, then $\mathcal{L}(X_n) \to \mathcal{L}(X)$.*

The tightness condition can be further made operational. It is proved in Billingsley (1968, p. 55) [see also Hall and Heyde (1980, p. 275)] that $\{X_n\}$ is tight if and only if

i) $P(|X_n(0)| > a) \to 0$ uniformly in n as $a \to \infty$;
ii) for each $\varepsilon > 0$, $P(\sup_{|s-t|<\delta} |X_n(s) - X_n(t)| > \varepsilon) \to 0$ uniformly in n as $\delta \to 0$.

Note that condition i) means tightness of $X_n(0)$. A sufficient moment condition for ii) to hold is also given in Billingsley (1968, p. 95).

In subsequent sections we consider various examples of $\{X_n\}$, for which weak convergence is discussed. To this end we will not go into details, but will only describe weak convergence results useful for later chapters. Details can be found in Chapter 2 of Billingsley (1968) and Chapter 4 of Hall and Heyde (1980).

3.3 THE FUNCTIONAL CENTRAL LIMIT THEOREM

As a sequel to the previous section, we continue to consider a sequence $\{X_n\}$ of stochastic processes in C. Here we take up a typical example of $\{X_n\}$. Suppose that $u_1, u_2, \ldots,$ be random variables on $(\Omega, \mathcal{F}, P)$ and define the partial sum

$$(3.2) \qquad S_j = S_{j-1} + u_j, \qquad (S_0 = 0),$$

$$= u_1 + \cdots + u_j, \quad (j = 1, \ldots, n).$$

We then construct, for $(j-1)/n \le t \le j/n$,

$$(3.3) \quad X_n(t) = \frac{1}{\sqrt{n}} S_{j-1} + n\left(t - \frac{j-1}{n}\right) \frac{1}{\sqrt{n}} u_j$$

$$= X_n\left(\frac{j-1}{n}\right) + n\left(t - \frac{j-1}{n}\right)\left(X_n\left(\frac{j}{n}\right) - X_n\left(\frac{j-1}{n}\right)\right)$$

$$= \frac{1}{\sqrt{n}} S_j + n\left(t - \frac{j}{n}\right) \frac{1}{\sqrt{n}} u_j$$

$$= X_n\left(\frac{j}{n}\right) + n\left(t - \frac{j}{n}\right)\left(X_n\left(\frac{j}{n}\right) - X_n\left(\frac{j-1}{n}\right)\right)$$

$$= \frac{1}{\sqrt{n}} \sum_{j=1}^{[nt]} u_j + (nt - [nt])\frac{1}{\sqrt{n}} u_{[nt]+1},$$

where $X_n(0) = 0$ and $X_n(1) = S_n/\sqrt{n}$. Note that, for each t, $X_n(t)$ is a random variable on $(\Omega, \mathcal{F}, P)$ and that $X_n(t)$ is a continuous function on $[0,1]$ for each $\omega \in \Omega$. Figure 3.1 explains how to construct $X_n(t)$ on $[(j-1)/n, j/n]$ for fixed $\omega \in \Omega$. In any case, $X_n = \{X_n(t)\}$ is a stochastic process in C and is called a *partial sum process*.

We now become more specific about random variables $u_1, u_2, \ldots,$ in (3.3) to obtain the *Functional Central Limit Theorem* (FCLT) or the *Invariance Principle* (IP) for $\{X_n\}$. Donsker (1951, 1952) provided the first general FCLT when $\{u_j\}$ follows i.i.d.$(0, \sigma^2)$, which we state below as Donsker's theorem. The proof starts by showing

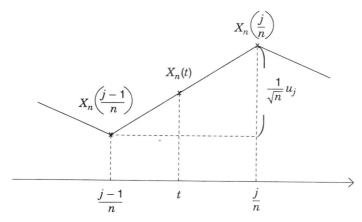

Figure 3.1.

that the finite-dimensional distributions converge weakly and then goes on proving that the sequence in question is tight [see, for details, Billingsley (1968, p. 68)].

Theorem 3.2 (**Donsker's Theorem**). *Suppose that the partial sum process* $\{X_n(t)\}$ *is defined in (3.3) with* $\{u_j\}$ *being* i.i.d.$(0, \sigma^2)$, *where* $\sigma^2 > 0$. *Then*

$$\mathcal{L}\left(\frac{X_n}{\sigma}\right) \to \mathcal{L}(w),$$

where $w = \{w(t)\}$ *is the one-dimensional standard Brownian motion on* $[0,1]$.

For later purposes we also consider the stochastic process of the following form:

(3.4) $$\bar{X}_n(t) = \frac{1}{\sqrt{n}} \sum_{j=1}^{[nt]} (u_j - \bar{u}) + (nt - [nt]) \frac{1}{\sqrt{n}} (u_{[nt]+1} - \bar{u})$$

$$= X_n(t) - tX_n(1),$$

where $\bar{u} = \sum_{j=1}^{n} u_j/n$. The process $\{\bar{X}_n(t)\}$ may be referred to as the *mean-corrected partial sum process*, for which we have the following result.

Corollary 3.1. *Suppose that the mean-corrected partial sum process* $\{\bar{X}_n(t)\}$ *is defined in (3.4) with* $\{u_j\}$ *being* i.i.d.$(0, \sigma^2)$, *where* $\sigma^2 > 0$. *Then*

$$\mathcal{L}\left(\frac{\bar{X}_n}{\sigma}\right) \to \mathcal{L}(\bar{w}),$$

where $\bar{w} = \{\bar{w}(t)\} = \{w(t) - tw(1)\}$ *is the one-dimensional Brownian bridge on* $[0,1]$.

Similarly, if we consider the demeaned partial sum process

$$(3.5) \quad \tilde{X}_n(t) = \frac{1}{\sqrt{n}} \sum_{j=1}^{[nt]} u_j - \frac{1}{n} \sum_{j=1}^{n} \left(\frac{1}{\sqrt{n}} \sum_{i=1}^{j} u_i \right) + \left(nt - [nt] \right) \frac{1}{\sqrt{n}} u_{[nt]+1}$$

$$= X_n(t) - \frac{1}{n} \sum_{j=1}^{n} X_n\left(\frac{j}{n} \right),$$

we have the following result.

Corollary 3.2. *Suppose that the demeaned partial sum process $\{\tilde{X}_n(t)\}$ is defined in (3.5) with $\{u_j\}$ being i.i.d.$(0, \sigma^2)$, where $\sigma^2 > 0$. Then*

$$\mathcal{L}\left(\frac{\tilde{X}_n}{\sigma} \right) \to \mathcal{L}(\tilde{w}),$$

where $\tilde{w} = \{\tilde{w}(t)\} = \{w(t) - \int_0^1 w(t)\, dt\}$ is the demeaned Brownian motion on $[0,1]$.

Note that the Brownian bridge and the demeaned Brownian motion are continuous functionals of the Brownian motion. Corollaries 3.1 and 3.2 may be proved from Donsker's theorem using the continuous mapping theorem discussed in the next section.

Donsker's FCLT was further developed in several directions where $\{u_j\}$ in (3.3) is a sequence of dependent random variables. Billingsley (1968, Chapter 4) established the FCLT under ϕ-*mixing conditions*, which was largely extended by McLeish (1975a,b, 1977) under the *mixingale conditions*. There is now a vast literature concerning mixing sequences, and Yoshihara (1992, 1993) gives excellent reviews of the literature. We, however, do not take this approach because it seems difficult to accommodate the limit theory on mixing sequences to linear processes discussed subsequently. In fact, not all linear processes satisfy *strong mixing conditions* [see, for example, Withers (1981) and Athreya and Pantula (1986)]. We will take an alternative approach that was advocated by Phillips and Solo (1992) and is especially designed for the case where $\{u_j\}$ follows a linear process.

3.4 CONTINUOUS MAPPINGS AND RELATED THEOREMS

In this section we present some useful theorems to establish the FCLT for linear processes dealt with subsequently. The first theorem, referred to as the *continuous mapping theorem*, is quite important [see, for the proof, Billingsley (1968, p.29)].

Theorem 3.3. *Let $h(x)$ be a continuous function defined on C. If $\mathcal{L}(X_n) \to \mathcal{L}(X)$, then $\mathcal{L}(h(X_n)) \to \mathcal{L}(h(X))$.*

This theorem is well known if the X_n are random variables [see Rao (1973, p. 124) and Problem 4.1].

To utilize fully the continuous mapping theorem, we need to define the Riemann integral in terms of convergence with probability 1. To see this consider

$$(3.6) \qquad h(X) = \int_0^1 X^2(t)\,dt,$$

where $X(t)$ belongs to C. Although we naturally require $X(t) \in L_2$ for every t, it does not necessarily imply $X^2(t) \in L_2$. Thus the integral in (3.6) cannot be defined as the m.s. Riemann integral unless $X^2(t) \in L_2$. In the present case, however, $X^2(t) = X^2(t, \omega)$ is a continuous function of t with $\omega \in \Omega$ fixed so that the Riemann integral $\int_0^1 X^2(t, \omega)\,dt$ is well defined at each ω. On the other hand, the collection of these integrals at all ω can be defined as limits of sequences of $\mathcal{F}$-measurable Riemann sums

$$\sum_j X^2(t'_j, \omega)(t_j - t_{j-1}), \qquad (t'_j \in [t_{j-1}, t_j)),$$

since a uniform sequence of partitions of $[0, 1]$ and uniform values t'_j can be used in obtaining $\int_0^1 X^2(t, \omega)\,dt$ at all ω. Thus $h(X)$ in (3.6) is $\mathcal{F}$-measurable and is independent of the sequences of partitions and the values t'_j involved in the limiting procedures. Thus the integral in (3.6) is well defined in the sense described above, which we call the *Riemann sample integral*. More details can be found in Soong (1973).

The continuity of $h(x)$ in (3.6) at a point $x \in C$ can be proved as follows. For x fixed consider

$$|h(y) - h(x)| = \left| \int_0^1 \left\{ (x(t) - y(t))^2 - 2x(t)(x(t) - y(t)) \right\} dt \right|$$

$$\leq \rho^2(x, y) + 2\rho(x, y) \int_0^1 |x(t)|\,dt,$$

which evidently tends to 0 as $y \to x$ ($\rho(x, y) \to 0$).

Three more examples of $h(x)$ follow (Problem 4.2).

$$(3.7) \quad h_1(x) = \sup_{0 \leq t \leq 1} x(t), \quad h_2(x) = \sup_{0 \leq t \leq 1} |x(t)|, \quad h_3(x) = (h_1(x), h_2(x)).$$

Note that $h(x)$ may be vector-valued, as in $h_3(x)$. As a special case of Theorem 3.3, suppose that $\mathcal{L}(X_n) \to \mathcal{L}(c)$, where c is a constant on $[0,1]$, which belongs to C. Then $\{X_n\}$ converges in probability to c (Problem 4.3) in the sense that $\rho(X_n, c) \to 0$ in probability, that is,

$$P(\rho(X_n, c) \geq \varepsilon) = P\left(\sup_{0 \leq t \leq 1} |X_n(t) - c| \geq \varepsilon \right) \to 0$$

for each positive ε. Therefore we have the following theorem.

Theorem 3.4. *Let $h(x)$ be a continuous function defined on C. If $\{X_n\}$ converges in probability to a constant c, then $h(X_n)$ converges in probability to $h(c)$.*

This theorem is also standard if the X_n are random variables [see Rao (1973, p. 124) and Problem 4.4]. The continuous mapping theorem can be extended to the case where h is not necessarily continuous. The following theorem is proved in Billingsley (1968, p. 31).

Theorem 3.5. *Let $h(x)$ be a measurable mapping of C into another metric space S' (with metric ρ' and σ-field $\mathcal{B}(S')$) and let D_h be the set of discontinuities of h. If $\mathcal{L}(X_n) \to \mathcal{L}(X)$ and $P(X \in D_h) = 0$, then $\mathcal{L}(h(X_n)) \to \mathcal{L}(h(X))$.*

As an example consider $h(w) = 1/\int_0^1 w^2(t)\,dt$, where $\{w(t)\}$ is the standard Brownian motion. Then $h(w)$ is measurable and $P(w = 0) = 0$ so that $P(w \in D_h) = 0$; hence the above theorem applies.

The next theorem relates convergence in probability on C to weak convergence. The proof is given in Billingsley (1968, p. 25).

Theorem 3.6. *If $\mathcal{L}(X_n) \to \mathcal{L}(X)$ and $\rho(X_n, Y_n) \to 0$ in probability, then $\mathcal{L}(Y_n) \to \mathcal{L}(X)$.*

This theorem is also well known if X_n and Y_n are random variables [see Rao (1973, p. 122) and Problem 4.5].

We now explore some applications of the theorems presented above. Let us consider a model

$$(3.8) \qquad y_j = \rho y_{j-1} + \varepsilon_j, \quad y_0 = 0, \quad (j = 1, \ldots, T),$$

where the true value of ρ is 1 and $\{\varepsilon_j\} \sim$ i.i.d.$(0, \sigma^2)$. The statistics dealt with here are

$$S_{1T} = \frac{1}{T^2} \sum_{j=1}^{T} y_j^2,$$

$$S_{2T} = \frac{1}{T^2} \sum_{j=1}^{T} (y_j - \bar{y})^2,$$

$$S_{3T} = T(\hat{\rho} - 1),$$

$$S_{4T} = \frac{\hat{\rho} - 1}{\hat{\sigma}\bigg/\sqrt{\sum_{j=2}^{T} y_{j-1}^2}},$$

where

$$\hat{\rho} = \sum_{j=2}^{T} y_{j-1} y_j \bigg/ \sum_{j=2}^{T} y_{j-1}^2 , \quad \hat{\sigma}^2 = \frac{1}{T-1} \sum_{j=2}^{T} (y_j - \hat{\rho} y_{j-1})^2 .$$

The above statistics S_{1T} through S_{4T} are functions of $\{y_j\}$. Noting that $y_j = \varepsilon_1 + \cdots + \varepsilon_j$, we define, for $(j-1)/T \le t \le j/T$,

(3.9) $$X_T(t) = \frac{1}{\sqrt{T}} \sum_{i=1}^{j} \varepsilon_i + T\left(t - \frac{j}{T}\right) \frac{1}{\sqrt{T}} \varepsilon_j$$

so that $X_T(j/T) = y_j/\sqrt{T}$ and $\mathcal{L}(X_T/\sigma) \to \mathcal{L}(w)$.
For S_{1T} we have

$$S_{1T} = \frac{1}{T} \sum_{j=1}^{T} X_T^2\left(\frac{j}{T}\right)$$
$$= h_1(X_T) + R_{1T} ,$$

where

$$h_1(x) = \int_0^1 x^2(t)\, dt , \quad x \in C ,$$

(3.10) $$R_{1T} = \frac{1}{T} \sum_{j=1}^{T} X_T^2\left(\frac{j}{T}\right) - \int_0^1 X_T^2(t)\, dt$$

$$= \sum_{j=1}^{T} \int_{(j-1)/T}^{j/T} \left[X_T^2\left(\frac{j}{T}\right) - X_T^2(t)\right] dt .$$

Since $\mathcal{L}(h_1(X_T/\sigma)) \to \mathcal{L}(h_1(w))$ by the continuous mapping theorem, $\mathcal{L}(S_{1T}/\sigma^2) \to \mathcal{L}(h_1(w))$ follows from Theorem 3.6 if R_{1T} converges in probability to 0. In fact, the integrand in (3.10) has the following bound:

$$\left| X_T^2\left(\frac{j}{T}\right) - X_T^2(t) \right| \le 2 \sup_{0 \le t \le 1} |X_T(t)| \max_{1 \le j \le T} \frac{|\varepsilon_j|}{\sqrt{T}} ,$$

where we have $\mathcal{L}(\sup_{0 \le t \le 1} |X_T(t)/\sigma|) \to \mathcal{L}(\sup_{0 \le t \le 1} |w(t)|)$. Here $P(\sup_{0 \le t \le 1} |w(t)| \ge b)$ is known as a boundary-crossing probability [Shorack and Wellner (1986, p. 34)]. We also have (Problem 4.6)

(3.11) $$\max_{1 \le j \le T} \frac{|\varepsilon_j|}{\sqrt{T}} \to 0 \quad \text{in probability.}$$

Thus it must hold (Problem 4.7) that

$$\sup_{0 \le t \le 1} |X_T(t)| \max_{1 \le j \le T} \frac{|\varepsilon_j|}{\sqrt{T}} \to 0 \quad \text{in probability.} \tag{3.12}$$

Therefore we obtain

$$\mathcal{L}\left(\frac{S_{1T}}{\sigma^2}\right) = \mathcal{L}\left(\frac{1}{T^2\sigma^2}\sum_{j=1}^{T} y_j^2\right) \to \mathcal{L}\left(\int_0^1 w^2(t)\,dt\right).$$

Note that this last integral expression is well defined both in the m.s. sense and in the sense of convergence with probability 1, while $h_1(X_T)$ is not necessarily m.s. integrable.

Similarly we can show (Problem 4.8) that

$$\mathcal{L}\left(\frac{S_{2T}}{\sigma^2}\right) = \mathcal{L}\left(\frac{1}{T^2\sigma^2}\sum_{j=1}^{T}(y_j - \bar{y})^2\right) \tag{3.13}$$

$$\to \mathcal{L}\left(\int_0^1 w^2(t)\,dt - \left(\int_0^1 w(t)\,dt\right)^2\right)$$

$$= \mathcal{L}\left(\int_0^1 (w(t) - tw(t))^2\,dt\right).$$

Establishing the last equality in (3.13), however, is not straightforward as far as $\{\varepsilon_j\} \sim$ i.i.d.$(0, \sigma^2)$. Under this assumption, weak convergence can be established as in the second line in (3.13). Thus we may consider a special case where $\{\varepsilon_j\} \sim \text{NID}(0, \sigma^2)$, from which the last expression in (3.13) results. This is a reason why Donsker's FCLT is also called the invariance principle. Once the weak convergence is established, we can find the limiting distribution in an easy special case.

We next consider

$$S_{3T} = T(\hat{\rho} - 1)$$

$$= \frac{1}{T}\sum_{j=2}^{T} y_{j-1}(y_j - y_{j-1}) \Big/ \left[\frac{1}{T^2}\sum_{j=2}^{T} y_{j-1}^2\right]$$

$$= U_T/V_T,$$

where

$$\text{(3.14)} \quad U_{T} = \frac{1}{T}\sum_{j=2}^{T} y_{j-1}(y_{j}-y_{j-1})$$

$$= \frac{1}{2}X_{T}^{2}(1) - \frac{1}{2T}\sum_{j=1}^{T}\varepsilon_{j}^{2},$$

$$\text{(3.15)} \quad V_{T} = \frac{1}{T^{2}}\sum_{j=2}^{T} y_{j-1}^{2}$$

$$= \frac{1}{T}\sum_{j=1}^{T} X_{T}^{2}\left(\frac{j}{T}\right) - \frac{1}{T^{2}} y_{T}^{2}.$$

Let us define a continuous function $h_3(x) = (h_{31}(x), h_{32}(x))$ for $x \in C$, where

$$h_{31}(x) = \frac{1}{2}x^{2}(1), \qquad h_{32}(x) = \int_{0}^{1} x^{2}(t)\,dt.$$

Then we have

$$U_{T} = h_{31}(X_{T}) - \frac{1}{2T}\sum_{j=1}^{T}\varepsilon_{j}^{2},$$

$$V_{T} = h_{32}(X_{T}) + R_{1T} - \frac{1}{T^{2}} y_{T}^{2},$$

where R_{1T} is defined in (3.10).

It is now easy to deduce that

$$\mathcal{L}\left(\frac{U_{T}}{\sigma^{2}}, \frac{V_{T}}{\sigma^{2}}\right) \to \mathcal{L}\left(h_{31}(w) - \frac{1}{2}, h_{32}(w)\right)$$

and Theorem 3.5 yields

$$\mathcal{L}(S_{3T}) = \mathcal{L}(T(\hat{\rho}-1))$$

$$\to \mathcal{L}\left(\frac{h_{31}(w) - \frac{1}{2}}{h_{32}(w)}\right)$$

$$= \mathcal{L}\left(\frac{\frac{1}{2}(w^{2}(1) - 1)}{\int_{0}^{1} w^{2}(t)\,dt}\right)$$

$$= \mathcal{L}\left(\frac{\int_{0}^{1} w(t)\,dw(t)}{\int_{0}^{1} w^{2}(t)\,dt}\right).$$

Finally we deal with the t-ratio-like statistic defined by

$$S_{4T} = \frac{\hat{\rho} - 1}{\hat{\sigma} \Big/ \sqrt{\sum_{j=2}^{T} y_{j-1}^2}} = \frac{U_T / V_T}{\hat{\sigma} / \sqrt{V_T}}$$

$$= \frac{U_T}{\hat{\sigma} \sqrt{V_T}},$$

where U_T and V_T are defined in (3.14) and (3.15), respectively. Since it can be shown (Problem 4.9) that

(3.16) $$\hat{\sigma}^2 = \frac{1}{T-1} \sum_{j=2}^{T} (y_j - \hat{\rho} y_{j-1})^2 \to \sigma^2$$

in probability, Theorem 3.5 again yields

$$\mathcal{L}(S_{4T}) \to \mathcal{L}\left(\frac{\int_0^1 w(t)\,dw(t)}{\sqrt{\int_0^1 w^2(t)\,dt}} \right).$$

As seen above, the FCLT combined with the continuous mapping theorem is powerful and plays an important role in deriving weak convergence results for various statistics. Since the present approach always starts with constructing a partial sum process in C, while our concern is a statistic, this approach may be referred to as the *stochastic process approach*. This terminology will also be used in Chapter 4 for another purpose. If the statistic under consideration is a quadratic form or the ratio of quadratic forms, we need not make such a detour as is involved in the present approach. An alternative approach will be presented in Section 5.6 of Chapter 5.

PROBLEMS

4.1 Prove Theorem 3.3 when the X_n are random variables.
4.2 Show that the functions $h_1(x)$, $h_2(x)$ and $h_3(x)$ in (3.7) are continuous in C.
4.3 Prove that $\mathcal{L}(X_n) \to \mathcal{L}(c)$ in C implies $\rho(X_n, c) \to 0$ in probability, where c is a constant.
4.4 Prove Theorem 3.4 when the X_n are random variables.
4.5 Prove Theorem 3.6 when X_n and Y_n are random variables.
4.6 Show that (3.11) holds.
4.7 Prove that, if $\mathcal{L}(X_n) \to \mathcal{L}(X)$ and $Y_n \to 0$ in probability, then $X_n Y_n \to 0$ in probability, where X, X_n and Y_n are random variables.
4.8 Derive the weak convergence results in (3.13).
4.9 Establish (3.16).

3.5 FCLT FOR LINEAR PROCESSES: CASE 1

In this section we consider a sequence $\{Y_n\}$ of stochastic processes in C defined by

$$(3.17) \quad Y_n(t) = \frac{1}{\sqrt{n}} \sum_{i=1}^{j} u_i + n\left(t - \frac{j}{n}\right) \frac{1}{\sqrt{n}} u_j, \quad \left(\frac{j-1}{n} \le t \le \frac{j}{n}\right),$$

where $\{u_j\}$ is assumed to be generated by

$$(3.18) \quad u_j = \sum_{l=0}^{\infty} \alpha_l \varepsilon_{j-l}, \quad \alpha_0 = 1.$$

Here $\{\varepsilon_j\}$ is a sequence of random variables defined on $(\Omega, \mathcal{F}, P)$, while $\{\alpha_l\}$ is a sequence of constants for which we assume

$$(3.19) \quad \sum_{l=0}^{\infty} l |\alpha_l| < \infty.$$

The condition (3.19) may be replaced, for example, by $\sum_{l=0}^{\infty} l^2 \alpha_l^2 < \infty$, but we assume the stronger condition (3.19) only for simplicity (Problem 5.1).

To establish the FCLT for $\{Y_n\}$ in (3.17), we need additional assumptions on $\{\varepsilon_j\}$. In this section we assume that $\{\varepsilon_j\} \sim$ i.i.d.$(0, \sigma^2)$ with $\sigma^2 > 0$ so that $\{u_j\}$ in (3.18) belongs to L_2 discussed in Chapter 2 and becomes ergodic and strictly stationary as well as second-order stationary [Hannan (1970, p. 204)]. Following Phillips and Solo (1992), let us decompose $\{u_j\}$ into

$$(3.20) \quad u_j = \alpha \varepsilon_j + \tilde{\varepsilon}_{j-1} - \tilde{\varepsilon}_j,$$

where it is easy to check (Problem 5.2) that

$$(3.21) \quad \alpha = \sum_{l=0}^{\infty} \alpha_l,$$

$$(3.22) \quad \tilde{\varepsilon}_j = \sum_{l=0}^{\infty} \tilde{\alpha}_l \varepsilon_{j-l}, \quad \tilde{\alpha}_l = \sum_{k=l+1}^{\infty} \alpha_k.$$

The sequence $\{\tilde{\varepsilon}_j\}$ also becomes stationary (Problem 5.3) with $|\alpha| < \infty$ and $\sum_{l=0}^{\infty} |\tilde{\alpha}_l| < \infty$. The decomposition (3.20) is known in the econometrics literature as the Beveridge–Nelson (1981) or BN decomposition. A similar decomposition had been used by Fuller (1976) for MA representations of finite order (Problem 5.4).

We now have

$$(3.23) \quad Y_n(t) = \alpha X_n(t) + R_n(t),$$

where, for $(j-1)/n \le t \le j/n$,

(3.24) $$X_n(t) = \frac{1}{\sqrt{n}} \sum_{i=1}^{j} \varepsilon_i + n\left(t - \frac{j}{n}\right) \frac{1}{\sqrt{n}} \varepsilon_j,$$

(3.25) $$R_n(t) = \frac{1}{\sqrt{n}} (\tilde{\varepsilon}_0 - \tilde{\varepsilon}_j) + n\left(t - \frac{j}{n}\right) \frac{1}{\sqrt{n}} (\tilde{\varepsilon}_{j-1} - \tilde{\varepsilon}_j).$$

By Donsker's theorem, we have $\mathcal{L}(X_n/\sigma) \to \mathcal{L}(w)$, where $w = \{w(t)\}$ is the standard Brownian motion, so that $\mathcal{L}(\alpha X_n/\sigma) \to \mathcal{L}(\alpha w)$ by the continuous mapping theorem. We now show that $\rho(Y_n, \alpha X_n) \to 0$ in probability so that, by Theorem 3.6, we have $\mathcal{L}(Y_n/\sigma) \to \mathcal{L}(\alpha w)$. Consider

(3.26) $$\rho(Y_n, \alpha X_n) = \sup_{0 \le t \le 1} |R_n(t)| \le \frac{4}{\sqrt{n}} \max_{0 \le j \le n} |\tilde{\varepsilon}_j|,$$

which converges in probability to 0 if

(3.27) $$\frac{1}{\sqrt{n}} \max_{0 \le j \le n} |\tilde{\varepsilon}_j| \to 0 \quad \text{in probability.}$$

This last condition is equivalent (Problem 5.5) to

(3.28) $$J_n = \frac{1}{n} \sum_{j=0}^{n} \tilde{\varepsilon}_j^2 I(\tilde{\varepsilon}_j^2 > n\delta) \to 0 \quad \text{in probability}$$

for any $\delta > 0$, where $I(A)$ is the indicator function of A. Since $E(J_n) \to 0$ because of strict and second-order stationarity of $\{\tilde{\varepsilon}_j\}$, (3.28) follows from Markov's inequality.

The above arguments are summarized in the following theorem.

Theorem 3.7. *Let $\{Y_n\}$ be defined by (3.17) with $\{u_j\}$ being generated by the linear process (3.18) under the summability condition (3.19). If $\{\varepsilon_j\}$ in (3.18) is an i.i.d.$(0, \sigma^2)$ sequence, then $\mathcal{L}(Y_n/\sigma) \to \mathcal{L}(\alpha w)$.*

As an application of this theorem, consider an integrated process

(3.29) $$y_j = y_{j-1} + u_j, \quad y_0 = 0, \quad u_j = \sum_{l=0}^{\infty} \alpha_l \varepsilon_{j-l}, \quad \alpha_0 = 1,$$

where $\{\varepsilon_j\} \sim$ i.i.d.$(0, 1)$ and $\{\alpha_l\}$ satisfies (3.19). Then we obtain (Problem 5.6)

(3.30) $$\mathcal{L}\left(\frac{1}{T^2} \sum_{j=1}^{T} y_j^2\right) \to \mathcal{L}\left(\alpha^2 \int_0^1 w^2(t)\, dt\right),$$

$$(3.31) \quad \mathcal{L}\left(\frac{1}{T^2}\sum_{j=1}^{T}(y_j - \bar{y})^2\right) \to \mathcal{L}\left(\alpha^2 \int_0^1 \left(w(t) - \int_0^1 w(s)\,ds\right)^2 dt\right),$$

where $\alpha = \sum_{l=0}^{\infty} \alpha_l$.

PROBLEMS

5.1 Show that $\sum_{l=0}^{\infty} l\,|\alpha_l| < \infty$ implies $\sum_{l=0}^{\infty} l^2 \alpha_l^2 < \infty$.

5.2 Derive the BN decomposition (3.20) from (3.18).

5.3 Show that $\{\tilde{\varepsilon}_j\}$ in (3.22) is second-order stationary.

5.4 Derive the BN decomposition as in (3.20) when the order of the MA representation in (3.18) is finite.

5.5 Prove that

$$P\left(\max_{1 \le j \le n} |Z_j| > \delta\right) = P\left(\sum_{j=1}^n Z_j^2 I(|Z_j| > \delta) > \delta^2\right)$$

so that (3.27) and (3.28) are equivalent.

5.6 Derive the weak convergence results (3.30) and (3.31).

3.6 FCLT FOR MARTINGALE DIFFERENCES

This section serves as a basis of discussions concerning FCLTs for linear processes presented in the next section. The FCLT that we have seen in Section 3.3 assumes the basic sequence that forms the partial sum as in (3.2) to be i.i.d., and this FCLT has been extended in Section 3.5 to the dependent case where the basic sequence is stationary and is represented by an infinite, weighted sum of i.i.d. random variables. In these FCLTs the sequence of stochastic processes has a constant variance, while covariances depend only on time differences. This is referred to as the *homogeneous case*. The present section deals with a *nonhomogeneous case*, where the basic sequence has nonconstant variances, while covariances are assumed to be zero. This last assumption is relaxed in the next section so that the basic sequence has nonconstant covariances.

For the above purpose, we assume $\{\varepsilon_j\}$ to be a sequence of *martingale differences*, that is, $E(|\varepsilon_j|) < \infty$ and $E(\varepsilon_j|\mathcal{F}_{j-1}) = 0$ (a.s.), where $\{\mathcal{F}_j\}$ is an increasing sequence of *sub-σ-fields* of $\mathcal{F}$. Note that $\{\varepsilon_j\}$ is defined on $(\Omega, \mathcal{F}, P)$ and that each ε_j is measurable with respect to $\mathcal{F}_j$. Then $\{\varepsilon_j\}$ is said to be *adapted* to $\{\mathcal{F}_j\}$. In subsequent discussions $\{\varepsilon_j\}$ is assumed to belong to L_2, that is, $E(\varepsilon_j^2) < \infty$, which is said to be *square integrable*. Note that square integrability of $\{\varepsilon_j\}$ does not necessarily imply $\sup_j E(\varepsilon_j^2) < \infty$, much less $E(\sup_j \varepsilon_j^2) < \infty$.

We now describe the FCLT due to Brown (1971) [see also Hall and Heyde (1980, p. 99)]. Let us define

$$(3.32) \quad \xi_n(t) = \frac{1}{s_n} \sum_{i=1}^{j-1} \varepsilon_i + \frac{ts_n^2 - s_{j-1}^2}{s_j^2 - s_{j-1}^2} \frac{1}{s_n} \varepsilon_j, \quad \left(\frac{s_{j-1}^2}{s_n^2} \le t \le \frac{s_j^2}{s_n^2}\right),$$

where

$$(3.33) \quad s_n^2 = \sum_{j=1}^n E(\varepsilon_j^2) = E\left[\left(\sum_{j=1}^n \varepsilon_j\right)^2\right].$$

It is noticeable that the construction of the partial sum process $\{\xi_n(t)\}$ is different from Donsker's. This is because of the nonhomogeneous nature of variances of $\{\varepsilon_j\}$. A geometrical interpretation of paths of $\xi_n(t)$ is that $\xi_n(t)$ in the interval $[s_{j-1}^2/s_n^2, s_j^2/s_n^2]$ is on the line joining $(s_{j-1}^2/s_n^2, \sum_{i=1}^{j-1} \varepsilon_i/s_n)$ and $(s_j^2/s_n^2, \sum_{i=1}^{j} \varepsilon_i/s_n)$ (Problem 6.1).

Theorem 3.8. *Let $\{\varepsilon_j\}$ be a sequence of square integrable martingale differences satisfying*

$$(3.34) \quad \frac{1}{s_n^2} \sum_{j=1}^n \varepsilon_j^2 \to 1 \quad \text{in probability,}$$

$$(3.35) \quad \frac{1}{s_n^2} \sum_{j=1}^n E\left[\varepsilon_j^2 I(|\varepsilon_j| > \delta s_n)\right] \to 0 \quad \text{for every } \delta > 0.$$

Then $\mathcal{L}(\xi_n) \to \mathcal{L}(w)$.

It is easy to check that the present theorem does imply Donsker's FCLT when $\{\varepsilon_j\}$ is i.i.d.$(0, \sigma^2)$ (Problem 6.2). A set of sufficient conditions for (3.34) and (3.35) to hold will be given later. The condition (3.35) is referred to as the *Lindeberg condition*, which implies (Problem 6.3) that

$$(3.36) \quad \max_{1 \le j \le n} \frac{E(\varepsilon_j^2)}{s_n^2} \to 0,$$

and thus, by Chebyshev's inequality,

$$(3.37) \quad \max_{1 \le j \le n} P\left(\frac{|\varepsilon_j|}{s_n} > \delta\right) \to 0 \quad \text{for every } \delta > 0.$$

Moreover, since it holds (Problem 6.4) that

$$(3.38) \quad P\left(\max_{1 \le j \le n} \frac{|\varepsilon_j|}{s_n} > \delta\right) = P\left(\frac{1}{s_n^2} \sum_{j=1}^n \varepsilon_j^2 I(|\varepsilon_j| > \delta s_n) > \delta^2\right),$$

the Lindeberg condition also implies

$$(3.39) \quad \max_{1 \le j \le n} \frac{|\varepsilon_j|}{s_n} \to 0 \quad \text{in probability.}$$

Phillips and Solo (1992) prove that, if (3.34) holds, (3.39) implies the Lindeberg condition (3.35). Therefore the conditions (3.35) and (3.39) are equivalent under (3.34).

We now give a set of sufficient conditions for Theorem 3.8 to hold. We first assume that there exists a random variable η with $E(\eta^2) < \infty$ such that

$$(3.40) \quad P(|\varepsilon_j| > x) \le cP(|\eta| > x)$$

for each $x \ge 0$, $j \ge 1$ and for some positive constant c. In general, the sequence $\{\varepsilon_j\}$ satisfying (3.40) with $E(|\eta|) < \infty$ is said to be *strongly uniformly integrable* (s.u.i.). Since we assume $E(\eta^2) < \infty$, $\{\varepsilon_j^2\}$ also becomes s.u.i., which implies that

$$E(\varepsilon_j^2) = \int_0^\infty P(\varepsilon_j^2 > x)\, dx$$
$$\le c \int_0^\infty P(\eta^2 > x)\, dx$$
$$= cE(\eta^2) < \infty$$

so that $\sup_j E(\varepsilon_j^2) < \infty$. It also implies *uniform integrability* of $\{\varepsilon_j^2\}$, that is

$$(3.41) \quad \lim_{\delta \to \infty} \sup_j E\left[\varepsilon_j^2 I(|\varepsilon_j| > \delta)\right] = 0,$$

since it holds (Problem 6.5) that

$$(3.42) \quad E\left[\varepsilon_j^2 I(|\varepsilon_j| > \delta)\right] = \delta^2 P(|\varepsilon_j| > \delta) + \int_{\delta^2}^\infty P\left(|\varepsilon_j| > \sqrt{x}\right) dx.$$

It can be shown (Problem 6.6) that uniform integrability of $\{\varepsilon_j^2\}$ implies

$$\sup_j E(\varepsilon_j^2) < \infty.$$

We next assume that

$$(3.43) \quad \frac{1}{n} \sum_{j=1}^n E\left(\varepsilon_j^2 | \mathcal{F}_{j-1}\right) \to \sigma_\varepsilon^2 \quad \text{in probability,}$$

where σ_ε^2 is a positive constant. Since Hall and Heyde (1980, p. 36) proved that, if $\{\varepsilon_j^2\}$ is s.u.i.,

$$\frac{1}{n}\sum_{j=1}^{n}\left[\varepsilon_j^2 - E\left(\varepsilon_j^2|\mathcal{F}_{j-1}\right)\right] \to 0 \quad \text{in probability,}$$

we necessarily have that

$$\frac{1}{n}\sum_{j=1}^{n}\varepsilon_j^2 \to \sigma_\varepsilon^2 \quad \text{in probability.}$$

Moreover, it is known [Chow and Teicher (1988, p. 102)] that $\{\sum_{j=1}^{n}\varepsilon_j^2/n\}$ is s.u.i. if $\{\varepsilon_j^2\}$ is. This fact, together with convergence in probability, implies [Chow and Teicher (1988, p.100)] that

$$\frac{1}{n}\sum_{j=1}^{n}E\left(\varepsilon_j^2\right) = \frac{s_n^2}{n} \to \sigma_\varepsilon^2.$$

Thus the first condition (3.34) in Theorem 3.8 is clearly satisfied if $\{\varepsilon_j^2\}$ is s.u.i. and (3.43) holds.

For the Lindeberg condition (3.35), we can deduce from (3.42) that, if $\{\varepsilon_j^2\}$ is s.u.i.,

$$E\left[\varepsilon_j^2 I\left(|\varepsilon_j| > \sqrt{n}\delta\sigma_\varepsilon\right)\right] \le cE\left[\eta^2 I(|\eta| > \sqrt{n}\delta\sigma_\varepsilon)\right].$$

Thus the Lindeberg condition (3.35) is ensured because $E(\eta^2) < \infty$.

We conclude the above arguments by the following corollary.

Corollary 3.3. *Let $\{\varepsilon_j\}$ be a sequence of square integrable martingale differences that satisfies (3.40) with $E(\eta^2) < \infty$ and (3.43) with $\sigma_\varepsilon^2 > 0$. Then the two conditions in Theorem 3.8 are satisfied so that $\mathcal{L}(\xi_n) \to \mathcal{L}(w)$.*

The strong uniform integrability condition plays an important role in the above corollary. That condition was also used by Hannan and Heyde (1972) in a different context. If we can impose quite a restrictive assumption

(3.44) $$E(\sup_j \varepsilon_j^2) < \infty,$$

we necessarily have $P(\varepsilon_j^2 > x) \le P(\sup_j \varepsilon_j^2 > x)$ so that (3.44) ensures strong uniform integrability of $\{\varepsilon_j^2\}$. Thus (3.44) implies the Lindeberg condition (3.35). Then, if (3.43) holds, the condition (3.34) in Theorem 3.8 is also satisfied.

The following corollary summarizes the above arguments.

Corollary 3.4. *Let $\{\varepsilon_j\}$ be a sequence of square integrable martingale differences such that*

i) $E(\sup_j \varepsilon_j^2) < \infty$;
ii) $1/n \sum_{j=1}^{n} E(\varepsilon_j^2 | \mathcal{F}_{j-1})$ *converges in probability to* $\sigma_\varepsilon^2 > 0$.

Then the two conditions in Theorem 3.8 are satisfied so that $\mathcal{L}(\xi_n) \to \mathcal{L}(w)$.

As a final remark to this section, we mention that the above results also apply to a triangular array $\{\varepsilon_{jn}, 1 \leq j \leq n, n \geq 1\}$ of square integrable martingale differences, where $\{\varepsilon_{jn}\}$ is adapted to $\{F_{jn}\}$, which is a triangular array of sub-σ-fields of $\mathcal{F}$ such that $\mathcal{F}_{j-1,n} \subset \mathcal{F}_{jn}$ for all n. This means that we may put $\varepsilon_{jn} = \varepsilon_j/s_n$ in Theorem 3.8. A simple example of $\{\varepsilon_{jn}\}$ is $\{j\eta_j/n\}$ with $\{\eta_j\}$ being i.i.d.$(0, \sigma^2)$. Here $\{j\eta_j/n\}$ is not identically distributed, although independent. It can be checked (Problem 6.7) that Theorem 3.8 still holds with ε_j replaced by $\varepsilon_{jn} = j\eta_j/n$.

PROBLEMS

6.1 Explain the geometrical interpretation of paths of $\xi_n(t)$ described after (3.33).
6.2 Check that the two conditions in Theorem 3.8 are satisfied if $\{\varepsilon_j\}$ is i.i.d.$(0, \sigma^2)$.
6.3 Derive (3.36) from the Lindeberg condition (3.35).
6.4 Prove the relation in (3.38).
6.5 Show that strong uniform integrability of $\{\varepsilon_j^2\}$ implies uniform integrability of $\{\varepsilon_j^2\}$, proving formula (3.42).
6.6 Prove that uniform integrability of $\{\varepsilon_j^2\}$ implies $\sup_j E(\varepsilon_j^2) < \infty$.
6.7 Show that the two conditions in Theorem 3.8 are satisfied if ε_j is replaced by $\varepsilon_{jn} = j\eta_j/n$ with $\{\eta_j\}$ being i.i.d.$(0, \sigma^2)$.

3.7 FCLT FOR LINEAR PROCESSES: CASE 2

This section deals with the linear process generated by a sequence $\{\varepsilon_j\}$ of square integrable martingale differences discussed in the previous section. More specifically, we consider

$$(3.45) \quad Y_n(t) = \frac{1}{s_n} \sum_{i=1}^{j-1} u_i + \frac{ts_n^2 - s_{j-1}^2}{s_j^2 - s_{j-1}^2} \frac{1}{s_n} u_j, \quad \left(\frac{s_{j-1}^2}{s_n^2} \leq t \leq \frac{s_j^2}{s_n^2} \right),$$

where

$$(3.46) \quad s_n^2 = \sum_{j=1}^{n} E(\varepsilon_j^2) = E\left[\left(\sum_{j=1}^{n} \varepsilon_j\right)^2\right],$$

$$(3.47) \quad u_j = \sum_{l=0}^{\infty} \alpha_l \varepsilon_{j-l}, \quad \alpha_0 = 1,$$

84 FUNCTIONAL CENTRAL LIMIT THEOREMS

$$\text{(3.48)} \qquad \sum_{l=0}^{\infty} l |\alpha_l| < \infty.$$

In Section 3.5 we saw that the FCLT for the linear process generated by an i.i.d.$(0, \sigma^2)$ sequence did not require any additional assumptions except (3.47) and (3.48). In the present case, it seems necessary to impose a slightly stronger moment condition on $\{\varepsilon_j\}$ than required in the previous section to establish the FCLT for $\{Y_n\}$ in (3.45), which we now discuss.

Using the BN decomposition as in Section 3.5 we obtain (Problem 7.1)

$$\text{(3.49)} \qquad Y_n(t) = \alpha \xi_n(t) + R_n(t),$$

where, for $s_{j-1}^2/s_n^2 \leq t \leq s_j^2/s_n^2$,

$$\text{(3.50)} \qquad \xi_n(t) = \frac{1}{s_n} \sum_{i=1}^{j-1} \varepsilon_i + \frac{ts_n^2 - s_{j-1}^2}{s_j^2 - s_{j-1}^2} \frac{1}{s_n} \varepsilon_j,$$

$$\text{(3.51)} \qquad R_n(t) = \frac{1}{s_n}(\tilde{\varepsilon}_0 - \tilde{\varepsilon}_{j-1}) + \frac{ts_n^2 - s_{j-1}^2}{s_j^2 - s_{j-1}^2} \frac{1}{s_n}(\tilde{\varepsilon}_{j-1} - \tilde{\varepsilon}_j),$$

$$\text{(3.52)} \qquad \alpha = \sum_{l=0}^{\infty} \alpha_l,$$

$$\text{(3.53)} \qquad \tilde{\varepsilon}_j = \sum_{l=0}^{\infty} \tilde{\alpha}_l \varepsilon_{j-l}, \qquad \tilde{\alpha}_l = \sum_{k=l+1}^{\infty} \alpha_k.$$

Suppose, for a moment, that $\{\varepsilon_j\}$ is a sequence of square integrable martingale differences satisfying that $\{\varepsilon_j^2\}$ is strongly uniformly integrable, that is, there exists η with $E(\eta^2) < \infty$ such that $P(|\varepsilon_j| > x) \leq cP(|\eta| > x)$ for each $x \geq 0$, each integer j, and some $c > 0$. We also assume that $\sum_{j=1}^n E(\varepsilon_j^2|\mathcal{F}_{j-1})/n$ converges in probability to $\sigma_\varepsilon^2 > 0$. Then it follows from Corollary 3.3 and the continuous mapping theorem that $\mathcal{L}(\alpha \xi_n) \to \mathcal{L}(\alpha w)$.

We next deal with the remainder term $R_n(t)$ defined in (3.51), for which we have (Problem 7.2)

$$\text{(3.54)} \qquad \sup_{0 \leq t \leq 1} |R_n(t)| = \rho(Y_n, \alpha \xi_n) \leq \frac{4}{s_n} \max_{0 \leq j \leq n} |\tilde{\varepsilon}_j|.$$

Note that $\{\tilde{\varepsilon}_j\}$ satisfies (Problem 7.3) that

$$\text{(3.55)} \qquad E(\tilde{\varepsilon}_j^2) \leq cE(\eta^2) \sum_{l=0}^{\infty} \tilde{\alpha}_l^2 < \infty.$$

As was explained in Section 3.5, the last quantity in (3.54) converges in probability to 0 if

(3.56) $$\frac{1}{s_n} \max_{0 \le j \le n} |\tilde{\varepsilon}_j| \to 0 \quad \text{in probability}$$

or, equivalently,

(3.57) $$J_n = \frac{1}{s_n^2} \sum_{j=0}^{n} \tilde{\varepsilon}_j^2 I\left(\tilde{\varepsilon}_j^2 > s_n^2 \delta\right) \to 0 \quad \text{in probability}$$

for any $\delta > 0$. The condition (3.57) holds if $E(J_n) \to 0$, which was automatically satisfied in Section 3.5 since $\{\tilde{\varepsilon}_j^2\}$ was uniformly integrable because of strict and second-order stationarity of $\{\tilde{\varepsilon}_j\}$. In the present case, however, $\{\tilde{\varepsilon}_j\}$ is not stationary. Nonetheless, we assume $\{\tilde{\varepsilon}_j^2\}$ to be uniformly integrable, as in Phillips and Solo (1992). A sufficient condition for this is $\sup_j E(|\tilde{\varepsilon}_j|^{2+\gamma}) < \infty$ for some $\gamma > 0$ [Billingsley (1968, p. 32) and Problem 7.4]. This, in turn, holds if $\sup_j E(|\varepsilon_j|^{2+\gamma}) < \infty$, which is implied by $E(|\eta|^{2+\gamma}) < \infty$ because of Hölder's inequality (Problem 7.5). We now have $\rho(Y_n, \alpha\xi_n) \to 0$ in probability so that $\mathcal{L}(Y_n) \to \mathcal{L}(\alpha w)$ by Theorem 3.6. The above arguments are summarized in the following theorem.

Theorem 3.9. *Let $\{Y_n\}$ be defined by (3.45) with $\{u_j\}$ being generated by the linear process (3.47) under the summability condition (3.48). If $\{\varepsilon_j\}$ in (3.47) is a sequence of square integrable martingale differences that satisfies (3.40) with $E(|\eta|^{2+\gamma}) < \infty$ for some $\gamma > 0$, and (3.43) with $\sigma_\varepsilon^2 > 0$, then $\mathcal{L}(Y_n) \to \mathcal{L}(\alpha w)$.*

Note that, in comparison with the condition $E(\eta^2) < \infty$ in the previous section, we have imposed a stronger moment condition $E(|\eta|^{2+\gamma}) < \infty$ for some $\gamma > 0$ so that $\sup_j E(|\varepsilon_j|^{2+\gamma}) < \infty$. This condition may be dispensed with if we can assume $E(\sup_j \varepsilon_j^2) < \infty$ (Problem 7.6). In fact, we have the following corollary.

Corollary 3.5. *Let $\{Y_n\}$ be defined as in Theorem 3.9. Suppose that $\{\varepsilon_j\}$ in (3.47) is a sequence of square integrable martingale differences such that*

i) $E(\sup_j \varepsilon_j^2) < \infty$;

ii) $1/n \sum_{j=1}^{n} E(\varepsilon_j^2 | \mathcal{F}_{j-1})$ *converges in probability to* $\sigma_\varepsilon^2 > 0$.

Then $\mathcal{L}(Y_n) \to \mathcal{L}(\alpha w)$.

PROBLEMS

7.1 Derive the expression in (3.49).

7.2 Prove the inequality in (3.54).

7.3 Show that $\{\tilde{\varepsilon}_j\}$ in (3.53) satisfies the relation (3.55) if $\{\varepsilon_j^2\}$ is s.u.i. and (3.48) holds.

7.4 Show that $\sup_j E(|\tilde{\varepsilon}_j|^{2+\gamma})$ for some $\gamma > 0$ implies uniform integrability of $\{\tilde{\varepsilon}_j^2\}$.

7.5 Prove that, if there exists η with $E(|\eta|^{2+\gamma}) < \infty$ for some $\gamma > 0$ such that $P(|\varepsilon_j| > x) \leq cP(|\eta| > x)$ for each $x \geq 0$, each integer j and some $c > 0$, then $\sup_j E(|\varepsilon_j|^{2+\gamma}) < \infty$ and $\sup_j E(|\tilde{\varepsilon}_j|^{2+\gamma}) < \infty$.

7.6 Prove that, if $E(\sup_j \varepsilon_j^2) < \infty$ and (3.48) holds, then (3.57) is ensured.

3.8 WEAK CONVERGENCE TO THE INTEGRATED BROWNIAN MOTION

In Section 2.4 of Chapter 2 we introduced the integrated Brownian motion and indicated that the so-called I(d) process is essentially the ($d - 1$)-fold integrated Brownian motion. In this section we refine this fact on the basis of results obtained so far in this chapter.

Let us first discuss weak convergence to the one-fold integrated Brownian motion $\{F_1(t)\}$ defined by

$$(3.58) \qquad F_1(t) = \int_0^t w(s)\, ds,$$

where $\{w(s)\}$ is the one-dimensional standard Brownian motion. Let us construct the I(2) process $\{y_j^{(2)}\}$ generated by

$$(3.59) \qquad (1 - L)^2 y_j^{(2)} = \varepsilon_j, \qquad y_{-1}^{(2)} = y_0^{(2)} = 0, \qquad (j = 1, \ldots, n),$$

where we assume, for the time being, that $\{\varepsilon_j\}$ is i.i.d.$(0, \sigma^2)$ with $\sigma^2 > 0$. Note that (3.59) can be rewritten (Problem 8.1) as

$$(3.60) \qquad y_j^{(2)} = y_{j-1}^{(2)} + y_j^{(1)} = y_1^{(1)} + \cdots + y_j^{(1)},$$

where $\{y_j^{(1)}\}$ is the I(1) process or the random walk following $y_j^{(1)} = y_{j-1}^{(1)} + \varepsilon_j$, $y_0^{(1)} = 0$. Define two sequences $\{Y_n^{(1)}\}$ and $\{Y_n^{(2)}\}$ of stochastic processes in C by

$$(3.61)\ Y_n^{(1)}(t) = \frac{1}{\sqrt{n}} y_{[nt]}^{(1)} + (nt - [nt]) \frac{1}{\sqrt{n}} \varepsilon_{[nt]+1}$$

$$= \frac{1}{\sqrt{n}} \sum_{i=1}^{j} \varepsilon_i + n\left(t - \frac{j}{n}\right) \frac{1}{\sqrt{n}} \varepsilon_j, \qquad \left(\frac{j-1}{n} \leq t \leq \frac{j}{n}\right),$$

$$(3.62) \quad Y_n^{(2)}(t) = \frac{1}{n\sqrt{n}} y_{[nt]}^{(2)} + (nt - [nt]) \frac{1}{n\sqrt{n}} y_{[nt]+1}^{(1)}$$

$$= \frac{1}{n} \sum_{i=1}^{j} Y_n^{(1)}\left(\frac{i}{n}\right) + n\left(t - \frac{j}{n}\right) \frac{1}{n\sqrt{n}} y_j^{(1)}, \quad \left(\frac{j-1}{n} \le t \le \frac{j}{n}\right).$$

It follows from Donsker's theorem that $\mathcal{L}(Y_n^{(1)}/\sigma) \to \mathcal{L}(w)$.

We now show that $\mathcal{L}(Y_n^{(2)}/\sigma) \to \mathcal{L}(F_1)$. For this purpose, define the integral version of (3.62) by

$$G_{1n}(t) = \int_0^t Y_n^{(1)}(s)\,ds.$$

Note that $\mathcal{L}(G_{1n}/\sigma) \to \mathcal{L}(F_1)$ by the continuous mapping theorem. Then it holds (Problem 8.2) that, for $(j-1)/n \le t \le j/n$,

$$|Y_n^{(2)}(t) - G_{1n}(t)| \le \left| \sum_{i=1}^{j} \int_{(i-1)/n}^{i/n} Y_n^{(1)}\left(\frac{i}{n}\right) ds - \int_0^t Y_n^{(1)}(s)\,ds \right| + \frac{1}{n\sqrt{n}} |y_j^{(1)}|$$

$$(3.63) \qquad \le \frac{2}{\sqrt{n}} \max_{1 \le j \le n} |\varepsilon_j| + \frac{1}{n} \sup_{0 \le t \le 1} |Y_n^{(1)}(t)|.$$

Now it can be shown (Problem 8.3) that

$$(3.64) \qquad \sup_{0 \le t \le 1} |Y_n^{(2)}(t) - G_{1n}(t)| \to 0 \quad \text{in probability.}$$

This fact together with $\mathcal{L}(G_{1n}/\sigma) \to \mathcal{L}(F_1)$ establishes that $\mathcal{L}(Y_n^{(2)}/\sigma) \to \mathcal{L}(F_1)$.

Weak convergence to the general g-fold integrated Brownian motion can be dealt with similarly. Define, for a positive integer g,

$$(3.65) \qquad F_g(t) = \int_0^t F_{g-1}(s)\,ds, \qquad F_0(s) = w(s),$$

and construct the I(d) process $\{y_j^{(d)}\}$ generated by

$$(3.66) \qquad (1-L)^d y_j^{(d)} = \varepsilon_j, \qquad (j = 1,\ldots,n),$$

with $y_{-(d-1)}^{(d)} = y_{-(d-2)}^{(d)} = \cdots = y_0^{(d)} = 0$ and $\{\varepsilon_j\}$ being i.i.d.$(0, \sigma^2)$. We have

$$(3.67) \qquad y_j^{(d)} = y_{j-1}^{(d)} + y_j^{(d-1)} = y_1^{(d-1)} + \cdots + y_j^{(d-1)}, \qquad y_j^{(0)} = \varepsilon_j$$

and put, for $d \geq 2$,

$$Y_n^{(d)}(t) = \frac{1}{n^{d-1/2}} y_{[nt]}^{(d)} + (nt - [nt]) \frac{1}{n^{d-1/2}} y_{[nt]+1}^{(d-1)}$$

$$= \frac{1}{n} \sum_{i=1}^{j} Y_n^{(d-1)}\left(\frac{i}{n}\right) + n\left(t - \frac{j}{n}\right) \frac{1}{n^{d-1/2}} y_j^{(d-1)}, \quad \left(\frac{j-1}{n} \leq t \leq \frac{j}{n}\right).$$

(3.68)

Define also the integral version of (3.68) by

$$G_{d-1,n}(t) = \int_0^t Y_n^{(d-1)}(s)\,ds.$$

We now prove by induction that $\mathcal{L}(Y_n^{(d)}/\sigma) \to \mathcal{L}(F_{d-1})$ for any $d \geq 2$. The case $d = 2$ was already established. Suppose that $\mathcal{L}(Y_n^{(k-1)}/\sigma) \to \mathcal{L}(F_{k-2})$ holds for some $k \geq 3$ so that $\mathcal{L}(G_{k-1,n}/\sigma) \to \mathcal{L}(F_{k-1})$ by the continuous mapping theorem. Then we have (Problem 8.4), for $(j-1)/n \leq t \leq j/n$,

$$(3.69) \quad |Y_n^{(k)}(t) - G_{k-1,n}(t)| \leq \left|\sum_{i=1}^{j} \int_{(i-1)/n}^{i/n} Y_n^{(k-1)}\left(\frac{i}{n}\right) ds - \int_0^t Y_n^{(k-1)}(s)\,ds\right|$$

$$+ \frac{1}{n^{k-1/2}} \left|y_j^{(k-1)}\right|$$

$$\leq \frac{2}{\sqrt{n}} \max_{1 \leq j \leq n} |\varepsilon_j| + \frac{1}{n} \sup_{0 \leq t \leq 1} \left|Y_n^{(k-1)}(t)\right|.$$

Thus it is seen that $\sup_{0 \leq t \leq 1} |Y_n^{(k)}(t) - G_{k-1,n}(t)|$ converges in probability to 0. Since $\mathcal{L}(G_{k-1,n}/\sigma) \to \mathcal{L}(F_{k-1})$ by assumption, (3.69) yields that $\mathcal{L}(Y_n^{(k)}/\sigma) \to \mathcal{L}(F_{k-1})$.

The above arguments can be easily extended to the case where the innovation sequence $\{\varepsilon_j\}$ follows a linear process. We state an extended result in the following theorem, whose proof is left as Problem 8.5.

Theorem 3.10. *Suppose that the* I(d) *process* $\{y_j^{(d)}\}$ *is generated by*

$$(1-L)^d y_j^{(d)} = u_j, \quad (d \geq 2, \ j = 1, \ldots, n),$$

where $y_{-(d-1)}^{(d)} = y_{-(d-2)}^{(d)} = \cdots = y_0^{(d)} = 0$ *and*

$$u_j = \sum_{l=0}^{\infty} \alpha_l \varepsilon_{j-l}, \quad \alpha_0 = 1, \quad \sum_{l=0}^{\infty} l|\alpha_l| < \infty,$$

with $\{\varepsilon_j\}$ *being i.i.d.*$(0, \sigma^2)$. *Define* $F_{d-1}(t)$ *and* $Y_n^{(d)}(t)$ *by* (3.65) *and* (3.68), *respectively. Then* $\mathcal{L}(Y_n^{(d)}/\sigma) \to \mathcal{L}(\alpha F_{d-1})$, *where* $\alpha = \sum_{l=0}^{\infty} \alpha_l$.

As the first application of Theorem 3.10, we establish the weak convergence of

$$(3.70) \qquad S_T = \frac{1}{T^{2d}} \sum_{j=1}^{T} \left(y_j^{(d)} \right)^2, \qquad (d \geq 2).$$

For this purpose we put

$$(3.71) \qquad h(x) = \int_0^1 x^2(t)\, dt, \qquad x \in C.$$

Using (3.68) and noting that $Y_T^{(d)}(j/T) = y_j^{(d)}/T^{d-1/2}$, we consider

$$S_T - h\left(Y_T^{(d)}\right) = \frac{1}{T} \sum_{j=1}^{T} \left(Y_T^{(d)} \left(\frac{j}{T} \right) \right)^2 - \int_0^1 \left(Y_T^{(d)}(t) \right)^2 dt$$

$$= \sum_{j=1}^{T} \int_{(j-1)/T}^{j/T} \left[\left(Y_T^{(d)} \left(\frac{j}{T} \right) \right)^2 - \left(Y_T^{(d)}(t) \right)^2 \right] dt,$$

where the integrand has the following bound:

$$(3.72) \qquad \left| \left(Y_T^{(d)} \left(\frac{j}{T} \right) \right)^2 - \left(Y_T^{(d)}(t) \right)^2 \right| \leq 2 \sup_{0 \leq t \leq 1} \left| Y_T^{(d)}(t) \right| \max_{1 \leq j \leq T} \frac{\left| y_j^{(d-1)} \right|}{T^{d-1/2}}.$$

Thus it can be shown (Problem 8.6) that

$$(3.73) \qquad S_T - h\left(Y_T^{(d)}\right) \to 0 \qquad \text{in probability.}$$

Since $\mathcal{L}(Y_T^{(d)}/\sigma) \to \mathcal{L}(\alpha F_{d-1})$ by Theorem 3.10 and

$$\mathcal{L}\left(h\left(Y_T^{(d)}/\sigma\right) \right) \to \mathcal{L}(h(\alpha F_{d-1}))$$

by the continuous mapping theorem, (3.71) and (3.73) lead us to

$$(3.74) \qquad \mathcal{L}\left(\frac{S_T}{\sigma^2}\right) \to \mathcal{L}\left(\alpha^2 \int_0^1 F_{d-1}^2(t)\, dt\right).$$

As the second application, we establish the weak convergence of

$$(3.75) \qquad U_T = \frac{1}{T^{2d-1}} \sum_{j=1}^{T} y_{j-1}^{(d)} \left(y_j^{(d)} - y_{j-1}^{(d)} \right)$$

$$= \frac{1}{2T^{2d-1}} \left(y_T^{(d)} \right)^2 - \frac{1}{2T^{2d-1}} \sum_{j=1}^{T} \left(y_j^{(d)} - y_{j-1}^{(d)} \right)^2.$$

Since $\{y_j^{(d)} - y_{j-1}^{(d)}\}$ is the I($d-1$) process because of (3.67), it is easy to see that the last term on the right side of (3.75) converges in probability to 0 for $d \geq 2$. Then we have, by Theorem 3.10 and the continuous mapping theorem,

$$\mathcal{L}\left(\frac{U_T}{\sigma^2}\right) \to \mathcal{L}\left(\frac{\alpha^2}{2} F_{d-1}^2(1)\right) = \mathcal{L}\left(\alpha^2 \int_0^1 F_{d-1}(t)\, dF_{d-1}(t)\right).$$

The equality above is due to $(d-1)$-times differentiability of $\{F_{d-1}(t)\}$. Note that the situation is completely different from the case $d = 1$, which will be discussed in Section 3.11.

PROBLEMS

8.1 Derive the recursive relations (3.60) and (3.67).

8.2 Prove the inequalities in (3.63).

8.3 Show that (3.64) holds.

8.4 Prove the inequalities in (3.69).

8.5 Prove Theorem 3.10.

8.6 Show that (3.73) holds.

3.9 WEAK CONVERGENCE TO THE ORNSTEIN–UHLENBECK PROCESS

We introduced in Section 2.7 of Chapter 2 the Ornstein–Uhlenbeck (O–U) process in C defined by

(3.76) $$dX(t) = -\beta X(t)\, dt + dw(t) \iff$$

$$X(t) = e^{-\beta t} X(0) + e^{-\beta t} \int_0^t e^{\beta s}\, dw(s),$$

where β is a constant. We will show in this section that the near random walk process

(3.77) $$y_j = \left(1 - \frac{\beta}{n}\right) y_{j-1} + \varepsilon_j, \qquad (j = 1, \ldots, n),$$

converges weakly to the O–U process in the sense described later, where $\{\varepsilon_j\}$ is assumed to be i.i.d.$(0, \sigma^2)$. Note that (3.77) may be rewritten (Problem 9.1) using Abel's transformation as

$$\text{(3.78)} \quad y_j = \rho_n^j y_0 + \rho_n^{j-1}\varepsilon_1 + \cdots + \rho_n \varepsilon_{j-1} + \varepsilon_j$$

$$= \rho_n^j y_0 + \sum_{i=1}^{j} \rho_n^{j-i}(S_i - S_{i-1})$$

$$= \rho_n^j y_0 + \rho_n^{-1} S_j - (1-\rho_n) \sum_{i=1}^{j} \rho_n^{j-i-1} S_i ,$$

where $\rho_n = 1 - (\beta/n)$ and

$$\text{(3.79)} \quad S_j = \varepsilon_1 + \cdots + \varepsilon_j, \qquad S_0 \equiv 0.$$

The last expression for y_j in (3.78) is quite useful for subsequent discussions. Note also in (3.78) that we retain the initial value y_0, which we assume, for the time being, to take the following form:

$$\text{(3.80)} \quad y_0 = \sqrt{n}\gamma\sigma,$$

where γ is a constant. We will consider later the case where y_0 is a random variable of stochastic order $\sqrt{n}$.

The present problem was studied to a large extent by Bobkoski (1983) for the case where y_0 is a constant or a random variable distributed independently of n. In that case, y_0 is asymptotically negligible. Let us define, for $(j-1)/n \le t \le j/n$,

$$\text{(3.81)} \quad X_n(t) = \frac{1}{\sqrt{n}} y_{j-1} + n\left(t - \frac{j-1}{n}\right)\frac{y_j - y_{j-1}}{\sqrt{n}}$$

$$= \frac{\rho_n^{j-1}}{\sqrt{n}} y_0 + \frac{\rho_n^{-1}}{\sqrt{n}} S_{j-1} - \frac{\beta}{n\sqrt{n}} \sum_{i=1}^{j-1} \rho_n^{j-i-2} S_i$$

$$+ n\left(t - \frac{j-1}{n}\right)\frac{y_j - y_{j-1}}{\sqrt{n}}$$

$$= \rho_n^{j-1}\gamma\sigma + \rho_n^{-1} Y_n\left(\frac{j-1}{n}\right) - \frac{\beta}{n}\sum_{i=1}^{j-1} \rho_n^{j-i-2} Y_n\left(\frac{i}{n}\right)$$

$$+ n\left(t - \frac{j-1}{n}\right)\frac{y_j - y_{j-1}}{\sqrt{n}},$$

where $X_n(0) = \gamma\sigma$ and

$$\text{(3.82)} \quad Y_n(t) = \frac{1}{\sqrt{n}} S_{j-1} + n\left(t - \frac{j-1}{n}\right)\frac{1}{\sqrt{n}}\varepsilon_j, \qquad \left(\frac{j-1}{n} \le t \le \frac{j}{n}\right).$$

Note that (3.78) has been applied to obtain the second expression in (3.81).

We also consider a function $h(y; \gamma)$ on C, whose value at t denoted as $h_t(y; \gamma)$ is defined by

$$h_t(y; \gamma) = e^{-\beta t}\gamma + y(t) - \beta e^{-\beta t} \int_0^t e^{\beta s} y(s)\, ds. \qquad (3.83)$$

It is easy to check (Problem 9.2) that h is a continuous mapping defined on C. We shall show that $\mathcal{L}(X_n/\sigma) \to \mathcal{L}(h(w; \gamma))$ so that $\mathcal{L}(X_n/\sigma) \to \mathcal{L}(X)$ with $X(0) = \gamma$ since it holds (Problem 9.3) that $X(t) = h_t(w; \gamma)$. For this purpose let us consider

$$|X_n(t) - h_t(Y_n; \gamma\sigma)| \le |\gamma|\sigma A_{jn} + B_{jn} + |\beta| C_{jn} + D_{jn}, \qquad (3.84)$$

where

$$A_{jn} = |\rho_n^{j-1} - e^{-\beta t}|,$$

$$B_{jn} = \left|\rho_n^{-1} Y_n\left(\frac{j-1}{n}\right) - Y_n(t)\right|,$$

$$C_{jn} = \left|\frac{1}{n}\sum_{i=1}^{j-1} \rho_n^{j-i-2} Y_n\left(\frac{i}{n}\right) - e^{-\beta t}\int_0^t e^{\beta s} Y_n(s)\, ds\right|,$$

$$D_{jn} = \frac{1}{\sqrt{n}} |y_j - y_{j-1}|.$$

It can be shown (Problem 9.4) that

$$A_{jn} \le \sup_{0 \le t \le 1} |\rho_n^{[nt]} - e^{-\beta t}| = O\left(\frac{1}{n}\right),$$

$$B_{jn} \le \frac{1}{\sqrt{n}} \max_{1 \le j \le n} |\varepsilon_j| + |\rho_n^{-1} - 1| \sup_{0 \le t \le 1} |Y_n(t)| = o_p(1),$$

$$C_{jn} \le \sum_{i=1}^{j-1} \int_{(i-1)/n}^{i/n} e^{-\beta(t-s)} \left|Y_n\left(\frac{i}{n}\right) - Y_n(s)\right| ds$$

$$+ \sum_{i=1}^{j-1} \int_{(i-1)/n}^{i/n} |\rho_n^{j-i-2} - e^{-\beta(t-s)}|\, ds \left|Y_n\left(\frac{i}{n}\right)\right|$$

$$+ \int_{[nt]/n}^{t} e^{-\beta(t-s)} |Y_n(s)|\, ds$$

$$\le \frac{O(1)}{\sqrt{n}} \max_{1 \le j \le n} |\varepsilon_j| + o(1) \sup_{0 \le t \le 1} |Y_n(t)| = o_p(1),$$

$$D_{jn} \le \frac{|\beta|}{n\sqrt{n}} \max_{1 \le j \le n} |y_j| + \frac{1}{\sqrt{n}} \max_{1 \le j \le n} |\varepsilon_j| = o_p(1).$$

Thus $\sup_{0 \le t \le 1} |X_n(t)/\sigma - h_t(Y_n/\sigma; \gamma)| \to 0$ in probability. Since $\mathcal{L}(Y_n/\sigma) \to \mathcal{L}(w)$, we have $\mathcal{L}(X_n/\sigma) \to \mathcal{L}(h(w; \gamma))$ by the continuous mapping theorem and it follows from Problem 9.3 that $\mathcal{L}(X_n/\sigma) \to \mathcal{L}(X)$.

The above arguments are summarized in the following theorem.

Theorem 3.11. *Let the near random walk $\{y_j\}$ be defined by (3.77). On the basis of $\{y_j\}$, construct the process $\{X_n(t)\}$ in C as in (3.81) with $y_0 = \sqrt{n}\gamma\sigma$. Then $\mathcal{L}(X_n/\sigma) \to \mathcal{L}(X)$, where $\{X(t)\}$ is the O–U process defined in (3.76) with $X(0) = \gamma$.*

The case of y_0 being a random variable needs some care. Suppose that

(3.85) $$y_0 = \sqrt{n}\sigma X(0),$$

where $X(0) \sim N(\gamma, \delta^2)$ and is independent of $\{\varepsilon_j\}$. Then we construct, as in (3.81),

(3.86) $$X_n(t) = \rho_n^{j-1}\sigma X(0) + \rho_n^{-1} Y_n\left(\frac{j-1}{n}\right) - \frac{\beta}{n}\sum_{i=1}^{j-1}\rho_n^{j-i-2} Y_n\left(\frac{i}{n}\right)$$
$$+ n\left(t - \frac{j-1}{n}\right)\frac{y_j - y_{j-1}}{\sqrt{n}},$$

which is composed of a random variable $X(0)$ and a stochastic process $\{Y_n(t)\}$ defined in (3.82). Let R be the real line, which is complete and separable under the Euclidean metric. Then the joint weak convergence of $(X(0), Y_n/\sigma)$ on $R \times C$ holds; that is, $\mathcal{L}(X(0), Y_n/\sigma) \to \mathcal{L}(X(0), w)$ [see Billingsley (1968, p. 224) and the next section]. Defining on $R \times C$

$$h_t(x, y) = e^{-\beta t}x + y(t) - \beta e^{-\beta t}\int_0^t e^{\beta s} y(s)\, ds,$$

we can obtain that $\sup_{0 \le t \le 1} |X_n(t)/\sigma - h_t(X(0), Y_n/\sigma)| \to 0$ in probability so that $\mathcal{L}(X_n/\sigma) \to \mathcal{L}(h(X(0), w)) = \mathcal{L}(X)$ with $X(0) \sim N(\gamma, \delta^2)$.

Theorem 3.12. *Assume the same conditions as in Theorem 3.11 except that $X_n(t)$ is defined in (3.86) with $y_0 = \sqrt{n}\sigma X(0)$ and $X(0) \sim N(\gamma, \delta^2)$. Then $\mathcal{L}(X_n/\sigma) \to \mathcal{L}(X)$ with $\{X(t)\}$ being the O–U process.*

In this theorem we may put $\delta = 0$. Then $y_0 = \sqrt{n}\sigma\gamma$ so that the theorem reduces to Theorem 3.11. If we assume that $\gamma = 0$ and $\delta^2 = 1/(2\beta)$ with $\beta > 0$, then $\{X(t)\}$ becomes stationary, as was indicated in (2.64).

As an application let us establish the weak convergence of

(3.87) $$V_T = \frac{1}{T^2}\sum_{j=1}^T y_j^2,$$

where $\{y_j\}$ is the near random walk defined in (3.77). As for y_0 we assume that $y_0 = \sqrt{T}\sigma X(0)$ with $X(0) \sim N(\gamma, \delta^2)$ and put

$$h(y) = \int_0^1 y^2(t)\,dt.$$

Using $X_T(t)$ defined in (3.86) and noting that $X_T(j/T) = y_j/\sqrt{T}$, it can be easily shown (Problem 9.5) that $V_T - h(X_T)$ converges in probability to 0. Since $h(X_T/\sigma) \to h(X)$ by Theorem 3.12 and the continuous mapping theorem, we have that

$$\mathcal{L}\left(\frac{V_T}{\sigma^2}\right) \to \mathcal{L}\left(\int_0^1 X^2(t)\,dt\right).$$

Extensions to near integrated processes seem straightforward. Consider the near integrated process $\{y_j\}$ defined by

(3.88) $$y_j = \left(1 - \frac{\beta}{n}\right) y_{j-1} + u_j,$$

(3.89) $$u_j = \sum_{l=0}^{\infty} \alpha_l \varepsilon_{j-l}, \quad \alpha_0 = 1, \quad \sum_{l=0}^{\infty} l|\alpha_l| < \infty,$$

where $\{\varepsilon_j\} \sim$ i.i.d.$(0, \sigma^2)$. The proof of the following theorem is left as Problem 9.6.

Theorem 3.13. *Assume that the near integrated process $\{y_j\}$ is defined by (3.88) and (3.89) with $y_0 = \sqrt{n}\alpha\sigma X(0)$, where $\alpha = \sum_{l=0}^{\infty} \alpha_l$ and $X(0) \sim N(\gamma, \delta^2)$. Define*

$$X_n(t) = \frac{1}{\sqrt{n}} y_{j-1} + n\left(t - \frac{j-1}{n}\right) \frac{y_j - y_{j-1}}{\sqrt{n}}, \quad \left(\frac{j-1}{n} \le t \le \frac{j}{n}\right).$$

Then $\mathcal{L}(X_n/\sigma) \to \mathcal{L}(\alpha X)$, where $\{X(t)\}$ is the O–U process with $X(0) \sim N(\gamma, \delta^2)$.

The extension to the case where $\{\varepsilon_j\}$ is a sequence of square integrable martingale differences is also straightforward. We do not pursue the matter here.

PROBLEMS

9.1 Establish the relations in (3.78) by showing that

$$\sum_{j=1}^n a_j(b_j - b_{j-1}) = a_{n+1}b_n - a_1 b_0 - \sum_{j=1}^n (a_{j+1} - a_j)b_j.$$

9.2 Prove that $h(y; \gamma)$ in (3.83) is a continuous mapping defined on C.

9.3 Show that $X(t) = h_t(w; \gamma)$ with $X(0) = \gamma$, where $X(t)$ and h_t are defined by (3.76) and (3.83), respectively.

9.4 Prove that $\sup_{0 \leq t \leq 1} |X_n(t) - h_t(Y_n; \gamma\sigma)|$ in (3.84) converges in probability to 0.

9.5 Establish the weak convergence of V_T in (3.87).

9.6 Prove Theorem 3.13.

3.10 WEAK CONVERGENCE OF VECTOR-VALUED STOCHASTIC PROCESSES

Our discussions have so far been concerned with weak convergence of scalar stochastic processes, although an exception is found in Section 3.9. In practice we need to deal with vector processes, whose FCLTs we describe here.

3.10.1 Space C^q

Let $(C^q, \mathcal{B}(C^q))$ be a measurable space, where $C^q = C[0,1] \times \cdots \times C[0,1]$ (q copies) and $\mathcal{B}(C^q)$ is the σ-field generated by the subsets of C^q that are open with respect to the metric ρ_q defined by

$$(3.90) \qquad \rho_q(x, y) = \max_{1 \leq i \leq q} \sup_{0 \leq t \leq 1} |x_i(t) - y_i(t)|$$

for $x = (x_1, \ldots, x_q)'$, $y = (y_1, \ldots, y_q)' \in C^q$. The space C^q is complete and separable under ρ_q. In particular, separability results from C being separable under the uniform metric and it holds that $\mathcal{B}(C^q) = \mathcal{B}(C) \times \cdots \times \mathcal{B}(C)$ (q copies) [Billingsley (1968, p. 224)]. Separability also implies that, for given probability measures $P^{(i)}(i = 1, \ldots, q)$ on $(C, \mathcal{B}(C))$, the product measure $P^{(1)} \times \cdots \times P^{(q)}$ is a probability measure on $\mathcal{B}(C^q)$ [Billingsley (1968, p. 21)].

Let $\{X_n\}$ be a sequence of q-dimensional stochastic processes in C^q and $\{Q_n\}$ the family of probability measures induced by $\{X_n\}$ as

$$Q_n(A) = P(X_n \in A) = P\left(X_n^{-1}(A)\right), \qquad A \in \mathcal{B}(C^q).$$

As in the scalar case, $\mathcal{L}(X_n) \to \mathcal{L}(X)$ if the finite-dimensional distributions of Q_n converge weakly to those of the probability measure Q induced by X and if $\{Q_n\}$ is tight. It is usually difficult to prove the joint weak convergence, but separability and independence facilitate its proof. Suppose that the q components of X_n are independent of each other; so are the q components of X. Suppose further that $\mathcal{L}(X_n(t_1), \ldots, X_n(t_k)) \to \mathcal{L}(X(t_1), \ldots, X(t_k))$ for each finite k and each collection $0 \leq t_1 < t_2 < \cdots < t_k \leq 1$. Then $\mathcal{L}(X_n) \to \mathcal{L}(X)$ if all the marginal probability measures of $\{Q_n\}$ are tight on the component spaces [Billingsley (1968, p. 41)].

In the next subsection we take up an example of $\{X_n\}$ and discuss its weak convergence, following the ideas described above.

3.10.2 Basic FCLT for Vector Processes

Let us now consider a sequence $\{X_n\}$ of stochastic processes in C^q defined by

$$(3.91) \qquad X_n(t) = \Sigma^{-1/2} \left[\frac{1}{\sqrt{n}} \sum_{j=1}^{[nt]} \varepsilon_j + (nt - [nt]) \frac{1}{\sqrt{n}} \varepsilon_{[nt]+1} \right],$$

where $\{\varepsilon_j\}$ is a sequence of q-dimensional i.i.d.$(0, \Sigma)$ random vectors on $(\Omega, \mathcal{F}, P)$ with $\Sigma > 0$. We shall show that $\mathcal{L}(X_n) \to \mathcal{L}(w)$, where $\{w(t)\}$ is the q-dimensional standard Brownian motion.

We first note that, for a single time point t,

$$\left\| X_n(t) - \frac{\Sigma^{-1/2}}{\sqrt{n}} \sum_{j=1}^{[nt]} \varepsilon_j \right\| \leq \|\Sigma^{-1/2}\| \frac{1}{\sqrt{n}} \|\varepsilon_{[nt]+1}\|,$$

where $\|M\| = [\operatorname{tr}(M'M)]^{1/2}$ for any matrix or vector M. Since

$$(3.92) \qquad \frac{1}{\sqrt{n}} \|\varepsilon_{[nt]+1}\| \to 0 \qquad \text{in probability}$$

by Chebyshev's inequality, and

$$\mathcal{L}\left(\frac{\Sigma^{-1/2}}{\sqrt{n}} \sum_{j=1}^{[nt]} \varepsilon_j \right) \to \mathcal{L}(w(t))$$

by the multivariate CLT [see, for example, Rao (1973, p.128)], it follows from the vector version of Theorem 3.6 that $\mathcal{L}(X_n(t)) \to \mathcal{L}(w(t))$.

Consider next two time points s and t with $s < t$. We are to prove $\mathcal{L}(X_n(s), X_n(t)) \to \mathcal{L}(w(s), w(t))$, which will follow by the continuous mapping theorem if we prove $\mathcal{L}(X_n(s), X_n(t) - X_n(s)) \to \mathcal{L}(w(s), w(t) - w(s))$. Because of (3.92) it is enough to prove

$$\mathcal{L}\left(\frac{\Sigma^{-1/2}}{\sqrt{n}} \sum_{j=1}^{[ns]} \varepsilon_j, \; \frac{\Sigma^{-1/2}}{\sqrt{n}} \left(\sum_{j=1}^{[nt]} \varepsilon_j - \sum_{j=1}^{[ns]} \varepsilon_j \right) \right) \to \mathcal{L}(w(s), w(t) - w(s)).$$

Since the two vectors on the left are independent, this follows by the multivariate CLT for each vector [Billingsley (1968, p. 26)]. A set of three or more time points can be treated in the same way, and hence the finite-dimensional distributions converge properly.

As was described in the previous subsection, tightness of the family $\{Q_n\}$ of probability measures induced by $\{X_n\}$ is ensured if all the marginal probability measures associated with each component $\{X_{in}\}$ of $\{X_n\}$ are tight on C. Since it does hold that $\mathcal{L}(X_{in}) \to \mathcal{L}(w)$, where $\{w(t)\}$ is the one-dimensional standard Brownian motion, the

WEAK CONVERGENCE OF VECTOR-VALUED STOCHASTIC PROCESSES   97

associated marginal probability measures must be relatively compact [Billingsley (1968, p. 35)]. Thus tightness results from completeness and separability of C under the uniform metric.

It is an immediate consequence of the above discussions and the continuous mapping theorem to obtain

$$(3.93) \qquad \mathcal{L}\left(\frac{1}{T^2}\sum_{j=1}^{T} y_j y_j'\right) \to \mathcal{L}\left(\Sigma^{1/2}\int_0^1 w(t)w'(t)\,dt\,\Sigma^{1/2}\right),$$

where $y_j = y_{j-1} + \varepsilon_j$, $y_0 = 0$, and $\{\varepsilon_j\} \sim$ i.i.d.$(0, \Sigma)$. It can also be shown (Problem 10.1) that

$$(3.94) \qquad \mathcal{L}\left(\frac{1}{T^2}\sum_{j=1}^{T} y_j' H' H y_j\right) \to \mathcal{L}\left(\int_0^1 w'(t)\Sigma^{1/2} H' H \Sigma^{1/2} w(t)\,dt\right)$$

for any $q \times q$ constant matrix H. In particular, suppose that any two components $\{y_{kj}\}$ and $\{y_{lj}\}$ ($k \neq l$) of $\{y_j\}$ are independent and $\Sigma = I_q$. It then follows from (3.94) that

$$\mathcal{L}\left(\frac{1}{T^2}\sum_{j=1}^{T} y_{kj} y_{lj}\right) \to \mathcal{L}\left(\int_0^1 w_k(t) w_l(t)\,dt\right).$$

A special case of this was dealt with in Section 1.4 of Chapter 1 together with the c.f. of the limiting distribution.

3.10.3 FCLT for Vector-Valued Linear Processes

Similar arguments can be applied to stochastic processes defined by

$$(3.95) \qquad Y_n(t) = A^{-1}\left[\frac{1}{\sqrt{n}}\sum_{j=1}^{[nt]} u_j + (nt - [nt])\frac{1}{\sqrt{n}} u_{[nt]+1}\right],$$

$$(3.96) \qquad u_j = \sum_{l=0}^{\infty} A_l \varepsilon_{j-l}, \qquad \sum_{l=0}^{\infty} l \, \|A_l\| < \infty,$$

$$(3.97) \qquad A = \sum_{l=0}^{\infty} A_l,$$

where $\{\varepsilon_j\} \sim$ i.i.d.$(0, I_q)$, $\|A_l\| = [\mathrm{tr}(A_l' A_l)]^{1/2}$ and A is nonsingular. Note that $V(\varepsilon_j) = I_q$ and we do not assume $A_0 = I_q$, but do assume A_0 to be nonsingular and block lower triangular.

The BN decomposition used before is also applied to decompose the vector u_j into

$$(3.98) \quad u_j = A\varepsilon_j + \tilde{\varepsilon}_{j-1} - \tilde{\varepsilon}_j, \qquad \tilde{\varepsilon}_j = \sum_{l=0}^{\infty} \tilde{A}_l \varepsilon_{j-l}, \qquad \tilde{A}_l = \sum_{k=l+1}^{\infty} A_k.$$

The sequence $\{\tilde{\varepsilon}_j\}$ also becomes stationary (Problem 10.2) with $0 < \|A\| < \infty$ and $0 < \sum_{l=0}^{\infty} \|\tilde{A}_l\| < \infty$. We now have

$$Y_n(t) = X_n(t) + R_n(t),$$

where

$$X_n(t) = \frac{1}{\sqrt{n}} \sum_{j=1}^{[nt]} \varepsilon_j + (nt - [nt]) \frac{1}{\sqrt{n}} \varepsilon_{[nt]+1},$$

$$R_n(t) = A^{-1} \left[\frac{1}{\sqrt{n}} (\tilde{\varepsilon}_0 - \tilde{\varepsilon}_{[nt]}) + (nt - [nt]) \frac{1}{\sqrt{n}} (\tilde{\varepsilon}_{[nt]} - \tilde{\varepsilon}_{[nt]+1}) \right].$$

It can be shown (Problem 10.3) that

$$(3.99) \quad |R_{in}(t)| \leq \|A^{-1}\| \left[\frac{1}{\sqrt{n}} \|\tilde{\varepsilon}_0\| + \frac{3}{\sqrt{n}} \max_{0 \leq j \leq n} \|\tilde{\varepsilon}_j\| \right] \to 0$$

in probability so that $\rho_q(Y_n, X_n) \to 0$ in probability, where $R_{in}(t)$ is the ith component of $R_n(t)$. Since $\mathcal{L}(X_n) \to \mathcal{L}(w)$, we establish the following theorem using the vector version of Theorem 3.6.

Theorem 3.14. *Let the sequence $\{Y_n\}$ of q-dimensional stochastic processes be defined by (3.95), (3.96) and (3.97), where $\{\varepsilon_j\} \sim$ i.i.d.$(0, I_q)$. Then $\mathcal{L}(Y_n) \to \mathcal{L}(w)$.*

As an application we consider the weak convergence of

$$(3.100) \quad V_T = \frac{1}{T^2} \sum_{j=1}^{T} y_j y_j',$$

where $y_j = y_{j-1} + u_j$, $y_0 = 0$ with $\{u_j\}$ being the linear process generated by (3.96) with $\{\varepsilon_j\} \sim$ i.i.d.$(0, I_q)$. Let us put

$$(3.101) \quad h(x) = \int_0^1 x(t) x'(t) \, dt, \qquad x \in C^q,$$

WEAK CONVERGENCE OF VECTOR-VALUED STOCHASTIC PROCESSES   99

which is a continuous function of x (Problem 10.4). Using $Y_T(t)$ defined in (3.95) and noting that $Y_T(j/T) = A^{-1}y_j/\sqrt{T}$ we have

$$A^{-1}V_T(A^{-1})' - h(Y_T) = \frac{1}{T^2}\sum_{j=1}^{T}(A^{-1}y_j)(A^{-1}y_j)' - \int_0^1 Y_T(t)Y_T'(t)\,dt$$

$$= \sum_{j=1}^{T}\int_{(j-1)/T}^{j/T}\left[Y_T\left(\frac{j}{T}\right)Y_T'\left(\frac{j}{T}\right) - Y_T(t)Y_T'(t)\right]dt.$$

The (k,l)th element of the integrand has the following bound (Problem 10.5):

(3.102)
$$\left|Y_{kT}\left(\frac{j}{T}\right)Y_{lT}\left(\frac{j}{T}\right) - Y_{kT}(t)Y_{lT}(t)\right|$$

$$\leq \left|\left(Y_{kT}\left(\frac{j}{T}\right) - Y_{kT}(t)\right)Y_{lT}\left(\frac{j}{T}\right)\right|$$

$$+ \left|\left(Y_{lT}\left(\frac{j}{T}\right) - Y_{lT}(t)\right)Y_{kT}(t)\right|$$

$$\leq 2\|A^{-1}\|\sup_{0\leq t\leq 1}\|Y(t)\|\max_{1\leq j\leq T}\frac{\|u_j\|}{\sqrt{T}},$$

which converges in probability to 0. Since $\mathcal{L}(h(Y_T)) \to \mathcal{L}(h(w))$ by the continuous mapping theorem, $\mathcal{L}(A^{-1}V_T(A^{-1})') \to \mathcal{L}(h(w))$ and, using again the continuous mapping theorem, we establish that

(3.103)
$$\mathcal{L}(V_T) \to \mathcal{L}(Ah(w)A')$$

$$= \mathcal{L}\left(A\int_0^1 w(t)w'(t)\,dtA'\right).$$

3.10.4 FCLT for the Vector-Valued Integrated Brownian Motion

In this subsection we extend the scalar I(d) ($d \geq 2$) processes discussed in Section 3.8 to vector processes, for which we establish the FCLT. Let us define

(3.104) $\quad (1-L)^d y_j^{(d)} = u_j, \quad y_{-(d-1)}^{(d)} = y_{-(d-2)}^{(d)} = \cdots = y_0^{(d)} = 0,$

(3.105) $\quad\quad u_j = \sum_{l=0}^{\infty} A_l\,\varepsilon_{j-l}, \quad \sum_{l=0}^{\infty} l\|A_l\| < \infty,$

(3.106) $\quad\quad A = \sum_{l=0}^{\infty} A_l,$

where $\{\varepsilon_j\}$ is i.i.d.$(0, I_q)$. Note that $y_j^{(d)} = y_{j-1}^{(d)} + y_j^{(d-1)}$ with $y_j^{(0)} = u_j$.

We now consider, as in Section 3.8, a sequence $\{Y_n^{(d)}\}$ of q-dimensional stochastic processes defined by

$$Y_n^{(d)}(t) = \frac{1}{n^{d-1/2}} y_{[nt]}^{(d)} + (nt - [nt]) \frac{1}{n^{d-1/2}} y_{[nt]+1}^{(d-1)}$$

$$= \frac{1}{n} \sum_{j=1}^{[nt]} Y_n^{(d-1)}\left(\frac{j}{n}\right) + (nt - [nt]) \frac{1}{n^{d-1/2}} y_{[nt]+1}^{(d-1)}.$$

It can be shown almost in the same way as in the scalar case (Problem 10.6) that

(3.107) $$\mathcal{L}(Y_n^{(2)}) \to \mathcal{L}(AF_1),$$

where $\{F_g(t)\}$ is the q-dimensional g-fold integrated Brownian motion defined by

$$F_g(t) = \int_0^t F_{g-1}(s)\,ds, \qquad F_0(t) = w(t).$$

For general d (≥ 3) we can prove (Problem 10.7) by induction that

(3.108) $$\mathcal{L}(Y_n^{(d)}) \to \mathcal{L}(AF_{d-1}).$$

As the first application let us consider the weak convergence of

$$S_T^{(d)} = \frac{1}{T^{2d}} \sum_{j=1}^{T} y_j^{(d)}(y_j^{(d)})'$$

$$= \frac{1}{T} \sum_{j=1}^{T} Y_T^{(d)}\left(\frac{j}{T}\right) \left(Y_T^{(d)}\left(\frac{j}{T}\right)\right)'.$$

For this purpose put

$$h(x) = \int_0^1 x(t)x'(t)\,dt, \qquad x \in C^q.$$

Then we have (Problem 10.8)

$$\|S_T^{(d)} - h(Y_T^{(d)})\|$$

$$\leq \sum_{j=1}^{T} \int_{(j-1)/T}^{j/T} \left\| Y_T^{(d)}\left(\frac{j}{T}\right)\left(Y_T^{(d)}\left(\frac{j}{T}\right)\right)' - Y_T^{(d)}(t)\left(Y_T^{(d)}(t)\right)' \right\| dt$$

$$\leq \frac{1}{T^{2d-1}} \max_{1 \leq j \leq T} \|y_j^{(d-1)}\|^2 + 2 \sup_{0 \leq t \leq 1} \|Y_T^{(d)}(t)\| \frac{1}{T^{d-1/2}} \max_{1 \leq j \leq T} \|y_j^{(d-1)}\|,$$

(3.109)

which evidently converges in probability to 0. Therefore we obtain, by the continuous mapping theorem,

$$\mathcal{L}(S_T^{(d)}) \to \mathcal{L}\big(h(AF_{d-1})\big)$$

$$= \mathcal{L}\left(A \int_0^1 F_{d-1}(t) F'_{d-1}(t)\, dt\, A'\right).$$

As the second application, we establish the weak convergence of

$$U_T^{(d)} = \frac{1}{T^{2d-1}} \sum_{j=1}^T y_{j-1}^{(d)} \left(y_j^{(d)} - y_{j-1}^{(d)}\right)'$$

$$= \sum_{j=1}^T Y_T^{(d)}\left(\frac{j-1}{T}\right) \left[Y_T^{(d)}\left(\frac{j}{T}\right) - Y_T^{(d)}\left(\frac{j-1}{T}\right)\right]'.$$

It can be shown (Problem 10.9) that

(3.110) $\quad \int_0^1 Y_T^{(d)}(t) \left(dY_T^{(d)}(t)\right)' = U_T^{(d)} + \frac{1}{2T^{2d-1}} \sum_{j=1}^T y_j^{(d-1)} \left(y_j^{(d-1)}\right)'.$

Here the last term on the right side converges in probability to the null matrix. Since $\mathcal{L}(Y_T^{(d)}) \to \mathcal{L}(AF_{d-1})$ and the limiting random vector is $(d-1)$-times continuously differentiable, we obtain, for $d \geq 2$,

$$\mathcal{L}(U_T^{(d)}) \to \mathcal{L}\left(A \int_0^1 F_{d-1}(t)\left(dF_{d-1}(t)\right)' A'\right)$$

$$= \mathcal{L}\left(A \int_0^1 F_{d-1}(t) F'_{d-2}(t)\, dt\, A'\right).$$

Because of the nonsymmetric nature of $U_T^{(d)}$, we cannot reduce the final expression above to a simple form. We, however, obtain (Problem 10.10)

(3.111) $\qquad \mathcal{L}\left(U_T^{(d)} + (U_T^{(d)})'\right) \to \mathcal{L}\left(AF_{d-1}(1) F'_{d-1}(1) A'\right)$

for $d \geq 2$.

Note that the above results do not hold for $d = 1$, which we discuss in the next section.

PROBLEMS

10.1 Establish the weak convergence in (3.94).

10.2 Prove that $\{\tilde{\varepsilon}_j\}$ defined in (3.98) is second-order stationary.

10.3 Prove the inequality in (3.99).

10.4 Prove that the function $h(x)$ defined in (3.101) is a continuous function of x.

10.5 Establish the weak convergence in (3.103) by proving that the right side of (3.102) converges in probability to 0.

10.6 Establish the weak convergence in (3.107).

10.7 Establish the weak convergence in (3.108) for $d \geq 3$.

10.8 Prove the inequalities in (3.109).

10.9 Show that the relation in (3.110) holds.

10.10 Establish the weak convergence in (3.111).

3.11 WEAK CONVERGENCE TO THE ITO INTEGRAL

As the final topic in this chapter, we discuss weak convergence to the Ito integral introduced in Sections 2.5 and 2.6 of Chapter 2. The difficulty consists in the fact that we cannot use the continuous mapping theorem because of the unbounded variation property of the Brownian motion.

Let us first deal with the scalar case and consider

$$S_T = \frac{1}{T} \sum_{j=1}^{T} x_{j-1}(x_j - x_{j-1}),$$

where $\{x_j\}$ is the near random walk defined by

$$x_j = \left(1 - \frac{\beta}{T}\right) x_{j-1} + \varepsilon_j, \qquad (j = 1, \ldots, T),$$

with $\{\varepsilon_j\} \sim$ i.i.d.$(0, 1)$. Suppose that $x_0 = \sqrt{T} X(0)$ with $X(0) \sim N(\gamma, \delta^2)$, which is independent of $\{\varepsilon_j\}$. Then we shall show

(3.112) $\quad \mathcal{L}(S_T) \to \mathcal{L}\left(\int_0^1 X(t)\, dX(t)\right) = \mathcal{L}\left(\frac{1}{2}(X^2(1) - X^2(0) - 1)\right),$

where $\{X(t)\}$ is the O–U process defined by

$$dX(t) = -\beta X(t)\, dt + dw(t) \iff X(t) = e^{-\beta t} X(0) + e^{-\beta t} \int_0^t e^{\beta s}\, dw(s).$$

To establish (3.112), it is convenient to rewrite S_T as

$$S_T = -\frac{1}{2T}\left[\sum_{j=1}^{T}(x_j - x_{j-1})^2 - \sum_{j=1}^{T}x_j^2 + \sum_{j=1}^{T}x_{j-1}^2\right]$$

$$= \frac{1}{2T}(x_T^2 - x_0^2) - \frac{1}{2T}\sum_{j=1}^{T}\left(-\frac{\beta}{T}x_{j-1} + \varepsilon_j\right)^2,$$

from which and Theorem 3.12 the latter expression on the right side of (3.112) evidently follows. The equivalence to the former is because of the Ito calculus described in (2.54).

If we follow arguments in Section 3.9, we are led to construct

$$X_T(t) = \frac{1}{\sqrt{T}}x_{j-1} + T\left(t - \frac{j-1}{T}\right)\frac{x_j - x_{j-1}}{\sqrt{T}}, \quad \left(\frac{j-1}{T} \leq t \leq \frac{j}{T}\right).$$

Then we have

$$S_T = \sum_{j=1}^{T}X_T\left(\frac{j-1}{T}\right)\left[X_T\left(\frac{j}{T}\right) - X_T\left(\frac{j-1}{T}\right)\right].$$

It can be shown (Problem 11.1) that

(3.113) $$\int_0^1 X_T(t)\,dX_T(t) = S_T + \frac{1}{2T}\sum_{j=1}^{T}\left(-\frac{\beta}{T}x_{j-1} + \varepsilon_j\right)^2.$$

Here the last term on the right side converges in probability to $\frac{1}{2}$ so that, although $\mathcal{L}(X_T) \to \mathcal{L}(X)$,

(3.114) $$\mathcal{L}\left(\int_0^1 X_T(t)\,dX_T(t)\right) \not\to \mathcal{L}\left(\int_0^1 X(t)\,dX(t)\right).$$

In fact it holds that

$$\mathcal{L}\left(\int_0^1 X_T(t)\,dX_T(t)\right) = \mathcal{L}\left(\frac{1}{2}(X_T^2(1) - X_T^2(0))\right)$$

$$\to \mathcal{L}\left(\frac{1}{2}(X^2(1) - X^2(0))\right)$$

$$= \mathcal{L}\left(\int_0^1 X(t)\,dX(t) + \frac{1}{2}\right).$$

The fact described in (3.114) is a consequence of the unbounded variation property of the Brownian motion. Note that $dX(t) = dw(t)$ when $\beta = 0$. The situation is certainly different if we consider the I(d) process with $d \geq 2$, as was discussed in Section 3.8 for the scalar case and in Section 3.10 for the vector case.

It is now easy to establish (Problem 11.2) that

$$(3.115) \qquad \mathcal{L}\left(\frac{1}{T}\sum_{j=1}^{T} x_{j-1}\varepsilon_j\right) \to \mathcal{L}\left(\int_0^1 X(t)\,dw(t)\right).$$

We can easily extend the above result to the case where $x_j = (1 - (\beta/T))\,x_{j-1} + u_j$ with $\{u_j\}$ being generated by

$$(3.116) \qquad u_j = \sum_{l=0}^{\infty} \alpha_l \varepsilon_{j-l}, \quad \sum_{l=0}^{\infty} l|\alpha_l| < \infty, \quad \{\varepsilon_j\} \sim \text{i.i.d.}(0,1).$$

The following theorem can be proved (Problem 11.3) by using Theorem 3.13 and the weak law of large numbers.

Theorem 3.15. *Suppose that $x_j = (1 - (\beta/T))x_{j-1} + u_j$, where $x_0 = \sqrt{T}\,\alpha X(0)$, $\alpha = \sum_{l=0}^{\infty} \alpha_l$ and $X(0) \sim N(\gamma, \delta^2)$ with $X(0)$ being independent of $\{u_j\}$ defined by (3.116). Then we have*

$$\mathcal{L}\left(\frac{1}{T}\sum_{j=1}^{T} x_{j-1}(x_j - x_{j-1})\right) \to \mathcal{L}\left(\frac{1}{2}\left(\alpha^2(X^2(1) - X^2(0)) - \sum_{l=0}^{\infty} \alpha_l^2\right)\right)$$

$$= \mathcal{L}\left(\alpha^2 \int_0^1 X(t)\,dX(t) + \frac{1}{2}\left(\alpha^2 - \sum_{l=0}^{\infty} \alpha_l^2\right)\right).$$

It is now an easy matter to establish (Problem 11.4) that

$$(3.117) \quad \mathcal{L}\left(\frac{1}{T}\sum_{j=1}^{T} x_{j-1}u_j\right) \to \mathcal{L}\left(\alpha^2 \int_0^1 X(t)\,dw(t) + \frac{1}{2}\left(\alpha^2 - \sum_{l=0}^{\infty} \alpha_l^2\right)\right).$$

Comparing (3.117) with (3.115) shows that the effect of the stationarity assumption on $\{u_j\}$ is not very simple. This is because of the presence of the Ito integral.

We next discuss the weak convergence of the random matrix defined by

$$(3.118) \qquad V_T = \frac{1}{T}\sum_{j=1}^{T} y_{j-1}(y_j - y_{j-1})',$$

where we assume, only for simplicity,

$$y_j = y_{j-1} + \varepsilon_j, \quad y_0 = 0,$$

$$\{\varepsilon_j\} \sim \text{i.i.d.}(0, I_q).$$

The present problem was solved by Chan and Wei (1988). We have

$$\mathcal{L}(V_T) \to \mathcal{L}\left(\int_0^1 w(t)\, dw'(t)\right), \tag{3.119}$$

where $\{w(t)\}$ is the q-dimensional standard Brownian motion. Using this result, it can be shown (Problem 11.5) that

$$\mathcal{L}\left(\frac{1}{T}\sum_{j=1}^{T} y'_j H \varepsilon_j\right) \to \mathcal{L}\left(\int_0^1 w'(t) H\, dw(t) + \text{tr}(H)\right) \tag{3.120}$$

for any $q \times q$ constant matrix H. A few special cases of (3.120) were discussed in Section 1.4 of Chapter 1, among which is Lévy's stochastic area.

The above situation was extended by Phillips (1988) to the case where

$$y_j = y_{j-1} + u_j, \quad y_0 = 0, \tag{3.121}$$

$$u_j = \sum_{l=0}^{\infty} A_l \varepsilon_{j-l}, \quad \sum_{l=0}^{\infty} l \|A_l\| < \infty, \tag{3.122}$$

$$A = \sum_{l=0}^{\infty} A_l, \tag{3.123}$$

with $\{\varepsilon_j\} \sim \text{i.i.d.}(0, I_q)$. The following is a simplified proof of Phillips (1988) for the weak convergence of V_T in (3.118). Using the BN decomposition, it can be shown (Problem 11.6) that

$$V_T = \frac{1}{T}\sum_{j=1}^{T}[Az_{j-1} + \tilde{\varepsilon}_0 - \tilde{\varepsilon}_{j-1}][A\varepsilon_j + \tilde{\varepsilon}_{j-1} - \tilde{\varepsilon}_j]' \tag{3.124}$$

$$= \frac{1}{T}\left[A\sum_{j=1}^{T} z_{j-1}\varepsilon'_j A' + A\sum_{j=1}^{T}\varepsilon_j\tilde{\varepsilon}'_j - \sum_{j=1}^{T}\tilde{\varepsilon}_{j-1}u'_j\right] + o_p(1),$$

where the term $o_p(1)$ is the matrix quantity that converges in probability to the null matrix, while

$$z_j = z_{j-1} + \varepsilon_j, \quad z_0 = 0,$$

$$\tilde{\varepsilon}_j = \sum_{l=0}^{\infty} \tilde{A}_l \varepsilon_{j-l}, \quad \tilde{A}_l = \sum_{k=l+1}^{\infty} A_k.$$

It holds (Problem 11.7) that

$$(3.125) \quad \frac{1}{T} A \sum_{j=1}^{T} \varepsilon_j \tilde{\varepsilon}_j' \to A(A - A_0)' \quad \text{in probability,}$$

$$(3.126) \quad \frac{1}{T} \sum_{j=1}^{T} \tilde{\varepsilon}_{j-1} u_j' \to \sum_{l=0}^{\infty} \left(\sum_{k=l+1}^{\infty} A_k \right) A_{l+1}' \quad \text{in probability,}$$

and the difference of these converges in probability to

$$(3.127) \quad A(A - A_0)' - \sum_{l=0}^{\infty} \left(\sum_{k=l+1}^{\infty} A_k \right) A_{l+1}' = \sum_{l=0}^{\infty} \sum_{m=l+1}^{\infty} A_l A_m'$$

$$= \sum_{l=0}^{\infty} \sum_{k=1}^{\infty} A_l A_{k+l}'$$

$$= \sum_{k=1}^{\infty} E(u_0 u_k').$$

Applying (3.119) to the first term on the right side of (3.124) and using (3.127), we can establish the following theorem.

Theorem 3.16. *Suppose that the q-dimensional I(1) process $\{y_j\}$ is generated by (3.121) with (3.122) and (3.123). Then we have*

$$(3.128) \quad \mathcal{L}\left(\frac{1}{T} \sum_{j=1}^{T} y_{j-1}(y_j - y_{j-1})' \right) \to \mathcal{L}\left(A \int_0^1 w(t) \, dw'(t) A' + \Lambda \right),$$

where

$$\Lambda = \sum_{k=1}^{\infty} E(u_0 u_k') = \sum_{k=1}^{\infty} \sum_{l=0}^{\infty} A_l A_{k+l}'.$$

It can be checked (Problem 11.8) that the present theorem reduces to Theorem 3.15 with $X(t) = w(t)$ when quantities are scalar.

Finally we indicate how to derive weak convergence results associated with statistics in the form of matrix quotients. For this purpose let us consider

$$(3.129) \quad \hat{B} = \sum_{j=2}^{T} y_j y_{j-1}' \left(\sum_{j=2}^{T} y_{j-1} y_{j-1}' \right)^{-1},$$

where $\{y_j\}$ is the q-dimensional I(1) process defined by (3.121) together with (3.122) and (3.123). We have $T(\hat{B} - I_q) = U_T V_T^{-1}$, where

$$U_T = \frac{1}{T} \sum_{j=2}^{T} u_j y'_{j-1}, \qquad V_T = \frac{1}{T^2} \sum_{j=2}^{T} y_{j-1} y'_{j-1}.$$

Then it holds that

$$\mathcal{L}(\theta_1 U_T + \theta_2 V_T) \to \mathcal{L}\big(\theta_1 h_1(w) + \theta_2 h_2(w)\big)$$

for any θ_1 and θ_2, where

$$h_1(w) = \left(A \int_0^1 w(t)\, dw'(t) A' + \Lambda\right)',$$

$$h_2(w) = A \int_0^1 w(t) w'(t)\, dt A'.$$

Therefore we can deduce that $\mathcal{L}(U_T, V_T) \to \mathcal{L}(h_1(w), h_2(w))$. Since $P(h_2(w) > 0) = 1$ if A is nonsingular, we finally obtain, by the continuous mapping theorem,

$$\mathcal{L}(U_T V_T^{-1}) = \mathcal{L}(T(\hat{B} - I_q)) \to \mathcal{L}(h_1(w) h_2^{-1}(w))$$

$$= \mathcal{L}\left(\left(A \int_0^1 w(t)\, dw'(t) A' + \Lambda\right)' \left(A \int_0^1 w(t) w'(t)\, dt A'\right)^{-1}\right).$$

(3.130)

A similar procedure can be used to derive weak convergence results for other kinds of matrix-valued statistics discussed in later chapters. A simple, scalar case corresponding to (3.130) was earlier discussed in Section 1.3 of Chapter 1.

PROBLEMS

11.1 Establish the relation in (3.113).
11.2 Establish the weak convergence in (3.115).
11.3 Prove Theorem 3.15.
11.4 Establish the weak convergence in (3.117).
11.5 Establish the weak convergence in (3.120).
11.6 Derive the expressions for V_T in (3.124).
11.7 Prove the convergence results (3.125) and (3.126).
11.8 Show that Theorem 3.16 reduces to Theorem 3.15 with $X(t) = w(t)$ when quantities in the former are scalar.

CHAPTER 4

The Stochastic Process Approach

We present a method for computing the c.f.s of quadratic or bilinear functionals of the Brownian motion, where functionals involve the single Riemann integral or the Ito integral. In doing so, we use a theorem concerning a transformation of measures induced by stochastic processes, from which the present approach called the *stochastic process approach* originates. It is recognized that the stochastic process approach does not require knowledge of eigenvalues, unlike the eigenvalue approach presented in Chapter 1. Advantages and disadvantages of the present approach are discussed in the last section.

4.1 GIRSANOV'S THEOREM: CASE 1

We have discussed some FCLTs in Chapter 3 and have indicated how to derive weak convergence results for various statistics. It is seen that the limiting random variables are functionals of the Brownian motion. In statistical applications, we need to compute distribution functions of those limiting random variables. In general, however, it is difficult to derive the distribution functions directly.

Here we present a method that we call the *stochastic process approach* for computing the c.f.s or m.g.f.s for quadratic and bilinear functionals of the Brownian motion. In this and the next sections, quadratic functionals are dealt with, while bilinear functionals are treated in Section 4.3. To illustrate the present methodology, we reconsider the statistics presented in Chapter 1, whose c.f.s were derived by the eigenvalue approach. It should be emphasized that the present approach does not require knowledge of eigenvalues, unlike the eigenvalue approach.

The stochastic process approach relies on *Girsanov's theorem* concerning a transformation of measures induced by stochastic processes. The idea is as follows: Suppose that $f(X)$ is a functional of a stochastic process $\{X(t)\}$ and that we would like to compute $E(f(X))$. Then, defining an auxiliary process $\{Y(t)\}$ that is equivalent in the sense of measures μ_X and μ_Y that the two processes induce, Girsanov's theorem yields $E(f(X)) = E(f(Y)d\mu_X(Y)/d\mu_Y)$, where the expectation on the left is taken with respect to μ_X, and that on the right with respect to μ_Y. An appropriate choice of $\{Y(t)\}$ may make the computation feasible.

As an example, let us take up

(4.1) $$S_1 = \int_0^1 X^2(t)\, dt,$$

where $\{X(t)\}$ is the O–U process defined by

(4.2) $$dX(t) = -\alpha X(t)\, dt + dw(t) \Leftrightarrow X(t) = e^{-\alpha t} X(0) + e^{-\alpha t} \int_0^t e^{\alpha s}\, dw(s)$$

with $\{w(t)\}$ being the one-dimensional standard Brownian motion. Note that $X(0)$ is assumed to be independent of increments of $\{w(t)\}$.

Our concern is to compute the c.f. $\phi_{S_1}(\theta) = E(e^{i\theta S_1})$. For this purpose, the following theorem due to Girsanov (1960) [see also Liptser and Shiryayev (1977, p. 277)] is useful.

Theorem 4.1. *Let $X = \{X(t) : 0 \leq t \leq 1\}$ and $Y = \{Y(t) : 0 \leq t \leq 1\}$ be the O–U processes on $C = C[0, 1]$ defined by*

(4.3) $$dX(t) = -\alpha X(t)\, dt + dw(t),$$
(4.4) $$dY(t) = -\beta Y(t)\, dt + dw(t), \quad X(0) = Y(0).$$

Let μ_X and μ_Y be probability measures on $(C, \mathcal{B}(C))$ induced by X and Y, respectively, by the relation

(4.5) $$\mu_X(A) = P(\omega : X \in A), \quad \mu_Y(A) = P(\omega : Y \in A), \quad A \in \mathcal{B}(C).$$

Then measures μ_X and μ_Y are equivalent and

(4.6) $$\frac{d\mu_X}{d\mu_Y}(x) = \exp\left[(\beta - \alpha) \int_0^1 x(t)\, dx(t) - \frac{\alpha^2 - \beta^2}{2} \int_0^1 x^2(t)\, dt\right],$$

where the left side is the Radon–Nikodym derivative evaluated at $x \in C$ with $x(0) = X(0)$.

Roughly speaking, the Radon–Nikodym derivative in (4.6) is an extended version of the likelihood ratio under contiguity. Suppose that $y_j = \rho_n(\alpha) y_{j-1} + \varepsilon_j$ ($j = 1, \ldots, n$) with $y_0 = 0$ and $\{\varepsilon_j\} \sim \text{NID}(0, 1)$. Let $l_n(\alpha)$ be the likelihood for $y_1, \ldots, y_n$ under $\rho_n(\alpha) = 1 - (\alpha/n)$. Then it holds (Problem 1.1 in this chapter) that

(4.7) $$\mathcal{L}\left(\left.\frac{l_n(\alpha)}{l_n(\beta)}\right|_\gamma\right) \to \mathcal{L}\left(\exp\left[(\beta - \alpha) \int_0^1 y(t)\, dy(t) - \frac{\alpha^2 - \beta^2}{2} \int_0^1 y^2(t)\, dt\right]\right),$$

where $dy(t) = -\gamma y(t)\, dt + dw(t)$ with $y(0) = 0$.

GIRSANOV'S THEOREM: CASE 1

On the basis of Theorem 4.1, it is an easy matter to compute $\phi_{S_1}(\theta)$. Let us first consider the case where $X(0) = Y(0) = \kappa$, a constant. Since

$$E(f(X)) = E(f(Y) d\mu_X(Y)/d\mu_Y),$$

we obtain [Liptser and Shiryayev (1978, p. 208) and Problem 1.2]

$$\begin{aligned}
(4.8)\ E(e^{\theta S_1}) &= E\left[\exp\left\{\theta \int_0^1 X^2(t)\,dt\right\}\right] \\
&= E\left[\exp\left\{\theta \int_0^1 Y^2(t)\,dt\right\} \frac{d\mu_X}{d\mu_Y}(Y)\right] \\
&= E\left[\exp\left\{\left(\theta - \frac{\alpha^2 - \beta^2}{2}\right)\int_0^1 Y^2(t)\,dt + (\beta - \alpha)\int_0^1 Y(t)\,dY(t)\right\}\right] \\
&= E\left[\exp\left\{\frac{\beta - \alpha}{2}(Y^2(1) - \kappa^2 - 1)\right\}\right] \\
&= \exp\left[\frac{\alpha}{2} + \frac{\kappa^2 \theta(\sinh\beta/\beta)}{\cosh\beta + \alpha(\sinh\beta/\beta)}\right]\left[\cosh\beta + \alpha\frac{\sinh\beta}{\beta}\right]^{-1/2},
\end{aligned}$$

where $\beta = \sqrt{\alpha^2 - 2\theta}$ and we have used the fact that $Y(1) \sim N(\kappa e^{-\beta}, (1 - e^{-2\beta})/(2\beta))$. Therefore we obtain

$$\begin{aligned}
(4.9)\quad \phi_{S_1}(\theta) &= E\left[\exp\left\{i\theta \int_0^1 X^2(t)\,dt\right\}\right] \\
&= \exp\left[\frac{\alpha}{2} + \frac{i\kappa^2 \theta(\sin\lambda/\lambda)}{\cos\lambda + \alpha(\sin\lambda/\lambda)}\right]\left[\cos\lambda + \alpha\frac{\sin\lambda}{\lambda}\right]^{-1/2},
\end{aligned}$$

where $\lambda = \sqrt{2i\theta - \alpha^2}$.

Consider a special case where $\alpha = 0$ so that

$$X(t) = \kappa + w(t) \iff dX(t) = dw(t),\quad X(0) = \kappa.$$

Then we have, from (4.9),

$$E\left[\exp\left\{i\theta \int_0^1 (\kappa + w(t))^2\,dt\right\}\right] = \exp\left\{i\kappa^2\theta\frac{\tan\sqrt{2i\theta}}{\sqrt{2i\theta}}\right\}\left(\cos\sqrt{2i\theta}\right)^{-1/2}.$$

A further special case of (4.9) with $\alpha = \kappa = 0$ so that $X(t) = w(t)$ leads us to

$$(4.10)\qquad E\left[\exp\left\{i\theta \int_0^1 w^2(t)\,dt\right\}\right] = (\cos\sqrt{2i\theta})^{-1/2},$$

which was formally presented in Section 1.1 of Chapter 1.

As the second example, we consider

$$S_2 = \int_0^1 \left\{ X(t) - \int_0^1 X(s)\,ds \right\}^2 dt = \int_0^1 X^2(t)\,dt - \left(\int_0^1 X(t)\,dt \right)^2,$$

where $\{X(t)\}$ is the O–U process defined in (4.2) with $X(0) = \kappa$. Proceeding in the same way as before we obtain (Problem 1.3)

$$E\left(e^{\theta S_2}\right) = E\left[\exp\left\{ \frac{\beta - \alpha}{2}(Y^2(1) - Y^2(0) - 1) - \theta \left(\int_0^1 Y(t)\,dt \right)^2 \right\}\right]$$

$$= \exp\left[\frac{1}{2}(\alpha - \beta)(\kappa^2 + 1)\right]$$

$$\times E\left[\exp\left\{ \frac{\beta - \alpha}{2} Y^2(1) - \theta \left(\int_0^1 Y(t)\,dt \right)^2 \right\}\right]$$

$$= \exp\left[\frac{\alpha}{2} + \frac{\alpha^2 \kappa^2 \theta}{g(\theta)} \left\{ \frac{1}{\beta^2} \frac{\sinh \beta}{\beta} - \frac{2}{\beta^4}(\cosh \beta - 1) \right\} \right] (g(\theta))^{-1/2},$$

(4.11)

where $\beta = \sqrt{\alpha^2 - 2\theta}$ and

$$g(\theta) = \frac{\alpha^3 - 2\theta}{\beta^2} \frac{\sinh \beta}{\beta} + \left(\frac{\alpha^2}{\beta^2} - \frac{4\alpha\theta}{\beta^4} \right) \cosh \beta + \frac{4\alpha\theta}{\beta^4}.$$

It can be checked that, when $\alpha = \kappa = 0$,

(4.12) $$E\left[\exp\left\{ i\theta \int_0^1 \left(w(t) - \int_0^1 w(s)\,ds \right)^2 dt \right\}\right] = \left(\frac{\sin \sqrt{2i\theta}}{\sqrt{2i\theta}} \right)^{-1/2},$$

which is the c.f. associated with the demeaned Brownian motion. In connection with (4.12) we can obtain (Problem 1.4) that

(4.13) $$E\left[\exp\left\{ i\theta \int_0^1 (w(t) - tw(1))^2 dt \right\}\right] = \left(\frac{\sin \sqrt{2i\theta}}{\sqrt{2i\theta}} \right)^{-1/2},$$

which is the c.f. associated with the Brownian bridge and was formally presented in Section 1.2 of Chapter 1. The equivalence of the left sides of (4.12) and (4.13) has now been established.

In the above arguments we have assumed the initial value $X(0)$ to be a constant. Suppose now that $X(0) \sim N(0, 1/(2\alpha))$ with $\alpha > 0$ so that $\{X(t)\}$ is stationary. Then

Theorem 4.1 yields (Problem 1.5)

$$(4.14) \quad E\left(e^{\theta S_1}\right) = E\left[\exp\left\{\frac{\beta-\alpha}{2}(Y^2(1) - Y^2(0) - 1)\right\}\right]$$

$$= e^{\alpha/2}\left[\cosh\beta + \frac{\alpha^2-\theta}{\alpha}\frac{\sinh\beta}{\beta}\right]^{-1/2},$$

$$(4.15) \quad E\left(e^{\theta S_2}\right) = E\left[\exp\left\{\frac{\beta-\alpha}{2}(Y^2(1) - Y^2(0) - 1) - \theta\left(\int_0^1 Y(t)\,dt\right)^2\right\}\right]$$

$$= e^{\alpha/2}\left[\left(1 + \frac{2\theta}{\beta^2} - \frac{2\alpha\theta}{\beta^4}\right)\cosh\beta\right.$$

$$\left. + \left(\alpha + \frac{(\alpha-2)\theta}{\beta^2}\right)\frac{\sinh\beta}{\beta} + \frac{2\alpha\theta}{\beta^4}\right]^{-1/2},$$

where $\beta = \sqrt{\alpha^2 - 2\theta}$. A much simpler way of deriving (4.14) and (4.15) is possible since we have already obtained $E(e^{\theta S_1})$ and $E(e^{\theta S_2})$ when $X(0) = \kappa$. Replacing κ in (4.8) by $X(0)$ and noting that $E\left[\exp\{\theta X^2(0)\}\right] = (1 - \theta/\alpha)^{-1/2}$, we have, in the present case,

$$E\left(e^{\theta S_1}\right) = E\left[E\left\{e^{\theta S_1} \mid X(0)\right\}\right]$$

$$= e^{\alpha/2}\left[\left(1 - \frac{\theta}{\alpha g(\theta)}\frac{\sinh\beta}{\beta}\right)g(\theta)\right]^{-1/2}$$

$$= e^{\alpha/2}\left[g(\theta) - \frac{\theta}{\alpha}\frac{\sinh\beta}{\beta}\right]^{-1/2},$$

where $g(\theta) = \cosh\beta + \alpha\sinh\beta/\beta$. This gives us (4.14). The derivation of (4.15) can be done similarly from (4.11).

The present approach can also be applied to obtain the c.f. corresponding to a random variable in ratio form. Let us consider

$$(4.16) \quad R_1 = \frac{\int_0^1 X(t)\,dX(t)}{\int_0^1 X^2(t)\,dt} = \frac{U}{S_1},$$

where $\{X(t)\}$ is the O–U process defined in (4.2) with $X(0) = \kappa$.

The statistic R_1 may be regarded as the MLE of $-\alpha$ for the O–U process if the likelihood $l(-\alpha)$ for $\{X(t)\}$ is interpreted as

$$l(-\alpha) = \exp\left[-\alpha\int_0^1 X(t)\,dX(t) - \frac{\alpha^2}{2}\int_0^1 X^2(t)\,dt\right] = \frac{d\mu_X}{d\mu_w}(X).$$

Since $P(R_1 \leq x) = P(xS_1 - U \geq 0)$, we are led to compute [Perron (1991a) and Problem 1.6]

$$(4.17) \quad E\left[\exp\{\theta(xS_1 - U)\}\right] = E\left[\exp\left\{\theta\left(x\int_0^1 X^2(t)\,dt - \int_0^1 X(t)\,dX(t)\right)\right\}\right]$$

$$= E\left[\exp\left\{\frac{\beta - \alpha - \theta}{2}(Y^2(1) - \kappa^2 - 1)\right\}\right]$$

$$= \exp\left[\frac{\alpha + \theta}{2} + \frac{\kappa^2\theta\left(\alpha + \frac{\theta}{2} + x\right)(\sinh\beta/\beta)}{\cosh\beta + (\alpha + \theta)(\sinh\beta/\beta)}\right]$$

$$\times \left[\cosh\beta + (\alpha + \theta)\frac{\sinh\beta}{\beta}\right]^{-1/2},$$

where $\beta = \sqrt{\alpha^2 - 2\theta x}$. The joint m.g.f. of U and S_1 can be easily obtained from (4.17) (Problem 1.7).

Moments of R_1 can also be computed following formula (1.39) as

$$E(R_1^k) = \frac{1}{(k-1)!}\int_0^\infty \theta_2^{k-1}\left.\frac{\partial^k \psi(\theta_1, -\theta_2)}{\partial\theta_1^k}\right|_{\theta_1=0} d\theta_2,$$

where $\psi(\theta_1, -\theta_2) = E[\exp\{\theta_1 U - \theta_2 S_1\}] = m(-\theta_1; \theta_2/\theta_1)$ with $m(\theta; x)$ being the m.g.f. of $xS_1 - U$ given in (4.17).

When $X(t) = w(t)$ so that $\alpha = \kappa = 0$, we have

$$(4.18) \quad E\left[\exp\{i\theta(xS_1 - U)\}\right] = E\left[\exp\left\{i\theta\left(x\int_0^1 w^2(t)\,dt - \int_0^1 w(t)\,dw(t)\right)\right\}\right]$$

$$= e^{i\theta/2}\left[\cos\sqrt{2i\theta x} + i\theta\frac{\sin\sqrt{2i\theta x}}{\sqrt{2i\theta x}}\right]^{-1/2},$$

which was first obtained by White (1958) and was discussed in Section 1.3 of Chapter 1.

If $X(0) \sim N(0, 1/(2\alpha))$ with $\alpha > 0$, we obtain (Problem 1.8), by the conditional argument described below (4.15),

$$E\left[\exp\{\theta(xS_1 - U)\}\right] = \exp\left(\frac{\alpha + \theta}{2}\right)\left[\cosh\beta + \frac{2\alpha^2 - \theta^2 - 2\theta x}{2\alpha}\frac{\sinh\beta}{\beta}\right]^{-1/2},$$

(4.19)

where $\beta = \sqrt{\alpha^2 - 2\theta x}$.

We note in passing (Problem 1.9) that the statistic R_1 in (4.16) naturally appears in

$$(4.20) \quad \mathcal{L}(T(\hat{\rho} - 1)) \to \mathcal{L}(R_1),$$

where $\hat{\rho}$ is the LSE of ρ in the near integrated model

(4.21)
$$y_j = \rho y_{j-1} + \varepsilon_j, \qquad (j = 1, \ldots, T),$$
$$\rho = 1 - \frac{\alpha}{T}, \quad \{\varepsilon_j\} \sim \text{i.i.d.}(0, 1),$$

with $y_0 = \sqrt{T}\kappa$ or $y_0 = \sqrt{T}X(0)$. Here we can recognize the usefulness of the present approach since it is quite complicated to obtain results like (4.17) and (4.19) by the eigenvalue approach. More general models than (4.21) will be considered in Chapter 7, where we also discuss approximations to the finite sample distribution of $\hat{\rho}$.

As another example of ratio statistics, we consider

(4.22)
$$R_2 = \frac{\int_0^1 X_1(t)\, dw_2(t)}{\int_0^1 X_1^2(t)\, dt} = \frac{V}{S_1},$$

where $\{X_1(t)\}$ is the O–U process defined by $dX_1(t) = -\alpha X_1(t)\, dt + dw_1(t)$ with $X_1(0) = \kappa$ and $\{w_1(t), w_2(t)\}$ is the two-dimensional standard Brownian motion. It is seen that, given $\{w_1(t)\}$ or S_1, R_2 is conditionally normal with the conditional mean 0 and the conditional variance S_1^{-1}. This fact may be expressed in a sophisticated way [Phillips (1989)] as

(4.23)
$$\mathcal{L}(R_2) = \mathcal{L}\left(\int_{S_1 > 0} N(0, S_1^{-1})\, dQ(S_1^{-1})\right),$$

where Q is the probability measure associated with S_1^{-1}. We also have that, for any real x, $xS_1 - V$ is conditionally normal with

$$E\left[xS_1 - V \mid \{w_1(t)\}\right] = xS_1, \quad V\left[xS_1 - V \mid \{w_1(t)\}\right] = S_1.$$

Thus, using (4.9), we can easily derive

(4.24)
$$E[\exp\{i\theta(xS_1 - V)\}] = E\left[\exp\left\{i\left(\theta x + \frac{i\theta^2}{2}\right)S_1\right\}\right]$$
$$= \exp\left[\frac{\alpha}{2} + \frac{i\kappa^2\theta\,(x + i\theta/2)\,(\sin\lambda/\lambda)}{\cos\lambda + \alpha(\sin\lambda/\lambda)}\right]$$
$$\times \left[\cos\lambda + \alpha\,\frac{\sin\lambda}{\lambda}\right]^{-1/2},$$

where $\lambda = \sqrt{2i\theta x - \theta^2 - \alpha^2}$.

Similarly, when $X_1(0) \sim N(0, 1/(2\alpha))$ with $\alpha > 0$, and $X_1(0)$ is independent of $\{w_2(t)\}$ and increments of $\{w_1(t)\}$, we obtain (Problem 1.10), using (4.14),

$$(4.25) \quad E\left[\exp\{i\theta(xS_1 - V)\}\right] = e^{\alpha/2} \left[\cos\lambda + \frac{\alpha^2 - i\theta\left(x + i\theta/2\right)}{\alpha} \frac{\sin\lambda}{\lambda}\right]^{-1/2},$$

where $\lambda = \sqrt{2i\theta x - \theta^2 - \alpha^2}$.

The statistic R_2 in (4.22) naturally appears in (Problem 1.11)

$$(4.26) \quad \mathcal{L}(T(\hat{\beta} - \beta)) \to \mathcal{L}(R_2),$$

where $\hat{\beta}$ is the LSE of β in the model

$$(4.27) \quad \begin{aligned} y_{2j} &= \beta y_{1j} + \varepsilon_{2j}, \\ y_{1j} &= \left(1 - \frac{\alpha}{T}\right) y_{1,j-1} + \varepsilon_{1j}, \quad (j = 1, \ldots, T), \end{aligned}$$

where $(\varepsilon_{1j}, \varepsilon_{2j})' \sim$ i.i.d.$(0, I_2)$ and $y_{10} = \sqrt{T}\kappa$ or $y_{10} \sim N(0, T/(2\alpha))$. In Section 1.5 of Chapter 1 we considered a simpler model with $\alpha = \kappa = 0$, which is a simplified version of the cointegrated system, and the c.f. corresponding to (4.24) was obtained in (1.83).

PROBLEMS

1.1 Establish the weak convergence result in (4.7).
1.2 Derive the m.g.f. in (4.8).
1.3 Derive the m.g.f. in (4.11) noting that

$$Y(t) = \kappa e^{-\beta t} + e^{-\beta t} \int_0^t e^{\beta s} dw(s),$$

$$\int_0^1 Y(t)\, dt = (1 - e^{-\beta})\frac{\kappa}{\beta} - \frac{1}{\beta}\int_0^1 \left(e^{-\beta(1-t)} - 1\right) dw(t).$$

1.4 Establish the result in (4.13).
1.5 Derive the m.g.f.s in (4.14) and (4.15).
1.6 Derive the m.g.f. in (4.17).
1.7 Derive the joint m.g.f. of U and S_1 defined in (4.16).
1.8 Derive the m.g.f. in (4.19) by the conditional argument described below (4.15).
1.9 Establish the weak convergence result in (4.20).
1.10 Derive the m.g.f. in (4.25).
1.11 Establish the weak convergence result in (4.26).

4.2 GIRSANOV'S THEOREM: CASE 2

In this section, we extend Girsanov's Theorem 4.1 to cover the case where $\{X(t)\}$ is the g-fold integrated Brownian motion ($g \geq 1$). Thus we consider

$$(4.28) \qquad F_g(t) = \int_0^t F_{g-1}(s)\,ds, \qquad F_0(t) = w(t)$$

and put

$$(4.29) \quad dY_g(t) = \beta Y_g(t)\,dt + dF_g(t) = \left(\beta Y_g(t) + F_{g-1}(t)\right)dt, \quad Y_g(0) = 0.$$

Our purpose here is to obtain the m.g.f. $m_g(\theta)$ of

$$(4.30) \qquad S(F_g) = \int_0^1 F_g^2(t)\,dt, \qquad (g \geq 1).$$

If it holds that $m_g(\theta) = E[\exp\{\theta S(Y_g)\,d\mu_{F_g}(Y_g)/d\mu_{Y_g}\}]$, where μ_{F_g} and μ_{Y_g} are measures induced by $\{F_g(t)\}$ and $\{Y_g(t)\}$, respectively, then the computation of $m_g(\theta)$ may be feasible. For this purpose we establish the following theorem.

Theorem 4.2. *Let $\{F_g(t)\}$ and $\{Y_g(t)\}$ be defined by (4.28) and (4.29), respectively. Then probability measures μ_{F_g} and μ_{Y_g} are equivalent and*

$$(4.31) \quad \frac{d\mu_{F_g}}{d\mu_{Y_g}}(y) = \exp\left[-\beta \int_0^1 \frac{d^g y(t)}{dt^g}\,d\left(\frac{d^g y(t)}{dt^g}\right) + \frac{\beta^2}{2}\int_0^1 \left(\frac{d^g y(t)}{dt^g}\right)^2 dt\right],$$

where $y(0) = 0$ and $y \in C^{(g)}$ — the space of g-times continuously differentiable functions on $[0,1]$.

Proof. It follows from (4.29) that

$$(4.32) \qquad Y_g(t) = e^{\beta t}\int_0^t e^{-\beta s}F_{g-1}(s)\,ds,$$

which is g-times continuously differentiable so that it holds (Problem 2.1) that

$$(4.33) \qquad Z(t) \equiv \frac{d^g Y_g(t)}{dt^g} = \beta \int_0^t Z(s)\,ds + w(t)$$

and thus

$$(4.34) \qquad dZ(t) = \beta Z(t)\,dt + dw(t), \qquad Z(0) = 0.$$

The measures μ_w and μ_Z are evidently equivalent by Theorem 4.1 and, by the same theorem, we have

$$(4.35) \qquad \rho(x) \equiv \frac{d\mu_w}{d\mu_Z}(x) = \exp\left[-\beta \int_0^1 x(t)\,dx(t) + \frac{\beta^2}{2}\int_0^1 x^2(t)\,dt\right]$$

for $x \in C$ with $x(0) = 0$. Noting that

$$\frac{d^g F_g(t)}{dt^g} = w(t), \qquad \frac{d^g Y_g(t)}{dt^g} = Z(t),$$

we may put $F_g(t) = \Phi_g(w)(t)$ and $Y_g(t) = \Phi_g(Z)(t)$, where

$$(4.36) \qquad \Phi_g(x)(t) = \int_0^t \Phi_{g-1}(x)(s)\,ds, \qquad \Phi_0(x)(t) = x(t), \qquad x \in C.$$

Since $\mu_{F_g}(A) = P(\omega : F_g \in A) = P(\omega : w \in \Phi_g^{-1}(A))$ for $A \in \mathcal{B}(C^{(g)})$, we have $\mu_{F_g} = \mu_w \Phi_g^{-1}$. Similarly we have $\mu_{Y_g} = \mu_Z \Phi_g^{-1}$. Thus measures μ_{F_g} and μ_{Y_g} are equivalent and

$$\frac{d\mu_{F_g}}{d\mu_{Y_g}}(y) = \rho\left(\Phi_g^{-1}(y)\right) = \frac{d\mu_w}{d\mu_Z}\left(\Phi_g^{-1}(y)\right), \qquad y \in C^{(g)},$$

which establishes the theorem since $\Phi_g^{-1}(y)(t) = d^g y(t)/dt^g$ because of (4.36). $\square$

A heuristic derivation of the above theorem follows. Consider the discrete-time processes defined by

$$(4.37) \qquad y_j = \left(1 + \frac{\beta}{T}\right) y_{j-1} + \frac{\varepsilon_j}{(1-L)^g},$$

$$(4.38) \qquad z_j = (1-L)^g y_j = \left(1 + \frac{\beta}{T}\right) z_{j-1} + \varepsilon_j, \quad (j = 1, \ldots, T),$$

where $\{\varepsilon_j\} \sim \text{NID}(0,1)$ and $y_j = z_j = 0$ for $j \leq 0$. Note that, when $\beta = 0$, $(1-L)^{g+1} y_j = \varepsilon_j$ and $(1-L) z_j = \varepsilon_j$. Let $l_T(0)$ and $l_T(\beta)$ be the likelihoods for $(y_1, \ldots, y_T)$ under $\beta = 0$ and $\beta \neq 0$, respectively. Then we can show (Problem 2.2) that

$$(4.39) \qquad \mathcal{L}\left(\left.\frac{l_T(0)}{l_T(\beta)}\right|_\beta\right) \to \mathcal{L}\left(\exp\left[-\beta \int_0^1 Z(t)\,dZ(t) + \frac{\beta^2}{2}\int_0^1 Z^2(t)\,dt\right]\right),$$

where $Z(t)$ is defined in (4.33).

GIRSANOV'S THEOREM: CASE 2

We now consider $E[\exp\{\theta S(F_g)\}] = E[\exp\{\theta S(Y_g)\,d\mu_{F_g}(Y_g)/d\mu_{Y_g}\}]$, which involves the computation of

$$(4.40) \quad Z(t) = \frac{d^g Y_g(t)}{dt^g} = \beta \frac{d^{g-1} Y_g(t)}{dt^{g-1}} + w(t)$$

$$= \beta^g Y_g(t) + \beta^{g-1} F_{g-1}(t) + \beta^{g-2} F_{g-2}(t) + \cdots + \beta F(t) + w(t).$$

The general case is evidently difficult to deal with. Let us restrict our attention to the case $g = 1$ and consider (Problem 2.3)

$$(4.41) \quad m_1(\theta) = E\left[\exp\left\{\theta \int_0^1 F_1^2(t)\,dt\right\}\right]$$

$$= E\left[\exp\left\{\theta \int_0^1 Y_1^2(t)\,dt\right\} \frac{d\mu_{F_1}}{d\mu_{Y_1}}(Y_1)\right]$$

$$= E\left[\exp\left\{\theta \int_0^1 Y_1^2(t)\,dt - \beta \int_0^1 \frac{dY_1(t)}{dt}\,d\left(\frac{dY_1(t)}{dt}\right)\right.\right.$$

$$\left.\left. + \frac{\beta^2}{2}\int_0^1 \left(\frac{dY_1(t)}{dt}\right)^2 dt\right\}\right]$$

$$= E\left[\exp\left\{\frac{\beta^2}{2}\int_0^1 w^2(t)\,dt - \frac{\beta}{2}w^2(1)\right.\right.$$

$$\left.\left. -\beta^2 w(1)e^\beta \int_0^1 e^{-\beta t} w(t)\,dt + \frac{\beta}{2}\right\}\right],$$

where $\beta = (2\theta)^{1/4}$. We are in the same situation as was discussed in the last section. Define $dX(t) = \gamma X(t)\,dt + dw(t)$ with $X(0) = 0$ and apply Theorem 4.1 to obtain (Problem 2.4)

$$(4.42) \quad m_1(\theta) = E\left[\exp\left\{-\frac{\beta+\gamma}{2}X^2(1) - \beta^2 X(1)e^\beta \int_0^1 e^{-\beta t}X(t)\,dt + \frac{\beta+\gamma}{2}\right\}\right]$$

$$= \left[\frac{1}{2}\left\{1 + \cos(2\theta)^{1/4}\cosh(2\theta)^{1/4}\right\}\right]^{-1/2},$$

where $\gamma = i\beta$.

The present approach can also be applied to derive the m.g.f.s associated with ratio statistics. Let us first consider

$$(4.43) \quad R_1 = \frac{\int_0^1 F_1(t)\,dF_1(t)}{\int_0^1 F_1^2(t)\,dt} = \frac{U}{S(F_1)}.$$

Then we obtain (Problem 2.5)

(4.44) $E[\exp\{\theta(xS(F_1) - U)\}]$

$$= E\left[\exp\left\{\frac{\beta^2}{2}\int_0^1 w^2(t)\,dt - \frac{\beta}{2}w^2(1) - \frac{\theta}{2}Y_1^2(1) - \beta^2 w(1)Y_1(1) + \frac{\beta}{2}\right\}\right]$$

$$= E\left[\exp\left\{-\frac{\beta + \gamma}{2}X^2(1) - \frac{\theta}{2}e^{2\beta}Z^2 - \beta^2 e^{\beta}X(1)Z + \frac{\beta + \gamma}{2}\right\}\right]$$

$$= \left[\frac{1}{2}(1 + \cos\beta\cosh\beta) + \frac{\beta^2}{4x}\left(\cosh\beta\frac{\sin\beta}{\beta} - \cos\beta\frac{\sinh\beta}{\beta}\right)\right]^{-1/2},$$

where $\beta = (2\theta x)^{1/4}$, $\gamma = i\beta$ and

(4.45) $$Z = \int_0^1 e^{-\beta t}X(t)\,dt.$$

The statistic R_1 in (4.43) arises in (Problem 2.6)

(4.46) $$\mathcal{L}(T(\hat{\rho} - 1)) \to \mathcal{L}(R_1),$$

where $\hat{\rho}$ is the LSE of $\rho\,(=1)$ in the model:

(4.47) $y_j = \rho y_{j-1} + v_j,$ $v_j = v_{j-1} + \varepsilon_j,$ $(j = 1, \ldots, T)$

with $v_0 = y_0 = 0$ and $\{\varepsilon_j\} \sim$ i.i.d.$(0, 1)$.

We also consider

(4.48) $$R_2 = \frac{\int_0^1 F_1(t)\,dw_2(t)}{\int_0^1 F_1^2(t)\,dt} = \frac{V}{S(F_1)},$$

where $\{w_2(t)\}$ is the standard Brownian motion independent of $\{F_1(t)\}$. Using the conditional argument given in the last section, we can easily obtain

(4.49) $$E[\exp\{\theta(xS(F_1) - V)\}] = \exp\left[\left(\theta x + \frac{\theta^2}{2}\right)S(F_1)\right]$$

$$= m_1\left(\theta x + \frac{\theta^2}{2}\right),$$

where $m_1(\theta)$ is defined in (4.42).

The statistic R_2 in (4.48) arises in (Problem 2.7)

(4.50) $$\mathcal{L}(T^2(\hat{\beta} - \beta)) \to \mathcal{L}(R_2),$$

where $\hat{\beta}$ is the LSE of β in the second-order cointegrated model:

(4.51)
$$y_{2j} = \beta y_{1j} + \varepsilon_{2j},$$
$$(1 - L)^2 y_{1j} = \varepsilon_{1j}, \quad (j = 1, \ldots, T),$$

with $(\varepsilon_{1j}, \varepsilon_{2j})' \sim$ i.i.d.$(0, I_2)$ and $y_{1,-1} = y_{10} = 0$.

The computation of the m.g.f.s or c.f.s for higher order integrated processes is much involved because of the complicated expression for $d^g Y_g(t)/dt^g$ given in (4.40). Even the case $g = 2$ turns out to be hard to deal with. In the next chapter we present another approach that makes the computation feasible by making use of computerized algebra.

PROBLEMS

2.1 Prove that the process $\{Y_g(t)\}$ defined in (4.32) satisfies the relation in (4.33).
2.2 Establish the weak convergence result in (4.39).
2.3 Derive the expressions in (4.41).
2.4 Obtain the m.g.f. $m_1(\theta)$ as in (4.42).
2.5 Derive the m.g.f. in (4.44).
2.6 Establish the weak convergence result in (4.46).
2.7 Establish the weak convergence result in (4.50).

4.3 GIRSANOV'S THEOREM: CASE 3

In this section we deal with the q-dimensional standard Brownian motion $\{w(t)\}$ and consider

(4.52) $$dX(t) = AX(t)\,dt + dw(t),$$
(4.53) $$dY(t) = BY(t)\,dt + dw(t), \quad X(0) = Y(0) = 0,$$

where A and B are $q \times q$ constant matrices. The processes $\{X(t)\}$ and $\{Y(t)\}$ are q-dimensional O–U processes, for which it is known [Arnold (1974, p. 129)] that

(4.54) $$X(t) = e^{At} \int_0^t e^{-As}\,dw(s), \quad Y(t) = e^{Bt} \int_0^t e^{-Bs}\,dw(s).$$

Note that e^{At} is a matrix-valued function of t defined by

(4.55) $$e^{At} = \sum_{n=0}^{\infty} \frac{t^n}{n!} A^n.$$

Later we shall also introduce matrix-valued functions $\cosh At = (e^{At} + e^{-At})/2$, $\sinh At = (e^{At} - e^{-At})/2$, $\tanh At = (\cosh At)^{-1} \sinh At$, and so on.

Girsanov's theorem still applies to the above situation and is stated as follows [Liptser and Shiryayev (1977, p. 279)].

Theorem 4.3. *Let μ_X and μ_Y be probability measures induced by $\{X(t)\}$ and $\{Y(t)\}$, respectively. Then μ_X and μ_Y are equivalent and*

$$\frac{d\mu_X}{d\mu_Y}(x) = \exp\left[\int_0^1 x'(t)(A-B)'\,dx(t) - \frac{1}{2}\int_0^1 x'(t)(A-B)'(A+B)x(t)\,dt\right],$$

(4.56)

where $x \in C^q$ with $x(0) = 0$.

As an application of this theorem, we derive the m.g.f. of

(4.57) $$S_1 = \int_0^1 w'(t)Hw(t)\,dt,$$

where H is a $q \times q$ symmetric matrix. Of course, it is easier in this case to use the result in (4.10) for the scalar case and the independence property of components of $\{w(t)\}$, which yields (Problem 3.1)

(4.58) $$E(e^{\theta S_1}) = \prod_{j=1}^{q}(\cos\sqrt{2\lambda_j\theta})^{-1/2},$$

where λ_js are the eigenvalues of H. Nonetheless we consider

$$E(e^{\theta S_1}) = E\left[\exp\left\{\theta\int_0^1 X'(t)HX(t)\,dt\right\}\frac{d\mu_w}{d\mu_X}(X)\right]$$

$$= E\left[\exp\left\{\int_0^1 X'(t)\left(\theta H + \frac{1}{2}A'A\right)X(t)\,dt - \int_0^1 X'(t)A'\,dX(t)\right\}\right].$$

Putting $A^2 = -2\theta H$ with A symmetric and using the matrix version of Ito's theorem [Arnold (1974, p. 143)]

$$d(X(t)X'(t)) = X(t)\,dX'(t) + dX(t)X'(t) + I_q\,dt,$$

we obtain (Problem 3.2), if A is symmetric,

(4.59) $$\int_0^1 X'(t)A'\,dX(t) = \int_0^1 X'(t)A\,dX(t)$$

$$= \frac{1}{2}[X'(1)AX(1) - \text{tr}(A)],$$

from which we derive (4.58) (Problem 3.3).

GIRSANOV'S THEOREM: CASE 3

It is an immediate consequence of the result in (4.58) to obtain, for example, the m.g.f. of

(4.60) $$S_2 = \int_0^1 w_1(t)w_2(t)\,dt = \int_0^1 w'(t)Hw(t)\,dt,$$

where H is the 2×2 matrix with $H_{11} = H_{22} = 0$ and $H_{12} = H_{21} = \frac{1}{2}$. We have (Problem 3.4)

(4.61) $$E(e^{\theta S_2}) = (\cos\sqrt{\theta}\cosh\sqrt{\theta})^{-1/2}.$$

The statistic S_2 in (4.60) was discussed in Section 1.4 of Chapter 1.

Another important and interesting statistic takes the following form:

(4.62) $$S_3 = \int_0^1 w'(t)G\,dw(t),$$

where G is any constant matrix. If G is symmetric, then we can use the relation in (4.59) and readily obtain the distribution of S_3 (Problem 3.5).

Two cases of G's being not symmetric were also discussed in Section 1.4 of Chapter 1. One was

(4.63) $$S_4 = \int_0^1 w_1(t)\,dw_2(t) = \int_0^1 w'(t)\begin{pmatrix} 0 & 1 \\ 0 & 0 \end{pmatrix} dw(t)$$

and it can be shown by the conditional argument (Problem 3.6) that $E(e^{\theta S_4}) = (\cos\theta)^{-1/2}$. The other was Lévy's stochastic area defined by

(4.64) $$S_5 = \frac{1}{2}\int_0^1 \left[w_1(t)\,dw_2(t) - w_2(t)\,dw_1(t)\right]$$

$$= \int_0^1 w'(t)\begin{pmatrix} 0 & \frac{1}{2} \\ -\frac{1}{2} & 0 \end{pmatrix} dw(t).$$

It holds (Problem 3.7) that $E[S_5\,|\,\{w_1(t)\}] = 0$ and

(4.65) $$V[S_5\,|\,\{w_1(t)\}] = \int_0^1 \left(w_1(t) - \frac{1}{2}w_1(1)\right)^2 dt.$$

We can now obtain (Problem 3.8) $E(e^{\theta S_5}) = (\cos(\theta/2))^{-1}$.

PROBLEMS

3.1 Derive the m.g.f. of S_1 in (4.57) using the result in Section 4.1 and the independence property of components of $\{w(t)\}$.

3.2 Establish the relation in (4.59).

3.3 Derive the m.g.f. of S_1 in (4.57) using the relation in (4.59).

3.4 Obtain the m.g.f. of S_2 as in (4.61).

3.5 Derive the m.g.f. of S_3 in (4.62) when G is symmetric.

3.6 Derive the m.g.f. of S_4 in (4.63).

3.7 Compute the conditional variance of S_5 in (4.65) given $\{w_1(t)\}$.

3.8 Derive the m.g.f. of S_5 in (4.64) using (4.65) and the relations described in Problem 1.3.

4.4 THE CAMERON–MARTIN FORMULA

Here we concentrate on quadratic functionals of the q-dimensional standard Brownian motion $\{w(t)\}$ and present a formula for computing the m.g.f.s especially designed for such functionals. The following result, known as the Cameron–Martin formula, is established in Liptser and Shiryayev (1977, p. 280).

Theorem 4.4. *Let $H(t)$ be a $q \times q$ symmetric nonnegative definite matrix whose elements $H_{jk}(t)$ are continuous and satisfy the condition*

$$\int_0^1 \sum_{j,k=1}^q |H_{jk}(t)|\, dt < \infty.$$

Then it holds that

$$(4.66) \quad E\left[\exp\left\{-\int_0^1 w'(t)H(t)w(t)\, dt\right\}\right] = \exp\left[\frac{1}{2}\int_0^1 \mathrm{tr}\bigl(G(t)\bigr) dt\right],$$

where $G(t)$ is a $q \times q$ symmetric nonpositive definite matrix, being a unique solution of the matrix-valued Riccati differential equation

$$(4.67) \quad \frac{dG(t)}{dt} = 2H(t) - G^2(t), \qquad G(1) = 0.$$

It is difficult, in general, to solve the matrix equation in (4.67). Suppose that $H(t)$ is a constant matrix so that we put $H(t) = H$. Then it is known [Bellman (1970, p. 323)] that the solution is given by

$$(4.68) \quad G(t) = D\tanh D(t-1) = D\bigl(\cosh D(t-1)\bigr)^{-1}\sinh D(t-1),$$

where D is a positive square root of $2H$, that is $D = (2H)^{1/2}$. Using the facts that

(4.69) $$\frac{d \log(\cosh D(t-1))}{dt} = D \tanh D(t-1),$$

(4.70) $$\exp\left[\frac{1}{2} \text{tr}\{\log(\cosh D)\}\right] = \prod_{j=1}^{q} \sqrt{\cosh \delta_j},$$

where δ_js are the eigenvalues of D, we can show that

(4.71) $$E\left[\exp\left\{\theta \int_0^1 w'(t)Hw(t)\,dt\right\}\right] = \prod_{j=1}^{q}(\cos\sqrt{2\lambda_j\theta})^{-1/2},$$

where λ_js are the eigenvalues of H. This result was already obtained in the last section. Evidently the assumption of positive definiteness of H is not necessary.

For our purpose, the usefulness of the Cameron–Martin formula depends crucially on the solvability of the matrix equation (4.67). Even for the scalar case, it cannot be solved explicitly, in general. Consider, for example

(4.72) $$S = \int_0^1 t^m w^2(t)\,dt.$$

Then Theorem 4.4 leads us to

$$E(e^{\theta S}) = \exp\left\{\frac{1}{2}\int_0^1 g(t)\,dt\right\},$$

where $g(t)$ is the solution to

$$\frac{dg(t)}{dt} = -2\theta t^m - g^2(t), \quad g(1) = 0.$$

The explicit solution may be obtained by quadrature for some ms after tedious efforts; then it must be integrated.

In the next chapter we consider another approach and obtain the m.g.f. of S in (4.72) for any $m\,(> -1)$.

4.5 ADVANTAGES AND DISADVANTAGES OF THE PRESENT APPROACH

The success of the stochastic process approach depends crucially on the computability of the expectation of a functional of the Brownian motion, where the expectation is taken with respect to the transformed measure given by Girsanov's theorem. As we have seen, the present approach is quite successful in dealing with quadratic functionals of the O–U process. In fact, it will be seen in the next chapter that the present approach is more suitable for the analysis of the O–U process than the approach introduced there.

In practice, however, we need to deal with other classes of functionals of the Brownian motion. As an example, let us consider the process

$$(4.73) \qquad dX(t) = \mu(X(t),t)\,dt + \sigma(X(t),t)\,dw(t),$$

which was introduced in Section 2.7 of Chapter 2 as the Ito stochastic differential equation. The existence of a unique solution to (4.73) was also discussed there. In connection with this process, we consider an auxiliary process

$$(4.74) \qquad dY(t) = m(Y(t),t)\,dt + \sigma(Y(t),t)\,dw(t), \quad Y(0) = X(0).$$

Then the two measures μ_X and μ_Y are equivalent under some suitable conditions and the Radon–Nikodym derivative is given in Liptser and Shiryayev (1977, p. 277) by

$$(4.75) \qquad \frac{d\mu_X}{d\mu_Y}(x) = \exp\left[\int_0^1 \frac{\mu(x(t),t) - m(x(t),t)}{\sigma^2(x(t),t)}\,dx(t) \right.$$
$$\left. - \frac{1}{2}\int_0^1 \frac{\mu^2(x(t),t) - m^2(x(t),t)}{\sigma^2(x(t),t)}\,dt\right],$$

where $x \in C$ with $x(0) = X(0)$. Then it might be thought to be possible to compute $E(f(X)) = E(f(Y)d\mu_X(Y)/d\mu_Y)$ for a functional $f(X)$, but the computation turns out to be difficult.

To see the difficulty, let us consider $dX(t) = t^k dw(t)$ with $X(0) = 0$, which is a special case of (4.73) with $\mu = 0$ and $\sigma = t^k$. Then, using (4.75), we shall have, for example,

$$(4.76) \quad E\left[\exp\left(\theta \int_0^1 X^2(t)\,dt\right)\right] = E\left[\exp\left\{-\sqrt{-2\theta}\int_0^1 \frac{Y(t)}{t^k}\,dY(t)\right\}\right],$$

where we have put $m(Y(t),t) = \sqrt{-2\theta}\,t^k Y(t)$. The stochastic process $\{Y(t)\}$ in the present case has the solution

$$Y(t) = \exp\left(\frac{\sqrt{-2\theta}\,t^{k+1}}{k+1}\right)\int_0^t \exp\left(-\frac{\sqrt{-2\theta}\,s^{k+1}}{k+1}\right) s^k\,dw(s),$$

but this does not help in computing the right side of (4.76) except for $k = 0$.

In the next chapter we consider another approach that overcomes the above difficulty. Since

$$dX(t) = t^k\,dw(t), \quad X(0) = 0 \iff X(t) = \int_0^t s^k\,dw(s),$$

ADVANTAGES AND DISADVANTAGES OF THE PRESENT APPROACH

the Riemann integral in (4.76) can be rewritten as

$$\int_0^1 X^2(t)\,dt = \int_0^1 \left\{ \int_0^t \int_0^t u^k v^k\, dw(u)\, dw(v) \right\} dt$$

$$= \int_0^1 \int_0^1 [1 - \max(s,t)]\, s^k t^k\, dw(s)\, dw(t).$$

This last expression enables us to obtain the c.f. by the approach presented in the next chapter.

We have also noted in Section 4.4 that the stochastic process approach is not suitable for obtaining the c.f. of the following form:

$$S = \int_0^1 t^{2k} w^2(t)\, dt$$

$$= \int_0^1 \int_0^1 \frac{1}{2k+1} \left[1 - (\max(s,t))^{2k+1} \right] dw(s)\, dw(t).$$

This last expression will again make the derivation of the c.f. possible.

CHAPTER 5

The Fredholm Approach

We present another method for computing the c.f.s of quadratic plus linear or bilinear functionals of the Brownian motion, where functionals are expressed by the Riemann–Stieltjes integral. This method requires some knowledge of the theory of integral equations of the Fredholm type, among which are the Fredholm determinant and the resolvent. We give an introductory discussion of these, together with various examples of how to derive them. We then indicate by some theorems and examples how to relate the c.f. to the Fredholm determinant and the resolvent. It turns out that the present approach enables us to deal with a wider class of functionals than do the stochastic process and eigenvalue approaches.

5.1 MOTIVATING EXAMPLES

In the first four sections of this chapter, we deal with a statistic expressed in the Riemann–Stieltjes integral

(5.1) $$S = \int_0^1 \int_0^1 K(s,t)\,dw(s)\,dw(t),$$

where $K(s,t)$ ($\neq 0$) is a function with some conditions imposed later, while $\{w(t)\}$ is the one-dimensional standard Brownian motion. The statistic S covers a wide class of quadratic functionals of the Brownian motion. As an example, let $X(t)$ be the O–U process defined by $dX(t) = -\beta X(t)\,dt + dw(t)$ with $X(0) = 0$. Then we have (Problem 1.1 in this chapter)

(5.2) $$\int_0^1 X^2(t)\,dt = \int_0^1 \int_0^1 \frac{e^{-\beta|s-t|} - e^{-\beta(2-s-t)}}{2\beta}\,dw(s)\,dw(t).$$

If $X(0)$ is a nonzero constant or a random variable, the integral on the right side of (5.2) contains a linear or bilinear functional of the Brownian motion as well. The quadratic, plus linear or bilinear forms, will be dealt with in Section 5.5.

As is seen above, the integral of the square of the O–U process with the initial value equal to 0 can always be expressed as in (5.1), and we have indicated in the last chapter that the c.f. of (5.2) can be easily obtained by the stochastic process approach. Thus, as far as the O–U process is concerned, there is no advantage in using the double integral expression as on the right side of (5.2). We, however, need to consider the other processes that do not fall into the O–U process. As an example, let us consider

$$(5.3) \qquad S_1 = \int_0^1 \left(g(t)w(t)\right)^2 dt = \int_0^1 \int_0^1 K(s,t) \, dw(s) \, dw(t),$$

where $g(t)$ is a nonstochastic, continuous function and

$$K(s,t) = \int_{\max(s,t)}^1 g^2(u) \, du.$$

Here $Y(t) = g(t)w(t)$ is not the O–U process, and it is difficult to obtain the c.f. of S_1 by the stochastic process approach, as was discussed in Section 4.5 of Chapter 4.

Some other examples that motivate the double integral expression follow. In Section 4.3 of Chapter 4 we dealt with Lévy's stochastic area defined by

$$(5.4) \qquad S_2 = \frac{1}{2} \int_0^1 \left[w_1(t) \, dw_2(t) - w_2(t) \, dw_1(t)\right],$$

where $w(t) = (w_1(t), w_2(t))'$ is the two-dimensional standard Brownian motion. The expression on the right side of (5.4) is far from the double integral expression, but we have seen that, given $\{w_1(t)\}$, S_2 is conditionally normal with the mean 0 and the variance given by the right side of (4.65), which can be rewritten (Problem 1.2) as

$$(5.5) \qquad V\left[S_2 \mid \{w_1(t)\}\right] = \int_0^1 \int_0^1 \frac{1}{4} \left[1 - 2|s - t|\right] dw_1(s) \, dw_1(t).$$

Then we can proceed to the computation of

$$E(e^{i\theta S_2}) = E\left[\exp\left\{-\frac{\theta^2}{2} \int_0^1 \int_0^1 \frac{1}{4} \left[1 - 2|s-t|\right] dw_1(s) \, dw_1(t)\right\}\right].$$

Returning to the one-dimensional standard Brownian motion $\{w(t)\}$, let us consider

$$(5.6) \qquad S_3 = \int_0^1 g(t) w(t) \, dw(t).$$

The conditional argument is not applicable here. Assuming $g(t)$ to be differentiable, we use Ito's theorem to obtain

$$d(g(t)w^2(t)) = 2g(t)w(t) \, dw(t) + \left(g'(t)w^2(t) + g(t)\right) dt$$

MOTIVATING EXAMPLES

so that we have (Problem 1.3)

(5.7) $$S_3 = \frac{1}{2}\int_0^1\int_0^1 g(\max(s,t))\,dw(s)\,dw(t) - \frac{1}{2}\int_0^1 g(t)\,dt.$$

We also consider the g-fold integrated Brownian motion $\{F_g(t)\}$ and put

(5.8) $$S_4 = \int_0^1 F_g^2(t)\,dt$$

$$= \int_0^1\int_0^1\left[\int_{\max(s,t)}^1 \frac{((u-s)(u-t))^g}{(g!)^2}\,du\right]dw(s)\,dw(t).$$

In Section 4.2 of Chapter 4, we have obtained the c.f. of S_4 for $g = 1$ by the stochastic process approach. For $g \geq 2$, however, that approach is very involved, as was mentioned there. We shall obtain the c.f. of S_4 for $g = 2$ using the last expression in (5.8).

The above examples deal directly with the Riemann–Stieltjes integrals. In some cases, these integrals naturally emerge from quadratic forms under finite samples. Here we take up just one example. Let us consider

(5.9) $$S_{T5} = \frac{1}{T^2}\varepsilon' \Sigma_T\, \varepsilon,$$

where $\varepsilon = (\varepsilon_1,\ldots,\varepsilon_T)'$ with $\{\varepsilon_j\} \sim$ i.i.d.$(0,1)$, while Σ_T is the $T \times T$ matrix with the (j,k)th element $\Sigma_T(j,k)$ being $\min(j,k) - (jk/T)$. Here it holds that, for $K(s,t) = \min(s,t) - st$,

$$\frac{1}{T}\Sigma_T(j,k) = K\left(\frac{j}{T},\frac{k}{T}\right) \quad \text{for all} \quad j,k.$$

Although we leave a rigorous treatment to Section 5.6, we can show that

$$\mathcal{L}(S_{T5}) = \mathcal{L}\left(\frac{1}{T}\sum_{j,k=1}^T K\left(\frac{j}{T},\frac{k}{T}\right)\varepsilon_j\varepsilon_k\right)$$

$$\to \mathcal{L}\left(\int_0^1\int_0^1 [\min(s,t) - st]\,dw(s)\,dw(t)\right).$$

In the present case, the limiting random variable may be expressed by the Riemann integral (Problem 1.4), but, given (5.9), the present expression seems more natural.

Thus we strongly feel the need of a study on how to compute the c.f. of the statistic that takes the form in (5.1). For this purpose, we introduce an approach, which we call the *Fredholm approach*, named after the Swedish mathematician E. I. Fredholm (1866–1927). To develop this approach, we require some knowledge of the theory of *integral equations of Fredholm type*, which we now describe briefly, following Courant and Hilbert (1953), Hochstadt (1973), and Whittaker and Watson (1958).

PROBLEMS

1.1 Establish the relation in (5.2).
1.2 Derive the expression in (5.5).
1.3 Establish the relation in (5.7).
1.4 Obtain the limiting form of $\mathcal{L}(S_{T5})$ in (5.9) by the FCLT discussed in Chapter 3.

5.2 THE FREDHOLM THEORY: THE HOMOGENEOUS CASE

Let us consider the integral equation for λ and $f(t)$

$$(5.10) \qquad f(t) = \lambda \int_0^1 K(s,t) f(s)\, ds,$$

where $K(s,t)$ is a known, continuous function on $[0,1] \times [0,1]$. A value λ (possibly complex) for which this integral equation possesses a nonvanishing continuous solution is called an *eigenvalue* of the *kernel* $K(s,t)$; the corresponding solution $f(t)$ is called an *eigenfunction* for the eigenvalue λ. The maximum number l of linearly independent solutions is called the *multiplicity* of λ.

It is usually difficult to solve for λ and $f(t)$ analytically. For our purpose, however, it is not necessary. We have only to obtain the *Fredholm determinant*, whose definition is given below.

The integral equation (5.10) may be approximated by the algebraic system

$$f\left(\frac{j}{T}\right) = \frac{\lambda}{T} \sum_{k=1}^{T} K\left(\frac{j}{T}, \frac{k}{T}\right) f\left(\frac{k}{T}\right) \qquad (j = 1, \ldots, T),$$

or, in matrix notation,

$$(5.11) \qquad f_T = \frac{\lambda}{T} K_T f_T,$$

where $f_T = [(f(j/T))]$ is a $T \times 1$ vector and $K_T = [(K(j/T, k/T))]$ is a $T \times T$ matrix. As in the theory of matrices, we can consider the characteristic equation that determines λ. In the present case, however, we study the asymptotic behavior, as $T \to \infty$, of

$$(5.12) \qquad D_T(\lambda) = \left| I_T - \frac{\lambda}{T} K_T \right|.$$

Clearly $D_T(\lambda)$ is a polynomial of degree T in λ. Thus we may put

$$(5.13) \qquad D_T(\lambda) = \sum_{l=0}^{T} \frac{a_l(T)}{l!} \lambda^l, \qquad a_l(T) = \left.\frac{d^l}{d\lambda^l} D_T(\lambda)\right|_{\lambda=0}.$$

THE FREDHOLM THEORY: THE HOMOGENEOUS CASE

Clearly $a_0(T) = 1$ and we have (Problem 2.1)

(5.14) $$a_1(T) = -\frac{1}{T}\sum_{j=1}^{T} K\left(\frac{j}{T},\frac{j}{T}\right) \to -\int_0^1 K(t,t)\,dt,$$

(5.15) $$a_2(T) = \frac{1}{T^2}\sum_{j,k=1}^{T} \begin{vmatrix} K\left(\frac{j}{T},\frac{j}{T}\right) & K\left(\frac{j}{T},\frac{k}{T}\right) \\ K\left(\frac{k}{T},\frac{j}{T}\right) & K\left(\frac{k}{T},\frac{k}{T}\right) \end{vmatrix}$$

$$\to \int_0^1\int_0^1 \begin{vmatrix} K(s,s) & K(s,t) \\ K(t,s) & K(t,t) \end{vmatrix} ds\,dt.$$

More generally it can be shown [Hochstadt (1973, p. 237)] that

$$a_l(T) = \frac{(-1)^l}{T^l}\sum_{j_1,\ldots,j_l} K\begin{pmatrix} \frac{j_1}{T} & \cdots & \frac{j_l}{T} \\ \frac{j_1}{T} & \cdots & \frac{j_l}{T} \end{pmatrix}$$

$$\to (-1)^l \int_0^1\cdots\int_0^1 K\begin{pmatrix} t_1 & \cdots & t_l \\ t_1 & \cdots & t_l \end{pmatrix} dt_1\cdots dt_l,$$

where

(5.16) $$K\begin{pmatrix} t_1 & \cdots & t_l \\ t_1 & \cdots & t_l \end{pmatrix} = \begin{vmatrix} K(t_1,t_1) & \cdots & K(t_1,t_l) \\ \vdots & & \vdots \\ K(t_l,t_1) & \cdots & K(t_l,t_l) \end{vmatrix}.$$

Then it can also be derived from (5.13) that

(5.17) $$D(\lambda) = \lim_{T\to\infty} D_T(\lambda)$$

$$= \sum_{n=0}^{\infty} \frac{(-1)^n \lambda^n}{n!}\int_0^1\cdots\int_0^1 K\begin{pmatrix} t_1 & \cdots & t_n \\ t_1 & \cdots & t_n \end{pmatrix} dt_1\cdots dt_n.$$

The function $D(\lambda)$ is called the *Fredholm determinant* (FD) of the kernel $K(s,t)$. It holds that the series in (5.17) converges for all λ, that is, $D(\lambda)$ is an *entire* or *integral* function with $D(0) = 1$. In fact, it holds [Hochstadt (1973, p. 239)] that

$$|D(\lambda)| \leq \sum_{n=0}^{\infty} \frac{(|\lambda|M\sqrt{n})^n}{n!},$$

where $M = \max|K(s,t)|$, and the series on the right side can be seen to converge for all λ by the ratio test.

Some properties of entire functions follow. Let $h(z)$ be an entire function of z that may be complex with $h(0) = 1$, let the zeros of $h(z)$ be at $a_1, a_2, \ldots$, where $\lim_{n\to\infty} |a_n| = \infty$, and let the zero at a_n be of order m_n. Then $h(z)$ can be expanded [Whittaker and Watson (1958, p. 139)] as

$$(5.18) \qquad h(z) = e^{G(z)} \prod_{n=1}^{\infty} \left[\left\{ \left(1 - \frac{z}{a_n}\right) e^{g_n(z)} \right\}^{m_n} \right],$$

where $G(z)$ is some entire function such that $G(0) = 0$, while

$$g_n(z) = \frac{z}{a_n} + \frac{1}{2}\left(\frac{z}{a_n}\right)^2 + \cdots + \frac{1}{k_n}\left(\frac{z}{a_n}\right)^{k_n}$$

with k_n being the smallest integer such that

$$\left| m_n \left(\frac{N}{a_n}\right)^{k_n+1} \right| < b_n, \qquad \sum_{n=1}^{\infty} b_n < \infty,$$

for a constant N. We put $g_n(z) = 0$ if $k_n = 0$. If it is possible to choose all the k_n equal to each other, then $k = k_n$ is called the *genus* associated with the infinite product.

In particular, if $D(\lambda)$ is the FD of a continuous kernel $K(s, t)$ with an infinite number of eigenvalues $\{\lambda_n\}$, the infinite product takes the form [Hochstadt (1973, p. 249)]

$$(5.19) \qquad D(\lambda) = \exp\left\{-\lambda \int_0^1 K(t,t)\, dt\right\} \prod_{n=1}^{\infty} \left[\left\{ \left(1 - \frac{\lambda}{\lambda_n}\right) \exp\left(\frac{\lambda}{\lambda_n}\right) \right\}^{m_n} \right],$$

where m_n is the order of the zero of $D(\lambda)$ at λ_n. Thus $D(\lambda)$ in the present case is an entire function of genus unity. Note that m_n is not necessarily equal to the multiplicity of λ_n. A much simpler representation for $D(\lambda)$ with genus zero and m_n equal to the multiplicity of λ_n will be obtained later by imposing some conditions on $K(s, t)$.

The following theorem holds because of (5.19) for the relationship between the zeros of $D(\lambda)$ and the eigenvalues of $K(s, t)$.

Theorem 5.1. *Every zero of $D(\lambda)$ is an eigenvalue of K, and in turn every eigenvalue of K is a zero of $D(\lambda)$.*

Note that zero is never an eigenvalue since $D(0) = 1 \neq 0$. It sometimes happens that $D(\lambda)$ never becomes zero so that there exists no eigenvalue. It is known, however, that, if $K(s, t)$ is symmetric as well as continuous, then there exists at least one eigenvalue insofar as $K(s, t)$ is not identically equal to zero.

In subsequent discussions, we assume $K(s, t)$ to be continuous and symmetric on $[0, 1] \times [0, 1]$. Then every eigenvalue is real. If there are an infinite number of eigenvalues, $K(s, t)$ is said to be *nondegenerate*; otherwise it is *degenerate*. When $K(s, t)$ is

THE FREDHOLM THEORY: THE HOMOGENEOUS CASE   135

nondegenerate, $\lambda = \infty$ is the only accumulation point of zeros. If all the eigenvalues of K have the same sign, then $K(s,t)$ is said to be *definite*. Alternatively, $K(s,t)$ is positive (negative) definite if $\int_0^1 \int_0^1 K(s,t)g(s)g(t)\,ds\,dt$ is nonnegative (nonpositive) for any continuous function $g(t)$ on $[0,1]$. If all but a finite number of eigenvalues have the same sign, $K(s,t)$ is said to be *nearly definite*. A necessary and sufficient condition for $K(s,t)$ to be nearly definite is that it can be expressed as the sum of a definite kernel and degenerate kernels.

Some examples of $K(s,t)$ follow. Consider the following functions:

(5.20) $K_1(s,t) = 1 - \max(s,t)$, $K_2(s,t) = \min(s,t) - st$, $K_3(s,t) = g(s)g(t)$,

where $g(t)$ is continuous. Then it can be shown (Problem 2.2) that these are all positive definite. It will be recognized later that K_3 is degenerate, while K_1 and K_2 are nondegenerate.

Suppose that $K(s,t)$ is nearly definite, which we shall also assume in subsequent discussions. Then the following theorem called *Mercer's theorem* holds [Hochstadt (1973, p. 91)].

Theorem 5.2. *Let $K(s,t)$ be continuous, symmetric, and nearly definite on $[0,1] \times [0,1]$. Then*

(5.21) $$K(s,t) = \sum_{n=1}^{\infty} \frac{1}{\lambda_n} f_n(s) f_n(t),$$

where $\{\lambda_n\}$ is a sequence of eigenvalues repeated as many times as their multiplicities, while $\{f_n(t)\}$ is an orthonormal sequence of eigenfunctions corresponding to eigenvalues λ_n and the series on the right side converges absolutely and uniformly to $K(s,t)$.

Mercer's theorem tells us (Problem 2.3) that

(5.22) $$\sum_{n=1}^{\infty} \frac{1}{\lambda_n} = \int_0^1 K(t,t)\,dt,$$

where λ_n is repeated as many times as its multiplicity. Mercer's theorem will also be effectively used for deriving the c.f. of a quadratic functional of the Brownian motion.

Recalling the infinite product representation for $D(\lambda)$ in (5.19) together with (5.22) leads us to the following theorem [Hochstadt (1973, p. 251)].

Theorem 5.3. *Suppose that $K(s,t)$ is continuous, symmetric, and nearly definite on $[0,1] \times [0,1]$. Then the FD of K can be expanded as*

(5.23) $$D(\lambda) = \prod_{n=1}^{\infty} \left(1 - \frac{\lambda}{\lambda_n}\right)^{l_n},$$

where λ_n is the eigenvalue of K with $\lambda_m \neq \lambda_n$ for $m \neq n$ and l_n is the multiplicity of λ_n.

Note that l_n in (5.23) is not only the order of the zero of $D(\lambda)$ at λ_n, but also the multiplicity of λ_n. The entire function $D(\lambda)$ in the present case is of genus zero.

PROBLEMS

2.1 Establish the convergence results in (5.14) and (5.15).

2.2 Show that the three functions in (5.20) are all positive definite.

2.3 Establish the relation in (5.22). Show also that

$$\sum_{n=1}^{\infty} \frac{1}{\lambda_n^2} = \int_0^1 \int_0^1 K^2(s,t)\, ds\, dt.$$

5.3 THE c.f. OF THE QUADRATIC BROWNIAN FUNCTIONAL

We now proceed to obtain the c.f. of

(5.24) $$S = \int_0^1 \int_0^1 K(s,t)\, dw(s)\, dw(t),$$

where $K(s,t)$ is symmetric, continuous, and nearly definite. For this purpose we can first show (Problem 3.1) that S is the same in the m.s. sense as

$$S = \sum_{n=1}^{\infty} \frac{1}{\lambda_n} \left\{ \int_0^1 f_n(t)\, dw(t) \right\}^2,$$

where $\{\lambda_n\}$ is a sequence of eigenvalues of K and $\{f_n(t)\}$ is an orthonormal sequence of eigenfunctions.

Then, noting that

$$\left\{ \int_0^1 f_n(t)\, dw(t) \right\} \sim \text{NID}(0,1),$$

and using the product expansion for $D(\lambda)$ in (5.23), we obtain the following theorem, which was first established by Anderson and Darling (1952).

Theorem 5.4. *Consider the statistic S in (5.24), where $K(s,t)$ is continuous, symmetric, and nearly definite on $[0,1] \times [0,1]$. Then we have*

$$
\begin{align}
(5.25) \qquad E(e^{i\theta S}) &= E\left[\exp\left\{i\theta \int_0^1 \int_0^1 K(s,t)\,dw(s)\,dw(t)\right\}\right] \\
&= E\left[\exp\left\{i\theta \sum_{n=1}^{\infty} \frac{1}{\lambda_n}\left(\int_0^1 f_n(t)\,dw(t)\right)^2\right\}\right] \\
&= \prod_{n=1}^{\infty}\left(1 - \frac{2i\theta}{\lambda_n}\right)^{-1/2} \\
&= (D(2i\theta))^{-1/2},
\end{align}
$$

where λ_n is the eigenvalue of K repeated as many times as its multiplicity, and $D(\lambda)$ is the FD of K.

As was mentioned before, it is usually difficult to obtain eigenvalues explicitly. Thus the next to last expression in (5.25) is not very useful. Note that this corresponds to the eigenvalue approach discussed in Chapter 1. Our concern here is the last expression in (5.25). The problem is how to obtain $D(\lambda)$, which is now discussed.

The function $D(\lambda)$ may be defined in several ways, among which are

$$
\begin{align}
(5.26) \qquad D(\lambda) &= \lim_{T\to\infty}\left|I_T - \frac{\lambda}{T} K_T\right| \\
&= \sum_{n=0}^{\infty} \frac{(-1)^n \lambda^n}{n!} \int_0^1 \cdots \int_0^1 K\begin{pmatrix} t_1 & \cdots & t_n \\ t_1 & \cdots & t_n \end{pmatrix} dt_1 \cdots dt_n \\
&= \prod_{n=1}^{\infty}\left(1 - \frac{\lambda}{\lambda_n}\right)^{l_n}.
\end{align}
$$

It is certainly true that there are cases where $D(\lambda)$ can be computed easily following one of the above formulas. One such example is the function $K_3(s,t)$ defined in (5.20). We obtain (Problem 3.2) the FD $D(\lambda)$ of K_3 as

$$
(5.27) \qquad D(\lambda) = 1 - \lambda \int_0^1 g^2(t)\,dt,
$$

which is also obtained by noting that

$$
E\left[\exp\left\{i\theta\left(\int_0^1 g(t)\,dw(t)\right)^2\right\}\right] = \left(1 - 2i\theta \int_0^1 g^2(t)\,dt\right)^{-1/2}.
$$

In general, however, the computation of $D(\lambda)$ via the above formulas is difficult.

An alternative method for obtaining the FD is demonstrated in Nabeya and Tanaka (1988, 1990a), which we now explain. We first present a set of sufficient conditions for a function of λ to be the FD.

Theorem 5.5. *Let $K(s,t)$ be continuous, symmetric, and nearly definite on $[0,1] \times [0,1]$ and $\{\lambda_n\}$ a sequence of eigenvalues of K. Suppose that $\tilde{D}(\lambda)$ is an entire function of λ with $\tilde{D}(0) = 1$. Then $\tilde{D}(\lambda)$ becomes the* FD *of K if*

i) *every zero of $\tilde{D}(\lambda)$ is an eigenvalue of K, and in turn every eigenvalue of K is a zero of $\tilde{D}(\lambda)$;*

ii) *$\tilde{D}(\lambda)$ can be expanded as*

$$\tilde{D}(\lambda) = \prod_{n=1}^{\infty} \left(1 - \frac{\lambda}{\lambda_n}\right)^{l_n}, \tag{5.28}$$

where l_n is equal to the multiplicity of λ_n.

A word may be in order. If $\tilde{D}(\lambda)$ is an entire function with $\tilde{D}(0) = 1$, so is $\tilde{D}^2(\lambda)$, for example. The zero of $\tilde{D}^2(\lambda)$ at λ_n, however, is of order $2l_n$, while the multiplicity of λ_n is l_n. Thus $\tilde{D}^2(\lambda)$ is not the FD of K.

To obtain a candidate $\tilde{D}(\lambda)$ for the FD of K, we work with a differential equation with some boundary conditions equivalent to the integral equation (5.10). As an illustration, let us take up $K(s,t) = 1 - \max(s,t)$, which is positive definite, and consider

$$f(t) = \lambda \int_0^1 \left[1 - \max(s,t)\right] f(s)\,ds \tag{5.29}$$

$$= \lambda \left[-t \int_0^t f(s)\,ds - \int_t^1 s f(s)\,ds + \int_0^1 f(s)\,ds\right].$$

By differentiation we have

$$f'(t) = -\lambda \int_0^t f(s)\,ds, \qquad f''(t) = -\lambda f(t).$$

Then it can be shown (Problem 3.3) that the integral equation (5.29) is equivalent to the following differential equation with two boundary conditions:

$$f''(t) + \lambda f(t) = 0, \qquad f(1) = f'(0) = 0. \tag{5.30}$$

Here the choice of boundary conditions is somewhat arbitrary insofar as they are linearly independent, but the simpler the better, as will be recognized shortly. The general solution to (5.30) is given by

$$f(t) = c_1 \cos\sqrt{\lambda}\,t + c_2 \sin\sqrt{\lambda}\,t, \qquad f(1) = f'(0) = 0, \tag{5.31}$$

where c_1 and c_2 are arbitrary constants. From the boundary conditions $f(1) = f'(0) = 0$, we have the following homogeneous equation on $c = (c_1, c_2)'$:

$$(5.32) \qquad \begin{pmatrix} \cos\sqrt{\lambda} & \sin\sqrt{\lambda} \\ 0 & \sqrt{\lambda} \end{pmatrix} \begin{pmatrix} c_1 \\ c_2 \end{pmatrix} = \begin{pmatrix} 0 \\ 0 \end{pmatrix} \Leftrightarrow M(\lambda)c = 0.$$

The eigenfunction $f(t)$ in (5.31) must be nonvanishing, which occurs only when $c \ne 0$. Thus (5.32) implies that $|M(\lambda)| = \sqrt{\lambda}\cos\sqrt{\lambda} = 0$. Then λ ($\ne 0$) is an eigenvalue if and only if $\cos\sqrt{\lambda} = 0$. We therefore obtain $\tilde{D}(\lambda) = \cos\sqrt{\lambda}$ with $\tilde{D}(0) = 1$ as a candidate for the FD of $K(s,t) = 1 - \max(s,t)$. Condition i) in Theorem 5.5 has now been established.

We proceed to establish ii) in the same theorem. From (5.32) we have $c_2 = 0$ so that $f(t) = c_1 \cos\sqrt{\lambda} t$ with $c_1 \ne 0$. Thus the multiplicity of every eigenvalue is unity. Since $\tilde{D}(\lambda) = \cos\sqrt{\lambda}$ can be expanded as in (5.28) with $\lambda_n = ((n-\frac{1}{2})\pi)^2$ and $l_n = 1$, every zero of $\tilde{D}(\lambda)$ is of order unity, which is equal to the multiplicity of each eigenvalue. Therefore $\tilde{D}(\lambda)$ is really the FD of K.

Theorem 5.4 now yields

$$(5.33) \qquad E\left[\exp\left\{i\theta \int_0^1 \int_0^1 (1 - \max(s,t))\, dw(s)\, dw(t)\right\}\right] = \left(\cos\sqrt{2i\theta}\right)^{-1/2},$$

which was formally presented in Section 1.1 of Chapter 1.

Similarly, if $K(s,t) = \min(s,t) - st$, we can show (Problem 3.4) that the associated FD is given by $\sin\sqrt{\lambda}/\sqrt{\lambda}$ so that $\lambda_n = n^2\pi^2$ and

$$(5.34) \qquad E\left[\exp\left\{i\theta \int_0^1 \int_0^1 (\min(s,t) - st)\, dw(s)\, dw(t)\right\}\right] = \left(\frac{\sin\sqrt{2i\theta}}{\sqrt{2i\theta}}\right)^{-1/2},$$

which was formally presented in Section 1.2 of Chapter 1.

In the above examples, it was quite easy to obtain the FDs following Theorem 5.5. The eigenvalues can also be given explicitly. In general, however, we need some effort in verifying condition ii) in the theorem together with the determination of the multiplicity l_n. For the multiplicity we have the following theorem, which describes nothing but the dimension of a null space in the theory of matrices.

Theorem 5.6. *Suppose that the integral equation (5.10) is equivalent to a differential equation with some boundary conditions. Suppose further that the latter is equivalent to*

$$f(t) = c_1\phi_1(t) + \cdots + c_r\phi_r(t),$$

$$M(\lambda)c = 0,$$

where $\phi_j s$ are linearly independent, continuous functions, while $M(\lambda)$ is the $r \times r$ coefficient matrix of the system of linear homogeneous equations in $c = (c_1, \ldots, c_r)'$. Then the multiplicity l_n of the eigenvalue λ_n is given by

$$l_n = r - \text{rank}(M(\lambda_n)).$$

As an application of Theorem 5.6, consider the statistic S in (5.24) with

(5.35) $$K(s,t) = \frac{1}{4}\left[1 - 2|s - t|\right],$$

which is shown to be positive definite (Problem 3.5). This kernel appears in connection with Lévy's stochastic area, as was shown in (5.5). We first have (Problem 3.6) that the integral equation (5.10) is equivalent to

(5.36) $$f''(t) + \lambda f(t) = 0, \quad f(0) + f(1) = 0, \quad f'(0) + f'(1) = 0,$$

so that we are led to the homogeneous equation $M(\lambda)c = 0$, where

$$|M(\lambda)| = \begin{vmatrix} 1 + \cos\sqrt{\lambda} & \sin\sqrt{\lambda} \\ -\sqrt{\lambda}\sin\sqrt{\lambda} & \sqrt{\lambda}(1 + \cos\sqrt{\lambda}) \end{vmatrix}$$

$$= 2\sqrt{\lambda}(1 + \cos\sqrt{\lambda}) = 4\sqrt{\lambda}\left(\cos\frac{\sqrt{\lambda}}{2}\right)^2.$$

Then $\lambda \ (\neq 0)$ is an eigenvalue if and only if $|M(\lambda)| = 0$. Thus we obtain $\tilde{D}(\lambda) = (\cos(\sqrt{\lambda}/2))^2$ as a candidate for the FD of K. Since $\text{rank}(M(\lambda_n)) = 0$ for every eigenvalue $\lambda_n = ((2n-1)\pi)^2$, $(n = 1, 2, \ldots)$, the multiplicity of each eigenvalue is two. In fact, for $K(s,t)$ in (5.35), we have

$$\int_0^1 K(t,t)\,dt = \frac{1}{4},$$

while

$$\sum_{n=1}^{\infty} \frac{1}{((2n-1)\pi)^2} = \frac{1}{8},$$

and thus the equality in (5.22) holds with the multiplicity two. Since it is known that

$$\left(\cos\frac{\sqrt{\lambda}}{2}\right)^2 = \prod_{n=1}^{\infty}\left(1 - \frac{\lambda}{((2n-1)\pi)^2}\right)^2,$$

THE c.f. OF THE QUADRATIC BROWNIAN FUNCTIONAL

$\tilde{D}(\lambda)$ is ensured to be the FD of K and

$$(5.37) \quad E(e^{i\theta S}) = E\left[\exp\left\{i\theta \int_0^1 \int_0^1 \frac{1}{4}\left[1 - 2|s - t|\right] dw(s) dw(t)\right\}\right]$$

$$= \left(\tilde{D}(2i\theta)\right)^{-1/2} = \left(\cos \frac{\sqrt{2i\theta}}{2}\right)^{-1}.$$

The c.f. of Lévy's stochastic area defined in (5.4) as

$$S_2 = \frac{1}{2} \int_0^1 \left[w_1(t) dw_2(t) - w_2(t) dw_1(t)\right]$$

can be easily derived from (5.37). Since

$$S_2 \mid \{w_1(t)\} \sim N\left(0, \int_0^1 \int_0^1 \frac{1}{4}\left[1 - 2|s - t|\right] dw_1(s) dw_1(t)\right),$$

(5.37) yields

$$(5.38) \quad E(e^{i\theta S_2}) = E\left[\exp\left\{-\frac{\theta^2}{2} \int_0^1 \int_0^1 \frac{1}{4}\left[1 - 2|s - t|\right] dw(s) dw(t)\right\}\right]$$

$$= \left(\tilde{D}(-\theta^2)\right)^{-1/2} = \left(\cosh \frac{\theta}{2}\right)^{-1},$$

which was earlier obtained in Section 1.4 of Chapter 1 by the eigenvalue approach and in Section 4.3 of Chapter 4 by the stochastic process approach.

Another case where the multiplicity of each eigenvalue is two can be found in Watson (1961), who suggested a goodness-of-fit test statistic on a circle with

$$(5.39) \quad K(s, t) = \min(s, t) - \frac{1}{2}(s + t) + \frac{1}{2}(s - t)^2 + \frac{1}{12}.$$

In fact, we can show (Problem 3.7) that the FD $D(\lambda)$ of K in (5.39) is given by

$$(5.40) \quad D(\lambda) = \left(\sin \frac{\sqrt{\lambda}}{2} \bigg/ \frac{\sqrt{\lambda}}{2}\right)^2.$$

Thus it holds that

$$E\left[\exp\left\{i\theta \int_0^1 \int_0^1 \left(\min(s, t) - \frac{1}{2}(s + t) + \frac{1}{2}(s - t)^2 + \frac{1}{12}\right) dw(s) dw(t)\right\}\right]$$

$$= \left(\sin \sqrt{\frac{i\theta}{2}} \bigg/ \sqrt{\frac{i\theta}{2}}\right)^{-1}.$$

Cases of multiplicities greater than unity may be rare. The most important is the case where the multiplicity is equal to unity for each eigenvalue. Then every zero of the candidate function $\tilde{D}(\lambda)$ must be of order unity. The infinite product expansion under such a circumstance is given by the following theorem [Whittaker and Watson (1958, p. 137)].

Theorem 5.7. *Let $h(z)$ be an entire function with $h(0) = 1$ and have simple zeros at the points $a_1, a_2, \ldots,$ where $\lim_{n\to\infty} |a_n| = \infty$. Suppose that there is a sequence $\{C_m\}$ of simple closed curves such that $h'(z)/h(z)$ is bounded on C_m as $m \to \infty$. Then $h(z)$ can be expanded as*

$$h(z) = \exp\{h'(0)z\} \prod_{n=1}^{\infty} \left\{ \left(1 - \frac{z}{a_n}\right) \exp\left(\frac{z}{a_n}\right) \right\}.$$

As an application of this theorem, let us consider a nearly definite kernel

(5.41) $$K(s,t) = 1 - \max(s,t) + b,$$

where b is any nonzero constant. We obtain

(5.42) $$f''(t) + \lambda f(t) = 0, \quad f(1) = \lambda b \int_0^1 f(s)\, ds, \quad f'(0) = 0,$$

the set of which is equivalent to the original integral equation (5.10) (Problem 3.8). Solving for $f(t)$ we have, from the two boundary conditions, the homogeneous equation $M(\lambda)c = 0$, where

$$M(\lambda) = \begin{pmatrix} \cos\sqrt{\lambda} - b\sqrt{\lambda}\sin\sqrt{\lambda} & \sin\sqrt{\lambda} + b\sqrt{\lambda}(\cos\sqrt{\lambda} - 1) \\ 0 & \sqrt{\lambda} \end{pmatrix}.$$

Thus we obtain, as a candidate for the FD,

(5.43) $$\tilde{D}(\lambda) = \cos\sqrt{\lambda} - b\sqrt{\lambda}\sin\sqrt{\lambda}, \quad \tilde{D}(0) = 1.$$

Evidently $\text{rank}(M(\lambda_n)) = 1$ for each eigenvalue λ_n so that, by Theorem 5.6, the multiplicity of λ_n is unity for all n.

Let us put $z = \sqrt{\lambda}$ and consider

(5.44) $$h(z) = \tilde{D}(\lambda) = \cos z - bz \sin z,$$

which is an even entire function with $h(0) = 1$ whose zeros are all simple (Problem 3.9). Then we can define the zeros of $h(z)$ by $\pm a_1, \pm a_2, \ldots,$ where $\lim_{n\to\infty} |a_n| = \infty$. Let C_m be the square in the complex plane with vertices $(2m + \frac{1}{2})\pi(\pm 1 \pm i)$, $m =$

1, 2, ... Then it is seen that

$$\frac{h'(z)}{h(z)} = \frac{-bz\cos z - (b+1)\sin z}{\cos z - bz\sin z}$$

is bounded on each side of C_m as $m \to \infty$. Note that $h'(z)/h(z)$ is not bounded on squares with vertices $m\pi(\pm 1 \pm i)$ since it takes the value $-bz$ at $z = m\pi$.
Using Theorem 5.7, we can expand the even function $h(z)$ with $h'(0) = 0$ as

$$h(z) = \prod_{n=1}^{\infty} \left\{ \left(1 - \frac{z}{a_n}\right) \exp\left(\frac{z}{a_n}\right) \left(1 + \frac{z}{a_n}\right) \exp\left(-\frac{z}{a_n}\right) \right\}$$

$$= \prod_{n=1}^{\infty} \left(1 - \frac{z^2}{a_n^2}\right) = \prod_{n=1}^{\infty} \left(1 - \frac{\lambda}{a_n^2}\right)$$

$$= \tilde{D}(\lambda),$$

from which we conclude that $\tilde{D}(\lambda)$ in (5.43) is the FD of K in (5.41). Note that $\{a_n^2\}$ is a sequence of eigenvalues of the symmetric kernel $K(s, t)$ so that each a_n^2 is real. It can be checked (Problem 3.10) that, when the value of b in (5.41) is nonnegative, every a_n^2 is positive, while all a_n^2s except one are positive when b is negative.

Some other examples of how to obtain the FDs of given kernels will be given in the next section.

PROBLEMS

3.1 Show that the following equality holds in the m.s. sense:

$$\int_0^1 \int_0^1 K(s,t)\, dw(s)\, dw(t) = \sum_{n=1}^{\infty} \frac{1}{\lambda_n} \left\{ \int_0^1 f_n(t)\, dw(t) \right\}^2,$$

where $K(s, t)$ is symmetric, continuous, and nearly definite, while $\{\lambda_n\}$ is a sequence of eigenvalues of K and $\{f_n(t)\}$ is an orthonormal sequence of eigenfunctions.

3.2 Derive the FD of K_3 defined in (5.20) following one of the definitions of $D(\lambda)$ in (5.26).

3.3 Show that the integral equation (5.10) is equivalent to (5.30) when $K(s, t) = 1 - \max(s, t)$.

3.4 Derive the c.f. of S in (5.1) when $K(s, t) = \min(s, t) - st$, following the Fredholm approach.

3.5 Show that the kernel defined in (5.35) is positive definite.

3.6 Show that the integral equation (5.10) is equivalent to (5.36) when $K(s, t) = \frac{1}{4}[1 - 2|s - t|]$.

3.7 Prove that the FD of $K(s,t)$ in (5.39) is given by (5.40).

3.8 Show that the integral equation (5.10) is equivalent to (5.42) when $K(s,t) = 1 - \max(s,t) + b$.

3.9 Show that the zeros of $h(z) = \cos z - bz \sin z$ are all simple.

3.10 Show that all the eigenvalues of $K(s,t)$ in (5.41) are positive when b is non-negative, while only one eigenvalue is negative when b is negative.

5.4 VARIOUS FREDHOLM DETERMINANTS

We continue to derive the c.f.s of various quadratic functionals of the Brownian motion. The examples considered here are thought to be those that cannot be dealt with easily by the stochastic process or eigenvalue approach.

Let us first consider the statistic S_1 defined in (5.3). In particular, we consider

$$(5.45) \quad U_1 = \int_0^1 (t^m w(t))^2 \, dt = \int_0^1 \int_0^1 \frac{1}{2m+1} \left[1 - (\max(s,t))^{2m+1}\right] dw(s)\, dw(t),$$

where $m > -\frac{1}{2}$. Let us show that

$$(5.46) \quad \mathcal{L}(U_1) = \mathcal{L}\left(\int_0^1 \int_0^1 s^m t^m \min(s,t)\, dw(s)\, dw(t)\right).$$

For this purpose, we first note that

$$V\left(\int_0^1 t^m w(t) g(t)\, dt\right) = \int_0^1 \int_0^1 s^m t^m \min(s,t) g(s) g(t)\, ds\, dt \geq 0$$

for any continuous function $g(t)$, which implies that $s^m t^m \min(s,t)$ is positive definite. Thus we can define

$$Z(t) = \sum_{n=1}^{\infty} \frac{f_n(t)}{\sqrt{\lambda_n}} Z_n,$$

where $\{Z_n\} \sim \text{NID}(0,1)$ and each $\lambda_n\,(>0)$ is the eigenvalue of $s^m t^m \min(s,t)$, while $f_n(t)$ is an orthonormal eigenfunction corresponding to λ_n. Since it can be easily checked that any finite dimensional distribution of $Z(t)$ is the same as $t^m w(t)$ and the finite-dimensional sets form a determining class (see Section 3.1 of Chapter 3), it follows from Mercer's theorem that

$$\mathcal{L}(U_1) = \mathcal{L}\left(\int_0^1 Z^2(t)\, dt\right)$$

$$= \mathcal{L}\left(\sum_{n=1}^{\infty} \frac{1}{\lambda_n} Z_n^2\right) = \mathcal{L}\left(\int_0^1 \int_0^1 s^m t^m \min(s,t)\, dw(s)\, dw(t)\right).$$

The above distributional equivalence for $m = 0$ was earlier presented in (1.11). Note that the kernel appearing in (5.45) is positive definite (Problem 4.1). The statistic U_1 was dealt with by MacNeill (1974) and Nabeya and Tanaka (1988) in connection with testing for parameter constancy. MacNeill (1974) assumed $m > -1$ rather than $m > -\frac{1}{2}$, although the kernel is not continuous when $-1 < m \leq -\frac{1}{2}$. Here we do not go into such complexity, but continue to assume that the kernel is continuous so that $m > -\frac{1}{2}$.

We can show (Problem 4.2) that the integral equation (5.10) with $K(s,t) = K_1(s,t) = (1 - (\max(s,t))^{2m+1})/(2m+1)$ is equivalent to

$$(5.47) \quad f''(t) - \frac{2m}{t} f'(t) + \lambda t^{2m} f(t) = 0, \quad \lim_{t \to 0} \frac{f'(t)}{t^{2m}} = 0, \quad f(1) = 0.$$

The differential equation in (5.47) is a special case of Bessel's equation

$$(5.48) \quad y''(x) - \frac{2\alpha - 1}{x} y'(x) + \left(\beta^2 \gamma^2 x^{2\gamma - 2} + \frac{\alpha^2 - \nu^2 \gamma^2}{x^2} \right) y(x) = 0.$$

It is known [Abramowitz and Stegun (1972)] that (5.48) has the general solution

$$(5.49) \quad y(x) = \begin{cases} x^\alpha \left(A J_\nu(\beta x^\gamma) + B Y_\nu(\beta x^\gamma) \right), & (\nu : \text{integer}), \\ x^\alpha \left(A J_\nu(\beta x^\gamma) + B J_{-\nu}(\beta x^\gamma) \right), & (\nu : \text{noninteger}), \end{cases}$$

where $J_\nu(z)$ is the Bessel function of the first kind defined by

$$(5.50) \quad J_\nu(z) = \left(\frac{z}{2} \right)^\nu \sum_{k=0}^{\infty} \frac{(-z^2/4)^k}{k! \, \Gamma(\nu + k + 1)},$$

while $Y_\nu(z)$ is the Bessel function of the second kind (also called *Weber's function*) defined by

$$(5.51) \quad Y_\nu(z) = \frac{J_\nu(z) \cos(\nu \pi) - J_{-\nu}(z)}{\sin(\nu \pi)}.$$

On the basis of (5.48) and (5.49) we obtain

$$f(t) = t^{(2m+1)/2} \left\{ c_1 J_\nu \left(\frac{\sqrt{\lambda}}{m+1} t^{m+1} \right) + c_2 J_{-\nu} \left(\frac{\sqrt{\lambda}}{m+1} t^{m+1} \right) \right\}$$

$$= c_1 \left(\frac{\sqrt{\lambda}}{2(m+1)} \right)^\nu t^{2m+1} \sum_{k=0}^{\infty} \frac{(-\lambda t^{2(m+1)}/4(m+1)^2)^k}{k! \, \Gamma(\nu + k + 1)}$$

$$+ c_2 \left(\frac{\sqrt{\lambda}}{2(m+1)} \right)^{-\nu} \sum_{k=0}^{\infty} \frac{(-\lambda t^{2(m+1)}/4(m+1)^2)^k}{k! \, \Gamma(-\nu + k + 1)},$$

where $\nu = (2m + 1)/(2(m + 1))$. Note that ν cannot be an integer when $m > -\frac{1}{2}$. From the two boundary conditions in (5.47), it follows (Problem 4.3) that $M(\lambda)c = 0$, where $c = (c_1, c_2)'$ and

$$(5.52) \quad M(\lambda) = \begin{pmatrix} \left(\dfrac{\sqrt{\lambda}}{2(m+1)}\right)^{\nu} \dfrac{2m+1}{\Gamma(\nu+1)} & 0 \\ J_{\nu}\left(\dfrac{\sqrt{\lambda}}{m+1}\right) & J_{-\nu}\left(\dfrac{\sqrt{\lambda}}{m+1}\right) \end{pmatrix}.$$

Then we obtain, from (5.50),

$$(5.53) \quad \tilde{D}_1(\lambda) = \Gamma(-\nu+1) J_{-\nu}\left(\dfrac{\sqrt{\lambda}}{m+1}\right) \bigg/ \left(\dfrac{\sqrt{\lambda}}{2(m+1)}\right)^{-\nu}$$

as a candidate for the FD of K_1 with $\tilde{D}_1(0) = 1$. It is clear that $\operatorname{rank}(M(\lambda_n)) = 1$ for each eigenvalue λ_n so that the multiplicity of every eigenvalue is unity. Since it is known [Watson (1958, p. 498)] that

$$J_{\nu}(z) = \dfrac{(z/2)^{\nu}}{\Gamma(\nu+1)} \prod_{n=1}^{\infty} \left(1 - \dfrac{z^2}{a_n^2}\right),$$

where $a_1 < a_2 < \cdots$ are the positive zeros of $J_{\nu}(z)$, $\tilde{D}_1(\lambda)$ can be expanded as

$$\tilde{D}_1(\lambda) = \prod_{n=1}^{\infty} \left(1 - \dfrac{\lambda}{(m+1)^2 a_n^2}\right),$$

which implies that all the zeros of $\tilde{D}_1(\lambda)$ are positive and simple. Thus we have verified that $\tilde{D}_1(\lambda)$ is the FD of K_1, and the c.f. of U_1 in (5.44) is given by $(\tilde{D}_1(2i\theta))^{-1/2}$.

It may be noted that, when $m = 0$ so that $\nu = \frac{1}{2}$, $\tilde{D}_1(\lambda)$ reduces to $\cos\sqrt{\lambda}$ since

$$J_{-1/2}(z) = \sqrt{\dfrac{2}{\pi z}} \cos z.$$

Closely related to the statistic U_1 is

$$(5.54) \quad U_2 = \int_0^1 t^m w(t)\, dw(t), \quad (m \geq 0),$$

which is a special case of S_3 defined in (5.6) with $g(t) = t^m$. It follows from (5.7) that

$$(5.55) \quad U_2 = \int_0^1 \int_0^1 \dfrac{1}{2}(\max(s,t))^m\, dw(s)\, dw(t) - \dfrac{1}{2(m+1)}.$$

Note here that the kernel $K_2(s,t) = (\max(s,t))^m/2$ is nearly definite since it is the sum of a negative definite kernel $-(1 - (\max(s,t))^m)/2$ and a degenerate kernel $\frac{1}{2}$.

It can be checked that

$$(5.56)\quad f''(t) - \frac{m-1}{t}f'(t) - \frac{\lambda m}{2}t^{m-1}f(t) = 0, \quad \lim_{t\to 0}\frac{f'(t)}{t^{m-1}} = 0, \quad f'(1) = mf(1)$$

are equivalent to the integral equation (5.10) with $K(s,t) = K_2(s,t)$. Then we obtain (Problem 4.4), as the FD of K_2,

$$(5.57)\quad D_2(\lambda) = \Gamma\left(\frac{1}{m+1}\right)\left(-\frac{m+1}{m}\right) J_{\nu-1}\left(\frac{\sqrt{-2\lambda m}}{m+1}\right) \Big/ \left(\frac{\sqrt{-2\lambda m}}{2(m+1)}\right)^{\nu-1}$$

$$= \Gamma(\nu) J_{\nu-1}\left(\frac{\sqrt{-2\lambda m}}{m+1}\right) \Big/ \left(\frac{\sqrt{-2\lambda m}}{2(m+1)}\right)^{\nu-1},$$

where $\nu = -m/(m+1)$. Thus the c.f. of U_2 in (5.55) is given by

$$(D_2(2i\theta))^{-1/2} \exp\{-i\theta/(2(m+1))\}.$$

The first expression in (5.57) is more suitable for the computational purpose. It may be noted (Problem 4.5) that, when $m = 0$, $D_2(\lambda)$ reduces to $1 - \frac{\lambda}{2}$, as should be.

We next consider a quadratic functional of the Brownian bridge. Let us put

$$(5.58)\quad U_3 = \int_0^1 \left[t^m\{w(t) - tw(1)\}\right]^2 dt$$

$$= \int_0^1\int_0^1 \left[\frac{1}{2m+1}(1 - (\max(s,t))^{2m+1}) - \frac{2 - s^{2m+2} - t^{2m+2}}{2m+2}\right.$$

$$\left. + \frac{1}{2m+3}\right] dw(s)\,dw(t),$$

where $m > -\frac{1}{2}$. Because of the same reasoning as in establishing (5.46) we also have

$$(5.59)\quad \mathcal{L}(U_3) = \mathcal{L}\left(\int_0^1\int_0^1 s^m t^m [\min(s,t) - st]\,dw(s)\,dw(t)\right).$$

The above distributional equivalence for $m = 0$ was earlier presented in (1.18). Proceeding in the same way as above, we obtain (Problem 4.6), as the FD of $K_3(s,t) = s^m t^m(\min(s,t) - st)$,

$$(5.60)\quad D_3(\lambda) = \Gamma\left(\frac{2m+3}{2(m+1)}\right) J_{1/(2(m+1))}\left(\frac{\sqrt{\lambda}}{m+1}\right) \Big/ \left(\frac{\sqrt{\lambda}}{2(m+1)}\right)^{1/(2(m+1))}.$$

It can be checked that, when $m = 0$, $D_3(\lambda)$ reduces to $D_3(\lambda) = \sin\sqrt{\lambda}/\sqrt{\lambda}$ since

$$J_{1/2}(z) = \sqrt{\frac{2}{\pi z}} \sin z.$$

It may also be noted that the statistic on the right side of (5.59) is a special case of

(5.61) $$W^2 = \int_0^1 \int_0^1 \sqrt{\psi(s)}\sqrt{\psi(t)}\left[\min(s,t) - st\right] dw(s)\,dw(t),$$

which was discussed by Anderson and Darling (1952) in connection with goodness of fit tests. The so-called Anderson–Darling statistic is the one with $\psi(t) = 1/(t(1-t))$, and the c.f. of W^2 in that case was obtained in Anderson and Darling (1952) by the eigenvalue approach, although the kernel is not continuous at $(s,t) = (0,0)$ or $(1,1)$. The statistic U_3 in (5.58) or (5.59) is W^2 in (5.61) with $\psi(t) = t^{2m}$. If we take m to be negative, then this may be an alternative to the Anderson–Darling statistic.

We also point out (Problem 4.7) that

(5.62) $$\mathcal{L}(U_3) = \mathcal{L}\left(\int_0^1 \int_0^1 \frac{1}{2m+1}\left[(\min(s,t))^{1/(2m+1)} - (st)^{1/(2m+1)}\right] dw(s)\,dw(t)\right).$$

A slightly modified version of U_3 is

(5.63) $$U_4 = \int_0^1 \left[t^m\left\{w(t) - (2m+1)\int_0^1 s^{2m}w(s)\,ds\right\}\right]^2 dt$$

$$= \int_0^1 \int_0^1 \frac{1}{2m+1}\left[(\min(s,t))^{2m+1} - (st)^{2m+1}\right] dw(s)\,dw(t),$$

where $m > -\frac{1}{2}$. It is now an easy matter to obtain (Problem 4.8) the FD of $K_4(s,t) = ((\min(s,t))^{2m+1} - (st)^{2m+1})/(2m+1)$ as

(5.64) $$D_4(\lambda) = \Gamma\left(\frac{4m+3}{2(m+1)}\right) J_{(2m+1)/2(m+1)}\left(\frac{\sqrt{\lambda}}{m+1}\right) \bigg/ \left(\frac{\sqrt{\lambda}}{2(m+1)}\right)^{(2m+1)/2(m+1)}.$$

The statistic U_4 in (5.63) was dealt with in Nabeya and Tanaka (1988).

Nabeya and Tanaka (1988) also suggested a method for computing c.f.s of quadratic functionals of the Brownian motion whose kernel is given by

(5.65) $$K(s,t) = \min(s,t) + \sum_{k=1}^{r} \xi_k(s)\psi_k(t),$$

where

i) $\xi_k(s)$ and $\psi_k(t)$ ($k = 1, \ldots, r$) are continuous and each set is linearly independent in the space $C[0,1]$;

VARIOUS FREDHOLM DETERMINANTS

ii) $\psi_k(t)$ $(k = 1,\ldots,r)$ are twice continuously differentiable and $\psi_k''(t)$ for $k = 1,\ldots,q$ are linearly independent, whereas $\psi_k''(t) = 0$ for $k = q+1,\ldots,r$ with $0 \le r - q \le 2$.

Condition ii) above allows for the space of linear combinations of $\psi_k(t)$ to contain a constant and/or a linear function.

We now consider

$$(5.66) \qquad f(t) = \lambda \int_0^1 \left[\min(s,t) + \sum_{k=1}^r \xi_k(s)\psi_k(t) \right] f(s)\,ds,$$

from which we obtain the nonhomogeneous differential equation

$$(5.67) \qquad f''(t) + \lambda f(t) = \lambda \sum_{k=1}^q a_k \psi_k''(t),$$

with the boundary conditions

$$(5.68) \qquad f(0) = \lambda \sum_{k=1}^r a_k \psi_k(0),$$

$$(5.69) \qquad f'(1) = \lambda \sum_{k=1}^r a_k \psi_k'(1),$$

where

$$(5.70) \qquad a_k = \int_0^1 \xi_k(s)f(s)\,ds, \qquad (k = 1,\ldots,r).$$

The associated FD is obtained as follows, by modifying the technique of Kac, Kiefer, and Wolfowitz (1955). For given $\lambda (\ne 0)$, the general solution to (5.67) is

$$(5.71) \qquad f(t) = c_1 \cos\sqrt{\lambda}\,t + c_2 \sin\sqrt{\lambda}\,t + \sum_{k=1}^q a_k g_k(t),$$

where $g_k(t)$ is a special solution of

$$(5.72) \qquad g_k''(t) + \lambda g_k(t) = \lambda \psi_k''(t).$$

Substituting $f(t)$ from (5.71), we regard (5.68), (5.69), and (5.70) as a system of $r+2$ linear homogeneous equations $M(\lambda)c = 0$ in $c = (a_1,\ldots,a_r, c_1, c_2)'$. Then it can be shown that $\lambda (\ne 0)$ is an eigenvalue of K if and only if the system has a nontrivial solution so that we are led to compute $|M(\lambda)|$ to obtain a candidate for the FD.

As an illustration, let us consider

$$(5.73) \quad U_5 = \int_0^1 \int_0^1 \left[\min(s,t) + \frac{t}{16}(9s^5 - 25s) - \frac{9}{16}t^5(s^5 - s) \right] dw(s)\,dw(t),$$

which is discussed in Nabeya and Tanaka (1988), where the kernel, denoted as $K_5(s,t)$, is positive definite and may be rewritten as in (5.65) with $q = 1$ and $r = 2$ by putting

$$\xi_1(s) = -s^5 + s, \qquad \xi_2(s) = 9s^5 - 25s,$$

$$\psi_1(t) = \frac{9}{16}t^5, \qquad \psi_2(t) = \frac{t}{16}.$$

We obtain

$$(5.74) \qquad f''(t) + \lambda f(t) = \frac{45}{4}\lambda a_1 t^3$$

with the boundary conditions

$$(5.75) \qquad f(0) = 0, \qquad f'(1) = \frac{\lambda}{16}(45a_1 + a_2),$$

where

$$(5.76) \qquad a_1 = \int_0^1 (-s^5 + s) f(s)\,ds, \qquad a_2 = \int_0^1 (9s^5 - 25s) f(s)\,ds.$$

The general solution to (5.74) is given by

$$(5.77) \qquad f(t) = c_1 \cos\sqrt{\lambda}\,t + c_2 \sin\sqrt{\lambda}\,t + \frac{45a_1}{4}\left(t^3 - \frac{6}{\lambda}t\right).$$

Substituting this into (5.75) and (5.76) yields $M(\lambda)c = 0$, where $c = (a_1, a_2, c_1, c_2)'$ and

$$(5.78) \quad M(\lambda) = \begin{pmatrix} 0 & 0 & 1 & 0 \\ M_{21}(\lambda) & -\dfrac{\lambda}{16} & -\sqrt{\lambda}\sin\sqrt{\lambda} & \sqrt{\lambda}\cos\sqrt{\lambda} \\ -\dfrac{90}{7\lambda} & 0 & M_{33}(\lambda) & M_{34}(\lambda) \\ M_{41}(\lambda) & -1 & M_{43}(\lambda) & M_{44}(\lambda) \end{pmatrix}$$

with

$$M_{21}(\lambda) = \frac{135}{4} - \frac{45}{16}\lambda - \frac{135}{2\lambda},$$

$$M_{33}(\lambda) = \frac{1}{\lambda^3}\left[-4(\lambda^2 - 15\lambda + 30)\cos\sqrt{\lambda} + 20\sqrt{\lambda}(\lambda - 6)\sin\sqrt{\lambda} + \lambda^2 - 120 \right],$$

$$M_{34}(\lambda) = \frac{4}{\lambda^3}\left[-5\sqrt{\lambda}(\lambda-1)\cos\sqrt{\lambda} - (\lambda^2 - 15\lambda + 30)\sin\sqrt{\lambda}\right],$$

$$M_{41}(\lambda) = -45 + \frac{3330}{7\lambda},$$

$$M_{43}(\lambda) = \frac{1}{\lambda^3}\Big[20(\lambda^2 - 27\lambda + 54)\cos\sqrt{\lambda}$$
$$- 4\sqrt{\lambda}(4\lambda^2 + 45\lambda - 270)\sin\sqrt{\lambda} - 25\lambda^2 + 1080\Big],$$

$$M_{44}(\lambda) = \frac{4}{\lambda^3}\left[\sqrt{\lambda}(4\lambda^2 + 45\lambda - 270)\cos\sqrt{\lambda} + 5(\lambda^2 - 27\lambda + 54)\sin\sqrt{\lambda}\right].$$

Making use of the computerized algebra REDUCE we obtain, as a candidate for the FD,

(5.79) $\quad\tilde{D}_5(\lambda) = |M(\lambda)|/\sqrt{\lambda}$

$$= \frac{1350}{\lambda^5}\left[\left(-\frac{\lambda^3}{6} + 2\lambda^2 - 6\lambda\right)\cos\sqrt{\lambda}\right.$$
$$\left. + \sqrt{\lambda}\left(-\frac{\lambda^3}{42} + \frac{7\lambda^2}{10} - 4\lambda + 6\right)\sin\sqrt{\lambda}\right].$$

Here every zero of $h(z) = \lambda^5\tilde{D}_5(\lambda)/1350$ with $z = \sqrt{\lambda}$ is real because nonzero $z^2 = \lambda$ is an eigenvalue of the positive definite kernel K_5. Let $a_1, a_2, \ldots$ be the positive zeros of $h(z)$. Then the rank of the 4×4 coefficient matrix $M(a_j^2)$ is 3, which implies that the multiplicity of every eigenvalue is unity. It can also be checked that every nonzero solution of $h(z) = 0$ is simple by showing that there exists no nonzero solution common to $h(z) = 0$ and $h'(z) = 0$. Moreover, the function $h^*(z) = 1350h(z)/z^{10} = \tilde{D}_5(\lambda)$ is even and analytic with $h^*(0) = 1$ and with the zeros $\pm a_1, \pm a_2, \ldots$. Since $h^{*\prime}(z)/h^*(z)$ is bounded on the square C_m with vertices $(2m + \frac{1}{2})\pi(\pm 1 \pm i)$, we can verify that $\tilde{D}_5(\lambda)$ in (5.79) is the FD of K_5.

In the derivation of $\tilde{D}_5(\lambda)$, we have used the four conditions given in (5.75) and (5.76), which enforces us to compute the determinant of the 4×4 matrix $M(\lambda)$ in (5.78). In the present case, however, it is more convenient to use the boundary condition $f(1) = 0$, instead of $f'(1) = \lambda(45a_1 + a_2)/16$ in (5.75). This enables us to dispense with the introduction of a_2 since the general solution $f(t)$ in (5.77) does not contain a_2. Then we can deal with a 3×3 matrix rather than the 4×4 matrix $M(\lambda)$, and we can arrive at the same result (Problem 4.9).

The same technique can be applied to obtain the c.f. of

(5.80) $\quad U_6 = \int_0^1\int_0^1\left[\min(s,t) - st - \frac{2}{\pi^2}\sin\pi s\sin\pi t\right]dw(s)\,dw(t).$

Denoting as $K_6(s,t)$ the kernel appearing in (5.80), we can show that the integral equation (5.10) with $K = K_6$ is equivalent to

(5.81) $\quad\quad\quad\quad\quad f''(t) + \lambda f(t) = 2\lambda a_1\sin\pi t$

with the boundary conditions $f(0) = f(1) = 1$ and

$$a_1 = \int_0^1 \sin \pi s f(s)\, ds.$$

When $\lambda \neq \pi^2$, the general solution to (5.81) is given by

$$f(t) = c_1 \cos \sqrt{\lambda}\, t + c_2 \sin \sqrt{\lambda}\, t - \frac{2a_1 \lambda}{\pi^2 - \lambda} \sin \pi t$$

and the above three conditions yield $M(\lambda)c = 0$, where $c = (a_1, c_1, c_2)'$ and

$$M(\lambda) = \begin{pmatrix} 0 & 1 & 0 \\ 0 & \cos \sqrt{\lambda} & \sin \sqrt{\lambda} \\ \dfrac{-\pi^2}{\pi^2 - \lambda} & \dfrac{\pi(\cos \sqrt{\lambda} + 1)}{\pi^2 - \lambda} & \dfrac{\pi \sin \sqrt{\lambda}}{\pi^2 - \lambda} \end{pmatrix},$$

(5.82) $$|M(\lambda)| = -\frac{\pi^2}{\pi^2 - \lambda} \sin \sqrt{\lambda}.$$

Suppose that $\lambda = \pi^2$. Then the general solution to (5.81) is

$$f(t) = c_1 \cos \pi t + c_2 \sin \pi t - a_1 \pi t \cos \pi t.$$

The above three conditions yield $Nc = 0$, where $c = (a_1, c_1, c_2)'$ and

$$N = \begin{pmatrix} 0 & 1 & 0 \\ \pi & -1 & 0 \\ -\tfrac{3}{4} & 0 & \tfrac{1}{2} \end{pmatrix}, \quad |N| = -\frac{\pi}{2} \neq 0.$$

Thus $\lambda = \pi^2$ is not an eigenvalue. Then we obtain, from (5.82),

(5.83) $$D_6(\lambda) = \frac{\sin \sqrt{\lambda}}{\sqrt{\lambda}} \bigg/ \left(1 - \frac{\lambda}{\pi^2}\right)$$

$$= \prod_{n=2}^{\infty} \left(1 - \frac{\lambda}{n^2 \pi^2}\right)$$

as the FD of $K_6(s, t) = \min(s, t) - st - 2 \sin \pi s \sin \pi t / \pi^2$; hence the c.f. of U_6 in (5.80) is given by $(D_6(2i\theta))^{-1/2}$.

The above result may be derived as follows. By Mercer's theorem (Theorem 5.2) it holds that

$$\min(s, t) - st = \sum_{n=1}^{\infty} \frac{2}{n^2 \pi^2} \sin n\pi s \sin n\pi t,$$

where $\{\sqrt{2}\sin n\pi t\}$ is an orthonormal sequence of eigenfunctions corresponding to $\lambda_n = n^2\pi^2$. Therefore

$$K_6(s,t) = \sum_{n=2}^{\infty} \frac{2}{n^2\pi^2} \sin n\pi s \sin n\pi t,$$

which yields $D_6(\lambda)$ in (5.83). The same reasoning can be applied to obtain the FD $D(\lambda)$ of

$$K(s,t) = \min(s,t) - st - \sum_{n=1}^{k} \frac{2}{n^2\pi^2} \sin n\pi s \sin n\pi t.$$

We have

$$D(\lambda) = \frac{\sin\sqrt{\lambda}}{\sqrt{\lambda}} \bigg/ \prod_{n=1}^{k}\left(1 - \frac{\lambda}{n^2\pi^2}\right)$$

$$= \prod_{n=k+1}^{\infty}\left(1 - \frac{\lambda}{n^2\pi^2}\right).$$

Let us consider a slightly different statistic

(5.84) $$U_7 = \int_0^1\int_0^1 \left[\min(s,t) - st - \frac{2}{\pi^2}\sin^2\pi s \sin^2\pi t\right] dw(s)\,dw(t).$$

Mercer's theorem is not helpful here, but, following the method demonstrated before, we can show [Nabeya (1992) and Problem 4.10] that $E(e^{i\theta U_7}) = (D_7(2i\theta))^{-1/2}$, where

(5.85) $$D_7(\lambda) = \left[\frac{\sin\sqrt{\lambda}}{\sqrt{\lambda}} + \frac{1-\cos\sqrt{\lambda}}{\pi^2}\bigg/\left(1 - \frac{\lambda}{4\pi^2}\right)\right]\bigg/\left(1 - \frac{\lambda}{4\pi^2}\right).$$

Nabeya (1992) points out that the corresponding result earlier obtained by Darling (1955) is incorrect. It can be checked that $\lambda = 4\pi^2$ is an eigenvalue although the factor $1 - (\lambda/(4\pi^2))$ appears in the denominator of $D_7(\lambda)$.

Similarly, if we consider an extended version

(5.86) $$U_8 = \int_0^1\int_0^1 \left[\min(s,t) - st - \frac{2}{\pi^2}\sin^2\pi s \sin^2\pi t\right.$$
$$\left. - \frac{1}{2\pi^2}\sin 2\pi s \sin 2\pi t\right] dw(s)\,dw(t),$$

we can show [Nabeya (1992) and Problem 4.11] that $E(e^{i\theta U_8}) = (D_8(2i\theta))^{-1/2}$, where

(5.87) $\quad D_8(\lambda) = D_7(\lambda) \bigg/ \left(1 - \dfrac{\lambda}{4\pi^2}\right)$

$\quad\quad\quad\quad = \left[\dfrac{\sin\sqrt{\lambda}}{\sqrt{\lambda}} + \dfrac{1 - \cos\sqrt{\lambda}}{\pi^2} \bigg/ \left(1 - \dfrac{\lambda}{4\pi^2}\right)\right] \bigg/ \left(1 - \dfrac{\lambda}{4\pi^2}\right)^2.$

It can be checked that $\lambda = 4\pi^2$ is not an eigenvalue.

As a final example, we deal with the integrated Brownian motion. Let us consider the statistic defined in (5.8). When $g = 1$, the statistic becomes

(5.88) $\quad\quad U_9 = \int_0^1 F_1^2(t)\,dt$

$\quad\quad\quad\quad = \int_0^1 \int_0^1 \left[\int_{\max(s,t)}^1 (u-s)(u-t)\,du\right] dw(s)\,dw(t),$

where $\{F_1(t)\}$ is the one-fold integrated Brownian motion defined by

$$F_1(t) = \int_0^t w(s)\,ds.$$

The kernel appearing in (5.88), which we denote as $K_9(s,t)$, has the expression

$$K_9(s,t) = \frac{1}{6}(1-t)^2(t+2-3s)$$

for $s \leq t$.

The integral equation (5.10) with $K = K_9$ is equivalent to

$$f^{(4)}(t) - \lambda f(t) = 0, \quad f(1) = f'(1) = f''(0) = f'''(0) = 0.$$

The differential equation has the general solution

$$f(t) = c_1 e^{At} + c_2 e^{-At} + c_3 e^{iAt} + c_4 e^{-iAt}, \quad A = \lambda^{1/4},$$

for arbitrary constants c_1 through c_4. The four boundary conditions above yield $M(\lambda)c = 0$, where $c = (c_1, c_2, c_3, c_4)'$ and

$$M(\lambda) = \begin{pmatrix} e^A & e^{-A} & e^{iA} & e^{-iA} \\ e^A & -e^{-A} & ie^{iA} & -ie^{-iA} \\ 1 & 1 & -1 & -1 \\ 1 & -1 & -i & i \end{pmatrix},$$

$|M(\lambda)| = 8i(1 + \cos A \cosh A)$.

Then we obtain, as the FD of K_9,

(5.89) $$D_9(\lambda) = \frac{1}{2}\left(1 + \cos\lambda^{1/4}\cosh\lambda^{1/4}\right).$$

The same result was obtained in Section 4.2 of Chapter 4 by the stochastic process approach.

It may be of interest to compute the eigenvalues of K_9, that is, the zeros of $D_9(\lambda)$. The first six eigenvalues $\lambda_n(1)$ ($n = 1, \ldots, 6$) are

$$\lambda_1(1) = 12.36236, \quad \lambda_2(1) = 485.5188, \quad \lambda_3(1) = 3806.546,$$
$$\lambda_4(1) = 14617.27, \quad \lambda_5(1) = 39943.83, \quad \lambda_6(1) = 89135.41.$$

Since $\mathcal{L}(U_9) = \mathcal{L}(\sum_{n=1}^{\infty} Z_n^2/\lambda_n(1))$, where $\{Z_n\} \sim \text{NID}(0, 1)$, and it holds that $E(U_9) = \frac{1}{12}$, only the first term $Z_1^2/\lambda_1(1)$ in the infinite sum expression for U_9 has quite a high relative weight. In fact,

$$E\left(Z_1^2/\lambda_1(1)\right)/E(U_9) = \frac{12}{12.36236} = 0.9707.$$

This observation leads us to approximate the distribution of U_9 just by $\chi^2(1)$ divided by $\lambda_1(1)$. Both distributions have monotone densities with an infinite peak at the origin. The graphical comparison will be made in the next chapter.

The case for the two-fold integrated Brownian motion can be dealt with similarly. The statistic is

(5.90) $$U_{10} = \int_0^1 F_2^2(t)\,dt$$
$$= \int_0^1 \int_0^1 \left[\int_{\max(s,t)}^1 \frac{1}{4}((u-s)(u-t))^2\,du\right] dw(s)\,dw(t),$$

where the kernel appearing above, denoted as $K_{10}(s, t)$, has the expression

$$K_{10}(s, t) = \frac{1}{120}(1-t)^3(t^2 + 3t + 6 - 5st + 10s^2 - 15s)$$

for $s \leq t$. We obtain

$$f^{(6)}(t) + \lambda f(t) = 0, \quad f(1) = f'(1) = f''(1) = f'''(0) = f^{(4)}(0) = f^{(5)}(0) = 0,$$

where the differential equation has the solution

$$f(t) = c_1 e^{i\alpha t} + c_2 e^{-i\alpha t} + c_3 e^{i\beta t} + c_4 e^{-i\beta t} + c_5 e^{i\gamma t} + c_6 e^{-i\gamma t}$$

for arbitrary constants c_1 through c_6 with $\alpha = \lambda^{1/6}$, $\beta = \lambda^{1/6}(1 - \sqrt{3}i)/2$, and $\gamma = \lambda^{1/6}(-1 - \sqrt{3}i)/2$. The six boundary conditions above yield $M(\lambda)c = 0$, where

$c = (c_1, c_2, c_3, c_4, c_5, c_6)'$ and

$$M(\lambda) = \begin{pmatrix} e^{i\alpha} & e^{-i\alpha} & e^{i\beta} & e^{-i\beta} & e^{i\gamma} & e^{-i\gamma} \\ \alpha e^{i\alpha} & -\alpha e^{-i\alpha} & \beta e^{i\beta} & -\beta e^{-i\beta} & \gamma e^{i\gamma} & -\gamma e^{-i\gamma} \\ \alpha^2 e^{i\alpha} & \alpha^2 e^{-i\alpha} & \beta^2 e^{i\beta} & \beta^2 e^{-i\beta} & \gamma^2 e^{i\gamma} & \gamma^2 e^{-i\gamma} \\ \alpha^3 & -\alpha^3 & \beta^3 & -\beta^3 & \gamma^3 & -\gamma^3 \\ \alpha^4 & \alpha^4 & \beta^4 & \beta^4 & \gamma^4 & \gamma^4 \\ \alpha^5 & -\alpha^5 & \beta^5 & -\beta^5 & \gamma^5 & -\gamma^5 \end{pmatrix},$$

$$|M(\lambda)| = 12i \left[\cos \alpha \left(\cos \alpha + 2 \cosh^2 \frac{\sqrt{3}\alpha}{2} + 3 \right) + 8 \cos \frac{\alpha}{2} \cosh \frac{\sqrt{3}\alpha}{2} + 4 \right].$$

The FD of K_{10} is found to be

$$(5.91) \quad D_{10}(\lambda) = \frac{1}{9} \left[2 \left(1 + \cos \lambda^{1/6} + \cos \lambda^{1/6} \omega + \cos \lambda^{1/6} \omega^2 \right) \right.$$
$$\left. + \cos \lambda^{1/6} \cos \lambda^{1/6} \omega \cos \lambda^{1/6} \omega^2 \right],$$

where $\omega = (1 + \sqrt{3}i)/2$.

The smallest eigenvalue $\lambda_1(2)$ of K_{10}, that is, the smallest zero of $D_{10}(\lambda)$ is 121.259, while $E(U_{10}) = 1/120$. Thus the first term in the expression $\mathcal{L}(U_{10}) = \mathcal{L}(\sum_{n=1}^{\infty} Z_n^2/\lambda_n(2))$ has the relative weight

$$E\left(Z_1^2/\lambda_1(2)\right)/E(U_{10}) = \frac{120}{121.259} = 0.9896.$$

The approximation of $\mathcal{L}(U_{10})$ by $\mathcal{L}(Z_1^2/\lambda_1(2))$ will also be graphically presented in the next chapter.

For the general g-integrated Brownian motion, the statistic takes the form as in (5.8), where the kernel has the expression

$$K(s,t) = \frac{1}{(g!)^2} \int_{\max(s,t)}^{1} ((u-s)(u-t))^g \, du$$

$$= \frac{1}{(g!)^2} \sum_{j,k=0}^{g} \binom{g}{j} \binom{g}{k} \frac{(-1)^{j+k}}{j+k+1} \left[1 - (\max(s,t))^{j+k+1} \right] s^{g-j} t^{g-k}.$$

As a function of t, $K(s,t)$ is a polynomial of degree $2g+1$ whose coefficient is

$$\frac{(-1)^{g+1}}{(g!)^2} \sum_{k=0}^{g} \binom{g}{k} \frac{(-1)^k}{g+k+1} = \frac{(-1)^{g+1}}{(g!)^2} \frac{g! \, \Gamma(g+1)}{\Gamma(2g+2)}$$

$$= \frac{(-1)^{g+1}}{(2g+1)!}.$$

THE FREDHOLM THEORY: THE NONHOMOGENEOUS CASE

The integral equation (5.10) with $K(s,t)$ given above yields

$$f^{(2g+2)}(t) + (-1)^g \lambda f(t) = 0,$$

$$f^{(j)}(1) = 0 \ (j = 0, 1, \ldots, g), \quad f^{(k)}(0) = 0 \ (k = g+1, g+2, \ldots, 2g+1).$$

In principle, we can solve the differential equation, and the $2g + 2$ boundary conditions will yield a homogeneous equation to obtain the FD.

PROBLEMS

4.1 Show that the kernel appearing in (5.45) is positive definite.

4.2 Show that (5.47) is equivalent to the integral equation (5.10) with $K(s,t) = (1 - (\max(s,t))^{2m+1})/(2m+1)$.

4.3 Show that the two boundary conditions in (5.47) imply $|M(\lambda)| = 0$, where $M(\lambda)$ is defined in (5.52).

4.4 Prove that $D_2(\lambda)$ in (5.57) is the FD of $K_2(s,t) = (\max(s,t))^m/2$, using the relation

$$J_\nu(z) = \frac{z}{2\nu}\{J_{\nu-1}(z) + J_{\nu+1}(z)\}.$$

4.5 Show that $D_2(\lambda)$ in (5.57) reduces to $1 - \frac{\lambda}{2}$ when $m = 0$.

4.6 Derive the c.f. of U_3 defined in (5.58).

4.7 Establish (5.62) by deriving the c.f. of the distribution on the right side.

4.8 Derive the c.f. of U_4 defined in (5.63).

4.9 Derive $\tilde{D}_5(\lambda)$ in (5.79) from the differential equation (5.74) with the boundary conditions $f(0) = f(1) = 0$ and the first condition in (5.76).

4.10 Derive the c.f. of U_7 defined in (5.84).

4.11 Derive the c.f. of U_8 defined in (5.86).

5.5 THE FREDHOLM THEORY: THE NONHOMOGENEOUS CASE

We have so far discussed how to obtain the c.f.s of purely quadratic functionals of the Brownian motion. In this section we deal with quadratic plus linear or bilinear functionals of the Brownian motion. More specifically we consider

(5.92) $$S_1 = \int_0^1 \int_0^1 K(s,t) \, dw(s) \, dw(t) + a \int_0^1 l(t) \, dw(t)$$

or

$$(5.93) \qquad S_2 = \int_0^1 \int_0^1 K(s,t)\,dw(s)\,dw(t) + aZ \int_0^1 l(t)\,dw(t) + bZ^2,$$

where $l(t)$ is continuous, a and b are constants, while Z follows $N(0,1)$ and is independent of $\{w(t)\}$. We continue to assume that $K(s,t)$ is continuous, symmetric, and nearly definite.

To derive the c.f.s of the above statistics, we need the Fredholm theory on nonhomogeneous integral equations, which we describe briefly. Let us consider

$$(5.94) \qquad f(t) = \lambda \int_0^1 K(s,t) f(s)\,ds + g(t),$$

where $g(t)$ is a continuous function on $[0, 1]$. The corresponding algebraic system is

$$(5.95) \qquad f_T = \frac{\lambda}{T} K_T f_T + g_T,$$

where f_T and K_T are defined in (5.11), while $g_T = [(g(j/T))]$ is a $T \times 1$ vector. Putting $D_T(\lambda) = |I_T - \lambda K_T/T|$ and assuming that $D_T(\lambda) \neq 0$, (5.95) can be solved to obtain

$$f_T = \frac{1}{D_T(\lambda)} G_T(\lambda) g_T,$$

where $G_T(\lambda)$ is the adjoint matrix of $I_T - \lambda K_T/T$. The jth component of this solution may be written as

$$(5.96) \qquad f\left(\frac{j}{T}\right) = \frac{1}{D_T(\lambda)} \left[G_T(j,j;\lambda) g\left(\frac{j}{T}\right) + \sum_{\substack{k=1 \\ k \neq j}}^T G_T(j,k;\lambda) g\left(\frac{k}{T}\right) \right],$$

where $G_T(j,k;\lambda)$ is the (j,k)th element of $G_T(\lambda)$. It is noted that $G_T(j,j;\lambda)$ is of the same type as $D_{T-1}(\lambda)$ so that

$$\lim_{T \to \infty} G_T(j,j;\lambda) = \lim_{T \to \infty} D_T(\lambda) = D(\lambda).$$

Then we take a limit of (5.96), which can be expressed, if $D(\lambda) \neq 0$, as

$$(5.97) \qquad f(t) = g(t) + \int_0^1 \frac{1}{D(\lambda)} G(t,s;\lambda) g(s)\,ds$$

$$= g(t) + \lambda \int_0^1 \Gamma(t,s;\lambda) g(s)\,ds,$$

THE FREDHOLM THEORY: THE NONHOMOGENEOUS CASE 159

where $G(s,t;\lambda)/\lambda$ is called the *Fredholm minor* and $\Gamma(s,t;\lambda) = G(s,t;\lambda)/(\lambda D(\lambda))$ the *Fredholm resolvent* or simply the resolvent of $K(s,t)$.

The above arguments are just formal, but can be made rigorous [see, for example, Courant and Hilbert (1953) and Hochstadt (1973)].

In subsequent discussions, the resolvent $\Gamma(s,t;\lambda)$ plays a fundamental role, for which various expressions are possible. We have

(5.98)
$$\Gamma(s,t;\lambda) = \sum_{j=1}^{\infty} \lambda^{j-1} K_{(j)}(s,t)$$

$$= \sum_{n=1}^{\infty} \frac{1}{\lambda_n - \lambda} f_n(s) f_n(t)$$

$$= K(s,t) + \lambda \int_0^1 \Gamma(s,u;\lambda) K(u,t)\, du\, ,$$

where $K_{(j)}(s,t)$ is the iterated kernel defined by

(5.99)
$$K_{(j)}(s,t) = \int_0^1 K(s,u) K_{(j-1)}(u,t)\, du$$

$$= \sum_{n=1}^{\infty} \frac{1}{\lambda_n^j} f_n(s) f_n(t)\, ,$$

with $K_{(1)}(s,t) = K(s,t)$, while $\{\lambda_n\}$ is a sequence of eigenvalues for the homogeneous integral equation (5.10), repeated as many times as their multiplicities, and $\{f_n(t)\}$ is an orthonormal sequence of eigenfunctions corresponding to $\{\lambda_n\}$. Note that $\Gamma(s,t;\lambda)$ is a symmetric function of s and t, and $\Gamma(s,t;0) = K(s,t)$.

It is known that the first expression in (5.98) is valid for $|\lambda| < 1/\max|K(s,t)|$. It follows from the second expression that $\Gamma(s,t;\lambda)$ is not an entire function of λ, unlike $D(\lambda)$, but is analytic except for simple poles at $\{\lambda_n\}$. Namely the resolvent is a meromorphic function of λ that possesses simple poles at the eigenvalues. The last expression may be most useful for obtaining $\Gamma(s,t;\lambda)$, although it is hard in general to obtain it for any s and t. For our purpose, however, it is not necessary, as will be shown below.

We now consider the statistic S_1 in (5.92). We first assume that $l(t) = K(0,t)$. The general case will be treated later. Thus we deal with

(5.100)
$$S = \int_0^1 \int_0^1 K(s,t)\, dw(s)\, dw(t) + a \int_0^1 K(0,t)\, dw(t)\, .$$

Using Mercer's theorem we have

(5.101)
$$\mathcal{L}(S) = \mathcal{L}\left(\sum_{n=1}^{\infty} \frac{1}{\lambda_n} (Z_n^2 + a f_n(0) Z_n)\right) ,$$

where $\{Z_n\} \sim \text{NID}(0, 1)$. It now follows (Problem 5.1) that

$$(5.102) \qquad E(e^{i\theta S}) = (D(2i\theta))^{-1/2} \exp\left[\frac{(ia\theta)^2}{2} \sum_{n=1}^{\infty} \frac{f_n^2(0)}{\lambda_n(\lambda_n - 2i\theta)}\right],$$

where $D(\lambda)$ is the FD of K. Moreover, using the definition of the resolvent $\Gamma(s, t; \lambda)$, we arrive at the following theorem [Tanaka (1990a) and Problem 5.2].

Theorem 5.8. *The c.f. $\phi(\theta)$ of S defined in (5.100) is given by*

$$\phi(\theta) = (D(2i\theta))^{-1/2} \exp\left[\frac{ia^2\theta}{4}\{\Gamma(0, 0; 2i\theta) - K(0, 0)\}\right],$$

where $D(\lambda)$ is the FD of K and $\Gamma(s, t; \lambda)$ is the resolvent of K.

To demonstrate how to obtain $\Gamma(0, 0; \lambda)$, let us consider

$$(5.103) \qquad U_1 = \int_0^1 t^{2m}(w(t) + \kappa)^2 \, dt$$

$$= \int_0^1 \int_0^1 \frac{1}{2m+1}\left[1 - (\max(s, t))^{2m+1}\right] dw(s) \, dw(t)$$

$$+ 2\kappa \int_0^1 \frac{1 - t^{2m+1}}{2m+1} dw(t) + \frac{\kappa^2}{2m+1},$$

where $m > -\frac{1}{2}$ and κ is a constant. Note that $(1 - t^{2m+1})/(2m + 1) = K_1(0, t)$, where $K_1(s, t) = (1 - (\max(s, t))^{2m+1})/(2m + 1)$. We already know the FD $D_1(\lambda)$ of K_1, which is given on the right side of (5.53). To obtain $\Gamma(0, 0; \lambda)$, put $h(t) = \Gamma(0, t; \lambda)$ and use the last relation in (5.98) to get

$$(5.104) \qquad h(t) = K_1(0, t) + \lambda \int_0^1 h(s)K_1(s, t) \, ds.$$

We can show (Problem 5.3) that (5.104) is equivalent to

$$(5.105) \qquad h''(t) - \frac{2m}{t}h'(t) + \lambda t^{2m}h(t) = 0, \quad \lim_{t \to 0} \frac{h'(t)}{t^{2m}} = -1, \quad h(1) = 0.$$

The general solution is given by

$$h(t) = t^{(2m+1)/2}\left\{c_1 J_\nu\left(\frac{\sqrt{\lambda}}{m+1} t^{m+1}\right) + c_2 J_{-\nu}\left(\frac{\sqrt{\lambda}}{m+1} t^{m+1}\right)\right\},$$

where $\nu = (2m + 1)/(2(m + 1))$. The two boundary conditions yield $M(\lambda)c = (-1, 0)'$, where $c = (c_1, c_2)'$ and $M(\lambda)$ is given in (5.52). Thus c_1 and c_2 can be uniquely determined as

$$c_1 = -\frac{\Gamma(\nu+1)}{2m+1}\left(\frac{\sqrt{\lambda}}{2(m+1)}\right)^{-\nu},$$

$$c_2 = \frac{J_\nu\left(\sqrt{\lambda}/(m+1)\right)}{J_{-\nu}\left(\sqrt{\lambda}/(m+1)\right)}\left(\frac{\sqrt{\lambda}}{2(m+1)}\right)^{-\nu}\frac{\Gamma(\nu+1)}{2m+1},$$

so that, by definition,

$$\Gamma(0,0;\lambda) = h(0)$$

$$= \frac{c_2}{\Gamma(-\nu+1)}\left(\frac{\sqrt{\lambda}}{2(m+1)}\right)^{-\nu}$$

$$= \frac{\Gamma(\nu+1)}{(2m+1)\Gamma(-\nu+1)}\frac{J_\nu\left(\sqrt{\lambda}/(m+1)\right)}{J_{-\nu}\left(\sqrt{\lambda}/(m+1)\right)}\left(\frac{\sqrt{\lambda}}{2(m+1)}\right)^{-2\nu}.$$

Therefore Theorem 5.8 yields

$$E\left(e^{i\theta U_1}\right) = (D_1(2i\theta))^{-1/2}\exp\left\{i\kappa^2\theta\left(\Gamma(0,0;2i\theta) - \frac{1}{2m+1}\right)\right\}e^{i\kappa^2\theta/(2m+1)}$$

$$= \left[\Gamma(-\nu+1)J_{-\nu}\left(\frac{\sqrt{2i\theta}}{m+1}\right)\bigg/\left(\frac{\sqrt{2i\theta}}{2(m+1)}\right)^{-\nu}\right]^{-1/2}$$

$$\times \exp\left\{\frac{i\kappa^2\theta\,\Gamma(\nu+1)}{(2m+1)\Gamma(-\nu+1)}\frac{J_\nu\left(\sqrt{2i\theta}/(m+1)\right)}{J_{-\nu}\left(\sqrt{2i\theta}/(m+1)\right)}\left(\frac{\sqrt{2i\theta}}{2(m+1)}\right)^{-2\nu}\right\}.$$

(5.106)

We note that, when $m = 0$, (5.106) reduces to

$$E\left(e^{i\theta U_1}\right) = (\cos\sqrt{2i\theta})^{-1/2}\exp\left\{i\kappa^2\theta\,\frac{\tan\sqrt{2i\theta}}{\sqrt{2i\theta}}\right\},$$

which was obtained in Section 4.1 of Chapter 4 by the stochastic process approach. In that case, U_1 reduces to $\int_0^1 X^2(t)\,dt$, where $\{X(t) = w(t) + \kappa\}$ follows a simple O–U process $dX(t) = dw(t)$ with $X(0) = \kappa$. As far as the O–U process is concerned, the stochastic process approach is preferable to the present approach. In fact, if $\{X(t)\}$ is the O–U process defined by $dX(t) = -\beta X(t)\,dt + dw(t)$ with $X(0) = \kappa$, and if we consider

$$(5.107) \quad U_2 = \int_0^1 X^2(t)\,dt$$

$$= \int_0^1 \left\{ e^{-\beta t}\int_0^t e^{\beta s}\,dw(s) + \kappa e^{-\beta t}\right\}^2 dt$$

$$= \int_0^1\int_0^1 \frac{e^{-\beta|s-t|} - e^{-\beta(2-s-t)}}{2\beta}\,dw(s)\,dw(t)$$

$$+ 2\kappa \int_0^1 \frac{e^{-\beta t} - e^{-\beta(2-t)}}{2\beta}\,dw(t) + \kappa^2 \frac{1 - e^{-2\beta}}{2\beta},$$

we can still follow the present approach noting that $(e^{-\beta t} - e^{-\beta(2-t)})/(2\beta) = K_2(0,t)$, where

$$K_2(s,t) = \frac{1}{2\beta}\left[e^{-\beta|s-t|} - e^{-\beta(2-s-t)}\right].$$

Since the FD of K_2 is shown to be

$$(5.108) \quad D_2(\lambda) = \left(\cos\sqrt{\lambda - \beta^2} + \beta\,\frac{\sin\sqrt{\lambda - \beta^2}}{\sqrt{\lambda - \beta^2}}\right)e^{-\beta},$$

which can also be deduced from (4.9), and it is also shown that the resolvent $\Gamma(s,t;\lambda)$ of K_2 evaluated at the origin is

$$\Gamma(0,0;\lambda) = \frac{\sin\sqrt{\lambda - \beta^2}}{\sqrt{\lambda - \beta^2}} \bigg/ \left[\cos\sqrt{\lambda - \beta^2} + \beta\,\frac{\sin\sqrt{\lambda - \beta^2}}{\sqrt{\lambda - \beta^2}}\right],$$

we obtain

$$(5.109) \quad E\left(e^{i\theta U_2}\right) = \left(\cos\mu + \beta\,\frac{\sin\mu}{\mu}\right)^{-1/2}\exp\left\{\frac{\beta}{2} + \frac{i\kappa^2\theta(\sin\mu/\mu)}{\cos\mu + \beta(\sin\mu/\mu)}\right\},$$

where $\mu = \sqrt{2i\theta - \beta^2}$. This last result, however, was obtained in (4.9) more easily by the stochastic process approach.

The present approach may be effectively used to obtain the c.f. of

$$(5.110) \quad U_3 = \int_0^1 t^m\left(w(t) + \kappa\right)dw(t)$$

$$= \int_0^1\int_0^1 \frac{1}{2}(\max(s,t))^m\,dw(s)\,dw(t) + 2\kappa\int_0^1 \frac{1}{2}t^m\,dw(t) - \frac{1}{2(m+1)}.$$

It follows (Problem 5.4) that

$$
(5.111) \quad E\left(e^{i\theta U_3}\right) = \left[\Gamma(\nu) J_{\nu-1}\left(\frac{\sqrt{-4i\theta m}}{m+1}\right) \Big/ \left(\frac{\sqrt{-4i\theta m}}{2(m+1)}\right)^{\nu-1}\right]^{-1/2}
$$

$$
\times \exp\left\{\frac{i\kappa^2 \theta}{2} \frac{\Gamma(-\nu+1)}{\Gamma(\nu+1)} \frac{J_{-\nu+1}\left(\sqrt{-4i\theta m}/(m+1)\right)}{J_{\nu-1}\left(\sqrt{-4i\theta m}/(m+1)\right)} \left(\frac{\sqrt{-4i\theta m}}{2(m+1)}\right)^{2\nu} - \frac{i\theta}{2(m+1)}\right\},
$$

where $\nu = -m/(m+1)$.

We have so far dealt with statistics that take the form as in (5.92) under the assumption that $l(t) = K(0,t)$. If this is not the case, Theorem 5.8 does not apply. For such cases Nabeya (1992) presented a solution, which we describe below. Let us consider

$$
(5.112) \quad S_Y = \int_0^1 \{Y(t) + m(t)\}^2 \, dt,
$$

where $\{Y(t)\}$ is a zero-mean Gaussian process with $\text{Cov}(Y(s), Y(t)) = K(s,t)$, while $m(t)$ is a continuous function.

We also define

$$
\frac{c_n}{\sqrt{\lambda_n}} = \int_0^1 m(t) f_n(t) \, dt, \quad q(t) = \sum_{n=1}^\infty \frac{c_n}{\sqrt{\lambda_n}} f_n(t), \quad r(t) = m(t) - q(t),
$$

where $\{\lambda_n\}$ is a sequence of eigenvalues of the positive definite kernel $K(s,t)$, repeated as many times as their multiplicities, while $\{f_n(t)\}$ is an orthonormal sequence of eigenfunctions corresponding to $\{\lambda_n\}$. Note that $c_n/\sqrt{\lambda_n}$ and $q(t)$ are the Fourier coefficients and Fourier series for $m(t)$, respectively, where the infinite series is assumed to converge uniformly. It holds (Problem 5.5) that

$$
(5.113) \quad \int_0^1 q(t) r(t) \, dt = 0
$$

so that

$$
\int_0^1 m(t) q(t) \, dt = \int_0^1 q^2(t) \, dt = \sum_{n=1}^\infty \frac{c_n^2}{\lambda_n}.
$$

Let $\{Z_n\}$ be a sequence of NID(0, 1) and define

$$
(5.114) \quad Z(t) = \sum_{n=1}^\infty \frac{f_n(t)}{\sqrt{\lambda_n}} Z_n.
$$

Then $\{Z(t)\}$ is also a zero-mean Gaussian process with

$$\mathrm{Cov}\bigl(Z(s),Z(t)\bigr) = \sum_{n=1}^{\infty} \frac{1}{\lambda_n} f_n(s) f_n(t) = K(s,t)$$

so that any finite-dimensional distribution of $\{Z(t) + m(t)\}$ is the same as that of $\{Y(t) + m(t)\}$. Therefore the statistic S_Y in (5.112) has the same c.f. as

(5.115)
$$\begin{aligned}
S_Z &= \int_0^1 \bigl(Z(t) + m(t)\bigr)^2 dt \\
&= \int_0^1 \left\{ \sum_{n=1}^{\infty} \frac{f_n(t)}{\sqrt{\lambda_n}} (Z_n + c_n) + r(t) \right\}^2 dt \\
&= \sum_{n=1}^{\infty} \frac{1}{\lambda_n} (Z_n + c_n)^2 + \int_0^1 r^2(t)\, dt\, .
\end{aligned}$$

We now obtain [Nabeya (1992) and Problem 5.6]

(5.116)
$$E\bigl(e^{i\theta S_Z}\bigr) = (D(2i\theta))^{-1/2} Q(\theta),$$

where $D(\lambda)$ is the FD of K and

$$\begin{aligned}
Q(\theta) &= \exp\left\{ i\theta \int_0^1 r^2(t)\, dt + \sum_{n=1}^{\infty} \frac{i\theta c_n^2}{\lambda_n - 2i\theta} \right\} \\
&= \exp\left\{ i\theta \int_0^1 m^2(t)\, dt - 2\theta^2 \sum_{n=1}^{\infty} \frac{c_n^2}{\lambda_n(\lambda_n - 2i\theta)} \right\}.
\end{aligned}$$

Using the second relation of the resolvent in (5.98) and the fact that

$$\frac{c_n^2}{\lambda_n} = \int_0^1 \int_0^1 m(s) m(t) f_n(s) f_n(t)\, ds\, dt\, ,$$

we can now establish the following theorem [Nabeya (1992)].

Theorem 5.9. *The c.f. of*

$$S_Y = \int_0^1 \{Y(t) + m(t)\}^2 dt$$

is given by

$$\phi(\theta) = (D(2i\theta))^{-1/2} \exp\left\{ i\theta \int_0^1 m^2(t)\, dt - 2\theta^2 \int_0^1 \int_0^1 \Gamma(s,t; 2i\theta) m(s) m(t)\, ds\, dt \right\},$$

THE FREDHOLM THEORY: THE NONHOMOGENEOUS CASE

where $D(\lambda)$ is the FD of $K(s,t) = \text{Cov}(Y(s), Y(t))$ and $\Gamma(s,t;\lambda)$ is the resolvent of K.

In the above theorem we do not need to derive $\Gamma(s,t;\lambda)$, as Nabeya (1992) demonstrates. Multiplying by $m(s)$ both sides of the last equation for $\Gamma(s,t;\lambda)$ in (5.98) and integrating with respect to s lead us to

$$h(t) = \int_0^1 K(s,t)m(s)\,ds + \lambda \int_0^1 K(s,t)h(s)\,ds, \tag{5.117}$$

where

$$h(t) = \int_0^1 \Gamma(s,t;\lambda)m(s)\,ds. \tag{5.118}$$

The nonhomogeneous equation (5.117) can be solved for $h(t)$ in the same way as before and it follows from (5.118) that

$$\int_0^1 \int_0^1 \Gamma(s,t;\lambda)m(s)m(t)\,ds\,dt = \int_0^1 h(t)m(t)\,dt. \tag{5.119}$$

As an example consider

$$U_4 = \int_0^1 (w(t) + a + bt)^2\,dt \tag{5.120}$$

$$= \int_0^1 \int_0^1 [1 - \max(s,t)]\,dw(s)\,dw(t)$$

$$+ 2\int_0^1 \left(a + \frac{b}{2} - at - \frac{bt^2}{2}\right)dw(t) + a^2 + ab + \frac{b^2}{3}.$$

Note that the statistic U_4 is not of the form given in (5.100) unless $b = 0$. We need to rely on Theorem 5.9 rather than Theorem 5.8. The integral equation (5.117) with $K(s,t) = \text{Cov}(w(s), w(t)) = \min(s,t)$ and $m(s) = a + bs$ is equivalent to

$$h''(t) + \lambda h(t) = -(a + bt), \qquad h(0) = h'(1) = 0,$$

where the general solution is given by

$$h(t) = c_1 \cos\sqrt{\lambda}\,t + c_2 \sin\sqrt{\lambda}\,t - \frac{1}{\lambda}(a + bt).$$

From the boundary conditions $h(0) = h'(1) = 0$, we can determine c_1 and c_2 uniquely as

$$c_1 = \frac{a+b}{\lambda \cos\sqrt{\lambda}} - \frac{b\tan\sqrt{\lambda}}{\lambda\sqrt{\lambda}}, \qquad c_2 = \frac{b}{\lambda\sqrt{\lambda}}.$$

Then Theorem 5.9 and (5.119) lead us to

$$E\left(e^{i\theta U_4}\right) = (\cos\sqrt{2i\theta})^{-1/2} \exp\left[\frac{1}{2}\left\{(a+b)^2\sqrt{2i\theta}\tan\sqrt{2i\theta}\right.\right.$$
$$\left.\left. + 2ab\left(1 - \frac{1}{\cos\sqrt{2i\theta}}\right) + b^2\left(\frac{\tan\sqrt{2i\theta}}{\sqrt{2i\theta}} - \frac{2}{\cos\sqrt{2i\theta}} + 1\right)\right\}\right].$$

(5.121)

A quite similar argument applies to derive the c.f. of

(5.122) $\quad U_5 = \int_0^1 (w(t) - tw(1) + a + bt)^2\, dt$

$$= \int_0^1 \int_0^1 \left[\frac{1}{3} - \max(s,t) + \frac{s^2 + t^2}{2}\right] dw(s)\, dw(t)$$
$$+ \int_0^1 \left(a(1 - 2t) + b\left(\frac{1}{3} - t^2\right)\right) dw(t) + a^2 + ab + \frac{1}{3}b^2.$$

We can show (Problem 5.7) that

$$E\left(e^{i\theta U_5}\right) = \left(\frac{\sin\sqrt{2i\theta}}{\sqrt{2i\theta}}\right)^{-1/2} \exp\left[\frac{1}{\cos\sqrt{2i\theta} + 1}\left\{a(a+b)\sqrt{2i\theta}\sin\sqrt{2i\theta}\right.\right.$$
$$\left.\left. + \frac{b^2}{2}\left(\frac{\sqrt{2i\theta}\cos\sqrt{2i\theta}\sin\sqrt{2i\theta}}{\cos\sqrt{2i\theta} - 1} + \cos\sqrt{2i\theta} + 1\right)\right\}\right].$$

(5.123)

Nabeya (1992) considers various statistics that take the form given in (5.112) arising in goodness-of-fit tests. Among them is the statistic

(5.124) $\quad U_6 = \int_0^1 \left(Y(t) + \frac{a}{\pi}\sin^2 \pi t\right)^2 dt,$

where

$$\mathrm{Cov}(Y(s), Y(t)) = \min(s,t) - st - \frac{1}{2\pi^2}\sin 2\pi s\, \sin 2\pi t$$
$$= K_6(s,t).$$

We already know from the discussions below (5.83) that the FD $D_6(\lambda)$ of K_6 is given by

$$D_6(\lambda) = \frac{\sin\sqrt{\lambda}}{\sqrt{\lambda}} \bigg/ \left(1 - \frac{\lambda}{4\pi^2}\right).$$

Thus, proceeding in the same way as above, we obtain [Nabeya (1992) and Problem 5.8]

$$(5.125)\quad E\left(e^{i\theta U_6}\right) = \left(D_6(2i\theta)\right)^{-1/2} \exp\left[a^2\left\{\frac{i\theta}{4(2\pi^2 - i\theta)}\right.\right.$$
$$\left.\left. + \frac{\pi^2\sqrt{2i\theta}}{(2\pi^2 - i\theta)^2}\frac{1 - \cos\sqrt{2i\theta}}{\sin\sqrt{2i\theta}}\right\}\right].$$

We next deal with the quadratic plus bilinear functionals of the Brownian motion given in (5.93). Assuming that $l(t) = K(0,t)$ in (5.93), let us consider

$$(5.126)\quad S = \int_0^1\int_0^1 K(s,t)\,dw(s)\,dw(t) + aZ\int_0^1 K(0,t)\,dw(t) + bZ^2,$$

where Z follows $N(0,1)$ and is independent of $\{w(t)\}$. Since

$$(5.127)\quad \mathcal{L}(S) = \mathcal{L}\left(\sum_{n=1}^\infty \frac{1}{\lambda_n}\left(Z_n^2 + af_n(0)Z_nZ\right) + bZ^2\right),$$

it follows (Problem 5.9) that

$$(5.128)\quad E\left(e^{i\theta S}\right) = \left[D(2i\theta)\left\{1 - 2ib\theta + a^2\theta^2\sum_{n=1}^\infty \frac{f_n^2(0)}{\lambda_n(\lambda_n - 2i\theta)}\right\}\right]^{-1/2},$$

where $D(\lambda)$ is the FD of K. We now arrive at the following theorem [Tanaka (1990a)].

Theorem 5.10. *The c.f. $\phi(\theta)$ of S defined in (5.126) is given by*

$$\phi(\theta) = \left[D(2i\theta)\left\{1 - 2ib\theta - \frac{ia^2\theta}{2}\left(\Gamma(0,0;2i\theta) - K(0,0)\right)\right\}\right]^{-1/2},$$

where $D(\lambda)$ is the FD of K and $\Gamma(s,t;\lambda)$ is the resolvent of K.

It is now an easy matter to obtain the c.f. of

$$(5.129)\quad U_7 = \int_0^1 t^{2m}\left(w(t) + \kappa Z\right)^2 dt$$
$$= \int_0^1\int_0^1 \frac{1}{2m+1}\left[1 - (\max(s,t))^{2m+1}\right]dw(s)\,dw(t)$$
$$+ 2\kappa Z\int_0^1 \frac{1 - t^{2m+1}}{2m+1}\,dw(t) + \frac{\kappa^2}{2m+1}Z^2,$$

where $m > -\frac{1}{2}$. Theorem 5.10 and (5.106) yield

$$E\left(e^{i\theta U_7}\right) = \left[\Gamma(-\nu+1) J_{-\nu}\left(\frac{\sqrt{2i\theta}}{m+1}\right) \bigg/ \left(\frac{\sqrt{2i\theta}}{2(m+1)}\right)^{-\nu}\right]^{-1/2}$$

$$\times \left[1 - \frac{2i\kappa^2 \theta \Gamma(\nu+1)}{(2m+1)\Gamma(-\nu+1)} \frac{J_\nu\left(\sqrt{2i\theta}/(m+1)\right)}{J_{-\nu}\left(\sqrt{2i\theta}/(m+1)\right)} \left(\frac{\sqrt{2i\theta}}{2(m+1)}\right)^{-2\nu}\right]^{-1/2},$$

(5.130)

where $\nu = (2m+1)/(2(m+1))$. A much simpler way of deriving (5.130) is to take the expectation of (5.106) with κ replaced by κZ by noting that $E\{\exp(i\theta Z^2)\} = (1 - 2i\theta)^{-1/2}$. This conditional argument was earlier presented in Section 4.1 of Chapter 4 [see arguments following (4.15)].

It is also an immediate consequence of (5.111) and the conditional argument that the c.f. of

$$(5.131) \quad U_8 = \int_0^1 t^m \left(w(t) + \kappa Z\right) dw(t)$$

$$= \int_0^1 \int_0^1 \frac{1}{2} \left(\max(s,t)\right)^m dw(s)\, dw(t) + 2\kappa Z \int_0^1 \frac{1}{2} t^m\, dw(t) - \frac{1}{2(m+1)}$$

is given by

$$E\left(e^{i\theta U_8}\right) = \left[\Gamma(\nu) J_{\nu-1}\left(\frac{\sqrt{-4i\theta m}}{m+1}\right) \bigg/ \left(\frac{\sqrt{-4i\theta m}}{2(m+1)}\right)^{\nu-1}\right]^{-1/2}$$

$$\times \left[1 - i\kappa^2 \theta \frac{\Gamma(-\nu+1)}{\Gamma(\nu+1)} \frac{J_{-\nu+1}\left(\sqrt{-4i\theta m}/(m+1)\right)}{J_{\nu-1}\left(\sqrt{-4i\theta m}/(m+1)\right)} \left(\frac{\sqrt{-4i\theta m}}{2(m+1)}\right)^{2\nu}\right]^{-1/2}$$

$$\times \exp\left\{-\frac{i\theta}{2(m+1)}\right\},$$

(5.132)

where $\nu = -m/(m+1)$.

If we consider the stationary O–U process $\{X(t)\}$ defined by $dX(t) = -\beta X(t)\, dt + dw(t)$ with $X(0) = Z/\sqrt{2\beta} \sim N(0, 1/(2\beta))$, we can also obtain the c.f. of

$$U_9 = \int_0^1 X^2(t)\,dt$$

$$= \int_0^1 \left\{ e^{-\beta t} \int_0^t e^{\beta s}\,dw(s) + \kappa Z e^{-\beta t} \right\}^2 dt$$

$$= \int_0^1 \int_0^1 K_2(s,t)\,dw(s)\,dw(t) + 2\kappa Z \int_0^1 K_2(0,t)\,dw(t) + \kappa^2 Z^2 \frac{1-e^{-2\beta}}{2\beta},$$

(5.133)

where $\kappa = 1/\sqrt{2\beta}$ and $K_2(s,t)$ is defined below (5.107). Since (5.109) is the c.f. of U_9 with κZ replaced by κ, we immediately obtain, by the conditional argument and (5.109),

(5.134) $$E\left(e^{i\theta U_9}\right) = e^{\beta/2} \left[\cos\mu + \left(\beta - \frac{i\theta}{\beta}\right) \frac{\sin\mu}{\mu} \right]^{-1/2},$$

where $\mu = \sqrt{2i\theta - \beta^2}$. The same result was earlier obtained in (4.14) by the stochastic process approach.

We now drop the assumption that $l(t) = K(0,t)$ in (5.93). Let us deal with

(5.135) $$S_Y = \int_0^1 \{Y(t) + m(t)Z\}^2\,dt,$$

where $\{Y(t)\}$ is a zero-mean Gaussian process with $\mathrm{Cov}(Y(s), Y(t)) = K(s,t)$, while Z follows $N(0,1)$ and is independent of $\{Y(t)\}$. Proceeding in the same way as before, we can show (Problem 5.10) that

(5.136) $$S_Z = \int_0^1 (Z(t) + m(t)Z)^2\,dt$$

$$= \sum_{n=1}^\infty \frac{1}{\lambda_n}(Z_n + c_n Z)^2 + Z^2 \int_0^1 r^2(t)\,dt$$

has the same c.f. as S_Y in (5.135), where $Z(t)$ is defined in (5.114), c_n and $r(t)$ are the same as those defined below (5.112), while $\{Z_n\}$ follows $NID(0,1)$ and is independent of Z.

Then we can establish the following theorem (Problem 5.11).

Theorem 5.11. *The c.f. of S_Y in (5.135) is given by*

(5.137) $$\phi(\theta) = \left[D(2i\theta) \left\{ 1 - 2i\theta \int_0^1 m^2(t)\,dt + 4\theta^2 \int_0^1 \int_0^1 m(s)m(t)\Gamma(s,t; 2i\theta)\,ds\,dt \right\} \right]^{-1/2},$$

where $D(\lambda)$ is the FD of $K(s,t) = \mathrm{Cov}(Y(s), Y(t))$ and $\Gamma(s,t;\lambda)$ is the resolvent of K.

The double integral in (5.137) can also be evaluated in the same way as before. As an example let us consider

$$U_{10} = \int_0^1 \{w(t) + (a + bt)Z\}^2 \, dt, \tag{5.138}$$

which is an extended version of U_4 given in (5.120). Comparing the expressions for $\phi(\theta)$ in Theorems 5.9 and 5.11, we obtain immediately, from (5.121),

$$E\left(e^{i\theta U_{10}}\right) = (\cos\sqrt{2i\theta})^{-1/2} \left[1 - (a+b)^2 \sqrt{2i\theta} \tan\sqrt{2i\theta} \right.$$
$$\left. - 2ab\left(1 - \frac{1}{\cos\sqrt{2i\theta}}\right) - b^2\left(\frac{\tan\sqrt{2i\theta}}{\sqrt{2i\theta}} - \frac{2}{\cos\sqrt{2i\theta}} + 1\right) \right]^{-1/2}.$$

(5.139)

PROBLEMS

5.1 Establish (5.102).
5.2 Prove Theorem 5.8 using (5.102), the second relation in (5.98), and Mercer's theorem.
5.3 Show that the nonhomogeneous integral equation (5.104) is equivalent to (5.105).
5.4 Prove that the c.f. of U_3 in (5.110) is given by (5.111).
5.5 Establish the relation in (5.113).
5.6 Show that the c.f. of S_Z in (5.115) is given by (5.116).
5.7 Show that the c.f. of U_5 in (5.122) is given by (5.123).
5.8 Show that the c.f. of U_6 in (5.124) is given by (5.125).
5.9 Derive (5.128) on the basis of (5.127).
5.10 Derive the second expression for S_Z in (5.136).
5.11 Prove Theorem 5.11.

5.6 WEAK CONVERGENCE OF QUADRATIC FORMS

In Chapter 3 we presented a set of FCLTs, and those theorems have been applied to establish weak convergence of various statistics. In doing so, we normally construct a partial sum process associated with the statistic under consideration, from which we deduce weak convergence making use of the continuous mapping theorem. This is a two-stage procedure, well accepted in the literature, with wide applicability.

Here we dispense with constructing partial sums. Rather we deal with statistics directly without relating them to underlying partial sum processes. The statistics to be considered are quadratic forms in an increasing number of random variables. It will be seen that the limiting random variable in the sense of weak convergence is expressed by a double integral with respect to the Brownian motion, which we have dealt with in this chapter. Thus the c.f.s of those random variables can be easily obtained by the Fredholm approach.

Following Nabeya and Tanaka (1988), let us consider

(5.140) $$S_T = \frac{1}{T} \sum_{j,k=1}^{T} B_T(j,k) \varepsilon_j \varepsilon_k = \frac{1}{T} \varepsilon' B_T \varepsilon,$$

where $\varepsilon = (\varepsilon_1, \ldots, \varepsilon_T)'$ and $B_T = [(B_T(j,k))]$ is a $T \times T$ real symmetric matrix. For the time being, we assume that $\{\varepsilon_j\} \sim$ i.i.d.$(0, 1)$, under which we would like to discuss the weak convergence of S_T as $T \to \infty$. Note that S_T is a quadratic form in an increasing number of i.i.d. random variables.

Let us assume that there exists a symmetric, continuous and nearly definite function $K(s,t)$ ($\not\equiv 0$) that satisfies

(5.141) $$\lim_{T \to \infty} \max_{j,k} \left| B_T(j,k) - K\left(\frac{j}{T}, \frac{k}{T}\right) \right| = 0.$$

This condition restricts the class of quadratic forms considered here. For example, $B_T(j,k) = \rho^{|j-k|}$ does not satisfy (5.141) insofar as ρ is constant with $-1 \leq \rho < 1$. The case of $\rho = 1$ is an exception, for which we find $K(s,t) \equiv 1$ that satisfies (5.141). Roughly speaking, it is necessary for (5.141) to hold that values of $B_T(j,k)$ for adjacent js and ks are close enough to each other. It can be shown (Problem 6.1) that, if $B_T(j,k) = (1 - (\beta/T))^{|j-k|}$ with β being fixed, then $K(s,t) = e^{-\beta|s-t|}$ is a positive definite kernel that satisfies (5.141).

We note in passing that, although quadratic forms are our concern, ratio statistics can be equally treated if the denominator has a positive definite kernel. In fact, $P(U_T/V_T < x) = P(xV_T - U_T > 0)$, where U_T and V_T are quadratic forms with $V_T > 0$. Then $xV_T - U_T$ has the same form as S_T in (5.140). The ratio statistics we shall deal with normally have a structure such that the kernel associated with U_T is degenerate, while that with V_T is positive definite; hence the kernel associated with $xV_T - U_T$ is nearly definite.

Under the assumption (5.141), we now discuss the weak convergence of $S_T = \varepsilon' B_T \varepsilon / T$. We first note [Nabeya and Tanaka (1988) and Problem 6.2] that

(5.142) $$R_T = \frac{1}{T} \varepsilon' B_T \varepsilon - \frac{1}{T} \sum_{j,k=1}^{T} K\left(\frac{j}{T}, \frac{k}{T}\right) \varepsilon_j \varepsilon_k$$

converges in probability to 0; hence it suffices to consider

$$\begin{aligned} S_T' &= \frac{1}{T} \sum_{j,k=1}^{T} K\left(\frac{j}{T}, \frac{k}{T}\right) \varepsilon_j \varepsilon_k \\ &= \frac{1}{T} \sum_{j,k=1}^{T} \left(\sum_{n=1}^{\infty} \frac{1}{\lambda_n} f_n\left(\frac{j}{T}\right) f_n\left(\frac{k}{T}\right) \right) \varepsilon_j \varepsilon_k \\ &= S_{1T}' + S_{2T}', \end{aligned}$$

where $\{\lambda_n\}$ is a sequence of eigenvalues of K repeated as many times as their multiplicities and $\{f_n(t)\}$ is an orthonormal sequence of eigenfunctions corresponding to $\{\lambda_n\}$, while

$$S_{1T}' = \sum_{n=1}^{M} \frac{1}{\lambda_n} \left(\frac{1}{\sqrt{T}} \sum_{j=1}^{T} f_n\left(\frac{j}{T}\right) \varepsilon_j \right)^2,$$

$$S_{2T}' = \sum_{n=M+1}^{\infty} \frac{1}{\lambda_n} \left(\frac{1}{\sqrt{T}} \sum_{j=1}^{T} f_n\left(\frac{j}{T}\right) \varepsilon_j \right)^2.$$

It is easy to see that, for M fixed,

$$\mathcal{L}\left(S_{1T}'\right) \to \mathcal{L}\left(\sum_{n=1}^{M} \frac{1}{\lambda_n} Z_n^2 \right) \qquad \text{as } T \to \infty,$$

where $\{Z_n\} \sim \text{NID}(0, 1)$, while, for every $\gamma > 0$ and $\delta > 0$,

$$P\left(|S_{2T}'| > \gamma\right) < \frac{1}{\gamma} \sum_{n=M+1}^{\infty} \frac{1}{\lambda_n} \frac{1}{T} \sum_{j=1}^{T} f_n^2\left(\frac{j}{T}\right)$$

$$< \delta,$$

for all T and sufficiently large M. Then, letting $M \to \infty$, we obtain the following theorem [Nabeya and Tanaka (1988)].

Theorem 5.12. *Let $S_T = \varepsilon' B_T \varepsilon / T$ be defined as in (5.140) with $\{\varepsilon_j\} \sim$ i.i.d.$(0, 1)$ and B_T satisfying (5.141). Then, as $T \to \infty$,*

$$\begin{aligned} (5.143) \qquad \mathcal{L}(S_T) &\to \mathcal{L}\left(\sum_{n=1}^{\infty} \frac{1}{\lambda_n} Z_n^2 \right) \\ &= \mathcal{L}\left(\int_0^1 \int_0^1 K(s,t) \, dw(s) \, dw(t) \right), \end{aligned}$$

where $\{Z_n\} \sim \text{NID}(0, 1)$, $\{\lambda_n\}$ is a sequence of eigenvalues of K and $\{w(t)\}$ is the one-dimensional standard Brownian motion.

This theorem tells us that the limiting distribution of $\varepsilon' B_T \varepsilon / T$ does not depend on the common distribution of ε as long as $\{\varepsilon_j\} \sim$ i.i.d.$(0, 1)$. This implies that the invariance principle holds in Donsker's sense. The expressions for the limiting random variables are now quite familiar to us. The second expression in (5.143) is more important for our purpose, but this has been obtained indirectly from the first as a consequence of Mercer's theorem. This theorem ensures our intuition that

$$\mathcal{L}(S_T) \sim \mathcal{L}\left(\frac{1}{T} \sum_{j,k=1}^{T} K\left(\frac{j}{T}, \frac{k}{T}\right) \varepsilon_j \varepsilon_k\right)$$

$$\sim \mathcal{L}\left(\sum_{j,k=1}^{T} K\left(\frac{j}{T}, \frac{k}{T}\right) \Delta w\left(\frac{j}{T}\right) \Delta w\left(\frac{k}{T}\right)\right)$$

$$\to \mathcal{L}\left(\int_0^1 \int_0^1 K(s, t) \, dw(s) \, dw(t)\right),$$

where $\Delta w(j/T) = w(j/T) - w((j-1)/T)$.

Some applications of Theorem 5.12 follow. Let us first consider

(5.144) $$S_{T1} = \frac{1}{T^2} y'y = \frac{1}{T^2} \varepsilon' C' C \varepsilon,$$

where $y = (y_1, \ldots, y_T)'$ with $y_j = y_{j-1} + \varepsilon_j$, $y_0 = 0$ and $\{\varepsilon_j\} \sim$ i.i.d.$(0, 1)$, while C is the random-walk generating matrix defined in (1.3). The statistic has already been discussed in Chapters 1 and 3 using the eigenvalue and stochastic process approaches. Noting that the (j, k)th element of $B_T = C'C/T$ is $(T + 1 - \max(j, k))/T$, we can readily find $K(s, t) = 1 - \max(s, t)$ that satisfies (5.141). It is much easier to work with Theorem 5.12 to establish the weak convergence of S_{T1}.

We next consider

(5.145) $$S_{T2} = \frac{1}{T^2} y' \Omega^{-2} y,$$

where $y_j = \varepsilon_j - \varepsilon_{j-1}$ with $\varepsilon_0, \varepsilon_1, \ldots \sim$ i.i.d.$(0, 1)$ and $\Omega = V(y)$ defined in Section 1.2 of Chapter 1. Since

$$\mathcal{L}(S_{T2}) = \mathcal{L}\left(\frac{1}{T^2} \varepsilon' \Omega^{-1} \varepsilon\right)$$

$$= \mathcal{L}\left(\frac{1}{T^2} \varepsilon' \left(CC' - \frac{1}{T+1} Cee'C'\right) \varepsilon\right),$$

where $e = (1, \ldots, 1)'$, and the (j, k)th element of $B_T = (CC' - Cee'C'/(T+1))/T$ is $(\min(j,k) - jk/(T+1))/T$, we find $K(s,t) = \min(s,t) - st$ that satisfies (5.141).

An extended version of the statistic S_{T1} in (5.144) is

(5.146) $$S_{T3} = \frac{1}{T^2} y'y = \frac{1}{T^2} \varepsilon' C'(\rho) C(\rho) \varepsilon,$$

where $y_j = \rho y_{j-1} + \varepsilon_j$ with $\rho = 1 - (\beta/T)$, $y_0 = 0$, and $\{\varepsilon_j\} \sim$ i.i.d.$(0, 1)$, while

(5.147) $$C(\rho) = \begin{pmatrix} 1 & & & & 0 \\ \rho & 1 & & & \\ \vdots & & \ddots & & \\ \vdots & & & \ddots & \\ \rho^{T-1} & \cdots & \cdots & \rho & 1 \end{pmatrix}.$$

The (j, k)th element of $B_T = C'(\rho)C(\rho)/T$ is $(\rho^{|j-k|} - \rho^{2T-j-k+2})/(T(1-\rho^2))$ and we can arrive at (Problem 6.3)

(5.148) $$\mathcal{L}(S_{T3}) \to \mathcal{L}\left(\int_0^1 \int_0^1 \frac{e^{-\beta|s-t|} - e^{-\beta(2-s-t)}}{2\beta} dw(s) dw(t) \right)$$
$$= \mathcal{L}\left(\int_0^1 X^2(t) dt \right),$$

where $\{X(t)\}$ is the O–U process defined by $dX(t) = -\beta X(t) dt + dw(t)$ with $X(0) = 0$. The equivalence of the above two distributions were established in (5.2) and the limiting c.f. is given by $(D(2i\theta))^{-1/2}$, where $D(\lambda)$ is defined by the right side of (5.108).

We also consider an extended version of S_{T2} in (5.145). Let us put

(5.149) $$S_{T4} = \frac{1}{T^2} y'\Omega^{-2} y + \frac{\gamma}{T^4} y'\Omega^{-3} y,$$

where $\{y_j\}$ follows an MA(1) process, as in (5.145). Since

$$\mathcal{L}(S_{T4}) = \mathcal{L}\left(\frac{1}{T^2} \varepsilon' \left(\Omega^{-1} + \frac{\gamma}{T^2} \Omega^{-2} \right) \varepsilon \right),$$

we shall have [Tanaka (1990b) and Problem 6.4]

(5.150) $$\mathcal{L}(S_{T4}) \to \mathcal{L}\left(\sum_{n=1}^{\infty} \left(\frac{1}{\lambda_n} + \frac{\gamma}{\lambda_n^2} \right) Z_n^2 \right)$$
$$= \mathcal{L}\left(\int_0^1 \int_0^1 (K(s,t) + \gamma K_{(2)}(s,t)) dw(s) dw(t) \right),$$

where $K(s,t) = \min(s,t) - st$ and the iterated kernel $K_{(2)}(s,t)$ is defined by

$$K_{(2)}(s,t) = \int_0^1 K(s,u)K(u,t)\,du$$

$$= \sum_{n=1}^{\infty} \frac{1}{\lambda_n^2} f_n(s) f_n(t).$$

The c.f. of the limiting distribution in (5.150) can also be derived. In a general setting we have the following theorem [Nabeya (1989), Tanaka (1990b), and Problem 6.5].

Theorem 5.13. *Suppose that the statistic S_T is defined by*

$$S_T = \frac{1}{T} \varepsilon' B_T \varepsilon + \frac{\gamma}{T^2} \varepsilon' B_T^2 \varepsilon,$$

where $\{\varepsilon_j\} \sim$ i.i.d.$(0,1)$ and B_T satisfies (5.141). Then

(5.151) $$\mathcal{L}(S_T) \to \mathcal{L}\left(\sum_{n=1}^{\infty} \left(\frac{1}{\lambda_n} + \frac{\gamma}{\lambda_n^2}\right) Z_n^2\right)$$

$$= \mathcal{L}\left(\int_0^1 \int_0^1 \left(K(s,t) + \gamma K_{(2)}(s,t)\right) dw(s)\,dw(t)\right),$$

where $\{Z_n\} \sim$ NID$(0,1)$ and $\{\lambda_n\}$ is a sequence of eigenvalues of K repeated as many times as their multiplicities. Moreover the c.f. of the limiting distribution is given by

$$\lim_{T\to\infty} E\left(e^{i\theta S_T}\right) = \prod_{n=1}^{\infty} \left[1 - 2i\theta\left(\frac{1}{\lambda_n} + \frac{\gamma}{\lambda_n^2}\right)\right]^{-1/2}$$

$$= \prod_{n=1}^{\infty} \left[1 - \frac{1}{\lambda_n}\left(i\theta + \sqrt{-\theta^2 + 2i\gamma\theta}\right)\right]^{-1/2}$$

$$\times \left[1 - \frac{1}{\lambda_n}\left(i\theta - \sqrt{-\theta^2 + 2i\gamma\theta}\right)\right]^{-1/2}$$

$$= \left[D\left(i\theta + \sqrt{-\theta^2 + 2i\gamma\theta}\right) D\left(i\theta - \sqrt{-\theta^2 + 2i\gamma\theta}\right)\right]^{-1/2},$$

(5.152)

where $D(\lambda)$ is the FD of K.

This theorem is useful for computing the limiting local power of test statistics for testing an MA unit root. In fact, the statistic S_{T4} in (5.149) is essentially the

LM statistic evaluated under the local alternative $H_1: \alpha = 1 - (\sqrt{\gamma}/T)$ for testing $H_0: \alpha = 1$ against H_1 in the MA(1) model $y_j = \varepsilon_j - \alpha \varepsilon_{j-1}$. The limiting local power of the test is then easily calculated by inverting the c.f. given in (5.152). We shall discuss more details in Chapter 10 extending the model (see also Chapter 8).

Here we extend the present discussion in two directions. Suppose first that we deal with a vector-valued error process $\{\varepsilon_j\}$ and our concern is a statistic

$$(5.153) \quad S_T^{(q)} = \frac{1}{T} \sum_{j,k=1}^{T} B_T(j,k) \varepsilon_j' H \varepsilon_k = \frac{1}{T} \varepsilon'(B_T \otimes H)\varepsilon,$$

where H is a nonzero $q \times q$ symmetric matrix with constant elements, $\{\varepsilon_j\} \sim$ i.i.d.$(0, I_q)$ and $\varepsilon = (\varepsilon_1', \ldots, \varepsilon_T')'$, while $B_T(j,k)$ is the same as before and has a uniform limit $K(s,t)$ as in (5.141) that is symmetric, continuous, and nearly definite. It is now an easy matter to establish (Problem 6.6) that

$$(5.154) \quad \mathcal{L}(S_T^{(q)}) \to \mathcal{L}\left(\int_0^1 \int_0^1 K(s,t) \, dw'(s) H \, dw(t)\right),$$

where $\{w(t)\}$ is the q-dimensional standard Brownian motion. The expression on the right side was discussed in Chapter 2 and it holds (Problem 6.7) that

$$(5.155) \quad \lim_{T \to \infty} E\left(e^{i\theta S_T^{(q)}}\right) = \prod_{j=1}^{q} \left(D(2i\delta_j \theta)\right)^{-1/2},$$

where $D(\lambda)$ is the FD of K while δ_js are the eigenvalues of H.

As an application of the above result, let us consider the model

$$(5.156) \quad y_j = \rho y_{j-m} + \varepsilon_j, \quad (j = 1, \ldots, T),$$

where $y_{1-m} = y_{2-m} = \cdots = y_0 = 0$ and $\{\varepsilon_j\} \sim$ i.i.d.$(0,1)$. The model may be referred to as a *seasonal AR model* with period m. Suppose that m is a divisor of T so that $N = T/m$ is an integer. Then we have

$$y = (C(\rho) \otimes I_m)\varepsilon,$$

where $y = (y_1, \ldots, y_T)'$, $\varepsilon = (\varepsilon_1, \ldots, \varepsilon_T)'$ and $C(\rho)$ is an $N \times N$ matrix defined in (5.147) with T replaced by N. Assuming that $\rho = 1 - (\beta/N)$ with β fixed, we can show (Problem 6.8) that, as $N \to \infty$,

$$(5.157) \quad \mathcal{L}\left(\frac{1}{N^2} y'y\right) \to \mathcal{L}\left(\int_0^1 \int_0^1 \frac{e^{-\beta|s-t|} - e^{-\beta(2-s-t)}}{2\beta} dw'(s) \, dw(t)\right),$$

where $\{w(t)\}$ is the m-dimensional standard Brownian motion. Thus we have, from (5.108) and (5.155),

$$\lim_{N \to \infty} E\left[\exp\left(\frac{i\theta}{N^2} y'y\right)\right] = \left(\cos\sqrt{2i\theta - \beta^2} + \beta \frac{\sin\sqrt{2i\theta - \beta^2}}{\sqrt{2i\theta - \beta^2}}\right)^{-m/2} e^{\beta m/2}.$$

The assumption that m is a divisor of T is not a restriction so far as the asymptotic result is concerned.

Another extension of the class of quadratic forms is to relax the i.i.d. assumption on the error process $\{\varepsilon_j\}$. For this purpose we consider

(5.158) $$V_T = \frac{1}{T} \sum_{j,k=1}^{T} B_T(j,k) u_j u_k = \frac{1}{T} u' B_T u,$$

where we assume that $\{u_j\}$ is generated by

(5.159) $$u_j = \sum_{l=0}^{\infty} \alpha_l \varepsilon_{j-l}, \quad \sum_{l=0}^{\infty} |\alpha_l| < \infty, \quad \alpha \equiv \sum_{l=0}^{\infty} \alpha_l \neq 0,$$

and $\{\varepsilon_j\} \sim$ i.i.d.$(0, 1)$. Note that $\{u_j\}$ is a stationary process with a slightly weaker condition than that imposed in Section 3.5 of Chapter 3.

We can show [Tanaka (1990a) and Problem 6.9] that

(5.160) $$R_T = V_T - \frac{1}{T} \sum_{j,k=1}^{T} K\left(\frac{j}{T}, \frac{k}{T}\right) u_j u_k$$

converges in probability to 0. Concentrating on the second term on the right side of (5.160), we can deduce (Problem 6.10) that

(5.161) $$\mathcal{L}(V_T) = \mathcal{L}\left(\frac{1}{T} \sum_{j,k=1}^{T} B_T(j,k) u_j u_k\right)$$
$$\to \mathcal{L}\left(\alpha^2 \int_0^1 \int_0^1 K(s,t) \, dw(s) \, dw(t)\right).$$

The existence of a factor α^2 is a consequence of the linear process assumption in (5.159), which we also discussed in Section 3.5 of Chapter 3.

A more general assumption on $\{u_j\}$ in (5.158) is possible [Tanaka (1990a)], to the extent that the weak convergence result in (5.161) holds. We do not pursue the matter here.

This section has discussed weak convergence of quadratic forms, but the discussion can be extended to the class of quadratic plus linear or bilinear forms such as

$$(5.162) \qquad S_{1T} = \frac{1}{T} \sum_{j,k=1}^{T} B_T(j,k) \varepsilon_j \varepsilon_k + \frac{a}{\sqrt{T}} \sum_{k=1}^{T} c_T(k) \varepsilon_k$$

or

$$(5.163) \qquad S_{2T} = \frac{1}{T} \sum_{j,k=1}^{T} B_T(j,k) \varepsilon_j \varepsilon_k + \frac{a}{\sqrt{T}} Z \sum_{k=1}^{T} c_T(k) \varepsilon_k + bZ^2,$$

where Z follows $N(0,1)$ and is independent of $\{\varepsilon_j\}$. Under the assumption (5.141) on $B_T(j,k)$ and a similar assumption on $c_T(k)$, the statistics S_{1T} and S_{2T} will converge in distribution to those random variables given in (5.92) and (5.93), respectively. Then the limiting c.f.s may be derived by the Fredholm approach.

In connection with time series problems, however, the linear or bilinear part in the above statistics arises as a result of nonnegligible influence of the initial value of an underlying process. As an example, let us consider

$$(5.164) \qquad y_j = y_{j-1} + \varepsilon_j, \qquad (j = 1, \ldots, T).$$

If $y_0 = \sqrt{T}\gamma$, then it holds that

$$\frac{1}{T^2} \sum_{j=1}^{T} y_j^2 = \frac{1}{T} \sum_{j,k=1}^{T} \frac{T + 1 - \max(j,k)}{T} \varepsilon_j \varepsilon_k + \frac{2\gamma}{\sqrt{T}} \sum_{k=1}^{T} \frac{T - k + 1}{T} \varepsilon_k + \gamma^2,$$

(5.165)

which has essentially the same form as S_{1T} in (5.162). We could use the Fredholm approach to derive the c.f. of the limiting distribution. A more general statistic is considered in (5.107). As was discussed there, we can take a much simpler route. We already know from Section 3.9 of Chapter 3 that

$$(5.166) \qquad \mathcal{L}\left(\frac{1}{T^2} \sum_{j=1}^{T} y_j^2\right) \to \mathcal{L}\left(\int_0^1 X^2(t)\, dt\right),$$

where $dX(t) = dw(t)$ with $X(0) = \gamma$ and the c.f. of the limiting distribution is available in (4.9) with $\alpha = 0$. Similarly, if $y_0 = \sqrt{T}\gamma Z$, $\sum_{j=1}^{T} y_j^2/T^2$ has the same form as the right side of (5.165) with γ replaced by γZ. That form is the same as S_{2T} in (5.163). Then we have the same weak convergence result as in (5.166) with $X(0)$ replaced by $X(0) = \gamma Z$, and the limiting c.f. is readily obtained by the stochastic process approach.

The initial value problem associated with a simple integrated model in (5.164) or, more generally, a near-integrated model can be solved by relating such discrete-time models to the O–U process. Thus we do not discuss more on the weak convergence of the statistics such as S_{1T} in (5.162) or S_{2T} in (5.163). We shall return to the initial value problem in Chapter 7.

PROBLEMS

6.1 Show that $B_T(j,k) = (1 - (\beta/T))^{|j-k|}$ with β fixed satisfies (5.141) with $K(s,t) = e^{-\beta|s-t|}$.

6.2 Prove that R_T defined in (5.142) converges in probability to 0.

6.3 Establish the weak convergence result in (5.148).

6.4 Establish the weak convergence result in (5.150).

6.5 Prove Theorem 5.13.

6.6 Deduce the weak convergence result in (5.154).

6.7 Show that the limiting c.f. of $S_T^{(q)}$ in (5.154) is given by (5.155).

6.8 Establish the weak convergence result in (5.157).

6.9 Prove that R_T defined in (5.160) converges in probability to 0.

6.10 Deduce the weak convergence result in (5.161).

CHAPTER 6

Numerical Integration

We discuss and demonstrate how to invert numerically the c.f.s to obtain the distribution functions and the probability densities of statistics presented so far. It should be emphasized that any computer package for integration cannot do the job properly if the integrand contains the square root of a complex-valued function, such as the c.f.s we have dealt with. We present a Fortran program for this purpose.

As an integration method, we use Simpson's rule, which proves to be successful for most cases. There are, however, some cases where the integrand is oscillating and converges to 0 rather slowly. For such cases, Euler's transformation of slowly convergent alternating series is employed to accelerate the convergence of integration based on Simpson's rule. Various graphs and Fortran programs are presented for demonstration purposes. We also discuss the saddlepoint method for computing approximate distribution functions and percent points.

6.1 INTRODUCTION

This chapter is concerned with computing numerically the distribution functions and the probability densities of statistics that are quadratic, plus linear or bilinear functionals of the Brownian motion. This necessarily entails inverting the c.f.s of those statistics. As we have seen, the c.f. is usually expressed as the square root of a complex-valued function, which a computer cannot compute properly. We first need to devise an algorithm for evaluating the square root correctly.

After obtaining the correct c.f., we proceed to examine the behavior of the integrand to find an effective interval of the integral together with a suitable method for integration. We shall find Simpson's rule successful in most cases, but we shall also use Euler's transformation of slowly convergent alternating series for cases where the integrand is oscillating and converges to 0 rather slowly. Since it is usually the case that the difficulty does not arise in the computation of distribution functions, but in the computation of probability densities, we shall also consider a practical method for computing the latter on the basis of the former.

The next section deals with statistics that take only nonnegative values with well-behaved c.f.s. The procedure for computing distribution functions and probability densities is explained in detail with Fortran programs that can be executed by a desktop computer. The oscillating case is discussed in Section 6.3, where Euler's transformation is introduced to make the computation more accurate and efficient. Section 6.4 deals with a general case where statistics take both positive and negative values. A practical method is considered here for computing probability densities. Section 6.5 discusses how to obtain percent points of distribution functions. A Fortran program is also presented to demonstrate the methodology. The last section describes briefly the saddlepoint method for computing approximate distribution functions and percent points.

6.2 NUMERICAL INTEGRATION: THE NONNEGATIVE CASE

Let S be a nonnegative statistic, for which we would like to compute the distribution function and the probability density. Lévy's inversion theorem for nonnegative random variables tells us that

$$(6.1) \qquad F(x) = P(S \le x)$$
$$= \frac{1}{\pi} \int_0^\infty \text{Re}\left[\frac{1 - e^{-i\theta x}}{i\theta} \phi(\theta)\right] d\theta,$$

where $\phi(\theta)$ is the c.f. of S.

If $F(x)$ is differentiable, we have the probability density of S given by

$$(6.2) \qquad f(x) = \frac{dF(x)}{dx}$$
$$= \frac{1}{\pi} \int_0^\infty \text{Re}\left[e^{-i\theta x} \phi(\theta)\right] d\theta.$$

Our purpose here is to devise an efficient method for numerical integration associated with (6.1) and (6.2). We can recognize, from the computational point of view, that the distribution function is handled more easily than the probability density since the integrand in the former converges to 0 more rapidly.

There is, however, a problem prior to numerical integration in (6.1) and (6.2). The c.f. $\phi(\theta)$ is usually the square root of a complex-valued function, which a computer cannot evaluate properly. As an example, let us consider

$$(6.3) \qquad \phi_1(\theta) = \left(\frac{\sin\sqrt{2i\theta}}{\sqrt{2i\theta}}\right)^{-1/2},$$

which is the c.f. of a nonnegative statistic defined by

$$S_1 = \int_0^1 \left(w(t) - \int_0^1 w(s)\,ds\right)^2 dt$$
$$= \int_0^1 \int_0^1 [\min(s,t) - st]\,dw(s)\,dw(t).$$

NUMERICAL INTEGRATION: THE NONNEGATIVE CASE

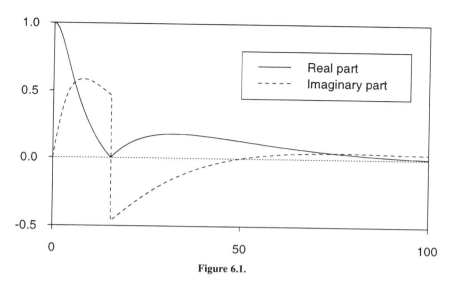

Figure 6.1.

Figure 6.1 draws the computer-generated graph of $\phi_1(\theta)$, $\tilde{\phi}_1(\theta)$, say. It is seen that there is a discontinuity point in $\text{Im}(\tilde{\phi}_1(\theta))$ at around $\theta = 15$ where $\text{Re}(\tilde{\phi}_1(\theta)) = 0$. Since $\phi_1(\theta)$ is continuous for all θ, this graph is not correct, although $|\tilde{\phi}_1(\theta)| = |\phi_1(\theta)|$. If we had replaced $\tilde{\phi}_1(\theta)$ by $-\tilde{\phi}_1(\theta)$ at the discontinuity point and the successive points, we would have obtained a correct graph.

To obtain the correct c.f., Nabeya and Tanaka (1988) used the following algorithm. We start computing a c.f. $\phi(\theta)$ at $\theta = \theta_0 = 0$, at which $\phi(\theta)$ is always unity. Then, for $\theta = \theta_1\ (>0)$ close to $\theta = \theta_0$, check if

(6.4) $$|\phi(\theta_0) + \tilde{\phi}(\theta_1)| \le |\phi(\theta_0) - \tilde{\phi}(\theta_1)|,$$

where $\tilde{\phi}(\theta_1)$ is the computer-generated value of $\phi(\theta_1)$. If (6.4) is true, it means that the computer has generated $\phi(\theta_1)$ with the wrong sign so that we put $\phi(\theta_1) = -\tilde{\phi}(\theta_1)$; otherwise we put $\phi(\theta_1) = \tilde{\phi}(\theta_1)$. Then we proceed to check if

(6.5) $$|\phi(\theta_1) + \tilde{\phi}(\theta_2)| \le |\phi(\theta_1) - \tilde{\phi}(\theta_2)|,$$

for $\theta_2\ (>\theta_1)$ close to θ_1. If (6.5) holds true, we put $\phi(\theta_2) = -\tilde{\phi}(\theta_2)$; otherwise we put $\phi(\theta_2) = \tilde{\phi}(\theta_2)$. Proceeding further in this way, we can compute the correct c.f. $\phi(\theta)$ for successive values of θ. Figure 6.2 shows the graph of $\phi_1(\theta)$ in (6.3) computed in the way described above.

We now proceed to the computation of $F(x)$ in (6.1). For this purpose, we need to examine the behavior of the integrand. Here we consider, by change of variables,

(6.6) $$F_1(x) = \frac{1}{\pi} \int_0^\infty \text{Re}\left[\frac{1 - e^{-i\theta x}}{i\theta} \phi_1(\theta)\right] d\theta$$

$$= \frac{1}{\pi} \int_0^\infty g_1(u; x)\, du,$$

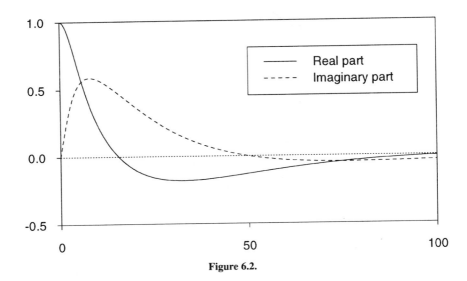

Figure 6.2.

where we have put $\theta = u^2$ and

$$(6.7) \qquad g_1(u; x) = \text{Re}\left[\frac{2(1 - e^{-iu^2 x})}{iu} \phi_1(u^2)\right].$$

We have transformed θ into $\theta = u^2$ because $\phi_1(\theta)$ in (6.3) involves $\sqrt{\theta}$. If the c.f. involves $\theta^{1/m}$ ($m > 1$), we shall consider $\theta = u^m$. The transformed integrand $g_1(u; x)$ in (6.7) reduces to 0 at $u = 0$. Thus we can dispense with computing the value of the integrand at the origin. This is especially advantageous to the numerical integration dealt with in Section 6.3, where such computation is complicated.

Figure 6.3 shows the graph of the transformed integrand $g_1(u; x)$ in (6.6) for $0 \le u \le 10$ and $x = 0.74346$. The present value of x is supposed to be the upper 1% point. The computation of percent points will be explained in Section 6.5. We need to determine an effective interval where the integrand is nonnegligible. It turns out that the interval $0 \le u < \infty$ of the integral in (6.6) may be replaced by $0 \le u \le 50$ for the present and other moderate values of x. Simpson's rule will do a proper job for numerical integration in (6.6), which will be discussed later by presenting a Fortran program.

The computation of the probability density $f(x)$ in (6.2) proceeds in much the same way as above, once the correct c.f. is obtained. We transform the integrand to get

$$(6.8) \qquad f_1(x) = \frac{1}{\pi} \int_0^\infty \text{Re}\left[e^{-i\theta x} \phi_1(\theta)\right] d\theta$$

$$= \frac{1}{\pi} \int_0^\infty h_1(u; x) \, du,$$

NUMERICAL INTEGRATION: THE NONNEGATIVE CASE

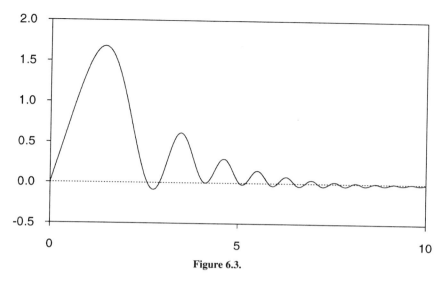

Figure 6.3.

where $\theta = u^2$ and

(6.9) $$h_1(u; x) = \text{Re}\left[2ue^{-iu^2 x}\phi_1(u^2)\right].$$

The transformed integrand $h_1(u; x)$ also vanishes at $u = 0$. The effective interval for which this integrand is nonnegligible is usually wider than that for the distribution function. Figure 6.4 shows the graph of $h_1(u; x)$ for $0 \le u \le 20$ and the same value of x as above. It is seen that, unlike the integrand in Figure 6.3, the present one

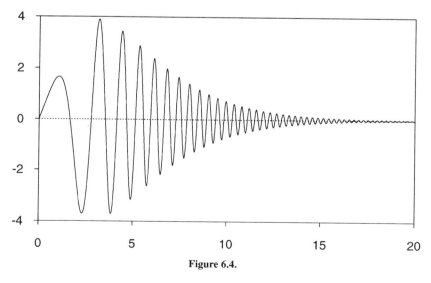

Figure 6.4.

is oscillating around zero and does not approach zero as rapidly. Simpson's rule, however, is found to be still applicable to the present situation, although we need a slightly wider effective interval for integration.

We are now ready to compute $F_1(x)$ in (6.6) and $f_1(x)$ in (6.8). *Simpson's rule* tells us that

$$(6.10) \quad I = \int_a^b f(u)\,du$$
$$= \frac{h}{3}\left[4\sum_{i=1}^{n} f(u_{2i-1}) + 2\sum_{i=1}^{n-1} f(u_{2i}) + f(a) + f(b)\right],$$

where $h = (b-a)/(2n)$ and $u_i = a + ih$. In the present case $a = 0$ and $f(u) = g_1(u; x)$ or $h_1(u; x)$. Note that $g_1(0; x) = h_1(0; x) = 0$ so that these terms do not contribute to the above sum.

Table 6.1 presents a Fortran program for computing $F_1(x)$ and $f_1(x)$ for $x = 0.01\,(+0.01)\,1.2$. In this program, we have chosen $b = M = 50$, $1/h = N = 50$ and $2n = L = 2{,}500$. It is desirable to try various values of $h = 1/N$ to ensure that the result does not depend on those values. The values of $f_1(x)$ may be used to draw the graph of the probability density, but some finer points of x will be necessary to get a correct graph. Such a graph has been already presented in Chapter 1 as part of Figure 1.2.

In Chapter 5, we presented various c.f.s for nonnegative statistics. The distribution functions and the probability densities of most of those statistics may be computed by the present method. The only exception is the statistic involving the integrated Brownian motion. A simple application of Simpson's rule does not accomplish the job properly since it is found that the associated integrand is oscillating and converges to 0 rather slowly. This will be discussed in the next section.

For c.f.s that contain Bessel functions together with gamma functions presented in Section 5.4 of Chapter 5, we can usually make use of a computer package for computing those functions. MacNeill (1974, 1978) and Nabeya and Tanaka (1988) dealt with such cases. Simpson's rule is found to be applicable in those cases.

6.3 NUMERICAL INTEGRATION: THE OSCILLATING CASE

Here we take up examples for which a simple application of Simpson's rule fails. We suggest using Euler's transformation to overcome the difficulty. Another practical remedy will be given in the next section.

As the first example, let us consider

$$(6.11) \quad S_2 = \int_0^1\int_0^1 \left[dw_1(s)\,dw_1(t) - dw_2(s)\,dw_2(t)\right]$$
$$= \int_0^1\int_0^1 dw'(s)\begin{pmatrix}1 & 0\\ 0 & -1\end{pmatrix} dw(t),$$

NUMERICAL INTEGRATION: THE OSCILLATING CASE

Table 6.1. Fortran Program for Computing $F_1(x)$ and $f_1(x)$

```
      PARAMETER(M=50,N=50,KX=120)
      IMPLICIT REAL*8(A-E,P-Z),COMPLEX*16(F)
      DIMENSION F(0:2),FDF(2),FPD(2)
      PI=1D0/3.141592653589793D0
      SN=DFLOAT(N)
      L=M*N
      DO 100 K=1,KX
      X=DFLOAT(K)*1D-2
      F(0)=1D0
      S1=0D0
      S2=0D0
      T1=0D0
      T2=0D0
      DO 1 I=1,L-1,2
      DO 2 J=1,2
      U=DFLOAT(I+J-1)/SN
      T=U*U
      CALL CHAR(T,F(J-1),F(J))
      FDF(J)=DCMPLX(DSIN(T*X),DCOS(T*X)-1D0)*F(J)*2D0/U
      FPD(J)=DCMPLX(DCOS(T*X),-DSIN(T*X))*F(J)*2D0*U
    2 CONTINUE
      S1=S1+DREAL(FDF(1))
      S2=S2+DREAL(FDF(2))
      T1=T1+DREAL(FPD(1))
      T2=T2+DREAL(FPD(2))
      F(0)=F(2)
    1 CONTINUE
      PROB=PI*(4D0*S1+2D0*S2-DREAL(FDF(2)))/(3D0*SN)
      PD=PI*(4D0*T1+2D0*T2-DREAL(FPD(2)))/(3D0*SN)
      WRITE(6,20)X,PROB,PD
   20 FORMAT(3F13.6)
  100 CONTINUE
      STOP
      END
      SUBROUTINE CHAR(T,FL,FP)
      IMPLICIT REAL*8(T),COMPLEX*16(F)
      F0=(0D0,1D0)
      F1=2D0*F0*T
      F2=CDSQRT(F1)
      FP=1D0/CDSQRT(CDSIN(F2)/F2)
      IF(CDABS(FL+FP).LE.CDABS(FL-FP))FP=-FP
      RETURN
      END
```

where $w(t) = (w_1(t), w_2(t))'$ is the two-dimensional standard Brownian motion. Of course, $\mathcal{L}(S_2) = \mathcal{L}(X^2 - Y^2)$, where $(X, Y)' \sim N(0, I_2)$. This observation leads us to take a more efficient method for computing the distribution function and the probability density of S_2, which will be considered later. We first stick to the method discussed so far.

The c.f. of S_2 is given by

(6.12) $$\phi_2(\theta) = (1 + 4\theta^2)^{-1/2}.$$

The distribution of S_2 is certainly symmetric around the origin since the c.f. is real. Thus we need not worry about computing the square root of a complex-valued function. Imhof's (1961) formula gives us

(6.13) $$F_2(x) = P(S_2 \leq x)$$
$$= \frac{1}{2} - \frac{1}{\pi} \int_0^\infty \frac{1}{\theta} \operatorname{Im}\left[e^{-i\theta x} \phi_2(\theta)\right] d\theta$$
$$= \frac{1}{2} + \frac{1}{\pi} \int_0^\infty \frac{\sin \theta x}{\theta \sqrt{1 + 4\theta^2}} d\theta,$$

(6.14) $$f_2(x) = \frac{dF_2(x)}{dx}$$
$$= \frac{1}{\pi} \int_0^\infty \frac{\cos \theta x}{\sqrt{1 + 4\theta^2}} d\theta.$$

The integrands above take a simple form and one might think that Simpson's rule would execute numerical integration properly. This is not the case, however. This is because the integrands are oscillating and converge to 0 quite slowly. In particular the integrand involved in (6.14) is of order $1/\theta$ as $\theta \to \infty$. Change of variables such as $\theta = u^2$ does not improve this situation.

Figure 6.5 shows a graph of

(6.15) $$g_2(\theta; x) = \frac{\cos \theta x}{\sqrt{1 + 4\theta^2}}$$

for $x = 3$. The function is oscillating around 0 and leaves rippling waves as far as $\theta = 100$ and beyond that point.

In the present case, however, we have another expression for $F_2(x)$ and $f_2(x)$. It can be shown (Problem 3.1 in this chapter) that, for $x \geq 0$,

(6.16) $$F_2(x) = P(S_2 \leq x) = P(X^2 - Y^2 \leq x)$$
$$= 1 - \frac{4}{\sqrt{2\pi}} \int_0^\infty \Phi(-\sqrt{x + \theta^2}) e^{-\theta^2/2} d\theta,$$

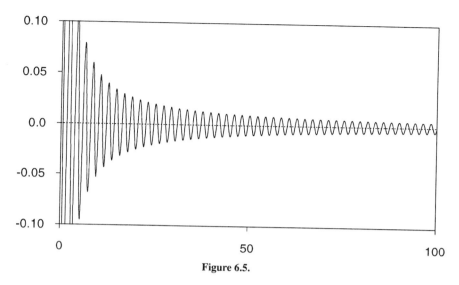

Figure 6.5.

and $F_2(x) = 1 - F_2(-x)$ for $x < 0$, where $(X, Y)' \sim N(0, I_2)$ and $\Phi(\cdot)$ is the distribution function of $N(0, 1)$. Thus, for $x \geq 0$,

(6.17) $$f_2(x) = \frac{1}{\pi} \int_0^\infty \frac{1}{\sqrt{x + \theta^2}} \exp\left(-\frac{x}{2} - \theta^2\right) d\theta,$$

and $f_2(x) = f_2(-x)$ for $x < 0$.

Figure 6.6 shows the graph of

(6.18) $$h_2(\theta; x) = \frac{1}{\sqrt{x + \theta^2}} \exp\left(-\frac{x}{2} - \theta^2\right)$$

for $x = 3$. It is to be recognized that the integration based on (6.17) gives a more efficient method for computing $f_2(x)$ than that based on (6.14). Simpson's rule or any other computer package can be used for numerical integration in (6.16) and (6.17). Note, however, that $f_2(x)$ diverges as $|x| \to 0$ so that, in drawing the graph of $f_2(x)$, the integration in (6.17) should be done for various values of x excluding $x = 0$. The probability density of $S_2/2$, that is, $f(x) = 2f_2(2x)$, was already presented in Figure 1.4.

In general it is not easy to find an alternative expression for integration that is computationally more efficient. As such an example, we consider the integral of the square of the one-fold integrated Brownian motion

(6.19) $$S_3 = \int_0^1 \left\{ \int_0^t w(s) \, ds \right\}^2 dt,$$

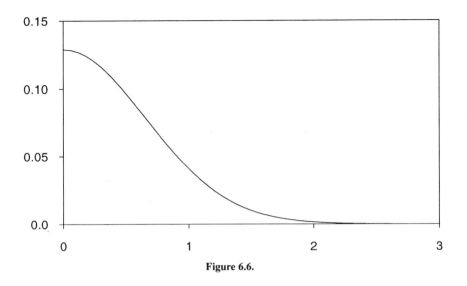

Figure 6.6.

whose c.f. is available from (4.42) or (5.89) as

$$(6.20) \qquad \phi_3(\theta) = \left[\frac{1}{2} \left\{ 1 + \cos(2i\theta)^{1/4} \cosh(2i\theta)^{1/4} \right\} \right]^{-1/2}.$$

Figure 6.7 draws $\phi_3(u^4)$ as a function of u.

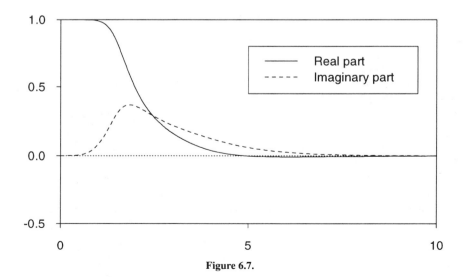

Figure 6.7.

Lévy's inversion formula (6.1) leads us (Problem 3.2) to

(6.21) $$F_3(x) = P(S_3 \le x)$$
$$= \frac{1}{\pi} \int_0^\infty g_3(u; x)\, du,$$

where we have put $\theta = u^4$ in (6.20) and

(6.22) $$g_3(u; x) = \frac{4}{u} \left[\text{Re}\left\{\phi_3(u^4)\right\} \sin u^4 x + \text{Im}\left\{\phi_3(u^4)\right\} (1 - \cos u^4 x) \right].$$

We also have the probability density given by

(6.23) $$f_3(x) = \frac{1}{\pi} \int_0^\infty h_3(u; x)\, du,$$

where

(6.24) $$h_3(u; x) = 4u^3 \left[\text{Re}\left\{\phi_3(u^4)\right\} \cos u^4 x + \text{Im}\left\{\phi_3(u^4)\right\} \sin u^4 x \right].$$

Figures 6.8 and 6.9 present graphs of $g_3(u; x)$ and $h_3(u; x)$, respectively, for $x = 0.3$. It turns out that Simpson's rule fails even in the computation of the former, much more the latter.

To overcome the difficulty, we use *Euler's transformation* of slowly convergent alternating series [Longman (1956)]. Suppose that the integrand $f(u)$ defined on

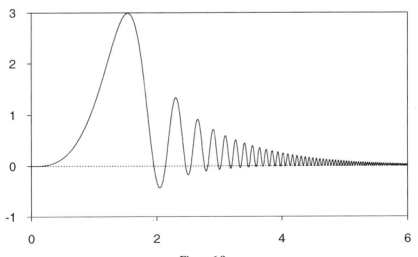

Figure 6.8.

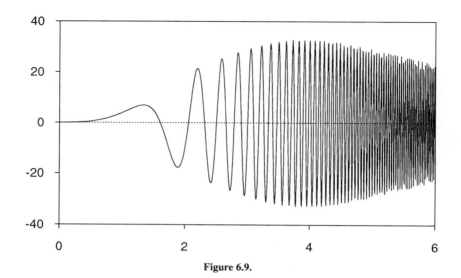

Figure 6.9.

$[0, \infty)$ has zeros at $u = u_k$ ($k = 1, 2, \ldots$). Then we have

(6.25)
$$I = \int_0^\infty f(u)\,du$$
$$= \int_0^{u_1} f(u)\,du + \int_{u_1}^{u_2} f(u)\,du + \ldots$$
$$= \sum_{k=0}^\infty (-1)^k V_k,$$

where

$$V_k = (-1)^k \int_{u_k}^{u_{k+1}} f(u)\,du, \qquad (u_0 = 0).$$

By definition $\{(-1)^k V_k\}$ is an alternating series. It can be shown (Problem 3.3) that

(6.26)
$$I = \sum_{k=0}^\infty (-1)^k V_k$$
$$= \sum_{k=0}^\infty (-1)^k \frac{(F-1)^k V_0}{2^{k+1}}$$
$$= \sum_{k=0}^{N-1} (-1)^k V_k + \sum_{k=0}^\infty (-1)^{k+N} \frac{(F-1)^k V_N}{2^{k+1}},$$

where F is the forward shift operator, that is, $FV_k = V_{k+1}$ and

(6.27) $\quad (F - 1)^k V_j = (F - 1)^{k-1}(V_{j+1} - V_j), \quad (F - 1)^0 V_j = V_j.$

Euler's transformation refers to the second relation in (6.26). The third may be referred to as *Euler's delayed transformation*. Since each V_k has the same sign and $\{|V_k|\}$ is supposed to be a decreasing sequence, it is expected that the k-fold forward difference $(F - 1)^k V_N$ of V_N divided by 2^{k+1} makes the convergence of the infinite series more rapid. Each V_k is easily computed by Simpson's rule.

Returning to the integral in (6.21), we need to find the zeros of the integrand $g_3(u; x)$. It is, however, not easy. Thus we split $g_3(u; x)$ into $g_3(u; x) = a(u; x) - b(u; x) + c(u)$ where

(6.28) $\quad a(u; x) = \dfrac{4}{u} \operatorname{Re}\left[\phi_3(u^4)\right] \sin u^4 x,$

(6.29) $\quad b(u; x) = \dfrac{4}{u} \operatorname{Im}\left[\phi_3(u^4)\right] \cos u^4 x,$

(6.30) $\quad c(u) = \dfrac{4}{u} \operatorname{Im}\left[\phi_3(u^4)\right].$

Figures 6.10, 6.11 and 6.12 present graphs of $a(u; x)$ and $b(u; x)$ for $x = 0.3$, and $c(u)$, respectively. It is expected that the integral of $c(u)$ is easily done by Simpson's rule. For the integrals of $a(u; x)$ and $b(u; x)$, we apply Euler's transformation (6.26) to the sequence of values of integrals computed by Simpson's rule, noting that the zeros of $a(u; x)$ are $u_k = (k\pi/x)^{1/4}$, while those of $b(u; x)$ are $\{(k + 1/2)\pi/x\}^{1/4}$ for $k = 0, 1, \ldots$, apart from the zeros of $\phi_3(u^4)$.

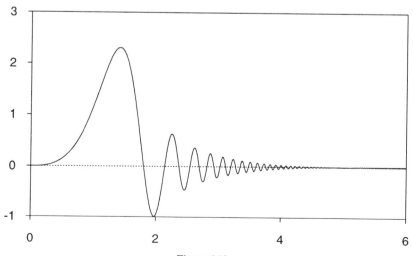

Figure 6.10.

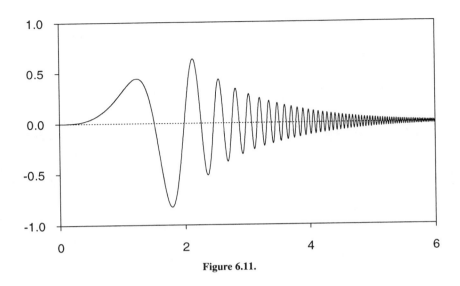

Figure 6.11.

Once the c.f. $\phi_3(\theta)$ is computed, the computation of the integrals of $a(u; x)$ and $b(u; x)$ in the way described above is easily programmed, which will be presented in Section 6.5 in connection with computing percent points.

Because of the nature of the function $h_3(u; x)$ presented in Figure 6.9, the computation of $f_3(x)$ in (6.23) by the above method is found to be not very accurate. Since we can compute $F_3(x)$ quite accurately, it is expected that $f_3(x)$ can be obtained, to a certain degree of accuracy, from numerical derivatives of $F_3(x)$. This will be a topic in the next section, where the graph of $f_3(x)$ computed in that way is presented.

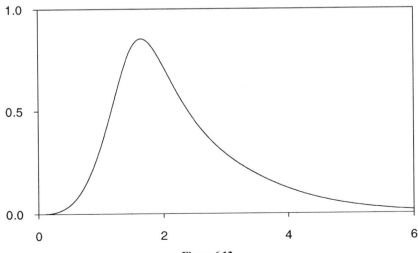

Figure 6.12.

NUMERICAL INTEGRATION: THE OSCILLATING CASE

Here we content ourselves with observing that the present method is also applicable to compute the distribution function of

$$(6.31) \qquad S_4 = \int_0^1 \left[\int_0^t \left\{ \int_0^s w(r)\,dr \right\} ds \right]^2 dt,$$

where the integrand is the square of the two-fold integrated Brownian motion. It follows from (5.91) that the c.f. of S_4 is given by

$$(6.32) \qquad \phi_4(\theta) = \left[\frac{1}{18} \left\{ \left(\cos\mu + 2\cosh^2 \frac{\sqrt{3}\mu}{2} + 3 \right) \cos\mu \right. \right.$$
$$\left. \left. + 8\cos\frac{\mu}{2} \cosh\frac{\sqrt{3}\mu}{2} + 4 \right\} \right]^{-1/2},$$

where $\mu = (2i\theta)^{1/6}$.

Figure 6.13 shows graphs of the distribution functions of S_3 in (6.19) and S_4 in (6.31). Percent points of these distributions will be tabulated in Section 6.5 together with a Fortran program.

PROBLEMS

3.1 Establish (6.16) when $(X, Y)' \sim N(0, I_2)$.
3.2 Derive the second equality in (6.21).
3.3 Derive Euler's transformation as in (6.26).

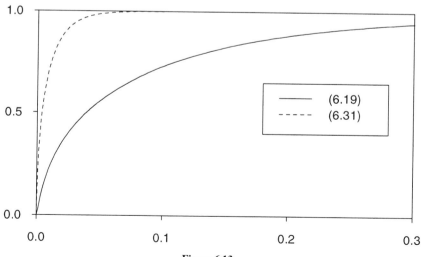

Figure 6.13.

6.4 NUMERICAL INTEGRATION: THE GENERAL CASE

In this section we deal with statistics that take both positive and negative values. As a simple example in terms of numerical integration, let us first consider Lévy's stochastic area defined by

$$(6.33) \qquad S_5 = \frac{1}{2} \int_0^1 \left[w_1(t) \, dw_2(t) - w_2(t) \, dw_1(t) \right],$$

where $w(t) = (w_1(t), w_2(t))'$ is the two-dimensional standard Brownian motion. The c.f. of S_5 is available from (1.57) or (5.38), which is

$$(6.34) \qquad \phi_5(\theta) = E\left(e^{i\theta S_5}\right) = \left(\cosh \frac{\theta}{2}\right)^{-1}.$$

The statistic S_5 has a symmetric distribution since $\phi_5(\theta)$ is real. Thus we need not worry about computing the square root of a complex-valued function. In fact, Imhof's formula described in (6.13) gives us

$$(6.35) \qquad F_5(x) = P(S_5 \leq x)$$

$$= \frac{1}{2} + \frac{1}{\pi} \int_0^\infty \frac{\sin \theta x}{\theta \cosh(\theta/2)} \, d\theta.$$

The integrand $\sin \theta x / (\theta \cosh(\theta/2))$ takes the value x at $\theta = 0$. Any computer package will compute the above integral fairly easily. The computation of the probability density given by

$$f(x) = \frac{1}{\pi} \int_0^\infty \frac{\cos \theta x}{\cosh(\theta/2)} \, d\theta$$

is also easy since the integrand approaches 0 exponentially. The graph of the probability density of $2 \times S_5$ was earlier presented in Figure 1.4 with percent points in Table 1.4.

We now deal with statistics for which numerical integration must be elaborated. Let S be such a statistic, which takes the form $S = U/V$, where $P(V > 0) = 1$. Imhof's formula for such a statistic gives us

$$(6.36) \qquad F(x) = P(S \leq x)$$

$$= P(xV - U \geq 0)$$

$$= \frac{1}{2} + \frac{1}{\pi} \int_0^\infty \frac{1}{\theta} \, \text{Im}\left[\phi(\theta; x)\right] d\theta,$$

NUMERICAL INTEGRATION: THE GENERAL CASE

where $\phi(\theta; x)$ is the c.f. of $xV - U$. In actual computation, we shall transform θ into another variable to make the integrand vanish at the origin, as was done in previous sections.

If $F(x)$ is differentiable, we have

(6.37)
$$f(x) = \frac{dF(x)}{dx}$$
$$= \frac{1}{\pi} \int_0^\infty \frac{1}{\theta} \operatorname{Im}\left[\frac{\partial \phi(\theta; x)}{\partial x}\right] d\theta.$$

Here it is usually the case that $\phi(\theta; x)$ is a complicated function of x, which makes the computation of $\partial \phi(\theta; x)/\partial x$ tedious. It is also the case, as was seen in the last section, that the integration for computing probability densities is more difficult than for distribution functions. Even if we use Euler's transformation, we require the values of θ for which $\partial \phi(\theta; x)/\partial x = 0$. It is also difficult in the present case. We, however, only need the values of $f(x)$ to draw its graph; hence very accurate values of $f(x)$ are not our concern.

The above discussions lead us to dispense with the computation of $f(x)$ that follows (6.37). Instead we proceed as follows. Let $F(x)$ and $F(x + \Delta x)$ be already computed by following (6.36), where Δx is a small number, 10^{-6}, say. Then we suggest computing $f(x)$ as

(6.38)
$$f(x) = \frac{F(x + \Delta x) - F(x)}{\Delta x}.$$

The right side above is a numerical derivative of $F(x)$. Computing $f(x)$ in this way, we can also avoid examining the behavior of the integrand associated with numerical integration in (6.37). We have only to concentrate on the computation of $F(x)$ discussed in the previous sections.

As an example, let us consider the statistic S_6 given by

(6.39)
$$S_6 = \frac{U_6}{V_6} = \frac{\int_0^1 w(t)\, dw(t)}{\int_0^1 w^2(t)\, dt},$$

which follows the AR(1) unit root distribution discussed in Section 1.3 of Chapter 1. An extended version of S_6 was discussed in Section 4.1 of Chapter 4. We have $P(S_6 \leq x) = P(xV_6 - U_6 \geq 0)$, and (1.35) or (4.17) yields

(6.40)
$$\phi_6(\theta; x) = E\left[\exp\{i\theta(xV_6 - U_6)\}\right]$$
$$= e^{i\theta/2}\left[\cos\sqrt{2i\theta x} + i\theta\,\frac{\sin\sqrt{2i\theta x}}{\sqrt{2i\theta x}}\right]^{-1/2}.$$

The c.f. $\phi_6(\theta; x)$ can be obtained (Problem 4.1) most easily by the stochastic process approach discussed in Chapter 4.

Putting $\theta = u^2$, we consider

(6.41)
$$F_6(x) = P(S_6 \le x)$$
$$= \frac{1}{2} + \frac{1}{\pi} \int_0^\infty g_6(u; x) \, du,$$

where

(6.42)
$$g_6(u; x) = \frac{2}{u} \operatorname{Im} \left[\phi_6(u^2; x) \right].$$

Note that $g_6(0; x) = 0$ (Problem 4.2) so that we can dispense with computing the value of the integrand at the origin. If we follow the untransformed formula (6.36), we need (Problem 4.3)

(6.43)
$$\lim_{\theta \to 0} \frac{1}{\theta} \operatorname{Im} \left[\phi_6(\theta; x) \right] = \frac{x}{2}.$$

Figure 6.14 gives the graph of the integrand $g_6(u; x)$ in (6.41) for $x = -8.03913$. The present value of x is supposed to be the 5% point of the distribution of S_6 in (6.39). It is expected that the numerical integration for (6.41) with the present value of x can be easily done by Simpson's rule. On the other hand, Figure 6.15 shows the graph of $g_6(u; x)$ for $x = 0.05$. This graph is quite different from that in Figure 6.14; the integrand is oscillating and converges slowly to 0. This is because $|x|$ is quite small. In fact, if $x = 0$, the integrand (6.42) becomes

$$g_6(u; 0) = \frac{2}{u} \operatorname{Im} \left[\frac{e^{iu^2/2}}{\sqrt{1 + iu^2}} \right],$$

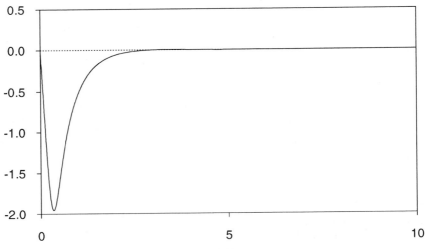

Figure 6.14.

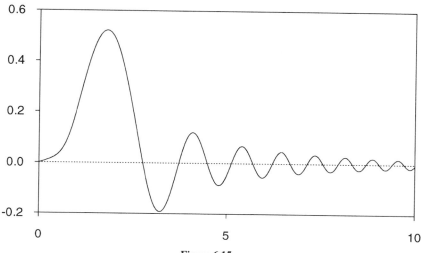

Figure 6.15.

which is of a similar nature to the integrand in (6.13). Of course, we can take another route to evaluate $F_6(0)$ more efficiently (Problem 4.4). For cases of $|x|$ ($\neq 0$) small we can still use Simpson's rule since the rate of convergence of the integrand to 0 is not very slow, although the effective interval of the integral in (6.41) becomes wider.

Table 6.2 presents a Fortran program for computing $F_6(x)$ and $f_6(x)$ for $x = -14.5 (+1) 2.5$, where the computation of $f_6(x)$ is done by numerical derivatives of $F_6(x)$ as in (6.38) with $\Delta x = 10^{-6}$. For the parameters used in Simpson's rule (6.10), we have chosen $b = 30$, $h = 0.01$, and $2n = 3{,}000$. The graph of $f_6(x)$ has been already presented in Chapter 1 as part of Figure 1.3.

As another example we take up

$$(6.44) \qquad S_7 = \frac{U_7}{V_7} = \frac{\int_0^1 \left\{ \int_0^t w_1(s)\,ds \right\} dw_2(t)}{\int_0^1 \left\{ \int_0^t w_1(s)\,ds \right\}^2 dt},$$

where $w(t) = (w_1(t), w_2(t))'$ is the two-dimensional standard Brownian motion. The statistic S_7 was earlier given in (4.48) and may be interpreted as the limit in distribution of the LSE arising from the second-order cointegrated process described in (4.51). It follows from (4.49) that the c.f. of $xV_7 - U_7$ is given by

$$(6.45) \qquad \phi_7(\theta; x) = \left[\frac{1}{2} \left\{ 1 + \cos(2i\theta x - \theta^2)^{1/4} \cosh(2i\theta x - \theta^2)^{1/4} \right\} \right]^{-1/2}.$$

Table 6.2. Fortran Program for Computing $F_6(x)$ and $f_6(x)$

```
      PARAMETER(M=30,N=100,KX=18)
      IMPLICIT REAL*8(A-E,P-Z),COMPLEX*16(F,G)
      DIMENSION F(0:2),G(0:2),FDF(2),GDF(2)
      DATA DINC/1D-6/
      PI=1D0/3.141592653589793D0
      SN=DFLOAT(N)
      L=M*N
      DO 100 K=1,KX
      X=DFLOAT(K)-15.5D0
      X1=X+DINC
      F(0)=1D0
      G(0)=1D0
      S1=0D0
      S2=0D0
      T1=0D0
      T2=0D0
      DO 1 I=1,L-1,2
      DO 2 J=1,2
      U=DFLOAT(I+J-1)/SN
      T=U*U
      CALL CHAR(X,T,U,F(J-1),F(J),FDF(J))
      CALL CHAR(X1,T,U,G(J-1),G(J),GDF(J))
    2 CONTINUE
      S1=S1+DIMAG(FDF(1))
      S2=S2+DIMAG(FDF(2))
      T1=T1+DIMAG(GDF(1))
      T2=T2+DIMAG(GDF(2))
      F(0)=F(2)
      G(0)=G(2)
    1 CONTINUE
      PROB=PI*(4D0*S1+2D0*S2-DIMAG(FDF(2)))/(3D0*SN)+5D-1
      PROB1=PI*(4D0*T1+2D0*T2-DIMAG(GDF(2)))/(3D0*SN)+5D-1
      PD=(PROB1-PROB)/DINC
      WRITE(6,20)X,PROB,PD
   20 FORMAT(3X,F8.4,F10.6,F8.4)
  100 CONTINUE
      STOP
      END
      SUBROUTINE CHAR(X,T,U,FL,FP,F)
      IMPLICIT REAL*8(X,T,U),COMPLEX*16(F)
      F0=(0D0,1D0)
      F1=CDEXP(F0*T/2D0)
      F2=2D0*F0*T*X
      F3=CDSQRT(F2)
      FP=F1/CDSQRT(CDCOS(F3)+F0*T*CDSIN(F3)/F3)
      IF(CDABS(FL+FP).LE.CDABS(FL-FP))FP=-FP
      F=FP*2D0/U
      RETURN
      END
```

NUMERICAL INTEGRATION: THE GENERAL CASE

Putting here $\theta = u^4$, we obtain

(6.46)
$$F_7(x) = P(S_7 \leq x)$$
$$= \frac{1}{2} + \frac{1}{\pi} \int_0^\infty g_7(u; x) \, du,$$

where

(6.47)
$$g_7(u; x) = \frac{4}{u} \operatorname{Im}\left[\phi_7(u^4; x)\right]$$

with $g_7(0; x) = 0$.

Figure 6.16 gives the graph of $g_7(u; x)$ for $x = -14.8468$. The present value of x is supposed to be the 5% point of the distribution of S_7 in (6.44). Although not presented here, the integrand $g_7(u; x)$ performs well even for $|x|$ small. In fact, if $x = 0$, $g_7(u; x)$ reduces to zero (Problem 4.5) so that $F_7(0) = \frac{1}{2}$. It can also be shown (Problem 4.6) that the distribution of S_7 is symmetric about the origin. The integration for (6.46) can be easily done by Simpson's rule for all moderate values of x.

Insofar as the computation of probability densities is based on numerical derivatives of distribution functions, we do not have to worry about the integration associated with probability densities. We have only to compute $F(x + \Delta x)$ as well as $F(x)$ for each x to obtain $f(x)$ as in (6.38). In subsequent chapters, we shall always use numerical derivatives whenever graphs of probability densities are presented. Here, as a sequel to the last section, the graphs of probability densities of S_3 in (6.19) and S_4 in (6.31) are presented in Figures 6.17 and 6.18, respectively. Also shown in each figure is an approximation by a constant multiple of $\chi^2(1)$ distribution discussed in

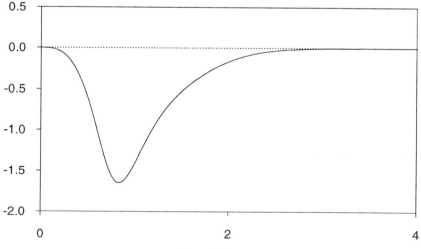

Figure 6.16.

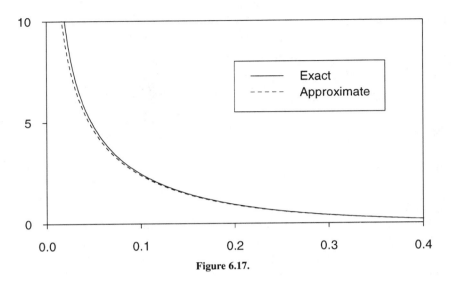

Figure 6.17.

Section 5.4 of Chapter 5, that is

$$\mathcal{L}(S_3) \sim \mathcal{L}\left(\frac{1}{12.36236} Z^2\right),$$

$$\mathcal{L}(S_4) \sim \mathcal{L}\left(\frac{1}{121.259} Z^2\right),$$

where $Z \sim N(0, 1)$.

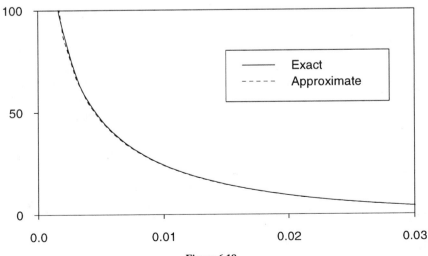

Figure 6.18.

PROBLEMS

4.1 Derive the c.f. of $xV_6 - U_6$ by the stochastic process approach, where U_6 and V_6 are defined in (6.39).

4.2 Show that $g_6(u; x)$ in (6.42) is equal to 0 when $u = 0$.

4.3 Establish the relation in (6.43).

4.4 Derive the easiest way of computing $F_6(x)$ in (6.41) when $x = 0$, and obtain its value.

4.5 Prove that $g_7(u; x)$ in (6.47) reduces to 0 if $x = 0$.

4.6 Show that the distribution of S_7 in (6.44) is symmetric about the origin.

6.5 COMPUTATION OF PERCENT POINTS

For testing purposes, it is necessary to obtain percent points. Since we have learned, to a large extent, how to compute distribution functions, the task of computing percent points is almost its by-product. Here we present two methods for computing percent points on the basis of distribution functions.

Suppose that we would like to find the $100\alpha\%$ point of $F(x)$, that is, the value x such that $F(x) = \alpha$. For this purpose we first consider *Newton's method of successive approximation*

(6.48) $$x_i = x_{i-1} - \frac{F(x_{i-1}) - \alpha}{f(x_{i-1})}, \quad (i = 1, 2, \ldots),$$

where x_0 is a starting value such that $F(x_0)$ is close to α, while $f(x) = dF(x)/dx$ or $f(x) = (F(x + \Delta x) - F(x))/\Delta x$ with Δx small, $\Delta x = 10^{-6}$, say. The above iteration may be terminated when $|x_i - x_{i-1}|$ becomes small for $i = n$, at which the $100\alpha\%$ point of $F(x)$ is obtained as x_n.

Another method for computing percent points is the *bisection method*. Let x_1 and x_2 be close to the solution to $F(x) = \alpha$, where $F(x_1) < \alpha$ and $F(x_2) > \alpha$. Then we compute $F(x)$ for $x = \bar{x} = (x_1 + x_2)/2$. If $F(\bar{x}) < \alpha$, then we replace x_1 by $\bar{x}$; otherwise we replace x_2 by $\bar{x}$. We again compute $F(x)$ for the mean of the newly defined values x_1 and x_2. This procedure is iterated until $|x_2 - x_1|$ becomes smaller than a preassigned level. In comparison with Newton's method, the present one needs less computation and avoids the computation of derivatives of $F(x)$, although the number of iterations required to attain convergence may be larger.

As an example, we take up again the distribution of S_3 in (6.19). A Fortran program that computes various percent points is presented in Table 6.3, where the bisection method is used together with Euler's transformation along the lines discussed in Section 6.3. Table 6.4 tabulates percent points of S_3 computed in this way together with those of S_4 in (6.31).

Various tables presented in Chapter 1 were also tabulated by either Newton's method or the bisection method. We shall present percent points of other distributions in later chapters.

Table 6.3. Fortran Program for Computing Percent Points of S_3 in (6.19)

```
      PARAMETER(M=50,N=50,II=50,IS=5,JK=7)
      IMPLICIT REAL*8(A-E,P-Z),COMPLEX*16(F,G)
      DIMENSION F(0:2),DF(2),XX(JK),YY(JK),PERPO(JK),
     * RA(N),RB(N),SA(0:II,0:II),SB(0:II,0:II)
      DATA XX/5D-4,17D-4,3D-3,39D-3,22D-2,31D-2,53D-2/
      DATA YY/6D-4,18D-4,4D-3,40D-3,23D-2,32D-2,54D-2/
      DATA PERPO/1D-2,5D-2,1D-1,5D-1,9D-1,95D-2,99D-2/
      PI=3.141592653589793D0
      SN=DFLOAT(N)
      L=M*N
      F(0)=1D0
      DO 1 I=1,L-1,2
      DO 2 J=1,2
      U=DFLOAT(I+J-1)/SN
      T=U*U*U*U
      CALL CHAR(T,F(J-1),F(J))
      DF(J)=DIMAG(F(J))*4D0/U
    2 CONTINUE
      S1=S1+DF(1)
      S2=S2+DF(2)
      F(0)=F(2)
    1 CONTINUE
      PROB0=(4D0*S1+2D0*S2-DF(2))/3D0/SN/PI
      DO 100 IK=1,JK
      X1=XX(IK)
      X2=YY(IK)
      IT=0
  222 IT=IT+1
      X=(X1+X2)/2D0
      FAL=1D0
      FBL=1D0
      AL=0D0
      BL=0D0
      DO 3 I=0,II
      DA=PI/X/2D0
      DB=PI/X
      AP=DSQRT(DSQRT(DFLOAT(2*I+1)*DA))
      BP=DSQRT(DSQRT(DFLOAT(I+1)*DB))
      HA=AP-AL
      HB=BP-BL
      DO 4 J=1,N
      UA=AL+HA*DFLOAT(J)/SN
      UB=BL+HB*DFLOAT(J)/SN
      TA=UA*UA*UA*UA
      TB=UB*UB*UB*UB
      CALL CHAR(TA,FAL,FAP)
      FAL=FAP
```

Table 6.3. (Continued)

```
      CALL CHAR(TB,FBL,FBP)
      FBL=FBP
      RA(J)=DCOS(TA*X)*DIMAG(FAP)*4D0/UA
      RB(J)=DSIN(TB*X)*DREAL(FBP)*4D0/UB
    4 CONTINUE
      AL=AP
      BL=BP
      S1=0D0
      S2=0D0
      T1=0D0
      T2=0D0
      DO 5 J=2,N,2
      S1=S1+RA(J-1)
      S2=S2+RA(J)
      T1=T1+RB(J-1)
      T2=T2+RB(J)
    5 CONTINUE
      VA=(4D0*S1+2D0*S2-RA(N))*HA
      VB=(4D0*T1+2D0*T2-RB(N))*HB
      IF(MOD(I,2).EQ.0)THEN
      SA(I,0)=VA
      SB(I,0)=VB
      ELSE
      SA(I,0)=-VA
      SB(I,0)=-VB
      END IF
    3 CONTINUE
      DO 6 K=1,II
      DO 7 J=IS,II-K
      SA(J,K)=(SA(J+1,K-1)-SA(J,K-1))/2D0
      SB(J,K)=(SB(J+1,K-1)-SB(J,K-1))/2D0
    7 CONTINUE
    6 CONTINUE
      SUMA=0D0
      SUMB=0D0
      SUMC=0D0
      SUMD=0D0
      DO 8 I=0,II
      IF(MOD(I,2).EQ.0.)THEN
      SGN=1D0
      ELSE
      SGN=-1D0
      END IF
      IF(I.LE.IS-1)THEN
      SUMA=SUMA+SGN*SA(I,0)
      SUMB=SUMB+SGN*SB(I,0)
```

Table 6.3. *(Continued)*

```
      ELSE
      SUMC=SUMC+SGN*SA(IS,I-IS)
      SUMD=SUMD+SGN*SB(IS,I-IS)
      END IF
    8 CONTINUE
      PROBA=(SUMA+SUMC/2D0)/3D0/SN/PI
      PROBB=(SUMB+SUMD/2D0)/3D0/SN/PI
      PROB=PROB0-PROBA+PROBB
      IF(PROB-PERPO(IK).LT.0D0)THEN
      X1=X
      ELSE
      X2=X
      END IF
      IF(X2-X1.LE.1D-10.OR.IT.EQ.30)GO TO 111
      WRITE(6,30)IT,X,PROB
      GO TO 222
  111 WRITE(6,30)IT,X,PROB
   30 FORMAT(I5,2F15.8)
  100 CONTINUE
      STOP
      END
      SUBROUTINE CHAR(T,FL,FP)
      IMPLICIT REAL*8(A-D,P-Z),COMPLEX*16(F)
      F0=(0D0,1D0)
      F1=2D0*F0*T
      F2=CDSQRT(F1)
      F3=CDSQRT(F2)
      F4=(1D0+CDCOS(F3)*CDCOS(F0*F3))/2D0
      FP=1D0/CDSQRT(F4)
      IF(CDABS(FL+FP).LE.CDABS(FL-FP))FP=-FP
      RETURN
      END
```

Table 6.4. Percent Points of S_3 in (6.19) and S_4 in (6.31)

Case	Probability of a Smaller Value						
	0.01	0.05	0.1	0.5	0.9	0.95	0.99
S_3	0.00056	0.00179	0.00347	0.0393	0.2213	0.3132	0.5391
S_4	0.000022	0.000099	0.000224	0.00384	0.02240	0.03176	0.05479

6.6 THE SADDLEPOINT APPROXIMATION

If our purpose is just to obtain approximations to distribution functions or percentiles, the *saddlepoint method* enables us to compute those without employing any numerical integration method. Daniels (1954) first introduced this method into statistics. For our purpose, we find the approach taken by Helstrom (1978) useful, which we now describe.

Let $\phi(\theta)$ be the c.f. of a random variable S and $F(x)$ its distribution function. To make the presentation simpler, we assume here that S takes only nonnegative values and is a quadratic functional of the Brownian motion. We then put

$$(6.49) \qquad h(z) = \phi(-iz) = E\left(e^{zS}\right),$$

where z is a complex variable. When z is real, $h(z)$ is the m.g.f. of S.

Helstrom (1978) recommends determining $F(x)$ following

$$(6.50) \qquad F(x) = \frac{1}{2\pi i} \int_{\bar{\theta}-i\infty}^{\bar{\theta}+i\infty} \left(-\frac{1}{z}\right) h(z) e^{-xz} \, dz, \qquad (\theta_L < \bar{\theta} < 0)$$

for the left-hand tail and

$$(6.51) \qquad 1 - F(x) = \frac{1}{2\pi i} \int_{\bar{\theta}-i\infty}^{\bar{\theta}+i\infty} \frac{1}{z} h(z) e^{-xz} dz, \qquad (0 < \bar{\theta} < \theta_R)$$

for the right-hand tail, where θ_L is the first singularity point of $h(\theta)$ to the left of the origin, while θ_R is that of $h(\theta)$ to the right of the origin. Note that the singularity points of $h(\theta)$ are all positive because of the assumption on S. Thus we may put $\theta_L = -\infty$ in the present case. If S is a nonpositive random variable, the role of θ_L and θ_R should be interchanged. If S takes both positive and negative values, the choice of $\bar{\theta}$ cannot be made in advance. This is the main reason why we assume S to be a nonnegative random variable. We cannot usually rely only on (6.50) or (6.51) to obtain good approximations, as is shown later by an illustrative example. For a given value of x, however, it is somewhat arbitrary whether we should use (6.50) or (6.51). A graphical solution will be given later. As the value of $\bar{\theta}$, we shall take the saddlepoint of the integrand in both of (6.50) and (6.51), which we now discuss.

Let us consider the logarithm of the integrand in (6.50) on the negative real axis

$$(6.52) \qquad \Psi_-(\theta; x) = \log h(\theta) - \theta x - \log(-\theta), \qquad (\theta < 0).$$

Since $\exp\{\Psi_-(\theta; x)\}$ is shown to be convex for any x, $\Psi_-(\theta; x)$ is expected to have a single minimum. If this is the case, the minimizer is a solution to

$$(6.53) \qquad \Psi_-^{(1)}(\theta; x) \equiv \frac{\partial \Psi_-(\theta; x)}{\partial \theta} = \frac{1}{h(\theta)} \frac{dh(\theta)}{d\theta} - x - \frac{1}{\theta} = 0, \qquad (\theta < 0).$$

The solution is called the *saddlepoint* of the complex-valued function $h(z)e^{-xz}/(-z)$, and is denoted by θ_-. The assumption that $\theta_L < \theta_- < 0$ is ensured in the present case.

Similarly, if we consider

(6.54) $$\Psi_+(\theta; x) = \log h(\theta) - \theta x - \log \theta, \qquad (\theta > 0),$$

the saddlepoint of the integrand $h(z)e^{-xz}/z$ in (6.51) is a solution to

(6.55) $$\Psi_+^{(1)}(\theta; x) \equiv \frac{\partial \Psi_+(\theta; x)}{\partial \theta} = \frac{1}{h(\theta)} \frac{dh(\theta)}{d\theta} - x - \frac{1}{\theta} = 0, \qquad (\theta > 0).$$

This solution is denoted by θ_+, for which we assume $0 < \theta_+ < \theta_R$.

Expanding $\exp\{\Psi_-(z; x)\}$ with $z = \theta_- + i\theta$ around θ_-, we have

$$\exp\{\Psi_-(z; x)\} = h(z)e^{-xz}/(-z)$$
$$\cong \exp\left\{\Psi_-(\theta_-; x) - \frac{1}{2}\Psi_-^{(2)}(\theta_-)\theta^2\right\},$$

where

$$\Psi_*^{(2)}(\theta_*) = \left.\frac{\partial^2 \Psi_*(\theta; x)}{\partial \theta^2}\right|_{\theta=\theta_*}.$$

Substituting this into (6.50) with $\bar{\theta} = \theta_-$, we obtain the saddlepoint approximation to $F(x)$

(6.56) $$F(x) \cong \frac{1}{2\pi} \exp\{\Psi_-(\theta_-; x)\} \int_{-\infty}^{\infty} \exp\left\{-\frac{1}{2}\Psi_-^{(2)}(\theta_-)\theta^2\right\} d\theta$$
$$= \frac{1}{\sqrt{2\pi \Psi_-^{(2)}(\theta_-)}} \exp\{\Psi_-(\theta_-; x)\}.$$

Similarly we obtain

(6.57) $$1 - F(x) \cong \frac{1}{\sqrt{2\pi \Psi_+^{(2)}(\theta_+)}} \exp\{\Psi_+(\theta_+; x)\}.$$

Once x is given, θ_- and θ_+ can be found by Newton's method. Then we can compute the saddlepoint approximate distribution following (6.56) or (6.57).

As an example, let us take up the c.f. $\phi(\theta) = (\sin \sqrt{2i\theta}/\sqrt{2i\theta})^{-1/2}$ so that

(6.58)
$$h(\theta) = \left(\frac{\sinh\sqrt{-2\theta}}{\sqrt{-2\theta}}\right)^{-1/2}, \quad (\theta < 0),$$

$$= \left(\frac{\sin\sqrt{2\theta}}{\sqrt{2\theta}}\right)^{-1/2}, \quad (\theta > 0).$$

It follows that $\theta_L = -\infty$ and $\theta_R = \pi^2/2 = 4.9348$.

Table 6.5 reports saddlepoints θ_- and θ_+ associated with (6.58) for some selected values of x. Also shown are approximate probabilities P_- and P_+ computed from (6.56) and (6.57), respectively, together with the exact probability P based on Simpson's rule. It is seen that the saddlepoints increase as x becomes large, but they are all smaller than θ_R. It is also observed that P_- behaves quite well on the left-hand tail, while P_+ does so on the right-hand tail. Neither P_- nor P_+ approximates P well for the whole range of x.

Figure 6.19 is a graphical version of Table 6.5. The two approximate distributions cross each other at $x = 0.26$. Thus we recommend in the present case that P_- be used for $x \le 0.26$ and P_+ for $x > 0.26$.

The saddlepoint approximate percentiles can be obtained as follows: Suppose that $F(x) = \alpha$ is given. Then, noting that

(6.59)
$$x = \frac{1}{h(\theta)}\frac{dh(\theta)}{d\theta} - \frac{1}{\theta}$$

at $\theta = \theta_-$ and substituting this into (6.52), it follows from (6.56) that θ_- must be approximately a solution to

(6.60) $\log \alpha = \log h(\theta) - \dfrac{\theta}{h(\theta)}\dfrac{dh(\theta)}{d\theta} + 1 - \log(-\theta) - \dfrac{1}{2}\log\{2\pi\Psi_-^{(2)}(\theta)\},$

where $\theta < 0$. Similarly, if $1 - F(x) = 1 - \beta$ is given, θ_+ must be approximately a solution to

(6.61) $\log(1 - \beta) = \log h(\theta) - \dfrac{\theta}{h(\theta)}\dfrac{dh(\theta)}{d\theta} + 1 - \log\theta - \dfrac{1}{2}\log\{2\pi\Psi_+^{(2)}(\theta)\},$

Table 6.5. Saddlepoint Approximations Associated with (6.58)

x	0.01	0.05	0.1	0.2	0.5	0.8	1
θ_-	-1395.97	-77.08	-25.27	-9.08	-2.68	-1.52	-1.17
θ_+	3.06	3.21	3.39	3.68	4.18	4.41	4.50
P_-	10^{-5}	0.12	0.42	0.75	1.00	1.05	1.06
P_+	0.50	0.57	0.65	0.79	0.96	0.99	1.00
P	10^{-5}	0.12	0.42	0.73	0.96	0.99	1.00

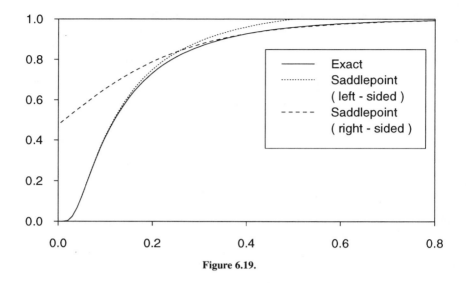

Figure 6.19.

where $\theta > 0$. The equations (6.60) and (6.61) can be most efficiently solved for θ by the *secant method*. Then percent points are obtained from (6.59) with θ replaced by the solution to (6.60) or (6.61).

The saddlepoints need not be computed accurately since (6.60) and (6.61) are just an approximation. The following examples also indicate that percent points computed from (6.59) are insensitive to a small departure from the solution to (6.60) or (6.61).

Let us take up two examples: one has the m.g.f. defined in (6.58) and the other is defined by

$$h(\theta) = \left[\frac{1}{2} + \frac{1}{4}\left\{\cos\sqrt{2}(-2\theta)^{1/4} + \cosh\sqrt{2}(-2\theta)^{1/4}\right\}\right]^{-1/2}, \quad (\theta < 0),$$

$$= \left[\frac{1}{2}\left\{1 + \cos(2\theta)^{1/4}\cosh(2\theta)^{1/4}\right\}\right]^{-1/2}, \quad (\theta > 0).$$

(6.62)

Note that a random variable with the m.g.f. (6.62) is given in (6.19). It follows that $\theta_L = -\infty$ and $\theta_R = 12.36236$.

Table 6.6 reports approximate percentiles associated with (6.58) and (6.62) for $\alpha = 0.05$ and $\beta = 0.95$ computed in the way described above, together with the exact percentiles based on numerical integration. The entries under the heading "θ" are those values around the solution to (6.60) or (6.61), while those under "odds" are differences of the left from the right side in (6.60) or (6.61). Those values of θ which have odds equal to 0 are approximate saddlepoints, and the corresponding values of x are approximate percentiles. We observe that the approximate percentiles

THE SADDLEPOINT APPROXIMATION

Table 6.6. Saddlepoint Approximate Percentiles Associated with (6.58) and (6.62)

				(6.58)			
	$\alpha = 0.05$			Exact			$x = 0.0366$
θ	-135		-133	-131.5921	-128		-126
odds	0.050		0.021	0	-0.054		-0.084
x	0.0360		0.0363	0.0365	0.0371		0.0374
	$\beta = 0.95$			Exact			$x = 0.4614$
θ	4.10		4.12	4.1454	4.16		4.18
odds	-0.205		-0.117	0	0.070		0.171
x	0.4288		0.4448	0.4661	0.4789		0.4972
				(6.62)			
	$\alpha = 0.05$			Exact			$x = 0.00179$
θ	-1330		-1310	-1295.4102	-1270		-1250
odds	0.027		0.011	0	-0.020		-0.036
x	0.00171		0.00173	0.00175	0.00178		0.00180
	$\beta = 0.95$			Exact			$x = 0.3132$
θ	5.15		5.17	5.1978	5.22		5.24
odds	-0.175		-0.103	0	0.087		0.168
x	0.2932		0.3035	0.3185	0.3311		0.3429

coincide with the exact ones up to the first two effective figures, as far as percentiles examined here are concerned. It is also seen that percentiles are insensitive to a small departure from approximate saddlepoints.

The saddlepoint method can be implemented to compute exact distribution functions and percent points. Helstrom (1995) also suggests that method, which we do not pursue in this book.

CHAPTER 7

Estimation Problems in Nonstationary Autoregressive Models

Estimation problems arising in nonstationary AR models are discussed. We start our discussion by considering regression models with the error term following a near integrated process. We shall then take account of seasonality and multiple unit roots as well as negative unit roots and complex roots on the unit circle. The weak convergence results on the estimators associated with these models are discussed, and the limiting distributions are computed and graphically presented.

7.1 NONSTATIONARY AUTOREGRESSIVE MODELS

The class of models that we first deal with is the regression model where the error term follows a nearly nonstationary process. More specifically, we deal with

(7.1)
$$y_j = x_j'\beta + \eta_j,$$
$$\eta_j = \rho\eta_{j-1} + u_j, \qquad (j = 1,\ldots,T),$$

where $\{x_j\}$ is a $p \times 1$ nonstochastic fixed sequence, ρ is any constant close to unity, and $\{u_j\}$ is a stationary process defined shortly. The above model may be put into the form

(7.2) $$y_j = \rho y_{j-1} + (x_j - \rho x_{j-1})'\beta + u_j$$

or

(7.3) $$\Delta y_j = (\rho - 1)y_{j-1} + (x_j - \rho x_{j-1})'\beta + u_j, \qquad \Delta = 1 - L.$$

213

Because of the representation of (7.2) the present model may be referred to as a *stochastic* (plus deterministic) *trend* model if $\rho \geq 1$. Since the model may also be rewritten as

$$(7.4) \qquad (1 - \rho L)(y_j - x_j'\beta) = u_j,$$

it is called a *trend stationary* model if $\rho < 1$. Note that we are mainly concerned with cases where ρ is close to unity.

The process $\{u_j\}$ in (7.1) is assumed to be generated by

$$(7.5) \qquad u_j = \sum_{l=0}^{\infty} \alpha_l \varepsilon_{j-l}, \quad \sum_{l=0}^{\infty} l|\alpha_l| < \infty, \quad \alpha \equiv \sum_{l=0}^{\infty} \alpha_l \neq 0.$$

Here we assume that $\{\varepsilon_j\}$ is a sequence of i.i.d.$(0, \sigma^2)$ random variables. This assumption can be relaxed, as was discussed in Chapter 3, to the extent that $\{\varepsilon_j\}$ is a sequence of martingale differences satisfying some additional conditions. We, however, assume $\{\varepsilon_j\}$ to be i.i.d.$(0, \sigma^2)$ for simplicity.

The stochastic nonstationarity aspect of the model (7.1) underlies the assumption that ρ is close to unity, which we express by

$$(7.6) \qquad \rho = 1 - \frac{c}{T},$$

where c is a fixed constant that may take any real value. The idea of a parameter depending on a sample size is similar to that in approximating a binomial distribution with parameters n, p by a Poisson distribution with parameter λ in such a way that $p = \lambda/n$.

The model described above is the one discussed in Nabeya and Tanaka (1990a), where the initial value η_0 to generate $\{\eta_j\}$ was assumed to be zero. Here we allow for the influence of η_0 in such a way that

$$(7.7) \qquad \eta_0 = \sqrt{T}\sigma_L Z,$$

where σ_L is a parameter involved in $\{u_j\}$, while Z is either a constant or a normal random variable independent of $\{u_j\}$.

The model (7.1) together with (7.5), (7.6), and (7.7) is fairly general and seems to cover various models studied in the literature. In particular, Fuller (1976) considered the following three types of models:

$$(7.8) \qquad y_j = \begin{cases} \rho y_{j-1} + \varepsilon_j, \\ \rho y_{j-1} + \alpha + \varepsilon_j, \\ \rho y_{j-1} + \beta + j\gamma + \varepsilon_j, \end{cases}$$

NONSTATIONARY AUTOREGRESSIVE MODELS

where the true values of (ρ, α, γ) are $(1, 0, 0)$, while β is arbitrary. Then he discussed the asymptotic properties of the ordinary LSEs of ρ applied to each model under $(\rho, \alpha, \gamma) = (1, 0, 0)$ and β's being arbitrary. It can be shown (Problem 1.1 in this chapter) that this curious assumption about parameters can be better understood for our model. On the other hand, another restricted model

(7.9) $$y_j = \rho y_{j-1} + j\gamma + \varepsilon_j,$$

where the true value of γ is 0, is not physically realizable from our model, although a closely related model is the one with $x_j = j$. The similarities and differences between (7.8), (7.9), and our model (7.1) will be made clear later.

The second and third models in (7.8) are quite special in the sense described above. It can be shown [Fuller (1985) and Problem 1.2] that, if the parameter α in the second model is a nonzero constant independent of T, then the ordinary LSE applied to that model tends to normality even if $\rho = 1$. The same is true if the parameter γ in the third model is a nonzero constant. The finite sample distribution, however, is not normal, and that case was considered by Evans and Savin (1984).

The present model does not take account of seasonality or multiple unit roots. We shall give a separate treatment of seasonality in Section 7.7, while the case of multiple unit roots will be discussed in Section 7.9 by considering a strongly nonstationary AR model. Negative unit roots and complex roots on the unit circle are also studied in Sections 7.3 and 7.8, respectively.

We now discuss the estimation problem associated with ρ in our model (7.1). It is quite natural to consider

(7.10) $$\hat{\rho} = \sum_{j=2}^{T} \hat{\eta}_{j-1} \hat{\eta}_j \bigg/ \sum_{j=2}^{T} \hat{\eta}_{j-1}^2 ,$$

where $\hat{\eta} = (\hat{\eta}_1, \ldots, \hat{\eta}_T)' = y - X\hat{\beta}$, $y = (y_1, \ldots, y_T)'$, $X = (x_1, \ldots, x_T)'$, and $\hat{\beta} = (X'X)^{-1}X'y$ with rank$(X) = p < T$.

In Section 1.3 of Chapter 1, we discussed the limiting distribution of $T(\hat{\rho} - 1)$ for the model $y_j = \rho y_{j-1} + \varepsilon_j$ with $y_0 = 0$ and $\rho = 1$, where we also considered the other estimators such as the Yule–Walker estimator. It was found that the limiting distributions were different from each other. Here we exclusively deal with the estimator $\hat{\rho}$ defined in (7.10), but we shall take up the Yule–Walker estimator as well in Chapter 9 when we consider AR unit root tests. Note also that Fuller's (1976) estimator applied to each of the models in (7.8) is slightly different from the corresponding estimator $\hat{\rho}$ in (7.10), but it will be found that the two estimators are asymptotically the same.

We have (Problem 1.3), from (7.10),

(7.11) $$T(\hat{\rho} - 1) = \frac{1}{T} \sum_{j=2}^{T} \hat{\eta}_{j-1}(\hat{\eta}_j - \hat{\eta}_{j-1}) \bigg/ \left[\frac{1}{T^2} \sum_{j=2}^{T} \hat{\eta}_{j-1}^2 \right]$$

$$= U_T / V_T ,$$

where

$$(7.12) \quad U_T = \frac{1}{2T}(\hat{\eta}_T^2 - \hat{\eta}_1^2) - \frac{1}{2T}\sum_{j=2}^{T}(\hat{\eta}_j - \hat{\eta}_{j-1})^2,$$

$$(7.13) \quad V_T = \frac{1}{T^2}\sum_{j=1}^{T}\hat{\eta}_j^2 - \frac{1}{T^2}\hat{\eta}_T^2.$$

The limiting distribution of $T(\hat{\rho} - 1) = U_T/V_T$ depends on the assumptions made concerning the regressor sequence $\{x_j\}$. Here we consider four simple cases by specifying $\{x_j\}$:

Model A: $y_j = \eta_j$,

Model B: $y_j = \beta_0 + \eta_j$,

Model C: $y_j = j\beta_1 + \eta_j$,

Model D: $y_j = \beta_0 + j\beta_1 + \eta_j$,

where $\eta_j = \rho\eta_{j-1} + u_j$ for all models.

White (1958) first obtained the m.g.f. associated with the limiting distribution of $T(\hat{\rho} - 1)$ for Model A with $\rho = 1$, $u_j = \varepsilon_j$, and $\eta_0 = 0$. Anderson (1959) showed that the corresponding limiting distribution depends on the common distribution of $\{\varepsilon_j\}$ if $|\rho| > 1$ and ρ is independent of T. It will be seen that this is not the case if $\rho = 1 - (c/T)$ with $c < 0$. Fuller (1976) gave percent points of $T(\hat{\rho} - 1)$ for Model A, Model B, and Model D with $\rho = 1$, $u_j = \varepsilon_j$, and $\eta_0 = 0$ on the basis of the Monte Carlo simulations conducted by Dickey (1976). Evans and Savin (1981a,b) inverted numerically the m.g.f. obtained by White (1958) to tabulate the limiting density and cumulative distribution. They also calculated under normality the finite sample distribution function of $T(\hat{\rho} - \rho)$ for ρ near unity and observed that the distribution function is very poorly approximated by the limiting distribution for moderately large values of T. Phillips (1977) also noted that his approximation is poor for $|\rho|$ close to 1. We shall give an alternative approximation in Section 7.6.

Bobkoski (1983) first developed asymptotic arguments analytically under $\rho = 1 - (c/T)$. He proved that, for Model A with $u_j = \varepsilon_j$ and $\eta_0 = 0$,

$$(7.14) \quad \mathcal{L}(T(\hat{\rho} - 1)) \to \mathcal{L}\left(\frac{\int_0^1 X(t)\,dX(t)}{\int_0^1 X^2(t)\,dt}\right)$$

as $T \to \infty$ under $\rho = 1 - (c/T)$, where $\{X(t)\}$ is the O–U process generated by $dX(t) = -cX(t)\,dt + dw(t)$ with $X(0) = 0$ and $\{w(t)\}$ being the standard Brownian

NONSTATIONARY AUTOREGRESSIVE MODELS

motion. Bobkoski (1983) also discussed the Yule–Walker estimator and obtained the associated c.f.'s to compute the distribution functions.

Phillips (1987a,b) attempted to weaken the assumption on $\{u_j\}$ in Model A with $\eta_0 = 0$ to the extent that $T(\hat{\rho} - 1)$ has a nondegenerate limiting distribution. Phillips (1987b) assumed mixing conditions on $\{u_j\}$, which allow for temporal dependence and heteroscedasticity in such a way that

$$(7.15) \quad \sigma_L^2 = \lim_{T \to \infty} E \left(\frac{1}{\sqrt{T}} \sum_{j=1}^{T} u_j \right)^2 \quad \text{and} \quad \sigma_S^2 = \lim_{T \to \infty} \frac{1}{T} \sum_{j=1}^{T} E(u_j^2)$$

exist, and he proved that, for Model A with $\eta_0 = 0$,

$$(7.16) \quad \mathcal{L}(T(\hat{\rho} - 1)) \to \mathcal{L} \left(\frac{\int_0^1 X(t)\, dX(t) + \frac{1}{2}\left(1 - \frac{\sigma_S^2}{\sigma_L^2}\right)}{\int_0^1 X^2(t)\, dt} \right)$$

as $T \to \infty$ under $\rho = \exp(-c/T)$.

In our model $\{u_j\}$ is assumed to be stationary and the two variances given in (7.15) reduce (Problem 1.4) to

$$(7.17) \quad \sigma_L^2 = \sigma^2 \left(\sum_{l=0}^{\infty} \alpha_l \right)^2 = \sum_{l=-\infty}^{\infty} \gamma_l \quad \text{and} \quad \sigma_S^2 = \sigma^2 \sum_{l=0}^{\infty} \alpha_l^2 = \gamma_0,$$

where γ_l is the lth order autocovariance of $\{u_j\}$. Since the spectral density $f(\omega)$ of $\{u_j\}$ is given by

$$f(\omega) = \frac{1}{2\pi} \sum_{l=-\infty}^{\infty} \gamma_l\, e^{il\omega} = \frac{\sigma^2}{2\pi} \left| \sum_{l=0}^{\infty} \alpha_l e^{il\omega} \right|^2$$

so that $\sigma_L^2 = 2\pi f(0)$, and $f(0)$ is related to the long-run variation of $\{u_j\}$, σ_L^2 is called the *long-run variance*. The variance σ_S^2, on the other hand, is $V(u_j)$, so it is called the *short-run variance*. We shall use $r = \sigma_S^2/\sigma_L^2 = 1/\sum_{l=-\infty}^{\infty} \rho_l$ in later discussions, where ρ_l is the lth order autocorrelation of $\{u_j\}$.

Nabeya and Tanaka (1990a) dealt with Models A through D assuming that $\eta_0 = 0$. For each model they obtained the c.f.s associated with the limiting distributions of $T(\hat{\rho} - 1)$ using the Fredholm approach and inverted the c.f.s to draw probability densities and tabulate percent points.

Perron (1991a,b) extended the estimation problem in the O–U process $\{X(t)\}$ with $X(0) = 0$ considered by Liptser and Shiryayev (1978) to the case where

$$(7.18) \quad dY(t) = -cY(t)\, dt + dw(t), \quad Y(0) = \gamma \quad \text{or} \quad Y(0) \sim N\left(0, \frac{1}{2c}\right),$$

with γ being a fixed constant and $c > 0$ for $Y(0) \sim N(0, 1/(2c))$. He derived the c.f. associated with the estimator

$$(7.19) \qquad -\tilde{c} = \frac{\int_0^1 Y(t)\,dY(t)}{\int_0^1 Y^2(t)\,dt}.$$

It is noticeable that $\mathcal{L}(-\tilde{c})$ reduces to the limiting distribution on the right side of (7.14) if $\gamma = 0$.

In subsequent sections, we shall verify that the limiting distributions of $T(\hat{\rho} - 1)$ for Models A through D with the exception of Model C are closely related to the distributions of estimators of $-c$ constructed from the O–U process (7.18).

PROBLEMS

1.1 Interpret the parameter restrictions imposed on the models (7.8) in terms of our model (7.1).

1.2 Prove that the ordinary LSE $\hat{\rho}$ of ρ in the second model in (7.8) tends to normality if $\rho = 1$ and α is a nonzero constant independent of the sample size. More specifically, show that $T\sqrt{T}(\hat{\rho} - 1) \to N(0, 12\sigma^2/\alpha^2)$.

1.3 Derive the expressions given in (7.11).

1.4 Show that the relations in (7.17) hold.

7.2 CONVERGENCE IN DISTRIBUTION OF LSES

The purpose of this section is to establish weak convergence results on the statistic $T(\hat{\rho} - 1)$ as $T \to \infty$ under $\rho = 1 - (c/T)$ for Models A through D described in the previous section. We first obtain the limiting expressions on the basis of a partial sum process constructed from the error term $\{\eta_j\}$, and then show that the expressions are closely related to those derived from estimators associated with an O–U process. The expressions are in terms of the single Riemann and Ito integrals. We could also derive the limiting expressions using the double Riemann–Stieltjes integral, which we do not pursue here since the former are more easily derived.

We now start discussions by constructing

$$(7.20) \quad Y_T(t) = \frac{1}{\sqrt{T}}\eta_{j-1} + T\left(t - \frac{j-1}{T}\right)\frac{\eta_j - \eta_{j-1}}{\sqrt{T}}, \quad \left(\frac{j-1}{T} \le t \le \frac{j}{T}\right),$$

where $Y_T(0) = \eta_0/\sqrt{T}$, $Y_T(1) = \eta_T/\sqrt{T}$ and

$$(7.21) \qquad \eta_j = \left(1 - \frac{c}{T}\right)\eta_{j-1} + u_j, \quad \eta_0 = \sqrt{T}\,\sigma_L Z.$$

Here $\{u_j\}$ is a stationary process defined in (7.5), $\sigma_L = \sqrt{\sigma_L^2}$ is the long-run standard deviation of $\{u_j\}$, while Z is an $N(\gamma, \delta^2)$ random variable independent of $\{u_j\}$. It follows from Theorem 3.13 that

$$\mathcal{L}(Y_T) \to \mathcal{L}(\sigma_L Y), \tag{7.22}$$

where $Y = \{Y(t)\}$ is the O–U process defined by

$$dY(t) = -cY(t)\,dt + dw(t), \quad Y(0) = Z \sim N(\gamma, \delta^2), \tag{7.23}$$

which is equivalent to

$$Y(t) = Y(0)e^{-ct} + e^{-ct} \int_0^t e^{cs}\,dw(s). \tag{7.24}$$

The FCLT in (7.22) associated with the partial sum process $\{Y_T(t)\}$ in (7.20) can be effectively used to establish weak convergence results on $T(\hat{\rho} - 1)$ as $T \to \infty$ under $\rho = 1 - (c/T)$. Since the limiting expression varies depending on the regressor sequence $\{x_j\}$, we need a separate treatment for each model, which we discuss in the following subsections.

7.2.1 Model A

The model dealt with here is

$$y_j = \eta_j = \rho \eta_{j-1} + u_j, \quad \eta_0 = \sqrt{T}\sigma_L Z, \tag{7.25}$$

so that (7.11) reduces to

$$T(\hat{\rho} - 1) = \frac{1}{2T}\left[\eta_T^2 - \eta_1^2 - \sum_{j=2}^{T}(\eta_j - \eta_{j-1})^2\right] \bigg/ \left[\frac{1}{T^2}\left(\sum_{j=1}^{T}\eta_j^2 - \eta_T^2\right)\right] \tag{7.26}$$

$$= U_{1T}/V_{1T}.$$

Since $\eta_T = \sqrt{T}Y_T(1) = O_p(\sqrt{T})$ because of (7.22), η_T^2/T^2 converges in probability to 0. Moreover it can be shown (Problem 2.1) that

$$\frac{1}{T}\sum_{j=2}^{T}(\eta_j - \eta_{j-1})^2 \to \sigma_S^2 \quad \text{in probability}, \tag{7.27}$$

$$\frac{\eta_1^2}{T} - \sigma_L^2 Z^2 \to 0 \quad \text{in probability}. \tag{7.28}$$

Thus we are led to consider, using the partial sum process $\{Y_T(t)\}$ in (7.20),

$$U_{1T} = \frac{1}{2}\left[\frac{1}{T}\eta_T^2 - \sigma_L^2 Z^2 - \sigma_S^2\right] + o_p(1)$$

$$= \frac{1}{2}\left[Y_T^2(1) - \sigma_L^2 Y^2(0) - \sigma_S^2\right] + o_p(1),$$

$$V_{1T} = \frac{1}{T^2}\sum_{j=1}^{T}\eta_j^2 + o_p(1)$$

$$= \frac{1}{T}\sum_{j=1}^{T}Y_T^2\left(\frac{j}{T}\right) + o_p(1).$$

Then it can be shown (Problem 2.2) that

(7.29) $$\mathcal{L}(U_{1T}, V_{1T}) \to \mathcal{L}(\sigma_L^2 U_1, \sigma_L^2 V_1),$$

where

(7.30) $$U_1 = \frac{1}{2}(Y^2(1) - Y^2(0) - r)$$

$$= \int_0^1 Y(t)\,dY(t) + \frac{1}{2}(1-r),$$

(7.31) $$V_1 = \int_0^1 Y^2(t)\,dt,$$

with $r = \sigma_S^2/\sigma_L^2$ being the ratio of the short-run to the long-run variances given in (7.17). Note that the second equality in (7.30) comes from the Ito calculus.

Finally, using the continuous mapping theorem, we obtain

Theorem 7.1. *For the LSE $\hat{\rho}$ for Model A in (7.25), it holds that, as $T \to \infty$ under $\rho = 1 - (c/T)$,*

(7.32) $$\mathcal{L}(T(\hat{\rho} - 1)) \to \mathcal{L}(U_1/V_1)$$

$$= \mathcal{L}\left(\frac{\int_0^1 Y(t)\,dY(t) + \frac{1}{2}(1-r)}{\int_0^1 Y^2(t)\,dt}\right).$$

The effect of r on the limiting distribution is clearly seen from (7.32); that is, the distribution will be shifted to the left as r becomes large. In particular, we have that $\mathcal{L}(U_1/(rV_1)) \to \mathcal{L}(-1/(2V_1))$ as $r \to \infty$. Note that the case of r large occurs when

CONVERGENCE IN DISTRIBUTION OF LSES 221

$u_j = \alpha(L)\varepsilon_j$ has a root close to unity, since $r = \sum_{l=0}^{\infty} \alpha_l^2/\alpha^2(1)$. It may also be noted that

(7.33)
$$\frac{1}{r} = \sum_{l=-\infty}^{\infty} \rho_l = 1 + 2\sum_{l=1}^{\infty} \rho_l,$$

where ρ_l is the lth order autocorrelation of $\{u_j\}$. Thus $r \to \infty$ is equivalent to $\sum_{l=1}^{\infty} \rho_l \to -\frac{1}{2}$.

The effect of $Y(0)$ can also be deduced if $Y(0)$ is a constant γ, say. Then we consider

$$\frac{U_1}{V_1} + c = \frac{\int_0^1 Y(t)\,dw(t) + \frac{1}{2}(1-r)}{\int_0^1 Y^2(t)\,dt}$$

$$= \frac{\int_0^1 \left[\gamma e^{-ct} + e^{-ct}\int_0^t e^{cs}\,dw(s)\right] dw(t) + \frac{1}{2}(1-r)}{\int_0^1 \left[\gamma e^{-ct} + e^{-ct}\int_0^t e^{cs}\,dw(s)\right]^2 dt},$$

which converges in probability to 0 as $|\gamma| \to \infty$ so that U_1/V_1 converges in probability to $-c$. Note that $(U_1/V_1) + c$ is the limit in distribution of $T(\hat{\rho} - \rho) = T(\hat{\rho} - 1) + c$. Moreover it can be shown (Problem 2.3) that, as $|\gamma| \to \infty$

(7.34)
$$\mathcal{L}\left(\gamma\left(\frac{U_1}{V_1} + c\right)\right) \to N\left(0, \frac{ce^c}{\sinh c}\right).$$

The effect of c as $|c| \to \infty$, however, is mixed up, depending on the value of γ, and is hard to deduce from (7.32). That problem will be discussed in Section 7.5 after the joint m.g.f. of U_1 and V_1 is derived in Section 7.4.

We note that, when $r = 1$, the limiting random variable in (7.32) coincides with the estimator $-\tilde{c}$ derived from the normal equation

(7.35)
$$\int_0^1 Y(t)\,dY(t) = -\tilde{c}\int_0^1 Y^2(t)\,dt$$

for the O–U process $dY(t) = -cY(t)\,dt + dw(t)$. Thus the estimator of $T(\rho - 1) = -c$ in the discrete-time AR(1) model $\Delta\eta_j = -c\eta_{j-1}/T + \varepsilon_j$ derived from (7.3) can be converted asymptotically into that of $-c$ in the continuous-time O–U process $dY(t) = -cY(t)\,dt + dw(t)$. In fact, the weak convergence result (7.29) implies that, when $r = 1$ or $u_j = \varepsilon_j$, the estimator $-\hat{c}$ in

$$\sum_{j=2}^{T} \eta_{j-1} \Delta \eta_j = (\hat{\rho} - 1) \sum_{j=2}^{T} \eta_{j-1}^2 = -\frac{\hat{c}}{T} \sum_{j=2}^{T} \eta_{j-1}^2$$

is carried over to $-\tilde{c}$ in (7.35) as $T \to \infty$.

7.2.2 Model B

We deal with the model

(7.36) $\quad y_j = \beta_0 + \eta_j, \quad \eta_j = \rho \eta_{j-1} + u_j, \quad \eta_0 = \sqrt{T} \sigma_L Z,$

so that $T(\hat{\rho} - 1) = U_{2T}/V_{2T}$, where

(7.37) $\quad U_{2T} = \frac{1}{2T}(\hat{\eta}_T^2 - \hat{\eta}_1^2) - \frac{1}{2T} \sum_{j=2}^{T}(\hat{\eta}_j - \hat{\eta}_{j-1})^2,$

(7.38) $\quad V_{2T} = \frac{1}{T^2} \sum_{j=1}^{T} \hat{\eta}_j^2 - \frac{1}{T^2} \hat{\eta}_T^2,$

$$\hat{\eta}_j = y_j - \bar{y} = \eta_j - \bar{\eta} = \eta_j - \frac{1}{T} \sum_{k=1}^{T} \eta_k.$$

As in the case of Model A, $\hat{\eta}_T^2 / T^2$ converges in probability to 0. It can also be shown (Problem 2.4) that

(7.39) $\quad \frac{1}{T} \sum_{j=2}^{T}(\hat{\eta}_j - \hat{\eta}_{j-1})^2 \quad \to \quad \sigma_S^2 \quad \text{in probability},$

(7.40) $\quad \frac{\hat{\eta}_1^2}{T} - \left(\sigma_L Y(0) - \frac{1}{T} \sum_{j=1}^{T} Y_T\left(\frac{j}{T}\right)\right)^2 \to 0 \quad \text{in probability}.$

We also have

$$\frac{\hat{\eta}_T^2}{T} = \left(Y_T(1) - \frac{1}{T} \sum_{j=1}^{T} Y_T\left(\frac{j}{T}\right)\right)^2,$$

$$\frac{1}{T^2} \sum_{j=1}^{T} \hat{\eta}_j^2 = \frac{1}{T} \sum_{j=1}^{T} \left(Y_T\left(\frac{j}{T}\right) - \frac{1}{T} \sum_{k=1}^{T} Y_T\left(\frac{k}{T}\right)\right)^2.$$

CONVERGENCE IN DISTRIBUTION OF LSES

Thus we may put

$$U_{2T} = \frac{1}{2}\left[\left(Y_T(1) - \frac{1}{T}\sum_{j=1}^{T}Y_T\left(\frac{j}{T}\right)\right)^2 - \left(\sigma_L Y(0) - \frac{1}{T}\sum_{j=1}^{T}Y_T\left(\frac{j}{T}\right)\right)^2 - \sigma_S^2\right] + o_p(1),$$

$$V_{2T} = \frac{1}{T}\sum_{j=1}^{T}\left(Y_T\left(\frac{j}{T}\right) - \frac{1}{T}\sum_{k=1}^{T}Y_T\left(\frac{k}{T}\right)\right)^2 + o_p(1).$$

Then it follows from (7.22) and the continuous mapping theorem that $\mathcal{L}(U_{2T}, V_{2T}) \to \mathcal{L}(\sigma_L^2 U_2, \sigma_L^2 V_2)$, where

$$U_2 = \frac{1}{2}\left[\left(Y(1) - \int_0^1 Y(t)\,dt\right)^2 - \left(Y(0) - \int_0^1 Y(t)\,dt\right)^2 - r\right]$$

$$= \int_0^1 Y(t)\,dY(t) - \int_0^1 dY(t)\int_0^1 Y(t)\,dt + \frac{1}{2}(1-r),$$

$$V_2 = \int_0^1 \left(Y(t) - \int_0^1 Y(s)\,ds\right)^2 dt$$

$$= \int_0^1 Y^2(t)\,dt - \left(\int_0^1 Y(t)\,dt\right)^2.$$

Therefore we obtain

Theorem 7.2. *For the LSE $\hat{\rho}$ for Model B in (7.36), it holds that, as $T \to \infty$ under $\rho = 1 - (c/T)$,*

(7.41) $\mathcal{L}(T(\hat{\rho} - 1)) \to \mathcal{L}(U_2/V_2)$

$$= \mathcal{L}\left(\frac{\int_0^1 Y(t)\,dY(t) - \int_0^1 dY(t)\int_0^1 Y(t)\,dt + \frac{1}{2}(1-r)}{\int_0^1 Y^2(t)\,dt - \left(\int_0^1 Y(t)\,dt\right)^2}\right).$$

As in Model A, $\mathcal{L}(U_2/(rV_2)) \to \mathcal{L}(-1/(2V_2))$ as $r \to \infty$. It is also deduced from (7.41) that $\mathcal{L}(U_2/V_2)$ does not depend on $Y(0)$ if $c = 0$. Suppose that $c \neq 0$ and $Y(0) = \gamma$ is a constant. Then U_2/V_2 converges in probability to $-c$ and $\gamma\{(U_2/V_2) + c\}$ converges to normal distribution as $|\gamma| \to \infty$ (Problem 2.5).

When $r = 1$, the limiting random variable in (7.41) may be obtained as follows: Let us construct a stochastic process

$$d\tilde{Y}(t) = (a - c\tilde{Y}(t))dt + dw(t), \quad \tilde{Y}(0) = Z \sim N(\gamma, \delta^2),$$

where the true value of a is 0 so that $\mathcal{L}(\tilde{Y}) = \mathcal{L}(Y)$. Then we obtain the normal equations for the estimators of a and $-c$ as

(7.42) $$\int_0^1 d\tilde{Y}(t) = \tilde{a} \int_0^1 dt + (-\tilde{c}) \int_0^1 \tilde{Y}(t) dt,$$

(7.43) $$\int_0^1 \tilde{Y}(t) d\tilde{Y}(t) = \tilde{a} \int_0^1 \tilde{Y}(t) dt + (-\tilde{c}) \int_0^1 \tilde{Y}^2(t) dt.$$

Solving for $-\tilde{c}$ we have

$$\left[\int_0^1 \tilde{Y}^2(t) dt - \left(\int_0^1 \tilde{Y}(t) dt\right)^2\right](-\tilde{c}) = \int_0^1 \tilde{Y}(t) d\tilde{Y}(t) - \int_0^1 d\tilde{Y}(t) \int_0^1 \tilde{Y}(t) dt.$$

(7.44)

Thus $-\tilde{c}$ coincides with the limiting random variable in (7.41) when $r = 1$ and $Y(t) = \tilde{Y}(t)$.

Finally we can point out that, when $r = 1$ or $u_j = \varepsilon_j$, the estimator $-\hat{c}$ in

(7.45) $$\sum_{j=2}^T \hat{\eta}_{j-1} \Delta \hat{\eta}_j = (\hat{\rho} - 1) \sum_{j=2}^T \hat{\eta}_{j-1}^2 = -\frac{\hat{c}}{T} \sum_{j=2}^T \hat{\eta}_{j-1}^2$$

is carried over to $-\tilde{c}$ in (7.44) asymptotically, as in Model A. It is seen that $-\hat{c}$ is asymptotically a solution to

(7.46) $$\sum_{j=2}^T \Delta \eta_j = \hat{a} - \frac{\hat{c}}{T} \sum_{j=2}^T \eta_{j-1},$$

(7.47) $$\sum_{j=2}^T \eta_{j-1} \Delta \eta_j = \frac{\hat{a}}{T} \sum_{j=2}^T \eta_{j-1} - \frac{\hat{c}}{T} \sum_{j=2}^T \eta_{j-1}^2,$$

and these equations have continuous analogs (7.42) and (7.43). We can also show (Problem 2.6) that Fuller's (1976) estimator of $T(\rho - 1) = -c$ has the same limiting distribution as $T(\hat{\rho} - 1) = -\hat{c}$.

7.2.3 Model C

The model dealt with here is

(7.48) $$y_j = j\beta_1 + \eta_j, \quad \eta_j = \rho \eta_{j-1} + u_j, \quad \eta_0 = \sqrt{T} \sigma_L Z,$$

CONVERGENCE IN DISTRIBUTION OF LSES

so that $T(\hat{\rho} - 1) = U_{3T}/V_{3T}$, where U_{3T} and V_{3T} are defined by the right sides of (7.37) and (7.38), respectively, with $\hat{\eta}_j$ replaced by

$$\hat{\eta}_j = y_j - j \sum_{k=1}^{T} k y_k \Big/ \sum_{k=1}^{T} k^2 = \eta_j - j \sum_{k=1}^{T} k \eta_k \Big/ \sum_{k=1}^{T} k^2.$$

It can be shown (Problem 2.7) that

(7.49) $$\frac{1}{T} \sum_{j=2}^{T} (\hat{\eta}_j - \hat{\eta}_{j-1})^2 \to \sigma_S^2 \quad \text{in probability},$$

(7.50) $$\frac{\hat{\eta}_1^2}{T} - \sigma_L^2 Y^2(0) \to 0 \quad \text{in probability}.$$

Noting that $\sum_{j=1}^{T} j^2 = T^3/3 + O(T^2)$, we also have

$$\frac{\hat{\eta}_T^2}{T} = \left(Y_T(1) - \frac{3}{T} \sum_{j=1}^{T} \left(\frac{j}{T}\right) Y_T \left(\frac{j}{T}\right) \right)^2 + o_p(1),$$

$$\frac{1}{T^2} \sum_{j=1}^{T} \hat{\eta}_j^2 = \frac{1}{T} \sum_{j=1}^{T} \left(Y_T\left(\frac{j}{T}\right) - \frac{3j}{T^2} \sum_{k=1}^{T} \left(\frac{k}{T}\right) Y_T\left(\frac{k}{T}\right) \right)^2 + o_p(1).$$

Thus we may put

$$U_{3T} = \frac{1}{2} \left[\left(Y_T(1) - \frac{3}{T} \sum_{j=1}^{T} \left(\frac{j}{T}\right) Y_T\left(\frac{j}{T}\right) \right)^2 - \sigma_L^2 Y^2(0) - \sigma_S^2 \right] + o_p(1),$$

$$V_{3T} = \frac{1}{T} \sum_{j=1}^{T} \left(Y_T\left(\frac{j}{T}\right) - \frac{3j}{T^2} \sum_{k=1}^{T} \left(\frac{k}{T}\right) Y_T\left(\frac{k}{T}\right) \right)^2 + o_p(1).$$

Then it can be shown that $\mathcal{L}(U_{3T}, V_{3T}) \to \mathcal{L}(\sigma_L^2 U_3, \sigma_L^2 V_3)$, where

(7.51) $$U_3 = \frac{1}{2} \left[\left(Y(1) - 3 \int_0^1 t Y(t)\, dt \right)^2 - Y^2(0) - r \right]$$

$$= \int_0^1 Y(t)\, dY(t) - \left(3Y(1) - \frac{9}{2} \int_0^1 t Y(t)\, dt \right) \int_0^1 t Y(t)\, dt + \frac{1}{2}(1 - r),$$

(7.52) $\quad V_3 = \int_0^1 \left(Y(t) - 3t \int_0^1 sY(s)\,ds\right)^2 dt$

$\qquad = \int_0^1 Y^2(t)\,dt - 3\left(\int_0^1 tY(t)\,dt\right)^2.$

Therefore we obtain

Theorem 7.3. *For the* LSE *$\hat{\rho}$ for Model C in (7.48), it holds that $\mathcal{L}(T(\hat{\rho}-1)) \to \mathcal{L}(U_3/V_3)$ as $T \to \infty$ under $\rho = 1 - (c/T)$, where U_3 and V_3 are defined in (7.51) and (7.52), respectively.*

As in Models A and B, we have that $\mathcal{L}(U_3/(rV_3)) \to \mathcal{L}(-1/(2V_3))$ as $r \to \infty$. Unlike in Model B, $\mathcal{L}(U_3/V_3)$ depends on $Y(0)$ even when $c = 0$. When $Y(0) = \gamma$ is a constant, U_3/V_3 converges in probability to a constant as $|\gamma| \to \infty$, whose value is $-3/2$ when $c = 0$ (Problem 2.8). Model C is quite different from the other models; it will be examined later.

It might be thought that, when $r = 1$, U_3/V_3 can be obtained as an estimator of $-c$ in the stochastic process

$$d\tilde{Y}(t) = \bigl(at - c\tilde{Y}(t)\bigr)dt + dw(t), \quad \tilde{Y}(0) = Z \sim N(\gamma, \delta^2),$$

where the true value of a is 0 so that $\mathcal{L}(\tilde{Y}) = \mathcal{L}(Y)$. This is, however, not the case. In fact, the normal equations are

$$\int_0^1 t\,d\tilde{Y}(t) = \tilde{a}\int_0^1 t^2\,dt + (-\tilde{c})\int_0^1 t\tilde{Y}(t)\,dt,$$

$$\int_0^1 \tilde{Y}(t)\,d\tilde{Y}(t) = \tilde{a}\int_0^1 t\tilde{Y}(t)\,dt + (-\tilde{c})\int_0^1 \tilde{Y}^2(t)\,dt$$

so that we have

$$\left[\int_0^1 \tilde{Y}^2(t)\,dt - 3\left(\int_0^1 t\tilde{Y}(t)\,dt\right)^2\right](-\tilde{c})$$

$$= \int_0^1 \tilde{Y}(t)\,d\tilde{Y}(t) - 3\int_0^1 t\tilde{Y}(t)\,dt \int_0^1 t\,d\tilde{Y}(t).$$

Thus $-\tilde{c}$ produces a different distribution from U_3/V_3 even if $r = 1$. Then it holds that, if $Y(0) = \gamma$ is a constant, $-\tilde{c}$ converges in probability to $-c$ and $\gamma(-\tilde{c} + c)$ tends to normality as $|\gamma| \to \infty$. It can also be shown (Problem 2.9) that the ordinary LSE of $T(\rho - 1) = -c$ applied to the model (7.9) reproduced here as

(7.53) $\qquad y_j = \rho y_{j-1} + j\gamma + \varepsilon_j,$

where the true value of γ is 0, has the same limiting distribution as $-\tilde{c}$.

7.2.4 Model D

Finally we deal with the model

(7.54) $\quad y_j = \beta_0 + j\beta_1 + \eta_j, \quad \eta_j = \rho\eta_{j-1} + u_j, \quad \eta_0 = \sqrt{T}\sigma_L Z,$

so that $T(\hat{\rho} - 1) = U_{4T}/V_{4T}$, where U_{4T} and V_{4T} are the same as the right sides of (7.37) and (7.38), respectively, except that

$$\hat{\eta}_j = y_j - \hat{\beta}_0 - j\hat{\beta}_1$$

$$= \eta_j - \frac{\left(\sum_{k=1}^{T} k^2 - j\sum_{k=1}^{T} k\right)\sum_{k=1}^{T}\eta_k + \left(jT - \sum_{k=1}^{T} k\right)\sum_{k=1}^{T} k\eta_k}{T\sum_{k=1}^{T} k^2 - \left(\sum_{k=1}^{T} k\right)^2}.$$

It can be shown (Problem 2.10) that

(7.55) $\quad \dfrac{1}{T}\sum_{j=2}^{T}(\hat{\eta}_j - \hat{\eta}_{j-1})^2 \to \sigma_S^2 \quad \text{in probability},$

(7.56) $\quad \dfrac{\hat{\eta}_1^2}{T} - \left(\sigma_L Y(0) - \dfrac{4}{T}\sum_{j=1}^{T}\dfrac{\eta_j}{\sqrt{T}} + \dfrac{6}{T}\sum_{j=1}^{T}\dfrac{j}{T}\dfrac{\eta_j}{\sqrt{T}}\right)^2 \to 0 \quad \text{in probability}.$

We also have

$$\frac{\hat{\eta}_T^2}{T} = \left(Y_T(1) + \frac{2}{T}\sum_{j=1}^{T} Y_T\left(\frac{j}{T}\right) - \frac{6}{T}\sum_{j=1}^{T}\frac{j}{T}Y_T\left(\frac{j}{T}\right)\right)^2 + o_p(1),$$

$$\frac{1}{T^2}\sum_{j=1}^{T}\hat{\eta}_j^2 = \frac{1}{T}\sum_{j=1}^{T}\left(Y_T\left(\frac{j}{T}\right) - \left(4 - \frac{6j}{T}\right)\frac{1}{T}\sum_{j=1}^{T} Y_T\left(\frac{j}{T}\right)\right.$$

$$\left. - \left(\frac{12j}{T} - 6\right)\frac{1}{T}\sum_{j=1}^{T}\frac{j}{T}Y_T\left(\frac{j}{T}\right)\right)^2 + o_p(1).$$

Thus we are led to consider

$$U_{4T} = \frac{1}{2}\left[\left(Y_T(1) + \frac{2}{T}\sum_{j=1}^{T} Y_T\left(\frac{j}{T}\right) - \frac{6}{T}\sum_{j=1}^{T}\frac{j}{T}Y_T\left(\frac{j}{T}\right)\right)^2\right.$$

$$-\left(\sigma_L Y(0) - \frac{4}{T}\sum_{j=1}^{T} Y_T\left(\frac{j}{T}\right) + \frac{6}{T}\sum_{j=1}^{T}\frac{j}{T}Y_T\left(\frac{j}{T}\right)\right)^2 - \sigma_S^2 \Bigg] + o_p(1),$$

$$V_{4T} = \frac{1}{T}\sum_{j=1}^{T}\left(Y_T\left(\frac{j}{T}\right) + \left(\frac{6j}{T} - 4\right)\frac{1}{T}\sum_{j=1}^{T}Y_T\left(\frac{j}{T}\right)\right.$$

$$\left. - \left(\frac{12j}{T} - 6\right)\frac{1}{T}\sum_{j=1}^{T}\frac{j}{T}Y_T\left(\frac{j}{T}\right)\right)^2 + o_p(1).$$

Then it can be shown that $\mathcal{L}(U_{4T}, V_{4T}) \to \mathcal{L}(\sigma_L^2 U_4, \sigma_L^2 V_4)$, where

(7.57) $\quad U_4 = \dfrac{1}{2}\Bigg[\left(Y(1) + 2\int_0^1 Y(t)\,dt - 6\int_0^1 tY(t)\,dt\right)^2$

$$- \left(Y(0) - 4\int_0^1 Y(t)\,dt + 6\int_0^1 tY(t)\,dt\right)^2 - r\Bigg]$$

$$= \int_0^1 Y(t)\,dY(t) - 6\left(\int_0^1 Y(t)\,dt\right)^2 + 2(Y(1) + 2Y(0))\int_0^1 Y(t)\,dt$$

$$-6(Y(1) + Y(0))\int_0^1 tY(t)\,dt + 12\int_0^1 Y(t)\,dt \int_0^1 tY(t)\,dt$$

$$+\frac{1}{2}(1 - r),$$

(7.58) $\quad V_4 = \displaystyle\int_0^1\left(Y(t) + (6t - 4)\int_0^1 Y(s)\,ds - (12t - 6)\int_0^1 sY(s)\,ds\right)^2 dt$

$$= \int_0^1 Y^2(t)\,dt - 4\left(\int_0^1 Y(t)\,dt\right)^2 - 12\left(\int_0^1 tY(t)\,dt\right)^2$$

$$+ 12\int_0^1 Y(t)\,dt \int_0^1 tY(t)\,dt.$$

Therefore we obtain

Theorem 7.4. *For the LSE $\hat{\rho}$ for Model D in (7.54), it holds that $\mathcal{L}(T(\hat{\rho} - 1)) \to \mathcal{L}(U_4/V_4)$ as $T \to \infty$ under $\rho = 1 - (c/T)$, where U_4 and V_4 are defined in (7.57) and (7.58), respectively.*

As in the previous models, we have that $\mathcal{L}(U_4/(rV_4)) \to \mathcal{L}(-1/(2V_4))$ as $r \to \infty$. We can also deduce that $\mathcal{L}(U_4/V_4)$ does not depend on $Y(0)$ when $c = 0$. If $c \neq 0$ and

$Y(0) = \gamma$ is a constant, U_4/V_4 converges in probability to $-c$ and $\gamma\{(U_4/V_4) + c\}$ tends to normality as $|\gamma| \to \infty$.

When $r = 1$, we can show (Problem 2.11) that $\mathcal{L}(U_4/V_4)$ coincides with the LSE $-\tilde{c}$ in the stochastic process

(7.59) $\quad d\tilde{Y}(t) = \big(a + bt - c\tilde{Y}(t)\big)\,dt + dw(t), \quad \tilde{Y}(0) = Z \sim N(\gamma, \delta^2),$

where the true values of a and b are both zero. It can also be shown that Fuller's estimator of $T(\rho - 1) = -c$ obtained from the discrete-time model

(7.60) $\quad\quad\quad\quad \Delta y_j = \beta + j\gamma - \dfrac{c}{T} y_{j-1} + u_j$

has the same limiting distribution as $T(\hat{\rho} - 1)$, where the true value of γ is 0.

Fuller (1976) tabulated, by simulations, percent points for the finite sample and limiting distributions of $T(\hat{\rho} - 1)$ for Models A, B, and D, assuming $\rho = 1$ and $\{u_j\} = \{\varepsilon_j\}$ to be NID(0, 1). In Section 7.5 we shall tabulate percent points of the limiting distributions under general conditions without resorting to simulations. For this purpose, we need to derive the c.f.'s associated with U_l and V_l ($l = 1, 2, 3, 4$), which we discuss in Section 7.4.

PROBLEMS

2.1 Prove (7.27) and (7.28).
2.2 Establish the weak convergence result in (7.29).
2.3 Show that (7.34) holds.
2.4 Prove (7.39) and (7.40).
2.5 Prove that $\gamma\{(U_2/V_2) + c\}$ tends to $N(0, 1/\tau^2)$ as $|\gamma| \to \infty$, where U_2/V_2 is given in (7.41) with $Y(0) = \gamma$ and

$$\tau^2 = \frac{c \sinh c - 2 \cosh c + 2}{c^2 e^c}.$$

2.6 Show that Fuller's estimator of $T(\rho - 1)$ applied to $y_j = \rho y_{j-1} + \alpha + u_j$ has the limiting distribution as given in (7.41), where the true value of α is 0.
2.7 Prove (7.49) and (7.50).
2.8 Show that, as $|\gamma| \to \infty$, U_3/V_3 converges in probability to $-3/2$ when $c = 0$, where U_3 and V_3 are defined in (7.51) and (7.52), respectively, with $Y(0) = \gamma$.
2.9 Derive the limiting distribution of the ordinary LSE of $T(\rho - 1)$ in the model (7.53).
2.10 Prove (7.55) and (7.56).
2.11 Show that an estimator of c obtained from (7.59) has the same distribution as that of U_4/V_4 with $r = 1$, where U_4 and V_4 are defined in (7.57) and (7.58), respectively.

7.3 THE NEGATIVE UNIT ROOT CASE

Although we are not concerned with the case where ρ is close to -1, the weak convergence of LSEs for this case may be of independent interest. We shall show, under a certain distributional condition, that the results can be easily derived from the near unit root case discussed in the previous section.

Suppose that our model is given by (7.1), where the true value of ρ is assumed to be $\rho = -1 + (c/T)$. Define the LSE $\hat{\rho}$ of ρ in the same way as before. Then we consider (Problem 3.1)

$$(7.61) \quad -T(\hat{\rho} + 1) = -T\left[\left\{\sum_{j=2}^{T} \hat{\eta}_{j-1}\hat{\eta}_j \bigg/ \sum_{j=2}^{T} \hat{\eta}_{j-1}^2\right\} + 1\right]$$

$$= W_T/X_T,$$

where $\hat{\eta}_j$ is defined in (7.10), while

$$(7.62) \quad W_T = \frac{1}{2T}\left(\hat{\eta}_T^2 - \hat{\eta}_1^2\right) - \frac{1}{2T}\sum_{j=2}^{T}\left(\hat{\eta}_j + \hat{\eta}_{j-1}\right)^2,$$

$$(7.63) \quad X_T = \frac{1}{T^2}\sum_{j=1}^{T}\hat{\eta}_j^2 - \frac{1}{T^2}\hat{\eta}_T^2.$$

We first consider Model A in (7.25) with $\rho = -1 + (c/T)$. Noting that $\hat{\eta}_j = \eta_j$, it is easy to see that

$$\frac{1}{T}\sum_{j=2}^{T}(\eta_j + \eta_{j-1})^2 \to \sigma_S^2 \quad \text{in probability},$$

$$\frac{\eta_1^2}{T} - \sigma_L^2 Z^2 \to 0 \quad \text{in probability}.$$

Moreover we have

$$(7.64) \quad \eta_j = \rho\eta_{j-1} + u_j$$

$$= \rho^j \eta_0 + \sum_{i=1}^{j} \rho^{j-i} u_i$$

$$= (-1)^j\left[\left(1 - \frac{c}{T}\right)^j \eta_0 + \sum_{i=1}^{j}\left(1 - \frac{c}{T}\right)^{j-i}\{(-1)^i u_i\}\right].$$

THE NEGATIVE UNIT ROOT CASE

Putting $v_i = (-1)^i u_i$, let us define $\xi_j = (-1)^j \eta_j$, where

(7.65)
$$\xi_j = \left(1 - \frac{c}{T}\right)^j \xi_0 + \sum_{i=1}^{j} \left(1 - \frac{c}{T}\right)^{j-i} v_i$$
$$= \left(1 - \frac{c}{T}\right) \xi_{j-1} + v_j.$$

Then it follows that, if $\{v_j\}$ has the same distribution as $\{u_j\}$, $\mathcal{L}(-T(\hat{\rho} + 1)) \to \mathcal{L}(U_1/V_1)$, where $\mathcal{L}(U_1/V_1)$ is the limiting distribution of $T(\hat{\rho} - 1)$ for Model A under $\rho = 1 - (c/T)$. Namely the limiting distribution has the mirror-image property

$$\lim_{T \to \infty} P\left(-T(\hat{\rho} + 1) \leq x \mid \rho = -1 + \frac{c}{T}\right) = \lim_{T \to \infty} P\left(T(\hat{\rho} - 1) \leq x \mid \rho = 1 - \frac{c}{T}\right).$$

This fact was first pointed out by Fuller (1976).

We next consider Model B in (7.36) under $\rho = -1 + (c/T)$. Then we have (Problem 3.2)

(7.66)
$$\bar{\eta} = \frac{1}{T} \sum_{j=1}^{T} \eta_j = \frac{1}{T} \sum_{j=1}^{T} (-1)^j \xi_j$$
$$= O_p\left(\frac{1}{\sqrt{T}}\right),$$

where ξ_j is defined in (7.65). Using (7.66), we can show (Problem 3.3) that, if $\mathcal{L}(\{(-1)^j u_j\}) = \mathcal{L}(\{u_j\})$,

(7.67)
$$\mathcal{L}\left(-T(\hat{\rho} + 1) \mid \rho = -1 + \frac{c}{T}\right) \to \mathcal{L}\left(\frac{U_1}{V_1}\right).$$

Note that the limit is not U_2/V_2, but U_1/V_1. This is because of the property of $\bar{\eta}$ described in (7.66), which was $O_p(\sqrt{T})$ under $\rho = 1 - (c/T)$.

For Model C we have (Problem 3.4)

$$\hat{\eta}_j = \eta_j - j \sum_{k=1}^{T} k \eta_k \bigg/ \sum_{k=1}^{T} k^2,$$

where

(7.68)
$$\sum_{k=1}^{T} k \eta_k = O_p(T\sqrt{T}).$$

Note that $\sum_{k=1}^{T} k \eta_k$ was of stochastic order $T^2 \sqrt{T}$ under $\rho = 1 - (c/T)$. Because of (7.68), we obtain the same weak convergence result (Problem 3.5) as in (7.67).

Model D is a mixture of Model B and Model C. Using the properties (7.66) and (7.68), we conclude (Problem 3.6) that $\mathcal{L}(-T(\hat{\rho}+1) \mid \rho = -1+(c/T)) \to \mathcal{L}(U_1/V_1)$ if $\mathcal{L}(\{(-1)^j u_j\}) = \mathcal{L}(\{u_j\})$.

The above arguments are summarized in the following theorem.

Theorem 7.5. *For Models A through D, assume that the distribution of $\{(-1)^j u_j\}$ is the same as that of $\{u_j\}$, where u_j is defined in (7.5). Then the LSEs $\hat{\rho}$ for these models have the same asymptotic distribution as $T \to \infty$ under $\rho = -1 + (c/T)$, and it holds that*

$$\mathcal{L}\left(-T(\hat{\rho}+1) \,\Big|\, \rho = -1 + \frac{c}{T}\right) \to \mathcal{L}\left(\frac{U_1}{V_1}\right),$$

where $\mathcal{L}(U_1/V_1)$ is the limiting distribution of $T(\hat{\rho} - 1)$ for Model A under $\rho = 1 - (c/T)$ described in Theorem 7.1.

PROBLEMS

3.1 Derive the expressions in (7.61).
3.2 Prove (7.66) under $\rho = -1 + (c/T)$.
3.3 Establish (7.67) for Model B under $\rho = -1 + (c/T)$.
3.4 Prove (7.68) under $\rho = -1 + (c/T)$.
3.5 Prove (7.67) for Model C under $\rho = -1 + (c/T)$.
3.6 Prove (7.67) for Model D under $\rho = -1 + (c/T)$.

7.4 THE c.f.s FOR THE LIMITING DISTRIBUTIONS OF LSES

In Section 7.2 we established weak convergence results on $T(\hat{\rho} - 1)$ under $\rho = 1 - (c/T)$ for Models A through D, where the limiting distributions were expressed as $\mathcal{L}(U_l/V_l)$ ($l = 1, 2, 3, 4$). For computational purposes we need to derive the associated c.f.s. Let $\phi_l(\theta)$ be the c.f. of $xV_l - U_l$. Then we have

(7.69)
$$F_l(x) = \lim_{T \to \infty} P(T(\hat{\rho} - 1) \leq x)$$
$$= P(xV_l - U_l \geq 0)$$
$$= \frac{1}{2} + \frac{1}{\pi} \int_0^\infty \frac{1}{\theta} \mathrm{Im}[\phi_l(\theta)] \, d\theta.$$

The purpose of this section is to derive $\phi_l(\theta)$ ($l = 1, 2, 3, 4$). Here we first obtain the joint m.g.f. $m_l(\theta_1, \theta_2) = E[\exp\{\theta_1 U_l + \theta_2 V_l\}]$ of U_l and V_l just because we require it for later discussions. Note that the c.f. $\phi_l(\theta)$ is recovered from $m_l(\theta_1, \theta_2)$ by putting $\phi_l(\theta) = m_l(-i\theta, i\theta x)$.

Let us define

(7.70) $$h_l(Y) = \exp\{\theta_1 U_l + \theta_2 V_l\}$$

so that $E(h_l(Y)) = m_l(\theta_1, \theta_2)$, where $Y = \{Y(t)\}$ is our basic process defined by

(7.71) $$dY(t) = -cY(t)\, dt + dw(t), \quad Y(0) \sim N(\gamma, \delta^2).$$

We also construct an auxiliary process

(7.72) $$dZ(t) = -\beta Z(t)\, dt + dw(t), \quad Z(0) = Y(0).$$

Then Girsanov's theorem ensures that the two measures μ_Y and μ_Z induced by $Y = \{Y(t)\}$ and $Z = \{Z(t)\}$, respectively, are equivalent and

(7.73) $$E(h_l(Y)) = E\left(h_l(Z)\frac{d\mu_Y}{d\mu_Z}(Z)\right),$$

where

$$\frac{d\mu_Y}{d\mu_Z}(Z) = \exp\left\{(\beta - c)\int_0^1 Z(t)\, dZ(t) + \frac{\beta^2 - c^2}{2}\int_0^1 Z^2(t)\, dt\right\}$$

$$= \exp\left\{\frac{\beta - c}{2}(Z^2(1) - Z^2(0) - 1) + \frac{\beta^2 - c^2}{2}\int_0^1 Z^2(t)\, dt\right\}.$$

In the following subsections, we compute (7.73) assuming either $Y(0) = \gamma$ or $Y(0) \sim N(0, 1/(2c))$ with $c > 0$, where $Y(0)$ is independent of increments of $\{w(t)\}$. This is because the expressions for the m.g.f.s become quite complicated if we just assume $Y(0) \sim N(\gamma, \delta^2)$. The former is referred to as the *fixed initial value case*, while the latter the *stationary case* because the process $\{Y(t)\}$ in (7.71) becomes stationary if $Y(0) \sim N(0, 1/(2c))$.

For later purposes we keep in mind that

(7.74) $$Z(t) = e^{-\beta t}Z(0) + e^{-\beta t}\int_0^t e^{\beta s}\, dw(s),$$

where $Z(0) = Y(0) = \gamma$ for the fixed initial value case, while $Z(0) \sim N(0, 1/(2c))$ for the stationary case. Note that $\{Z(t)\}$ is not stationary even for the stationary case unless $\beta = c$.

7.4.1 The Fixed Initial Value Case

Here we discuss the fixed initial value case where $Y(0) = Z(0) = \gamma$. Let $m_{l1}(\theta_1, \theta_2)$ be the joint m.g.f. of U_l and V_l ($l = 1, 2, 3, 4$) for this case, where U_l and V_l are given in Section 7.2.

We first obtain a general expression for $m_{l1}(\theta_1, \theta_2)$. Using (7.73) and (7.74) with $Y(0) = Z(0) = \gamma$, we can show (Problem 4.1) that

(7.75) $\quad m_{l1}(\theta_1, \theta_2) = E[\exp\{\theta_1 U_l + \theta_2 V_l\}]$

$$= \exp\left\{-\frac{a}{2}(\gamma^2 + 1) + \frac{\theta_1}{2}(1 - r)\right\}$$

$$\times E\left[\exp\left\{\frac{1}{2}W_l' A_l W_l + \gamma h_l' W_l\right\}\right],$$

where $a = \beta + \theta_1 - c$, $\beta = \sqrt{c^2 - 2\theta_2}$ and

$$W_1 = Z(1), \qquad h_1 = 0, \qquad A_1 = a,$$

$$W_2 = \begin{pmatrix} Z(1) \\ \int_0^1 Z(t)dt \end{pmatrix}, \quad h_2 = \begin{pmatrix} 0 \\ \theta_1 \end{pmatrix}, \quad A_2 = \begin{pmatrix} a & -\theta_1 \\ -\theta_1 & -2\theta_2 \end{pmatrix},$$

$$W_3 = \begin{pmatrix} Z(1) \\ \int_0^1 tZ(t)dt \end{pmatrix}, \quad h_3 = \begin{pmatrix} 0 \\ 0 \end{pmatrix}, \quad A_3 = \begin{pmatrix} a & -3\theta_1 \\ -3\theta_1 & 9\theta_1 - 6\theta_2 \end{pmatrix},$$

$$W_4 = \begin{pmatrix} Z(1) \\ \int_0^1 Z(t)dt \\ \int_0^1 tZ(t)dt \end{pmatrix}, \quad h_4 = \begin{pmatrix} 0 \\ 4\theta_1 \\ -6\theta_1 \end{pmatrix},$$

$$A_4 = \begin{pmatrix} a & 2\theta_1 & -6\theta_1 \\ 2\theta_1 & -12\theta_1 - 8\theta_2 & 12\theta_1 + 12\theta_2 \\ -6\theta_1 & 12\theta_1 + 12\theta_2 & -24\theta_2 \end{pmatrix}.$$

Here each W_l follows $N(\gamma \kappa_l, \Omega_l)$, where

$$\kappa_1 = e^{-\beta}, \quad \Omega_1 = \frac{1 - e^{-2\beta}}{2\beta},$$

$$\kappa_2 = \frac{1}{\beta}\begin{pmatrix} \beta e^{-\beta} \\ 1 - e^{-\beta} \end{pmatrix}, \quad \Omega_2 = \frac{1}{2\beta^3}\begin{pmatrix} \beta^2(1 - e^{-2\beta}) & \beta(1 - e^{-\beta})^2 \\ \beta(1 - e^{-\beta})^2 & 2\beta - (1 - e^{-\beta})(3 - e^{-\beta}) \end{pmatrix},$$

$$\kappa_3 = \frac{1}{\beta^2}\begin{pmatrix} \beta^2 e^{-\beta} \\ 1 - (\beta + 1)e^{-\beta} \end{pmatrix},$$

$$\Omega_3 = \frac{1}{6\beta^5} \begin{pmatrix} 3\beta^4(1-e^{-2\beta}) & 3\beta^2(\beta-1+(\beta+1)e^{-2\beta}) \\ 3\beta^2(\beta-1+(\beta+1)e^{-2\beta}) & 2\beta^3-3\beta^2+3-3(\beta+1)^2e^{-2\beta} \end{pmatrix},$$

$$\kappa_4 = \frac{1}{\beta^2} \begin{pmatrix} \beta^2 e^{-\beta} \\ \beta(1-e^{-\beta}) \\ 1-(\beta+1)e^{-\beta} \end{pmatrix},$$

$$\Omega_4 = \begin{pmatrix} \Omega_2 & & \frac{\beta-1+(\beta+1)e^{-2\beta}}{2\beta^3} \\ & & \frac{\beta^2-(\beta+1)(1-e^{-\beta})^2}{2\beta^4} \\ \frac{\beta-1+(\beta+1)e^{-2\beta}}{2\beta^3} & \frac{\beta^2-(\beta+1)(1-e^{-\beta})^2}{2\beta^4} & \frac{2\beta^3-3\beta^2+3-3(\beta+1)^2e^{-2\beta}}{6\beta^5} \end{pmatrix}.$$

Thus (7.75) yields (Problem 4.2)

(7.76) $m_{l1}(\theta_1, \theta_2)$

$$= \exp\left[\frac{1}{2}\left\{-a(\gamma^2+1)+\theta_1(1-r)+\gamma^2\left(\Omega_l^{-1}\kappa_l+h_l\right)'\left(\Omega_l^{-1}-A_l\right)^{-1}\right.\right.$$
$$\left.\left.\times\left(\Omega_l^{-1}\kappa_l+h_l\right)-\gamma^2\kappa_l'\Omega_l^{-1}\kappa_l\right\}\right]|B_l-\Omega_l A_l|^{-1/2}$$

$$= \exp\left[\frac{1}{2}\left\{c-r\theta_1+\gamma^2\left(-a+2h_l'\kappa_l+h_l'\Omega_l h_l+(\kappa_l+\Omega_l h_l)'A_l\right.\right.\right.$$
$$\left.\left.\left.\times(B_l-\Omega_l A_l)^{-1}(\kappa_l+\Omega_l h_l)\right)\right\}\right]\left[e^{\beta}|B_l-\Omega_l A_l|\right]^{-1/2},$$

where B_l is the identity matrix with the same dimension as Ω_l. This last expression attempts to reduce the computational burden of matrix inversion, where it is noticed that m_{l1} depends, not on γ, but on $|\gamma|$.

The computation of the last expression in (7.76) can be done easily by any computerized algebra. Putting $\mu = \sqrt{2\theta_2-c^2}$, we obtain (Problem 4.3)

$$m_{11}(\theta_1,\theta_2) = \exp\left[\frac{c-r\theta_1}{2}+\frac{\gamma^2(\theta_1^2-2c\theta_1+2\theta_2)}{2H_1(\theta_1,\theta_2)}\frac{\sin\mu}{\mu}\right][H_1(\theta_1,\theta_2)]^{-1/2},$$

(7.77)

$m_{21}(\theta_1,\theta_2)$

$$= \exp\left[\frac{c-r\theta_1}{2}+\frac{c^2\gamma^2(\theta_1^2-2c\theta_1+2\theta_2)}{2H_2(\theta_1,\theta_2)}\left(-\frac{\sin\mu}{\mu^3}-\frac{2}{\mu^4}(\cos\mu-1)\right)\right]$$
$$\times [H_2(\theta_1,\theta_2)]^{-1/2},$$

(7.78)

$m_{31}(\theta_1, \theta_2)$

$$= \exp\left[\frac{c - r\theta_1}{2} + \frac{\gamma^2}{2H_3(\theta_1, \theta_2)}\right]$$

$$\times \left\{\left(\frac{2c^3\theta_1 - (c^2 + 3c + 3)\left(\theta_1^2 - 3\theta_1 + 2\theta_2\right) + 3c^2\theta_1}{\mu^2}\right.\right.$$

$$\left. - \frac{3\left(2c^3\theta_1 - c^2\left(\theta_1^2 - 3\theta_1 + 2\theta_2\right) - 3c\theta_1^2 - 3\theta_1^2\right)}{\mu^4}\right)\frac{\sin\mu}{\mu}$$

$$+ \frac{3\left(2c^3\theta_1 - c^2\left(\theta_1^2 - 5\theta_1 + 2\theta_2\right) - (c+1)\left(3\theta_1^2 - 6\theta_1 + 4\theta_2\right)\right)}{\mu^4}\cos\mu$$

$$\left.- \frac{6}{\mu^4}\left(c^2\theta_1 + (c+1)(3\theta_1 - 2\theta_2)\right)\right\}\left[H_3(\theta_1, \theta_2)\right]^{-1/2},$$

(7.79)

$$m_{41}(\theta_1, \theta_2) = \exp\left[\frac{c - r\theta_1}{2} + \frac{c^4\gamma^2\left(\theta_1^2 - 2c\theta_1 + 2\theta_2\right)}{2H_4(\theta_1, \theta_2)}\right]\left\{\left(\frac{1}{\mu^4} - \frac{24}{\mu^6}\right)\frac{\sin\mu}{\mu}\right.$$

$$\left.+ 8\left(\frac{1}{\mu^6} - \frac{3}{\mu^8}\right)\cos\mu + 4\left(\frac{1}{\mu^6} + \frac{6}{\mu^8}\right)\right\}\left[H_4(\theta_1, \theta_2)\right]^{-1/2},$$

(7.80)

where

$$H_1(\theta_1, \theta_2) = \cos\mu + (c - \theta_1)\frac{\sin\mu}{\mu},$$

$$H_2(\theta_1, \theta_2) = \frac{\theta_1^2 + c^2\theta_1 - c^3 + 2\theta_2}{\mu^2}\frac{\sin\mu}{\mu} - \frac{c^2}{\mu^2}\cos\mu$$

$$+ \frac{2(\theta_1^2 + c^2\theta_1 - 2c\theta_2)}{\mu^4}(\cos\mu - 1),$$

$$H_3(\theta_1, \theta_2) = \frac{\theta_1\left(c^2 + 3c + 3\right) - c^3}{\mu^2}\frac{\sin\mu}{\mu} - \frac{c^2}{\mu^2}\cos\mu$$

$$- \frac{3\theta_1\left(c^2 + 3c + 3\right) - 6\theta_2(c+1)}{\mu^4}\left(\frac{\sin\mu}{\mu} - \cos\mu\right),$$

$$H_4(\theta_1, \theta_2) = \frac{c^5 - c^4\theta_1 - 4\theta_1^2\left(c^2 + 3c + 27\right) - 8\theta_2\left(c^2 - 3c - 3\right)}{\mu^4}\frac{\sin\mu}{\mu}$$

$$+ \frac{24\left(c^4\theta_1 + 8\theta_1^2\theta_2 + 4(c+1)\left(3\theta_1^2 - \theta_2^2\right)\right)}{\mu^6}\left(\frac{\sin\mu}{\mu} + \frac{\cos\mu}{\mu^2} - \frac{1}{\mu^2}\right)$$

$$+ \left(\frac{c^4}{\mu^4} - \frac{8\left(c^3(c\theta_1 - 2\theta_2) + 4\theta_1^2\left(c^2 + 3c + 6\right)\right)}{\mu^6} \right) \cos\mu$$

$$- \frac{4\left(c^4\theta_1 + 4\theta_1^2\left(c^2 + 3c - 3\right) - 2c^2\theta_2(c+3)\right)}{\mu^6}.$$

The c.f. $\phi_{l1}(\theta)$ of $xV_l - U_l$ is given by $\phi_{l1}(\theta) = m_{l1}(-i\theta, i\theta x)$ for $l = 1, 2, 3, 4$. The c.f. $\phi_{11}(\theta)$ for the case where $c = 0$, $r = 1$ and $\gamma = 0$ was earlier obtained in White (1958) by the eigenvalue approach, while $\phi_{11}(\theta)$ with $c \neq 0$, $r = 1$, and $\gamma = 0$ was first treated by Bobkoski (1983) on the basis of a method from the theory of stochastic differential equations. Phillips (1987b) dealt with $\phi_{11}(\theta)$ for the case where $c \neq 0$, $r \neq 1$, and $\gamma = 0$, following White (1958). Perron (1991a) derived $\phi_{11}(\theta)$ with $c \neq 0$, $r = 1$, and $\gamma \neq 0$ by the stochastic process approach. Nabeya and Tanaka (1990a) obtained $\phi_{l1}(\theta)$ ($l = 1, 2, 3, 4$) for the case where $c \neq 0$, $r \neq 1$, and $\gamma = 0$ by the Fredholm approach, while Nabeya and Sørensen (1994) dealt with $\phi_{l1}(\theta_1, \theta_2)$ ($l = 1, 2, 4$) for the case where $c \neq 0$, $r \neq 1$, and $\gamma \neq 0$ by the Fredholm approach. We find the stochastic process approach most convenient for the present problem.

An interesting feature associated with Models B and D is that the distribution is independent of γ when $c = 0$, as was mentioned in Section 7.2, where we also discussed the effect of the parameters r and γ. The effect of c as $|c| \to \infty$ will be examined in the next section by using the joint m.g.f.s obtained above.

7.4.2 The Stationary Case

Here we assume that $Y(0) = Z(0) \sim N(0, 1/(2c))$, which is independent of increments of the Brownian motion $\{w(t)\}$. We recognize that the m.g.f.s in the present case can be readily obtained from those in the fixed initial value case by the conditional argument given below (4.15).

Let $m_{l2}(\theta_1, \theta_2)$ be the joint m.g.f. of U_l and V_l ($l = 1, 2, 3, 4$) for the present case. Using (7.73), (7.74), (7.75), and (7.76) with γ replaced by $Y(0) = Z(0) \sim N(0, 1/(2c))$, we can show (Problem 4.4) that

(7.81) $\quad m_{l2}(\theta_1, \theta_2) = E\left[E\left[\exp\{\theta_1 U_l + \theta_2 V_l\} | Z(0)\right]\right]$

$$= E\left[\exp\left\{-\frac{a}{2}(Z^2(0) + 1) + \frac{\theta_1}{2}(1-r)\right\}\right.$$

$$\left. \times E\left[\exp\left\{\frac{1}{2}W_l'A_l W_l + Z(0)h_l'W_l\right\} \Big| Z(0)\right]\right]$$

$$= \exp\left(\frac{c - r\theta_1}{2}\right)\left[1 - \frac{1}{2c}\{-a + 2h_l'\kappa_l + h_l'\Omega_l h_l \right.$$

$$\left. + (\kappa_l + \Omega_l h_l)'A_l(B_l - \Omega_l A_l)^{-1}(\kappa_l + \Omega_l h_l)\}\right]^{-1/2}$$

$$\times \left[e^\beta \, | B_I - \Omega_I A_I | \right]^{-1/2}$$

$$= E\left[m_{l1}(\theta_1, \theta_2) \right],$$

where the quantities appearing above are defined below (7.75). The expression $E[m_{l1}(\theta_1, \theta_2)]$ means that the expectation is taken with respect to $\gamma^2 = Z^2(0) \sim \chi^2(1)/(2c)$.

On the basis of (7.77) through (7.80), it is now an easy matter to obtain

(7.82) $\quad m_{12}(\theta_1, \theta_2) = \exp\left(\dfrac{c - r\theta_1}{2} \right) \left[\cos \mu + \dfrac{2c^2 - \theta_1^2 - 2\theta_2}{2c} \dfrac{\sin \mu}{\mu} \right]^{-1/2},$

(7.83) $\quad m_{22}(\theta_1, \theta_2)$

$$= \exp\left(\dfrac{c - r\theta_1}{2} \right) \left[\left(1 - \dfrac{2\theta_2}{\mu^2} \right) \cos \mu + \dfrac{(c+2)\theta_1^2 - 2c\theta_2}{\mu^4} (\cos \mu - 1) \right.$$

$$\left. + \left(c + \dfrac{c\theta_1^2}{2\mu^2} + \dfrac{\theta_1^2 - (c-2)\theta_2}{\mu^2} \right) \dfrac{\sin \mu}{\mu} \right]^{-1/2},$$

$m_{32}(\theta_1, \theta_2)$

$$= \exp\left(\dfrac{c - r\theta_1}{2} \right) \left[\left\{ \dfrac{1}{2\mu^2} \left(-2c^3 + c\left(\theta_1^2 + 2\theta_2\right) + 3\left(\theta_1^2 - \theta_1 + 2\theta_2\right) \right. \right. \right.$$

$$+ \dfrac{3}{c} \left(\theta_1^2 - 3\theta_1 + 2\theta_2 \right) \bigg) - \dfrac{3}{2\mu^4} \left(c\left(\theta_1^2 + 3\theta_1 - 2\theta_2\right) + 2\left(3\theta_1 - 2\theta_2\right) \right.$$

$$\left. + 3\theta_1^2 + \dfrac{3}{c} \theta_1^2 \right) \bigg\} \dfrac{\sin \mu}{\mu} + \bigg\{ -\dfrac{c^2}{\mu^2} + \dfrac{3}{2\mu^4} \left(c\left(\theta_1^2 + \theta_1 - 2\theta_2\right) + 3\theta_1^2 \right.$$

$$\left. + \dfrac{1}{c} \left(3\theta_1^2 - 6\theta_1 + 4\theta_2 \right) \right) \bigg\} \cos \mu + \dfrac{3}{\mu^4} \bigg\{ c\theta_1 + \left(1 + \dfrac{1}{c} \right)(3\theta_1 - 2\theta_2) \bigg\} \bigg]^{-1/2},$$

(7.84)

$m_{42}(\theta_1, \theta_2) = \exp\left(\dfrac{c - r\theta_1}{2} \right)$

$$\times \left[\left\{ -\dfrac{c^3}{\mu^2} - \dfrac{c^3\left(\theta_1^2 - 2\theta_2\right) + 8\theta_1^2\left(c^2 + 3c + 27\right) + 16\theta_2\left(c^2 - 3c - 3\right)}{2\mu^4} \right\} \right.$$

$$\times \dfrac{\sin \mu}{\mu} + \left\{ \dfrac{c^4}{\mu^4} - \dfrac{4\left(c^3\left(\theta_1^2 - 2\theta_2\right) + 8\theta_1^2\left(c^2 + 3c + 6\right)\right)}{\mu^6} \right\} \cos \mu$$

$$+ \frac{12\left(c^3\left(\theta_1^2+2\theta_2\right)+16\theta_1^2\theta_2+8(c+1)\left(3\theta_1^2-\theta_2^2\right)\right)}{\mu^6}$$

$$\times\left(\frac{\sin\mu}{\mu}+\frac{\cos\mu}{\mu^2}-\frac{1}{\mu^2}\right)$$

$$-\left.\frac{2\left(c^3\left(\theta_1^2-2\theta_2\right)+8\theta_1^2\left(c^2+3c-3\right)-12c^2\theta_2\right)}{\mu^6}\right]^{-1/2},$$

(7.85)

where $\mu = \sqrt{2\theta_2 - c^2}$.

The c.f. $\phi_{l2}(\theta)$ of $xV_l - U_l$ in the present case is obtained as $\phi_{l2}(\theta) = m_{l2}(-i\theta, i\theta x)$ ($l = 1, 2, 3, 4$). Perron (1991b) discussed in detail the case of $l = 1$. These c.f.s will be used in Section 7.6 to approximate the exact distribution of $T(\hat{\rho} - 1)$, where $\{y_j\}$ follows a zero-mean or nonzero-mean Gaussian stationary AR(1) process with the coefficient parameter close to unity. It is also of interest to examine the effect of c on the limiting distributions of $T(\hat{\rho} - 1)$, which will be discussed in the next section.

PROBLEMS

4.1 Derive the last expression in (7.75) for $l = 2$.
4.2 Derive the expressions given in (7.76).
4.3 Derive the m.g.f. $m_{11}(\theta_1, \theta_2)$ as in (7.77).
4.4 Establish the relations in (7.81).

7.5 TABLES AND FIGURES OF LIMITING DISTRIBUTIONS

This section presents percent points and probability densities of the limiting distributions of $T(\hat{\rho} - 1)$. The computation follows the same lines as given in Chapter 6. Moments of the distributions are also computed as a by-product.

To save space we consider only a few combinations of the parameter values of c, r, and γ. In particular, we set r at $r = 1$. Thus we are not concerned with the effect of r on the limiting distributions. Note that the distributions will be shifted to the left as r becomes large. Table 7.1 tabulates percent points for the limiting distributions of $T(\hat{\rho} - 1)$ in Model A together with the mean and standard deviation. Note that the case $(c, r, \gamma) = (0, 1, 0)$ for the fixed initial value case corresponds to a classical unit root distribution. Percent points were obtained by Newton's method of successive approximation described in Section 6.5 of Chapter 6, while moments were computed following the formula in (1.39). Figure 7.1 shows the corresponding probability densities for the fixed initial value case where $c = 0$, while Figure 7.2 shows those for the stationary case.

Table 7.1. Limiting Distributions of $T(\hat{\rho} - 1)$ in Model A

Case	\multicolumn{7}{c}{Probability of a Smaller Value}	Mean	SD
	0.01	0.05	0.1	0.5	0.9	0.95	0.99		
\multicolumn{10}{c}{The Fixed Initial Value Case ($r = 1$)}
$c = -2$									
$\gamma = 0$	−10.054	−4.559	−2.343	1.689	2.519	2.758	3.355	0.749	2.644
$\gamma = 0.5$	−7.254	−2.845	−1.020	1.824	2.376	2.540	2.968	1.155	1.982
$\gamma = 1$	−2.876	0.260	1.122	1.937	2.249	2.336	2.529	1.701	0.959
$c = 0$									
$\gamma = 0$	−13.695	−8.039	−5.714	−0.853	0.928	1.285	2.033	−1.781	3.180
$\gamma = 0.5$	−10.956	−6.431	−4.571	−0.679	0.764	1.048	1.634	−1.418	2.549
$\gamma = 1$	−6.848	−4.020	−2.857	−0.391	0.588	0.778	1.130	−0.844	1.622
$c = 2$									
$\gamma = 0$	−17.039	−11.138	−8.641	−2.941	−0.538	−0.082	0.785	−3.930	3.680
$\gamma = 0.5$	−14.276	−9.504	−7.472	−2.759	−0.695	−0.316	0.392	−3.554	3.044
$\gamma = 1$	−10.102	−7.026	−5.695	−2.482	−0.919	−0.631	−0.141	−2.979	2.086
\multicolumn{10}{c}{The Stationary Case ($r = 1$)}
$c = 0.1$	−8.265	−4.006	−2.505	−0.279	0.395	0.627	1.211	−0.788	1.766
$c = 0.5$	−11.138	−6.370	−4.521	−0.999	0.270	0.574	1.241	−1.698	2.427
$c = 1$	−12.850	−7.831	−5.818	−1.651	−0.029	0.322	1.043	−2.431	2.781
$c = 2$	−15.260	−9.929	−7.717	−2.775	−0.669	−0.262	0.533	−3.633	3.234
$c = 5$	−20.737	−14.808	−12.232	−5.888	−2.620	−2.029	−1.095	−6.821	4.144
$c = 10$	−28.470	−21.834	−18.846	−10.939	−6.206	−5.271	−3.846	−11.904	5.246

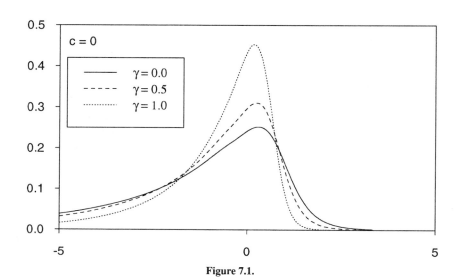

Figure 7.1.

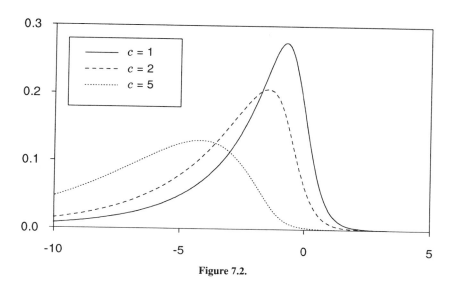
Figure 7.2.

It is seen from Table 7.1 that

i) the distributions for both cases are shifted to the left and become more dispersed as c becomes large;

ii) the distributions for the fixed initial value case become more concentrated as $|\gamma|$ becomes large;

iii) the mean is smaller than the median in both cases, implying that the distribution has a heavy left-hand tail;

iv) the distribution for the stationary case with $c = 2$ is comparable to that for the fixed initial value case with $c = 2$ and $\gamma = 0.5$.

The above conclusions are partly seen in Figures 7.1 and 7.2. Moreover, it is observed from Figure 7.1 that the distributions tend to become symmetric around $x = c\ (=0)$ as γ becomes large, while Figure 7.2 tells us that the distributions tend to become symmetric around $x = -c$ as c becomes large. The behavior of distributions as $|\gamma|$ becomes large has been already examined in Section 7.2. The effect of c as $|c| \to \infty$ will be discussed later.

Table 7.2 and Figures 7.3 and 7.4 are concerned with Model B. Conclusions similar to those of Model A can be drawn, although the distributions have smaller means and are more dispersed. Note, as was mentioned before, that the distribution for the fixed initial value case with $c = 0$ does not depend on γ.

Table 7.3 and Figures 7.5 and 7.6 are concerned with Model C and Model D, where we have only presented the fixed initial value case with $\gamma = 0$. It is seen that the distributions are further located to the left of those in Models A and B and are more dispersed. The distributions are shifted to the left as c becomes large, but the degree of shift is not as large as in Models A and B. This fact will explain the lower

Table 7.2. Limiting Distributions of $T(\hat{\rho} - 1)$ in Model B

Case	\multicolumn{6}{c}{Probability of a Smaller Value}	Mean	SD
	0.01	0.05	0.1	0.5	0.9	0.95	0.99		

Case	0.01	0.05	0.1	0.5	0.9	0.95	0.99	Mean	SD
\multicolumn{10}{c}{The Fixed Initial Value Case ($r = 1$)}									
$c = -2$									
$\gamma = 0$	−16.182	−9.439	−6.423	0.313	1.809	2.127	2.831	−1.187	3.969
$\gamma = 0.5$	−14.339	−7.601	−4.558	1.097	2.129	2.421	3.107	−0.201	3.504
$\gamma = 1$	−8.963	−2.132	0.325	1.704	2.388	2.650	3.296	1.254	2.117
$c = 0$									
Any γ	−20.626	−14.094	−11.251	−4.357	−0.845	−0.143	1.054	−5.379	4.511
$c = 2$									
$\gamma = 0$	−23.682	−16.896	−13.891	−6.308	−2.277	−1.487	−0.135	−7.374	4.951
$\gamma = 0.5$	−22.605	−15.998	−13.095	−5.858	−2.021	−1.246	0.079	−6.881	4.744
$\gamma = 1$	−19.687	−13.672	−11.092	−4.862	−1.553	−0.856	0.353	−5.751	4.139
\multicolumn{10}{c}{The Stationary Case ($r = 1$)}									
$c = 0.1$	−20.717	−14.181	−11.335	−4.314	−0.898	−0.189	1.017	−5.448	4.522
$c = 0.5$	−21.108	−14.551	−11.291	−4.714	−1.121	−0.390	0.846	−5.736	4.569
$c = 1$	−21.644	−15.054	−12.172	−5.091	−1.419	−0.668	0.600	−6.118	4.636
$c = 2$	−22.811	−16.143	−13.209	−5.894	−2.401	−1.269	0.048	−6.935	4.786
$c = 5$	−26.640	−19.704	−16.601	−8.558	−4.016	−3.143	−1.740	−9.615	5.284
$c = 10$	−33.267	−25.871	−22.492	−13.334	−7.640	−6.488	−4.712	−14.385	6.103

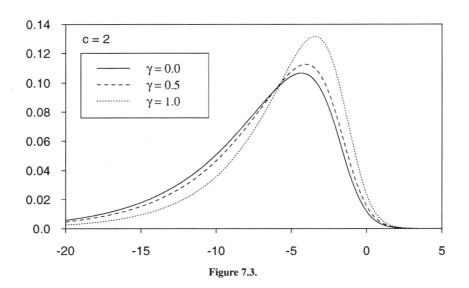

Figure 7.3.

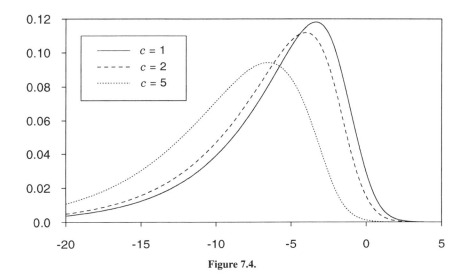

Figure 7.4.

power of unit root tests associated with Models C and D. That topic will be discussed in Chapter 9.

We now examine the effect of the parameter c on the limiting distributions $\mathcal{L}(U_l/V_l)$ of $T(\hat{\rho} - 1)$. We first deal with the situation where $c \to \infty$. For this purpose, let us consider $U_l/(cV_l)$. The joint m.g.f. of U_l and cV_l is given by $m_{l1}(\theta_1, c\theta_2)$ or $m_{l2}(\theta_1, c\theta_2)$, where m_{l1} and m_{l2} are the joint m.g.f.s of U_l and V_l obtained in the last section. As an example, we consider

$$m_{l1}(\theta_1, c\theta_2) = \exp\left[\frac{c - r\theta_1}{2} + \frac{\gamma^2(\theta_1^2 - 2c\theta_1 + 2c\theta_2)}{2H_1(\theta_1, c\theta_2)} \frac{\sinh \nu}{\nu}\right] [H_1(\theta_1, c\theta_2)]^{-1/2},$$

Table 7.3. Limiting Distributions of $T(\hat{\rho} - 1)$ in Models C and D for the Fixed Initial Value Case

	Probability of a Smaller Value								
c	0.01	0.05	0.1	0.5	0.9	0.95	0.99	Mean	SD
		Model C		($r = 1$	$\gamma = 0$)				
-2	-21.838	-14.696	-11.507	-3.266	1.374	2.175	3.479	-4.348	5.414
0	-23.736	-16.636	-13.479	-5.511	-1.383	-0.602	0.686	-6.661	5.163
2	-24.699	-17.566	-14.386	-6.301	-2.043	-1.249	0.068	-7.448	5.247
		Model D		($r = 1$	$\gamma = 0$)				
-2	-27.347	-19.621	-16.097	-6.563	-0.622	0.475	2.173	-7.655	6.354
0	-29.359	-21.711	-18.245	-9.103	-3.767	-2.673	-0.854	-10.246	6.033
2	-30.266	-22.576	-19.083	-9.815	-4.367	-3.269	-1.447	-10.962	6.116

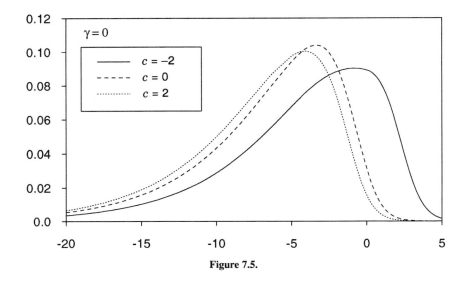

Figure 7.5.

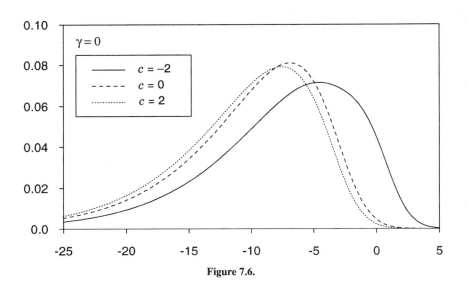

Figure 7.6.

where

$$\nu = (c^2 - 2c\theta_2)^{1/2}$$
$$= c - \theta_2 + O\left(\frac{1}{c}\right),$$
$$H_1(\theta_1, c\theta_2) = \cosh \nu + (c - \theta_1) \frac{\sinh \nu}{\nu}$$
$$= e^\nu + (c - \theta_1 - \nu) \frac{\sinh \nu}{\nu}$$
$$= e^\nu \left(1 + O\left(\frac{1}{c}\right)\right).$$

Then we have

$$m_{11}(\theta_1, c\theta_2) = \exp\left[\frac{c - r\theta_1}{2} + \frac{\gamma^2(\theta_1^2 - 2c\theta_1 + 2c\theta_2)}{2e^\nu \left(1 + O\left(c^{-1}\right)\right)} \frac{e^\nu - e^{-\nu}}{2\nu} - \frac{\nu}{2}\right]$$
$$\times \left(1 + O\left(\frac{1}{c}\right)\right)$$
$$\to \exp\left[-\frac{r + \gamma^2}{2} \theta_1 + \frac{1 + \gamma^2}{2} \theta_2\right].$$

Thus, for the fixed initial value case, $U_1/(cV_1)$ converges in probability to $-(r + \gamma^2)/(1 + \gamma^2) = -\delta$ as $c \to \infty$. Similarly we can show that $U_1/(cV_1)$ for the stationary case has the probability limit $-r$. Moreover $U_l/(cV_l)$ for $l = 2, 3, 4$ are shown to have the same probability limit as in the case for $l = 1$. Figures 7.2 and 7.4 reflect this fact to some extent.

To deduce the distributional property as $c \to \infty$, we transform U_l/V_l into

$$\frac{1}{\sqrt{2c}} \left(\frac{U_l}{V_l} + \frac{c(r + \gamma^2)}{1 + \gamma^2}\right) = \sqrt{\frac{c}{2}} (U_l + c\delta V_l) \bigg/ (cV_l)$$

extending the idea of Phillips (1987b). The normalization here is complicated, but may be interpreted in the usual way. To see this, note first that $U_l/(cV_l)$ converges in probability to $-\delta$ so that, treating c as a sample size, we can expect that

$$\sqrt{c} \left(\frac{U_l}{cV_l} + \delta\right) = \frac{1}{\sqrt{c}} \left(\frac{U_l}{V_l} + c\delta\right)$$

converges in distribution. The factor $1/\sqrt{2}$ makes the limiting distribution standardized when $r = 1$ and $\gamma = 0$.

As an example let us consider

$$m_{11}\left(\sqrt{\frac{c}{2}}\theta_1,\ c\sqrt{\frac{c}{2}}\delta\theta_1 + c\theta_2\right)$$
$$= \exp\left[\frac{1}{2}\left(c - r\sqrt{\frac{c}{2}}\theta_1\right) + \frac{\gamma^2\left(\frac{c}{2}\theta_1^2 - 2c\sqrt{\frac{c}{2}}\theta_1 + 2c\sqrt{\frac{c}{2}}\delta\theta_1 + 2c\theta_2\right)}{2H_1}\right.$$
$$\left. \times \frac{\sinh \nu}{\nu}\right](H_1)^{-1/2},$$

where

$$\nu = \left(c^2 - 2c\sqrt{\frac{c}{2}}\delta\theta_1 - 2c\theta_2\right)^{1/2}$$
$$= c - \sqrt{\frac{c}{2}}\delta\theta_1 - \theta_2 - \frac{\delta^2\theta_1^2}{4} + O\left(\frac{1}{\sqrt{c}}\right),$$
$$\frac{1}{\nu} = \frac{1}{c}\left(1 + \frac{\delta\theta_1}{\sqrt{2c}} + O\left(\frac{1}{c}\right)\right),$$
$$H_1 = \cosh\nu + \left(c - \sqrt{\frac{c}{2}}\theta_1\right)\frac{\sinh\nu}{\nu}$$
$$= e^\nu\left(1 + \frac{(\delta - 1)\theta_1}{2\sqrt{2c}} + O\left(\frac{1}{c}\right)\right).$$

Then we obtain

$$m_{11}\left(\sqrt{\frac{c}{2}}\theta_1,\ c\sqrt{\frac{c}{2}}\delta\theta_1 + c\theta_2\right) \to \exp\left[\frac{\theta_1^2\,\delta^2\,(1+\gamma^2)}{2\ \ \ \ 4} + \theta_2\frac{1+\gamma^2}{2}\right]$$

so that cV_1 converges in probability to $(1+\gamma^2)/2$, while

$$\mathcal{L}\left(\sqrt{\frac{c}{2}}(U_1 + c\delta V_1)\right) \to N\left(0,\ \frac{\delta^2(1+\gamma^2)}{4}\right),$$

so that $(U_1 + c\delta V_1)/(\sqrt{2c}V_1)$ tends to $N(0, \delta^2/(1+\gamma^2))$ as $c \to \infty$. Similarly we can show that

(7.86) $$\mathcal{L}\left(\frac{1}{\sqrt{2c}}\left(\frac{U_l}{V_l} + c\delta\right)\right) \to N\left(0,\ \frac{\delta^2}{1+\gamma^2}\right)$$

as $c \to \infty$ for $l = 2, 3,$ and 4.

TABLES AND FIGURES OF LIMITING DISTRIBUTIONS

For the stationary case where the m.g.f. is given by $m_{12}(\theta_1, \theta_2)$, the same result as in (7.86) holds with δ and γ replaced by r and 0, respectively (Problem 5.1). Asymptotic normality as $c \to \infty$ is also recognized, to some extent, in Figures 7.2, 7.4, 7.5, and 7.6.

We next consider the situation where $c \to -\infty$. The situation is pertinent only to the fixed initial value case. We shall show that the asymptotic behavior is quite different between $\gamma = 0$ and $\gamma \neq 0$. It will also be found that Model C is quite different from the other models.

Suppose first that $\gamma = 0$ and consider

$$(7.87) \qquad \frac{1}{2ce^c}\left(\frac{U_l}{V_l} + c\right) = \frac{ce^c(U_l + cV_l)}{2c^2 e^{2c} V_l}.$$

It can be shown (Problem 5.2) that, for $l = 1, 2, 3$, and 4, $-4ce^{2c}U_l$ and $4c^2 e^{2c} V_l$ both converge in distribution to the same $\chi^2(1)$ random variable as $c \to -\infty$ so that $(U_l + cV_l)/(cV_l)$ converges in probability to 0. It is expected that $e^{-c}(U_l + cV_l)/(cV_l)$ will have a nondegenerate limiting distribution, from which the expression in (7.87) arises. Phillips (1987b) has shown that (7.87) for $l = 1$ with $\gamma = 0$ converges to standard Cauchy distribution as $c \to -\infty$. The same result also holds for $l = 2$ and 4 with $\gamma = 0$ (Problem 5.3). The case of $l = 3$ is exceptional since $\mathcal{L}\left(-4c^3 e^{2c}(U_3 + cV_3)/3\right)$ converges to $\chi^2(1)$ so that (7.87) for $l = 3$ diverges to $-\infty$ (Problem 5.4).

When $\gamma \neq 0$, we consider the limiting distribution as $c \to -\infty$ of

$$(7.88) \qquad \frac{-1}{\sqrt{-c}\,e^c}\left(\frac{U_l}{V_l} + c\right) = \frac{\sqrt{-c}\,e^c(U_l + cV_l)}{ce^{2c} V_l}.$$

It can be shown that $e^{2c}U_l$ and $ce^{2c}V_l$ converge in probability to $\gamma^2/2$ and $-\gamma^2/2$, respectively, for all l as $c \to -\infty$ so that $(U_l + cV_l)/(cV_l)$ converges in probability to 0. It can also be checked that $e^c(U_l + cV_l)$ for $l = 1, 2$, and 4 converges in probability to 0. Thus it is expected that (7.88) for $l = 1, 2$, and 4 will have a nondegenerate limiting distribution. The case of $l = 3$ is exceptional. In fact, it holds that $c^2 e^{2c}(U_3 + cV_3)$ converges in probability to $3\gamma^2/2$ (Problem 5.5). Thus (7.88) for $l = 3$ diverges to $-\infty$.

For $l = 1$ we are led to consider

$$(7.89) \qquad m_{11}\left(\sqrt{-c}\,e^c \theta_1, c\sqrt{-c}\,e^c \theta_1 + ce^{2c}\theta_2\right)$$

$$= \exp\left[\frac{c - r\sqrt{-c}\,e^c \theta_1}{2} - \frac{\gamma^2 ce^{2c}(\theta_1^2 - 2\theta_2)}{2H_1} \frac{\sinh \nu}{\nu}\right](H_1)^{-1/2},$$

where

$$v = \left(c^2 - 2c\sqrt{-c}\, e^c \theta_1 - 2ce^{2c}\theta_2\right)^{1/2} = -c + O(\sqrt{-c}\, e^c),$$

$$\frac{c}{v} = -1 + \frac{e^c}{\sqrt{-c}}\theta_1 + O\left(\frac{e^{2c}}{c}\right),$$

$$H_1 = \cosh v + \left(c - \sqrt{-c}\, e^c \theta_1\right)\frac{\sinh v}{v} = e^{-v} + O\left(\frac{e^c}{c}\right).$$

Since we have

$$ce^{2c}\frac{\sinh v}{v} = -\frac{1}{2}e^c + O\left(\sqrt{-c}\, e^{2c}\right),$$

it is seen that (7.89) converges to $\exp\{\gamma^2(\theta_1^2 - 2\theta_2)/4\}$ as $c \to -\infty$. Thus $ce^{2c}V_1$ converges in probability to $-\gamma^2/2$, while

$$\mathcal{L}\left(\sqrt{-c}\, e^c(U_1 + cV_1)\right) \to \mathrm{N}\left(0, \frac{\gamma^2}{2}\right),$$

so that (7.88) for $l = 1$ with $\gamma \neq 0$ converges to $\mathrm{N}(0, 2/\gamma^2)$ as $c \to -\infty$. The same result also holds for $l = 2$ and 4 with $\gamma \neq 0$ (Problem 5.6).

The fact that the asymptotic distribution of $(U_l/V_l) + c$ as $c \to -\infty$ depends crucially on the initial value γ is also discussed in a different context by Anderson (1959).

PROBLEMS

5.1 Prove that, for the stationary case, (7.86) holds for $l = 1$ with δ and γ replaced by r and 0, respectively.

5.2 Show that $-4ce^{2c}U_1$ and $4c^2e^{2c}V_1$ both converge in distribution to the same $\chi^2(1)$ random variable as $c \to -\infty$, where U_1 and V_1 are defined in (7.30) and (7.31), respectively, with $Y(0) = \gamma = 0$.

5.3 Prove that the random variable in (7.87) for $l = 4$ with $\gamma = 0$ converges to standard Cauchy distribution as $c \to -\infty$.

5.4 Show that $-4c^3e^{2c}(U_3 + cV_3)/3$ converges in distribution to $\chi^2(1)$ as $c \to -\infty$, where U_3 and V_3 are defined in (7.51) and (7.52), respectively, with $Y(0) = \gamma = 0$.

5.5 Show that $c^2e^{2c}(U_3 + cV_3)$ converges in probability to $3\gamma^2/2$ as $c \to -\infty$, where U_3 and V_3 are defined in (7.51) and (7.52), respectively, with $Y(0) = \gamma \neq 0$.

5.6 Prove that the random variable in (7.88) for $l = 2$ with $\gamma \neq 0$ converges to $\mathrm{N}(0, 2/\gamma^2)$ as $c \to -\infty$.

7.6 APPROXIMATIONS TO THE DISTRIBUTIONS OF THE LSES

In this section we consider approximations to the finite sample distributions of the LSEs for stationary, but nearly nonstationary AR(1) models. For this purpose we take up a simplified version of Models A and B:

(7.90) $\qquad y_j = \rho y_{j-1} + \varepsilon_j,$

(7.91) $\qquad y_j = \mu(1 - \rho) + \rho y_{j-1} + \varepsilon_j, \qquad (j = 1, 2, \ldots, T),$

where $|\rho| < 1$ and $\{\varepsilon_j\} \sim \text{NID}(0, \sigma^2)$. We assume that $y_0 \sim N(0, \sigma^2/(1 - \rho^2))$ for (7.90) and $y_0 \sim N(\mu, \sigma^2/(1 - \rho^2))$ for (7.91), where y_0 is independent of $\{\varepsilon_j\}$ so that $\{y_j\}$ is a Gaussian stationary AR(1) process.

As an estimator of ρ in (7.90), we consider

$$\tilde{\rho}_1 = \sum_{j=1}^{T} y_{j-1} y_j \bigg/ \sum_{j=1}^{T} y_{j-1}^2,$$

while we define an estimator of ρ in (7.91) by

$$\tilde{\rho}_2 = \left[\sum_{j=1}^{T} y_{j-1} y_j - \frac{1}{T} \sum_{j=1}^{T} y_{j-1} \sum_{j=1}^{T} y_j\right] \bigg/ \left[\sum_{j=1}^{T} y_{j-1}^2 - \frac{1}{T} \left(\sum_{j=1}^{T} y_{j-1}\right)^2\right].$$

Suppose that we are concerned with the distributions of $T(\tilde{\rho}_l - \rho)$ ($l = 1, 2$). Following Imhof (1961), we can show (Problem 6.1) that the exact distributions of $T(\tilde{\rho}_l - \rho)$ can be computed as

(7.92) $\qquad P(T(\tilde{\rho}_l - \rho) \leq x) = \frac{1}{2} + \frac{1}{\pi} \int_0^{\infty} \frac{\sin\left\{\frac{1}{2} \sum_{j=1}^{T+1} \tan^{-1}(\lambda_{jl} u)\right\}}{u \prod_{j=1}^{T+1} (1 + \lambda_{jl}^2 u^2)^{1/4}} du,$

where λ_{jl}s ($j = 1, \ldots, T + 1; l = 1, 2$) are the eigenvalues of $xB_l - A_l$ with

$$A_1 = \frac{1}{2T}\left[\begin{pmatrix} \mathbf{0}' & 0 \\ C_{11} & \mathbf{0} \end{pmatrix} + \begin{pmatrix} \mathbf{0} & C_{11}' \\ 0 & \mathbf{0}' \end{pmatrix}\right],$$

$$A_2 = A_1 - \frac{1}{2T^2}\left[\begin{pmatrix} \mathbf{0}' & 0 \\ ee'C_{11} & \mathbf{0} \end{pmatrix} + \begin{pmatrix} \mathbf{0} & C_{11}'ee' \\ 0 & \mathbf{0}' \end{pmatrix}\right],$$

$$B_1 = \frac{1}{T^2} \begin{pmatrix} C_{11}'C_{11} & \mathbf{0} \\ \mathbf{0}' & 0 \end{pmatrix},$$

$$B_2 = B_1 - \frac{1}{T^3} \begin{pmatrix} C_{11}ee'C_{11}' & \mathbf{0} \\ \mathbf{0}' & 0 \end{pmatrix},$$

$$C_{11} = \begin{pmatrix} 0 & & & & & \\ 1 & 0 & & & & \\ h(\rho) & \rho & 1 & \cdot & & \mathbf{0} \\ & \cdot & \cdot & \cdot & \cdot & \\ & \cdot & & \cdot & \cdot & \\ & \cdot & & & \cdot & 0 \\ & \rho^{T-2} & & & \rho & 1 \end{pmatrix},$$

$$h(\rho) = \frac{1}{\sqrt{1-\rho^2}} (1, \rho, \ldots, \rho^{T-1})',$$

$$e = (1, \ldots, 1)' : T \times 1.$$

Numerical integration in (7.92) can be done easily by using any computer package.

Since $\{y_j\}$ is stationary, the Edgeworth-type approximation may be used to approximate the distributions of $T(\tilde{\rho}_l - \rho)$. Phillips (1978) obtained

$$(7.93)\; P\left(T(\tilde{\rho}_1 - \rho) \leq x\right) \\ \sim \Phi(y) + \phi(y) \left[\frac{1}{\sqrt{T}} \frac{\rho(1+y^2)}{\sqrt{1-\rho^2}} + \frac{1}{4T} \frac{(1-\rho^2)y + (1+\rho^2)y^3 - 2\rho^2 y^5}{1-\rho^2} \right],$$

while Tanaka (1983c) obtained

$$(7.94)\qquad P\left(T(\tilde{\rho}_2 - \rho) \leq x\right) \sim \Phi(y) + \phi(y) \left[\frac{1}{\sqrt{T}} \frac{2\rho + 1 + \rho y^2}{\sqrt{1-\rho^2}} \right. \\ \left. + \frac{1}{4T} \frac{-(1+\rho)(1+7\rho)y + (1-4\rho - 3\rho^2)y^3 - 2\rho^2 y^5}{1-\rho^2} \right],$$

where $y = x/\sqrt{T(1-\rho^2)}$, and $\Phi(y)$ and $\phi(y)$ are the N(0, 1) distribution function and probability density, respectively.

On the other hand, if ρ is close to 1, we may approximate $P(T(\tilde{\rho}_l - \rho) \leq x)$ under $\rho = 1 - (c/T)$ by

$$(7.95)\qquad \lim_{T \to \infty} P\left(T(\tilde{\rho}_l - \rho) \leq x\right) = \lim_{T \to \infty} P\left(T(\tilde{\rho}_l - 1) \leq z\right) \\ = \frac{1}{2} + \frac{1}{\pi} \int_0^\infty \frac{1}{\theta} \text{Im}\left[m_{l2}(-i\theta, i\theta z)\right] d\theta,$$

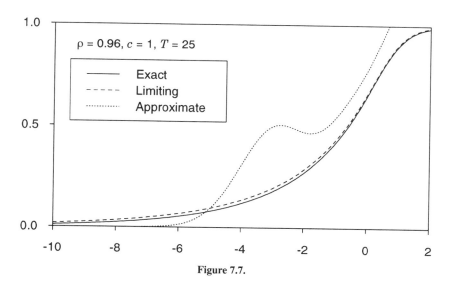

Figure 7.7.

where $z = x - c$, and m_{12} and m_{22} are defined in (7.82) and (7.83), respectively, by putting $r = 1$ (Problem 6.2).

Figures 7.7 and 7.8 show distribution functions of $T(\tilde{p}_1 - \rho)$ based on (7.92)—Exact, (7.93)—Approximate and (7.95)—Limiting, where the former figure is Case 1 with $\rho = 0.96$ and $T = 25$ so that $c = 1$, while the latter is Case 2 with $\rho = 0.6$ and $T = 25$ so that $c = 10$. Figures 7.9 and 7.10 show the corresponding results for $T(\tilde{p}_2 - \rho)$. It is seen from these figures that the distributions of $T(\tilde{p}_l - \rho)$ can be well approximated by the limiting distributions when ρ is close to 1, even if T is small.

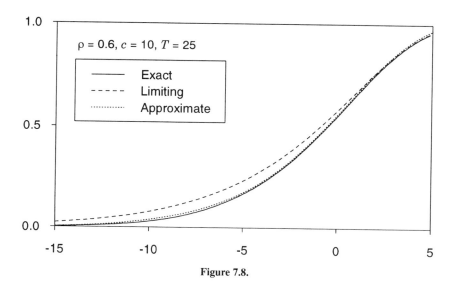

Figure 7.8.

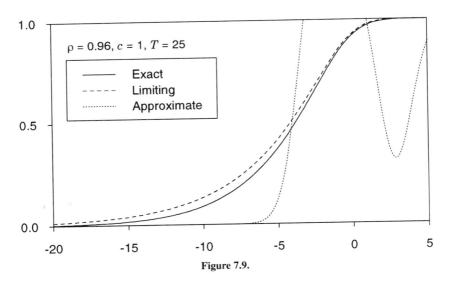

Figure 7.9.

The Edgeworth-type approximation fails to be monotone for ρ close to 1, although it tends to show superiority over the approximation by the limiting distribution as ρ gets away from unity.

PROBLEMS

6.1 Establish (7.92) using Imhof's formula.

6.2 Prove (7.95) when $\rho = 1 - (c/T)$ with c fixed.

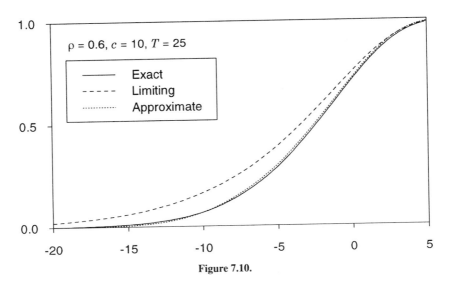

Figure 7.10.

7.7 NEARLY NONSTATIONARY SEASONAL AR MODELS

In this and the next two sections, we extend the models dealt with so far in three directions. An extension to seasonal models is discussed in this section, while complex roots on the unit circle and higher order nonstationarity are studied in the next two sections.

Nonstationary seasonal AR models were first treated by Dickey, Hasza, and Fuller (1984), which we generalize in the sense described later. Let us consider the regression model

$$y_j = x_j'\beta + \eta_j, \tag{7.96}$$

which is the same as before, except that the error term $\{\eta_j\}$ follows a seasonal AR process

$$\eta_j = \rho_m \eta_{j-m} + u_j, \quad (j = 1, \ldots, T). \tag{7.97}$$

Here m is a positive integer, while $\{u_j\}$ is a stationary process generated by

$$u_j = \sum_{l=0}^{\infty} \alpha_l \varepsilon_{j-l}, \quad \sum_{l=0}^{\infty} l|\alpha_l| < \infty, \quad \alpha \equiv \sum_{l=0}^{\infty} \alpha_l \neq 0,$$

where $\{\varepsilon_j\} \sim$ i.i.d.$(0, \sigma^2)$. The parameter ρ_m and the initial values $\eta_{1-m}, \eta_{2-m}, \ldots, \eta_0$ to generate (7.97) will be specified later, together with the regressor sequence $\{x_j\}$.

In subsequent discussions, we assume that m is a divisor of T so that $N = T/m$ is an integer. As far as asymptotic arguments developed below are concerned, this is not restrictive. Then (7.96) and (7.97) may be put into the following form:

$$\mathbf{y}_j = X_j \beta + \boldsymbol{\eta}_j, \tag{7.98}$$
$$\boldsymbol{\eta}_j = \rho_m \boldsymbol{\eta}_{j-1} + \mathbf{u}_j, \quad (j = 1, \ldots, N), \tag{7.99}$$

where $\mathbf{y}_j = (y_{(j-1)m+1}, \ldots, y_{jm})'$: $m \times 1$, and $\boldsymbol{\eta}_j$ and $\mathbf{u}_j$ are $m \times 1$ vectors defined similarly, while

$$X_j = \begin{pmatrix} x'_{(j-1)m+1} \\ \vdots \\ x'_{jm} \end{pmatrix} : m \times p.$$

As in the previous sections, we consider four models specifying the regressor sequence $\{X_j\}$ in (7.98), which are

Model A: $\mathbf{y}_j = \boldsymbol{\eta}_j$,

Model B: $\mathbf{y}_j = \boldsymbol{\beta}_0 + \boldsymbol{\eta}_j$, $X_j = I_m = 1 \otimes I_m$,

Model C: $\quad y_j = j\boldsymbol{\beta}_1 + \boldsymbol{\eta}_j, \quad X_j = jI_m = j \otimes I_m,$

Model D: $\quad y_j = \boldsymbol{\beta}_0 + j\boldsymbol{\beta}_1 + \boldsymbol{\eta}_j, \quad X_j = (1,j) \otimes I_m,$

where $\boldsymbol{\eta}_j = \rho_m \boldsymbol{\eta}_{j-1} + \boldsymbol{u}_j$ ($j = 1, \ldots, N$) for all models.

For the above models, we assume that the initial vector $\boldsymbol{\eta}_0 = 0$ for simplicity of presentation. For the parameter ρ_m, we put

$$\rho_m = 1 - \frac{c}{N} = 1 - \frac{cm}{T} \tag{7.100}$$

with c fixed.

A word may be in order. The Kronecker structure of regressors such as $X_j = Z_j' \otimes I_m$ in Models B, C, and D means that each seasonal has different regression coefficients. If the regression coefficients are common among seasonals, for example, $y_j = (1, \ldots, 1)'\beta_0 + \boldsymbol{\eta}_j$ in Model B with β_0 scalar, the asymptotic results developed below will be different. Here we do not impose such a restriction.

The LSEs of ρ_m for the above models are defined in the same way as before. To study the asymptotic behavior, it seems more convenient to make combined use of the stochastic process and Fredholm approaches. For that purpose we rewrite the above models (Problem 7.1) as

$$y = (\bar{X} \otimes I_m)\beta + \eta, \tag{7.101}$$

$$\eta = (C(\rho_m) \otimes I_m)u, \tag{7.102}$$

where $y = (y_1, \ldots, y_T)' = (y_1', \ldots, y_N')'$, and η and u are defined similarly, while

$$\bar{X} = \begin{cases} e = (1, \ldots, 1)' & : \quad N \times 1 \quad \text{for Model B} \\ d = (1, \ldots, N)' & : \quad N \times 1 \quad \text{for Model C} \\ (e, d) & : \quad N \times 2 \quad \text{for Model D}, \end{cases}$$

$$C(\rho) = \begin{pmatrix} 1 & & & & \\ \rho & 1 & & 0 & \\ \cdot & \cdot & \cdot & & \\ \cdot & & \cdot & \cdot & \\ \cdot & & & \cdot & \\ \rho^{N-1} & \cdot & \cdot & \rho & 1 \end{pmatrix} \quad : N \times N.$$

Then we first obtain the LSE of β as

$$\hat{\beta} = ((\bar{X}'\bar{X})^{-1} \otimes I_m)(\bar{X}' \otimes I_m)y = \beta + ((\bar{X}'\bar{X})^{-1}\bar{X}' \otimes I_m)\eta$$

so that

$$\hat{\eta} = y - (\bar{X} \otimes I_m)\hat{\beta} = (\bar{M} \otimes I_m)\eta = (\bar{M}C(\rho_m) \otimes I_m)u,$$

NEARLY NONSTATIONARY SEASONAL AR MODELS

where $\bar{M} = I_N - \bar{X}(\bar{X}'\bar{X})^{-1}\bar{X}'$. The LSE of ρ_m is now expressed as

$$\hat{\rho}_m = \sum_{j=m+1}^{T} \hat{\eta}_{j-m} \hat{\eta}_j \Big/ \sum_{j=m+1}^{T} \hat{\eta}_{j-m}^2$$

$$= \sum_{j=2}^{N} \hat{\boldsymbol{\eta}}'_{j-1} \hat{\boldsymbol{\eta}}_j \Big/ \sum_{j=2}^{N} \hat{\boldsymbol{\eta}}'_{j-1} \hat{\boldsymbol{\eta}}_{j-1} \ .$$

It then follows (Problem 7.2) that

(7.103) $$N(\hat{\rho}_m - 1) = U_N/V_N \ ,$$

where

(7.104) $$U_N = \frac{1}{N} \sum_{j=2}^{N} \hat{\boldsymbol{\eta}}'_{j-1} \left(\hat{\boldsymbol{\eta}}_j - \hat{\boldsymbol{\eta}}_{j-1} \right)$$

$$= \frac{1}{2N} \hat{\boldsymbol{\eta}}'_N \hat{\boldsymbol{\eta}}_N - \frac{1}{2N} \sum_{j=1}^{N} \left(\hat{\boldsymbol{\eta}}_j - \hat{\boldsymbol{\eta}}_{j-1} \right)' \left(\hat{\boldsymbol{\eta}}_j - \hat{\boldsymbol{\eta}}_{j-1} \right)$$

$$= \frac{1}{2N} \left[\left(e'_N \otimes I_m \right) \hat{\eta} \right]' \left(e'_N \otimes I_m \right) \hat{\eta}$$

$$- \frac{1}{2N} \sum_{j=1}^{N} \left(\hat{\boldsymbol{\eta}}_j - \hat{\boldsymbol{\eta}}_{j-1} \right)' \left(\hat{\boldsymbol{\eta}}_j - \hat{\boldsymbol{\eta}}_{j-1} \right)$$

$$= \frac{1}{N} \sum_{j,k=1}^{N} K_{1N}(j,k) \boldsymbol{u}'_j \boldsymbol{u}_k - \frac{1}{2N} \sum_{j=1}^{N} \left(\hat{\boldsymbol{\eta}}_j - \hat{\boldsymbol{\eta}}_{j-1} \right)' \left(\hat{\boldsymbol{\eta}}_j - \hat{\boldsymbol{\eta}}_{j-1} \right) \ ,$$

(7.105) $$V_N = \frac{1}{N^2} \sum_{j=2}^{N} \hat{\boldsymbol{\eta}}'_{j-1} \hat{\boldsymbol{\eta}}_{j-1}$$

$$= \frac{1}{N^2} \hat{\eta}' \hat{\eta} - \frac{1}{N^2} \hat{\boldsymbol{\eta}}'_N \hat{\boldsymbol{\eta}}_N$$

$$= \frac{1}{N} \sum_{j,k=1}^{N} K_{2N}(j,k) \boldsymbol{u}'_j \boldsymbol{u}_k - \frac{1}{N^2} \hat{\boldsymbol{\eta}}'_N \hat{\boldsymbol{\eta}}_N \ ,$$

with $\hat{\boldsymbol{\eta}}_0 = 0$, $\boldsymbol{e}_N = (0, \ldots, 0, 1)' : N \times 1$, and

$$K_{1N} = \left[(K_{1N}(j,k)) \right] = \frac{1}{2} C'(\rho_m) \bar{M} \boldsymbol{e}_N \boldsymbol{e}'_N \bar{M} C(\rho_m) \ ,$$

$$K_{2N} = \left[(K_{2N}(j,k)) \right] = \frac{1}{N} C'(\rho_m) \bar{M} C(\rho_m) \ .$$

Here it can be shown that $\hat{\boldsymbol{\eta}}_N' \hat{\boldsymbol{\eta}}_N/N$ has a nondegenerate limiting distribution so that

$$\frac{1}{N^2} \hat{\boldsymbol{\eta}}_N' \hat{\boldsymbol{\eta}}_N \to 0 \quad \text{in probability}.$$

Moreover it holds (Problem 7.3) that, for Models A, B, C, and D,

(7.106) $\quad \dfrac{1}{N} \displaystyle\sum_{j=1}^{N} (\hat{\boldsymbol{\eta}}_j - \hat{\boldsymbol{\eta}}_{j-1})' (\hat{\boldsymbol{\eta}}_j - \hat{\boldsymbol{\eta}}_{j-1}) \to m\sigma^2 \displaystyle\sum_{l=0}^{\infty} \alpha_l^2 \quad$ in probability.

Invoking arguments developed in Section 5.6 of Chapter 5, we now deduce (Problem 7.4) that, for Models A, B, C, and D

(7.107) $\mathcal{L}\left(\dfrac{1}{\alpha^2 \sigma^2}(xV_N - U_N)\right) \to \mathcal{L}\left(\displaystyle\int_0^1 \int_0^1 K(s,t)\, d\mathbf{w}'(s)\, d\mathbf{w}(t) + \dfrac{mr}{2}\right),$

where $r = \sum_{l=0}^{\infty} \alpha_l^2/\alpha^2 = \sigma_S^2/\sigma_L^2$ and $\{\mathbf{w}(t)\}$ is the m-dimensional standard Brownian motion, while $K(s,t)$ is the uniform limit of $xK_{2N} - K_{1N}$ that satisfies

$$\lim_{N\to\infty} \max_{j,k} \left| xK_{2N}(j,k) - K_{1N}(j,k) - K\left(\frac{j}{N}, \frac{k}{N}\right) \right| = 0.$$

In particular, for Model A with $c = 0$, we have

(7.108) $\quad \mathcal{L}\big(N(\hat{\rho}_m - 1)\big) = \mathcal{L}(U_N/V_N)$

$$\to \mathcal{L}\left(\frac{\displaystyle\int_0^1 \mathbf{w}'(t)\, d\mathbf{w}(t) + \frac{m}{2}(1-r)}{\displaystyle\int_0^1 \mathbf{w}'(t)\mathbf{w}(t)\, dt} \right).$$

Thus the limiting distribution of $N(\hat{\rho}_m - 1)$ has a convolutional property and the c.f. $\phi_l(\theta; m)$ of the limiting distribution in (7.107) for Model A ($l = 1$), B ($l = 2$), C ($l = 3$), and D ($l = 4$) is given by

(7.109) $\quad\quad\quad\quad \phi_l(\theta; m) = \big[m_{l1}(-i\theta, i\theta x)\big]^m,$

where $m_{l1}(\theta_1, \theta_2)$ ($l = 1, 2, 3, 4$) are available from Section 7.4 with γ replaced by 0. Numerical inversion of $\phi_l(\theta; m)$ will be given later.

The convolutional property of the limiting distribution associated with nonstationary seasonal models may be best seen if we consider the statistic

(7.110) $\quad\quad\quad\quad S_T^{(m)} = \dfrac{1}{T^2} \displaystyle\sum_{j=1}^{T} y_j^2, \quad y_j = y_{j-m} + \varepsilon_j,$

where $y_{1-m} = y_{2-m} = \cdots = y_0 = 0$ and $\{\varepsilon_j\} \sim$ i.i.d.$(0, \sigma^2)$. Since

$$\sum_{j=1}^{T} y_j^2 = \varepsilon' \left[C'(1)C(1) \otimes I_m \right] \varepsilon$$

$$= \sum_{j,k=1}^{N} \min(N+1-j, N+1-k) \varepsilon_j' \varepsilon_k,$$

it is easy to deduce that

(7.111) $\quad \mathcal{L} \left(\dfrac{m^2}{\sigma^2} S_T^{(m)} \right) \to \mathcal{L} \left(\int_0^1 \int_0^1 [1 - \max(s,t)] \, d\boldsymbol{w}'(s) \, d\boldsymbol{w}(t) \right)$

$$= \mathcal{L} \left(\int_0^1 \boldsymbol{w}'(t) \boldsymbol{w}(t) \, dt \right)$$

and it holds [Sen and Srivastava (1973) and Problem 7.5] that

(7.112) $\quad P \left(\displaystyle\int_0^1 \boldsymbol{w}'(t) \boldsymbol{w}(t) \, dt \leq x \right) = 2^{(m+2)/2} \sum_{k=0}^{\infty} \binom{-\tfrac{m}{2}}{k} \Phi \left(-\dfrac{2k + m/2}{\sqrt{x}} \right),$

where $\Phi(x)$ is the distribution function of $N(0,1)$. Note that the result (7.112) with $m = 1$ was earlier presented in (1.10).

A spurious property of covariances and regressions, as was discussed in Section 1.4 of Chapter 1, can also be seen. Suppose that $\{y_j\}$ is generated as in (7.110) so that

$$\boldsymbol{y}_j = \boldsymbol{y}_{j-1} + \boldsymbol{\varepsilon}_j, \quad \{\boldsymbol{\varepsilon}_j\} \sim \text{i.i.d.}(0, \sigma^2 I_m).$$

Because of the assumption on $\{\boldsymbol{\varepsilon}_j\}$, the m component processes of $\{\boldsymbol{y}_j\}$ are independent of each other. Nonetheless the covariance

(7.113) $\quad S_T^{(m)}(l) = \dfrac{1}{T^2} \displaystyle\sum_{j=l+1}^{T} \boldsymbol{y}_{j-l} \boldsymbol{y}_j \quad (l = 0, 1, \ldots)$

does have a nondegenerate limiting distribution, even for $l \neq 0$.

As an example, let us consider $S_T^{(m)}(1)$. Defining by e_k the $m \times 1$ vector with all components equal to 0 except the kth component, which is 1, we obtain (Problem 7.6)

(7.114) $\quad \displaystyle\sum_{j=2}^{T} \boldsymbol{y}_{j-1} \boldsymbol{y}_j = \sum_{j=1}^{N} \sum_{k=1}^{m-1} \boldsymbol{y}_j' e_k e_{k+1}' \boldsymbol{y}_j + \sum_{j=2}^{N} \boldsymbol{y}_{j-1}' e_m e_1' \boldsymbol{y}_j$

$$= \sum_{j,k=1}^{N} \min(N+1-j, N+1-k) \boldsymbol{\varepsilon}_j' H_1 \boldsymbol{\varepsilon}_k - \sum_{j=1}^{N} \boldsymbol{\varepsilon}_j' e_m e_1' \boldsymbol{y}_j,$$

where

$$H_1 = \frac{1}{2}\sum_{k=1}^{m-1}(e_k e'_{k+1} + e_{k+1} e'_k) + \frac{1}{2}(e_1 e'_m + e_m e'_1).$$

Since it can be shown (Problem 7.7) that

(7.115) $$\frac{1}{N^2}\sum_{j=1}^{N}\varepsilon'_j e_m e'_1 y_j \to 0 \quad \text{in probability},$$

we conclude that

(7.116) $$\mathcal{L}\left(\frac{m^2}{\sigma^2} S_T^{(m)}(1)\right) \to \mathcal{L}\left(\int_0^1 \int_0^1 [1 - \max(s,t)] \, dw'(s) H_1 \, dw(t)\right)$$

$$= \mathcal{L}\left(\int_0^1 w'(t) H_1 w(t) \, dt\right).$$

For general l, the statistic $S_T^{(m)}(l)$ in (7.113) has the following limiting distribution (Problem 7.8):

(7.117) $$\mathcal{L}\left(\frac{m^2}{\sigma^2} S_T^{(m)}(l)\right) \to \mathcal{L}\left(\int_0^1 \int_0^1 [1 - \max(s,t)] \, dw'(s) H_l \, dw(t)\right),$$

where

$$H_l = \frac{1}{2}\sum_{k=1}^{m}(e_k e'_{k+l} + e_{k+l} e'_k), \quad e_{k+im} = e_k, \quad (i = 0, 1, \ldots).$$

The fact that regressions of $\{y_j\}$ on $\{y_{j-l}\}$ for $l \neq im$ $(i = 0, 1, \ldots)$ are spurious can also be verified easily. In fact, it holds that, if $l \neq im$,

$$\mathcal{L}\left(\frac{\sum_{j=l+1}^{T} y_{j-l} y_j}{\sum_{j=l+1}^{T} y_{j-l}^2}\right) \to \mathcal{L}\left(\frac{\int_0^1 \int_0^1 [1 - \max(s,t)] \, dw'(s) H_l \, dw(t)}{\int_0^1 \int_0^1 [1 - \max(s,t)] \, dw'(s) \, dw(t)}\right)$$

$$= \mathcal{L}(W),$$

and

$$P(W \leq x) = \frac{1}{2} + \frac{1}{\pi}\int_0^\infty \frac{1}{\theta} \operatorname{Im}\left[\prod_{j=1}^{m}\left(\cos\sqrt{2i\delta_j\theta}\right)^{-1/2}\right] d\theta,$$

where δ_js are the eigenvalues of $xI_m - H_l$.

NEARLY NONSTATIONARY SEASONAL AR MODELS

Extensions to nearly nonstationary seasonal models with any period m are quite straightforward. Nabeya and Perron (1994) discussed such an extension for $m = 2$. The above analysis can also be extended to higher order nonstationary seasonal models like

$$(1 - L^m)^D y_j = u_j,$$

where D is a positive integer greater than unity [Chan and Wei (1988)].

Another extension somewhat related to the present model is to consider

(7.118) $\qquad y_{hj} = \rho y_{h(j-1)} + \varepsilon_{hj}, \qquad (j = 1, \ldots, T),$

where h is the sampling interval, $\{\varepsilon_{hj}\} \sim$ i.i.d.$(0, \sigma^2 h)$, and $\rho = 1 - ch$ with c a fixed constant. Here we assume that $y_0 = 0$ for simplicity and that $M = hT$ is a fixed sampling span, while $h \to 0$ and $T \to \infty$. The present situation may be applied to fields where a near-continuous record of data is available. The associated analysis is referred to as *continuous record asymptotics* [Phillips (1987a)].

Let us construct a partial sum process

$$Y_T(t) = y_{hj} + T\left(t - \frac{j}{T}\right)(y_{hj} - y_{h(j-1)}), \qquad \left(\frac{j-1}{T} \leq t \leq \frac{j}{T}\right),$$

$$= \sqrt{M}\left[\frac{1}{\sqrt{T}}\sum_{i=1}^{j}\rho^{j-i}\xi_i + T\left(t - \frac{j}{T}\right)\frac{y_{hj} - y_{h(j-1)}}{\sqrt{M}}\right],$$

where $\xi_j = \varepsilon_{hj}/\sqrt{h} = \sqrt{T}\varepsilon_{hj}/\sqrt{M}$ so that $\{\xi_j\} \sim$ i.i.d.$(0, \sigma^2)$. Noting that $\rho = 1 - ch = 1 - (cM/T)$, we can deduce that $\mathcal{L}(Y_T/\sigma) \to \mathcal{L}(\sqrt{M}Y)$, where $Y = \{Y(t)\}$ is the O–U process defined by

(7.119) $\qquad dY(t) = -cMY(t)\,dt + dw(t), \quad Y(0) = 0, \quad (0 \leq t \leq 1),$

which is equivalent to

$$Y(t) = e^{-cMt}\int_0^t e^{cMs}\,dw(s), \quad (0 \leq t \leq 1).$$

It follows that, when $M = \rho = \sigma = 1$,

$$\mathcal{L}(y_{h[Tt]}) \to \mathcal{L}(w(t))$$

for each t. Thus $\{y_{hj}\}$ itself behaves much like the standard Brownian motion as $h \to 0$. It is now an easy matter to obtain

(7.120) $$\mathcal{L}(T(\hat{\rho}-1)) \to \mathcal{L}\left(\frac{\int_0^1 Y(t)\,dY(t)}{\int_0^1 Y^2(t)\,dt}\right),$$

where

$$\hat{\rho} = \sum_{j=2}^T y_{h(j-1)} y_{hj} \Big/ \sum_{j=2}^T y_{h(j-1)}^2.$$

The limiting distribution in (7.120) is given in Section 7.4 by replacing c by cM. In particular we have

(7.121) $$E\left[\exp\left\{\theta_1 \int_0^1 Y(t)\,dY(t) + \theta_2 \int_0^1 Y^2(t)\,dt\right\}\right]$$

$$= \exp\left(\frac{cM - \theta_1}{2}\right)\left[\cos\mu + (cM - \theta_1)\frac{\sin\mu}{\mu}\right]^{-1/2},$$

where $\mu = \sqrt{2\theta_2 - c^2 M^2}$.

It is of some interest to relate the limiting distribution of $T(\hat{\rho}-1)$ with an estimator in the extended O–U process

(7.122) $\quad dZ_M(u) = -cZ_M(u)\,du + dw(u), \quad Z_M(0) = 0, \quad (0 \le u \le M).$

Note here that the Brownian motion $\{w(u)\}$ is now defined on $[0, M]$. It holds that $\mathcal{L}[\{Z_M(Mt) : 0 \le t \le 1\}] = \mathcal{L}[\{\sqrt{M}Y(t) : 0 \le t \le 1\}]$. The LSE or MLE $-\hat{c}$ of $-c$ is given by

$$-\hat{c} = \frac{\int_0^M Z_M(u)\,dZ_M(u)}{\int_0^M Z_M^2(u)\,du},$$

where

$$\int_0^M Z_M(u)\,dZ_M(u) = \frac{1}{2}\left(Z_M^2(M) - Z_M^2(0) - M\right)$$

$$= \frac{1}{2}\left(Z_M^2(M) - M\right)$$

because of the Ito calculus

$$d\left(Z_M^2(u)\right) = 2Z_M(u)\,dZ_M(u) + du, \quad (0 \le u \le M).$$

Then we can show (Problem 7.9) that

(7.123) $$\mathcal{L}(T(\hat{\rho} - 1)) \to \mathcal{L}(-M\hat{c}).$$

Equivalently we have $\mathcal{L}(h^{-1}(\hat{\rho} - 1)) \to \mathcal{L}(-\hat{c})$. It follows from (7.123) that

(7.124) $$\mathcal{L}\left(\frac{\frac{1}{M}\int_0^M Z_M(u)\,dZ_M(u)}{\frac{1}{M^2}\int_0^M Z_M^2(u)\,du}\right) = \mathcal{L}\left(\frac{\int_0^1 Y(t)\,dY(t)}{\int_0^1 Y^2(t)\,dt}\right).$$

The above results (7.123) and (7.124) may also be established more easily by noting that $\mathcal{L}[\{Z_M(Mt) : 0 \le t \le 1\}] = \mathcal{L}[\{\sqrt{M}Y(t) : 0 \le t \le 1\}]$.

If we allow the sampling span M to vary, asymptotics concerning M emerge. When $c > 0$, it can be shown (Problem 7.10) that

(7.125) $$-\hat{c} = \frac{\int_0^M Z_M(u)\,dZ_M(u)}{\int_0^M Z_M^2(u)\,du} \to -c \quad \text{in probability as } M \to \infty,$$

(7.126) $$\sqrt{M}(\hat{c} - c) \to N(0, 2c) \quad \text{as } M \to \infty.$$

When $c = 0$, $\mathcal{L}(-M\hat{c})$ is independent of M and

$$\mathcal{L}(-M\hat{c}) = \mathcal{L}\left(\frac{\int_0^1 w(t)\,dw(t)}{\int_0^1 w^2(t)\,dt}\right)$$

for any M. When $c < 0$, it holds (Problem 7.11) that

(7.127) $$\mathcal{L}\left(\frac{1}{2ce^{cM}}(-\hat{c} + c)\right) = \mathcal{L}\left(\frac{ce^{cM}\int_0^M Z_M(u)\,dw(u)}{2c^2 e^{2cM}\int_0^M Z_M^2(u)\,du}\right)$$

$\to$ standard Cauchy as $M \to \infty$.

Continuous record asymptotics can also be extended to deal with regression models $y_{hj} = x'_{hj}\beta + \eta_{hj}$, where $\eta_{hj} = (1 - ch)\eta_{h(j-1)} + u_{hj}$ and $\{u_{hj}\}$ follows a stationary process, and to allow for nonnegligible influence of η_0. Such an extension is discussed in Perron (1991a). We do not pursue the matter here.

We return to the normalized LSEs $N(\hat{\rho}_m - 1)$ given in (7.103), whose limiting distributions are computed as

$$F_l(x; m) = \lim_{N \to \infty} P\left(N(\hat{\rho}_m - 1) \le x\right)$$

$$= \frac{1}{2} + \frac{1}{\pi} \int_0^\infty \frac{1}{\theta} \operatorname{Im}\left[\{m_{l1}(-i\theta, i\theta x)\}^m\right] d\theta,$$

where $m_{l1}(\theta_1, \theta_2)$ ($l = 1, 2, 3, 4$) are defined in (7.109). Table 7.4 tabulates percent points for the limiting distributions of $N(\hat{\rho}_m - 1)$ in Model A together with the mean and standard deviation. We have set $r = \sigma_S^2/\sigma_L^2 = 1$ and the parameter values of m examined are $m = 2, 4$, and 12, while those of c are $c = -2, 0$, and 2. The corresponding percent points in Models B, C, and D are presented in Tables 7.5, 7.6, and 7.7, respectively. Dickey, Hasza, and Fuller (1984) obtained percent points, by simulations, for the finite sample and limiting distributions of $T(\hat{\rho}_m - 1) = mN(\hat{\rho}_m - 1)$ in Models A and B with $r = 1$ and $c = 0$ ($\rho_m = 1$). The entries in our tables should be multiplied by m when compared with those in their tables.

PROBLEMS

7.1 Show that the original model (7.96) with (7.97) can be rewritten as (7.101) and (7.102).

7.2 Prove that $N(\hat{\rho}_m - 1)$ in (7.103) can be expressed as U_N/V_N, where U_N and V_N are defined in (7.104) and (7.105), respectively.

Table 7.4. Limiting Distributions of $N(\hat{\rho}_m - 1)$ in Model A

	Probability of a Smaller Value								
c	0.01	0.05	0.1	0.5	0.9	0.95	0.99	Mean	SD
			$m = 2$,	$r = 1$					
-2	-3.322	-0.560	0.452	1.876	2.395	2.540	2.864	1.566	1.149
0	-7.058	-4.195	-3.005	-0.387	0.770	1.014	1.478	-0.832	1.730
2	-10.255	-7.142	-5.789	-2.472	-0.756	-0.414	0.176	-2.955	2.175
			$m = 4$,	$r = 1$					
-2	-0.007	1.002	1.328	1.946	2.294	2.388	2.578	1.850	0.493
0	-3.735	-2.268	-1.648	-0.180	0.615	0.791	1.107	-0.386	0.988
2	-6.738	-5.033	-4.265	-2.235	-0.989	-0.724	-0.280	-2.473	1.360
			$m = 12$,	$r = 1$					
-2	1.399	1.627	1.724	1.984	2.181	2.234	2.331	1.964	0.190
0	-1.503	-0.965	-0.722	-0.057	0.410	0.521	0.715	-0.118	0.463
2	-4.160	-3.436	-3.091	-2.078	-1.321	-1.142	-0.838	-2.156	0.708

Table 7.5. Limiting Distributions of $N(\hat{\rho}_m - 1)$ in Model B

c	\multicolumn{7}{c	}{Probability of a Smaller Value}	Mean	SD
	0.01	0.05	0.1	0.5	0.9	0.95	0.99		
\multicolumn{10}{c}{$m = 2, \quad r = 1$}
−2	−7.982	−3.954	−2.205	0.723	1.650	1.860	2.281	0.108	2.004
0	−13.106	−9.507	−7.894	−3.693	−1.190	−0.664	0.214	−4.207	2.800
2	−15.975	−12.154	−10.407	−5.689	−2.739	−2.111	−1.071	−6.223	3.157
\multicolumn{10}{c}{$m = 4, \quad r = 1$}
−2	−3.074	−1.122	−0.431	0.876	1.506	1.653	1.926	0.650	0.965
0	−9.023	−6.961	−6.006	−3.344	−1.547	−1.145	−0.471	−3.606	1.813
2	−11.713	−9.470	−8.411	−5.368	−3.205	−2.710	−1.883	−5.635	2.094
\multicolumn{10}{c}{$m = 12, \quad r = 1$}
−2	−0.271	0.224	0.430	0.962	1.327	1.417	1.576	0.909	0.378
0	−5.857	−4.929	−4.479	−3.113	−2.038	−1.776	−1.322	−3.201	0.968
2	−8.314	−7.268	−6.753	−5.148	−3.837	−3.510	−2.943	−5.237	1.150

Table 7.6. Limiting Distributions of $N(\hat{\rho}_m - 1)$ in Model C

c	\multicolumn{7}{c	}{Probability of a Smaller Value}	Mean	SD
	0.01	0.05	0.1	0.5	0.9	0.95	0.99		
\multicolumn{10}{c}{$m = 2, \quad r = 1$}
−2	−13.561	−9.440	−7.529	−2.205	0.992	1.601	2.577	−2.839	3.492
0	−15.625	−11.594	−9.749	−4.783	−1.780	−1.171	−0.191	−5.374	3.289
2	−16.586	−12.527	−10.662	−5.605	−2.492	−1.856	−0.834	−6.191	3.360
\multicolumn{10}{c}{$m = 4, \quad r = 1$}
−2	−8.817	−6.318	−5.120	−1.665	0.586	1.060	1.829	−2.023	2.300
0	−11.092	−8.715	−7.594	−4.396	−2.205	−1.725	−0.940	−4.700	2.173
2	−12.045	−9.644	−8.508	−5.238	−2.956	−2.450	−1.622	−5.537	2.232
\multicolumn{10}{c}{$m = 12, \quad r = 1$}
−2	−4.863	−3.671	−3.088	−1.327	0.012	0.329	0.865	−1.454	1.230
0	−7.454	−6.347	−5.804	−4.132	−2.805	−2.483	−1.932	−4.237	1.185
2	−8.388	−7.261	−6.706	−4.985	−3.600	−3.260	−2.678	−5.087	1.226

Table 7.7. Limiting Distributions of $N(\hat{\rho}_m - 1)$ in Model D

	Probability of a Smaller Value								
c	0.01	0.05	0.1	0.5	0.9	0.95	0.99	Mean	SD
			$m = 2,$	$r = 1$					
−2	−18.482	−13.898	−11.725	−5.397	−1.158	−0.312	1.016	−6.025	4.246
0	−20.709	−16.257	−14.178	−8.336	−4.410	−3.543	−2.113	−8.914	3.961
2	−21.597	−17.109	−15.007	−9.074	−5.065	−4.183	−2.730	−9.655	4.025
			$m = 4,$	$r = 1$					
−2	−13.267	−10.405	−9.004	−4.761	−1.730	−1.069	0.000	−5.123	2.887
0	−15.760	−13.067	−11.772	−7.928	−5.049	−4.372	−3.233	−8.221	2.680
2	−16.627	−13.905	−12.593	−8.684	−5.742	−5.050	−3.883	−8.977	2.728
			$m = 12,$	$r = 1$					
−2	−8.753	−7.316	−6.597	−4.343	−2.534	−2.095	−1.346	−4.477	1.598
0	−11.663	−10.360	−9.709	−7.645	−5.905	−5.465	−4.693	−7.744	1.498
2	−12.500	−11.178	−10.517	−8.414	−6.636	−6.184	−5.392	−8.513	1.526

7.3 Establish (7.106) for Model C.
7.4 Establish (7.107) for Model A.
7.5 Derive the relation in (7.112) using the fact that the inverse Laplace transform of $e^{-c\sqrt{\theta}}$ is $c\exp(-c^2/(4x))/(2\sqrt{\pi x^3})$ for $c > 0$.
7.6 Prove that the relation in (7.114) holds.
7.7 Establish (7.115).
7.8 Establish the weak convergence result in (7.117).
7.9 Prove (7.123) by deriving the joint m.g.f. of

$$\frac{1}{M}\int_0^M Z_M(u)\,dZ_M(u) \quad \text{and} \quad \frac{1}{M^2}\int_0^M Z_M^2(u)\,du.$$

7.10 Show that (7.125) and (7.126) hold when $c > 0$.
7.11 Prove (7.127) when $c < 0$.

7.8 COMPLEX ROOTS ON THE UNIT CIRCLE

This section deals with AR models that have complex roots on the unit circle. Let us consider the simplest model

$$(7.128) \quad \left(1 - e^{i\theta}L\right)\left(1 - e^{-i\theta}L\right)y_j = \varepsilon_j \iff y_j = 2y_{j-1}\cos\theta - y_{j-2} + \varepsilon_j,$$

where $0 < \theta < \pi$, $y_{-1} = y_0 = 0$, and $\{\varepsilon_j\} \sim$ i.i.d.$(0, \sigma^2)$. Ahtola and Tiao (1987) discussed the distributions of the LSEs associated with (7.128), which Chan and Wei

(1988) extended to higher order cases. Here we stick to the model (7.128) and study the asymptotic distributions of the LSEs following the idea of Chan and Wei (1988).

We first note (Problem 8.1) that (7.128) can be rewritten as

$$(7.129) \quad y_j = \frac{1}{\sin\theta} \sum_{k=1}^{j} \varepsilon_k \sin(j-k+1)\theta$$

$$= \frac{1}{\sin\theta} [X_j \sin(j+1)\theta - Y_j \cos(j+1)\theta],$$

where

$$(7.130) \quad X_j = \sum_{k=1}^{j} \varepsilon_k \cos k\theta, \quad Y_j = \sum_{k=1}^{j} \varepsilon_k \sin k\theta.$$

It may be noted that $X_j = X_{j-1} + \varepsilon_j \cos j\theta$ and $Y_j = Y_{j-1} + \varepsilon_j \sin j\theta$ with $X_0 = Y_0 = 0$. Thus $\{X_j\}$ and $\{Y_j\}$ are random walks with martingale difference innovations. If we define a partial sum process $\{Z_T(t)\}$ in $C^2 = C[0,1] \times C[0,1]$ by

$$Z_T(t) = \frac{\sqrt{2}}{\sqrt{T}\sigma} \left[\begin{pmatrix} X_j \\ Y_j \end{pmatrix} + T\left(t - \frac{j}{T}\right) \begin{pmatrix} \cos j\theta \\ \sin j\theta \end{pmatrix} \varepsilon_j \right], \quad \left(\frac{j-1}{T} \leq t \leq \frac{j}{T}\right),$$

it follows from Helland (1982) and Chan and Wei (1988) that $\mathcal{L}(Z_T) \to \mathcal{L}(w)$, where $w = \{w(t)\}$ is the two-dimensional standard Brownian motion.

We now consider the LSE $\hat{\beta} = (\hat{\beta}_1, \hat{\beta}_2)'$ of $\beta = (\beta_1, \beta_2)' = (2\cos\theta, -1)'$ for the model (7.128). We have

$$(7.131) \quad T(\hat{\beta} - \beta) = A_T^{-1} b_T,$$

where

$$A_T = \frac{1}{T^2} \sum_{j=3}^{T} \begin{pmatrix} y_{j-1}^2 & y_{j-1}y_{j-2} \\ y_{j-1}y_{j-2} & y_{j-2}^2 \end{pmatrix}, \quad b_T = \frac{1}{T} \sum_{j=3}^{T} \begin{pmatrix} y_{j-1}\varepsilon_j \\ y_{j-2}\varepsilon_j \end{pmatrix}.$$

Following Chan and Wei (1988), we obtain

$$\frac{1}{T^2} \sum_{j=1}^{T} y_j^2 = \frac{1}{T^2 \sin^2\theta} \sum_{j=1}^{T} [X_j \sin(j+1)\theta - Y_j \cos(j+1)\theta]^2$$

$$= \frac{\sigma^2}{4T \sin^2\theta} \sum_{j=1}^{T} Z_T'\left(\frac{j}{T}\right) Z_T\left(\frac{j}{T}\right)$$

$$- \frac{1}{2T^2 \sin^2\theta} \sum_{j=1}^{T} [(X_j^2 - Y_j^2) \cos 2(j+1)\theta + 2X_j Y_j \sin 2(j+1)\theta].$$

Here Chan and Wei (1988) proved that, if $e^{i\theta} \neq 1$, then

(7.132) $$\sup_{1 \leq j \leq T} \left| \sum_{k=1}^{j} e^{ik\theta} W_k \right| = o_p(T^2),$$

where $W_k = X_k^2$ or Y_k^2 or $X_k Y_k$. Thus it follows that

$$\mathcal{L}\left(\frac{1}{T^2} \sum_{j=1}^{T} y_j^2 \right) \to \mathcal{L}\left(\frac{\sigma^2}{4 \sin^2 \theta} \int_0^1 w'(t) w(t)\, dt \right).$$

Using (7.129) and (7.132), we obtain (Problem 8.2)

(7.133) $$\frac{1}{T^2} \sum_{j=2}^{T} y_{j-1} y_j = \frac{\sigma^2 \cos\theta}{4T \sin^2\theta} \sum_{j=1}^{T} Z_T'\left(\frac{j}{T}\right) Z_T\left(\frac{j}{T}\right) + o_p(1),$$

(7.134) $$\frac{1}{T} \sum_{j=2}^{T} y_{j-1} \varepsilon_j = \frac{\sigma^2}{2 \sin\theta} \sum_{j=1}^{T} \left[X_T\left(\frac{j-1}{T}\right) \Delta Y_T\left(\frac{j}{T}\right) \right.$$
$$\left. - Y_T\left(\frac{j-1}{T}\right) \Delta X_T\left(\frac{j}{T}\right) \right],$$

$$\frac{1}{T} \sum_{j=3}^{T} y_{j-2} \varepsilon_j$$

$$= \frac{\sigma^2}{2 \sin\theta} \sum_{j=1}^{T} \left[\cos\theta \left\{ X_T\left(\frac{j-1}{T}\right) \Delta Y_T\left(\frac{j}{T}\right) - Y_T\left(\frac{j-1}{T}\right) \Delta X_T\left(\frac{j}{T}\right) \right\} \right.$$
$$\left. - \sin\theta \left\{ X_T\left(\frac{j-1}{T}\right) \Delta X_T\left(\frac{j}{T}\right) + Y_T\left(\frac{j-1}{T}\right) \Delta Y_T\left(\frac{j}{T}\right) \right\} \right] + o_p(1),$$

(7.135)

where $Z_T(j/T) = (X_T(j/T), Y_T(j/T))'$ and $\Delta X_T(j/T) = X_T(j/T) - X_T((j-1)/T)$.

The joint weak convergence of the above quantities leads us to establish that $\mathcal{L}(T(\hat{\beta} - \beta)) = \mathcal{L}(T(\hat{\beta}_1 - 2\cos\theta, \hat{\beta}_2 + 1)') \to \mathcal{L}(Z)$, where $Z = (Z_1, Z_2)'$ and

(7.136) $$Z_1 = \frac{2 \int_0^1 \left[\{w_1(t)\, dw_2(t) - w_2(t)\, dw_1(t)\} \sin\theta + w'(t)\, dw(t) \cos\theta \right]}{\int_0^1 w'(t) w(t)\, dt},$$

Table 7.8. Limiting Distribution of $T(\hat{\beta}_2 + 1)$ in (7.137)

		Probability of a Smaller Value						
0.01	0.05	0.1	0.5	0.9	0.95	.099	Mean	SD
−2.956	−2.028	−1.539	0.775	6.010	8.389	14.115	1.664	3.460

$$(7.137) \qquad Z_2 = \frac{-2 \int_0^1 w'(t) \, dw(t)}{\int_0^1 w'(t) w(t) \, dt}.$$

The limiting random variable Z_1 depends on θ and involves Lévy's stochastic area as well as the Ito integral, while Z_2 is independent of θ and is closely related to the seasonal unit root statistic discussed in the last section. In fact, the $100\alpha\%$ point $x(\alpha)$ of Z_2 is equal to $-2y(1-\alpha)$, where $y(1-\alpha)$ is the upper $100\alpha\%$ point of the limiting distribution in (7.108) with $m = 2$ and $r = 1$. Table 7.8 gives percent points of Z_2 constructed from the entries in Table 7.4 for $c = 0$, $m = 2$, and $r = 1$. Ahtola and Tiao (1987) presented a similar table based on simulations.

We emphasize here that the weak convergence results in (7.136) and (7.137) follow under the assumption that $0 < \theta < \pi$. Thus, $\sin\theta \neq 0$ and $\cos\theta \neq \pm 1$. If $\theta = 0$ or $\theta = \pi$, then the asymptotic distribution of $\hat{\beta}$ will be different, which we discuss in the next section.

PROBLEMS

8.1 Show that $\{y_j\}$ defined in (7.128) can be rewritten as in (7.129).

8.2 Establish (7.133) through (7.135) using (7.132).

7.9 AUTOREGRESSIVE MODELS WITH MULTIPLE UNIT ROOTS

Here we deal with the AR(p) model

$$(7.138) \qquad y_j = \delta_1 y_{j-1} + \cdots + \delta_p y_{j-p} + \varepsilon_j = \delta' \mathbf{y}_{j-1} + \varepsilon_j,$$

where $\delta = (\delta_1, \ldots, \delta_p)'$, $\mathbf{y}_{j-1} = (y_{j-1}, \ldots, y_{j-p})'$ and $\{\varepsilon_j\} \sim$ i.i.d.$(0, \sigma^2)$. It is assumed that

$$(7.139) \qquad \delta(L) = \beta(L)(1 - L)^d,$$

where $\delta(L) = 1 - \delta_1 L - \cdots - \delta_p L^p$, $\beta(L) = 1 - \beta_1 L - \cdots - \beta_q L^q$ with $q = p - d$ and $\beta(x) \neq 0$ for $|x| \leq 1$. We also assume that the initial values y_j ($j = -p + 1, -p + 2, \ldots, 0$) to generate $y_1, y_2 \ldots$ are set at 0. Thus the process $\{y_j\}$ is the I(d) process with the error term given by $\varepsilon_j / \beta(L)$.

Our purpose here is to study the asymptotic distribution of the LSE of δ obtained from (7.138). One difficulty arises because the regressor $\{y_{j-1}\}$ in (7.138) follows (Problem 9.1)

$$(7.140) \qquad \mathcal{L}\left(\frac{1}{T^{2d}\sigma^2}\sum_{j=p+1}^{T} y_{j-1}y'_{j-1}\right) \to \mathcal{L}\left(\alpha^2\int_0^1 F_{d-1}^2(t)\,dt\, ee'\right),$$

where $e = (1,\ldots,1)' : p \times 1$ and $\{F_g(t)\}$ is the g-fold integrated Brownian motion, while $\alpha = \sum_{l=0}^{\infty} \alpha_l$ with α_l being the coefficient of L^l in the expansion of $1/\beta(L)$. Thus $\{y_{j-1}\}$ is collinear in the stochastic sense or cointegrated.

Various attempts have been made to cope with the above difficulty by a suitable transformation of the original model (7.138) and the parameter vector δ. Fuller (1976) gave a representation

$$(7.141) \qquad y_j = \sum_{i=1}^{d} \phi_i \Delta^{i-1} y_{j-1} + \sum_{i=1}^{q} \phi_{i+d} \Delta^d y_{j-i} + \varepsilon_j$$

$$= \phi' M y_{j-1} + \varepsilon_j,$$

where $\Delta = 1 - L$, $\phi = (\phi_1, \ldots, \phi_p)'$, $M = (M_1', M_2')'$, and

$$M_1 = \begin{pmatrix} 1 & 0 & 0 & \cdot & \cdot & \cdot & 0 \\ 1 & -1 & & & & & \cdot \\ 1 & -2 & 1 & & & & 0 \\ \cdot & \cdot & & & & & \cdot \\ \cdot & \cdot & & & & & \cdot \\ \cdot & \cdot & & & & & \cdot \\ 1 & (-1)\binom{d-1}{1} & \cdot & \cdot & \cdot & (-1)^{d-1} \end{pmatrix} : d \times p,$$

$$M_2 = \begin{pmatrix} 1 & \gamma_1 & \cdot & \cdot & \cdot & \gamma_d & & 0 \\ \cdot & \cdot & & & & \cdot & & \\ 0 & & & \cdot & & & \cdot & \\ & & 1 & \gamma_1 & \cdot & \cdot & & \gamma_d \end{pmatrix} : q \times p,$$

with $\gamma_k = (-1)^k \binom{d}{k}$. It follows from (7.138) and (7.141) that $\delta = M'\phi$. Thus it can be shown (Problem 9.2) that, when $d = 1$,

$$(7.142) \qquad \phi_1 = \sum_{i=1}^{q+1} \delta_i, \quad \phi_k = -\sum_{i=k}^{q+1} \delta_i, \quad (k = 2, \ldots, q+1),$$

while, when $d = 2$,

(7.143) $$\phi_1 = \sum_{i=1}^{q+2} \delta_i, \quad \phi_2 = -\sum_{i=2}^{q+2}(i-1)\delta_i, \quad \phi_k = \sum_{i=k}^{q+2}(i-k+1)\delta_i, \quad (k = 3, \ldots, q+2).$$

Note that $\phi_1 = 1$ for $d = 1$ and $\phi_1 = \phi_2 = 1$ for $d = 2$. In general $\phi_1 = \cdots = \phi_d = 1$ in (7.141). Since it holds that

(7.144) $$\Delta^d y_j = \beta_1 \Delta^d y_{j-1} + \cdots + \beta_q \Delta^d y_{j-q} + \varepsilon_j,$$
$$\Delta^d = \Delta \qquad (d = 1),$$
$$= \Delta - L\left(\Delta + \cdots + \Delta^{d-1}\right) \qquad (d \geq 2),$$

we can deduce that

$$\phi_i = 1 \quad (i = 1, \ldots, d), \qquad \phi_{i+d} = \beta_i \quad (i = 1, \ldots, q).$$

The d regressors $\{\Delta^{i-1} y_{j-1}\}$ $(i = 1, \ldots, d)$ in (7.141) follow I(d), I($d-1$), ..., I(1) processes, respectively, while the q regressors $\{\Delta^d y_{j-i}\} = \{\varepsilon_{j-i}/\beta(L)\}$ $(i = 1, \ldots, q)$ follow stationary processes. It then holds (Problem 9.3) that

(7.145) $$\mathcal{L}\left(D_T^{-1} M \sum_{j=p+1}^{T} y_{j-1} y'_{j-1} M' D_T^{-1}\right)$$
$$= \mathcal{L}\left(D_T^{-1} \sum_{j=p+1}^{T} \begin{pmatrix} x_{j-1} \\ u_{j-1} \end{pmatrix} (x'_{j-1}, u'_{j-1}) D_T^{-1}\right)$$
$$\to \mathcal{L}\left(\begin{pmatrix} \alpha^2 \sigma^2 F & 0 \\ 0 & \Gamma \end{pmatrix}\right),$$

where

$$D_T = \text{diag}(T^d, \ldots, T, \sqrt{T}, \ldots, \sqrt{T}) : p \times p,$$
$$x_{j-1} = (y_{j-1}, \Delta y_{j-1}, \ldots, \Delta^{d-1} y_{j-1})' : d \times 1,$$
$$u_{j-1} = (\Delta^d y_{j-1}, \ldots, \Delta^d y_{j-q})' = \left(\frac{\varepsilon_{j-1}}{\beta(L)}, \ldots, \frac{\varepsilon_{j-q}}{\beta(L)}\right)' : q \times 1,$$
$$F = [(F_{kl})] = \left[\left(\int_0^1 F_{d-k}(t) F_{d-l}(t)\, dt\right)\right] : d \times d,$$
$$\Gamma = [(\Gamma_{kl})] = E(u_j u'_j) = \left[\left(\frac{\sigma^2}{2\pi} \int_0^1 \frac{e^{i(k-l)\lambda}}{|\beta(e^{i\lambda})|^2}\, d\lambda\right)\right] : q \times q.$$

It is seen that Fuller's representation avoids the singularity problem arising from the original model (7.138).

Chan and Wei (1988) developed, to a large extent, asymptotic arguments concerning the LSEs for nonstationary AR models that contain, not only unit roots, but also negative unit roots and complex roots on the unit circle, which Jeganathan (1991) further extended to cases where roots are near the unit circle. Chan and Wei's transformation is somewhat hard to interpret, as will be seen shortly, but their way in asymptotic arguments which we follow in subsequent discussions, is much more convenient than Fuller's.

Let $\hat{\phi}$ be the LSE of ϕ in (7.141), which is given by

$$(7.146) \quad \hat{\phi} = \left(M \sum_{j=p+1}^{T} y_{j-1} y'_{j-1} M' \right)^{-1} M \sum_{j=p+1}^{T} y_{j-1} y_j$$

$$= \phi + \left(\sum_j \binom{x_{j-1}}{u_{j-1}} (x'_{j-1}, u'_{j-1}) \right)^{-1} \sum_j \binom{x_{j-1}}{u_{j-1}} \varepsilon_j$$

so that

$$D_T(\hat{\phi} - \phi) = \left(D_T^{-1} \sum_j \binom{x_{j-1}}{u_{j-1}} (x'_{j-1}, u'_{j-1}) D_T^{-1} \right)^{-1} D_T^{-1} \sum_j \binom{x_{j-1}}{u_{j-1}} \varepsilon_j.$$

Following Chan and Wei (1988), it may be proved that

$$\mathcal{L}\left(D_T^{-1} \sum_j \binom{x_{j-1}}{u_{j-1}} \varepsilon_j \right) \to \mathcal{L}\left(\binom{\alpha \sigma^2 \xi}{\eta} \right),$$

where η follows $N(0, \sigma^2 \Gamma)$ and is independent of

$$\xi = [(\xi_k)] = \left[\left(\int_0^1 F_{d-k}(t) dw(t) \right) \right] : d \times 1.$$

It follows from (7.145) and the continuous mapping theorem that

$$(7.147) \quad \mathcal{L}\left(D_T(\hat{\phi} - \phi) \right) = \mathcal{L}\left((M' D_T^{-1})^{-1} (\hat{\delta} - \delta) \right)$$

$$\to \mathcal{L}\left(\binom{\frac{1}{\alpha} F^{-1} \xi}{\Gamma^{-1} \eta} \right).$$

This last result appeals to intuition since ϕ_i ($i = 1, \ldots, d$) is the coefficient of the $I(d - i + 1)$ process $\{\Delta^{i-1} y_{j-1}\}$, while ϕ_{i+d} ($i = 1, \ldots, q$) is the coefficient of the stationary process $\{\Delta^d y_{j-i}\} = \{\varepsilon_{j-i}/\beta(L)\}$.

Chan and Wei (1988) considered the limiting distribution of $(\bar{M}' D_T^{-1})^{-1} (\hat{\delta} - \delta)$, where

$$\bar{M} = \begin{pmatrix} \bar{M}_1 Q_1 \\ M_2 \end{pmatrix} : p \times p,$$

with

$$\bar{M}_1 = M_1 \begin{pmatrix} I_d \\ 0 \end{pmatrix} : d \times d,$$

$$Q_1 = \begin{pmatrix} 1 & -\beta_1 & \cdot & \cdot & \cdot & -\beta_q & & 0 \\ & \ddots & \ddots & & & & \ddots & \\ 0 & & 1 & -\beta_1 & \cdot & \cdot & \cdot & -\beta_q \end{pmatrix} : d \times p.$$

It can be established (Problem 9.4) that

(7.148) $(\bar{M}' D_T^{-1})^{-1} (\hat{\delta} - \delta) = \left(D_T^{-1} \bar{M} \sum_j y_{j-1} y_{j-1}' \bar{M}' D_T^{-1} \right)^{-1} D_T^{-1} \bar{M} \sum_j y_{j-1} \varepsilon_j$

$$= \left(D_T^{-1} \sum_j \begin{pmatrix} z_{j-1} \\ u_{j-1} \end{pmatrix} (z_{j-1}', u_{j-1}') D_T^{-1} \right)^{-1}$$

$$\times D_T^{-1} \sum_j \begin{pmatrix} z_{j-1} \\ u_{j-1} \end{pmatrix} \varepsilon_j,$$

where

$$z_{j-1} = \left(\frac{\varepsilon_{j-1}}{(1-L)^d}, \ldots, \frac{\varepsilon_{j-1}}{1-L} \right)' : d \times 1,$$

$$u_{j-1} = \left(\frac{\varepsilon_{j-1}}{\beta(L)}, \ldots, \frac{\varepsilon_{j-q}}{\beta(L)} \right)' : q \times 1.$$

Then Chan and Wei (1988) proved that

$$\mathcal{L}\left((\bar{M}' D_T^{-1})^{-1} (\hat{\delta} - \delta) \right) \to \mathcal{L}\left(\begin{pmatrix} F^{-1} \xi \\ \Gamma^{-1} \eta \end{pmatrix} \right).$$

Comparing this expression with (7.147), it is seen that a complicated transformation with $(\bar{M}'D_T^{-1})^{-1}$ makes the limiting distribution independent of the factor $1/\alpha$, but that transformation, in turn, introduces a regressor z_{j-1} constructed from the matrix Q_1 which depends on β_is. In this sense, Fuller's representation (7.141) seems more appealing.

We have seen that some portions of the LSEs have a nonnormal limiting distribution under the transformations defined above. On the other hand, it follows from (7.147) that

$$\mathcal{L}\left(\sqrt{T}\left(\hat{\phi} - \phi\right)\right) \to \mathcal{L}\left(\begin{pmatrix} 0 \\ \Gamma^{-1}\eta \end{pmatrix}\right),$$

which yields

(7.149)
$$\mathcal{L}\left(\sqrt{T}\left(\hat{\delta} - \delta\right)\right) = \mathcal{L}\left(\sqrt{T}M'\left(\hat{\phi} - \phi\right)\right)$$
$$\to \mathcal{L}\left(M_2'\Gamma^{-1}\eta\right)$$
$$= N\left(0, \sigma^2 M_2'\Gamma^{-1}M_2\right),$$

where M_2 and Γ are defined in (7.141) and (7.145), respectively. The same asymptotic result arises if we first obtain the LSEs of β_i ($i = 1, \ldots, q$) from (7.144) and then substitute these into (7.139). It also follows from (7.145) that

$$\left(\frac{1}{T}\sum_{j=p+1}^{T} y_{j-1}y_{j-1}'\right)^{-1} \to M_2'\Gamma^{-1}M_2 \quad \text{in probability,}$$

which contrasts with (7.140). Since M_2 is a $q \times p$ matrix with rank q ($< p$) and Γ is a $q \times q$ nonsingular matrix, the $p \times p$ matrix $M_2'\Gamma^{-1}M_2$ is singular with rank q. Note that the rank decreases by the number of unit roots. Sims, Stock, and Watson (1990) also discussed a similar situation. Choi (1993) argues that a joint hypothesis such as $\sum_{i=1}^{p+d} \delta_i = 1$ is impossible to perform because of singularity of the covariance matrix, but it is still possible to conduct tests on individual coefficients. We shall consider this problem in Chapter 9.

Asymptotic normality is always ensured, regardless of the order q, if we consider the regression of y_j on $y_{j-1}, \ldots, y_{j-d-1}$ when $\{y_j\} \sim I(d)$. Suppose, for simplicity, that $q = 0$ so that $y_j = \delta'y_{j-1} + \varepsilon_j$, where $\delta = (\delta_1, \ldots, \delta_d)'$ and $y_{j-1} = (y_{j-1}, \ldots, y_{j-d})'$. Then we consider the regression relation

$$y_j = \hat{\delta}'y_{j-1} + \hat{\kappa}y_{j-d-1} + e_j = \delta'y_{j-1} + \kappa y_{j-d-1} + \varepsilon_j,$$

where the true value of κ is 0. We obtain

AUTOREGRESSIVE MODELS WITH MULTIPLE UNIT ROOTS

$$\begin{pmatrix} \hat{\delta} \\ \hat{\kappa} \end{pmatrix} = \left(\sum_j \begin{pmatrix} y_{j-1} \\ y_{j-d-1} \end{pmatrix} (y'_{j-1}, y_{j-d-1}) \right)^{-1} \sum_j \begin{pmatrix} y_{j-1} \\ y_{j-d-1} \end{pmatrix} y_j$$

$$= \begin{pmatrix} \delta \\ \kappa \end{pmatrix} + H' \left(H \sum_j \begin{pmatrix} y_{j-1} \\ y_{j-d-1} \end{pmatrix} (y_{j-1}, y_{j-d-1}) H' \right)^{-1}$$

$$\times H \sum_j \begin{pmatrix} y_{j-1} \\ y_{j-d-1} \end{pmatrix} \varepsilon_j,$$

where

$$H = \begin{pmatrix} H_1 \\ H_2 \end{pmatrix} = \begin{pmatrix} H_1 \\ 1 \;\; (-1)\binom{d}{1} \;\; (-1)^2\binom{d}{2} \cdots (-1)^d \end{pmatrix} \;:\; (d+1) \times (d+1)$$

with H_1 being the $d \times (d+1)$ matrix constructed from the first $d+1$ columns of M_1 in (7.141). Noting that

$$H \begin{pmatrix} y_{j-1} \\ y_{j-d-1} \end{pmatrix} = \left(\frac{\varepsilon_{j-1}}{(1-L)^d}, \ldots, \frac{\varepsilon_{j-1}}{1-L}, \varepsilon_{j-1} \right)' \;:\; (d+1) \times 1,$$

and using the same reasoning as before, we have

$$\sqrt{T} \begin{pmatrix} \hat{\delta} - \delta \\ \hat{\kappa} \end{pmatrix} = \left(\frac{1}{T} \sum_j \varepsilon_{j-1}^2 \right)^{-1} \frac{1}{\sqrt{T}} \sum_j \varepsilon_{j-1} \varepsilon_j H'_2 + o_p(1)$$

so that

$$\mathcal{L}\left(\sqrt{T} \begin{pmatrix} \hat{\delta} - \delta \\ \hat{\kappa} \end{pmatrix} \right) \to N(0, H'_2 H_2).$$

Note that the $(d+1) \times (d+1)$ covariance matrix $H'_2 H_2$ is of rank unity. In particular, if $d = 1$, we have

$$H'_2 H_2 = \begin{pmatrix} 1 & -1 \\ -1 & 1 \end{pmatrix}.$$

The above discussions have assumed that the true orders d and q are known. It may be of some interest to examine effects resulting from misspecifying these orders. Suppose first that the order q is misspecified to be $q = 0$. Then we attempt to regress

y_j on $\mathbf{y}_{j-1} = (y_{j-1}, \ldots, y_{j-d})'$ to obtain

$$\hat{\delta} = \delta + \left(\sum_j \mathbf{y}_{j-1}\mathbf{y}'_{j-1}\right)^{-1} \sum_j \mathbf{y}_{j-1}u_j,$$

where δ is now the parameter in $y_j = \delta' \mathbf{y}_{j-1} + u_j$ with $u_j = \varepsilon_j/\beta(L)$. It now follows (Problem 9.5) that

(7.150) $$\mathcal{L}\left(\left(\bar{M}'_1 G_T^{-1}\right)^{-1}(\hat{\delta} - \delta)\right) \to \mathcal{L}\left(F^{-1}\left(\xi + \frac{1-r}{2}e_d\right)\right),$$

where $\bar{M}_1$ is the $d \times d$ matrix constructed from the first d columns of M_1 in (7.141), $G_T = \text{diag}(T^d, \ldots, T)$, $e_d = (0, \ldots, 0, 1)' : d \times 1$, and $r = \sum_{l=0}^{\infty} \alpha_l^2/\alpha^2$. Note that the case of $d = 1$ was discussed in previous sections. In the present case, there exists no transformation that makes the distribution of $\hat{\delta} - \delta$ tend to normality.

We finally examine the effect when d is misspecified downward. Let us assume that the true value of d is greater than unity and q is arbitrary. Then suppose that we only consider

$$\hat{\delta} = \sum_{j=2}^{T} y_{j-1}y_j \bigg/ \sum_{j=2}^{T} y_{j-1}^2 = 1 + \sum_{j=2}^{T} y_{j-1}(y_j - y_{j-1}) \bigg/ \sum_{j=2}^{T} y_{j-1}^2.$$

It can be shown (Problem 9.6) that

(7.151) $$\mathcal{L}\left(T(\hat{\delta} - 1)\right) \to \mathcal{L}\left(\frac{1}{2}F_{d-1}^2(1) \bigg/ \int_0^1 F_{d-1}^2(t)\, dt\right).$$

This result implies that the LSE of δ applied to

$$y_j = \delta y_{j-1} + \eta_j, \quad (1-L)^{d-1}\eta_j = u_j, \quad \delta = 1,$$

is still consistent even if the error term η_j follows an I($d-1$) process for any integer $d\ (>1)$. Moreover, the limit of $T(\hat{\delta} - 1)$ is positive and does not depend on the parameters in $\{u_j\} = \{\varepsilon_j/\beta(L)\}$. Nabeya and Perron (1994) discussed in detail the case of $d = 2$ and obtained the c.f. associated with the limiting distribution.

The c.f. $\phi(\theta; d)$ of

$$Y_d = x \int_0^1 F_{d-1}^2(t)\, dt - \frac{1}{2}F_{d-1}^2(1)$$

$$= \int_0^1 \int_0^1 K_d(s, t)\, dw(s)\, dw(t)$$

may be most conveniently obtained by the Fredholm approach, where

$$K_d(s,t) = \frac{1}{((d-1)!)^2}\left[x\int_{\max(s,t)}^1 ((u-s)(u-t))^{d-1}\,du\right.$$

$$\left.-\frac{1}{2}((1-s)(1-t))^{d-1}\right].$$

We shall have $\phi(\theta;d) = (D_d(2i\theta))^{-1/2}$, where $D_d(\lambda)$ is the FD associated with $K_d(s,t)$. For $d = 2$ and 3 we obtain, using the computerized algebra REDUCE,

$$D_2(\lambda) = \frac{1}{2}(1 + \cos A \cos iA) + \frac{1}{4x}(A\cos iA \sin A + iA\cos A \sin iA),$$

$$D_3(\lambda) = \frac{1}{18}\left[4\left(1 + \cos B + \cos B\omega + \cos B\omega^2\right) + 2\cos B\cos B\omega\cos B\omega^2\right.$$

$$+\frac{1}{x}\left\{\frac{1}{2}\left(B\sin 2B + B\omega\sin 2B\omega + B\omega^2\sin 2B\omega^2\right)\right.$$

$$\left.\left.+2\left(B\sin B + B\omega\sin B\omega + B\omega^2\sin B\omega^2\right)\right\}\right],$$

where $A = (\lambda x)^{1/4}$, $B = (\lambda x)^{1/6}$ and $\omega = (1 + \sqrt{3}i)/2$. The stochastic process approach may be used to confirm the above results, at least, for $d = 2$. In fact, if we define $dX(t) = (\beta X(t) + w(t))\,dt$ with $X(0) = 0$ and $dZ(t) = \gamma Z(t)\,dt + dw(t)$ with $Z(0) = 0$, the two-stage application of Girsanov's theorem yields (Problem 9.7)

(7.152) $E\left(e^{\theta Y_2}\right) = E\left[\exp\left\{\frac{\beta^2}{2}\int_0^1 w^2(t)dt - \frac{\beta}{2}w^2(1) - \frac{\theta}{2}h^2(w)\right.\right.$

$$\left.\left.-\beta^2 w(1)h(w) + \frac{\beta}{2}\right\}\right]$$

$$= E\left[\exp\left\{-\frac{\beta+\gamma}{2}(Z^2(1) - 1) - \frac{\theta}{2}h^2(Z) - \beta^2 Z(1)h(Z)\right\}\right]$$

$$= |I_2 + A\Sigma|^{-1/2}\exp\left(\frac{\beta+\gamma}{2}\right),$$

where $\beta = (2\theta x)^{1/4}$, $\gamma = i\beta$ and

$$A = \begin{pmatrix} \beta+\gamma & \beta^2 \\ \beta^2 & \theta \end{pmatrix}, \quad \Sigma = E\left\{\begin{pmatrix} Z(1) \\ h(Z) \end{pmatrix}(Z(1), h(Z))\right\},$$

$$Z(1) = \int_0^1 e^{\gamma(1-t)}\,dw(t), \quad h(Z) = \int_0^1 e^{\beta(1-t)}Z(t)\,dt.$$

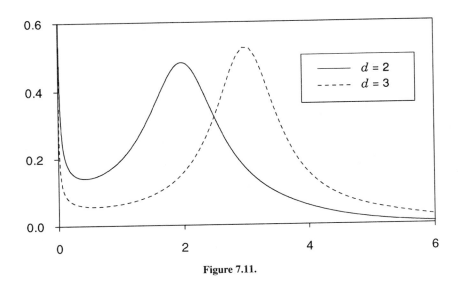

Figure 7.11.

Computing the last expression in (7.152) by any computerized algebra will give us $(D_2(2\theta))^{-1/2}$.

Figure 7.11 shows the graphs of limiting densities in (7.151) for $d = 2$ and 3. Both densities are bimodal. One peak is at the origin; the other is at about $x = 2$ for $d = 2$ and $x = 3$ for $d = 3$. Table 7.9 presents percent points of these two distributions together with the mean and standard deviation. It is interesting to note that

$$(7.153) \qquad E\left(\frac{1}{2}F_{d-1}^2(1) \bigg/ \int_0^1 F_{d-1}^2(t)\,dt\right) = d$$

for $d = 2, 3$. We have shown in (1.40) that (7.153) holds for $d = 1$. Then we naturally conjecture that (7.153) holds for any integer d (≥ 1).

Table 7.9. Limiting Unit Root Distributions in (7.151)

	Probability of a Smaller Value								
d	0.01	0.05	0.1	0.5	0.9	0.95	.099	Mean	SD
2	0.0075	0.168	0.510	1.946	3.384	4.038	5.528	2	1.146
3	0.0331	0.607	1.376	2.984	4.505	5.312	6.897	3	1.310

PROBLEMS

9.1 Establish the weak convergence result in (7.140).
9.2 Prove the relationships between δ and ϕ given in (7.142) and (7.143).
9.3 Establish the weak convergence result in (7.145).
9.4 Show that the relations as given in (7.148) hold.
9.5 Establish the weak convergence result in (7.150).
9.6 Prove that (7.151) holds.
9.7 Show that the relations as given in (7.152) hold.

CHAPTER 8

Estimation Problems in Noninvertible Moving Average Models

Dealing first with the noninvertible MA(1) model, we explore various asymptotic properties of ML estimators of the coefficient parameter, where ML estimation is restricted to local because of the complicated nature of global estimation. It turns out that the behavior of the estimator depends crucially on the assumption made concerning the initial value of the error process. We establish the consistency of the local MLE and approximate its asymptotic distribution. The discussion is directly extended to noninvertible seasonal MA models. We also extend the discussion to noninvertible MA models of infinite order, for which the pseudolocal MLE is found to be not necessarily consistent. Finally we describe a close relationship of the present problem with a problem arising in the state space model introduced in Chapter 1.

8.1 NONINVERTIBLE MOVING AVERAGE MODELS

This chapter is concerned with linear time series models with a root of the MA part on or near the unit circle. The phenomenon is usually incurred by overdifferencing. The model that we first deal with is

(8.1) $$y_j = \varepsilon_j - \alpha \varepsilon_{j-1}, \quad (j = 1, \ldots, T),$$

where the initial value ε_0 to generate $\{y_j\}$ is assumed to be either a random variable or a fixed constant. The difference turns out to be quite important. When ε_0 is random, it is assumed that $\varepsilon_0, \varepsilon_1 \ldots$ are i.i.d.$(0, \sigma^2)$ random variables so that $\{y_j\}$ is stationary. Then we naturally require that $|\alpha| \leq 1$ because of the identifiability condition. This case is referred to as the *stationary case*. When ε_0 is constant, we assume that $\varepsilon_0 = 0$. The zero initial condition may be replaced by any other constant independent of T as far as asymptotic arguments discussed below are concerned. The process $\{y_j\}$ is now not stationary. This is referred to as the *conditional case*. In this case, the parameter α

may take any value for the model (8.1) to be identifiable (Problem 1.1 in this chapter), but we assume, as in the stationary case, that $|\alpha| \leq 1$.

Let α_0 be the true value of α, which is assumed to take the form

$$(8.2) \qquad \alpha_0 = 1 - \frac{c}{T},$$

where c is a nonnegative constant. Note that $c \geq 0$ because $\alpha_0 \leq 1$. We say that the MA(1) model (8.1) is *noninvertible* when $\alpha_0 = 1$ and is *nearly noninvertible* when α_0 is close to unity.

With the above setting, our purpose is to study the asymptotic properties of estimators of α_0. In doing so, we assume that $\varepsilon_0, \varepsilon_1 \ldots \sim \text{NID}(0, \sigma^2)$ in the stationary case, while $\varepsilon_0 = 0$ and $\varepsilon_1, \varepsilon_2 \ldots \sim \text{NID}(0, \sigma^2)$ in the conditional case. The normality assumption is required only to construct likelihoods, but may be dropped for asymptotic arguments. The log-likelihood $l_{T1}(\alpha, \sigma^2)$ for the former is given by

$$(8.3) \qquad l_{T1}(\alpha, \sigma^2) = -\frac{T}{2}\log(2\pi\sigma^2) - \frac{1}{2}\log|\Omega(\alpha)| - \frac{1}{2\sigma^2} y'\Omega^{-1}(\alpha)y,$$

while the log-likelihood $l_{T2}(\alpha, \sigma^2)$ for the latter (Problem 1.2) is

$$(8.4) \qquad l_{T2}(\alpha, \sigma^2) = -\frac{T}{2}\log(2\pi\sigma^2) - \frac{1}{2\sigma^2} y'\Phi^{-1}(\alpha)y,$$

where $y = (y_1, \ldots, y_T)'$ and

$$(8.5) \quad \Omega(\alpha) = \begin{pmatrix} 1+\alpha^2 & -\alpha & & & 0 \\ -\alpha & 1+\alpha^2 & \cdot & & \\ & \cdot & \cdot & \cdot & \\ & & \cdot & \cdot & -\alpha \\ 0 & & & -\alpha & 1+\alpha^2 \end{pmatrix} : T \times T,$$

$$(8.6) \qquad \Phi(\alpha) = \Omega(\alpha) - \alpha^2 e_1 e_1' : T \times T,$$

with $e_1 = (1, 0, \ldots, 0)' : T \times 1$.

The analysis associated with $l_{T1}(\alpha, \sigma^2)$ in (8.3) is easier. We first have [Anderson (1971, p. 292) and Problem 1.3] that the eigenvalues λ_{jT} of $\Omega(\alpha)$ are given by

$$(8.7) \qquad \lambda_{jT} = 1 + \alpha^2 - 2\alpha \cos\frac{j\pi}{T+1}, \quad (j = 1, \ldots, T).$$

It then holds (Problem 1.4) that

$$(8.8) \qquad |\Omega(\alpha)| = \prod_{j=1}^{T} \lambda_{jT} = \sum_{j=0}^{T} \alpha^{2j}.$$

Moreover $\Omega(\alpha)$ and $d\Omega(\alpha)/d\alpha$ commute and the eigenvalues of $d\Omega(\alpha)/d\alpha$ can be explicitly given. These facts enable us to proceed to the analysis relying on the eigenvalue approach.

It is, however, not the case with $l_{T2}(\alpha, \sigma^2)$ in (8.4). The eigenvalues of $\Phi(\alpha)$ cannot be obtained explicitly except when $\alpha = 1$. Moreover $\Phi(\alpha)$ and $d\Phi(\alpha)/d\alpha$ do not commute. We cannot take the eigenvalue approach in this case, for which the Fredholm approach proves useful.

The estimator of α_0 considered in subsequent discussions is mainly the local MLE that attains the local maximum of $l_{T1}(\alpha, \sigma^2)$ or $l_{T2}(\alpha, \sigma^2)$ at a point closest to unity. It is certainly true that the study of the global MLE is important, but it seems quite hard to establish any meaningful asymptotic results on the global MLE.

The first simulation study of the global MLE under finite samples and α_0 close to unity seems to be that by Kang (1975), who noticed a tendency for the MLE to occur at $\alpha = 1$ even when $\alpha_0 < 1$. It was also argued that the MLE is quite different from the LSE that minimizes $y'\Omega^{-1}(\alpha)y$ in (8.3) when α_0 is close to unity. Cryer and Ledolter (1981) gave a theoretical explanation for the above fact and computed a bounding probability that the global MLE occurs at $\alpha = 1$. Anderson and Takemura (1986) extended the discussions to stationary autoregressive-moving average (ARMA) models with multiple unit roots on the MA part.

Sargan and Bhargava (1983) discussed the asymptotic properties of the local MLE in regression models with the error term given by a stationary, noninvertible MA(1) model. It was proved that there exists a local MLE $\hat{\alpha}$ such that $\hat{\alpha} - 1 = O_p(T^{-1})$, and the limiting probability of the local maximum at $\alpha = 1$ was computed. Tanaka and Satchell (1989) were more concerned with the asymptotic distribution of the local MLE for both the stationary and conditional cases, when $\alpha_0 = 1$. The existence of a consistent estimator was established for the conditional case by the Fredholm approach and they gave some ad hoc approximate distributions for both cases. Davis and Dunsmuir (1996) considered the stationary case when $\alpha_0 = 1 - (c/T)$, and studied the asymptotic distribution of the local MLE, pointing out the difference from the global MLE.

In Section 8.2 we deal with the stationary case with $\alpha_0 = 1 - (c/T)$, where the local MLE is examined in detail. The limiting distribution of the normalized local MLE, although not given explicitly, is approximated and is compared with the finite sample distributions and the Edgeworth-type approximation obtained in Tanaka (1984). Section 8.3 discusses the local MLE in the conditional case by the Fredholm approach taken in Tanaka and Satchell (1989). The analysis turns out to be more difficult than in the stationary case; hence we mainly deal with the situation where $\alpha_0 = 1$.

The above discussions are extended in two directions. One is oriented toward incorporating seasonality, which is quite parallel to discussions in Chapter 7. Continuous record asymptotics are also discussed as a related topic. These will be dealt with in Section 8.4. The other extension, which is discussed in Section 8.5, considers the situation where $\{\varepsilon_j\}$ is a dependent process. Nonetheless we act as if the log-likelihood (8.3) or (8.4) were true. It turns out that the consistency of the local MLE is not always ensured, but depends on the degree of dependence. This sharply contrasts with the consistency result in the AR unit root case, where the MLE or LSE of the AR(1) coefficient close to unity is always consistent even if the error terms are dependent.

The probability that the local MLE of α occurs at $\alpha = 1$ is of independent interest. In Section 8.6 we shall compute the limiting as well as finite sample probabilities for both cases. In Section 8.7 we point out a close relationship of the local MLEs considered so far with estimators in a state space model, whose simplest model was introduced in Section 1.1 of Chapter 1.

PROBLEMS

1.1 Show that the MA(1) model (8.1) is always identifiable when $\varepsilon_0 = 0$.
1.2 Derive the log-likelihood $l_{T2}(\alpha, \sigma^2)$ as in (8.4).
1.3 Derive the eigenvalues of $\Omega(\alpha)$ defined in (8.5).
1.4 Prove the second equality in (8.8).

8.2 THE LOCAL MLE IN THE STATIONARY CASE

Since we are mainly concerned with the estimation of α_0, our analysis is based on the concentrated log-likelihood

$$(8.9) \quad l_{T1}(\alpha) = -\frac{T}{2} \log y'\Omega^{-1}(\alpha)y - \frac{1}{2} \log |\Omega(\alpha)| - \frac{T}{2} \log \frac{2\pi}{T} - \frac{T}{2}.$$

The local MLE in the present case is equivalent to the one that attains the local maximum of $g_{T1}(\alpha)$ closest to unity, where

$$(8.10) \quad g_{T1}(\alpha) = l_{T1}(\alpha) - l_{T1}(1)$$

$$= -\frac{T}{2} \log \frac{y'\Omega^{-1}(\alpha)y}{y'\Omega^{-1}(1)y} - \frac{1}{2} \log \frac{|\Omega(\alpha)|}{|\Omega(1)|}$$

$$= -\frac{T}{2} \log \left[1 + \frac{1}{T} \frac{y'\left(\Omega^{-1}(\alpha) - \Omega^{-1}(1)\right) y}{\frac{1}{T} y'\Omega^{-1}(1)y} \right] - \frac{1}{2} \log \frac{\sum_{j=0}^{T} \alpha^{2j}}{T+1}.$$

The advantage of dealing with $g_{T1}(\alpha)$ rather than $l_{T1}(\alpha)$ will be made clear shortly. The function $g_{T1}(\alpha)$ is the logarithm of the likelihood ratio at the point α relative to $\alpha = 1$. Then it is intuitively clear that $g_{T1}(\alpha)$ as $T \to \infty$ will serve as the logarithm of the Radon-Nikodym derivative introduced in Chapter 4. It is also recognized that $g_{T1}(\alpha)$ serves as a test statistic for testing an MA unit root against a simple alternative, which will be discussed in Chapter 10.

THE LOCAL MLE IN THE STATIONARY CASE

Let us consider maximizing $g_{T1}(\alpha)$ in (8.10) over $-1 \leq \alpha \leq 1$. Suppose that T is large enough. Then we shall claim later that the maximization may be done over the values of α such that

$$(8.11) \qquad \alpha = 1 - \frac{\theta}{T}, \qquad 0 \leq \theta \leq \theta_1,$$

where θ_1 is some positive constant. We now proceed to consider $\mathcal{L}(g_{T1}(\alpha))$ as $T \to \infty$ under $\alpha = 1 - (\theta/T)$ and $\alpha_0 = 1 - (c/T)$. We can first show (Problem 2.1) that

$$(8.12) \qquad \frac{1}{T} y' \Omega^{-1}(1) y \to \sigma^2 \qquad \text{in probability},$$

$$(8.13) \qquad \frac{1}{T+1} \sum_{j=0}^{T} \alpha^{2j} \to \frac{\sinh \theta}{\theta e^{\theta}}.$$

We next consider

$$S_{T1} = \frac{1}{\sigma^2} y' \left(\Omega^{-1}(\alpha) - \Omega^{-1}(1) \right) y$$

$$= \sum_{j=1}^{T} \left(1 + \alpha_0^2 - 2\alpha_0 \delta_{jT}\right) \left(\frac{1}{1 + \alpha^2 - 2\alpha \delta_{jT}} - \frac{1}{2 - 2\delta_{jT}} \right) Z_j^2,$$

where $\delta_{jT} = \cos(j\pi/(T+1))$ and $\{Z_j\} \sim \text{NID}(0, 1)$. Noting that

$$1 - \delta_{jT} = 2 s_{jT}^2, \qquad \left(s_{jT} = \sin \frac{j\pi}{2(T+1)} \right),$$

$$1 + \alpha^2 - 2\alpha \delta_{jT} = (1 - \alpha)^2 + 2\alpha(1 - \delta_{jT})$$

$$= \frac{\theta^2}{T^2} + 4 \left(1 - \frac{\theta}{T} \right) s_{jT}^2,$$

we obtain

$$(8.14) \qquad S_{T1} = \frac{1}{T^2} \sum_{j=1}^{T} \frac{\left(c^2 + 4(T^2 - cT) s_{jT}^2\right)\left(-\theta^2 + 4\theta T s_{jT}^2\right)}{4 s_{jT}^2 \left(\theta^2 + 4(T^2 - \theta T) s_{jT}^2\right)} Z_j^2$$

$$= -\theta^2 \sum_{j=1}^{T} \frac{c^2 + 4(T^2 - cT) s_{jT}^2}{4T^2 s_{jT}^2 \left(\theta^2 + 4(T^2 - \theta T) s_{jT}^2\right)} Z_j^2 + \frac{\theta}{T} \sum_{j=1}^{T} Z_j^2$$

$$+ \frac{1}{T} \sum_{j=1}^{T} \left[\frac{\theta \left(c^2 + 4(T^2 - cT) s_{jT}^2\right)}{\theta^2 + 4(T^2 - \theta T) s_{jT}^2} - \theta \right] Z_j^2.$$

Then we can show [Davis and Dunsmuir (1996) and Problem 2.2] that

(8.15) $\quad \mathcal{L}(S_{T1}) \to \mathcal{L}\left(-\theta^2 \sum_{n=1}^{\infty} \frac{n^2\pi^2 + c^2}{n^2\pi^2(n^2\pi^2 + \theta^2)} Z_n^2 + \theta\right)$

so that

$$\mathcal{L}(g_{T1}(\alpha)) = \mathcal{L}\left(-\frac{1}{2}S_{T1} - \frac{1}{2}\log\frac{\sinh\theta}{\theta e^\theta} + o_p(1)\right)$$
$$\to \mathcal{L}(X_1(\theta)),$$

where

(8.16) $\quad X_1(\theta) = \frac{\theta^2}{2}\sum_{n=1}^{\infty} \frac{n^2\pi^2 + c^2}{n^2\pi^2(n^2\pi^2 + \theta^2)} Z_n^2 - \frac{1}{2}\log\frac{\sinh\theta}{\theta}.$

The limiting expression such as $X_1(\theta)$ was first dealt with in Tanaka and Satchell (1989), when $c = 0$. Davis and Dunsmuir (1996) obtained the present expression to argue that the MLE—global or local—is asymptotically the maximizer of $X_1(\theta)$. Although the expression $X_1(\theta)$ is more useful for present purposes, another equivalent expression based on the Fredholm approach is necessary for later purposes. We can show (Problem 2.3) that

(8.17) $\quad \mathcal{L}(X_1(\theta)) = \mathcal{L}\left(\int_0^1 \int_0^1 K_1(s, t; \theta) \, dw(s) \, dw(t) - \frac{1}{2}\log\frac{\sinh\theta}{\theta}\right),$

where $\{w(t)\}$ is the one-dimensional standard Brownian motion, while $K_1(s, t; \theta)$ is a symmetric kernel defined for $s \le t$ by

$$K_1(s, t; \theta) = -\frac{1}{2}\bigg[1 + c - c(s+t) - c^2 s(1-t) - (\theta - c)e^{-\theta(t-s)}$$
$$+ \frac{(\theta - c)^2}{\theta} e^{-\theta(1-s)} \sinh\theta(1-t)$$
$$- \frac{\theta e^{-\theta}}{\sinh\theta}\left(\cosh\theta(1-s) + \frac{c}{\theta}\sinh\theta(1-s)\right)$$
$$\times \left(\cosh\theta(1-t) + \frac{c}{\theta}\sinh\theta(1-t)\right)\bigg].$$

We can deduce from (8.16) that $K_1(s, t; \theta)$ with $\theta \neq 0$ is positive definite and its eigenvalues are given by $2n^2\pi^2(n^2\pi^2 + \theta^2)/(\theta^2(n^2\pi^2 + c^2))$. We can also obtain (Problem 2.4) the c.f. $\phi_1(u)$ of $X_1(\theta)$ in (8.16) as

(8.18) $\quad \phi_1(u) = \left[\frac{\sin\sqrt{a(u) + b(u)}}{\sqrt{a(u) + b(u)}} \frac{\sin\sqrt{a(u) - b(u)}}{\sqrt{a(u) - b(u)}} \bigg/ \frac{\sinh\theta}{\theta}\right]^{-1/2}$

$$\times \exp\left[-\frac{iu}{2}\log\frac{\sinh\theta}{\theta}\right],$$

where

$$a(u) = \frac{\theta^2(iu-1)}{2}, \quad b(u) = \frac{\theta\sqrt{\theta^2(1-iu)^2 + 4ic^2u}}{2}.$$

Let $\hat{\alpha}$ be the local MLE, which is defined to be the value of α closest to unity that attains the local maximum of $g_{T1}(\alpha)$ in (8.10). Since $g_{T1}(\alpha)$ is a continuously differentiable function of α on $-1 \leq \alpha \leq 1$, and it holds (Problem 2.5) that

(8.19) $$\left.\frac{dg_{T1}(\alpha)}{d\alpha}\right|_{\alpha=\pm 1} = 0,$$

$\hat{\alpha}$ is always a solution to $dg_{T1}(\alpha)/d\alpha = 0$. It follows that

$$\hat{\alpha} = \sup\left\{\alpha : |\alpha| \leq 1, \ \frac{dg_{T1}(\alpha)}{d\alpha} = 0, \ \frac{d^2 g_{T1}(\alpha)}{d\alpha^2} < 0\right\}.$$

The above definition of $\hat{\alpha}$ may be alternatively given in terms of $\hat{\theta} = T(1-\hat{\alpha})$ by introducing

$$h_{T1}(\theta) = l_{T1}(\alpha(\theta)) - l_{T1}(1) = g_{T1}(\alpha(\theta)),$$

where $\alpha(\theta) = 1 - (\theta/T)$. Then we can define the local MLE $\hat{\theta}$ of θ by

(8.20) $$\hat{\theta} = \inf\left\{\theta : 0 \leq \theta \leq 2T, \ \frac{dh_{T1}(\theta)}{d\theta} = 0, \ \frac{d^2 h_{T1}(\theta)}{d\theta^2} < 0\right\},$$

where

$$\frac{d^l h_{T1}(\theta)}{d\theta^l} = \left(-\frac{1}{T}\right)^l \frac{d^l l_{T1}(\alpha)}{d\alpha^l} = \left(-\frac{1}{T}\right)^l \frac{d^l g_{T1}(\alpha)}{d\alpha^l}.$$

We are now led to consider the weak convergence of

$$\frac{dh_{T1}(\theta)}{d\theta} = \frac{1}{2} \frac{y' \dfrac{d\Omega^{-1}(\alpha)}{d\alpha} y}{y'\Omega^{-1}(\alpha)y} + \frac{1}{2T}\frac{d}{d\alpha} \log|\Omega(\alpha)|,$$

$$\frac{d^2 h_{T1}(\theta)}{d\theta^2} = -\frac{1}{2T} \frac{y' \dfrac{d^2\Omega^{-1}(\alpha)}{d\alpha^2} y \cdot y'\Omega^{-1}(\alpha)y - \left(y' \dfrac{d\Omega^{-1}(\alpha)}{d\alpha} y\right)^2}{(y'\Omega^{-1}(\alpha)y)^2}$$

$$- \frac{1}{2T^2}\frac{d^2}{d\alpha^2} \log|\Omega(\alpha)|.$$

We have already shown that

$$\mathcal{L}(h_{T1}(\theta)) \to \mathcal{L}(X_1(\theta)),$$

where $X_1(\theta)$ is given in (8.16). We can further show (Problem 2.6) that

$$\mathcal{L}\left(\frac{d^l h_{T1}(\theta)}{d\theta^l}\right) \to \mathcal{L}\left(\frac{d^l X_1(\theta)}{d\theta^l}\right), \quad (l = 1, 2), \tag{8.21}$$

where

$$\frac{dX_1(\theta)}{d\theta} = \sum_{n=1}^{\infty} \frac{\theta(n^2\pi^2 + c^2)}{(n^2\pi^2 + \theta^2)^2} Z_n^2 - \frac{1}{2}b_1(\theta), \tag{8.22}$$

$$\frac{d^2 X_1(\theta)}{d\theta^2} = \sum_{n=1}^{\infty} \frac{(n^2\pi^2 + c^2)(n^2\pi^2 - 3\theta^2)}{(n^2\pi^2 + \theta^2)^3} Z_n^2 - \frac{1}{2}b_2(\theta), \tag{8.23}$$

with

$$b_1(\theta) = \frac{d}{d\theta} \log \frac{\sinh \theta}{\theta}$$

$$= \coth \theta - \frac{1}{\theta} = \sum_{n=1}^{\infty} \frac{2\theta}{n^2\pi^2 + \theta^2},$$

$$b_2(\theta) = \frac{d^2}{d\theta^2} \log \frac{\sinh \theta}{\theta}$$

$$= -\operatorname{cosech}^2 \theta + \frac{1}{\theta^2} = \sum_{n=1}^{\infty} \frac{2(n^2\pi^2 - \theta^2)}{(n^2\pi^2 + \theta^2)^2}.$$

Note here that $b_1(0) = 0$ and $b_2(0) = \frac{1}{3}$ so that

$$\left.\frac{dX_1(\theta)}{d\theta}\right|_{\theta=0} = 0,$$

$$\left.\frac{d^2 X_1(\theta)}{d\theta^2}\right|_{\theta=0} = \sum_{n=1}^{\infty} \left(\frac{1}{n^2\pi^2} + \frac{c^2}{n^4\pi^4}\right) Z_n^2 - \frac{1}{6}. \tag{8.24}$$

It is an immediate consequence of the above results that the limiting probability of $\hat{\alpha}$ occurring at $\alpha = 1$ is given (Problem 2.7) by

$$\lim_{T\to\infty} P(\hat{\alpha} = 1) = \lim_{T\to\infty} P\left(\left.\frac{d^2 h_{T1}(\theta)}{d\theta^2}\right|_{\theta=0} < 0\right) \tag{8.25}$$

$$= P\left(\left.\frac{d^2 X_1(\theta)}{d\theta^2}\right|_{\theta=0} < 0\right)$$

$$= \frac{1}{\pi} \int_0^{\infty} \operatorname{Re}\left[\frac{1 - e^{-i\theta/6}}{i\theta} \phi(\theta)\right] d\theta,$$

where $\phi(\theta)$ is the c.f. of $d^2X_1(\theta)/d\theta^2|_{\theta=0} + \frac{1}{6}$ given by

$$(8.26) \quad \phi(\theta) = \left[\frac{\sin\sqrt{i\theta + a(\theta)}}{\sqrt{i\theta + a(\theta)}} \frac{\sin\sqrt{i\theta - a(\theta)}}{\sqrt{i\theta - a(\theta)}}\right]^{-1/2},$$

$$a(\theta) = \sqrt{-\theta^2 + 2ic^2\theta}.$$

Note that $\lim_{T\to\infty} P(\hat{\alpha} = 1)$ is clearly a decreasing function of c. The limiting probability will be computed in Section 8.6. The c.f. $\phi(\theta)$ defined above will also emerge when we discuss limiting local powers of MA unit root tests in Chapter 10.

The weak convergence result in (8.21) may also be described by the Fredholm approach, that is,

$$\mathcal{L}\left(\frac{d^l h_{T1}(\theta)}{d\theta^l}\right) \to \mathcal{L}\left(\int_0^1 \int_0^1 \frac{\partial^l K_1(s,t;\theta)}{\partial \theta^l} dw(s)\,dw(t) - \frac{1}{2}\frac{d^l}{d\theta^l}\log\frac{\sinh\theta}{\theta}\right)$$

for $l = 1, 2$. By means of the Cramér–Wold device [Billingsley (1968, p. 48)], it is ensured that the joint weak convergence of $dh_{T1}(\theta)/d\theta$ and $d^2h_{T1}(\theta)/d\theta^2$ holds.

To describe the asymptotic distribution of the local MLE $\hat{\alpha}$, we first show that, for any $\delta > 0$ and for all $T \geq T_1$ with T_1 fixed, there exists a positive constant θ_1 such that

$$(8.27) \quad P\left(1 - \frac{\theta_1}{T} \leq \hat{\alpha} \leq 1\right) = P(0 \leq \hat{\theta} \leq \theta_1) \geq 1 - \delta,$$

that is,

$$\hat{\alpha} - 1 = O_p\left(\frac{1}{T}\right) \quad \text{or} \quad \hat{\theta} = T(1 - \hat{\alpha}) = O_p(1).$$

The case of $\alpha_0 = 1$ was first proved by Sargan and Bhargava (1983), following which we consider

$$(8.28) \quad \mu(\theta) = E\left(\frac{dX_1(\theta)}{d\theta}\right) = -\sum_{n=1}^{\infty} \frac{\theta(\theta^2 - c^2)}{(n^2\pi^2 + \theta^2)^2} < 0 \quad \text{for} \quad \theta > c,$$

$$(8.29) \quad \sigma^2(\theta) = V\left(\frac{dX_1(\theta)}{d\theta}\right) = \sum_{n=1}^{\infty} \frac{2\theta^2(n^2\pi^2 + c^2)^2}{(n^2\pi^2 + \theta^2)^4}$$

$$\leq \sum_{n=1}^{\infty} \frac{2\theta^2}{(n^2\pi^2 + \theta^2)^2} \quad \text{for} \quad \theta > c.$$

We have

$$\frac{\sigma^2(\theta)}{\mu^2(\theta)} \leq \frac{2}{(\theta^2 - c^2)^2} \frac{1}{\sum_{n=1}^{\infty} \frac{1}{(n^2\pi^2 + \theta^2)^2}} \quad \text{for} \quad \theta > c.$$

Since it holds (Problem 2.8) that

$$\sum_{n=1}^{\infty} \frac{\theta^3}{(n^2\pi^2 + \theta^2)^2} = \frac{1}{4} - \frac{1}{2\theta} + \frac{1}{2(e^{2\theta} - 1)} + \frac{\theta e^{2\theta}}{(e^{2\theta} - 1)^2}, \quad (8.30)$$

and the right side converges monotonically to $\frac{1}{4}$ from below, there exists constants A and B such that

$$\frac{\sigma^2(\theta)}{\mu^2(\theta)} \leq \frac{A}{\theta} \quad \text{for} \quad \theta \geq B > c.$$

Then, from Chebyshev's inequality,

$$\lim_{T \to \infty} P\left(\frac{dh_{T1}(\theta)}{d\theta} \geq 0\right) = P\left(\frac{dX_1(\theta)}{d\theta} - \mu(\theta) \geq -\mu(\theta)\right)$$

$$\leq \frac{\sigma^2(\theta)}{\mu^2(\theta)} \leq \frac{A}{\theta},$$

which can be made smaller than δ by taking $\theta = \theta_1 = \max(B, A/\delta)$. Then T_1 can be chosen so that (8.27) holds for all $T \geq T_1$.

We can now move on to study the limiting distribution of $T(\hat{\alpha} - 1) = -\hat{\theta}$. Unfortunately, $T(\hat{\alpha} - 1)$ cannot be expressed explicitly, unlike the AR unit root case. In fact, even when $\alpha_0 = 1$ so that $c = 0$, (8.16) and (8.21) will lead us to

$$\mathcal{L}\left(\left.\frac{d^l h_{T1}(\theta)}{d\theta^l}\right|_{\theta=0}\right) \to \mathcal{L}\left(\left.\frac{d^l X_1(\theta)}{d\theta^l}\right|_{\theta=0}\right)$$

$$= \begin{cases} \mathcal{L}(0) & (l = 2k-1), \\ \mathcal{L}\left((-1)^{k-1}(2k-1)! \sum_{n=1}^{\infty} \frac{kZ_n^2 - 1}{(n\pi)^{2k}}\right) & (l = 2k). \end{cases}$$

Thus an expansion such as

$$0 = \left.\frac{dX_1(\theta)}{d\theta}\right|_{\theta=\hat{\theta}} = \hat{\theta}\left.\frac{d^2X_1(\theta)}{d\theta^2}\right|_{\theta=0} + \frac{\hat{\theta}^3}{6}\left.\frac{d^4X_1(\theta)}{d\theta^4}\right|_{\theta=0} + \cdots$$

is not helpful here. We need to return to (8.20) and must content ourselves with the following result, whose proof is given in Davis and Dunsmuir (1996).

Theorem 8.1. *Suppose that $y_j = \varepsilon_j - \alpha\varepsilon_{j-1}$, where $\varepsilon_0, \varepsilon_1, \ldots, \sim \text{NID}(0, \sigma^2)$. Then it holds that, as $T \to \infty$ under $\alpha = 1 - (\theta/T)$ and $\alpha_0 = 1 - (c/T)$,*

$$\mathcal{L}(T(\hat{\alpha} - 1)) = \mathcal{L}(-\hat{\theta}) \to \mathcal{L}(-\theta(c)),$$

where $\hat{\alpha}$ is the local MLE and

$$\theta(c) = \inf\left\{\theta \,:\, \theta \geq 0, \; \frac{dX_1(\theta)}{d\theta} = 0, \; \frac{d^2X_1(\theta)}{d\theta^2} < 0\right\}.$$

Unlike the AR unit root case, the limiting distribution of $T(\hat{\alpha} - 1)$ or the distribution of $-\theta(c)$ cannot be exactly computed, so approximations are inevitable. One way is to truncate the infinite series for $X_1(\theta)$ in (8.16) and to find a local maximizer of this truncated series. A number of replications of this procedure will yield an approximation, which was employed by Davis and Dunsmuir (1996). The approximation should be essentially the same as the one based on the likelihood under large samples.

Another intuitively appealing, although ad hoc, approximation was used by Shephard (1993) when $c = 0$, based on an idea of Huber's (1964). Suppose that a local maximum of an objective function $f(\theta)$ defined on $[0, \infty)$ occurs at $\bar{\theta}$. If $f(\theta)$ is continuously differentiable and $df(\theta)/d\theta$ is strictly decreasing, the maximum $\bar{\theta}$ is unique and

(8.31) $$P(\bar{\theta} \geq x) = P\left(\left.\frac{df(\theta)}{d\theta}\right|_{\theta=x} \geq 0\right).$$

Our objective function is $X_1(\theta)$ defined in (8.16), for which we cannot expect monotonicity of $dX_1(\theta)/d\theta$, so (8.31) can only be used as an approximation in the following way:

$$\lim_{T \to \infty} P(T(\hat{\alpha} - 1) \leq x) = P(-\theta(c) \leq x)$$

$$\cong P\left(\left.\frac{dX_1(\theta)}{d\theta}\right|_{\theta=-x} \geq 0\right), \quad (x < 0).$$

It may be argued that this approximation formula can be equally well used to approximate the limiting distribution of the global MLE. The approximate probability need not be simulated, but can be computed (Problem 2.9) for $x < 0$ as

(8.32) $$F_1(x) = P\left(\left.\frac{dX_1(\theta)}{d\theta}\right|_{\theta=-x} \geq 0\right)$$

$$= P\left(-x \sum_{n=1}^{\infty} \frac{n^2\pi^2 + c^2}{(n^2\pi^2 + x^2)^2} Z_n^2 \geq \frac{1}{2}\left(\frac{1}{x} - \coth x\right)\right)$$

$$= 1 - \frac{1}{\pi} \int_0^{\infty} \mathrm{Re}\left[\frac{1 - \exp\left\{\frac{i\theta}{2x}\left(\frac{1}{x} - \coth x\right)\right\}}{i\theta} \phi(\theta; x)\right] d\theta,$$

where

(8.33) $$\phi(\theta; x) = E\left[\exp\left\{i\theta \sum_{n=1}^{\infty} \frac{n^2\pi^2 + c^2}{(n^2\pi^2 + x^2)^2} Z_n^2\right\}\right]$$

$$= \frac{\sinh x}{x}\left[\frac{\sin\sqrt{c(\theta)+d(\theta)}}{\sqrt{c(\theta)+d(\theta)}} \frac{\sin\sqrt{c(\theta)-d(\theta)}}{\sqrt{c(\theta)-d(\theta)}}\right]^{-1/2},$$

with

$$c(\theta) = i\theta - x^2, \quad d(\theta) = \sqrt{2i\theta(c^2 - x^2) - \theta^2}.$$

Figure 8.1 shows graphs of $F_1(x)$ in (8.32) for $(\alpha_0, c) = (1, 0)$ and the distribution function of $T(\hat{\alpha} - 1)$ for the same (α_0, c) and $T = 50$. This latter distribution was computed from 50,000 replications. It is seen that $F_1(x)$ gives a reasonably good approximation, although the left-hand tail is slightly heavier. This implies that the difference between the global and local MLEs is slight. In fact, Davis and Dunsmuir (1996) report on the basis of simulations that, when $c = 0$, the global maximum of $X_1(\theta)$ and the local maximum closest to zero differ in about 1% of the realizations of the process $X_1(\theta)$. Figures 8.2, 8.3, and 8.4 present the corresponding graphs for $(\alpha_0, c) = (0.9, 5), (0.8, 10)$, and $(0.6, 20)$, respectively, with $T = 50$. In Figure 8.4 we have also presented the Edgeworth-type approximation obtained in Tanaka (1984):

$$P(T(\hat{\alpha} - 1) \leq x) \sim \Phi(y) - \frac{\alpha_0}{\sqrt{T(1-\alpha_0^2)}} \phi(y),$$

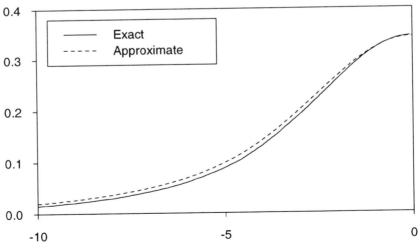

Figure 8.1.

THE LOCAL MLE IN THE STATIONARY CASE

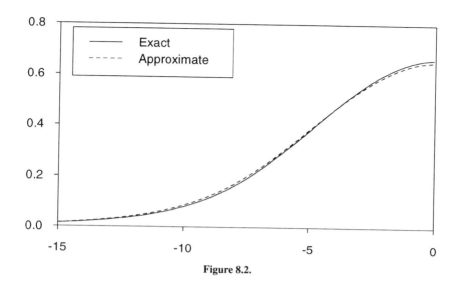

Figure 8.2.

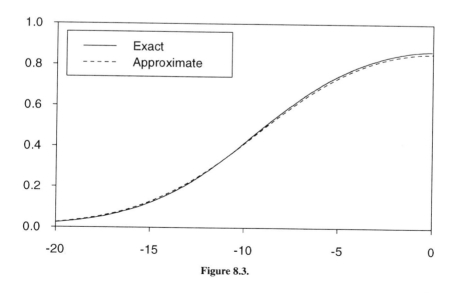

Figure 8.3.

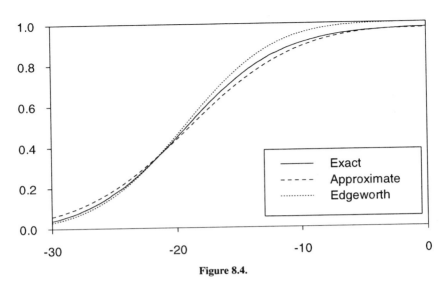

Figure 8.4.

where $y = x/\sqrt{T} + \sqrt{T}(1 - \alpha_0)$, $\Phi(y)$ and $\phi(y)$ are the N(0, 1) distribution function and probability density, respectively. The approximation by $F_1(x)$ performs fairly well for $(\alpha_0, c) = (0.9, 5)$ and $(0.8, 10)$. Even for $(\alpha_0, c) = (0.6, 20)$, it performs better than the Edgeworth approximation.

Figure 8.5 presents graphs of $dF_1(x)/dx$ for $c = 0, 5$, and 10, which are supposed to approximate probability densities of $-\theta(c)$ for $x < 0$. It is seen that the area under the curve gets large with c. It is also seen that the shape tends to be symmetric around $x = -c$ as c becomes large. In fact, as is shown in Figure 8.6, $dF_1(x)/dx$ with $c = 20$ is reasonably well approximated by the density of $N(-c, 2c)$.

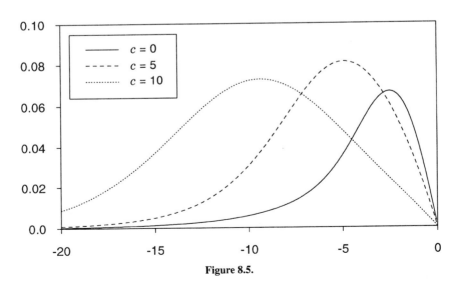

Figure 8.5.

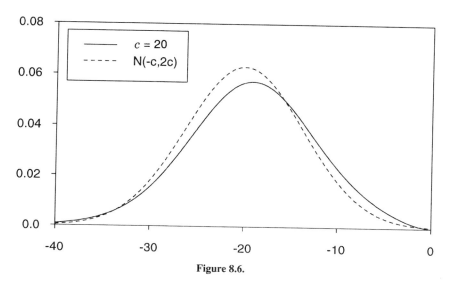

Figure 8.6.

Table 8.1 tabulates percent points of $F_1(x)$ in (8.32) for $c = 0, 5, 10$, and 15, which serve as an approximation to those of the limiting distribution of $T(\hat{\alpha} - 1)$. The percent points obtained by simulations in Davis and Dunsmuir (1996) are also presented. It is seen that the discrepancy between the two approximations becomes smaller as c gets large.

PROBLEMS

2.1 Prove (8.12) and (8.13).
2.2 Show that (8.15) holds.
2.3 Prove (8.17).
2.4 Show that the c.f. of $X_1(\theta)$ defined in (8.16) is given by (8.18).

Table 8.1. Approximate Percent Points of $\lim_{T \to \infty} P(T(\hat{\alpha} - 1) \leq x)$

| Case | Method | Probability of a Smaller Value | | | | |
		0.01	0.05	0.1	0.2	0.3
$c = 0$	F_1	-12.309	-6.970	-4.950	-3.002	-1.430
	D–D[a]	-11.253	-6.522	-4.736	-2.892	
$c = 5$	F_1	-16.118	-11.507	-9.569	-7.486	-6.051
	D–D[a]	-15.913	-11.245	-9.373	-7.240	
$c = 10$	F_1	-22.599	-17.943	-15.787	-13.345	-11.639
	D–D[a]	-22.704	-17.972	-15.699	-13.226	
$c = 15$	F_1	-29.282	-24.299	-21.885	-19.091	-17.124
	D–D[a]	-29.407	-24.472	-21.897	-19.019	

[a]Percent points obtained by simulations in Davis and Dunsmuir (1996).

2.5 Prove (8.19).

2.6 Establish (8.21).

2.7 Show that the c.f. of $d^2X_1(\theta)/d\theta^2|_{\theta=0} + \frac{1}{6}$ is given by (8.26), where $d^2X_1(\theta)/d\theta^2|_{\theta=0}$ is defined in (8.24).

2.8 Prove that (8.30) holds using

$$\frac{\sinh\theta}{\theta} = \prod_{n=1}^{\infty}\left(1 + \frac{\theta^2}{n^2\pi^2}\right).$$

2.9 Derive the c.f. $\phi(\theta; x)$ as in (8.33).

8.3 THE LOCAL MLE IN THE CONDITIONAL CASE

As in the stationary case, we consider the concentrated log-likelihood

$$(8.34)\qquad l_{T2}(\alpha) = -\frac{T}{2}\log y'\Phi^{-1}(\alpha)y - \frac{T}{2}\log\frac{2\pi}{T} - \frac{T}{2}$$

and define

$$(8.35)\qquad g_{T2}(\alpha) = l_{T2}(\alpha) - l_{T2}(1)$$

$$= -\frac{T}{2}\log\left[1 + \frac{1}{T}\frac{y'\left(\Phi^{-1}(\alpha) - \Phi^{-1}(1)\right)y}{\frac{1}{T}y'\Phi^{-1}(1)y}\right],$$

where $\Phi^{-1}(\alpha) = C'(\alpha)C(\alpha)$ and

$$(8.36)\qquad C(\alpha) = \begin{pmatrix} 1 & & & & 0 \\ \alpha & 1 & & & \\ \cdot & \cdot & \cdot & & \\ \cdot & & \cdot & \cdot & \\ \cdot & & & \cdot & \cdot \\ \alpha^{T-1} & \cdot & \cdot & \alpha & 1 \end{pmatrix} : T\times T.$$

Let us derive $\mathcal{L}(g_{T2}(\alpha))$ as $T\to\infty$ under $\alpha = 1 - (\theta/T)$ and $\alpha_0 = 1 - (c/T)$. Noting that $y = C^{-1}(\alpha_0)\varepsilon$, it is easy to show (Problem 3.1) that

$$(8.37)\qquad \frac{1}{T}y'\Phi^{-1}(1)y = \frac{1}{T}\varepsilon'\left(C^{-1}(\alpha_0)\right)'C'(1)C(1)C^{-1}(\alpha_0)\varepsilon \to \sigma^2$$

THE LOCAL MLE IN THE CONDITIONAL CASE

in probability. Then we consider

$$S_{T2} = \frac{1}{\sigma^2} y' \left(\Phi^{-1}(\alpha) - \Phi^{-1}(1) \right) y$$

$$= \frac{1}{T} Z'(B_T - \theta I_T) Z + \frac{\theta}{T} Z'Z,$$

where $Z \sim N(0, I_T)$ and

$$B_T = T \left[(C^{-1}(\alpha_0))' \{ C'(\alpha)C(\alpha) - C'(1)C(1) \} C^{-1}(\alpha_0) \right].$$

Since the matrices in the square brackets do not commute, the eigenvalue approach does not apply here. Using the Fredholm approach, it can be shown (Problem 3.2) that

(8.38) $$\mathcal{L}(g_{T2}(\alpha)) = \mathcal{L}\left(-\frac{1}{2} S_{T2} + o_p(1) \right)$$
$$\to \mathcal{L}(X_2(\theta)),$$

where

(8.39) $$X_2(\theta) = \int_0^1 \int_0^1 K_2(s, t; \theta) \, dw(s) \, dw(t) - \frac{\theta}{2},$$

and $K_2(s, t; \theta)$ is a symmetric kernel defined for $s \leq t$ by

(8.40) $$K_2(s, t; \theta) = -\frac{1}{2} \Bigg[-c - c^2(1 - t) - (\theta - c) e^{-\theta(t-s)}$$
$$+ \frac{(\theta - c)^2}{\theta} e^{-\theta(1-s)} \sinh \theta(1 - t) \Bigg].$$

The MLE—global or local—will be asymptotically the maximum of $X_2(\theta)$, as in the stationary case.

The distribution of $X_2(\theta)$ may be obtained by computing its c.f., although much is involved. Suppose, for simplicity, that $c = 0$. Then the kernel defined in (8.40) for $s \leq t$ reduces to

(8.41) $$K_2(s, t; \theta) = \frac{\theta}{2} e^{-\theta(1-s)} \cosh \theta(1 - t).$$

We can show (Problem 3.3) that the FD of this kernel is given by

(8.42) $$D_2(\lambda) = \left[\cos \theta \sqrt{\frac{\lambda}{2} - 1} - \sqrt{\frac{\lambda}{2} - 1} \sin \theta \sqrt{\frac{\lambda}{2} - 1} \right] \Big/ e^{\theta}.$$

Thus the c.f. $\phi_2(u)$ of $X_2(\theta)$ when $c = 0$ is given by $(D_2(2iu))^{-1/2} \exp(-iu\theta/2)$. It is easily verified that the zeros of $D_2(\lambda)$ are all positive and simple, and its nth smallest zero or eigenvalue λ_n lies in the interval

$$2\left(\left(\frac{(n-1)\pi}{\theta}\right)^2 + 1\right) < \lambda_n < 2\left(\left(\frac{(n-\frac{1}{2})\pi}{\theta}\right)^2 + 1\right).$$

Then we have, when $c = 0$,

(8.43) $$X_2(\theta) = \sum_{n=1}^{\infty} \frac{1}{\lambda_n} Z_n^2 - \frac{\theta}{2},$$

where $\{Z_n\} \sim \text{NID}(0, 1)$.

Let $\hat{\alpha}$ be the local MLE, that is, the value of α closest to unity that attains the local maximum of $g_{T2}(\alpha)$ in (8.35). Note that $dg_{T2}(\alpha)/d\alpha|_{\alpha=\pm 1} \neq 0$, unlike the stationary case. It follows that $\hat{\alpha} = 1$ if $dg_{T2}(\alpha)/d\alpha|_{\alpha=1} > 0$; otherwise

$$\hat{\alpha} = \sup\left\{\alpha : |\alpha| < 1, \frac{dg_{T2}(\alpha)}{d\alpha} = 0, \frac{d^2 g_{T2}(\alpha)}{d\alpha^2} < 0\right\}.$$

If there exists no such $\hat{\alpha}$, we put $\hat{\alpha} = -1$.

Let us define

$$h_{T2}(\theta) = l_{T2}(\alpha(\theta)) - l_{T2}(1) = g_{T2}(\alpha(\theta)),$$

where $\alpha(\theta) = 1 - (\theta/T)$. Then, in terms of the local MLE $\hat{\theta} = T(1 - \hat{\alpha})$, we have that $\hat{\theta} = 0$ if $dh_{T2}(\theta)/d\theta|_{\theta=0} < 0$; otherwise

$$\hat{\theta} = \inf\left\{\theta : 0 < \theta < 2T, \frac{dh_{T2}(\theta)}{d\theta} = 0, \frac{d^2 h_{T2}(\theta)}{d\theta^2} < 0\right\},$$

where

$$\frac{d^l h_{T2}(\theta)}{d\theta^l} = \left(-\frac{1}{T}\right)^l \frac{d^l l_{T2}(\alpha)}{d\alpha^l} = \left(-\frac{1}{T}\right)^l \frac{d^l g_{T2}(\alpha)}{d\alpha^l}.$$

We put $\hat{\theta} = 2T$ if the above $\hat{\theta}$ does not exist.

To discuss the asymptotic behavior of $\hat{\theta}$, we need to establish that, as $T \to \infty$ under $\alpha = 1 - (\theta/T)$ and $\alpha_0 = 1 - (c/T)$,

(8.44) $$\mathcal{L}\left(\frac{d^l h_{T2}(\theta)}{d\theta^l}\right) = \mathcal{L}\left(\left(-\frac{1}{T}\right)^l \frac{d^l g_{T2}(\alpha)}{d\alpha^l}\right)$$

$$\to \mathcal{L}\left(\frac{d^l X_2(\theta)}{d\theta^l}\right)$$

THE LOCAL MLE IN THE CONDITIONAL CASE

for $l = 0$ and 1, where $X_2(\theta)$ is given in (8.39), while

(8.45) $\quad \dfrac{dX_2(\theta)}{d\theta} = \displaystyle\int_0^1 \int_0^1 \dfrac{\partial K_2(s,t;\theta)}{\partial \theta} \, dw(s) \, dw(t) - \dfrac{1}{2},$

(8.46) $\quad \dfrac{\partial K_2(s,t;\theta)}{\partial \theta} = \dfrac{1}{4}\left[e^{-\theta|s-t|}\left\{1 + \dfrac{c^2}{\theta^2} - \left(\theta - \dfrac{c^2}{\theta}\right)|s-t|\right\}\right.$

$\qquad\qquad\qquad\qquad \left. + e^{-\theta(2-s-t)}\left\{1 - \dfrac{c^2}{\theta^2} - \dfrac{(\theta-c)^2}{\theta}(2-s-t)\right\}\right].$

Note that the case of $l = 0$ in (8.44) was already established in (8.38). The case of $l = 1$ can be similarly verified [Tanaka and Satchell (1989) and Problem 3.4].

We now obtain

$\qquad \left.\dfrac{dX_2(\theta)}{d\theta}\right|_{\theta=0} = \displaystyle\int_0^1 \int_0^1 \left.\dfrac{\partial K_2(s,t;\theta)}{\partial \theta}\right|_{\theta=0} dw(s)\, dw(t) - \dfrac{1}{2},$

(8.47) $\quad \left.\dfrac{\partial K_2(s,t;\theta)}{\partial \theta}\right|_{\theta=0} = \dfrac{1}{2} + c - \dfrac{c}{2}(s+t) + \dfrac{c^2}{2}(1-s)(1-t).$

It can be shown (Problem 3.5) by finding the FD of (8.47) as $1 - \lambda(c^2 + 3c + 3)/6$ that

(8.48) $\quad \mathcal{L}\left(\left.\dfrac{dX_2(\theta)}{d\theta}\right|_{\theta=0} + \dfrac{1}{2}\right) \sim \dfrac{c^2 + 3c + 3}{6}\chi^2(1).$

Thus we can deduce that

$\qquad \lim_{T\to\infty} P(\hat{\alpha} = 1) = \lim_{T\to\infty} P\left(\left.\dfrac{dh_{T2}(\theta)}{d\theta}\right|_{\theta=0} < 0\right)$

$\qquad\qquad\qquad\qquad = P\left(\left.\dfrac{dX_2(\theta)}{d\theta}\right|_{\theta=0} < 0\right)$

$\qquad\qquad\qquad\qquad = P\left(Z^2 < \dfrac{3}{c^2 + 3c + 3}\right),$

where $Z \sim N(0,1)$. Note that $\lim_{T\to\infty} P(\hat{\alpha} = 1)$ is a decreasing function of c.

We next show that there exists the local MLE $\hat{\alpha}$ such that $\hat{\alpha} - 1 = O_p(T^{-1})$ as $T \to \infty$. The proof is quite similar to that for the stationary case. Suppose that $c = 0$. Then we have

$\qquad E\left(\dfrac{dX_2(\theta)}{d\theta}\right) = \displaystyle\int_0^1 \dfrac{\partial K_2(t,t;\theta)}{\partial \theta}\, dt - \dfrac{1}{2}$

$$= -\frac{1}{4}(1 - e^{-2\theta}) < 0,$$

$$V\left(\frac{dX_2(\theta)}{d\theta}\right) = 2\int_0^1\int_0^1 \left(\frac{\partial K_2(s,t;\theta)}{\partial \theta}\right)^2 ds\,dt$$

$$= \frac{1}{16}\left[\frac{1}{\theta} - \frac{1}{4\theta^2} + e^{-2\theta}\left(4 - \frac{8\theta}{3}\right) + e^{-4\theta}\left(2 + \frac{1}{4\theta^2}\right)\right]$$

$$= \frac{1}{16\theta} + O\left(\frac{1}{\theta^2}\right).$$

The same reasoning as before ensures the existence of the local MLE $\hat{\alpha}$ such that $\hat{\alpha} - 1 = O_p(T^{-1})$. The case of c positive can be dealt with similarly, although the computation of the above moments is much involved.

The limiting distribution of $T(\hat{\alpha} - 1)$, even for $c = 0$, is hard to simulate in the present case if we attempt to truncate the infinite series for $X_2(\theta)$ in (8.43), since the eigenvalues $\{\lambda_n\}$ are unknown. For the ad hoc approximation introduced before, we have

$$\lim_{T\to\infty} P(T(\hat{\alpha} - 1) \leq x) \cong P\left(\left.\frac{dX_2(\theta)}{d\theta}\right|_{\theta=-x} \geq 0\right)$$

$$= P\left(\int_0^1\int_0^1 \left.\frac{\partial K_2(s,t;\theta)}{\partial \theta}\right|_{\theta=-x} dw(s)\,dw(t) - \frac{1}{2} \geq 0\right).$$

Let us consider only the case of $c = 0$. Then

$$\left.\frac{\partial K_2(s,t;\theta)}{\partial \theta}\right|_{\theta=-x} = \frac{1}{4}\left[e^{x|s-t|}(1 + x|s-t|) + e^{x(2-s-t)}(1 + x(2-s-t))\right]$$

(8.49)

and it can be shown (Problem 3.6) that the FD of (8.49) is given by

(8.50) $\quad D(\lambda) = e^{2x}\left[\cosh A \cosh B\right.$

$$-\frac{1}{4x+\lambda}\left\{4x^2 + 2\lambda x + \frac{\lambda^2}{4} + \frac{\lambda+6x}{4}\mu\right\}\cosh A \frac{\sinh B}{B}$$

$$-\frac{1}{4x+\lambda}\left\{4x^2 + 2\lambda x + \frac{\lambda^2}{4} - \frac{\lambda+6x}{4}\mu\right\}\frac{\sinh A}{A}\cosh B$$

$$\left. + x^2 \frac{\sinh A}{A}\frac{\sinh B}{B}\right],$$

where $\mu = \sqrt{\lambda^2 + 4\lambda x}$ and

$$A = \sqrt{x^2 + \frac{(\lambda - \mu)x}{2}}, \qquad B = \sqrt{x^2 + \frac{(\lambda + \mu)x}{2}}.$$

Therefore we have that, when $c = 0$,

$$(8.51) \quad F_2(x) = P\left(\int_0^1 \int_0^1 \left.\frac{\partial K_2(s,t;\theta)}{\partial \theta}\right|_{\theta=-x} dw(s)\, dw(t) - \frac{1}{2} \geq 0\right)$$

$$= \frac{1}{2} + \frac{1}{\pi} \int_0^\infty \frac{1}{\theta} \operatorname{Im}\left[\left(e^{i\theta} D(2i\theta)\right)^{-1/2}\right] d\theta.$$

Figure 8.7 shows graphs of $F_2(x)$ in (8.51) and the distribution function of $T(\hat{\alpha} - 1)$ for $(\alpha_0, T) = (1, 50)$. This latter distribution was computed from 50,000 replications. Comparison with Figure 8.1 tells us that the distributions of the MLEs in the stationary and conditional cases are quite different. The probability density in the latter seems to be monotone for $x < 0$. The difference from the approximate distribution is also bigger in the conditional case, which will imply more frequent occurrence of multiple local maxima.

PROBLEMS

3.1 Prove (8.37).
3.2 Prove (8.38).
3.3 Derive the FD of $K_2(s, t; \theta)$ in (8.41) as in (8.42).

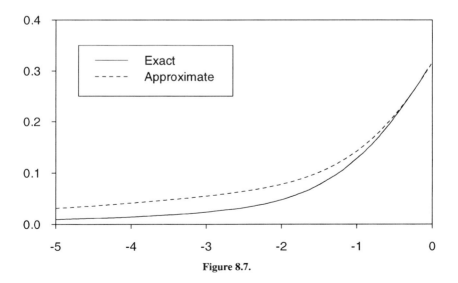

Figure 8.7.

3.4 Establish (8.44) for $l = 1$.
3.5 Prove (8.48).
3.6 Show that the FD of (8.49) is given by (8.50).

8.4 NONINVERTIBLE SEASONAL MODELS

The model first considered here is

(8.52) $\qquad y_j = \varepsilon_j - \alpha_m \varepsilon_{j-m}, \qquad (j = 1, \ldots, T),$

where m is a positive integer assumed to be a divisor of T so that $N = T/m$ is an integer. Then (8.52) may be rewritten as

(8.53) $\qquad \boldsymbol{y}_j = \boldsymbol{\varepsilon}_j - \alpha_m \boldsymbol{\varepsilon}_{j-1}, \qquad (j = 1, \ldots, N),$

where $\boldsymbol{y}_j = (y_{(j-1)m+1}, \ldots, y_{jm})'$ and $\boldsymbol{\varepsilon}_j = (\varepsilon_{(j-1)m+1}, \ldots, \varepsilon_{jm})'$. For $\{\boldsymbol{\varepsilon}_j\}$ we consider two cases in the same way as before. One is the stationary case where $\boldsymbol{\varepsilon}_0, \boldsymbol{\varepsilon}_1, \ldots, \sim \text{NID}(0, \sigma^2 I_m)$ and the other the conditional case where $\boldsymbol{\varepsilon}_0 = 0$ and $\boldsymbol{\varepsilon}_1, \boldsymbol{\varepsilon}_2, \ldots, \sim \text{NID}(0, \sigma^2 I_m)$. Note that

$$y \sim N\left(0, \sigma^2 \Omega(\alpha_m) \otimes I_m\right)$$

for the stationary case, while

$$y \sim N\left(0, \sigma^2 \Phi(\alpha_m) \otimes I_m\right)$$

for the conditional case, where $\Omega(\alpha_m)$ and $\Phi(\alpha_m)$ are $N \times N$ matrices defined in (8.5) and (8.6), respectively, with T replaced by N.

In the following subsections we explore the asymptotic properties of the local MLE $\hat{\alpha}_m$ of α_{m0} for the above two cases when the true value α_{m0} of α_m takes the form:

(8.54) $\qquad \alpha_{m0} = 1 - \dfrac{c}{N} = 1 - \dfrac{cm}{T}, \qquad (c \geq 0),$

while the parameter α_m takes the form

(8.55) $\qquad \alpha_m = 1 - \dfrac{\theta}{N} = 1 - \dfrac{\theta m}{T}, \qquad (\theta \geq 0).$

As a by-product of the above discussions, we also deal with continuous record asymptotics as the sampling interval approaches zero. For that purpose we shall consider

(8.56) $\qquad y_{hj} = \varepsilon_{hj} - \alpha \varepsilon_{h(j-1)}, \qquad (j = 1, \ldots, T),$

where $\{y_{hj}\}$ is observed on the fixed sampling span $[0, M]$ with the sampling interval $h = M/T$. We deal with the stationary and conditional cases concerning the distribution of $\{\varepsilon_{hj}\}$ $(j = 0, 1 \ldots)$, assuming either $\varepsilon_0, \varepsilon_h \ldots \sim \text{NID}(0, \sigma^2 h)$ or $\varepsilon_h, \varepsilon_{2h} \ldots \sim \text{NID}(0, \sigma^2 h)$ with $\varepsilon_0 = 0$.

8.4.1 The Stationary Case

The concentrated log-likelihood for α_m in the present case is given (Problem 4.1) by

(8.57) $\quad l_{T1}(\alpha_m) = -\dfrac{T}{2} \log y'(\Omega^{-1}(\alpha_m) \otimes I_m) y - \dfrac{m}{2} \log |\Omega(\alpha_m)| - \dfrac{T}{2} \log \dfrac{2\pi}{T} - \dfrac{T}{2}.$

We first have that, as $N \to \infty$ under $\alpha_m = 1 - (\theta/N)$ and $\alpha_{m0} = 1 - (c/N)$,

$$\dfrac{1}{T} y'(\Omega^{-1}(\alpha_m) \otimes I_m) y \to \sigma^2 \quad \text{in probability},$$

$$\dfrac{|\Omega(\alpha_m)|}{|\Omega(1)|} = \dfrac{1}{N+1} \sum_{j=0}^{N} \alpha_m^{2j} \to \dfrac{\sinh \theta}{\theta e^\theta}.$$

Let us next consider

$$S_{T1} = \dfrac{1}{\sigma^2} y' \left[(\Omega^{-1}(\alpha_m) - \Omega^{-1}(1)) \otimes I_m \right] y$$

$$= Z' \left[\left\{ \Omega^{1/2}(\alpha_{m0}) \left(\Omega^{-1}(\alpha_m) - \Omega^{-1}(1) \right) \Omega^{1/2}(\alpha_{m0}) \right\} \otimes I_m \right] Z$$

$$= \sum_{j=1}^{N} (1 + \alpha_{m0}^2 - 2\alpha_{m0} \delta_{jN}) \left(\dfrac{1}{1 + \alpha_m^2 - 2\alpha_m \delta_{jN}} - \dfrac{1}{2 - 2\delta_{jN}} \right) Z_j' Z_j,$$

where $\delta_{jN} = \cos(j\pi/(N+1))$, $\{Z_j\} \sim \text{NID}(0, I_m)$, and $Z = (Z_1', \ldots, Z_N')'$. It follows from (8.14) and (8.15) that

(8.58) $\quad \mathcal{L}(S_{T1}) \to \mathcal{L} \left(-\theta^2 \sum_{n=1}^{\infty} \dfrac{n^2 \pi^2 + c^2}{n^2 \pi^2 (n^2 \pi^2 + \theta^2)} Z_n' Z_n + m\theta \right).$

Thus we obtain

(8.59) $\quad \mathcal{L}(g_{T1}(\alpha_m)) = \mathcal{L}(l_{T1}(\alpha_m) - l_{T1}(1))$

$$= \mathcal{L} \left(-\dfrac{T}{2} \log \dfrac{y'(\Omega^{-1}(\alpha_m) \otimes I_m) y}{y'(\Omega^{-1}(1) \otimes I_m) y} - \dfrac{m}{2} \log \dfrac{|\Omega(\alpha_m)|}{|\Omega(1)|} \right)$$

$$\to \mathcal{L}(X_{m1}(\theta)),$$

where

$$(8.60) \quad X_{m1}(\theta) = \frac{\theta^2}{2} \sum_{n=1}^{\infty} \frac{n^2\pi^2 + c^2}{n^2\pi^2(n^2\pi^2 + \theta^2)} Z'_n Z_n - \frac{m}{2} \log \frac{\sinh \theta}{\theta}.$$

The asymptotic properties of the local MLE $\hat{\alpha}_m = 1 - (\hat{\theta}/N)$ can be studied on the basis of $X_{m1}(\theta)$ in (8.60). We first have that

$$(8.61) \quad \lim_{N \to \infty} P(\hat{\alpha}_m = 1) = P\left(\left.\frac{d^2 X_{m1}(\theta)}{d\theta^2}\right|_{\theta=0} < 0\right)$$

$$= P\left(\sum_{n=1}^{\infty} \left(\frac{1}{n^2\pi^2} + \frac{c^2}{n^4\pi^4}\right) Z'_n Z_n < \frac{m}{6}\right)$$

$$= \frac{1}{\pi} \int_0^{\infty} \mathrm{Re}\left[\frac{1 - e^{-im\theta/6}}{i\theta}(\phi(\theta))^m\right] d\theta,$$

where $(\phi(\theta))^m$ is the c.f. of $d^2 X_{m1}(\theta)/d\theta^2|_{\theta=0} + m/6$ with $\phi(\theta)$ defined in (8.26). In Section 8.6 we shall compute $\lim_{N \to \infty} P(\hat{\alpha}_m = 1)$ for various values of c and m and find that the probability is a decreasing function of c and m.

The existence of the local MLE $\hat{\alpha}_m$ such that $\hat{\alpha}_m - 1 = O_p(N^{-1})$ can be proved easily. In fact, it holds that

$$E\left(\frac{dX_{m1}(\theta)}{d\theta}\right) = m\mu(\theta), \quad V\left(\frac{dX_{m1}(\theta)}{d\theta}\right) = m\sigma^2(\theta),$$

where $\mu(\theta)$ and $\sigma^2(\theta)$ are defined in (8.28) and (8.29), respectively. Thus the consistency proof for $m = 1$ can be carried over to the present case.

The limiting distribution of $N(\hat{\alpha}_m - 1) = -\hat{\theta}$ may be approximated in the same way as before. We have (Problem 4.2)

$$(8.62) \quad \lim_{N \to \infty} P\left(N(\hat{\alpha}_m - 1) \leq x\right) \cong F_{m1}(x), \quad (x < 0),$$

where

$$(8.63) \quad F_{m1}(x) = P\left(\left.\frac{dX_{m1}(\theta)}{d\theta}\right|_{\theta=-x} \geq 0\right)$$

$$= P\left(-x \sum_{n=1}^{\infty} \frac{n^2\pi^2 + c^2}{(n^2\pi^2 + x^2)^2} Z'_n Z_n \geq \frac{m}{2}\left(\frac{1}{x} - \coth x\right)\right)$$

$$= 1 - \frac{1}{\pi} \int_0^{\infty} \mathrm{Re}\left[\frac{1 - e^{-iam\theta}}{i\theta}(\phi(\theta;x))^m\right] d\theta$$

with $a = (\coth x - 1/x)/(2x)$ and $\phi(\theta;x)$ is defined in (8.33).

8.4.2 The Conditional Case

The concentrated log-likelihood for α_m is given by

(8.64) $\quad l_{T2}(\alpha_m) = -\dfrac{T}{2} \log y' \left(\Phi^{-1}(\alpha_m) \otimes I_m \right) y - \dfrac{T}{2} \log \dfrac{2\pi}{T} - \dfrac{T}{2}.$

As $N \to \infty$ under $\alpha_m = 1 - (\theta/N)$ and $\alpha_{m0} = 1 - (c/N)$, we have

$$\dfrac{1}{T} y' \left(\Phi^{-1}(\alpha_m) \otimes I_m \right) y \to \sigma^2 \quad \text{in probability}.$$

Noting that $y = \left(C^{-1}(\alpha_{m0}) \otimes I_m \right) \varepsilon$, let us consider

$$\begin{aligned}
S_{T2} &= \dfrac{1}{\sigma^2} y' \left[\left(\Phi^{-1}(\alpha_m) - \Phi^{-1}(1) \right) \otimes I_m \right] y \\
&= \dfrac{1}{\sigma^2} \varepsilon' \left[\left\{ \left(C^{-1}(\alpha_{m0}) \right)' \left(\Phi^{-1}(\alpha_m) - \Phi^{-1}(1) \right) C^{-1}(\alpha_{m0}) \right\} \otimes I_m \right] \varepsilon \\
&= \dfrac{1}{N} Z' \{ (B_N - \theta I_N) \otimes I_m \} Z + \dfrac{\theta}{N} Z'Z ,
\end{aligned}$$

where $Z \sim N(0, I_T)$ and

$$B_N = N \left[\left(C^{-1}(\alpha_{m0}) \right)' \left(\Phi^{-1}(\alpha_m) - \Phi^{-1}(1) \right) C^{-1}(\alpha_{m0}) \right].$$

Then it can be shown (Problem 4.3) that

(8.65) $\quad \mathcal{L}(S_{T2}) \to \mathcal{L} \left(\int_0^1 \int_0^1 \bar{K}_2(s, t; \theta) \, d\mathbf{w}'(s) \, d\mathbf{w}(t) + m\theta \right),$

where $\bar{K}_2 = -2K_2$ with K_2 defined in (8.40) and $\{\mathbf{w}(t)\}$ is the m-dimensional standard Brownian motion.

It now follows that

(8.66) $\quad \begin{aligned} \mathcal{L}(g_{T2}(\alpha_m)) &= \mathcal{L}(l_{T2}(\alpha_m) - l_{T2}(1)) \\ &= \mathcal{L} \left(-\dfrac{T}{2} \log \dfrac{y' \left(\Phi^{-1}(\alpha_m) \otimes I_m \right) y}{y' \left(\Phi^{-1}(1) \otimes I_m \right) y} \right) \\ &\to \mathcal{L}(X_{m2}(\theta)), \end{aligned}$

where

(8.67) $\quad X_{m2}(\theta) = \displaystyle\int_0^1 \int_0^1 K_2(s, t; \theta) \, d\mathbf{w}'(s) \, d\mathbf{w}(t) - \dfrac{m\theta}{2}.$

As in the stationary case, the asymptotic properties of the local MLE $\hat{\alpha}_m = 1 - (\hat{\theta}/N)$ can be studied on the basis of $X_{m2}(\theta)$.

Recalling the definition of $\partial K_2(s, t; \theta)/\partial\theta|_{\theta=0}$ given in (8.47), we obtain (Problem 4.4)

$$(8.68) \quad \mathcal{L}\left(\left.\frac{dX_{m2}(\theta)}{d\theta}\right|_{\theta=0} + \frac{m}{2}\right) = \mathcal{L}\left(\int_0^1\int_0^1 \left.\frac{\partial K_2(s,t;\theta)}{\partial\theta}\right|_{\theta=0} dw'(s)\,dw(t)\right)$$

$$\sim \frac{c^2 + 3c + 3}{6}\chi^2(m).$$

Thus we can deduce that

$$(8.69) \quad \lim_{N\to\infty} P(\hat{\alpha}_m = 1) = P\left(\left.\frac{dX_{m2}(\theta)}{d\theta}\right|_{\theta=0} < 0\right)$$

$$= P\left(\chi_m^2 < \frac{3m}{c^2 + 3c + 3}\right),$$

where χ_m^2 follows a χ^2 distribution with m degrees of freedom. It will be verified in Section 8.6 that $\lim_{N\to\infty} P(\hat{\alpha}_m = 1)$ is a decreasing function of c and m.

The existence of the local MLE $\hat{\alpha}_m$ such that $\hat{\alpha}_m - 1 = O_p(N^{-1})$ can be proved in the same way as before (Problem 4.5). For the limiting distribution of $N(\hat{\alpha}_m - 1) = -\hat{\theta}$, we consider only the case of $\alpha_{m0} = 1$, that is, $c = 0$. An approximation may be given (Problem 4.6) as

$$(8.70) \quad \lim_{N\to\infty} P\left(N(\hat{\alpha}_m - 1) \leq x\right) \cong F_{m2}(x), \quad (x < 0),$$

where

$$(8.71) \quad F_{m2}(x) = P\left(\left.\frac{dX_{m2}(\theta)}{d\theta}\right|_{\theta=-x} \geq 0\right)$$

$$= P\left(\int_0^1\int_0^1 \left.\frac{\partial K_2(s,t;\theta)}{\partial\theta}\right|_{\theta=-x} dw'(s)\,dw(t) - \frac{m}{2} \geq 0\right)$$

$$= \frac{1}{2} + \frac{1}{\pi}\int_0^\infty \frac{1}{\theta} \text{Im}\left[\left(e^{i\theta}D(2i\theta)\right)^{-m/2}\right] d\theta.$$

Here $D(\lambda)$ is the FD of $\partial K_2(s,t;\theta)/\partial\theta|_{\theta=-x}$ defined in (8.50).

8.4.3 Continuous Record Asymptotics

The model considered here is somewhat related to (8.52) and is given by

$$(8.72) \quad y_{hj} = \varepsilon_{hj} - \alpha\varepsilon_{h(j-1)}, \quad (j = 1,\ldots,T),$$

where $\{y_{hj}\}$ is observed on the fixed sampling span $[0, M]$ with the sampling interval $h = M/T$. We study the situation where $h \to 0$ with M fixed, while the true value α_0 of α takes the form

$$(8.73) \quad \alpha_0 = 1 - ch = 1 - \frac{cM}{T}.$$

For $\{\varepsilon_{hj}\}$ $(j = 0, 1, \ldots)$ we assume either

(8.74)
$$\varepsilon_0, \varepsilon_h, \varepsilon_{2h}, \ldots \sim \text{NID}(0, \sigma^2 h)$$

or

(8.75)
$$\varepsilon_h, \varepsilon_{2h}, \ldots \sim \text{NID}(0, \sigma^2 h) \quad \text{and} \quad \varepsilon_0 = 0.$$

Then the log-likelihood $l_{T1}(\alpha, \sigma^2)$, when (8.74) is assumed, is

$$l_{T1}(\alpha, \sigma^2) = -\frac{T}{2}\log(2\pi\sigma^2 h) - \frac{1}{2\sigma^2 h} y'\Omega^{-1}(\alpha)y - \frac{1}{2}\log|\Omega(\alpha)|,$$

while the corresponding $l_{T2}(\alpha, \sigma^2)$, when (8.75) is assumed, is

$$l_{T2}(\alpha, \sigma^2) = -\frac{T}{2}\log(2\pi\sigma^2 h) - \frac{1}{2\sigma^2 h} y'\Phi^{-1}(\alpha)y.$$

Thus it can be checked easily that the concentrated log-likelihoods for α are the same as those given in previous sections, except that $\alpha_0 = 1 - (c/T)$ is replaced by $\alpha_0 = 1 - ch = 1 - (cM/T)$. This observation leads us to justify that the parameter α may take the form

(8.76)
$$\alpha = 1 - \theta h = 1 - \frac{\theta M}{T}.$$

It is now an easy matter to deduce that the asymptotic properties of the local MLE $\hat{\alpha} = 1 - \hat{\theta}h = 1 - (\hat{\theta}M/T)$ can be studied in terms of

$$X_{M1}(\theta) = \frac{\theta^2 M^2}{2} \sum_{n=1}^{\infty} \frac{n^2\pi^2 + c^2 M^2}{n^2\pi^2(n^2\pi^2 + \theta^2 M^2)} Z_n^2 - \frac{1}{2}\log\frac{\sinh \theta M}{\theta M}$$

for the stationary case (8.74) and

$$X_{M2}(\theta) = \int_0^1 \int_0^1 K_{M2}(s, t; \theta) \, dw(s) \, dw(t) - \frac{\theta M}{2}$$

for the conditional case (8.75), where

$$K_{M2}(s, t; \theta) = -\frac{1}{2}\left[-cM - c^2 M^2(1-t) - M(\theta - c)e^{-\theta M(t-s)}\right.$$
$$\left. + \frac{M(\theta - c)^2}{\theta}e^{-\theta M(1-s)}\sinh \theta M(1-t)\right], \quad (s \le t).$$

Thus the existence of the local MLE $\hat{\alpha}$ such that $\hat{\alpha} - 1 = O_p(h)$ is ensured for both cases. Moreover, it can be verified (Problem 4.8) that

$$(8.77) \quad \lim_{h \to 0} P(\hat{\alpha} = 1) = P\left(\left. \frac{d^2 X_{M1}(\theta)}{d\theta^2} \right|_{\theta=0} < 0 \right)$$

$$= \frac{1}{\pi} \int_0^\infty \operatorname{Re}\left[\frac{1 - e^{-i\theta/6}}{i\theta} \tilde{\phi}(\theta) \right] d\theta$$

for the stationary case, where $\tilde{\phi}(\theta)$ is defined by $\phi(\theta)$ in (8.26) with c replaced by cM, while

$$(8.78) \quad \lim_{h \to 0} P(\hat{\alpha} = 1) = P\left(\left. \frac{d X_{M2}(\theta)}{d\theta} \right|_{\theta=0} < 0 \right)$$

$$= P\left(Z^2 < \frac{3}{c^2 M^2 + 3cM + 3} \right)$$

for the nonstationary case, where $Z \sim N(0, 1)$. Note that the limiting probabilities (8.77) and (8.78) are decreasing functions of both c and M.

The asymptotic distribution of $\hat{\alpha}$ can be discussed similarly. Noting that

$$T(\alpha_0 - 1) = -cM, \qquad T(\hat{\alpha} - 1) = -\hat{\theta}M,$$

we can deduce that the limiting distributions of $T(\hat{\alpha} - 1)$ in continuous record asymptotics are the same as those in usual asymptotics with c replaced by cM. Equivalently, the limiting distributions of $(\hat{\alpha} - 1)/h$ in the former coincides with those of $T(\hat{\alpha} - 1)$ in the latter. Of course, if we use the parameterizations $\alpha_0 = 1 - (ch/M) = 1 - (c/T)$ and $\alpha = 1 - (\theta h/M) = 1 - (\theta/T)$, instead of $\alpha_0 = 1 - ch$ and $\alpha = 1 - \theta h$, then continuous record asymptotics including the limiting probabilities (8.77) and (8.78) are completely the same as usual asymptotics.

PROBLEMS

4.1 Show that the concentrated log-likelihood for α_m in the stationary case is given by (8.57).

4.2 Prove (8.63).

4.3 Prove the weak convergence result in (8.65).

4.4 Prove the distributional result in (8.68).

4.5 Show that there exists the local MLE $\hat{\alpha}_m$ such that $\hat{\alpha}_m - 1 = O_p(N^{-1})$ for the conditional case.

4.6 Prove (8.71).

4.7 Show that (8.77) and (8.78) hold.

8.5 THE PSEUDOLOCAL MLE

The MA(1) model in (8.1) may be naturally extended to

(8.79) $$y_j = u_j - \alpha u_{j-1}, \quad (j = 1, \ldots, T),$$

where $\{u_j\}$ is now not independent. In the same way as before, we consider two cases concerning the initial value u_0. One is the stationary case where $u_0, u_1, \ldots$ are generated by

(8.80) $$u_j = \sum_{l=0}^{\infty} \phi_l \varepsilon_{j-l}, \quad \sum_{l=0}^{\infty} l|\phi_l| < \infty, \quad \phi \equiv \sum_{l=0}^{\infty} \phi_l \neq 0,$$

with $\{\varepsilon_j\} \sim \text{NID}(0, \sigma^2)$. The other is the conditional case where $u_0 = 0$ and $u_1, u_2, \ldots$ follow (8.80).

Suppose that we misspecify the above model as the pure MA(1) so that $\{u_j\} = \{\varepsilon_j\}$ and proceed to estimate $\alpha_0 = 1 - (c/T)$ using the same likelihoods as before. Our purpose here is to study the asymptotic properties of the estimators of α_0 obtained from such pseudolikelihoods.

8.5.1 The Stationary Case

The concentrated pseudolog-likelihood for α is defined by $l_{T1}(\alpha)$ in (8.9), which gives, under $\alpha = 1 - (\theta/T)$ and $\alpha_0 = 1 - (c/T)$,

$$g_{T1}(\alpha) = l_{T1}(\alpha) - l_{T1}(1)$$

$$= -\frac{T}{2} \log \left[1 + \frac{1}{T} \frac{y'(\Omega^{-1}(\alpha) - \Omega^{-1}(1))y}{\frac{1}{T} y'\Omega^{-1}(1)y} \right] + \frac{\theta}{2} - \frac{1}{2} \log \frac{\sinh \theta}{\theta} + o(1),$$

where $y = D(\alpha_0)u^*$ with $u^* = (u_0, u_1, \ldots, u_T)'$ and

$$D(\alpha_0) = (-\alpha_0 e_1, C^{-1}(\alpha_0)) : T \times (T+1),$$

$$e_1 = (1, 0, \ldots, 0)' : T \times 1.$$

Note that $C(\alpha)$ is defined in (8.36). Note also that $dg_{T1}(\alpha)/d\alpha|_{\alpha=\pm 1} = 0$ so that the pseudolocal MLE $\hat{\alpha}$ is always a solution to $dg_{T1}(\alpha)/d\alpha = 0$.

The eigenvalue approach is not helpful for obtaining $\mathcal{L}(g_{T1}(\alpha))$ as $T \to \infty$ under $\alpha = 1 - (\theta/T)$ and $\alpha_0 = 1 - (c/T)$. The Fredholm approach yields (Problem 5.1)

$$\frac{1}{T} y'\Omega^{-1}(1)y = \frac{1}{T} u^{*'} D'(\alpha_0) \left[C'(1)C(1) - \frac{1}{T+1} C'(1)ee'C(1) \right] D(\alpha_0) u^*$$

$$= \frac{1}{T} u'(C^{-1}(\alpha_0))' C'(1) \left[I_T - \frac{1}{T+1} ee' \right] C(1) C^{-1}(\alpha_0) u + o_p(1)$$

$$\to \sigma^2 \sum_{l=0}^{\infty} \phi_l^2 \quad \text{in probability},$$

(8.81)

where $u = (u_1, \ldots, u_T)'$ and $e = (1, \ldots, 1)' : T \times 1$. Similarly we can show (Problem 5.2) that

(8.82) $\mathcal{L}\left(y' \left(\Omega^{-1}(\alpha) - \Omega^{-1}(1) \right) y \right)$

$$= \mathcal{L}\left(\frac{1}{T} u' \left[T \left(C^{-1}(\alpha_0) \right)' \left(\Omega^{-1}(\alpha) - \Omega^{-1}(1) \right) C^{-1}(\alpha_0) - \theta I_T \right] u \right.$$

$$\left. + \frac{\theta}{T} u'u + o_p(1) \right)$$

$$\to \mathcal{L}\left(\phi^2 \sigma^2 \int_0^1 \int_0^1 \bar{K}_1(s,t;\theta) \, dw(s) \, dw(t) + \theta \sigma^2 \sum_{l=0}^{\infty} \phi_l^2 \right),$$

where $\bar{K}_1 = -2K_1$ with K_1 given in (8.17). Therefore we simply obtain, from (8.17),

(8.83) $\mathcal{L}(g_{T1}(\alpha)) \to \mathcal{L}\left(\frac{1}{r} \int_0^1 \int_0^1 K_1(s,t;\theta) \, dw(s) \, dw(t) - \frac{1}{2} \log \frac{\sinh \theta}{\theta} \right)$

$$= \mathcal{L}(Y_1(\theta;r)),$$

where $r = \sum_{l=0}^{\infty} \phi_l^2 / \phi^2$ and

$$Y_1(\theta;r) = \frac{1}{2r} \left[\theta^2 \sum_{n=1}^{\infty} \frac{n^2\pi^2 + c^2}{n^2\pi^2(n^2\pi^2 + \theta^2)} Z_n^2 - r \log \frac{\sinh \theta}{\theta} \right].$$

$$= \frac{1}{r} \left[X_1(\theta) + \frac{1-r}{2} \log \frac{\sinh \theta}{\theta} \right]$$

with $X_1(\theta)$ defined in (8.16). Moreover, it can be verified by putting $g_{T1}(\alpha) = g_{T1}(\alpha(\theta))$ with $\alpha(\theta) = 1 - (\theta/T)$ that

$$\mathcal{L}\left(\frac{dg_{T1}(\alpha(\theta))}{d\theta} \right) \to \mathcal{L}\left(\frac{dY_1(\theta;r)}{d\theta} \right),$$

$$\mathcal{L}\left(\frac{d^2 g_{T1}(\alpha(\theta))}{d\theta^2} \right) \to \mathcal{L}\left(\frac{d^2 Y_1(\theta;r)}{d\theta^2} \right),$$

where

$$\frac{dY_1(\theta;r)}{d\theta} = \frac{1}{r}\left[\frac{dX_1(\theta)}{d\theta} + \frac{1-r}{2}\left(\coth\theta - \frac{1}{\theta}\right)\right],$$

$$\frac{d^2Y_1(\theta;r)}{d\theta^2} = \frac{1}{r}\left[\frac{d^2X_1(\theta)}{d\theta^2} + \frac{1-r}{2}\left(-\operatorname{cosech}^2\theta + \frac{1}{\theta^2}\right)\right].$$

It holds that $dY_1(\theta;r)/d\theta|_{\theta=0} = 0$. It also follows from (8.24) and (8.25) that

(8.84)
$$\lim_{T\to\infty} P(\hat{\alpha} = 1) = P\left(\frac{d^2Y_1(\theta;r)}{d\theta^2}\bigg|_{\theta=0} < 0\right)$$

$$= P\left(\frac{d^2X_1(\theta)}{d\theta^2}\bigg|_{\theta=0} < \frac{r-1}{6}\right)$$

$$= \frac{1}{\pi}\int_0^\infty \operatorname{Re}\left[\frac{1-e^{-ir\theta/6}}{i\theta}\phi(\theta)\right]d\theta,$$

where $\phi(\theta)$ is the c.f. of $d^2X_1(\theta)/d\theta^2|_{\theta=0} + \frac{1}{6}$ defined in (8.26). It is clear that $\lim_{T\to\infty} P(\hat{\alpha} = 1)$ is an increasing function of r.

Let us next consider the consistency problem of the pseudolocal MLE. For this purpose we have

(8.85) $$E\left(\frac{dY_1(\theta;r)}{d\theta}\right) = \frac{1}{r}\left[\mu(\theta) + \frac{1-r}{2}\left(\coth\theta - \frac{1}{\theta}\right)\right],$$

(8.86) $$V\left(\frac{dY_1(\theta;r)}{d\theta}\right) = \frac{\sigma^2(\theta)}{r^2} = O\left(\frac{1}{\theta}\right),$$

where $\mu(\theta) = E(dX_1(\theta)/d\theta)$ and $\sigma^2(\theta) = V(dX_1(\theta)/d\theta)$ are defined in (8.28) and (8.29), respectively. It can be verified that $E(dY_1(\theta;r)/d\theta)$ is continuous and uniformly bounded for $\theta \geq 0$. Thus, for sufficiently large θ, $E(dY_1(\theta;r)/d\theta)$ is negative if and only if

$$\lim_{\theta\to\infty} E\left(\frac{dY_1(\theta;r)}{d\theta}\right) = \frac{1}{r}\left(-\frac{1}{4} + \frac{1-r}{2}\right) < 0,$$

that is, $r > \frac{1}{2}$. Then Chebyshev's inequality and (8.86) ensure the existence of the local maximum $\hat{\alpha}$ such that $\hat{\alpha} - 1 = O_p(T^{-1})$. When $0 < r \leq \frac{1}{2}$, we cannot always find θ such that $E(dY_1(\theta;r)/d\theta) < 0$. Note that the consistency condition $r > \frac{1}{2}$ is equivalent to $\sum_{l=0}^\infty \rho_l < \frac{1}{2}$, where ρ_l is the lth order autocorrelation of $\{u_j\}$. Thus $\{u_j\}$ must not be positively highly correlated for the consistency of $\hat{\alpha}$. For instance, if $u_j = \beta u_{j-1} + \varepsilon_j$ so that $\rho_l = \beta^{|l|}$, the consistency condition is $\beta < \frac{1}{3}$.

Pötscher (1991) proved the above fact by a different approach. He also discussed the pseudoglobal MLE, whose consistency was shown to require a stronger condition.

8.5.2 The Conditional Case

The concentrated log-likelihood $l_{T2}(\alpha)$ defined in (8.34) gives us

$$g_{T2}(\alpha) = -\frac{T}{2}\log\left[1 + \frac{1}{T}\frac{y'\left(\Phi^{-1}(\alpha) - \Phi^{-1}(1)\right)y}{\frac{1}{T}y'\Phi^{-1}(1)y}\right],$$

where $y = C^{-1}(\alpha_0)u$. The Fredholm approach yields (Problem 5.3)

(8.87) $$\frac{1}{T}y'\Phi^{-1}(1)y = \frac{1}{T}u'\left(C^{-1}(\alpha_0)\right)'C'(1)C(1)C^{-1}(\alpha_0)u$$

$$\to \sigma^2 \sum_{l=0}^{\infty} \phi_l^2 \quad \text{in probability},$$

while

(8.88) $$\mathcal{L}\left(y'\left(\Phi^{-1}(\alpha) - \Phi^{-1}(1)\right)y\right)$$

$$\to \mathcal{L}\left(\sigma^2\phi^2\left(\int_0^1\int_0^1 \bar{K}_2(s,t;\theta)\,dw(s)\,dw(t) + r\theta\right)\right),$$

where $\bar{K}_2 = -2K_2$ with K_2 given in (8.40). Thus

(8.89) $$\mathcal{L}(g_{T2}(\alpha)) \to \mathcal{L}\left(\frac{1}{r}\int_0^1\int_0^1 K_2(s,t;\theta)\,dw(s)\,dw(t) - \frac{\theta}{2}\right)$$

$$= \mathcal{L}(Y_2(\theta;r)),$$

where

$$Y_2(\theta;r) = \frac{1}{r}\left[X_2(\theta) + \frac{1-r}{2}\theta\right]$$

with $X_2(\theta)$ defined in (8.39). Putting $g_{T2}(\alpha) = g_{T2}(\alpha(\theta))$ with $\alpha(\theta) = 1 - (\theta/T)$, we also have

$$\mathcal{L}\left(\frac{dg_{T2}(\alpha(\theta))}{d\theta}\right) \to \mathcal{L}\left(\frac{dY_2(\theta;r)}{d\theta}\right)$$

$$= \mathcal{L}\left(\frac{1}{r}\left(\frac{dX_2(\theta)}{d\theta} + \frac{1-r}{2}\right)\right).$$

It now follows from (8.89) and (8.48) that

$$\lim_{T\to\infty} P(\hat{\alpha} = 1) = P\left(\left.\frac{dY_2(\theta;r)}{d\theta}\right|_{\theta=0} < 0\right) \tag{8.90}$$

$$= P\left(\left.\frac{dX_2(\theta)}{d\theta}\right|_{\theta=0} < \frac{r-1}{2}\right)$$

$$= P\left(Z^2 < \frac{3r}{c^2 + 3c + 3}\right),$$

where $Z \sim N(0, 1)$. It is clear that $\lim_{T\to\infty} P(\hat{\alpha} = 1)$ is an increasing function of r, as in the stationary case.

For the consistency problem of the pseudolocal MLE, we only consider the case of $c = 0$ for simplicity of presentation. The previous arguments in Section 8.3 give us

$$E\left(\frac{dY_2(\theta;r)}{d\theta}\right) = \frac{1}{r}\left[E\left(\frac{dX_2(\theta)}{d\theta}\right) + \frac{1-r}{2}\right]$$

$$= \frac{1}{r}\left[-\frac{1 - e^{-2\theta}}{4} + \frac{1-r}{2}\right],$$

$$V\left(\frac{dY_2(\theta;r)}{d\theta}\right) = \frac{1}{r^2} V\left(\frac{dX_2(\theta)}{d\theta}\right)$$

$$= O\left(\frac{1}{\theta}\right).$$

Thus we derive the same consistency condition $r > \frac{1}{2}$ for the existence of the pseudolocal MLE such that $\hat{\alpha} - 1 = O_p(T^{-1})$.

PROBLEMS

5.1 Prove (8.81).
5.2 Prove the weak convergence result in (8.82).
5.3 Prove (8.87) and (8.88).

8.6 PROBABILITY OF THE LOCAL MLE AT UNITY

The finite sample and limiting probabilities of the local MLE $\hat{\alpha}$ occurring at $\alpha = 1$ are evaluated here for the seasonal MA model discussed in Section 8.4:

$$y_j = \varepsilon_j - \alpha \varepsilon_{j-m}, \quad (j = 1, \ldots, T).$$

Computational results shown below are based on Tanaka and Satchell (1989).

As for the finite sample probabilities in the stationary case we compute (Problem 6.1)

$$P_1(\alpha_0, m, N) = P\left(\left.\frac{d^2 l_{T1}(\alpha)}{d\alpha^2}\right|_{\alpha=1} < 0\right)$$

$$= P\left(\sum_{j=1}^{N} \frac{1 + \alpha_0^2 - 2\alpha_0 \delta_{jN}}{1 - \delta_{jN}}\left(\frac{N+2}{3} - \frac{1}{1 - \delta_{jN}}\right) Z_j' Z_j > 0\right),$$

(8.91)

where $l_{T1}(\alpha)$ is defined in (8.57), α_0 is the true value of α, $\{Z_j\} \sim \text{NID}(0, I_m)$, and $\delta_{jN} = \cos(j\pi/(N+1))$. The corresponding probabilities in the conditional case are computed (Problem 6.2) as

(8.92) $$P_2(\alpha_0, m, N) = P\left(\left.\frac{dl_{T2}(\alpha)}{d\alpha}\right|_{\alpha=1} > 0\right)$$

$$= P\left(\sum_{j,k=1}^{N} A_{jk}(\alpha_0) Z_j' Z_k > 0\right),$$

where $l_{T2}(\alpha)$ is defined in (8.64) and

$$A_{jk}(\alpha) = \begin{cases} 1 + (1-\alpha)^2(N-j) - ((N-j)(1-\alpha) + 1)^2, & j = k, \\ 1 - \alpha + (1-\alpha)^2(N-k) \\ \quad - ((N-j)(1-\alpha) + 1)((N-k)(1-\alpha) + 1), & j < k. \end{cases}$$

The computation of (8.91) and (8.92) may be done by Imhof's formula. Table 8.2 reports values of P_1 and P_2 for various values of α_0, m, and N. It is seen from Table 8.2 that

i) P_1 and P_2 decrease as α_0 gets away from 1 with m and N fixed;
ii) P_1 and P_2 decrease as N becomes large with α_0 (< 1) and m fixed. When $\alpha_0 = 1$, P_1 and P_2 increase as N becomes large with m fixed;
iii) P_1 and P_2 decrease as m becomes large with α_0 and N fixed;
iv) $P_1 < P_2$ for the same m and N when $\alpha_0 = 1$, while the inequality is reversed as α_0 gets away from 1.

We now turn to the limiting probabilities. In Section 8.4 we have obtained, under $\alpha_0 = 1 - (c/N)$,

(8.93) $$\lim_{N \to \infty} P_1(\alpha_0, m, N) = \frac{1}{\pi} \int_0^\infty \text{Re}\left[\frac{1 - e^{-im\theta/6}}{i\theta}(\phi(\theta))^m\right] d\theta,$$

Table 8.2. Finite Sample Probabilities of the Local MLE at Unity

Period	Case	N	$\alpha_0 = 0$	0.5	0.8	0.9	1
$m = 1$	P_1	50	0.001	0.013	0.130	0.333	0.653
		100	0.000	0.001	0.027	0.136	0.655
	P_2	50	0.059	0.070	0.122	0.206	0.678
		100	0.041	0.046	0.071	0.121	0.680
$m = 4$	P_1	10	0.001	0.041	0.350	0.509	0.572
		25	0.000	0.000	0.062	0.299	0.579
	P_2	10	0.002	0.009	0.074	0.207	0.580
		25	0.000	0.001	0.009	0.049	0.588
$m = 12$	P_1	5	0.000	0.052	0.408	0.503	0.533
		10	0.000	0.001	0.185	0.431	0.542
	P_2	5	0.000	0.001	0.048	0.188	0.537
		10	0.000	0.000	0.003	0.047	0.546

(8.94) $$\lim_{N \to \infty} P_2(\alpha_0, m, N) = P\left(\chi_m^2 < \frac{3m}{c^2 + 3c + 3}\right),$$

where $\phi(\theta)$ is defined in (8.26).

Table 8.3 shows the limiting probabilities (8.93) and (8.94) for various values of c and m. It is seen that both probabilities are decreasing with c and m. It is also noticeable that $P_1 < P_2$ when $c = 0$, while $P_1 > P_2$ as c becomes large. These are consistent with the results for finite sample probabilities. The entries in Table 8.3 may be used to approximate $P_1(\alpha_0, m, N)$ and $P_2(\alpha_0, m, N)$ when α_0 is close to unity and mN is large. For example, $P_1(0.9, 1, 50) = 0.333$ in Table 8.2 may be approximated as 0.3474, which is the value corresponding to P_1 with $m = 1$ and $c = N(1 - \alpha_0) = 50 \times 0.1 = 5$ in Table 8.3. As another example, $P_2(0.9, 4, 10) = 0.207$ may be approximated as 0.2119.

Table 8.3. Limiting Probabilities of the Local MLE at Unity

Period	Case	$c = 0$	0.05	1	2	5	10
$m = 1$	P_1	0.6574	0.6510	0.6326	0.5698	0.3474	0.1431
	P_2	0.6827	0.5732	0.4873	0.3690	0.2083	0.1194
$m = 2$	P_1	0.6167	0.6071	0.5795	0.4864	0.2016	0.0389
	P_2	0.6321	0.4682	0.3486	0.2061	0.0674	0.0223
$m = 4$	P_1	0.5841	0.5700	0.5296	0.3970	0.0814	0.0036
	P_2	0.5940	0.3601	0.2119	0.0788	0.0089	0.0010
$m = 12$	P_1	0.5489	0.5238	0.4528	0.2405	0.0037	0.0000
	P_2	0.5543	0.1829	0.0470	0.0030	0.0000	0.0000

PROBLEMS

6.1 Derive the expression for $P_1(\alpha_0, m, N)$ in (8.91).
6.2 Derive the expression for $P_2(\alpha_0, m, N)$ in (8.92).

8.7 THE RELATIONSHIP WITH THE STATE SPACE MODEL

In Section 1.1 of Chapter 1 we presented the so-called state space model

$$y_j = \mu_j + \varepsilon_j,$$
(8.95)
$$\mu_j = \mu_{j-1} + \xi_j, \quad \mu_0 = 0, \quad (j = 1, \ldots, T),$$

where $\{\varepsilon_j\}$ and $\{\xi_j\}$ are sequences of NID$(0, \sigma_\varepsilon^2)$ $(\sigma_\varepsilon^2 > 0)$ and NID$(0, \sigma_\xi^2)$ $(\sigma_\xi^2 \geq 0)$, respectively, and are independent of each other. We define the signal-to-noise ratio by $\rho = \sigma_\xi^2/\sigma_\varepsilon^2$. Here we shall show that the asymptotic properties of estimators of ρ when ρ is close to zero can be easily derived from the results obtained in previous sections. This is intuitively clear since

(8.96)
$$\Delta y_j = \xi_j + \Delta \varepsilon_j, \quad (j = 2, \ldots, T)$$

so that $\{\Delta y_j\}$ $(j \geq 2)$ follows a nearly noninvertible MA(1) process when ρ is close to zero.

Let us put $y = (y_1, \ldots, y_T)'$ and define ρ_0 to be the true value of ρ. Then

(8.97)
$$y \sim N(0, \sigma_\varepsilon^2(I_T + \rho_0 CC')),$$

where C is the random walk generating matrix. The concentrated log-likelihood $k_{T1}(\rho)$ is given, except for constants, by

(8.98)
$$k_{T1}(\rho) = -\frac{T}{2} \log y'(I_T + \rho CC')^{-1} y - \frac{1}{2} \log |I_T + \rho CC'|.$$

We now consider the local MLE $\hat{\rho}$ of ρ that is the local maximum of (8.98) closest to zero. For this purpose we put

(8.99)
$$\sqrt{\rho_0} = \frac{c}{T} \quad \text{and} \quad \sqrt{\rho} = \frac{\theta}{T},$$

where c and θ are arbitrary positive constants independent of T. Recalling that the eigenvalues λ_{jT}s of CC' are given by

$$\lambda_{jT} = \frac{1}{4}\left(\sin\frac{j - \frac{1}{2}}{2T + 1}\pi\right)^{-2},$$

THE RELATIONSHIP WITH THE STATE SPACE MODEL

we have

$$|I_T + \rho CC'| = \prod_{j=1}^{T}(1 + \rho \lambda_{jT})$$

$$= \prod_{j=1}^{T}\left\{1 + \frac{\theta^2}{T^2} \times \frac{1}{4}\left(\sin \frac{j - \frac{1}{2}}{2T + 1}\pi\right)^{-2}\right\}$$

$$\to \prod_{n=1}^{\infty}\left(1 + \frac{\theta^2}{\left(n - \frac{1}{2}\right)^2 \pi^2}\right)$$

$$= \cosh \theta.$$

Moreover, we obtain (Problem 7.1), as $T \to \infty$ under $\rho = \theta^2/T^2$ and $\rho_0 = c^2/T^2$, that $y'y/T \to \sigma_\varepsilon^2$ in probability, while

(8.100) $\quad \mathcal{L}\left(\frac{1}{\sigma_\varepsilon^2} y'\left((I_T + \rho CC')^{-1} - I_T\right) y\right)$

$$= \mathcal{L}\left(Z'(I_T + \rho_0 CC')\left((I_T + \rho CC')^{-1} - I_T\right)Z\right)$$

$$= \mathcal{L}\left(\sum_{j=1}^{T}(1 + \rho_0 \lambda_{jT})\left(\frac{1}{1 + \rho \lambda_{jT}} - 1\right)Z_j^2\right)$$

$$\to \mathcal{L}\left(-\theta^2 \sum_{n=1}^{\infty} \frac{\left(n - \frac{1}{2}\right)^2 \pi^2 + c^2}{\left(n - \frac{1}{2}\right)^2 \pi^2 \left(\left(n - \frac{1}{2}\right)^2 \pi^2 + \theta^2\right)} Z_n^2\right),$$

where $Z \sim N(0, I_T)$ and $\{Z_n\} \sim NID(0, 1)$.

The above arguments yield

$$\mathcal{L}\bigl(k_{T1}(\rho) - k_{T1}(0)\bigr) \to \mathcal{L}\bigl(Y_1(\theta)\bigr),$$

where

(8.101) $\quad Y_1(\theta) = \frac{\theta^2}{2} \sum_{n=1}^{\infty} \frac{\left(n - \frac{1}{2}\right)^2 \pi^2 + c^2}{\left(n - \frac{1}{2}\right)^2 \pi^2 \left(\left(n - \frac{1}{2}\right)^2 \pi^2 + \theta^2\right)} Z_n^2 - \frac{1}{2} \log(\cosh \theta).$

Let us put $\kappa = \sqrt{\rho}$ and $\kappa_0 = \sqrt{\rho_0}$. Then we can show (Problem 7.2) completely in the same way as before that there exists the local maximum $\hat{\kappa}$ of κ such that $\hat{\kappa} = O_p(T^{-1})$. This ensures the existence of the local maximum $\hat{\rho}$ of ρ such that $\hat{\rho} = O(T^{-2})$.

It can also be deduced (Problem 7.3) that

$$(8.102) \quad \lim_{T\to\infty} P(\hat{\rho} = 0) = P\left(\left.\frac{d^2 Y_1(\theta)}{d\theta^2}\right|_{\theta=0} < 0\right)$$

$$= \frac{1}{\pi}\int_0^\infty \mathrm{Re}\left[\frac{1 - e^{-i\theta/2}}{i\theta}\phi_1(\theta)\right] d\theta,$$

where

$$(8.103) \quad \left.\frac{d^2 Y_1(\theta)}{d\theta^2}\right|_{\theta=0} = \sum_{n=1}^\infty \left(\frac{1}{\left(n - \frac{1}{2}\right)^2 \pi^2} + \frac{c^2}{\left(n - \frac{1}{2}\right)^4 \pi^4}\right) Z_n^2 - \frac{1}{2},$$

and $\phi_1(\theta)$ is the c.f. of $d^2 Y_1(\theta)/d\theta^2|_{\theta=0} + \frac{1}{2}$ given by

$$(8.104) \quad \phi_1(\theta) = \left[\cos\sqrt{i\theta + a(\theta)}\cos\sqrt{i\theta - a(\theta)}\right]^{-1/2}, \quad a(\theta) = \sqrt{-\theta^2 + 2ic^2\theta}.$$

Shephard and Harvey (1990) obtained (8.102) for $c = 0$. The limiting probability (8.102) will be computed later.

The limiting distribution of $\hat{\theta} = T\hat{\kappa} = T\sqrt{\hat{\rho}}$ may be approximated in the same way as before. We suggest (Problem 7.4)

$$(8.105) \quad \lim_{T\to\infty} P(\hat{\theta} \le x) \cong G_1(x) = P\left(\left.\frac{dY_1(\theta)}{d\theta}\right|_{\theta=x} \le 0\right)$$

$$= P\left(\sum_{n=1}^\infty \frac{x\left(\left(n - \frac{1}{2}\right)^2 \pi^2 + c^2\right)}{\left(\left(n - \frac{1}{2}\right)^2 \pi^2 + x^2\right)^2} Z_n^2 \le \frac{1}{2}\tanh x\right)$$

$$= \frac{1}{\pi}\int_0^\infty \mathrm{Re}\left[\frac{1 - \exp(-i\theta \tanh x/(2x))}{i\theta}\psi_1(\theta)\right] d\theta,$$

where

$$(8.106) \quad \psi_1(\theta) = \cosh x \left[\cos\sqrt{a(\theta) + b(\theta)}\cos\sqrt{a(\theta) - b(\theta)}\right]^{-1/2},$$

$$a(\theta) = i\theta - x^2, \quad b(\theta) = \sqrt{2i\theta(c^2 - x^2) - \theta^2}.$$

The above approximation and the exact distribution will be graphically presented later.

We next show that the estimation of ρ is quite sensitive to the specification of the initial value or the likelihood used, as in the MA(1) model. To this end, let us consider

THE RELATIONSHIP WITH THE STATE SPACE MODEL

first differences of $\{y_j\}$ described in (8.96), which may be expressed as

$$(8.107) \qquad \Delta y = \xi + \Delta \varepsilon,$$

where $\Delta y = (y_2 - y_1, \ldots, y_T - y_{T-1})'$, $\xi = (\xi_2, \ldots, \xi_T)'$, and $\Delta \varepsilon = (\varepsilon_2 - \varepsilon_1, \ldots, \varepsilon_T - \varepsilon_{T-1})'$. Note that $\Delta y \sim N(0, \sigma_\varepsilon^2(\Omega + \rho_0 I_{T-1}))$, where Ω is the matrix $\Omega(1)$ defined in (8.5) with T replaced by $T - 1$.

The concentrated log-likelihood $k_{T2}(\rho)$ is given, except for constants, by

$$(8.108) \quad k_{T2}(\rho) = -\frac{T-1}{2} \log \Delta y'(\Omega + \rho I_{T-1})^{-1}\Delta y - \frac{1}{2} \log |\Omega + \rho I_{T-1}|.$$

This case was first treated by Shephard (1993) when $\rho_0 = 0$. It can be shown (Problem 7.5) that, as $T \to \infty$ under $\rho = \theta^2/T^2$ and $\rho_0 = c^2/T^2$, $\Delta y' \Omega^{-1} \Delta y / T \to \sigma_\varepsilon^2$ in probability and

$$\mathcal{L}\left(\frac{1}{\sigma_\varepsilon^2}\Delta y'\left((\Omega + \rho I_{T-1})^{-1} - \Omega^{-1}\right)\Delta y\right) \to \mathcal{L}\left(-\theta^2 \sum_{n=1}^{\infty} \frac{n^2\pi^2 + c^2}{n^2\pi^2(n^2\pi^2 + \theta^2)} Z_n^2\right).$$

(8.109)

We also have (Problem 7.6)

$$(8.110) \qquad \frac{|\Omega + \rho I_{T-1}|}{|\Omega|} \to \frac{\sinh \theta}{\theta}.$$

The above arguments yield

$$\mathcal{L}\big(k_{T2}(\rho) - k_{T2}(0)\big) \to \mathcal{L}\big(Y_2(\theta)\big),$$

where

$$(8.111) \qquad Y_2(\theta) = \frac{\theta^2}{2} \sum_{n=1}^{\infty} \frac{n^2\pi^2 + c^2}{n^2\pi^2(n^2\pi^2 + \theta^2)} Z_n^2 - \frac{1}{2}\log\frac{\sinh \theta}{\theta}.$$

It is noticeable that $Y_2(\theta)$ is identical with $X_1(\theta)$ defined in (8.16). Thus it follows from the previous result that there exists the local maximum $\hat{\kappa} = \sqrt{\hat{\rho}}$ of κ such that $\hat{\kappa} = O_p(T^{-1})$, ensuring the existence of $\hat{\rho}$ such that $\hat{\rho} = O_p(T^{-2})$.

We have already calculated

$$(8.112) \qquad \lim_{T\to\infty} P(\hat{\rho} = 0) = P\left(\frac{d^2 Y_2(\theta)}{d\theta^2}\bigg|_{\theta=0} < 0\right)$$

$$= P\left(\sum_{n=1}^{\infty}\left(\frac{1}{n^2\pi^2} + \frac{c^2}{n^4\pi^4}\right)Z_n^2 < \frac{1}{6}\right).$$

Table 8.4. Limiting Probabilities (8.102) and (8.112)

Case	$c = 0$	0.05	1	5	10	20	40
(8.102)	0.6778	0.6543	0.5952	0.2126	0.0830	0.0196	0.0021
(8.112)	0.6574	0.6510	0.6326	0.3474	0.1431	0.0290	0.0020

The limiting distribution of $\hat{\theta} = T\hat{\kappa} = T\sqrt{\hat{\rho}}$ in the present case may be approximated as

$$(8.113)\ \lim_{T\to\infty} P(\hat{\theta} \le x) \cong G_2(x) = P\left(\left. \frac{dY_2(\theta)}{d\theta} \right|_{\theta=x} \le 0 \right)$$

$$= P\left(\sum_{n=1}^{\infty} \frac{x(n^2\pi^2 + c^2)}{(n^2\pi^2 + x^2)^2} Z_n^2 \le \frac{1}{2}\left(\coth x - \frac{1}{x}\right) \right)$$

$$= \frac{1}{\pi}\int_0^{\infty} \text{Re}\left[\frac{1 - \exp\left\{-\frac{i\theta}{2x}\left(\coth x - \frac{1}{x}\right)\right\}}{i\theta} \psi_2(\theta) \right] d\theta,$$

where $\psi_2(\theta) = \phi(\theta; x)$ with $\phi(\theta; x)$ defined in (8.33).

Table 8.4 reports the limiting probabilities (8.102) and (8.112) for some values of c. It is seen that the former is slightly larger when c is small, while the latter is slightly larger as c becomes large. Of course, both probabilities tend to 0 as $c \to \infty$. Figure 8.8 presents graphs of $G_1(x)$ in (8.105) and the distribution function of the local MLE $\hat{\theta} = T\sqrt{\hat{\rho}}$ based on the likelihood $k_{T1}(\rho)$ in (8.98), where $(\rho_0, T) = (0, 50)$. Figure

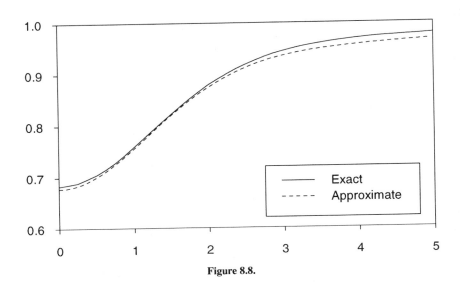

Figure 8.8.

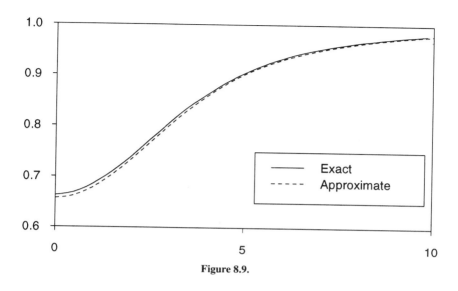

Figure 8.9.

8.9 shows the corresponding graphs of $G_2(x)$ in (8.113) and $\mathcal{L}(\hat{\theta})$ based on $k_{T2}(\rho)$ in (8.108). Note that $G_2(x)$ is a mirror image of $F_1(x)$ shown in Figure 8.1; that is, $G_2(x) = 1 - F_1(-x)$.

PROBLEMS

7.1 Prove the weak convergence result in (8.100).
7.2 Prove the existence of the local maximum $\hat{\rho}$ of $k_{T1}(\rho)$ in (8.98) such that $\hat{\rho} = O_p(T^{-2})$.
7.3 Derive the c.f. $\phi_1(\theta)$ given in (8.104).
7.4 Derive the c.f. $\psi_1(\theta)$ given in (8.106).
7.5 Prove the weak convergence result in (8.109).
7.6 Establish (8.110).

CHAPTER 9

Unit Root Tests in Autoregressive Models

The so-called unit root tests are discussed. Unlike the literature, we start the discussion with devising tests that have certain optimal properties. We then derive the limiting powers of those optimal tests, which are compared with those of ad hoc tests suggested in the literature. To make an absolute comparison, the limiting power envelope is computed, and the point optimal test is conducted to improve the power performance. We also consider reversing the null and alternative hypotheses of unit root tests. It emerges that the reversed unit root tests are closely related to MA unit root tests dealt with in the next chapter.

9.1 INTRODUCTION

This chapter is concerned with testing for a unit root in the AR part of linear time series models. The model we first deal with is

$$(9.1) \quad \begin{aligned} y_j &= x_j'\beta + \eta_j, \\ \eta_j &= \rho\eta_{j-1} + \varepsilon_j, \end{aligned} \quad (j = 1, \ldots, T),$$

where $\{x_j\}$ is a $p \times 1$ nonstochastic fixed sequence and $\{\varepsilon_j\}$ is an i.i.d.$(0, \sigma^2)$ sequence. For the initial value η_0 we assume in subsequent discussions that $\eta_0 = 0$. Note that the present model is a simplified version of the model dealt with in Chapter 7 and may be rewritten as

$$(9.2) \quad y = X\beta + \eta, \quad \eta = C(\rho)\varepsilon,$$

where $C(\rho)$ is defined in (8.36) with ρ replaced by α. We also put $V(\eta) = \sigma^2 \Phi(\rho) = \sigma^2 C(\rho)C'(\rho)$.

The problem we consider here is to test either

(9.3) $$H_0 : \rho = 1 \quad \text{versus} \quad H_1 : \rho < 1$$

or

(9.4) $$H_0 : \rho = 1 \quad \text{versus} \quad H : \rho \neq 1,$$

which we call the *AR unit root testing problem*. We recall that $\{y_j\}$ is trend stationary if $|\rho| < 1$, while it is of stochastic trend if $\rho \geq 1$. The distinction between the two cases is quite important. Hamilton (1994) and Hatanaka (1996) give extensive arguments concerning its implications for economic time series.

Fuller (1976) and Dickey and Fuller (1979) first discussed the present problem. They suggested various unit root tests and obtained finite sample powers of those tests by simulation experiments, which Evans and Savin (1981b, 1984) did by numerical integration. It is desirable, whenever any test is devised, to give optimality arguments for the test suggested. In Section 9.2 we do that job, which equally well applies to the MA unit root testing problem discussed in the next chapter. Since the test statistics depend on the regressor sequence $\{x_j\}$, we consider, as in Chapter 7, four simple models by specifying $\{x_j\}$:

Model A: $\quad y = \eta$,

Model B: $\quad y = \beta_0 e + \eta, \quad e = (1, \ldots, 1)' : T \times 1$,

Model C: $\quad y = \beta_1 d + \eta, \quad d = (1, \ldots, T)' : T \times 1$,

Model D: $\quad y = \beta_0 e + \beta_1 d + \eta$,

where $\eta = C(\rho)\varepsilon$ for all models. Under the normality assumption on η, locally optimal tests are suggested for these models following Nabeya and Tanaka (1990b).

In Section 9.3 we prove the equivalence of the optimal test with the LM test that can be derived more easily. Section 9.4 deals with various unit root tests suggested in the literature, among which are the Dickey–Fuller and Durbin–Watson tests. Integral expressions for the limiting powers of those tests as $T \to \infty$ under $\rho = 1 - (c/T)$ are derived in Section 9.5, where c is a fixed constant. To make an absolute comparison among the unit root tests, we derive the limiting power envelopes in Section 9.6. We also consider the point optimal test to improve the power performance. The numerical computation of the limiting powers of preceding unit root tests together with power envelopes is done in Section 9.7.

The rest of the present chapter is concerned with extending and modifying the tests to allow for seasonality and to cope with the situation where $\{\varepsilon_j\}$ in (9.1) is dependent. We also consider reversed unit root tests in the sense that the null and alternative in (9.3) or (9.4) are reversed. The resulting tests turn out to be those for the MA unit root.

9.2 OPTIMAL TESTS

Before moving on to discussions of the unit root tests, we give a general treatment concerning a derivation of a test that has certain optimal properties. For this purpose, let us consider a general regression model

(9.5) $$y = X\beta + u, \quad E(u) = 0, \quad V(u) = \sigma^2 \Sigma(\theta),$$

where X is a $T \times p$ nonstochastic matrix of rank $p (< T)$, while θ is a scalar parameter to be tested. We assume that $\Sigma(\theta)$ is positive definite and is a twice continuously differentiable function of θ. The parameters β and σ^2 are regarded as nuisance parameters. The testing problem is now formulated as either

(9.6) $$H_0 : \theta = \theta_0 \quad \text{versus} \quad H_1 : \theta < \theta_0,$$

or

(9.7) $$H_0 : \theta = \theta_0 \quad \text{versus} \quad H_1 : \theta \neq \theta_0.$$

To derive an optimal test, we assume normality on u in (9.5) so that $y \sim N(X\beta, \sigma^2 \Sigma(\theta))$. Then it is seen that the testing problems (9.6) and (9.7) are invariant under the group of transformations

$$y \rightarrow ay + Xb \quad \text{and} \quad (\theta, \beta, \sigma^2) \rightarrow (\theta, a\beta + b, a^2 \sigma^2),$$

where $0 < a < \infty$ and b is a $p \times 1$ vector. Following Kariya (1980), let us define

(9.8) $$M = I_T - X(X'X)^{-1}X'$$

and choose a $T \times (T - p)$ matrix H such that $H'H = I_{T-p}$ and $HH' = M$. Noting that $MX = HH'X = 0$ and thus $H'X = 0$, we have

(9.9) $$H'y \sim N(0, \sigma^2 H' \Sigma(\theta) H).$$

The distribution of the statistic $H'y$ is independent of the nuisance parameter β. It might be argued that the matrix H must be found explicitly. This is, however, not the case for our purpose. We can put $H = QF$ with a $T \times (T - p)$ matrix Q to be specified and a $(T - p) \times (T - p)$ nonsingular matrix F to be left unspecified. Then we have only to find the matrix Q such that $Q'X = 0$, whose usefulness will be demonstrated later.

We now concentrate on test statistics that are functions of $H'y$. The statistic $v = H'y/\sqrt{y'HH'y}$ is known to be a maximal invariant under the group of transformations

$$H'y \rightarrow aH'y \quad \text{and} \quad (\theta, \sigma^2) \rightarrow (\theta, a^2 \sigma^2),$$

where $0 < a < \infty$. Thus invariant tests may be constrained to be those based on v only. We can now assume that $\sigma^2 = 1$. It is known [Kariya (1980) and King (1980)] that the probability density $f(v \mid \theta)$ of v is given by

$$f(v|\theta) = \frac{1}{2}\Gamma\left(\frac{T-p}{2}\right)\pi^{-(T-p)/2}|H'\Sigma(\theta)H|^{-1/2}\left(v'(H'\Sigma(\theta)H)^{-1}v\right)^{-(T-p)/2}.$$
(9.10)

Note that $v'v = 1$; hence $f(v \mid \theta)$ is the density with respect to the uniform measure on the surface of the unit $T - p$ sphere.

The probability density $f(v \mid \theta)$ in (9.10) plays an important role in deriving invariant tests with certain optimal properties. Among those are *locally best invariant* (LBI), and LBI and *unbiased* (LBIU) tests. We shall also consider the Lagrange multiplier (LM) or Rao's score test introduced in Chapter 1 and show that it is equivalent to the LBI or LBIU test. Note here that the optimality is only local because there exists no unit root test that is uniformly best.

9.2.1 The LBI Test

Let us consider the testing problem (9.6). Then the LBI test is derived from Ferguson (1967, p. 235) as rejecting $H_0 : \theta = \theta_0$ when

$$\left.\frac{\partial \log f(v \mid \theta)}{\partial \theta}\right|_{\theta=\theta_0} < c,$$

where c is some constant, while $f(v \mid \theta)$ is defined in (9.10). We now have the following theorem.

Theorem 9.1. *Consider the regression model* $y = X\beta + u$ *with* $u \sim N(0, \sigma^2\Sigma(\theta))$. *Then the* LBI *test for the testing problem (9.6) rejects the null hypothesis when*

(9.11) $$\frac{y'H(H'\Sigma(\theta_0)H)^{-1}H'\left.\frac{d\Sigma(\theta)}{d\theta}\right|_{\theta=\theta_0}H(H'\Sigma(\theta_0)H)^{-1}H'y}{y'H(H'\Sigma(\theta_0)H)^{-1}H'y} < c.$$

Let us deal with Models A through D introduced in Section 9.1, and consider the testing problem (9.3). For this purpose we assume normality on $\eta = C(\rho)\varepsilon$, and put $\theta = \rho$, $\theta_0 = 1$, and $\Sigma(\theta) = \Phi(\rho) = C(\rho)C'(\rho)$. We obtain (Problem 2.1 in this chapter)

(9.12) $$\left.\frac{d\Phi(\rho)}{d\rho}\right|_{\rho=1} = dd' - \Phi(1),$$

where $d = (1, 2, \ldots, T)'$. Thus it follows from (9.11) that the LBI test in the present case rejects $H_0 : \rho = 1$ when

(9.13) $$\frac{y'H(H'\Phi(1)H)^{-1}H'dd'H(H'\Phi(1)H)^{-1}H'y}{y'H(H'\Phi(1)H)^{-1}H'y} < c.$$

OPTIMAL TESTS

It is noted here that the left side vanishes when $H'd = 0$. This occurs if d belongs to the column space of X since $H'X = 0$. Thus we cannot devise LBI tests for Models C and D. In such cases we consider LBIU tests, which we discuss in the next subsection. Here we deal only with Models A and B.

Let us consider Model A. Since $y = \eta = C(\rho)\varepsilon$, we may put $H = I_T$ so that the LBI statistic for the testing problem (9.3) is defined as

$$(9.14) \qquad S_{T1} = \frac{y'\Phi^{-1}(1)dd'\Phi^{-1}(1)y}{y'\Phi^{-1}(1)y} = \frac{y_T^2}{\sum_{j=1}^{T}(y_j - y_{j-1})^2}.$$

It can be shown (Problem 2.2) that, under H_0, $S_{T1}/T \sim \text{Be}(1/2, (T-1)/2)$, where $\text{Be}(\alpha, \beta)$ denotes beta distribution with shape parameters α and β. The limiting powers of the test based on S_{T1} and the other tests suggested below will be computed in Section 9.7.

For Model B, where $y = \beta_0 e + \eta$ with $e = (1, \ldots, 1)' : T \times 1$ and $\eta = C(\rho)\varepsilon$, we need to specify the $T \times (T-1)$ matrix H, which we take as $H = Q_2 F$, where

$$(9.15) \qquad Q_2' = \begin{pmatrix} 1 & -1 & & & & 0 \\ & \cdot & \cdot & & & \\ & & \cdot & \cdot & & \\ & 0 & & \cdot & \cdot & \\ & & & & 1 & -1 \end{pmatrix} : (T-1) \times T$$

and F is a $(T-1) \times (T-1)$ nonsingular matrix to be chosen in such a way that $F'Q_2'Q_2F = I_{T-1}$ and $Q_2FF'Q_2' = M = I_T - ee'/T$. It is seen from (9.13) that the LBI statistic does not depend on the matrix F; hence we need not specify F explicitly. We have only to find Q_2' as in (9.15) ensuring that $Q_2'e = 0$ because of the requirement $H'X = 0$.

The LBI statistic is now obtained (Problem 2.3) as

$$(9.16)\ S_{T2} = \frac{y'Q_2(Q_2'\Phi(1)Q_2)^{-1}Q_2'dd'Q_2(Q_2'\Phi(1)Q_2)^{-1}Q_2'y}{y'Q_2(Q_2'\Phi(1)Q_2)^{-1}Q_2'y} = \frac{(y_T - y_1)^2}{\sum_{j=2}^{T}(y_j - y_{j-1})^2}.$$

It holds (Problem 2.4) that $S_{T2}/(T-1) \sim \text{Be}(1/2, (T-2)/2)$ under H_0.

9.2.2 The LBIU Test

Here we consider the testing problem (9.7). Then the LBIU test is derived from Ferguson (1967, p.237) as rejecting H_0 when

$$\left.\frac{\partial^2 \log f(v \mid \theta)}{\partial \theta^2}\right|_{\theta=\theta_0} + \left(\left.\frac{\partial \log f(v \mid \theta)}{\partial \theta}\right|_{\theta=\theta_0}\right)^2$$

$$> c_1 + c_2 \left.\frac{\partial \log f(v \mid \theta)}{\partial \theta}\right|_{\theta=\theta_0}$$

for some constants c_1 and c_2. Suppose that the first derivative above reduces to a constant, which occurs, on account of (9.11), if $H'd\Sigma(\theta)/d\theta|_{\theta=\theta_0}H$ is a constant multiple of $H'\Sigma(\theta_0)H$. Then the following theorem holds.

Theorem 9.2. *Consider the regression model $y = X\beta + u$ with $u \sim N(0, \sigma^2\Sigma(\theta))$. Let H be a $T \times (T-p)$ matrix such that $H'H = I_{T-p}$ and $HH' = I_T - X(X'X)^{-1}X'$. Suppose that $H'd\Sigma(\theta)/d\theta|_{\theta=\theta_0}H$ is a constant multiple of $H'\Sigma(\theta_0)H$. Then the LBIU test for the testing problem (9.7) rejects the null hypothesis when*

(9.17) $$\frac{y'H(H'\Sigma(\theta_0)H)^{-1}H'\left.\frac{d^2\Sigma(\theta)}{d\theta^2}\right|_{\theta=\theta_0}H(H'\Sigma(\theta_0)H)^{-1}H'y}{y'H(H'\Sigma(\theta_0)H)^{-1}H'y} > c.$$

Let us consider the unit root testing problem (9.4). Since $H'd\Phi(\rho)/d\rho|_{\rho=1}H = H'(dd' - \Phi(1))H = -H'\Phi(1)H$ for Models C and D, the LBIU tests for these models can be derived from Theorem 9.2. We can first show (Problem 2.5) that the (j,k)th element of $d^2\Phi(\rho)/d\rho^2|_{\rho=1}$, A_{jk} say, is given by

(9.18) $$A_{jk} = \begin{cases} jk(k-3) + \dfrac{j(j^2+5)}{3}, & (j \le k) \\ jk(j-3) + \dfrac{k(k^2+5)}{3}, & (j > k). \end{cases}$$

Let us consider Model C, where $y = \beta_1 d + \eta$ with $\eta = C(\rho)\varepsilon$. We choose the $T \times (T-1)$ matrix H as $H = Q_3 F$, where F is a $(T-1) \times (T-1)$ nonsingular matrix, while

(9.19) $$Q_3' = \begin{pmatrix} -2 & 1 & & & & \\ 1 & -2 & 1 & & 0 & \\ & \cdot & \cdot & \cdot & & \\ & & \cdot & \cdot & \cdot & \\ & 0 & & \cdot & \cdot & \cdot \\ & & & 1 & -2 & 1 \end{pmatrix} : (T-1) \times T.$$

Note that $Q_3'd = 0$. Then it can be checked after some algebra that

(9.20) $$Q_3' \left.\frac{d^2\Phi(\rho)}{d\rho^2}\right|_{\rho=1} Q_3 = 2Q_3'\Phi(1)Q_3 - 2I_{T-1}.$$

Here the (j,k)th element of $(Q_3'\Phi(1)Q_3)^{-1}$ is $\min(j,k) - jk/T$, $(j,k = 1,\ldots,T-1)$. Then it follows from (9.17) that the LBIU test for (9.4) rejects H_0 when

$$(9.21) \qquad \frac{y'Q_3(Q_3'\Phi(1)Q_3)^{-2}Q_3'y}{y'Q_3(Q_3'\Phi(1)Q_3)^{-1}Q_3'y} < c.$$

We can now define (Problem 2.6) the LBIU statistic by

$$(9.22) \quad S_{T3} = \frac{1}{T}\frac{y'Q_3(Q_3'\Phi(1)Q_3)^{-2}Q_3'y}{y'Q_3(Q_3'\Phi(1)Q_3)^{-1}Q_3'y} = \frac{1}{T}\frac{\sum_{j=1}^{T}\left(y_j - \frac{j}{T}y_T\right)^2}{\sum_{j=1}^{T}(y_j - y_{j-1})^2 - \frac{1}{T}y_T^2}.$$

The factor $1/T$ was placed here to make the limiting distribution of S_{T3} nondegenerate. In fact it can be shown (Problem 2.7) that, under H_0,

$$(9.23) \qquad \mathcal{L}(S_{T3}) \to \mathcal{L}\left(\int_0^1\int_0^1 (\min(s,t) - st)\,dw(s)\,dw(t)\right),$$

where $\{w(t)\}$ is the one-dimensional standard Brownian motion.

Model D can be dealt with similarly, where $y = \beta_0 e + \beta_1 d + \eta$ with $\eta = C(\rho)\varepsilon$. We define

$$(9.24) \qquad Q_4' = \begin{pmatrix} 1 & -2 & 1 & & & & 0 \\ & \cdot & \cdot & \cdot & & & \\ & & \cdot & \cdot & \cdot & & \\ 0 & & & \cdot & \cdot & \cdot & \\ & & & & 1 & -2 & 1 \end{pmatrix} : (T-2) \times T$$

to ensure $Q_4'(e,d) = 0$. Then it holds that

$$(9.25) \qquad Q_4'\left.\frac{d^2\Phi(\rho)}{d\rho^2}\right|_{\rho=1} Q_4 = 2Q_4'\Phi(1)Q_4 - 2I_{T-2},$$

where the (j,k)th element of $(Q_4'\Phi(1)Q_4)^{-1}$ is $\min(j,k) - jk/(T-1)$, $(j,k = 1,\ldots,T-2)$. Thus the LBIU test has the same form as in (9.21) with Q_3 replaced by Q_4; hence we obtain (Problem 2.8) the rejection region of the LBIU test as

$$(9.26) \qquad S_{T4} = \frac{1}{T}\frac{y'Q_4(Q_4'\Phi(1)Q_4)^{-2}Q_4'y}{y'Q_4(Q_4'\Phi(1)Q_4)^{-1}Q_4'y}$$

$$= \frac{1}{T} \frac{\sum_{j=2}^{T} \left(y_j - y_1 - \frac{j-1}{T-1}(y_T - y_1)\right)^2}{\sum_{j=2}^{T}(y_j - y_{j-1})^2 - \frac{1}{T-1}(y_T - y_1)^2} < c.$$

It can be shown (Problem 2.9) that, under H_0, S_{T4} has the same limiting distribution as S_{T3} described in (9.23).

The last and present subsections have discussed how to derive LBI and LBIU tests and have applied these to the unit root testing problem. It is recognized that the computations of $Q'_k d^2 \Phi(\rho)/d\rho^2 Q_k$ and $(Q'_k \Phi(1) Q_k)^{-1}$ for $k = 3, 4$ are troublesome. In the next section we show that the LM test yields completely the same statistic without these matrix computations.

PROBLEMS

2.1 Show that the relation in (9.12) holds.
2.2 Prove that the statistic S_{T1}/T with S_{T1} defined in (9.14) follows Be$(1/2, (T-1)/2)$ under H_0.
2.3 Prove that S_{T2} in (9.16) is the LBI statistic for Model B.
2.4 Prove that the statistic $S_{T2}/(T-1)$ with S_{T2} defined in (9.16) follows Be$(1/2, (T-2)/2)$ under H_0.
2.5 Show that the (j, k)th element of $d^2 \Phi(\rho)/d\rho^2|_{\rho=1}$ is given by (9.18).
2.6 Prove that S_{T3} in (9.22) is the LBIU statistic for Model C.
2.7 Prove the weak convergence result in (9.23).
2.8 Prove that S_{T4} in (9.26) is the LBIU statistic for Model D.
2.9 Derive the limiting distribution of S_{T4} in (9.26) under H_0.

9.3 EQUIVALENCE OF THE LM TEST WITH THE LBI OR LBIU TEST

In this section we deal with the general regression model described in (9.5):

(9.27) $$y = X\beta + u, \quad u \sim N(0, \sigma^2 \Sigma(\theta)),$$

where $\Sigma(\theta)$ is assumed to be positive definite and twice continuously differentiable. For the above model, we consider the easily derivable LM test for the testing problem

(9.28) $$H_0: \theta = \theta_0 \quad \text{versus} \quad H_1: \theta < \theta_0$$

or

(9.29) $$H_0: \theta = \theta_0 \quad \text{versus} \quad H_1: \theta \neq \theta_0.$$

The above model (9.27) covers the AR unit root model discussed in this chapter and the MA unit root model treated in the next chapter.

Our purpose here is to show that the LM test for (9.28) is equivalent to the LBI test, while, under a certain condition, the LM test for (9.29) is equivalent to the LBIU test. These problems are discussed in King and Hillier (1985) and Tanaka (1995).

9.3.1 Equivalence with the LBI Test

The LM statistic is easy to derive. We have only to compute the slope of the log-likelihood evaluated at the MLE under H_0. Let us put

$$(9.30) \quad l(\theta, \beta, \sigma^2) = -\frac{T}{2}\log(2\pi\sigma^2) - \frac{1}{2}\log|\Sigma(\theta)|$$
$$- \frac{1}{2\sigma^2}(y - X\beta)'\Sigma^{-1}(\theta)(y - X\beta),$$

so that

$$\frac{\partial l(\theta, \beta, \sigma^2)}{\partial \theta} = K + \frac{1}{2\sigma^2}(y - X\beta)'\Sigma^{-1}(\theta)\frac{d\Sigma(\theta)}{d\theta}\Sigma^{-1}(\theta)(y - X\beta),$$

where K is a constant. Let $\tilde{\theta}$, $\tilde{\beta}$ and $\tilde{\sigma}^2$ be the MLEs under $H_0: \theta = \theta_0$. Then $\tilde{\theta} = \theta_0$ and

$$\tilde{\beta} = \left(X'\Sigma^{-1}(\theta_0)X\right)^{-1} X'\Sigma^{-1}(\theta_0)y, \quad \tilde{\sigma}^2 = \frac{1}{T}\tilde{\eta}'\Sigma^{-1}(\theta_0)\tilde{\eta},$$

where $\tilde{\eta} = y - X\tilde{\beta} = \tilde{M}y$ with

$$(9.31) \quad \tilde{M} = I_T - X\left(X'\Sigma^{-1}(\theta_0)X\right)^{-1} X'\Sigma^{-1}(\theta_0).$$

We now obtain the LM test for (9.28) as rejecting H_0 when

$$\partial l(\theta, \beta, \sigma^2)/\partial\theta\big|_{\theta=\theta_0, \beta=\tilde{\beta}, \sigma^2=\tilde{\sigma}^2}$$

takes small values, that is, when

$$(9.32) \quad \mathrm{LM}_1 = \frac{y'\tilde{M}'\Sigma^{-1}(\theta_0)\dfrac{d\Sigma(\theta)}{d\theta}\bigg|_{\theta=\theta_0}\Sigma^{-1}(\theta_0)\tilde{M}y}{y'\tilde{M}'\Sigma^{-1}(\theta_0)\tilde{M}y} < c.$$

The LBI test in the present case has been derived in Theorem 9.1, which rejects H_0 when

$$(9.33) \quad \frac{y'H(H'\Sigma(\theta_0)H)^{-1}H'\dfrac{d\Sigma(\theta)}{d\theta}\bigg|_{\theta=\theta_0} H(H'\Sigma(\theta_0)H)^{-1}H'y}{y'H(H'\Sigma(\theta_0)H)^{-1}H'y} < c.$$

We show

(9.34) $$\tilde{M}'\Sigma^{-1}(\theta_0)\tilde{M} = H(H'\Sigma(\theta_0)H)^{-1}H',$$

(9.35) $$\tilde{M}'\Sigma^{-1}(\theta_0)\left.\frac{d\Sigma(\theta)}{d\theta}\right|_{\theta=\theta_0}\Sigma^{-1}(\theta_0)\tilde{M}$$
$$= H(H'\Sigma(\theta_0)H)^{-1}H'\left.\frac{d\Sigma(\theta)}{d\theta}\right|_{\theta=\theta_0}H(H'\Sigma(\theta_0)H)^{-1}H',$$

to conclude that the LM test is identical with the LBI test. For this purpose we have [Rao (1973, p. 77)]

Lemma 9.1. *Let P and Q be $T \times (T - p)$ and $T \times p$ matrices such that $R = (P \quad Q)$ is orthogonal. Then, for any $T \times T$ nonsingular symmetrix matrix A, it holds that*

(9.36) $$\begin{aligned}P(P'AP)^{-1}P' &= A^{-1} - A^{-1}Q(Q'A^{-1}Q)^{-1}Q'A^{-1} \\ &= A^{-1}\bar{M} = \bar{M}'A^{-1} = \bar{M}'A^{-1}\bar{M},\end{aligned}$$

where $\bar{M} = I_T - Q(Q'A^{-1}Q)^{-1}Q'A^{-1}$.

Proof. Let us define the $T \times T$ matrix G by

$$G = \begin{pmatrix} P'A \\ Q' \end{pmatrix} : T \times T,$$

which is nonsingular since

$$GR = \begin{pmatrix} P'A \\ Q' \end{pmatrix}(P \quad Q) = \begin{pmatrix} P'AP & P'AQ \\ 0 & I_p \end{pmatrix}$$

is nonsingular. Then the first equality in (9.36) is established if we show

$$G\left[P(P'AP)^{-1}P' - \left(A^{-1} - A^{-1}Q(Q'A^{-1}Q)^{-1}Q'A^{-1}\right)\right] = 0,$$

which is certainly true because $P'Q = 0$. The rest of (9.36) can be proved easily.

It is now an easy matter to prove (9.34) and (9.35) (Problem 3.1). Thus we can establish the following theorem.

EQUIVALENCE OF THE LM TEST WITH THE LBI OR LBIU TEST

Theorem 9.3. *Consider the regression model $y = X\beta + u$ with $u \sim N(0, \sigma^2 \Sigma(\theta))$. Then the LM and LBI tests for the testing problem (9.28) are equivalent to each other, where the former is given in (9.32), while the latter in (9.33).*

Because of Theorem 9.3, the LBI statistic can be derived just by computing the statistic LM_1 in (9.32) without being worried about finding the matrix H and computing $(H'\Sigma(\theta_0)H)^{-1}$. For the unit root tests, it is more convenient to rewrite (Problem 3.2) LM_1 in (9.32) as

$$(9.37) \quad LM_1' = -\frac{y'\tilde{M}' \left.\frac{d\Phi^{-1}(\rho)}{d\rho}\right|_{\rho=1} \tilde{M}y}{y'\tilde{M}'\Phi^{-1}(1)\tilde{M}y} = \frac{\tilde{\eta}_T^2}{\sum_{j=1}^{T}(\tilde{\eta}_j - \tilde{\eta}_{j-1})^2} - 1,$$

where $\tilde{\eta}_0 \equiv 0$ and $\tilde{\eta} = (\tilde{\eta}_1, \ldots, \tilde{\eta}_T)' = \tilde{M}y$. For Model B we have $\tilde{\eta}_j = y_j - y_1$ so that we obtain $LM_1 = S_{T2} - 1$, where the discrepancy -1 arises from an abbreviated construction of S_{T2}. We can also deduce from (9.37) that the LM statistic reduces to a constant if $\tilde{\eta}_T = 0$, which occurs if $d = (1, \ldots, T)'$ belongs to the column space of X (Problem 3.3).

9.3.2 Equivalence with the LBIU Test

We consider the situation where the statistic LM_1 in (9.32) becomes a constant, which occurs if

$$(9.38) \quad \left.\frac{d\Sigma(\theta)}{d\theta}\right|_{\theta=\theta_0} \Sigma^{-1}(\theta_0)\tilde{M} = c \times \tilde{M}$$

for some constant c. In terms of the likelihood, the above situation occurs if the slope of the likelihood evaluated at the MLE under H_0 becomes a constant.

Under the above situation, we suggest the LM test for (9.29) as rejecting H_0 when $\partial^2 l(\theta, \beta, \sigma^2)/\partial\theta^2|_{\theta=\theta_0, \beta=\tilde{\beta}, \sigma^2=\tilde{\sigma}^2}$ takes large values. It follows from (9.30) and (9.38) that the LM test has the rejection region:

$$(9.39) \quad LM_2 = \frac{y'\tilde{M}'\Sigma^{-1}(\theta_0) \left.\frac{d^2\Sigma(\theta)}{d\theta^2}\right|_{\theta=\theta_0} \Sigma^{-1}(\theta_0)\tilde{M}y}{y'\tilde{M}'\Sigma^{-1}(\theta_0)\tilde{M}y} > c.$$

The corresponding LBIU test has been derived in Theorem 9.2, which rejects H_0 when

$$(9.40) \quad \frac{y'H(H'\Sigma(\theta_0)H)^{-1}H' \left.\frac{d^2\Sigma(\theta)}{d\theta^2}\right|_{\theta=\theta_0} H(H'\Sigma(\theta_0)H)^{-1}H'y}{y'H(H'\Sigma(\theta_0)H)^{-1}H'y} > c.$$

The equivalence of the LM and LBIU tests can be proved easily by using Lemma 9.1. Thus we can establish the following theorem.

Theorem 9.4. *Consider the regression model $y = X\beta + u$ with $u \sim N(0, \sigma^2\Sigma(\theta))$, and suppose that (9.38) holds. Then the LM and LBIU tests for the testing problem (9.29) are equivalent to each other, where the former is given in (9.39), while the latter in (9.40).*

For the unit root tests, the condition (9.38) is satisfied for Models C and D, but the computation of the LM statistics following (9.39) is much involved. Since $d^2\Phi^{-1}(\rho)/d\rho^2|_{\rho=1} = 2(I_T - e_T e_T')$ with $e_T = (0, \ldots, 0, 1)' : T \times 1$, it is more convenient in the present case to compute

$$\begin{aligned}
S_T &= -\frac{1}{2\sigma^2}(y - X\beta)' \frac{d^2\Omega^{-1}(\rho)}{d\rho^2}(y - X\beta)\Big|_{\rho=1, \beta=\tilde{\beta}, \sigma^2=\tilde{\sigma}^2} \\
&= -\frac{1}{2\tilde{\sigma}^2}\tilde{\eta}'(2I_T - 2e_T e_T')\tilde{\eta} \\
&= -\frac{\tilde{\eta}'\tilde{\eta}}{\tilde{\sigma}^2} + \frac{\tilde{\eta}_T^2}{\tilde{\sigma}^2} \\
&= -T \frac{\sum_{j=1}^{T}\tilde{\eta}_j^2}{\sum_{j=1}^{T}(\tilde{\eta}_j - \tilde{\eta}_{j-1})^2},
\end{aligned}$$

where $\tilde{\eta}_0 \equiv 0$ and we have assumed $\tilde{\eta}_T = 0$, which occurs if $d = (1, \ldots, T)'$ belongs to the column space of X, as was mentioned before. Since we reject H_0 when S_T becomes large, the rejection region of the LBIU unit root test is given by

$$(9.41) \qquad \text{LM}_2' = \frac{1}{T} \frac{\sum_{j=1}^{T}\tilde{\eta}_j^2}{\sum_{j=1}^{T}(\tilde{\eta}_j - \tilde{\eta}_{j-1})^2} < c.$$

It can now be easily checked (Problem 3.4) that the statistic LM_2' in (9.41) does produce the LBIU statistics S_{T3} in (9.22) and S_{T4} in (9.26).

PROBLEMS

3.1 Prove (9.34) and (9.35) using Lemma 9.1.

3.2 Derive the expressions for LM_1' in (9.37) from (9.32).

3.3 Prove that the LM statistic LM_1' in (9.37) reduces to a constant if $d = (1, \ldots, T)'$ belongs to the column space of X.

3.4 Prove that the LM statistic LM_2' in (9.41) yields the LBIU statistics S_{T3} in (9.22) and S_{T4} in (9.26).

9.4 VARIOUS UNIT ROOT TESTS

In this section we take up some other unit root tests, suggested in the literature, whose power properties are examined in the next three sections. We continue to consider the time series regression model

$$(9.42) \qquad y = X\beta + \eta, \qquad \eta \sim N(0, \sigma^2 \Phi(\rho)).$$

The testing problem discussed here is the unit root test against the left-sided alternative

$$(9.43) \qquad H_0 : \rho = 1 \quad \text{versus} \quad H_1 : \rho < 1.$$

To describe test statistics we define

$$(9.44) \qquad \hat{\eta} = y - X\hat{\beta}, \qquad \hat{\beta} = (X'X)^{-1}X'y,$$

$$(9.45) \qquad \tilde{\eta} = y - X\tilde{\beta}, \qquad \tilde{\beta} = \left(X'\Phi^{-1}(1)X\right)^{-1} X'\Phi^{-1}(1)y.$$

Note that $\hat{\beta}$ is the ordinary LSE which yields ordinary least squares (OLS) residuals $\hat{\eta}$, while $\tilde{\beta}$ is the generalized LSE under H_0 defined in the last section together with generalized least squares (GLS) residuals $\tilde{\eta}$.

The test statistics R_1 through R_6 defined below are from Nabeya and Tanaka (1990b), and are all based on $\hat{\eta}$ or $\tilde{\eta}$. In the definitions we put $\hat{\eta}_0 = \tilde{\eta}_0 = 0$ and $\bar{\tilde{\eta}} = \sum_{j=1}^{T} \tilde{\eta}_j / T$. Each statistic R_k is defined in such a way that the null hypothesis is rejected when R_k takes small values and its limiting null distribution as $T \to \infty$ is nondegenerate.

(i) LBI statistic:

$$R_1 = \tilde{\eta}_T^2 \bigg/ \sum_{j=1}^{T} (\tilde{\eta}_j - \tilde{\eta}_{j-1})^2 ,$$

(ii) LBIU statistic:

$$R_2 = \frac{1}{T} \sum_{j=1}^{T} \tilde{\eta}_j^2 \bigg/ \sum_{j=1}^{T} (\tilde{\eta}_j - \tilde{\eta}_{j-1})^2 ,$$

(iii) von Neumann ratio:

$$R_3 = \frac{1}{T}\sum_{j=1}^{T}(\tilde{\eta}_j - \bar{\tilde{\eta}})^2 \Big/ \sum_{j=1}^{T}(\tilde{\eta}_j - \tilde{\eta}_{j-1})^2,$$

(iv) Dickey–Fuller statistic:

$$R_4 = T(\hat{\rho} - 1), \quad \hat{\rho} = \sum_{j=2}^{T}\hat{\eta}_{j-1}\hat{\eta}_j \Big/ \sum_{j=2}^{T}\hat{\eta}_{j-1}^2,$$

(v) Durbin–Watson statistic:

$$R_5 = \frac{1}{T}\sum_{j=1}^{T}\hat{\eta}_j^2 \Big/ \sum_{j=2}^{T}(\hat{\eta}_j - \hat{\eta}_{j-1})^2,$$

(vi) Modified Durbin–Watson statistic:

$$R_6 = \frac{1}{T}\sum_{j=1}^{T}\hat{\eta}_j^2 \Big/ \sum_{j=1}^{T}(\hat{\eta}_j - \hat{\eta}_{j-1})^2.$$

The test statistics R_1 through R_3 are the ratio of quadratic forms in GLS residuals $\tilde{\eta}$, and R_4 through R_6 in OLS residuals $\hat{\eta}$, which implies that these are invariant under the group of transformations $y \to ay + Xb$ for any nonzero a and any $p \times 1$ vector b. A brief explanation of each test follows.

i) R_1: This is LBI, as was shown in the last section. This test cannot be applicable when $\tilde{\eta}_T = 0$, that is, when $d = (1,\ldots,T)'$ belongs to the column space of X.

ii) R_2: This is LBIU against $H_1 : \rho \neq 1$ if d belongs to the column space of X. Bhargava (1986) also suggested this test, claiming that it is LBI if the Anderson approximation $\Phi^{-1}(\rho) \cong (1-\rho)^2 I_T + \rho\Phi^{-1}(1)$ is used. Note that $\Phi^{-1}(\rho) = (1-\rho)^2 I_T + \rho\Phi^{-1}(1) + \rho(1-\rho)e_T e_T'$, where $e_T = (0,\ldots,0,1)' : T \times 1$.

iii) R_3: This is equivalent to the von Neumann ratio applied to $\{\tilde{\eta}_j\}$. Note that $\bar{\tilde{\eta}}$ is nonzero in general so that it is of interest to compare it with the LBIU statistic R_2, which is only slightly different in appearance.

iv) R_4: This is most frequently used because the statistic is easily derived from the ordinary LSE of ρ. We could suggest a statistic based on the Yule–Walker estimator, which will be taken up in Section 9.7 to compare power properties.

v) R_5: This is equivalent to the Durbin–Watson statistic, as is seen by considering $1/(TR_5)$. The present form arises because of deriving the nondegenerate limiting null distribution and rejecting H_0 for small values.

vi) R_6: This is a modification of R_5, the only difference being the range of summation in the denominator. This difference yields a different asymptotic distribution in some cases.

A frequently used test statistic not considered here is the t-ratio statistic. For the simplest model $y_j = \rho y_{j-1} + \varepsilon_j$ with $y_0 = 0$, it takes the form

$$t_{\hat{\rho}} = \frac{\hat{\rho} - 1}{\hat{\sigma} \Big/ \sqrt{\sum_{j=2}^{T} y_{j-1}^2}}, \qquad \hat{\sigma}^2 = \frac{1}{T-1} \sum_{j=2}^{T} (y_j - \hat{\rho} y_{j-1})^2.$$

The corresponding test is shown to be equivalent to the likelihood ratio (LR) test. We do not deal with the t-ratio because it is not the ratio of quadratic forms, and Dickey and Fuller (1979) found by simulations that the t-ratio test is inferior to the test based on $R_4 = T(\hat{\rho} - 1)$ against the left-sided alternative $H_1 : \rho < 1$. Note that the latter test is the Wald test, while the test based on R_1 or R_2 is the LM test, as was shown in the last section. Here we just mention that the LM, LR and Wald tests do produce different, limiting power properties.

One might argue that we should also consider a test based on $T(\tilde{\rho} - 1)$, where the statistic is defined in the same way as R_4 with $\hat{\eta}$ replaced by $\tilde{\eta}$. For Model A this test is nothing but the R_4 test. We can also deduce that the tests for Models A and B are asymptotically the same. Moreover, the tests for Models C and D turn out to be asymptotically equivalent to the corresponding R_2 tests. Thus we shall exclude the test based on $T(\tilde{\rho} - 1)$ from consideration.

9.5 INTEGRAL EXPRESSIONS FOR THE LIMITING POWERS

Limiting powers of the six tests presented in the last section are derived here by using integral expressions. For this purpose, we consider Models A through D introduced in Section 9.1. For these models we derive the limiting distributions of R_1 through R_6 as $T \to \infty$ under $\rho = 1 - (c/T)$, where c is a fixed constant. When $c = 0$, that is, $\rho = 1$, these give limiting null distributions. Results described in subsequent discussions are based on Nabeya and Tanaka (1990b).

We first have (Problem 5.1) the following lemma.

Lemma 9.2. *Under $\rho = 1 - (c/T)$, it holds for Models A through D that*

(9.46) $$\frac{1}{T\sigma^2} \sum_{j=2}^{T} (\hat{\eta}_j - \hat{\eta}_{j-1})^2 \to 1 \quad \text{in probability},$$

(9.47) $$\frac{1}{T\sigma^2} \sum_{j=1}^{T} (\tilde{\eta}_j - \tilde{\eta}_{j-1})^2 \to 1 \quad \text{in probability}.$$

Note the difference of the ranges of summation in (9.46) and (9.47). It turns out that $\hat{\eta}_1^2/T$ is nondegenerate for Models B and D, though it is negligible for Models A and C. This fact may be deduced from discussions in Chapter 7 [see (7.28), (7.40), (7.50), and (7.56) with $Z = Y(0) = 0$].

We derive in the following subsections integral expressions for the limiting powers of the tests based on R_k ($k = 1, \ldots, 6$) applied to Models A through D. As in Chapter 7, we first establish the weak convergence results on R_k. For this purpose we define the O–U process on [0,1]:

(9.48) $$dY(t) = -cY(t)\,dt + dw(t), \qquad Y(0) = 0.$$

9.5.1 Model A

Since $\hat{\eta}_j = \tilde{\eta}_j = \eta_j = \rho\eta_{j-1} + \varepsilon_j$ and $\bar{\hat{\eta}} = \bar{\eta}$, the following theorem can be easily established (Problem 5.2).

Theorem 9.5. *For the test statistics R_k ($k = 1, \ldots, 6$) applied to Model A, it holds that, as $T \to \infty$ under $\rho = 1 - (c/T)$,*

$$\mathcal{L}(R_1) \to \frac{1 - e^{-2c}}{2c}\chi^2(1),$$

$$\mathcal{L}(R_2), \mathcal{L}(R_5), \mathcal{L}(R_6) \to \mathcal{L}\left(\int_0^1 Y^2(t)\,dt\right),$$

$$\mathcal{L}(R_3) \to \mathcal{L}\left(\int_0^1 \left(Y(t) - \int_0^1 Y(s)\,ds\right)^2 dt\right),$$

$$\mathcal{L}(R_4) \to \mathcal{L}\left(\frac{\int_0^1 Y(t)\,dY(t)}{\int_0^1 Y^2(t)\,dt}\right),$$

where $\chi^2(q)$ denotes a χ^2 distribution with q degrees of freedom, while $\{Y(t)\}$ is the O–U process defined in (9.48).

The limiting null distributions can be easily derived from this theorem by putting $c = 0$; so can the significance points. Note that $(1 - e^{-2c})/(2c) = 1$ and $Y(t) = w(t)$ when $c = 0$. Note also that the limiting distribution of R_4, not only for Model A but also for the other models, is available from Chapter 7.

The expressions for the limiting powers of the six tests can now be obtained (Problem 5.3) in the following theorem.

Theorem 9.6. *Under Model A, let $x_k(\alpha)$ be the $100\alpha\%$ point of the limiting null distribution of R_k and $\beta_k(\alpha)$ the limiting power of the test based on R_k at the $100\alpha\%$ significance level. Then it holds that*

INTEGRAL EXPRESSIONS FOR THE LIMITING POWERS   337

$$\beta_1(\alpha) = P\left(Z^2 \leq \frac{2c}{1-e^{-2c}} x_1(\alpha)\right),$$

$$\beta_k(\alpha) = \frac{1}{\pi} \int_0^\infty \mathrm{Re}\left[\frac{1-\exp\{-i\theta x_k(\alpha)\}}{i\theta} e^{c/2} \left(\cos\nu + c\frac{\sin\nu}{\nu}\right)^{-1/2}\right] d\theta, (k=2,5,6),$$

$$\beta_3(\alpha) = \frac{1}{\pi} \int_0^\infty \mathrm{Re}\left[\frac{1-\exp\{-i\theta x_3(\alpha)\}}{i\theta} \phi_1(\theta)\right] d\theta,$$

$$\beta_4(\alpha) = \frac{1}{2} + \frac{1}{\pi} \int_0^\infty \frac{1}{\theta} \mathrm{Im}\left[m_{11}(-i\theta, i\theta x_4(\alpha))\right] d\theta,$$

where $Z \sim N(0,1)$, $\nu = \sqrt{2i\theta - c^2}$, and $m_{11}(-i\theta, i\theta x_4(\alpha))$ is defined by (7.77) with r and γ replaced by $r=1$ and $\gamma=0$, respectively, while $\phi_1(\theta)$ is defined by

$$(9.49) \quad \phi_1(\theta) = e^{c/2} \left[\frac{2i\theta - c^3}{\nu^2} \frac{\sin\nu}{\nu} - \frac{c^2\nu^2 + 4ic\theta}{\nu^4} \cos\nu + \frac{4ic\theta}{\nu^4}\right]^{-1/2}.$$

The expressions for $\beta_2(\alpha) (=\beta_5(\alpha)=\beta_6(\alpha))$ and $\beta_3(\alpha)$ may be deduced from (4.9) and (4.11), respectively, and $\beta_4(\alpha)$ from (7.77). The limiting powers depend on the value of c and will be computed later. At this stage we confirm (Problem 5.4) that the above tests applied to Model A are all consistent in the sense that $\beta_k(\alpha) \to 1$ as $c \to \infty$ for $k=1,\ldots,6$. The same is shown to be true for Models B, C, and D discussed subsequently.

9.5.2 Model B

For Model B we have $\tilde{\eta}_j = \eta_j - \eta_1 = \eta_j + O_p(1)$ so that it can be deduced that $\tilde{\eta}_T^2/T = \eta_T^2/T + o_p(1)$ and

$$\frac{1}{T^2} \sum_{j=1}^T \tilde{\eta}_j^2 = \frac{1}{T^2} \sum_{j=1}^T \eta_j^2 + o_p(1),$$

$$\frac{1}{T^2} \sum_{j=1}^T (\tilde{\eta}_j - \bar{\tilde{\eta}})^2 = \frac{1}{T^2} \sum_{j=1}^T (\eta_j - \bar{\eta})^2.$$

Thus the tests based on R_1 through R_3 have the same limiting properties as in Model A. Since $\hat{\eta}_j = \eta_j - \bar{\eta}$, the other tests will have different properties.

We can now establish the following theorem (Problem 5.5).

Theorem 9.7. *For Model B, the test statistics $R_1, R_2,$ and R_3 have the same limiting distributions as the corresponding statistics for Model A as $T \to \infty$ under $\rho = 1 -$*

(c/T), while it holds that

$$\mathcal{L}(R_4) \to \mathcal{L}\left(\frac{\int_0^1 Y(t)\,dY(t) - \int_0^1 dY(t)\int_0^1 Y(t)\,dt}{\int_0^1 \left(Y(t) - \int_0^1 Y(s)\,ds\right)^2 dt}\right),$$

$$\mathcal{L}(R_5) \to \mathcal{L}\left(\int_0^1 \left(Y(t) - \int_0^1 Y(s)\,ds\right)^2 dt\right),$$

$$\mathcal{L}(R_6) \to \mathcal{L}\left(\frac{\int_0^1 \left(Y(t) - \int_0^1 Y(s)\,ds\right)^2 dt}{\left(\int_0^1 Y(t)\,dt\right)^2 + 1}\right).$$

Using the same notations as in Theorem 9.6, we obtain the expressions for the limiting powers of the six tests in the following theorem (Problem 5.6).

Theorem 9.8. *Under Model B, the limiting powers of the tests based on $R_1, R_2,$ and R_3 are the same as the corresponding tests under Model A. The limiting powers of the other tests at the $100\alpha\%$ significance level can be expressed as follows:*

$$\beta_4(\alpha) = \frac{1}{2} + \frac{1}{\pi}\int_0^\infty \frac{1}{\theta}\,\text{Im}\left[m_{21}(-i\theta, i\theta x_4(\alpha))\right]d\theta,$$

$$\beta_5(\alpha) = \frac{1}{\pi}\int_0^\infty \text{Re}\left[\frac{1 - \exp\{-i\theta x_5(\alpha)\}}{i\theta}\phi_1(\theta)\right]d\theta,$$

$$\beta_6(\alpha) = \frac{1}{2} + \frac{1}{\pi}\int_0^\infty \frac{1}{\theta}\,\text{Im}\left[\exp\{i\theta x_6(\alpha)\}\psi_1(\theta; x_6(\alpha))\right]d\theta,$$

where $m_{21}(-i\theta, i\theta x_4(\alpha))$ is defined by (7.78) with r and γ replaced by $r = 1$ and $\gamma = 0$, respectively, while $\phi_1(\theta)$ is defined in (9.49) and

$$\psi_1(\theta; x) = e^{c/2}\left[-\frac{c^3 + 2i\theta(x + 1 - cx)}{\mu^2}\frac{\sin\mu}{\mu} - \frac{c^2 - 2i\theta x}{\mu^2}\cos\mu \right.$$
$$\left. + \frac{4ic\theta(x + 1)}{\mu^4}(\cos\mu - 1)\right]^{-1/2}$$

with $\mu = \sqrt{-2i\theta - c^2}$.

9.5.3 Model C

It holds for Model C that

$$\hat{\eta}_j = \eta_j - j\sum_{i=1}^{T} i\eta_i \Big/ \sum_{i=1}^{T} i^2, \qquad \tilde{\eta}_j = \eta_j - \frac{j}{T}\eta_T.$$

Thus the test statistics will have different distributions from those for Models A and B. Since $\tilde{\eta}_T \equiv 0$, we leave the LBI statistic R_1 out of consideration.
It is easy to deduce that

$$\mathcal{L}\left(\frac{1}{T^2\sigma^2}\sum_{j=1}^{T}\tilde{\eta}_j^2\right) = \mathcal{L}\left(\frac{1}{T^2\sigma^2}\sum_{j=1}^{T}\left(\eta_j - \frac{j}{T}\eta_T\right)^2\right)$$

$$\to \mathcal{L}\left(\int_0^1 (Y(t) - tY(1))^2 dt\right),$$

which gives $\mathcal{L}(R_2)$ as $T \to \infty$ because of Lemma 9.2. The limiting distribution of R_4 is available from Theorem 7.3, from which $\mathcal{L}(R_5)$ and $\mathcal{L}(R_6)$ as $T \to \infty$ can be easily derived. Leaving the expression for the limiting distribution of R_3 as Problem 5.7, we can establish the following theorem.

Theorem 9.9. *For Model C, the test statistics R_2 through R_6 have the following limiting distributions as $T \to \infty$ under $\rho = 1 - (c/T)$:*

$$\mathcal{L}(R_2) \to \mathcal{L}\left(\int_0^1 (Y(t) - tY(1))^2 dt\right),$$

$$\mathcal{L}(R_3) \to \mathcal{L}\left(\int_0^1 \left(Y(t) - \int_0^1 Y(s)\,ds - \left(t - \frac{1}{2}\right)Y(1)\right)^2 dt\right),$$

$$\mathcal{L}(R_4) \to \mathcal{L}(U_3/V_3),$$

$$\mathcal{L}(R_5), \mathcal{L}(R_6) \to \mathcal{L}(V_3),$$

where U_3 and V_3 are defined by (7.51) and (7.52), respectively, with r and $Y(0)$ replaced by $r = 1$ and $Y(0) = 0$, respectively.

The limiting powers of the five tests based on R_2 through R_6 can be evaluated using the following theorem (Problem 5.8).

Theorem 9.10. *Under Model C, the limiting powers of the tests based on R_2 through R_6 at the $100\alpha\%$ significance level can be expressed as follows:*

$$\beta_k(\alpha) = \frac{1}{\pi}\int_0^\infty \mathrm{Re}\left[\frac{1-\exp\{-i\theta x_k(\alpha)\}}{i\theta}\phi_k(\theta)\right]d\theta, \quad (k=2,3,5,6),$$

$$\beta_4(\alpha) = \frac{1}{2} + \frac{1}{\pi}\int_0^\infty \frac{1}{\theta}\mathrm{Im}\left[m_{31}(-i\theta, i\theta x_4(\alpha))\right]d\theta,$$

where $m_{31}(-i\theta, i\theta x_4(\alpha))$ is defined by (7.79) with r and γ replaced by $r=1$ and $\gamma=0$, respectively, while

$$\phi_2(\theta) = e^{c/2}\left[\frac{c^4}{\nu^4}\cos\nu - \left(\frac{3c^3-2i\theta(c^2+3c+3)}{3\nu^2} + \frac{2ic^2\theta}{\nu^4}\right)\frac{\sin\nu}{\nu}\right]^{-1/2},$$

$$\phi_3(\theta) = e^{c/2}\left[\left(\frac{12c^5-2ic^2\theta(c^2+12c+12)}{12\nu^4} + \frac{4ic^4\theta}{\nu^6}\right)\frac{\sin\nu}{\nu} - \frac{c^6}{\nu^6}\cos\nu\right.$$
$$\left. + \left(\frac{6ic^3\theta(c+2)+8\theta^2(c^2+3c+3)}{3\nu^6} - \frac{8c^2\theta^2}{\nu^8}\right)(\cos\nu-1)\right]^{-1/2},$$

$$\phi_5(\theta) = e^{c/2}\left[\left(-\frac{c^3}{\nu^2}+\frac{6i\theta(c+1)}{\nu^4}\right)\frac{\sin\nu}{\nu} - \left(\frac{c^2}{\nu^2}+\frac{6i\theta(c+1)}{\nu^4}\right)\cos\nu\right]^{-1/2}$$

$$= \phi_6(\theta)$$

with $\nu = \sqrt{2i\theta - c^2}$.

9.5.4 Model D

We exclude the LBI statistic R_1, as in Model C. Since it holds for Model D that

$$\tilde{\eta}_j = \eta_j - \eta_1 - \frac{j-1}{T-1}(\eta_T - \eta_1) = \eta_j - \frac{j}{T}\eta_T + O_p(1),$$

it can be shown that the tests based on R_2 and R_3 have the same limiting properties as in Model C. The limiting properties of the other tests can be easily deduced from the limiting distribution of $T(\hat{\rho}-1)$ for Model D obtained in Chapter 7.

We can first establish the following theorem (Problem 5.9).

Theorem 9.11. *For Model D, the test statistics R_2 and R_3 have the same limiting distributions as the corresponding statistics for Model C as $T \to \infty$ under $\rho = 1 - (c/T)$, while it holds that*

$$\mathcal{L}(R_4) \to \mathcal{L}(U_4/V_4),$$

$$\mathcal{L}(R_5) \to \mathcal{L}(V_4),$$

INTEGRAL EXPRESSIONS FOR THE LIMITING POWERS

$$\mathcal{L}(R_6) \to \mathcal{L}\left(\frac{V_4}{\left(4\int_0^1 Y(t)\,dt - 6\int_0^1 tY(t)\,dt\right)^2 + 1}\right),$$

where U_4 and V_4 are defined by (7.57) and (7.58), respectively, with r and $Y(0)$ replaced by $r = 1$ and $Y(0) = 0$, respectively.

We now derive the integral expressions for limiting powers of the five tests based on R_2 through R_6 (Problem 5.10).

Theorem 9.12. *Under Model D, the limiting powers of the tests based on R_2 and R_3 are the same as the corresponding tests under Model C. The limiting powers of the other tests at the $100\alpha\%$ significance level can be expressed as follows:*

$$\beta_4(\alpha) = \frac{1}{2} + \frac{1}{\pi}\int_0^\infty \frac{1}{\theta}\operatorname{Im}\left[m_{41}\left(-i\theta, i\theta x_4(\alpha)\right)\right]d\theta,$$

$$\beta_5(\alpha) = \frac{1}{\pi}\int_0^\infty \operatorname{Re}\left[\frac{1 - \exp\{-i\theta x_5(\alpha)\}}{i\theta}\phi_7(\theta)\right]d\theta,$$

$$\beta_6(\alpha) = \frac{1}{2} + \int_0^\infty \frac{1}{\theta}\operatorname{Im}\left[\exp\{i\theta x_6(\alpha)\}\,\psi_2\left(\theta; x_6(\alpha)\right)\right]d\theta,$$

where $m_{41}(-i\theta, i\theta x_4(\alpha))$ is defined by (7.80) with r and γ replaced by $r = 1$ and $\gamma = 0$, respectively, while

$$\phi_7(\theta) = e^{c/2}\left[\left(\frac{c^5 - 8ic^2\theta}{\nu^4} - \frac{24i\theta(c+1)(c^2+2i\theta)}{\nu^6}\right)\frac{\sin\nu}{\nu}\right.$$
$$+ \left(\frac{c^4}{\nu^4} + \frac{16ic^3\theta}{\nu^6} + \frac{96\theta^2(c+1)}{\nu^8}\right)\cos\nu$$
$$\left. + \left(\frac{8ic^2\theta(c+3)}{\nu^6} - \frac{96\theta^2(c+1)}{\nu^8}\right)\right]^{-1/2},$$

$$\psi_2(\theta; x) = e^{c/2}\left[\left(\frac{c^5 + 8i\theta\left((c^2 - 3c - 3)(3x + 1) - c^2(c-1)x\right)}{\mu^4}\right.\right.$$
$$\left. + \frac{96\theta^2(c+1)(4x+1)}{\mu^6}\right)\frac{\sin\mu}{\mu}$$
$$+ \left(\frac{c^4 - 8i\theta\left((c^2 - 5c + 3)x - 2c\right)}{\mu^4} - \frac{32c\theta^2(4x+1)}{\mu^6}\right.$$

$$+ \frac{96\theta^2(c+1)(4x+1)}{\mu^8} \bigg) \cos\mu$$

$$- \frac{8ic^2\theta(c+3)(4x+1)}{\mu^6} - \frac{96\theta^2(c+1)(4x+1)}{\mu^8} \bigg]^{-1/2},$$

with $\nu = \sqrt{2i\theta - c^2}$ and $\mu = \sqrt{-2i\theta - c^2}$.

PROBLEMS

5.1 Prove Lemma 9.2.

5.2 Prove the weak convergence results on R_1 and R_3 in Theorem 9.5.

5.3 Derive the expressions for the limiting powers $\beta_2(\alpha)$ and $\beta_3(\alpha)$ in Theorem 9.6.

5.4 Prove that the limiting powers $\beta_k(\alpha)$ ($k = 1, \ldots, 6$) defined in Theorem 9.6 all converge to 1 as $c \to \infty$.

5.5 Prove the weak convergence result on R_6 in Theorem 9.7.

5.6 Derive the expression for the limiting power $\beta_6(\alpha)$ in Theorem 9.8.

5.7 Prove the weak convergence result on R_3 in Theorem 9.9.

5.8 Derive the expression for the limiting power $\beta_3(\alpha)$ in Theorem 9.10.

5.9 Prove the weak convergence result on R_6 in Theorem 9.11.

5.10 Derive the expression for the limiting power $\beta_6(\alpha)$ in Theorem 9.12.

9.6 LIMITING POWER ENVELOPES AND POINT OPTIMAL TESTS

In this section we first derive the limiting power envelope of all the invariant unit root tests applied to the model (9.42). We then consider the so-called point optimal test, whose limiting power is tangent to the limiting power envelope at the point the test is conducted.

Let us first consider the testing problem

$$(9.50) \qquad H_0 : \rho = 1 \quad \text{versus} \quad H_1 : \rho = 1 - \frac{\theta}{T},$$

where θ is a given positive constant. The Neyman–Pearson lemma tells us (Problem 6.1) that the test applied to Model M = A, B, C, and D, which rejects H_0 when

$$(9.51) \quad V_T^{(M)}(\theta) = T \, \frac{\sum_{j=1}^{T} \left(\tilde{\eta}_j^{(0)} - \tilde{\eta}_{j-1}^{(0)} \right)^2 - \sum_{j=1}^{T} \left(\tilde{\eta}_j^{(1)} - \left(1 - \frac{\theta}{T}\right) \tilde{\eta}_{j-1}^{(1)} \right)^2}{\sum_{j=1}^{T} \left(\tilde{\eta}_j^{(0)} - \tilde{\eta}_{j-1}^{(0)} \right)^2}$$

takes large values, is the most powerful invariant (MPI), where $\tilde{\eta}^{(0)} = \left(\tilde{\eta}_1^{(0)}, \ldots, \tilde{\eta}_T^{(0)}\right)' = y - X\tilde{\beta}^{(0)}$, $\tilde{\eta}^{(1)} = \left(\tilde{\eta}_1^{(1)}, \ldots, \tilde{\eta}_T^{(1)}\right)' = y - X\tilde{\beta}^{(1)}$ with

$$\tilde{\beta}^{(0)} = \left(X'\left(C(1)C'(1)\right)^{-1}X\right)^{-1}X'\left(C(1)C'(1)\right)^{-1}y,$$

$$\tilde{\beta}^{(1)} = \left(X'\left(C\left(1-\frac{\theta}{T}\right)C'\left(1-\frac{\theta}{T}\right)\right)^{-1}X\right)^{-1}$$

$$\times X'\left(C\left(1-\frac{\theta}{T}\right)C'\left(1-\frac{\theta}{T}\right)\right)^{-1}y.$$

Note that $\tilde{\eta}^{(0)}$ and $\tilde{\eta}^{(1)}$ are the GLS residuals under H_0 and H_1, respectively.

In practice, however, the parameter space of ρ is an interval. Thus the above test fails to be optimal in general. Suppose that the true value of ρ, ρ_0 say, is given by $\rho_0 = 1 - (c/T)$ with c a nonnegative constant. Then we shall have $\mathcal{L}\left(V_T^{(M)}(\theta)\right) \to \mathcal{L}\left(V^{(M)}(c, \theta)\right)$ as $T \to \infty$ under $\rho_0 = 1 - (c/T)$ and $\rho = 1 - (\theta/T)$. Imagine now that the test based on $V_T^{(M)}(\theta)$ can be conducted for each $\theta = c > 0$. Then the limiting power envelope at the $100\alpha\%$ level is defined simply as $P\left(V^{(M)}(c, c) \geq x_{c,\alpha}^{(M)}\right)$, where $x_{c,\alpha}^{(M)}$ is the upper $100\alpha\%$ point of $\mathcal{L}\left(V^{(M)}(0, c)\right)$.

To be more specific, let us consider Model A, which yields

$$\mathcal{L}\left(V_T^{(A)}(\theta)\right) = \mathcal{L}\left(\frac{\sum_{j=1}^T (y_j - y_{j-1})^2 - \sum_{j=1}^T \left(y_j - \left(1-\frac{\theta}{T}\right)y_{j-1}\right)^2}{\frac{1}{T}\sum_{j=1}^T (y_j - y_{j-1})^2}\right)$$

$$= \mathcal{L}\left(\frac{-\frac{\theta^2}{T^2}\sum_{j=2}^T y_{j-1}^2 - \frac{2\theta}{T}\sum_{j=2}^T y_{j-1}(y_j - y_{j-1})}{\frac{1}{T}\sum_{j=1}^T (y_j - y_{j-1})^2}\right)$$

$$\to \mathcal{L}\left(V^{(A)}(c, \theta)\right),$$

where

(9.52) $$V^{(A)}(c, \theta) = -\theta^2 \int_0^1 Y^2(t)\,dt - 2\theta \int_0^1 Y(t)\,dY(t),$$

$$dY(t) = -cY(t)\,dt + dw(t), \quad Y(0) = 0.$$

We can also deduce (Problem 6.2) that $\mathcal{L}(V^{(A)}(c,\theta)) = \mathcal{L}(V^{(B)}(c,\theta))$ and $\mathcal{L}(V^{(C)}(c,\theta)) = \mathcal{L}(V^{(D)}(c,\theta))$, where

$$(9.53) \quad V^{(C)}(c,\theta) = -\theta^2 \left[\int_0^1 Y^2(t)\,dt - \frac{2(\theta+1)}{\delta} Y(1) \int_0^1 tY(t)\,dt \right.$$

$$\left. + \frac{\theta+1}{3\delta} Y^2(1) - \frac{\theta^2}{\delta} \left(\int_0^1 tY(t)\,dt \right)^2 \right] + \theta$$

with $\delta = (\theta^2 + 3\theta + 3)/3$. Thus the limiting power envelopes for Models A and B are the same; so are those for Models C and D. Elliott, Rothenberg, and Stock (1992) also derived the same expressions for $V^{(M)}(c,\theta)$, but the computation of the limiting power envelopes was done by simulations.

Here we compute the limiting power envelopes using the following theorem (Problem 6.3).

Theorem 9.13. *The limiting power envelope at the $100\alpha\%$ level of all the invariant unit root tests applied to Model M is computed as*

$$(9.54) \quad P\left(V^{(M)}(c,c) \geq x_{c,\alpha}^{(M)}\right) = \frac{1}{\pi} \int_0^\infty \mathrm{Re}\left[\frac{1-e^{-iau}}{iu} \phi^{(M)}(u;c)\right] du,$$

where $a = (c - x_{c,\alpha}^{(M)})/c^2$ and $\phi^{(M)}(u;c)$ is the c.f. of $(c - V^{(M)}(c,c))/c^2$ given by

$$(9.55) \quad \phi^{(M)}(u;c) = \left(\frac{\cos\mu - \frac{\mu}{c}\sin\mu}{e^c} \right)^{-1/2} \qquad (M = A, B),$$

$$= \left\{ \frac{-\frac{c^2}{\mu^2}\cos\mu + \left(c + 1 + \frac{c^2}{\mu^2}\right)\frac{\sin\mu}{\mu}}{e^c} \right\}^{-1/2} \qquad (M = C, D)$$

with $\mu = \sqrt{2iu - c^2}$. Moreover $x_{c,\alpha}^{(M)}$ in (9.54) is determined from

$$(9.56) \quad \alpha = P\left(V^{(M)}(0,c) \geq x_{c,\alpha}^{(M)}\right)$$

$$= \frac{1}{\pi} \int_0^\infty \mathrm{Re}\left[\frac{1-e^{-iau}}{iu} \phi^{(M)}(u)\right] du,$$

where $\phi^{(M)}(u)$ is the c.f. of $(c - V^{(M)}(0,c))/c^2$ given by

$$\phi^{(M)}(u) = \left[\cos \nu - \frac{\nu}{c} \sin \nu\right]^{-1/2} \qquad (M = A, B),$$

$$= \left[\frac{1}{\gamma}\left\{-\frac{c^2}{\nu^2}\cos \nu + \left(c + 1 + \frac{c^2}{\nu^2}\right)\frac{\sin \nu}{\nu}\right\}\right]^{-1/2} \qquad (M = C, D)$$

with $\gamma = (c^2 + 3c + 3)/3$ and $\nu = \sqrt{2iu}$.

The above theorem enables us to compute the limiting power envelopes accurately, which will be reported in the next section.

The test based on $V_T^{(M)}(\theta)$ in (9.51) is not uniformly MPI if the alternative is given by $H_1 : \rho < 1$. Nonetheless, we can still use $V_T^{(M)}(\theta)$ as a test statistic by suitably choosing the value of θ. The limiting power of this test will be tangent to the limiting power envelope at $\theta = c$. In this sense, the test is MPI at that point and is referred to as the point optimal invariant (POI) test [King (1988) and Elliott, Rothenberg, and Stock (1992)].

The limiting powers of the POI tests applied to Models A through D can be easily computed by deriving the c.f.s of $V^{(M)}(c, \theta)$. In fact, we can establish the following theorem (Problem 6.4).

Theorem 9.14. *For the testing problem $H_0 : \rho = 1$ versus $H_1 : \rho < 1$, the limiting powers at the $100\alpha\%$ level of the POI tests, which reject H_0 for large values of $V_T^{(M)}(\theta)$ in (9.51), can be computed as*

$$(9.57) \quad P\left(V^{(M)}(c, \theta) \geq x_{\theta,\alpha}^{(M)}\right) = \frac{1}{\pi}\int_0^\infty \text{Re}\left[\frac{1 - e^{-ibu}}{iu}\phi^{(M)}(u; c, \theta)\right] du,$$

where $x_{\theta,\alpha}^{(M)}$ is the upper $100\alpha\%$ point of $\mathcal{L}(V^{(M)}(0, \theta))$ and $b = (\theta - x_{\theta,\alpha}^{(M)})/\theta^2$. Moreover $\phi^{(M)}(u; c, \theta)$ is the c.f. of $(\theta - V^{(M)}(c, \theta))/\theta^2$ given by

$$\phi^{(M)}(u; c, \theta) = \left[\left\{\cos \mu + \left(c - \frac{2iu}{\theta}\right)\frac{\sin \mu}{\mu}\right\}\bigg/ e^c\right]^{-1/2}$$

for $M = A$ and B, while

$$\phi^{(M)}(u; c, \theta) = e^{c/2}\left[\left\{-\frac{c^2}{\mu^2} + \frac{2iu}{\delta\mu^4}(c - \theta)(c + c\theta + \theta)\right\}\cos \mu\right.$$

$$+ \left\{c + \frac{2iu}{3\delta\mu^2}\left((c^2 + 6)(\theta + 1) - c\theta^2\right)\right.$$

$$\left.\left. + \frac{2iu}{\delta\mu^4}\left((c + 1)\theta^2 - 2iu(\theta + 1)\right)\right\}\frac{\sin \mu}{\mu}\right]^{-1/2},$$

for $M = C$ and D, where $\mu = \sqrt{2iu - c^2}$ and $\delta = (\theta^2 + 3\theta + 3)/3$.

Note that, when $\theta = c$, $\phi^{(M)}(u; c, \theta)$ in (9.57) reduces to $\phi^{(M)}(u; c)$ in (9.54). The difficulty with conducting the POI test is the choice of the value of θ. One reasonable value of θ may be the one at which the limiting power envelope attains a power of 50%. The power performance of the POI tests will be examined in the next section.

PROBLEMS

6.1 Show that the test that rejects H_0 in (9.50) for large values of $V_T^{(M)}(\theta)$ in (9.51) is MPI.

6.2 Establish the weak convergence results on $V_T^{(M)}(\theta)$ in (9.51) for M = A, B, C and D as $T \to \infty$ under $\rho_0 = 1 - (c/T)$ and $\rho = 1 - (\theta/T)$.

6.3 Prove Theorem 9.13.

6.4 Prove Theorem 9.14.

9.7 COMPUTATION OF THE LIMITING POWERS

The limiting powers of the unit root tests and POI tests introduced in previous sections can now be easily computed by numerical integration. The computation of the limiting power envelopes, although more involved, can be done similarly.

We first need percent points of the limiting null distributions of the test statistics R_k ($k = 1, \ldots, 6$). Table 9.1 gives 1 and 5% points for Models A through D. Note that percent points of R_1 for Models C and D are not available for the reason given

Table 9.1. Percent Points of the Limiting Null Distributions of $R_k (k = 1, \ldots, 6)$

Level	R_1	R_2	R_3	R_4	R_5	R_6
			Model A			
1%	0.000157	0.0345	0.0248	−13.695	0.0345	0.0345
5%	0.00393	0.0565	0.0366	−8.039	0.0565	0.0565
			Model B			
1%	0.000157	0.0345	0.0248	−20.626	0.0248	0.0233
5%	0.00393	0.0565	0.0366	−14.094	0.0366	0.0336
			Model C			
1%		0.0248	0.0197	−23.736	0.0203	0.0203
5%		0.0366	0.0274	−16.636	0.0286	0.0286
			Model D			
1%		0.0248	0.0197	−29.359	0.0173	0.0166
5%		0.0366	0.0274	−21.711	0.0234	0.0222

before. It is seen from Table 9.1 that the distributions are shifted to the left as the model complexity increases, although those of R_1, R_2, and R_3 remain unchanged for transitions from A to B and C to D.

Table 9.2 reports the limiting percent powers of the above tests at the 5% significance level. Note that our testing problem is of the form:

$$H_0 : \rho = 1 \quad \text{versus} \quad H_1 : \rho = 1 - \frac{c}{T} < 1$$

so that the limiting powers are computed for various positive values of c.

We can first draw the following general conclusions from Table 9.2.

a) There exists no uniformly best test for each model.
b) The powers decrease as the model contains more trending regressors. Especially for Models C and D, the powers are considerably lower even at $c = 10$.

Table 9.2. Limiting Percent Powers of the Tests Based on $R_k(k = 1, \ldots, 6)$ at the 5% Significance Level

c	R_1	R_2	R_3	R_4	R_5	R_6
			Model A			
0.2	5.507	5.474	5.449	5.492	5.474	5.474
0.5	6.286	6.253	6.162	6.302	6.253	6.253
1	7.598	7.741	7.452	7.858	7.741	7.741
10	22.085	73.115	52.350	75.570	73.115	73.115
20	30.833	99.830	96.439	99.882	99.830	99.830
			Model B			
0.5	6.286	6.253	6.162	6.074	6.162	6.051
1	7.598	7.741	7.452	7.187	7.452	7.144
10	22.085	73.115	52.350	45.925	52.350	44.863
20	30.833	99.830	96.439	94.367	96.439	93.699
			Model C			
0.5		5.084	5.074	5.081	5.078	5.078
1		5.319	5.283	5.309	5.298	5.298
10		29.781	26.909	30.660	29.722	29.722
20		76.536	84.603	85.403	84.112	84.112
			Model D			
0.5		5.084	5.074	5.070	5.073	5.067
1		5.319	5.283	5.265	5.279	5.257
10		29.781	26.909	23.255	25.520	22.852
20		76.536	84.603	70.703	74.639	69.533

The following comments are specific to each test.

i) R_1: The LBI nature is seen only when c is very small. Its percent power is 30.833 even when $c = 20$. The optimality of the present LBI test is really local.
ii) R_2: This test may be recommended for Models B and D. Note, in particular, the higher power for Model B when $c = 10$. The LBIU property possessed by Models C and D, however, is also local.
iii) R_3: This test, called the *von Neumann ratio test*, is dominated in all models except in Model D.
iv) R_4: The most frequently used Dickey–Fuller test is recommended for Models A and C. The test, however, is dominated in Models B and D.
v) R_5: The Durbin–Watson test is dominated in all models; hence this is of little use.
vi) R_6: The modified Durbin–Watson test makes the power performance worse. This is dominated in all models by the Durbin–Watson test.

Table 9.3 tabulates the limiting power envelopes at the 5% level together with the upper 5% points of $V^{(M)}(0, c)$ under the heading $x^{(M)}_{c, 0.05}$ and the 5% points of $(c - V^{(M)}(0, c))/c^2$ under the heading $z^{(M)}_{c, 0.05}$ $(= (c - x^{(M)}_{c, 0.05})/c^2)$, whose computation is based on Theorem 9.13. Also shown are the limiting powers of the POI tests at the 5% level conducted at several values of θ, whose computation is based on Theorem 9.14.

The powers of the POI tests become higher for large values of c as θ increases, by sacrificing, to some extent, the powers for small values of c. This is because the

Table 9.3. Limiting Percent Power Envelopes and Limiting Percent Powers of the POI Tests at the 5% Level

Test	$c = 1$	5	7	10	13	20
			Models A and B			
$x^{(M)}_{c, 0.05}$	0.9061	3.2391	3.6945	3.5135	2.3043	−4.4922
$z^{(M)}_{c, 0.05}$	0.0939	0.0704	0.0675	0.0649	0.0633	0.0612
Envelope	7.972	31.948	49.940	75.818	91.975	99.885
$\theta = 5$	7.931	31.948	49.815	74.948	90.461	99.384
$\theta = 7$	7.912	31.892	49.940	75.613	91.433	99.700
$\theta = 10$	7.890	31.730	49.825	75.818	91.895	99.834
			Models C and D			
$x^{(M)}_{c, 0.05}$	0.9641	4.1677	5.4033	6.8138	7.6988	7.7389
$z^{(M)}_{c, 0.05}$	0.0359	0.0333	0.0326	0.0319	0.0314	0.0307
Envelope	5.319	11.485	17.600	31.005	48.367	85.526
$\theta = 5$	5.318	11.485	17.588	30.850	47.763	83.314
$\theta = 10$	5.315	11.463	17.585	31.005	48.322	85.034
$\theta = 13$	5.314	11.443	17.555	30.983	48.367	85.352

powers are tangent from below to the power envelope at $\theta = c$. It is observed in Table 9.3 that tangency at about 50% power occurs at $\theta = 7$ for Models A and B, while it occurs at $\theta = 13$ for Models C and D. Comparison with Table 9.2 shows that the Dickey–Fuller (R_4) tests applied to Models A and C almost attain the upper bound of attainable powers. For the other models, there is some room for improving the power performance by conducting the POI tests.

Figure 9.1 shows graphs of the limiting power functions of the LBIU (R_2) and POI ($\theta = 5$) tests for Model A, together with the power envelope, when the significance level is 5%. Note that the LBIU test is equivalent to the Durbin–Watson (R_5) and its modified (R_6) tests in the present case. Since the power of the Dickey–Fuller test is so close to the power envelope, the corresponding graph is not drawn here. For the same reason, the power function of the POI ($\theta = 7$) test, which achieves 50% power at $c = 7$, is also not drawn here. Figures 9.2, 9.3, and 9.4 present similar graphs for Models B, C, and D. In Figure 9.3 the power function of the Dickey–Fuller test is not shown for the same reason as above. In Figures 9.3 and 9.4 we still continue to present the power of the POI ($\theta = 5$) test, although larger values of θ yield more powerful tests for large values of c.

In conclusion we recommend the following tests for each model:

> Model A: Dickey–Fuller test,
>
> Model B: POI test at $\theta = 7$,
>
> Model C: Dickey–Fuller test,
>
> Model D: POI test at $\theta = 13$.

The limiting powers computed above may be used as a good approximation to the powers under finite samples of any moderate size. For demonstration purposes, we

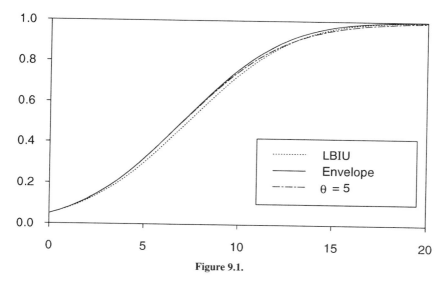

Figure 9.1.

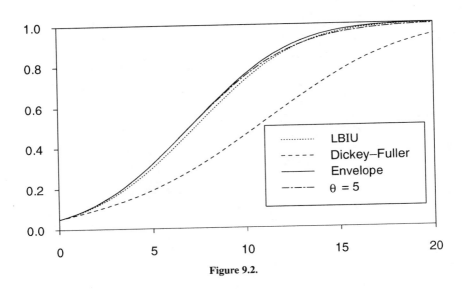

Figure 9.2.

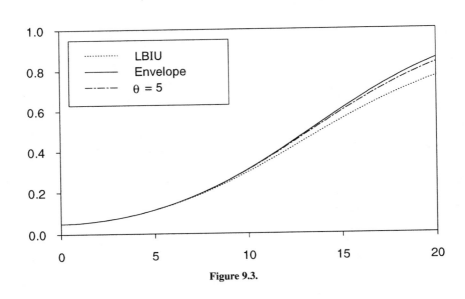

Figure 9.3.

COMPUTATION OF THE LIMITING POWERS

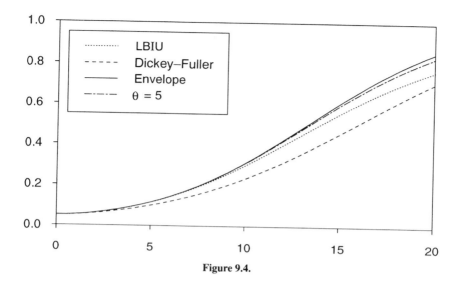

Figure 9.4.

take up the Dickey–Fuller (R_4) statistic only. We first compute the 5% points of the exact null distributions of R_4 for four models with $T = 25, 50, 100$. These are shown in Table 9.4, together with the limiting 5% points approached from above. Note that the limiting points are available from Tables 7.1, 7.2, and 7.3.

Then we obtain, for various values of ρ, the finite sample powers on the basis of simulations of 10,000 replications. The results are reported in Table 9.5. The entries are listed in the order of $c = T(1 - \rho) = 1, 10, 20$ and the limiting powers at these values of c are also reproduced from Table 9.2. A general feature observed in Table 9.5 is that the finite sample powers at $c = T(1 - \rho)$ approach the limiting value at the same value of c from above, although it is not very clear when $c = 1$.

A question of the possibility of devising a better unit root test other than the POI test is of interest. One might argue that the unit root tests shoud not be constrained to be those based on the OLS residuals $\hat{\eta}$ or the GLS residuals $\tilde{\eta}$. Here we just explore one such possibility by modifying the Dickey–Fuller test based on $T(\hat{\rho} - 1)$, where $\hat{\rho}$ is the ordinary LSE of ρ.

Table 9.4. 5% Points of the Null Distributilons of the Dickey–Fuller (R_4) Statistic

		Model		
T	A	B	C	D
25	−7.371	−12.527	−14.210	−18.047
50	−7.692	−13.278	−15.356	−19.776
100	−7.862	−13.677	−15.978	−20.716
∞	−8.039	−14.094	−16.636	−21.711

Table 9.5. Percent Powers of the Dickey–Fuller (R_4) Test at the 5% Significance Level

			Model			
c	T	ρ	A	B	C	D
1	25	0.96	7.7	6.9	5.2	5.1
1	50	0.98	8.0	7.3	5.4	5.5
1	100	0.99	7.6	7.1	5.2	5.6
1	∞		7.9	7.2	5.3	5.3
10	25	0.6	81.4	51.1	36.3	26.0
10	50	0.8	78.0	47.9	32.9	24.8
10	100	0.9	76.4	47.6	32.6	24.8
10	∞		75.6	45.9	30.7	23.3
20	25	0.2	99.9	97.2	93.6	83.3
20	50	0.6	99.9	96.1	90.4	77.5
20	100	0.8	99.9	95.4	88.3	73.1
20	∞		99.9	94.4	85.4	70.7

Let us take up Model A. Then we can suggest using the test statistic $T(\hat{\rho}(\delta) - 1)$, where

$$\text{(9.58)} \qquad \hat{\rho}(\delta) = \frac{\sum_{j=2}^{T} y_{j-1} y_j}{\sum_{j=2}^{T} y_{j-1}^2 + \delta y_T^2}.$$

Note that $\hat{\rho}(0)$ corresponds to the ordinary LSE, while $\hat{\rho}(1)$ to the Yule–Walker estimator. By specifying the value of δ, various estimators of ρ are possible.

The limiting distribution of $T(\hat{\rho}(\delta) - 1)$ as $T \to \infty$ under $\rho = 1$ was obtained in Section 1.3 of Chapter 1 by the eigenvalue approach. The stochastic process approach now yields the following theorem (Problem 7.1).

Theorem 9.15. *Let the estimator $\hat{\rho}(\delta)$ of ρ for Model A be defined by (9.58). Then it holds that, under $\rho = 1 - (c/T)$,*

$$\text{(9.59)} \qquad \lim_{T \to \infty} P(T(\hat{\rho}(\delta) - 1) \leq x) = P\left(\frac{\int_0^1 Y(t)\, dY(t) - \delta Y^2(1)}{\int_0^1 Y^2(t)\, dt} \leq x \right)$$

$$= \frac{1}{2} + \frac{1}{\pi} \int_0^\infty \frac{1}{\theta} \operatorname{Im}\left[\phi_\delta(\theta; x) \right] d\theta,$$

COMPUTATION OF THE LIMITING POWERS

Table 9.6. Limiting Percent Powers of the $\hat{\rho}(\delta)$-Test at the 5% Significance Level

δ	5% Point	$c = 0.2$	0.5	1	5	10	20
−10	−4.275	5.51	6.36	7.96	28.4	54.2	77.9
−5	−5.274	5.51	6.36	7.97	30.4	63.5	89.5
−2	−6.400	5.51	6.35	7.96	31.7	71.9	97.5
0	−8.039	5.49	6.30	7.86	31.4	75.6	99.9
0.5	−8.856	5.47	6.25	7.74	30.0	73.1	99.8
1	−10.107	5.43	6.12	7.44	26.6	66.5	99.5
2	−13.942	5.23	5.62	6.34	16.9	44.4	94.7
5	−29.262	4.94	4.87	4.83	6.0	10.3	34.6

where $\{Y(t)\}$ is the O–U process defined in (9.48) and

$$(9.60) \quad \phi_\delta(\theta; x) = \exp\left(\frac{c + i\theta}{2}\right) \left[\cos\mu + (c - i\theta(2\delta - 1))\frac{\sin\mu}{\mu}\right]^{-1/2}$$

with $\mu = \sqrt{2i\theta x - c^2}$.

The limiting powers of the test based on $T(\hat{\rho}(\delta) - 1)$ can be easily computed from (9.59). Table 9.6 gives the results at the 5% significance level, which are taken from Tanaka (1990a). It is seen that the power is higher, though not appreciable, for δ small when the alternative is close to the null. In fact the $\hat{\rho}(\delta)$-test with δ negative is slightly better than the $\hat{\rho}(0)$-test, that is, the Dickey–Fuller test when c is small. The Dickey–Fuller test, however, dominates as c gets large. The $\hat{\rho}(1)$-test, which is the Yule–Walker test, is uniformly worse than the Dickey–Fuller test. It seems that, for $\delta \geq 0$, the power decreases as δ gets large. Figure 9.5 shows limiting power functions of the $\hat{\rho}(\delta)$-test for $\delta = -2, 0, 2$.

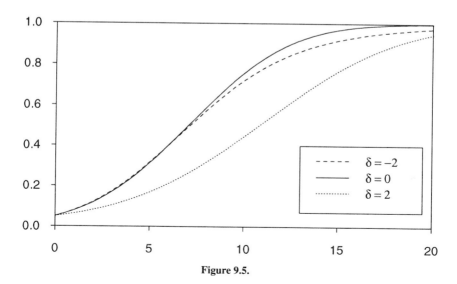

Figure 9.5.

We can also devise a test from a different viewpoint. We have seen in Section 7.9 of Chapter 7 that asymptotic normality arises from an estimator of ρ obtained from an augmented regression relation. More specifically, let us take up Model A and consider the regression relation:

$$(9.61) \qquad y_j = \hat{\rho}_1 y_{j-1} + \hat{\rho}_2 y_{j-2} + \hat{\varepsilon}_j,$$

where $\hat{\rho}_1$ and $\hat{\rho}_2$ are the ordinary LSEs of $\rho_1 = \rho$ and $\rho_2 = 0$, respectively. We have shown that, under $\rho = 1$,

$$\mathcal{L}\left(\sqrt{T}\begin{pmatrix}\hat{\rho}_1 - \rho_1 \\ \hat{\rho}_2 - \rho_2\end{pmatrix}\right) \to N\left(0, \begin{pmatrix}1 & -1 \\ -1 & 1\end{pmatrix}\right).$$

Suppose now that we use $\sqrt{T}(\hat{\rho}_1 - 1)$ as a unit root test statistic for Model A. Then we can establish (Problem 7.2) the following theorem concerning the finite sample powers of this normal test.

Theorem 9.16. *Let $\hat{\rho}_1$ be the LSE of ρ obtained from the regression relation (9.61), where $\{y_j\}$ follows Model A. Then the power of the test based on $\sqrt{T}(\hat{\rho}_1 - 1)$ against a fixed alternative $|\rho| < 1$ can be approximated as*

$$(9.62) \qquad P\left(\sqrt{T}(\hat{\rho}_1 - 1) \leq x\right) \cong \Phi\left(x + \sqrt{T}(1 - \rho)\right),$$

where $\Phi(\cdot)$ is the distribution function of $N(0,1)$.

Table 9.7 compares the sampling powers of the normal and Dickey–Fuller tests under $T = 100$, which are based on simulations of 10,000 replications. We have also reported the approximated powers of both tests, where the former was computed from (9.62), while the latter was computed from $\beta_4(\alpha)$ in Theorem 9.6 by putting $c = T(1 - \rho)$. The significance points used are all based on the limiting null dis-

Table 9.7. Percent Powers of the Normal and Dickey–Fuller Tests for Model A ($T = 100$)

Nominal Level	1%			5%			10%		
ρ	1	0.9	0.8	1	0.95	0.85	1	0.95	0.9
				Normal Test					
Sampling	1.3	11.7	40.3	6.4	15.4	48.5	12.6	25.2	42.7
Approximated	1.0	9.2	37.2	5.0	12.6	44.2	10.0	21.7	38.9
				Dickey–Fuller Test					
Sampling	0.9	30.5	91.5	4.9	31.0	97.1	9.8	52.8	92.8
Approximated	1.0	30.9	91.1	5.0	31.4	96.9	10.0	52.8	92.5

tributions; hence the significance levels are just nominal. It is deduced from the entries under the heading $\rho = 1$ that the sampling distribution of $\sqrt{T}(\hat{\rho}_1 - 1)$ has a heavier left-hand tail than its limiting distribution, which results in upward bias of the sampling power. The sampling distribution of $T(\hat{\rho} - 1)$, on the other hand, is well approximated by its limiting distribution, as was seen before. In any case, the performance of the normal test is surprisingly poor. Inference based on nonstandard distributions is worthy of pursuit.

PROBLEMS

7.1 Prove Theorem 9.15.
7.2 Prove Theorem 9.16.

9.8 SEASONAL UNIT ROOT TESTS

The unit root tests considered so far can be easily extended to deal with the seasonal unit root testing problem. Such an extension was earlier considered by Dickey, Hasza, and Fuller (1984).

Let us consider the seasonal regression model

(9.63)
$$y_j = x_j'\beta + \eta_j,$$
$$\eta_j = \rho_m \eta_{j-m} + \varepsilon_j, \quad (j = 1, \ldots, T),$$

where m is a positive integer and is assumed to be a divisor of T so that $N = T/m$ is an integer. As for the regressor sequence $\{x_j\}$ we assume the Kronecker structure, as in Chapter 7. Thus, assuming the m initial values $\eta_{1-m}, \eta_{2-m}, \ldots, \eta_0$ to be all zero, we may rewrite (9.63) as

(9.64)
$$y = (\bar{X} \otimes I_m)\beta + \eta,$$
$$\eta = (C(\rho_m) \otimes I_m)\varepsilon,$$

where $C(\rho_m)$ is the $N \times N$ matrix defined in (8.36) with ρ_m replaced by α, while $\bar{X}$ is an $N \times p$ matrix with rank $p(<N)$. As a direct extension of nonseasonal models we consider four models by specifying $\bar{X}$:

Model A: $y = \eta,$
Model B: $y = (e \otimes I_m)\beta_0 + \eta,$
Model C: $y = (d \otimes I_m)\beta_1 + \eta,$
Model D: $y = ((e, d) \otimes I_m)\beta + \eta,$

where $e = (1, \ldots, 1)' : N \times 1$, $d = (1, \ldots, N)'$, and $\eta = (C(\rho_m) \otimes I_m)\varepsilon$ for all models.

Our purpose here is to test either

(9.65) $$H_0 : \rho_m = 1 \quad \text{versus} \quad H_1 : \rho_m < 1$$

or

(9.66) $$H_0 : \rho_m = 1 \quad \text{versus} \quad H_1 : \rho_m \neq 1.$$

The LBI test for (9.65) and the LBIU test for (9.66) can be derived in the same way as before under the assumption that $y \sim N((\bar{X} \otimes I_m)\beta, \sigma^2(\Phi(\rho_m) \otimes I_m))$, where $\Phi(\rho_m) = C(\rho_m)C'(\rho_m)$. In fact, we can easily verify the equivalence of the LM test with the LBI or LBIU test in the present case. Then we can first establish (Problem 8.1) the following theorem concerning the LBI test.

Theorem 9.17. *Consider the regression model (9.64) with $\eta \sim N(0, \sigma^2(\Phi(\rho_m) \otimes I_m))$. Then the LBI test for the testing problem (9.65) rejects the null hypothesis when*

(9.67) $$\text{LM}_1 = \frac{\sum_{j=1}^{m} \tilde{\eta}_{T-m+j}^2}{\sum_{j=1}^{T} \left(\tilde{\eta}_j - \tilde{\eta}_{j-m}\right)^2} < c,$$

where $\tilde{\eta}_i \equiv 0$ for $i = 1 - m, 2 - m, \ldots, 0$ and $\tilde{\eta} = (\tilde{\eta}_1, \ldots, \tilde{\eta}_T)' = (\tilde{M} \otimes I_m)y$ with $\tilde{M} = I_N - \bar{X}(\bar{X}'\Phi^{-1}(1)\bar{X})^{-1}\bar{X}'\Phi^{-1}(1)$.

We can easily deduce from this theorem that the LBI test for (9.65) applied to Model A rejects H_0 when

(9.68) $$R_{A1} = m \times \frac{\sum_{j=1}^{m} y_{T-m+j}^2}{\sum_{j=1}^{T} \left(y_j - y_{j-m}\right)^2} < c.$$

Note that $R_{A1}/T \sim \text{Be}(m/2, (T-m)/2)$ under H_0. Similarly, the LBI test for (9.65) applied to Model B rejects H_0 when

(9.69) $$R_{B1} = m \times \frac{\sum_{j=1}^{m} (y_{T-m+j} - y_j)^2}{\sum_{j=m+1}^{T} \left(y_j - y_{j-m}\right)^2} < c.$$

It holds that $R_{B1}/(T-m) \sim \text{Be}(m/2, (T-2m)/2)$ under H_0.

There are some cases where the LBI statistic in (9.67) reduces to 0, as in non-seasonal models. It can be shown (Problem 8.2) that this occurs if $d = (1,\ldots,N)'$ belongs to the column space of $\bar{X}$. In those cases we devise the LBIU test, which is described in the following theorem (Problem 8.3.)

Theorem 9.18. *Consider the regression model (9.64) with $\eta \sim N(0, \sigma^2(\Phi(\rho_m) \otimes I_m))$, and suppose that $d = (1,\ldots,N)'$ belongs to the column space of $\bar{X}$. Then the LBIU test for the testing problem (9.66) rejects the null hypothesis when*

$$\text{(9.70)} \qquad \text{LM}_2 = \frac{\sum_{j=1}^{T} \tilde{\eta}_j^2}{\sum_{j=1}^{T} (\tilde{\eta}_j - \tilde{\eta}_{j-m})^2} < c.$$

It follows (Problem 8.4) from Theorem 9.18 that the LBIU test for (9.66) applied to Model C rejects H_0 when

$$\text{(9.71)} \qquad R_{C2} = \frac{m}{N} \times \frac{\sum_{j=1}^{N} \left(y_j - \frac{j}{N} y_N\right)' \left(y_j - \frac{j}{N} y_N\right)}{\sum_{j=1}^{N} (y_j - y_{j-1})' (y_j - y_{j-1}) - \frac{1}{N} y_N' y_N} < c,$$

where $y_j = (y_{(j-1)m+1}, \ldots, y_{jm})' : m \times 1$ with $y_0 = 0$. The corresponding test applied to Model D is found (Problem 8.5) to reject H_0 when

$$R_{D2} = \frac{m}{N} \times \frac{\sum_{j=2}^{N} \left(y_j - y_1 - \frac{j-1}{N-1}(y_N - y_1)\right)' \left(y_j - y_1 - \frac{j-1}{N-1}(y_N - y_1)\right)}{\sum_{j=2}^{N} (y_j - y_{j-1})'(y_j - y_{j-1}) - \frac{1}{N-1}(y_N - y_1)'(y_N - y_1)}$$

$$< c.$$
(9.72)

We can also consider the other seasonal unit root tests, as in previous sections. Here we only deal with the Dickey–Fuller test for the testing problem (9.65), which rejects H_0 when $N(\hat{\rho}_m - 1)$ takes small values, where

$$\text{(9.73)} \qquad \hat{\rho}_m = \frac{\sum_{j=m+1}^{T} \hat{\eta}_{j-m} \hat{\eta}_j}{\sum_{j=m+1}^{T} \hat{\eta}_{j-m}^2}$$

with $\hat{\eta} = (\hat{\eta}_1, \ldots, \hat{\eta}_T)' = (\bar{M} \otimes I_m) y$ and $\bar{M} = I_N - \bar{X}(\bar{X}'\bar{X})^{-1}\bar{X}'$. The Dickey–Fuller statistics for Models A through D can be easily produced from (9.73). Let the resulting statistics be R_{A3}, R_{B3}, R_{C3}, and R_{D3}.

The limiting powers of the above tests can be easily computed as $T \to \infty$ under

$$\rho_m = 1 - \frac{c}{N} = 1 - \frac{cm}{T}.$$

For that purpose we can establish the following theorem (Problem 8.6).

Theorem 9.19. *The test statistics defined above have the following limiting distributions as $T \to \infty$ under $\rho_m = 1 - (c/N)$:*

$$P(R_{A1} \leq x), P(R_{B1} \leq x) \to P\left(\chi_m^2 \leq \frac{2c}{1 - e^{-2c}} x\right),$$

$$P(R_{C2} \leq x), P(R_{D2} \leq x) \to \frac{1}{\pi} \int_0^\infty \mathrm{Re}\left[\frac{1 - e^{-i\theta x}}{i\theta}(\phi_2(\theta))^m\right] d\theta,$$

$$P(R_{M3} \leq x) \to \frac{1}{2} + \frac{1}{\pi} \int_0^\infty \frac{1}{\theta} \mathrm{Im}\left[\{m_{l1}(-i\theta, i\theta x)\}^m\right] d\theta,$$

where $\chi_m^2 \sim \chi^2(m)$, $\phi_2(\theta)$ is defined in Theorem 9.10 and $(M, l) = (A, 1), (B, 2), (C, 3), (D, 4)$ with m_{l1} ($l = 1, 2, 3, 4$) given in Theorems 9.6, 9.8, 9.10, and 9.12, respectively.

Note that the limiting expressions for $P(R_{M3} \leq x)$, $(M = A, B, C, D)$, were earlier obtained in Section 7.7 of Chapter 7.

The limiting power envelope of all the invariant seasonal unit root tests can also be computed by numerical integration. We first note that the test which rejects $H_0 : \rho_m = 1$ in favor of $H_1 : \rho_m = 1 - (\theta/N) = 1 - (\theta m/T)$, when $V_{Tm}^{(M)}(\theta)$ is large, is MPI, where

$$(9.74) \quad V_{Tm}^{(M)}(\theta) = T \frac{\sum_{j=1}^{T}\left(\tilde{\eta}_j^{(0)} - \tilde{\eta}_{j-m}^{(0)}\right)^2 - \sum_{j=1}^{T}\left(\tilde{\eta}_j^{(1)} - \left(1 - \frac{\theta}{N}\right)\tilde{\eta}_{j-m}^{(1)}\right)^2}{\sum_{j=1}^{T}\left(\tilde{\eta}_j^{(0)} - \tilde{\eta}_{j-m}^{(0)}\right)^2}$$

for M=A, B, C, and D. Then it is easy to establish the following theorem.

Theorem 9.20. *Suppose that the true value of ρ_m is given by $1 - (c/N) = 1 - (cm/T)$. Let $\beta^{(M)}(\alpha; c, m)$ be the limiting power envelope at the $100\alpha\%$ level of all the invariant seasonal unit root tests applied to Model M with period m. Then*

$$\beta^{(M)}(\alpha; c, m) = \frac{1}{\pi} \int_0^\infty \mathrm{Re}\left[\frac{1 - e^{-iau}}{iu} \{\phi^{(M)}(u; c)\}^m\right] du,$$

SEASONAL UNIT ROOT TESTS

where $a = (cm - x_{cm,\alpha}^{(M)})/c^2$ and $\phi^{(M)}(u; c)$ is defined in Theorem 9.13. Here $x_{cm,\alpha}^{(M)}$ is determined from

$$\alpha = \frac{1}{\pi} \int_0^\infty \text{Re}\left[\frac{1 - e^{-iau}}{iu} \left\{\phi^{(M)}(u)\right\}^m\right] du,$$

where $\phi^{(M)}(u)$ is given in Theorem 9.13.

Note that $\{\phi^{(M)}(u; c)\}^m$ is the c.f. of the m-fold convolution of $(c - V^{(M)}(c, c))/c^2$, where $V^{(M)}(c, \theta)$ is defined in (9.52) for M = A, B and in (9.53) for M = C, D.

We can also conduct the seasonal POI test, which rejects $H_0 : \rho_m = 1$ in favor of $H_1 : \rho_m < 1$ when $V_{Tm}^{(M)}(\theta)$ in (9.74) is large. The following theorem gives a formula for computing its limiting power.

Theorem 9.21. *Suppose that the seasonal POI test is conducted at θ when the true value of ρ_m is given by $1 - (c/N) = 1 - (cm/T)$. Let $\beta_{POI}^{(M)}(\alpha; c, \theta, m)$ be the limiting power of the test at the $100\alpha\%$ level applied to Model M with period m. Then*

$$\beta_{POI}^{(M)}(\alpha; c, \theta, m) = \frac{1}{\pi} \int_0^\infty \text{Re}\left[\frac{1 - e^{-ibu}}{iu} \left\{\phi^{(M)}(u; c, \theta)\right\}^m\right] du,$$

where $b = (\theta m - x_{\theta m,\alpha}^{(M)})/\theta^2$ and $\phi^{(M)}(u; c, \theta)$ is defined in Theorem 9.14.

Table 9.8 reports the limiting powers of the LBI, LBIU, and Dickey–Fuller tests at the 5% significance level, when $m = 4$ and 12. The 5% significance point of

Table 9.8. Limiting Percent Powers of Seasonal Unit Root Tests at the 5% Significance Level

	$m = 4$				$m = 12$			
Test	5% point	$c = 1$	5	10	5% point	$c = 0.5$	1	5
				Model A				
R_{A1}	0.711	19.9	87.0	99.3	5.226	23.6	56.1	100.0
R_{A3}	−2.268	20.0	99.2	100.0	−0.965	23.3	57.7	100.0
				Model B				
R_{B1}	0.711	19.9	87.0	99.3	5.226	23.6	56.1	100.0
R_{B3}	−6.961	13.1	66.7	99.6	−4.929	14.6	28.1	99.4
				Model C				
R_{C2}	0.301	5.99	34.2	89.6	1.271	5.53	7.21	80.4
R_{C3}	−8.715	5.94	34.8	94.4	−6.347	5.51	7.11	82.4
				Model D				
R_{D2}	0.301	5.99	34.2	89.6	1.271	5.53	7.21	80.4
R_{D3}	−13.067	5.74	23.8	76.7	−10.360	5.39	6.55	56.8

each test statistic is also reported. The following conclusions may be drawn from Table 9.8.

a) The LBI test based on R_{A1} or R_{B1} is not as bad as in the nonseasonal case. In fact, this test for Model B performs better than the Dickey–Fuller test based on R_{B3}, especially when $m = 12$.

b) The LBIU test based on R_{D2} is better than the Dickey–Fuller test, as in the nonseasonal case, although the latter test performs slightly better for Model C.

Table 9.9 tabulates, for each model M ($=$ A, B, C, D) with period m ($= 4, 12$), the limiting percent power envelopes and the limiting percent powers of POI tests conducted at various values of θ, when the significance level is 5%. Also shown are the 5% points of the m-fold convolution of $(c - V^{(M)}(0, c))/c^2$ under the heading $z_{cm, 0.05}^{(M)}$.

Comparing Table 9.9 with Table 9.8, we can observe that the Dickey–Fuller tests applied to Models A and C almost achieve the upper bound of attainable powers, while there is some room for improving the power performance by conducting the POI test for Models B and D. The situation is quite similar to the nonseasonal case. It is found that tangency at about 50% power occurs at $\theta = 2$ when $m = 4$, and $\theta = 0.9$ when $m = 12$ for Models A and B, while it occurs at $\theta = 6$ when $m = 4$ and $\theta = 3.5$ when $m = 12$ for Models C and D.

Figure 9.6 presents graphs of the limiting power functions of the LBI, Dickey–Fuller and POI ($\theta = 2$) tests, together with the power envelope for Model B with $m = 4$ when the significance level is 5%. The corresponding graphs for Model D with $m = 4$ are shown in Figure 9.7, where the POI test is conducted at $\theta = 6$, whose

Table 9.9. Limiting Percent Power Envelopes and Limiting Percent Powers of the POI Tests at the 5% Level: The Seasonal Case

Test	$m = 4$				$m = 12$			
	$c = 1$	2	5	10	$c = 0.5$	1	2	5
Models A and B								
$z_{cm, 0.05}^{(M)}$	1.532	1.132	0.862	0.759	14.105	8.816	6.118	4.435
Envelope	20.84	49.98	99.19	100.00	23.85	58.30	97.89	100.00
$\theta = 1$	20.84	49.16	97.08	100.00	23.73	58.30	97.63	100.00
$\theta = 1.5$	20.77	49.83	98.14	100.00	23.50	58.10	97.84	100.00
$\theta = 2$	20.65	49.98	98.65	100.00	23.25	57.65	97.89	100.00
Models C and D								
$z_{cm, 0.05}^{(M)}$	0.290	0.278	0.254	0.238	1.250	1.214	1.147	1.029
Envelope	5.99	9.01	35.48	94.53	5.53	7.21	15.46	83.04
$\theta = 4$	5.98	8.99	35.45	93.74	5.53	7.19	15.40	82.98
$\theta = 5$	5.98	8.98	35.48	94.08	5.52	7.17	15.34	83.04
$\theta = 6$	5.97	8.96	35.46	94.29	5.52	7.16	15.28	83.00

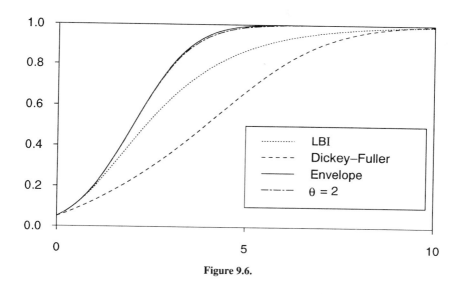

Figure 9.6.

power is found to be very close to the power envelope. The POI tests in both figures really improve the power performance.

In conclusion, we recommend that the Dickey–Fuller test be used for Models A and C and the POI test for Models B and D with the value of θ mentioned above.

Though not reported here, the limiting powers at c may be used as a good approximation to the finite sample powers at $\rho_m = 1 - (c/N)$. Dickey, Hasza, and Fuller (1984) reported the finite sample powers of various tests for Models A and B on the basis of simulations.

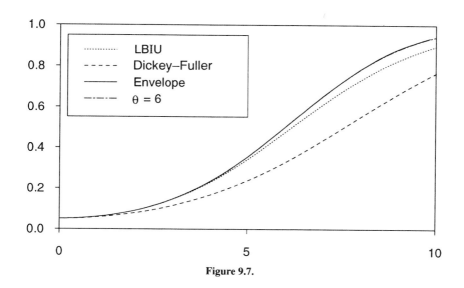

Figure 9.7.

PROBLEMS

8.1 Prove Theorem 9.17.
8.2 Show that the statistic LM_1 in (9.67) reduces to 0 if $d = (1, \ldots, N)'$ belongs to the column space of $\bar{X}$, where $\bar{X}$ is defined in (9.64).
8.3 Prove Theorem 9.18.
8.4 Prove that the LBIU test for Model C rejects H_0 when (9.71) holds.
8.5 Prove that the LBIU test for Model D rejects H_0 when (9.72) holds.
8.6 Establish the limiting expression for $P(R_{C2} \leq x)$ given in Theorem 9.19.

9.9 UNIT ROOT TESTS IN THE DEPENDENT CASE

In this section we extend the time series regression model dealt with in previous sections to

$$(9.75) \quad \begin{aligned} y_j &= x_j'\beta + \eta_j, \\ \eta_j &= \rho \eta_{j-1} + u_j, \quad (j = 1, \ldots, T), \end{aligned}$$

where $\{u_j\}$ is a dependent process generated by

$$(9.76) \quad u_j = \sum_{l=0}^{\infty} \alpha_l \varepsilon_{j-l}, \quad \sum_{l=0}^{\infty} l |\alpha_l| < \infty, \quad \sum_{l=0}^{\infty} \alpha_l \neq 0,$$

with $\{\varepsilon_j\} \sim \text{NID}(0, \sigma^2)$. Assuming that $\eta_0 = 0$, we may rewrite the above model as

$$(9.77) \quad y = X\beta + \eta, \quad \eta = C(\rho)u, \quad y \sim N\left(X\beta, \sigma^2 C(\rho)\Lambda(\theta)C'(\rho)\right),$$

where $V(u) = \sigma^2 \Lambda(\theta)$ and $\theta = (\alpha_1, \alpha_2, \cdots)'$. Our testing problem is the same as before; that is, we test $H_0 : \rho = 1$ against either $H_1 : \rho < 1$ or $H_1 : \rho \neq 1$.

It is impossible in the present case to devise an optimal test unless some restrictions are imposed on the sequence $\{u_j\}$ or the matrix $\Lambda(\theta)$. When there is no such restriction, one way to go on is just to act as if $\{u_j\} = \{\varepsilon_j\}$ or $\Lambda(\theta) = I_T$. Then we shall have the LBI and LBIU tests together with the other tests introduced before. The distributions of the corresponding statistics, however, will change. To examine this, let us consider Models A through D in the same way as before by specifying $\{x_j\}$ in (9.75). We have shown in Section 9.5 that, when $\{u_j\} = \{\varepsilon_j\}$, the limiting distributions of the statistics R_1 through R_6 as $T \to \infty$ under $\rho = 1 - (c/T)$ take the following forms:

$$\mathcal{L}(R_k) \to \mathcal{L}(W_k) \quad \text{for all models and } k = 1, 2, 3, 5,$$

$$\mathcal{L}(R_4) \to \mathcal{L}(U/V) \quad \text{for all models},$$

$$\mathcal{L}(R_6) \to \begin{cases} \mathcal{L}(W_6) & \text{for Models A and C}, \\ \mathcal{L}\left(\dfrac{W_6}{X_6 + 1}\right) & \text{for Models B and D}, \end{cases}$$

UNIT ROOT TESTS IN THE DEPENDENT CASE

where $W_1 = 0$ for Models C and D, and the other expressions for W_ks together with U, V, and X_6 vary among the models.

Suppose that the statistics R_1 through R_6 are constructed from each of Models A through D based on (9.75) in the same way as before. Then we can establish (Problem 9.1) the following theorem.

Theorem 9.22. *Let the statistics R_1 through R_6 be constructed as described above. Then it holds that, as $T \to \infty$ under $\rho = 1 - (c/T)$,*

$$\mathcal{L}(R_k) \to \mathcal{L}\left(\frac{1}{r} W_k\right) \quad \text{for all models and } k = 1, 2, 3, 5,$$

$$\mathcal{L}(R_4) \to \mathcal{L}\left(\frac{U}{V} + \frac{1-r}{2V}\right) \quad \text{for all models,}$$

$$\mathcal{L}(R_6) \to \begin{cases} \mathcal{L}\left(\dfrac{1}{r} W_6\right) & \text{for Models A and C,} \\ \mathcal{L}\left(\dfrac{W_6}{X_6 + r}\right) & \text{for Models B and D,} \end{cases}$$

where r is the ratio of the short-run to the long-run variances of $\{u_j\}$ defined by

$$(9.78) \qquad r = \frac{\sigma_S^2}{\sigma_L^2}, \qquad \sigma_S^2 = \sigma^2 \sum_{l=0}^{\infty} \alpha_l^2, \qquad \sigma_L^2 = \sigma^2 \left(\sum_{l=0}^{\infty} \alpha_l\right)^2.$$

The above theorem implies that, if we can find a consistent estimator of r, $\tilde{r}$ say, we can also find suitable transformations of the statistics so that the limiting distributions of the transformed statistics are the same as those of the untransformed statistics derived under the assumption that $\{u_j\} = \{\varepsilon_j\}$. To be specific, we have the following theorem [Nabeya and Tanaka (1990b) and Problem 9.2].

Theorem 9.23. *Let $\tilde{r}$ be an estimator of r that takes the form $\tilde{r} = \tilde{\sigma}_S^2/\tilde{\sigma}_L^2$, where $\tilde{\sigma}_S^2$ and $\tilde{\sigma}_L^2$ are the estimators of σ_S^2 and σ_L^2 such that $\tilde{\sigma}_S^2 \to \sigma_S^2$ and $\tilde{\sigma}_L^2 \to \sigma_L^2$ in probability as $T \to \infty$ under $\rho = 1 - (c/T)$. Then it holds that, as $T \to \infty$ under $\rho = 1 - (c/T)$,*

$$\mathcal{L}(\tilde{r} R_k) \to \mathcal{L}(W_k) \quad \text{for all models and } k = 1, 2, 3, 5,$$

$$\mathcal{L}\left(R_4 + \frac{\tilde{\sigma}_S^2 - \tilde{\sigma}_L^2}{T}\bigg/ 2\sum_{j=2}^{T} \hat{\eta}_{j-1}^2 \bigg/ T^2\right) \to \mathcal{L}\left(\frac{U}{V}\right) \quad \text{for all models,}$$

$$\mathcal{L}(\tilde{r} R_6) \to \mathcal{L}(W_6) \quad \text{for Models A and C,}$$

$$\mathcal{L}\left(\left(\frac{1}{R_6} + \frac{\tilde{\sigma}_L^2 - \tilde{\sigma}_S^2}{\sum_{j=1}^{T}\hat{\eta}_j^2 \Big/ T^2}\right)^{-1}\right) \to \mathcal{L}\left(\frac{W_6}{X_6 + 1}\right) \quad \text{for Models B and D},$$

where $\hat{\eta}_j$ is the OLS residual obtained from the regression $y_j = x_j'\beta + \eta_j$.

The extended test associated with R_4 was first suggested in Phillips (1987a) and was further extended in Phillips and Perron (1988), who indicated how to construct $\tilde{\sigma}_S^2$ and $\tilde{\sigma}_L^2$. In our context we consider

$$(9.79) \qquad \tilde{\sigma}_S^2 = \frac{1}{T}\sum_{j=1}^{T}\left(\tilde{\eta}_j - \tilde{\eta}_{j-1}\right)^2,$$

where $\tilde{\eta}_j$ is the GLS residual defined in the same way as before.

The estimator $\tilde{\sigma}_L^2$ may be defined as

$$(9.80) \quad \tilde{\sigma}_L^2 = \tilde{\sigma}_S^2 + \frac{2}{T}\sum_{i=1}^{l}\left(1 - \frac{i}{l+1}\right)\sum_{j=i+1}^{T}(\tilde{\eta}_j - \tilde{\eta}_{j-1})(\tilde{\eta}_{j-i} - \tilde{\eta}_{j-i-1}),$$

where l is a lag truncation number depending on T in such a way that $l \to \infty$ and $l/T \to 0$ as $T \to \infty$. This is the modified Bartlett estimator of $2\pi f_u(0)$, where $f_u(\omega)$ is the spectrum of $\{u_j\}$. The other estimators of σ_L^2 are possible [Anderson (1971)], but those are all constructed by truncating a weighted sum of sample covariances or averaging the periodogram over the neighboring frequencies around zero.

Phillips and Perron (1988) did some simulations to assess the adequacy of the extended Dickey–Fuller (R_4) test based on Theorem 9.23. As a dependent process $\{u_j\}$, they exclusively considered the MA(1) process $u_j = \varepsilon_j - \alpha\varepsilon_{j-1}$. It was found that the extended Dickey–Fuller statistic's distribution is not sensitive to moderate choices of l, but does suffer serious size distortions when α is close to unity. As α approaches unity, the null distribution is continually shifted to the left of the distribution for $\alpha = 0$. Thus the extended Dickey–Fuller test is too liberal in such cases.

Similar results apply to the other statistics introduced in Theorem 9.23. Figure 9.8 shows the sampling null distributions of $\tilde{r}R_2$ for $T = 100$, where three cases of Model A, $y_j = \rho y_{j-1} + u_j, u_j = \varepsilon_j - \alpha\varepsilon_{j-1}$ with $\alpha = -0.8, 0, 0.8$ are examined. The value of l used is 4. The limiting null distribution is also shown. It is seen from Figure 9.8 that the distribution with $\alpha = 0$ is quite close to the limiting distribution, as should be. The distribution with $\alpha = 0.8$, however, is seriously distorted. Note that, in the present case,

$$r = \frac{1 + \alpha^2}{(1 - \alpha)^2} = 1 + \frac{2\alpha}{(1 - \alpha)^2}$$

UNIT ROOT TESTS IN THE DEPENDENT CASE

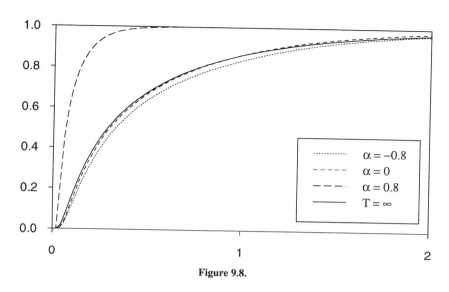

Figure 9.8.

so that $r = 41$ when $\alpha = 0.8$. It is found that $\tilde{r}$ produces a large downward bias when r is large, which forces $\tilde{r}R_2$ to take smaller values. It is also found that $\tilde{r}$ tends to produce upward bias when r is smaller than unity, which makes the distribution shift to the right. Note that $r = 0.51$ when $\alpha = -0.8$. It is seen that the distribution with $\alpha = -0.8$ is located to the right of the limiting distribution, but the degree of shift is slight since r is not far away from unity. Figure 9.9 shows the corresponding distributions when the value of l used is 8. As was mentioned before, we can confirm that the distributions are not sensitive to moderate choices of l.

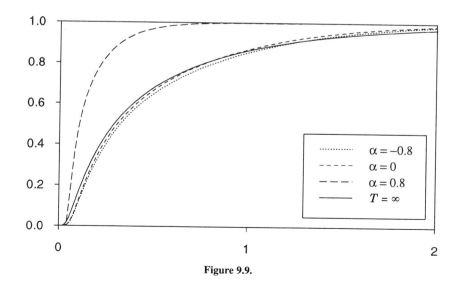

Figure 9.9.

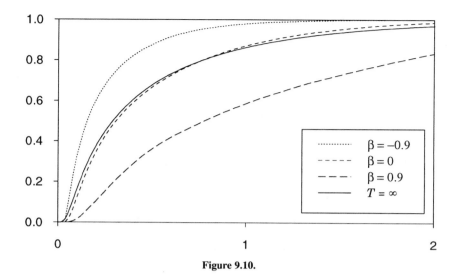

Figure 9.10.

Figure 9.10 shows the corresponding distributions when $u_j = \beta u_{j-1} + \varepsilon_j$ with $\beta = -0.9, 0, 0.9$. The value of l is 8. Note that, in the present case,

$$r = \frac{1-\beta}{1+\beta}$$

so that $r = 19$ for $\beta = -0.9$ and $r = 0.053$ for $\beta = 0.9$. It is seen that distortions are more serious when $\beta = 0.9$, since r is very small.

Size distortions may be explained by the difficulty in estimating $\sigma_L^2 = 2\pi f_u(0)$ when $f_u(0)$ is quite large or small compared with $f_u(\omega)$ at the neighboring frequencies. Hatanaka (1996) gives more explanations.

The above discussions apply equally to the extended POI test. Recalling the POI statistic $V_T^{(M)}(\theta)$ in (9.51), we deduce (Problem 9.3) that

(9.81) $$\mathcal{L}\left(V_T^{(M)}(\theta)\right) \to \mathcal{L}\left(\frac{1}{r}V^{(M)}(c,\theta) + \left(1-\frac{1}{r}\right)\theta\right),$$

where $V^{(A)}(c,\theta) = V^{(B)}(c,\theta)$ and $V^{(C)}(c,\theta) = V^{(D)}(c,\theta)$ are defined in (9.52) and (9.53), respectively. Thus we can conduct the extended POI test using $\tilde{r}V_T^{(M)}(\theta) - (\tilde{r}-1)\theta$, although this statistic also suffers size distortions in cases mentioned above.

Various attempts have been made to overcome size distortions under finite samples. The usual remedy is to assume $\{u_j\}$ to follow an ARMA(p,q) process. In fact, Fuller (1976) considered the unit root test $H_0 : \rho = 1$ when $\{y_j\}$ follows AR$(p+1)$, using Fuller's representation in (7.141)

(9.82) $$y_j = \rho y_{j-1} + \sum_{i=1}^{p} \beta_i \Delta y_{j-i} + \varepsilon_j.$$

It follows from (7.147) that, under H_0,

$$\mathcal{L}\left(T(\hat{\rho}-1)\right) \to \mathcal{L}\left((1-\beta_1-\cdots-\beta_p)\frac{\int_0^1 w(t)\,dw(t)}{\int_0^1 w^2(t)\,dt}\right),$$

where $\hat{\rho}$ is the ordinary LSE obtained from (9.82). Then we can easily devise a transformation of $T(\hat{\rho}-1)$ which eliminates the nuisance parameters $\beta_1,\ldots,\beta_p$.

Dickey and Fuller (1981) devised a likelihood ratio test for the model (9.82) with the constant and trend terms. Solo (1984) devised the LM-like test when $\{y_j\}$ follows ARMA(p,q), while Said and Dickey (1984) considered a t-test using (9.82) with p tending to ∞ as $T \to \infty$. Hall (1989) suggested a test based on $T(\tilde{\rho}_I - 1)$ when $\{u_j\}$ follows MA(q), where $\tilde{\rho}_I$ is an instrumental variable estimator of ρ. The test statistics for these tests eliminate nuisance parameters asymptotically in a parametric way. Size distortions, however, still remain under finite samples [Phillips and Perron (1988)]. Finite sample powers of some of the above tests are reported in Dickey and Fuller (1981), Dickey, Bell, and Miller (1986), Phillips and Perron (1988), and Schwert (1989).

PROBLEMS

9.1 Prove the weak convergence results on R_2 and R_6 in Theorem 9.22.

9.2 Prove the weak convergence result associated with R_6 in Theorem 9.23.

9.3 Show that (9.81) holds for Model A.

9.10 THE UNIT ROOT TESTING PROBLEM REVISITED

We have so far considered unit root tests for testing $H_0 : \rho = 1$ versus $H_1 : \rho < 1$. This may be regarded as testing nonstationarity against stationarity. If we allow the time series under consideration to contain a linear time trend under both H_0 and H_1, it may be interpreted as testing difference stationarity against trend stationarity. Then it seems more natural to take trend stationarity as the null hypothesis. In this section we attempt to conduct unit root tests in this reversed direction.

For the present purpose we consider first the state space model

(9.83) $$\begin{aligned} y_j &= \mu + \beta_j + j\gamma + u_j, \\ \beta_j &= \beta_{j-1} + \xi_j, \quad \beta_0 = 0, \quad (j=1,\ldots,T), \end{aligned}$$

where $\{\xi_j\} \sim \text{NID}(0,\sigma_\xi^2)$ and is independent of $\{u_j\}$ generated by

(9.84) $$u_j = \sum_{l=0}^{\infty} \alpha_l \varepsilon_{j-l}, \qquad \sum_{l=0}^{\infty} l|\alpha_l| < \infty, \qquad \{\varepsilon_j\} \sim \text{NID}\left(0,\sigma_\varepsilon^2\right).$$

The model (9.83) is essentially the same as the one considered in Kwiatkowski, Phillips, Schmidt, and Shin (1992), which we abbreviate as KPSS hereafter. The reversed unit root test suggested by KPSS is just to test

(9.85) $$H_0 : \sigma_\xi^2 = 0 \quad \text{versus} \quad H_1 : \sigma_\xi^2 > 0.$$

Note that $\{y_j\}$ is trend stationary under H_0, while it is difference stationary under H_1.

There exists no exact test that has any optimality because of $\{u_j\}$ being dependent upon an infinite number of nuisance parameters. Suppose that $\{u_j\} = \{\varepsilon_j\} \sim$ NID$(0, \sigma_\varepsilon^2)$. Then it follows from (9.83) that

(9.86) $$y = \mu e + \gamma d + \varepsilon + C\xi \sim \text{N}\left(\mu e + \gamma d, \sigma_\varepsilon^2(I_T + \rho CC')\right),$$

where $e = (1,\ldots,1)' : T \times 1$, $d = (1,\ldots,T)'$, $\rho = \sigma_\xi^2/\sigma_\varepsilon^2$, and C is the random walk generating matrix. The testing problem may now be regarded as that for the constancy of a regression coefficient discussed in Nabeya and Tanaka (1988), where the LBI test for (9.85) applied to the model (9.86) is also treated. In any case, the model (9.86) and the testing problem (9.85) are seen to fit a general framework discussed in Sections 9.2 and 9.3. It is now an easy matter to show (Problem 10.1) that the LBI test rejects H_0 when

(9.87) $$U_T = \frac{1}{T} \frac{y'MCC'My}{y'My}$$

takes large values, where $M = I_T - X(X'X)^{-1}X'$ and $X = (e, d)$.

The limiting power of the KPSS test based on U_T as $T \to \infty$ under $\rho = c^2/T^2$ can be easily obtained. We first have (Problem 10.2) that

(9.88) $$\frac{1}{T} y'My \to \sigma_\varepsilon^2 \quad \text{in probability.}$$

We next consider

(9.89) $$\mathcal{L}\left(\frac{1}{T^2\sigma_\varepsilon^2} y'MCC'My\right) = \mathcal{L}\left(\frac{1}{T^2\sigma_\varepsilon^2}(\varepsilon + C\xi)'MCC'M(\varepsilon + C\xi)\right)$$
$$= \mathcal{L}\left(\frac{1}{T^2} Z'(I_T + \rho CC')^{1/2} MCC'M(I_T + \rho CC')^{1/2} Z\right)$$
$$= \mathcal{L}\left(\frac{1}{T^2} Z'C'M\left(I_T + \frac{c^2}{T^2}CC'\right)MCZ\right),$$

where $Z \sim \text{N}(0, I_T)$. Then the weak convergence of this last random variable can be established by the Fredholm approach, which yields (Problem 10.3)

(9.90) $$\mathcal{L}(U_T) \to \mathcal{L}\left(\int_0^1 \int_0^1 \left[K(s,t) + c^2 K_{(2)}(s,t)\right] dw(s)\, dw(t)\right),$$

where $\{w(t)\}$ is the one-dimensional standard Brownian motion, while

(9.91) $$K(s,t) = \min(s,t) - 4st + 3st(s+t) - 3s^2t^2,$$

(9.92) $$K_{(2)}(s,t) = \int_0^1 K(s,u)K(u,t)\,du.$$

It follows (Problem 10.4) from Theorem 5.13 that

(9.93) $$\lim_{T\to\infty} P(U_T \geq x) = 1 - \frac{1}{\pi}\int_0^\infty \mathrm{Re}\left[\frac{1-e^{-i\theta x}}{i\theta}\phi(\theta)\right]d\theta,$$

where

$$\phi(\theta) = \left[D\left(i\theta + \sqrt{-\theta^2 + 2ic^2\theta}\right)D\left(i\theta - \sqrt{-\theta^2 + 2ic^2\theta}\right)\right]^{-1/2}$$

and $D(\lambda)$ is the FD of $K(s,t)$ given by

(9.94) $$D(\lambda) = \frac{12}{\lambda^2}\left(2 - \sqrt{\lambda}\sin\sqrt{\lambda} - 2\cos\sqrt{\lambda}\right).$$

The limiting power under $\rho = c^2/T^2$ can be easily computed from (9.93). Since the present problem is also closely related to the MA unit root, as is seen shortly, we postpone the detailed discussions until the next chapter.

The LBI statistic U_T in (9.87) was obtained under the assumption that $\{u_j\} = \{\varepsilon_j\} \sim \mathrm{NID}(0, \sigma_\varepsilon^2)$. Since the true process $\{u_j\}$ is generated as in (9.84), U_T depends on the parameters involved in $\{u_j\}$. As was suggested in KPSS, we can consider $\tilde{r}U_T$ as a test statistic for the original testing problem, where $\tilde{r}$ is the nonparametric estimator of $r = \sum_{l=0}^\infty \alpha_l^2/(\sum_{l=0}^\infty \alpha_l)^2$ constructed in the last section. Some simulation experiments conducted by KPSS exhibit size distortions especially when r is large or small, as in the unit root statistics dealt with in the last section.

Another model useful for the reversed unit root test is

(9.95) $$\begin{aligned}y_1 &= \mu + \gamma + v_1,\\ \Delta y_j &= \gamma + v_j - \alpha v_{j-1}, \quad (j=2,\ldots,T),\end{aligned}$$

where $\{v_j\}$ is a stationary process to be specified below and α is a parameter lying on $[-1,1]$. Then we consider testing

(9.96) $$H_0 : \alpha = 1 \quad \text{versus} \quad H_1 : \alpha < 1,$$

which tests the unit root in the MA part of $\{\Delta y_j\}$. Note that $\{y_j\}$ itself becomes trend stationary under H_0, while it is difference stationary under H_1; hence the present test can be regarded as a reversed unit root test. This idea is due to Saikkonen and

Luukkonen (1993a), which we refer to as S–L hereafter. The idea may be easily implemented to deal with multiple unit roots.

Suppose that $\{v_j\} = \{\varepsilon_j\} \sim \text{NID}(0, \sigma_\varepsilon^2)$, whose case with $\gamma = 0$ was already introduced in (1.19). The model (9.95) with $\{v_j\} = \{\varepsilon_j\}$ may be written as

$$(9.97) \quad y = \mu e + \gamma d + CC^{-1}(\alpha)\varepsilon \sim \text{N}\left(\mu e + \gamma d, \sigma_\varepsilon^2 C(C'(\alpha)C(\alpha))^{-1}C'\right),$$

where $C(\alpha)$ is defined in (8.36). The present model and the testing problem (9.96) also fit a general framework discussed in Sections 9.2 and 9.3. It follows (Problem 10.5) that the LBIU test for (9.96) rejects H_0 when

$$(9.98) \quad V_T = \frac{1}{T} \frac{y'MCC'My}{y'My}$$

takes large values, where $M = I_T - X(X'X)^{-1}X'$ with $X = (e, d)$. It is seen that V_T has the same form as U_T in (9.87), although the processes $\{y_j\}$ appearing in V_T and U_T are different and follow (9.97) and (9.86), respectively. It, however, holds that the distributions of U_T under $\rho = 0$ and V_T under $\alpha = 1$ are the same.

For the distribution of V_T under the alternative, let us put $\alpha = 1 - (c/T)$ with c a positive constant. Then it holds (Problem 10.6) that

$$(9.99) \quad \frac{1}{T} y'My \to \sigma_\varepsilon^2 \quad \text{in probability.}$$

We can also show (Problem 10.7) that

$$(9.100) \quad \mathcal{L}\left(\frac{1}{T^2\sigma_\varepsilon^2} y'MCC'My\right)$$

$$= \mathcal{L}\left(\frac{1}{T^2} Z'C'MC(C'(\alpha)C(\alpha))^{-1}C'MCZ\right),$$

$$\to \mathcal{L}\left(\int_0^1 \int_0^1 \left[K(s,t) + c^2 K_{(2)}(s,t)\right] dw(s)\,dw(t)\right),$$

where $Z \sim \text{N}(0, I_T)$, while K and $K_{(2)}$ are defined in (9.91) and (9.92), respectively. Thus we can conclude that U_T in (9.87) under $\rho = c^2/T^2$ and V_T in (9.98) under $\alpha = 1 - (c/T)$ have the same limiting distribution given on the right side of (9.90). Thus the limiting power of the S–L test under $\alpha = 1 - (c/T)$ can also be computed from (9.93).

We need to modify the statistic V_T to allow for $\{v_j\}$ being dependent. The modified S–L test dispenses with a nonparametric correction to eliminate the nuisance parameters involved in $\{v_j\}$. Instead $\{v_j\}$ is assumed to follow ARMA(p, q). By taking the second derivative of the log-likelihood evaluated at the MLEs under H_0 and retaining the highest order term, the modified S–L test rejects H_0 when

$$\tilde{V}_T = \frac{y'\tilde{M}'\tilde{\Sigma}^{-1}C\tilde{\Sigma}C'\tilde{\Sigma}^{-1}\tilde{M}y}{y'\tilde{M}'\tilde{\Sigma}^{-1}\tilde{M}y}$$

takes large values, where $\tilde{\Sigma}$ is the MLE of $\Sigma = V(v)$ and

$$\tilde{M} = I_T - X\left(X'\tilde{\Sigma}^{-1}X\right)^{-1}X'\tilde{\Sigma}^{-1}, \quad X = (e, d).$$

Empirical sizes and powers are examined in S–L by using the model (9.95) with $\gamma = 0$, where it is reported that the modified S–L test performs quite well. An efficient method of computing $\tilde{\Sigma}^{-1}$ is also discussed.

The reversed unit root tests considered here are closely related to MA unit root tests, as was mentioned before. Further discussions are given in the next chapter.

PROBLEMS

10.1 Show that the LBI test for (9.85) applied to the model (9.86) rejects H_0 when U_T takes large values, where U_T is defined in (9.87).

10.2 Prove (9.88).

10.3 Establish the weak convergence result in (9.90).

10.4 Prove (9.93).

10.5 Show that the LBIU test for (9.96) applied to the model (9.97) rejects H_0 when V_T takes large values, where V_T is defined in (9.98).

10.6 Prove (9.99).

10.7 Establish the weak convergence result in (9.100).

CHAPTER 10

Unit Root Tests in Moving Average Models

Another unit root testing problem, which occurs in the MA part of linear time series models, is discussed. We allow for two cases concerning the initial value that generates the MA(1) process. The LBI and LBIU tests are suggested for these cases, and the power properties are examined. The relationship with other testing problems is also explored.

10.1 INTRODUCTION

This chapter deals with another unit root testing problem, which occurs in the MA part of linear time series models. The model we first deal with is

(10.1) $\qquad y_j = x_j'\beta + \varepsilon_j - \alpha\varepsilon_{j-1}, \qquad (j = 1, \ldots, T),$

where $\{x_j\}$ is a $p \times 1$ nonstochastic fixed sequence and $\{\varepsilon_j\}$ $(j = 1, 2, \ldots)$ is an i.i.d.$(0, \sigma^2)$ sequence. For the initial value ε_0 we consider two cases as in Chapter 8; one is the *conditional case* where $\varepsilon_0 = 0$, and the other is the *stationary case* where $\varepsilon_0, \varepsilon_1 \ldots \sim$ i.i.d.$(0, \sigma^2)$. For both cases we assume that the parameter space of α is restricted to be $-1 \leq \alpha \leq 1$.

Our purpose here is to test

(10.2) $\qquad H_0 : \alpha = 1 \qquad \text{versus} \qquad H_1 : \alpha < 1,$

which we call the *MA unit root testing problem*. For this purpose we assume that $\varepsilon_1, \varepsilon_2, \ldots \sim \text{NID}(0, \sigma^2)$ for the conditional case and $\varepsilon_0, \varepsilon_1, \ldots \sim \text{NID}(0, \sigma^2)$ for the stationary case. It follows that

(10.3) $\qquad y \sim N\left(X\beta, \sigma^2 \left(C'(\alpha)C(\alpha)\right)^{-1}\right) : \quad$ conditional case,

(10.4) $\qquad y \sim N\left(X\beta, \sigma^2 \Omega(\alpha)\right) : \qquad$ stationary case, where $C(\alpha)$ and $\Omega(\alpha)$ are defined in (8.36) and (8.5), respectively. The testing problem (10.2) and the model (10.3) or (10.4) are seen to fit a general framework for deriving the LBI or LBIU test discussed in the last chapter. Thus the LM principle yields the LBI or LBIU test, which we deal with in Section 10.2. We also establish the weak convergence results on the test statistics for both the conditional and stationary cases as $T \to \infty$ under $\alpha = 1 - (c/T)$ with c a nonnegative constant. Since the limiting random variables depend on the regressor sequence $\{x_j\}$, we consider, as in previous chapters, four simple models by specifying $\{x_j\}$:

Model A: $\quad y_j = \varepsilon_j - \alpha \varepsilon_{j-1}$,

Model B: $\quad y_j = \beta_0 + \varepsilon_j - \alpha \varepsilon_{j-1}$,

Model C: $\quad y_j = j\beta_1 + \varepsilon_j - \alpha \varepsilon_{j-1}$,

Model D: $\quad y_j = \beta_0 + j\beta_1 + \varepsilon_j - \alpha \varepsilon_{j-1}$.

In Section 10.3 we consider a model where the first differences of observations follow MA(1). The MA unit root test applied to this model may be interpreted as a test for stationarity against nonstationarity, as was discussed in the last chapter. We explore the relationship of such a test with our MA unit root test.

The limiting powers of the LBI and LBIU unit root tests are computed in Section 10.4 by numerical integration. We examine the power properties in detail, concentrating on Model A in the conditional and stationary cases. To see how powerful the suggested tests are, we compute the limiting power envelopes of all the invariant MA unit root tests, as in the AR unit root case. The POI test is also conducted for Model A in the stationary case, and its limiting powers are computed.

In Section 10.5 we extend the model to allow for seasonality, while in Section 10.6 we modify the tests to cope with the situation where $\{\varepsilon_j\}$ is dependent. Section 10.7, the last section of the present chapter, discusses the relationship with a testing problem in the state space model, where the similarity and difference of the two testing problems are discussed.

10.2 THE LBI AND LBIU TESTS

It is an easy matter to derive the LM test for the testing problem (10.2) applied to the model (10.1). In the next subsection we deal with the conditional case, where the LM test is shown to be LBI. We then deal with the stationary case, where the LM test is found to be LBIU. The limiting powers of the tests are derived in both cases for Models A through D described in the last section.

10.2.1 The Conditional Case

The model (10.1) may be rewritten as

$$(10.5) \qquad y = X\beta + C^{-1}(\alpha)\varepsilon \sim \mathrm{N}\left(X\beta, \sigma^2 \left(C'(\alpha)C(\alpha)\right)^{-1}\right).$$

Using the relationship between the LM and LBI tests established in the last chapter, we can show (Problem 2.1 in this chapter) that the LBI test rejects $H_0 : \alpha = 1$ when

$$(10.6) \qquad S_{T1} = \frac{y'C'\tilde{M}ee'\tilde{M}Cy}{y'C'\tilde{M}Cy} > c,$$

where $e = (1, \ldots, 1)' : T \times 1$ and $C = C(1)$ is the random walk generating matrix, while

$$\tilde{M} = I_T - CX(X'C'CX)^{-1}X'C' = \tilde{M}^2.$$

Let us first derive the null distribution of S_{T1}. Noting that $\tilde{M}Cy = \tilde{M}\varepsilon$ under H_0, we consider

$$S_{T1}(H_0) = \frac{\varepsilon'\tilde{M}ee'\tilde{M}\varepsilon}{\varepsilon'\tilde{M}\varepsilon}$$

$$= e'\tilde{M}e \times \frac{\varepsilon'A\varepsilon}{\varepsilon'A\varepsilon + \varepsilon'B\varepsilon},$$

where

$$A = \tilde{M}ee'\tilde{M}/e'\tilde{M}e = A^2,$$
$$B = \tilde{M} - A = B^2.$$

Since it holds that $AB = 0$, $\text{rank}(A) = 1$ and $\text{rank}(B) = T - p - 1$, we can deduce that

$$\mathcal{L}\left(S_{T1}(H_0)/e'\tilde{M}e\right) \sim \text{Be}\left(\frac{1}{2}, \frac{T - p - 1}{2}\right),$$

where

$$\begin{aligned}
e'\tilde{M}e &= T & \text{for Model A}, \\
&= \tfrac{1}{4}T + O(1) & \text{for Model B}, \\
&= \tfrac{4}{9}T + O(1) & \text{for Model C}, \\
&= \tfrac{1}{9}T + O(1) & \text{for Model D}.
\end{aligned}$$

We now derive the limiting distribution of S_{T1} in (10.6) when the true value of α, α_0 say, takes the form

$$\alpha_0 = 1 - \frac{c}{T},$$

where c is a positive constant. Then we can show (Problem 2.2) that

(10.7) $$\frac{1}{T} y'C'\tilde{M}Cy \to \sigma^2$$

in probability for Models A through D as $T \to \infty$. Concentrating on the numerator of S_{T1} in (10.6) we can establish (Problem 2.3) the following theorem.

Theorem 10.1. *Let $S_{T1}^{(M)}$ be the LBI statistic for Model $M = A, B, C, D$, where*

$$S_{T1}^{(A)} = S_{T1}, \quad S_{T1}^{(B)} = 4S_{T1}, \quad S_{T1}^{(C)} = 9S_{T1}/4, \quad S_{T1}^{(D)} = 9S_{T1},$$

with S_{T1} defined in (10.6). Then it holds that, as $T \to \infty$ under $\alpha_0 = 1 - (c/T)$,

$$\mathcal{L}\left(S_{T1}^{(A)}\right) \to \frac{c^2 + 3c + 3}{3} \chi^2(1),$$

$$\mathcal{L}\left(S_{T1}^{(B)}\right) \to \frac{2c^2 + 15c + 60}{60} \chi^2(1),$$

$$\mathcal{L}\left(S_{T1}^{(C)}\right) \to \frac{9c^2 + 56c + 126}{126} \chi^2(1),$$

$$\mathcal{L}\left(S_{T1}^{(D)}\right) \to \frac{3c^2 + 35c + 315}{315} \chi^2(1).$$

Note that the scale transformation of S_{T1} makes the limiting distribution follow $\chi^2(1)$ under H_0. The limiting powers can be easily computed in the present case and will be reported in Section 10.4.

To see how powerful the present test is, we consider the MPI test

(10.8) $$H_0 : \alpha = 1 \quad \text{versus} \quad H_1 : \alpha = 1 - \frac{\theta}{T},$$

where θ is a given positive constant. For this purpose we concentrate on Model A. Then the Neyman–Pearson lemma leads us to conclude that the MPI test for (10.8) applied to Model A rejects H_0 when

(10.9) $$V_{T1}(\theta) = T \frac{y' \left[C'C - C'(\alpha)C(\alpha)\right] y}{y'C'Cy} > c,$$

where $\theta = T(1 - \alpha)$.

It now follows from the arguments in Section 8.3 of Chapter 8 that, as $T \to \infty$ under $\alpha = 1 - (\theta/T)$ and $\alpha_0 = 1 - (c/T)$,

(10.10) $$\mathcal{L}(V_{T1}(\theta)) \to \mathcal{L}\left(\int_0^1 \int_0^1 K(s, t; \theta)\, dw(s)\, dw(t) - \theta\right),$$

where $K(s, t; \theta) = 2K_2(s, t; \theta)$ with K_2 defined in (8.40). In particular, the limiting null distribution of $V_{T1}(\theta)$ as $T \to \infty$ under $\alpha = 1 - (\theta/T)$ and $\alpha_0 = 1$ can be computed (Problem 2.4) as

$$(10.11) \quad \lim_{T\to\infty} P(V_{T1}(\theta) \leq x)$$

$$= \frac{1}{\pi} \int_0^\infty \mathrm{Re} \left[\frac{1 - \exp\left\{-\frac{iu(\theta + x)}{\theta^2}\right\}}{iu} \left(\frac{\cos \mu - \frac{\mu}{\theta} \sin \mu}{e^\theta} \right)^{-1/2} \right] du,$$

where $\mu = \sqrt{2iu - \theta^2}$.

Imagine that the MPI test above can be conducted for each $\theta = c > 0$. Then it holds (Problem 2.5) that, as $T \to \infty$ under $\alpha = \alpha_0 = 1 - (c/T)$,

$$(10.12) \quad \lim_{T\to\infty} P\left(V_{T1}(c) \geq x\right)$$

$$= 1 - \frac{1}{\pi} \int_0^\infty \mathrm{Re} \left[\frac{1 - \exp\left\{-\frac{iu(c + x)}{c^2}\right\}}{iu} \left(\cos \nu - \frac{\nu}{c} \sin \nu \right)^{-1/2} \right] du,$$

where $\nu = \sqrt{2iu}$. If x is the upper $100\gamma\%$ point of the limiting distribution in (10.11) with $\theta = c$, the quantity on the right side of (10.12) gives us the limiting power envelope at the $100\gamma\%$ level obtained from the class of all the invariant MA unit root tests, which will also be computed in Section 10.4.

10.2.2 The Stationary Case

In the present case the model (10.1) may be rewritten as

$$(10.13) \qquad y = X\beta + \eta, \qquad \eta = B(\alpha)\varepsilon^* \sim \mathrm{N}\left(0, \sigma^2 \Omega(\alpha)\right),$$

where $\varepsilon^* = (\varepsilon_0, \varepsilon')' : (T + 1) \times 1$ and $B(\alpha) = \left(-\alpha e_1, C^{-1}(\alpha)\right) : T \times (T + 1)$ with $e_1 = (1, 0, \ldots, 0)' : T \times 1$. It can be easily checked that

$$\left. \frac{\partial L(\alpha, \theta)}{\partial \alpha} \right|_{\alpha=1, \theta=\hat{\theta}} = 0,$$

where $\theta = (\beta', \sigma^2)'$, $\hat{\theta} = (\hat{\beta}', \hat{\sigma}^2)'$, $\hat{\beta} = (X'\Omega^{-1}(1)X)^{-1}X'\Omega^{-1}(1)y$, $\hat{\sigma}^2 = (y - X\hat{\beta})'\Omega^{-1}(1)(y - X\hat{\beta})/T$, and

$$L(\alpha, \theta) = -\frac{T}{2} \log(2\pi\sigma^2) - \frac{1}{2} \log |\Omega(\alpha)| - \frac{1}{2\sigma^2}(y - X\beta)'\Omega^{-1}(\alpha)(y - X\beta).$$

(10.14)

Thus we are led to consider

(10.15) $$\left.\frac{\partial^2 L(\alpha, \theta)}{\partial \alpha^2}\right|_{\alpha=1, \theta=\hat{\theta}} = T \frac{\tilde{\eta}' \Omega^{-2} \tilde{\eta}}{\tilde{\eta}' \Omega^{-1} \tilde{\eta}} - \frac{T(T+5)}{6},$$

where $\Omega = \Omega(1)$, $\tilde{\eta} = \tilde{N}y$ and

(10.16) $$\tilde{N} = I_T - X(X'\Omega^{-1}X)^{-1}X'\Omega^{-1}.$$

We can now derive the LBIU test, which rejects H_0 when

(10.17) $$S_{T2} = \frac{1}{T} \frac{\tilde{\eta}' \Omega^{-2} \tilde{\eta}}{\tilde{\eta}' \Omega^{-1} \tilde{\eta}}$$

$$= \frac{1}{T} \frac{\tilde{\eta}' \left\{ C' \left(I_T - \frac{1}{T+1}ee' \right) C \right\}^2 \tilde{\eta}}{\tilde{\eta}' C' \left(I_T - \frac{1}{T+1}ee' \right) C \tilde{\eta}}$$

$$= \frac{1}{T} \frac{\tilde{\eta}'(H'H)^2 \tilde{\eta}}{\tilde{\eta}' H'H \tilde{\eta}} > c,$$

where $\Omega^{-1} = C'(I_T - ee'/(T+1))C = H'H$ with

(10.18) $$H = \begin{pmatrix} \frac{1}{\sqrt{2}} & & & & \\ \frac{1}{\sqrt{6}} & \frac{2}{\sqrt{6}} & & 0 & \\ \cdot & \cdot & \cdot & & \cdot \\ \cdot & \cdot & & & \\ \cdot & \cdot & & & \\ \frac{1}{\sqrt{T(T+1)}} & \frac{2}{\sqrt{T(T+1)}} & \cdot & \cdot & \frac{T}{\sqrt{T(T+1)}} \end{pmatrix}$$

The matrix H gives the Cholesky decomposition of Ω^{-1} as $\Omega^{-1} = H'H$, which can be naturally found from the *Kalman filter algorithm*. To see this, consider the noninvertible MA(1) model $y_j = \varepsilon_j - \varepsilon_{j-1}$, for which we have the equivalent state space model:

$$y_j = (1 \ 0) \begin{pmatrix} \beta_{1j} \\ \beta_{2j} \end{pmatrix} = a'\beta_j,$$

$$\beta_j = \begin{pmatrix} 0 & -1 \\ 0 & 0 \end{pmatrix} \beta_{j-1} + \begin{pmatrix} 1 \\ 1 \end{pmatrix} \varepsilon_j = G\beta_{j-1} + b\varepsilon_j.$$

Let $\beta(j|k) = E(\beta_j | y_k, y_{k-1}, \ldots)$ and $P(j|k) = V(\beta_j | y_k, y_{k-1}, \ldots)$. Then it is known [Jazwinski (1970)] that the Kalman filter algorithm evolves as

$$\beta(j|j-1) = G\beta(j-1|j-1),$$

$$\beta(j|j) = \beta(j|j-1) + \frac{P(j|j-1)a}{a'P(j|j-1)a}(y_j - a'\beta(j|j-1)),$$

$$P(j|j-1) = GP(j-1|j-1)G' + bb'\sigma^2,$$

$$P(j|j) = P(j|j-1) - \frac{P(j|j-1)aa'P(j|j-1)}{a'P(j|j-1)a},$$

where we set the initial value $\beta(0|0)$ to generate the above recursions at 0.

Since it is also known [Schweppe (1965)] that the log-likelihood for $y_j = \varepsilon_j - \varepsilon_{j-1}$, $j = 1, \ldots, T$, is given, except for constants, by

$$(10.19) \quad l(\sigma^2) = -\frac{T}{2}\log\sigma^2 - \frac{1}{2}\log|\Omega| - \frac{1}{2\sigma^2}y'\Omega^{-1}y$$

$$= -\frac{1}{2}\sum_{j=1}^{T}\log(a'P(j|j-1)a) - \frac{1}{2}\sum_{j=1}^{T}\frac{(y_j - a'\beta(j|j-1))^2}{a'P(j|j-1)a},$$

the MLEs of σ^2 obtained from (10.19) and (10.14) with $y \sim N(0, \sigma^2\Omega)$ should produce the same value. It can be shown (Problem 2.6) that the MLE $\tilde{\sigma}^2$ is given by

$$(10.20) \quad \tilde{\sigma}^2 = \frac{1}{T}y'\Omega^{-1}y$$

$$= \frac{1}{T}\sum_{j=1}^{T}\frac{j}{j+1}(y_j - a'\beta(j|j-1))^2$$

$$= \frac{1}{T}\sum_{j=1}^{T}\frac{1}{j(j+1)}(y_1 + 2y_2 + \cdots + jy_j)^2$$

$$= \frac{1}{T}y'H'Hy,$$

which yields the Cholesky decomposition $\Omega^{-1} = H'H$. The matrix H will prove useful in Section 10.6.

The statistic S_{T2} in (10.17) can be interpreted as follows. Let us put $v = H\tilde{\eta}$. Then we have

$$S_{T2} = \frac{1}{T}\frac{v'\Omega^{-1}v}{v'v} = \frac{1}{T}\frac{\sum_{j=1}^{T}w_j^2 - \frac{1}{T+1}\left(\sum_{j=1}^{T}w_j\right)^2}{\sum_{j=1}^{T}v_j^2},$$

where $w_j = w_{j-1} + v_j$ with $w_0 = 0$. It follows that S_{T2} is essentially the ratio of the corrected sum of squares of an integrated process $\{w_j\}$ to the sum of squares of the innovation sequence $\{v_j\}$ which drives $\{w_j\}$.

Let us derive the limiting distribution of S_{T2} as $T \to \infty$ under $\alpha = 1 - (c/T)$. We first note that

$$\tilde{\eta}'\Omega^{-1}\tilde{\eta} = \eta'\tilde{N}'\Omega^{-1}\tilde{N}\eta$$
$$= \varepsilon^{*\prime}B'(\alpha)\left[\Omega^{-1} - \Omega^{-1}X\left(X'\Omega^{-1}X\right)^{-1}X'\Omega^{-1}\right]B(\alpha)\varepsilon^*.$$

Then it can be checked (Problem 2.7) that

(10.21) $$\frac{1}{T}\tilde{\eta}'\Omega^{-1}\tilde{\eta} \to \sigma^2$$

in probability for Models A through D. We next note that

(10.22) $$\mathcal{L}\left(\tilde{\eta}'\Omega^{-2}\tilde{\eta}\right) = \mathcal{L}\left(\varepsilon^{*\prime}B'(\alpha)\tilde{N}'\Omega^{-2}\tilde{N}B(\alpha)\varepsilon^*\right)$$
$$= \mathcal{L}\left(\varepsilon'\Omega^{-1}\tilde{N}\Omega(\alpha)\tilde{N}'\Omega^{-1}\varepsilon\right)$$
$$= \mathcal{L}\left(\alpha\varepsilon'A_T\,\varepsilon + \frac{c^2}{T^2}\varepsilon'A_T^2\,\varepsilon\right),$$

where

$$A_T = \Omega^{-1}\tilde{N} = \Omega^{-1} - \Omega^{-1}X\left(X'\Omega^{-1}X\right)^{-1}X'\Omega^{-1}$$

and we have used the facts that

$$\Omega(\alpha) = B(\alpha)B'(\alpha) = \alpha\Omega + (1-\alpha)^2 I_T$$
$$= \alpha\Omega + \frac{c^2}{T^2}I_T,$$
$$\Omega^{-1}\tilde{N} = \tilde{N}'\Omega^{-1} = \Omega^{-1}\tilde{N}\Omega\tilde{N}'\Omega^{-1} = A_T.$$

If we can find a symmetric, continuous, and nearly definite function $K(s,t)$ such that

(10.23) $$\lim_{T\to\infty}\max_{j,k}\left|\frac{1}{T}A_T(j,k) - K\left(\frac{j}{T},\frac{k}{T}\right)\right| = 0,$$

it follows from (10.21), (10.22), and Theorem 5.13 that

(10.24) $$\mathcal{L}(S_{T2}) \to \mathcal{L}\left(\int_0^1\int_0^1 \left(K(s,t) + c^2 K_{(2)}(s,t)\right)dw(s)\,dw(t)\right),$$

THE LBI AND LBIU TESTS

where

$$K_{(2)}(s,t) = \int_0^1 K(s,u)K(u,t)\,du.$$

Moreover the c.f. of the limiting distribution of S_{T2} is given by

$$(10.25) \quad \phi(\theta) = \left[D\left(i\theta + \sqrt{-\theta^2 + 2ic^2\theta}\right) D\left(i\theta - \sqrt{-\theta^2 + 2ic^2\theta}\right)\right]^{-1/2},$$

where $D(\lambda)$ is the FD of $K(s,t)$.

We can easily find the function $K(s,t)$ for each model of A through D. The corresponding FD can also be obtained after some algebra by the Fredholm approach. In fact, the following theorem holds [Tanaka (1990, 1995) and Problem 2.8].

Theorem 10.2. *Let $S_{T2}^{(M)}$ be the LBIU statistic for Model $M = A, B, C, D$ constructed from S_{T2} in (10.17). Then each $S_{T2}^{(M)}$ follows the weak convergence as described in (10.24) with K replaced by K_M, where*

$$K_A(s,t) = \min(s,t) - st,$$
$$K_B(s,t) = K_A(s,t) - 3st(1-s)(1-t),$$
$$K_C(s,t) = K_A(s,t) - \tfrac{5}{4}st(1-s^2)(1-t^2),$$
$$K_D(s,t) = K_A(s,t) - 2st(1-s)(1-t)(4 - 5s - 5t + 10st).$$

Moreover the FD $D_M(\lambda)$ of $K_M(s,t)$ is given by

$$D_A(\lambda) = \frac{\sin\sqrt{\lambda}}{\sqrt{\lambda}},$$

$$D_B(\lambda) = \frac{12}{\lambda^2}\left(2 - \sqrt{\lambda}\sin\sqrt{\lambda} - 2\cos\sqrt{\lambda}\right),$$

$$D_C(\lambda) = \frac{45}{\lambda^3}\left(\sqrt{\lambda}\left(1 - \frac{\lambda}{3}\right)\sin\sqrt{\lambda} - \lambda\cos\sqrt{\lambda}\right),$$

$$D_D(\lambda) = \frac{8640}{\lambda^4}\left(2 + \frac{\lambda}{3} - \sqrt{\lambda}\left(2 - \frac{\lambda}{12}\right)\sin\sqrt{\lambda} - \left(2 - \frac{2\lambda}{3}\right)\cos\sqrt{\lambda}\right).$$

The FDs appearing in this theorem were earlier obtained in Nabeya and Tanaka (1988) in connection with a testing problem in the state space model. This last topic will be discussed in Section 10.7. It is now an easy matter to compute the limiting powers of the LBIU tests, which will be reported in Section 10.4.

The limiting power envelope specific to Model A can be obtained in the same way as in the conditional case. The MPI test for (10.8) applied to Model A rejects H_0

when

(10.26) $$V_{T2}(\theta) = T\frac{y'[\Omega^{-1} - \Omega^{-1}(\alpha)]y}{y'\Omega^{-1}y} > c,$$

where $\theta = T(1-\alpha)$. It follows from the arguments in Section 8.2 of Chapter 8 that, as $T \to \infty$ under $\alpha = 1 - (\theta/T)$ and $\alpha_0 = 1 - (c/T)$,

(10.27) $$\mathcal{L}(V_{T2}(\theta)) \to \mathcal{L}\left(\theta^2 \sum_{n=1}^{\infty} \frac{n^2\pi^2 + c^2}{n^2\pi^2(n^2\pi^2 + \theta^2)} Z_n^2 - \theta\right),$$

where $\{Z_n\} \sim \text{NID}(0, 1)$. This result was also obtained in Saikkonen and Luukkonen (1993b). Then we can show (Problem 2.9) that, as $T \to \infty$ under $\alpha = 1 - (\theta/T)$ and $\alpha_0 = 1$,

(10.28) $$P(V_{T2}(\theta) \le x) \to \frac{1}{\pi}\int_0^\infty \text{Re}\left[\frac{1 - e^{-iau}}{iu}\left(\frac{\sin\mu}{\mu}\bigg/\frac{\sinh\theta}{\theta}\right)^{-1/2}\right] du,$$

where $a = (\theta + x)/\theta^2$ and $\mu = \sqrt{2iu - \theta^2}$.

The limiting power envelope at the $100\gamma\%$ level is given (Problem 2.10) by

(10.29) $$\lim_{T\to\infty} P(V_{T2}(c) \ge x) = 1 - \frac{1}{\pi}\int_0^\infty \text{Re}\left[\frac{1 - e^{-ibu}}{iu}\left(\frac{\sin\sqrt{2iu}}{\sqrt{2iu}}\right)^{-1/2}\right] du,$$

where $b = (c + x)/c^2$ and the limit is taken under $\alpha = \alpha_0 = 1 - (c/T)$, while x is the upper $100\gamma\%$ point of the limiting distribution in (10.28) with $\theta = c$.

PROBLEMS

2.1 Show that S_{T1} in (10.6) is the LBI statistic for the model (10.5).
2.2 Prove (10.7).
2.3 Prove Theorem 10.1.
2.4 Show that the limiting distribution of $V_{T1}(\theta)$ in (10.9) as $T \to \infty$ under $\alpha = 1 - (\theta/T)$ and $\alpha_0 = 1$ can be computed from (10.11).
2.5 Establish (10.12).
2.6 Establish the relations in (10.20).
2.7 Show that (10.21) holds.
2.8 Prove Theorem 10.2.
2.9 Show that the limiting distribution of $V_{T2}(\theta)$ in (10.26) as $T \to \infty$ under $\alpha = 1 - (\theta/T)$ and $\alpha_0 = 1$ is given by (10.28).
2.10 Establish (10.29).

10.3 THE RELATIONSHIP WITH THE TEST STATISTICS IN DIFFERENCED FORM

Saikkonen and Luukkonen (1993a), referred to as S–L hereafter, suggested LBI and LBIU tests for the MA(1) model in differenced form. An extended version of their model takes the following form:

$$
\begin{aligned}
(10.30) \quad y_1 &= \mu + \gamma + \delta + \varepsilon_1, \\
\Delta y_j &= \gamma + (2j-1)\delta + \varepsilon_j - \alpha \varepsilon_{j-1}, \quad (j = 2, \ldots, T),
\end{aligned}
$$

where $\{\varepsilon_j\} \sim \text{NID}(0, \sigma^2)$. Various test statistics for testing $H_0 : \alpha = 1$ versus $H_1 : \alpha < 1$ arise depending on whether the parameters μ, γ and δ are assumed to be known or unknown. We shall show that the LBI and LBIU tests for variants of the above model have completely the same asymptotic properties as those derived in the last section. For this purpose, we rewrite the model (10.30) as

$$
(10.31) \qquad C^{-1} y = \mu e_1 + \gamma e + \delta f_1 + C^{-1}(\alpha)\varepsilon,
$$

where $e_1 = (1, 0, \ldots, 0)' : T \times 1$, $e = (1, \ldots, 1)' : T \times 1$, and $f_1 = (1, 3, \ldots, 2T-1)'$, or equivalently as

$$
\begin{aligned}
(10.32) \qquad y &= \mu e + \gamma d + \delta f + CC^{-1}(\alpha)\varepsilon \\
&= X\beta + CC^{-1}(\alpha)\varepsilon,
\end{aligned}
$$

where $d = (1, \ldots, T)'$, $f = (1^2, 2^2, \ldots, T^2)'$, $X = (e, d, f)$, and $\beta = (\mu, \gamma, \delta)'$.

Suppose first that the parameter μ is known to be 0. Note that μ appears only in the first equation of the model (10.30). Then we derive the LBI test which rejects H_0 when

$$
(10.33) \qquad \frac{y' \bar{M} e e' \bar{M} y}{y' \bar{M} y} > c,
$$

where $\bar{M} = I_T - \bar{X}(\bar{X}'\bar{X})^{-1}\bar{X}'$ and $\bar{X} = (d, f)$. The LBI test in (10.33) yields four variants by assuming some knowledge of the parameters γ and δ, which are

$$
(10.34) \qquad \text{SL}_{T1}^{(k)} = \frac{y' \bar{M}_k e e' \bar{M}_k y}{y' \bar{M}_k y} > c, \quad (k = 1, 2, 3, 4),
$$

where

$$
\bar{M}_1 = I_T, \qquad \bar{M}_2 = I_T - dd'/d'd,
$$
$$
\bar{M}_3 = I_T - ff'/f'f, \qquad \bar{M}_4 = I_T - \bar{X}(\bar{X}'\bar{X})^{-1}\bar{X}'.
$$

Note that $SL_{T1}^{(1)}$ arises from the restriction $\mu = \gamma = \delta = 0$, which is the LBI statistic suggested by S–L. The statistic $SL_{T1}^{(2)}$ assumes $\mu = \delta = 0$, while $SL_{T1}^{(3)}$ assumes $\mu = \gamma = 0$. The statistic $SL_{T1}^{(4)}$ imposes no restriction except $\mu = 0$.

It is easy to see that, under H_0,

$$SL_{T1}^{(1)}/T \sim Be\left(\frac{1}{2}, \frac{T-1}{2}\right),$$

$$SL_{T1}^{(k)}/e'\bar{M}_k e \sim Be\left(\frac{1}{2}, \frac{T-2}{2}\right) \quad \text{for} \quad k = 2, 3,$$

$$SL_{T1}^{(4)}/e'\bar{M}_4 e \sim Be\left(\frac{1}{2}, \frac{T-3}{2}\right),$$

where $e'\bar{M}_2 e = T/4 + O(1)$, $e'\bar{M}_3 e = 4T/9 + O(1)$ and $e'\bar{M}_4 e = T/9 + O(1)$. Thus the LBI statistics $SL_{T1}^{(k)}$ ($k = 1, 2, 3, 4$) suitably normalized have the same null distributions as $S_{T1}^{(M)}$ ($M = A, B, C, D$) in the conditional case. Moreover the same relationships are shown to be carried over to the limiting distributions as $T \to \infty$ under $\alpha_0 = 1 - (c/T)$.

Theorem 10.3. *Let us assume the model in differenced form given in (10.30) with $\mu = 0$, and consider the testing problem $H_0 : \alpha = 1$ versus $H_1 : \alpha < 1$. Then the LBI statistics $SL_{T1}^{(1)}$, $4SL_{T1}^{(2)}$, $9SL_{T1}^{(3)}/4$, and $9SL_{T1}^{(4)}$ with $SL_{T1}^{(k)}$ defined in (10.34) have the same limiting distributions as $S_{T1}^{(M)}$ ($M = A, B, C, D$), respectively, as $T \to \infty$ under $\alpha_0 = 1 - (c/T)$, where the limiting distributions of the latter are described in Theorem 10.1.*

Suppose next that the parameter μ is unknown. Then it follows from (9.98) that the LBIU test, when μ is unknown, rejects H_0 when

$$(10.35) \qquad \frac{y'MCC'My}{y'My} > c,$$

where $M = I_T - X(X'X)^{-1}X'$. The LBIU test in (10.35) yields four variants, as in the case of $\mu = 0$, by assuming some knowledge of the parameters γ and δ, which are

$$(10.36) \qquad SL_{T2}^{(k)} = \frac{1}{T} \frac{y'M_k CC'M_k y}{y'M_k y} > c, \quad (k = 1, 2, 3, 4),$$

where $M_k = I_T - X_k(X_k'X_k)^{-1}X_k'$ and

$$X_1 = e, \quad X_2 = (e, d), \quad X_3 = (e, f), \quad X_4 = (e, d, f).$$

Let us examine each of the LBIU statistics $SL_{T2}^{(k)}$ ($k = 1, 2, 3, 4$). The statistic $SL_{T2}^{(1)}$ restricts γ and δ to be 0, which corresponds to the LBIU test suggested by S–L. In

Section 1.2 of Chapter 1 we proved that

$$\mathrm{SL}_{T2}^{(1)} = \frac{1}{T} \frac{z'\Omega_*^{-2}z}{z'\Omega_*^{-1}z},$$

where $z = (y_2 - y_1, \ldots, y_T - y_{T-1})' : (T - 1) \times 1$ and Ω_* is the first $(T - 1) \times (T - 1)$ submatrix of $\Omega = \Omega(1)$ with $\Omega(\alpha)$ defined in (8.5). Since it holds in the present model that

$$z \sim \mathrm{N}\left(0, \sigma^2 \Omega_*(\alpha)\right),$$

where $\Omega_*(\alpha)$ is the first $(T-1) \times (T-1)$ submatrix of $\Omega(\alpha)$, we can deduce from (10.13) and (10.17) that

$$\mathcal{L}\left(\frac{T}{T-1}\mathrm{SL}_{T2}^{(1)}\right) = \mathcal{L}\left(S_{T-1,2}^{(A)}\right),$$

where $S_{T2}^{(A)}$ is the LBIU statistic applied to Model A in the stationary case under the sample of size T. Thus we can immediately obtain $\mathcal{L}\left(\mathrm{SL}_{T2}^{(1)}\right)$ as $T \to \infty$ under $\alpha_0 = 1 - (c/T)$ and its limiting c.f. from Theorem 10.2.

The statistic $\mathrm{SL}_{T2}^{(2)}$ restricts δ to be 0 and was discussed in Section 9.10 of Chapter 9. The arguments there imply that $\mathrm{SL}_{T2}^{(2)}$ has the same limiting distribution as $S_{T2}^{(B)}$ described in Theorem 10.2. The statistic $\mathrm{SL}_{T2}^{(3)}$ restricts γ to be 0, while $\mathrm{SL}_{T2}^{(4)}$ does not impose any restriction. We can show that $\mathrm{SL}_{T2}^{(3)}$ and $\mathrm{SL}_{T2}^{(4)}$ have the same limiting distributions as $S_{T2}^{(C)}$ and $S_{T2}^{(D)}$, respectively, described in Theorem 10.2.

The above arguments are summarized in the following theorem.

Theorem 10.4. *Let us assume the model in differenced form given in (10.30) with μ unknown and consider the testing problem $H_0 : \alpha = 1$ versus $H_1 : \alpha < 1$. Then the LBIU statistics $\mathrm{SL}_{T2}^{(k)}$ ($k = 1, 2, 3, 4$) in (10.36) have the same limiting distributions as $S_{T2}^{(M)}$ ($M = A, B, C, D$), respectively, as $T \to \infty$ under $\alpha_0 = 1 - (c/T)$, where the limiting distributions of the latter are described in Theorem 10.2.*

10.4 PERFORMANCE OF THE LBI AND LBIU TESTS

We first compute the limiting powers of the LBI tests devised in the conditional case. The limiting power envelope is also computed for Model A. These are followed by the similar computation for the LBIU tests devised in the stationary case.

10.4.1 The Conditional Case

Let $S_{T1}^{(M)}$ be the LBI statistic for Model M = A, B, C, D, where $S_{T1}^{(A)} = S_{T1}$, $S_{T1}^{(B)} = 4S_{T1}$, $S_{T1}^{(C)} = 9S_{T1}/4$ and $S_{T1}^{(D)} = 9S_{T1}$ with S_{T1} defined in (10.6). Then it follows from Theorem 10.1 that the limiting powers of the LBI test at the $100\gamma\%$ level can be computed as

$$\beta_{M1}(\gamma) = \lim_{T \to \infty} P\left(S_{T1}^{(M)} \geq x_\gamma\right) \tag{10.37}$$

$$= P\left(Z^2 \geq \frac{x_\gamma}{a(M)}\right),$$

where x_γ is the upper $100\gamma\%$ point of $\chi^2(1)$, $Z^2 \sim \chi^2(1)$ and

$$a(A) = \frac{c^2 + 3c + 3}{3}, \qquad a(B) = \frac{2c^2 + 15c + 60}{60},$$

$$a(C) = \frac{9c^2 + 56c + 126}{126}, \qquad a(D) = \frac{3c^2 + 35c + 315}{315}.$$

Note that the tests are all consistent since $a(M) \to \infty$ as $c \to \infty$ so that $\beta_{M1}(\gamma) \to 1$. It can also be easily checked that, when $c > 0$,

$$a(A) > a(C) > a(B) > a(D).$$

Since the significance point x_γ is common to Models A through D, we can deduce that $\beta_{A1}(\gamma) > \beta_{C1}(\gamma) > \beta_{B1}(\gamma) > \beta_{D1}(\gamma)$.

Table 10.1 reports the limiting percent powers at the significance levels $\gamma = 0.01$, 0.05, and 0.1 for various values of c. The corresponding significance points are also

Table 10.1. Limiting Percent Powers of the LBI Tests

Model	x	$c = 1$	5	10	20	50	60
			$\gamma = 0.01$				
A	6.6349	9.17	49.63	69.89	83.57	93.10	94.22
B	6.6349	2.30	14.24	32.44	55.80	79.35	82.52
C	6.6349	3.64	24.97	46.78	67.79	85.61	87.87
D	6.6349	1.50	5.44	14.11	33.14	64.02	69.09
			$\gamma = 0.05$				
A	3.8415	19.95	60.47	76.85	87.46	94.75	95.60
B	3.8415	8.36	26.43	45.34	65.58	84.21	86.65
C	3.8415	11.14	38.11	58.07	75.20	89.03	90.75
D	3.8415	6.41	14.33	26.28	45.98	72.21	76.22
			$\gamma = 0.1$				
A	2.7055	28.16	66.40	80.49	89.47	95.59	96.31
B	2.7055	14.65	34.89	52.92	70.83	86.72	88.78
C	2.7055	18.16	46.23	64.29	79.08	90.78	92.24
D	2.7055	12.02	21.94	34.73	53.51	76.53	79.95

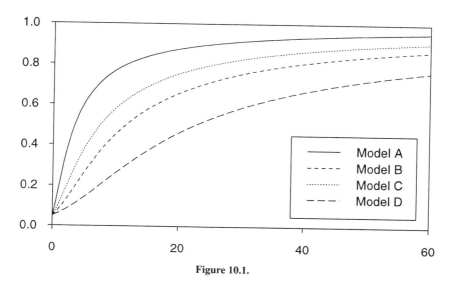

Figure 10.1.

reported under the heading x. Figure 10.1 shows graphs of the limiting powers at the 5% level. The LBI test in the present case performs better than the corresponding test for an AR unit root, as is seen from comparing Table 10.1 with Table 9.2. The relationship of the limiting powers among the four models, that is, $\beta_{A1}(\gamma) > \beta_{C1}(\gamma) > \beta_{B1}(\gamma) > \beta_{D1}(\gamma)$, is clearly seen in Figure 10.1.

Let us concentrate on Model A. In Figure 10.2 we draw the limiting power envelope at the 5% level computed from (10.12) together with the limiting power of the LBI test applied to Model A. Table 10.2 reports these two values for various values of c.

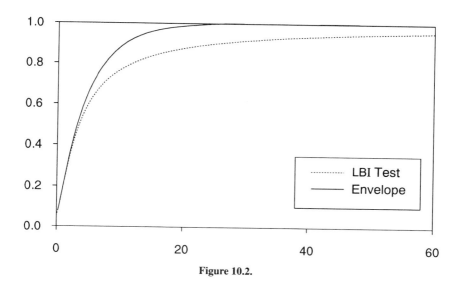

Figure 10.2.

Table 10.2. Limiting Percent Power Envelope at the 5% Level in Model A: Conditional Case

Test	c = 1	5	10	20	50	60
x	1.3577	1.0434	−0.3360	−3.7760	−15.7197	−19.9299
Envelope	20.16	66.31	88.29	98.59	100.00	100.00
LBI	19.95	60.47	76.85	87.46	94.75	95.60

The upper 5% points computed from (10.11) with θ replaced by c are also reported under the heading x. It is seen from Figure 10.2 that the LBI test performs quite well when c is small, but eventually becomes worse. There is some room for devising an invariant test that possesses higher powers than the LBI test when c is large. The same situation will emerge in the stationary case dealt with in the next subsection, where we consider the POI test with higher powers than the LBIU test when c is large.

10.4.2 The Stationary Case

The limiting powers of the LBIU test at the $100\gamma\%$ level can be computed as

$$(10.38) \quad \beta_{M2}(\gamma) = \lim_{T \to \infty} P\left(S_{T2}^{(M)} \geq x_\gamma^{(M)}\right)$$

$$= 1 - \frac{1}{\pi} \int_0^\infty \text{Re}\left[\frac{1 - \exp\{-i\theta x_\gamma^{(M)}\}}{i\theta} \phi_M(\theta)\right] d\theta,$$

where $S_{T2}^{(M)}$ is the LBIU statistic for Model M = A, B, C, D and $x_\gamma^{(M)}$ is the upper $100\gamma\%$ point of the limiting null distribution of $S_{T2}^{(M)}$, while $\phi_M(\theta)$ is the limiting c.f. of $S_{T2}^{(M)}$ as $T \to \infty$ under $\alpha_0 = 1 - (c/T)$. Note that $\phi_M(\theta)$ is defined by the right side of (10.25) with D replaced by D_M, where D_M is given in Theorem 10.2.

Figure 10.3 draws the limiting probability densities of $S_{T2}^{(M)}$ under H_0 for M = B, C, and D. The corresponding density of $S_{T2}^{(A)}$ is omitted here since it is already shown in Figure 1.2. It is noticed that the limiting null distributions of $S_{T2}^{(B)}$ and $S_{T2}^{(C)}$ are quite close to each other. Figure 10.4 draws the limiting power functions of the LBIU tests for Models A through D at the 5% level, while Table 10.3 reports the limiting percent powers at the significance levels $\gamma = 0.01, 0.05,$ and 0.1 for the same values of c as in Table 10.1. The corresponding percent points are also reported under the heading x. It is seen from Figure 10.4 that the limiting powers at the 5% level are highest for Model A and are lowest for Model D. The same is true at the other significance levels, as is seen from Table 10.3. The limiting powers for Models B and C are almost the same, unlike the conditional case. Hence we have $\beta_{A2}(\gamma) > \beta_{B2}(\gamma) \cong \beta_{C2}(\gamma) > \beta_{D2}(\gamma)$ in the present case.

Let us examine the power properties of the LBIU test, concentrating on Model A. Table 10.4 reports the finite sample percent powers at the 5% level for $T = 25$ and

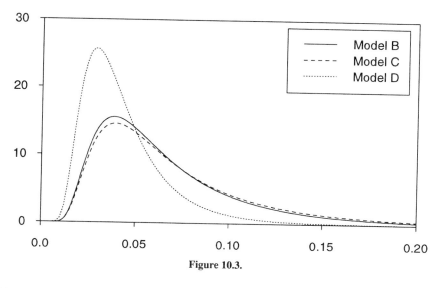

Figure 10.3.

50. These are computed from

$$P\left(S_{T2}^{(A)} \geq x\right) = P\left(\frac{1}{T}\frac{y'\Omega^{-2}y}{y'\Omega^{-1}y} \geq x\right)$$

$$= P\left(\sum_{j=1}^{T}\frac{1+\alpha^2-2\alpha\delta_j}{2-2\delta_j}\left(\frac{\frac{1}{T^2}}{2-2\delta_j} - \frac{x}{T}\right)Z_j^2 \geq 0\right),$$

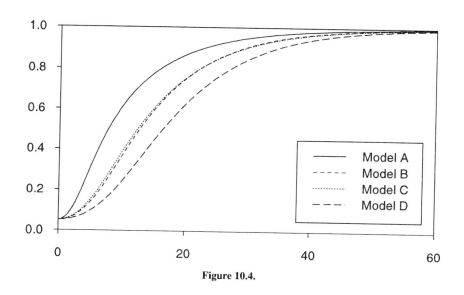

Figure 10.4.

Table 10.3. Limiting Percent Powers of the LBIU Tests

Model	x	c = 1	5	10	20	50	60
			$\gamma = 0.01$				
A	0.7435	1.41	17.72	47.42	78.10	97.68	98.80
B	0.2177	1.10	4.76	20.89	60.27	96.60	98.48
C	0.2472	1.12	5.49	22.95	60.85	96.22	98.24
D	0.1205	1.05	2.65	10.85	44.52	94.45	97.58
			$\gamma = 0.05$				
A	0.4614	6.20	31.10	61.06	86.65	99.09	99.59
B	0.1479	5.31	13.69	36.71	74.77	98.73	99.51
C	0.1642	5.36	14.86	38.62	75.08	98.59	99.44
D	0.0860	5.16	9.46	24.15	62.17	97.82	99.19
			$\gamma = 0.1$				
A	0.3473	11.74	40.19	68.80	90.61	99.53	99.80
B	0.1192	10.48	21.69	46.95	81.62	99.33	99.76
C	0.1305	10.55	23.01	48.68	81.90	99.26	99.73
D	0.0715	10.25	16.39	34.11	71.27	98.81	99.60

where $\delta_j = \cos(j\pi/(T+1))$ and x is the upper 5% point of the null distribution of $S_{T2}^{(A)}$. The value x is 0.48053 for $T = 25$ and 0.47121 for $T = 50$. We also report in Table 10.4 the finite sample powers of the MPI test computed from (10.26) with θ replaced by c, that is, the finite sample envelope. The corresponding upper 5% points of the null distributions of the MPI statistics $V_{T2}(\theta)$ in (10.26) are also shown under the heading x (Problem 4.1). It is seen from Table 10.4 that the LBIU test is as good as the MPI test only when α is close to 1. It is also observed that the powers of each test with the same values of $T(1-\alpha)$ are close to each other. In fact the powers depend only on $c = T(1-\alpha)$ asymptotically, as the results in Section 10.2 indicate.

Moving on to the limiting powers of the LBIU and MPI tests for Model A, we show in Figure 10.5 the limiting power function of the LBIU test and the limiting

Table 10.4. Finite Sample Percent Powers of the LBIU and MPI Tests for Model A at the 5% Level: Stationary Case

	T = 25				T = 50			
α	$T(1-\alpha)$	LBIU	MPI	x	$T(1-\alpha)$	LBIU	MPI	x
0.95	1.25	6.91	6.92	−0.5946	2.5	12.93	12.97	−0.5224
0.9	2.5	13.12	13.16	−0.4726	5	31.99	33.20	−0.5267
0.8	5	32.86	34.24	−0.4251	10	61.78	69.80	−1.6047
0.7	7.5	50.13	55.60	−0.8350	15	77.27	88.55	−3.3017
0.6	10	62.03	72.11	−1.5001	20	85.50	96.16	−5.3992
0.5	12.5	70.14	83.48	−2.3536	25	90.04	98.84	−7.8642

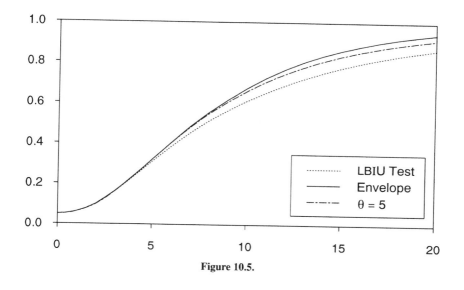

Figure 10.5.

power envelope at the 5% level. Also shown is the limiting power function of the POI test conducted at $\theta = 5$. Note that the POI test is not MPI for $\theta \neq c$; nonetheless the test conducted at $\theta = 5$ performs, as a whole, better than the LBIU test. Table 10.5 tabulates Figure 10.5 for selected values of c. The 5% significance points for the computation of the limiting power envelope are also reported under the heading x (Problem 4.2), together with the limiting percent powers of the POI tests conducted at $\theta = 7$ and 10, as well as at $\theta = 5$ (Problem 4.3). It is seen that the power of the POI ($\theta = 5$) test is tangent to the power envelope at a power of about 32%. We also observe that tangency at a power of 50% occurs at $\theta \doteq 7$. As was mentioned in the last chapter, the powers of the POI test become higher for large values of c as θ increases, by sacrificing, to some extent, the powers for small values of c. In comparison with Table 10.4, it is seen that the limiting powers can be used as a good approximation to the finite sample powers with the same value of $c = T(1 - \alpha)$.

Table 10.5. Limiting Percent Power Envelope at the 5% Level in Model A: Stationary Case

Test	$c = 1$	5	7	10	20	50
x	−0.5747	−0.6218	−0.9533	−1.6857	−4.9261	−16.7111
Envelope	6.20	32.13	48.18	67.39	94.47	99.99
LBIU	6.20	31.10	45.11	61.06	86.65	99.09
$\theta = 5$	6.15	32.13	47.87	65.91	91.65	99.79
$\theta = 7$	6.08	31.89	48.18	66.96	92.84	99.88
$\theta = 10$	5.97	31.00	47.76	67.39	93.76	99.93

PROBLEMS

4.1 Indicate how to obtain the percent points x reported in Table 10.4.
4.2 Describe the procedure for computing the power envelope reported in Table 10.5.
4.3 Explain how to compute the limiting powers of the POI test reported in Table 10.5.

10.5 SEASONAL UNIT ROOT TESTS

As in the AR unit root tests, the MA unit root tests can be easily extended to deal with the seasonal MA unit root testing problem. Let us consider the seasonal regression model:

$$(10.39) \qquad y_j = x_j'\beta + \varepsilon_j - \alpha_m \varepsilon_{j-m}, \qquad (j = 1, \ldots, T),$$

where α_m is a parameter on $[-1, 1]$, while m is a positive integer and is assumed to be a divisor of T so that $N = T/m$ is an integer. By specifying the regressor sequence $\{x_j\}$ we consider, as before, four simple models:

Model A: $\quad \mathbf{y}_j = \boldsymbol{\varepsilon}_j - \alpha_m \boldsymbol{\varepsilon}_{j-1},$

Model B: $\quad \mathbf{y}_j = \beta_0 + \boldsymbol{\varepsilon}_j - \alpha_m \boldsymbol{\varepsilon}_{j-1},$

Model C: $\quad \mathbf{y}_j = j\beta_1 + \boldsymbol{\varepsilon}_j - \alpha_m \boldsymbol{\varepsilon}_{j-1},$

Model D: $\quad \mathbf{y}_j = \beta_0 + j\beta_1 + \boldsymbol{\varepsilon}_j - \alpha_m \boldsymbol{\varepsilon}_{j-1},$

where $\mathbf{y}_j = (y_{(j-1)m+1}, y_{(j-1)m+2}, \ldots, y_{jm})' : m \times 1$ and $\boldsymbol{\varepsilon}_j$ is the $m \times 1$ vector defined similarly, while β_0 and β_1 are $m \times 1$ coefficient vectors.

In the following two subsections we derive the seasonal MA unit root tests for

$$(10.40) \qquad H_0 : \alpha_m = 1 \quad \text{versus} \quad H_1 : \alpha_m < 1,$$

allowing for two cases concerning the initial vector $\boldsymbol{\varepsilon}_0 = (\varepsilon_{1-m}, \varepsilon_{2-m}, \ldots, \varepsilon_0)' : m \times 1$. In the third subsection we compute the limiting powers of the tests together with the limiting power envelope specific to Model A in the two cases.

10.5.1 The Conditional Case

Assume that $\boldsymbol{\varepsilon}_0 = 0$ and $\boldsymbol{\varepsilon}_1, \boldsymbol{\varepsilon}_2, \ldots, \sim \text{NID}(0, \sigma^2 I_m)$. Then the seasonal model (10.39) may be rewritten as

$$(10.41) \qquad y = (\bar{X} \otimes I_m)\beta + \left(C^{-1}(\alpha_m) \otimes I_m\right)\varepsilon,$$

where $\bar{X}$ is an $N \times p$ matrix with rank p and $C(\alpha_m)$ is the $N \times N$ matrix defined in (8.36) with α_m replaced by α. Note that $\bar{X}$ does not enter into Model A, while

SEASONAL UNIT ROOT TESTS

$\bar{X} = e$ for Model B, $\bar{X} = d$ for Model C and $\bar{X} = (e, d)$ for Model D, where $e = (1, \ldots, 1)' : N \times 1$ and $d = (1, \ldots, N)'$.

The LM principle leads us (Problem 5.1) to derive the LBI test which rejects H_0 when

$$(10.42) \qquad S_{N1} = \frac{y'\{(C'\tilde{M}ee'\tilde{M}C) \otimes I_m\}y}{y'\{(C'\tilde{M}C) \otimes I_m\}y} > c,$$

where $C = C(1)$ is the $N \times N$ random walk generating matrix and

$$(10.43) \qquad \tilde{M} = I_N - C\bar{X}(\bar{X}'C'C\bar{X})^{-1}\bar{X}'C' = \tilde{M}^2.$$

It holds that

$$\mathcal{L}\left(S_{N1}/e'\tilde{M}e\right) \sim \mathrm{Be}\left(\frac{m}{2}, \frac{(N-p-1)m}{2}\right)$$

under H_0.

Let us consider the limiting distribution of S_{N1} as $N \to \infty$ under $\alpha_{m0} = 1 - (c/N) = 1 - (cm/T)$, where α_{m0} is the true value of α_m. For this purpose we rewrite S_{N1} as

$$S_{N1} = \frac{\varepsilon'\left[\left\{(C^{-1}(\alpha_{m0}))'C'\tilde{M}ee'\tilde{M}CC^{-1}(\alpha_{m0})\right\} \otimes I_m\right]\varepsilon}{\varepsilon'\left[\left\{(C^{-1}(\alpha_{m0}))'C'\tilde{M}CC^{-1}(\alpha_{m0})\right\} \otimes I_m\right]\varepsilon}$$

$$= \frac{\sum_{j,k=1}^{N} A_{jk}\varepsilon_j'\varepsilon_k}{\sum_{j,k=1}^{N} B_{jk}\varepsilon_j'\varepsilon_k},$$

where A_{jk} and B_{jk} are defined in a self-evident manner. Since A_{jk} and B_{jk} are invariant with respect to m, the following theorem is readily established from the weak convergence results for the case of $m = 1$.

Theorem 10.5. *Let $S_{N1}^{(M)}$ be the LBI statistic for Model $M = A, B, C, D$, where*

$$S_{N1}^{(A)} = mS_{N1}, \qquad S_{N1}^{(B)} = 4mS_{N1},$$
$$S_{N1}^{(C)} = 9mS_{N1}/4, \qquad S_{N1}^{(D)} = 9mS_{N1},$$

with S_{N1} defined in (10.42). Then it holds that, as $N \to \infty$ under $\alpha_{m0} = 1 - (c/N) = 1 - (cm/T)$,

$$\mathcal{L}\left(S_{N1}^{(A)}\right) \to \frac{c^2 + 3c + 3}{3}\chi^2(m),$$

$$\mathcal{L}\left(S_{N1}^{(B)}\right) \to \frac{2c^2 + 15c + 60}{60} \chi^2(m),$$

$$\mathcal{L}\left(S_{N1}^{(C)}\right) \to \frac{9c^2 + 56c + 126}{126} \chi^2(m),$$

$$\mathcal{L}\left(S_{N1}^{(D)}\right) \to \frac{3c^2 + 35c + 315}{315} \chi^2(m).$$

Let us concentrate on Model A and derive the limiting power envelope on the basis of the MPI test:

(10.44) $\quad H_0 : \alpha_m = 1 \quad$ versus $\quad H_1 : \alpha_m = 1 - \dfrac{\theta}{N} = 1 - \dfrac{m\theta}{T}.$

The MPI test rejects H_0 when

(10.45) $\quad V_{N1}(\theta) = T \dfrac{y' \left[\{C'C - C'(\alpha_m)C(\alpha_m)\} \otimes I_m\right] y}{y' \{(C'C) \otimes I_m\} y} > c,$

where $\theta = N(1 - \alpha_m)$. It follows from Section 8.4 of Chapter 8 that, as $N \to \infty$ under $\alpha_m = 1 - (\theta/N)$ and $\alpha_{m0} = 1 - (c/N)$,

(10.46) $\quad \mathcal{L}(V_{N1}(\theta)) \to \mathcal{L}\left(\int_0^1 \int_0^1 K(s,t;\theta) \, d\mathbf{w}'(s) \, d\mathbf{w}(t) - m\theta\right),$

where $K(s,t;\theta) = 2K_2(s,t;\theta)$ with K_2 defined in (8.40) and $\{\mathbf{w}(t)\}$ is the m-dimensional standard Brownian motion. In particular, we have (Problem 5.2), under $\alpha_m = 1 - (\theta/N)$ and $\alpha_{m0} = 1$,

$$\lim_{N \to \infty} P(V_{N1}(\theta) \le x) = \frac{1}{\pi} \int_0^\infty \operatorname{Re}\left[\frac{1 - e^{-iau}}{iu} \left(\cos\mu - \frac{\mu}{\theta} \sin\mu\right)^{-m/2} e^\theta\right] du,$$

(10.47)

where $a = (m\theta + x)/\theta^2$ and $\mu = \sqrt{2iu - \theta^2}$.

The limiting power envelope at the $100\gamma\%$ level is now obtained (Problem 5.3) as

$$\lim_{N \to \infty} P(V_{N1}(c) \ge x) = 1 - \frac{1}{\pi} \int_0^\infty \operatorname{Re}\left[\frac{1 - e^{-ibu}}{iu} \left(\cos\nu - \frac{\nu}{c} \sin\nu\right)^{-m/2}\right] du,$$

(10.48)

where the limit is taken under $\alpha_m = \alpha_{m0} = 1 - (c/N)$, $b = (cm + x)/c^2$, $\nu = \sqrt{2iu}$, and x is the upper $100\gamma\%$ point of the limiting distribution in (10.47) with $\theta = c$.

10.5.2 The Stationary Case

Here we assume that $\varepsilon_0, \varepsilon_1, \ldots, \sim \text{NID}(0, \sigma^2 I_m)$. Then the seasonal model (10.39) may be rewritten as

$$(10.49) \qquad y = (\bar{X} \otimes I_m)\beta + (B(\alpha_m) \otimes I_m)\varepsilon^*,$$

where $\varepsilon^* = (\varepsilon_0', \varepsilon')' : (T+m) \times 1$ and $B(\alpha_m) = (-\alpha_m e_1, C^{-1}(\alpha_m)) : N \times (N+1)$ with $e_1 = (1, 0, \ldots, 0)' : N \times 1$.

We can show (Problem 5.4) that the test which rejects $H_0 : \alpha_m = 1$ when

$$(10.50) \qquad S_{N2} = \frac{m}{N} \frac{y'\{(\tilde{N}'\Omega^{-2}\tilde{N}) \otimes I_m\}y}{y'\{(\tilde{N}'\Omega^{-1}\tilde{N}) \otimes I_m\}y} > c$$

is LBIU, where $\Omega = \Omega(1)$ with $\Omega(\alpha) = B(\alpha)B'(\alpha)$ and

$$(10.51) \qquad \tilde{N} = I_N - \bar{X}(\bar{X}'\Omega^{-1}\bar{X})^{-1}\bar{X}'\Omega^{-1}.$$

Let us derive the limiting distribution of S_{N2} as $N \to \infty$ under $\alpha_{m0} = 1 - (c/N) = 1 - (cm/T)$. We first have

$$\frac{1}{T} y' \left[(\tilde{N}'\Omega^{-1}\tilde{N}) \otimes I_m\right] y \to \sigma^2$$

in probability. We also have

$$\mathcal{L}\left(y' \left[(\tilde{N}'\Omega^{-2}\tilde{N}) \otimes I_m\right] y\right) = \mathcal{L}\left(\varepsilon^{*'} \left[\{B'(\alpha_m)\tilde{N}'\Omega^{-2}\tilde{N}B(\alpha_m)\} \otimes I_m\right] \varepsilon^*\right)$$

$$= \mathcal{L}\left(\varepsilon' \left[\{\Omega^{-1}\tilde{N}\Omega(\alpha_m)\tilde{N}'\Omega^{-1}\} \otimes I_m\right] \varepsilon\right).$$

Then, using the decomposition described in (10.22), we can establish the following theorem.

Theorem 10.6. *Let $S_{N2}^{(M)}$ be the LBIU statistic for Model $M = A, B, C, D$ constructed from S_{N2} in (10.50). Then it holds that, as $N \to \infty$ under $\alpha_{m0} = 1 - (c/N) = 1 - (cm/T)$,*

$$\mathcal{L}(S_{N2}^{(M)}) \to \mathcal{L}\left(\int_0^1 \int_0^1 \left(K_M(s,t) + c^2 K_{M(2)}(s,t)\right) d\mathbf{w}'(s) d\mathbf{w}(t)\right),$$

where $\{\mathbf{w}(t)\}$ is the m-dimensional standard Brownian motion, while the kernels K_M ($M = A, B, C, D$) are defined in Theorem 10.2.

The limiting power envelope associated with Model A can be derived easily, as in the conditional case. The MPI test for $H_0 : \alpha_m = 1$ versus $H_1 : \alpha_m = 1 - (\theta/N)$

rejects H_0 when

$$(10.52) \qquad V_{N2}(\theta) = T \frac{y' \left[\{\Omega^{-1} - \Omega^{-1}(\alpha_m)\} \otimes I_m \right] y}{y' \left(\Omega^{-1} \otimes I_m \right) y} > c,$$

where $\theta = N(1 - \alpha_m)$. It follows from Section 8.4 of Chapter 8 that, as $N \to \infty$ under $\alpha_m = 1 - (\theta/N)$ and $\alpha_{m0} = 1 - (c/N)$,

$$(10.53) \qquad \mathcal{L}(V_{N2}(\theta)) \to \mathcal{L} \left(\theta^2 \sum_{n=1}^{\infty} \frac{n^2 \pi^2 + c^2}{n^2 \pi^2 (n^2 \pi^2 + \theta^2)} Z'_n Z_n - m\theta \right),$$

where $\{Z_n\} \sim \text{NID}(0, I_m)$. In particular we have (Problem 5.5), under $\alpha_m = 1 - (\theta/N)$ and $\alpha_{m0} = 1$,

$$\lim_{N \to \infty} P\left(V_{N2}(\theta) \leq x \right) = \frac{1}{\pi} \int_0^{\infty} \text{Re} \left[\frac{1 - e^{-iau}}{iu} \left(\frac{\sin \mu}{\mu} \Big/ \frac{\sinh \theta}{\theta} \right)^{-m/2} \right] du,$$

(10.54)

where $a = (m\theta + x)/\theta^2$ and $\mu = \sqrt{2iu - \theta^2}$.

The limiting power envelope at the $100\gamma\%$ level is now obtained (Problem 5.6) as

$$\lim_{N \to \infty} P\left(V_{N2}(c) \geq x \right) = 1 - \frac{1}{\pi} \int_0^{\infty} \text{Re} \left[\frac{1 - e^{-ibu}}{iu} \left(\frac{\sin \sqrt{2iu}}{\sqrt{2iu}} \right)^{-m/2} \right] du,$$

(10.55)

where the limit is taken under $\alpha_m = \alpha_{m0} = 1 - (c/N)$, $b = (cm + x)/c^2$, and x is the upper $100\gamma\%$ point of the limiting distribution in (10.54) with $\theta = c$.

10.5.3 Power Properties

Here we first compute the limiting powers of the LBI tests for Models A through D when the period m is 4 or 12. We then compute the limiting power envelope specific to Model A. The computations associated with the LBIU tests follow similarly.

Let $\beta_{M1}^{(m)}(\gamma)$ be the limiting power of the LBI test at the $100\gamma\%$ level applied to Model M ($=$ A, B, C, D) with period m. Then it follows from Theorem 10.5 that

$$(10.56) \qquad \beta_{M1}^{(m)}(\gamma) = P\left(\chi_m^2 \geq \frac{x_\gamma(m)}{a(M)} \right),$$

where $x_\gamma(m)$ is the upper $100\gamma\%$ point of $\chi^2(m)$, $\chi_m^2 \sim \chi^2(m)$ and $a(M)$ is defined in (10.37). We can deduce from the arguments below (10.37) that $\beta_{A1}^{(m)}(\gamma) > \beta_{C1}^{(m)}(\gamma) > \beta_{B1}^{(m)}(\gamma) > \beta_{D1}^{(m)}(\gamma)$ for each m.

Table 10.6. Limiting Percent Powers of the LBI Tests at the 5% Level for $m = 4, 12$

Model	$c = 1$	2	$m = 4$ 5	10	15	20
A	39.71	70.10	95.59	99.47	99.87	99.95
B	11.65	21.39	54.50	84.62	94.18	97.44
C	18.06	35.91	75.51	94.45	98.25	99.30
D	7.59	11.05	25.89	54.17	74.08	85.30

Model	$c = 0.5$	1	$m = 12$ 2	5	8	10
A	34.90	70.20	96.27	99.99	100.00	100.00
B	9.99	17.43	37.83	86.93	98.16	99.50
C	15.13	30.90	64.50	97.96	99.88	99.98
D	6.95	9.44	16.19	46.82	75.19	86.65

Table 10.6 reports $\beta_{M1}^{(m)}(\gamma)$ at the level $\gamma = 0.05$, where the 5% significance points are 9.4877 for $m = 4$ and 21.0261 for $m = 12$. Figure 10.6 shows the limiting power functions at $\gamma = 0.05$ and $m = 4$, Figure 10.7 at $\gamma = 0.05$ and $m = 12$. Comparing Table 10.6 with Table 10.1 for $m = 1$, we observe that the powers become higher for each model as m increases. In particular, the power function for Model A approaches unity very quickly when m is large, as Figures 10.6 and 10.7 demonstrate. Figure 10.8 draws the power envelope as well as the power function of the LBI test at the 5% level applied to Model A with $m = 4$. It is seen that the LBI test attains nearly the highest possible power. This feature is more clearly seen in Figure 10.9, where the corresponding graphs for $m = 12$ are drawn.

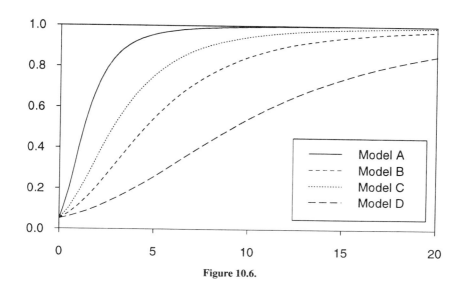

Figure 10.6.

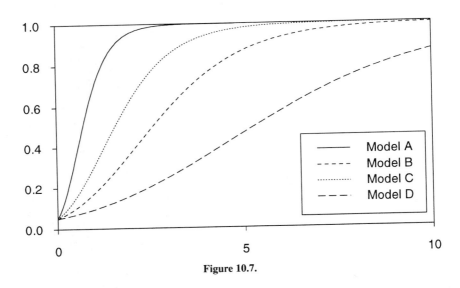

Figure 10.7.

We next deal with the LBIU test. Let $\beta_{M2}^{(m)}(\gamma)$ be the limiting power at the $100\gamma\%$ level applied to Model M ($=$ A, B, C, D) with period m. Then it follows from Theorem 10.6 that

$$(10.57) \quad \beta_{M2}^{(m)}(\gamma) = 1 - \frac{1}{\pi} \int_0^\infty \text{Re}\left[\frac{1 - \exp\left\{-i\theta x_\gamma^{(M)}(m)\right\}}{i\theta} (\phi_M(\theta))^m\right] d\theta,$$

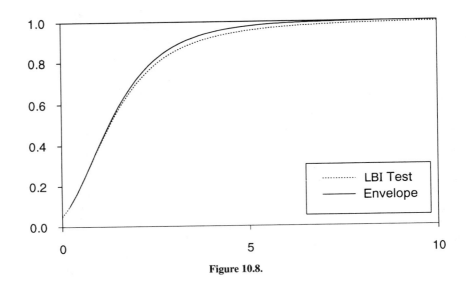

Figure 10.8.

SEASONAL UNIT ROOT TESTS

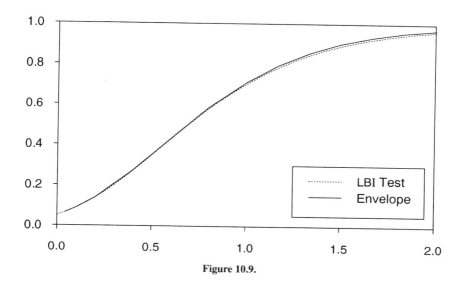

Figure 10.9.

where $x_\gamma^{(M)}(m)$ is the upper $100\gamma\%$ point of the limiting null distribution of the LBIU statistic in (10.50) for Model M with period m, while $\phi_M(\theta)$ is defined by the right side of (10.25) with D replaced by D_M given in Theorem 10.2.

Figure 10.10 draws the limiting probability densities of the LBIU statistic under H_0 for Model A with $m = 4$ and 12. The density is continually shifted to the right as m increases. A similar feature is observed for the other models though not shown here. Figure 10.11 draws the limiting power functions at the 5% level when $m = 4$, while the corresponding functions for $m = 12$ are presented in Figure 10.12. The

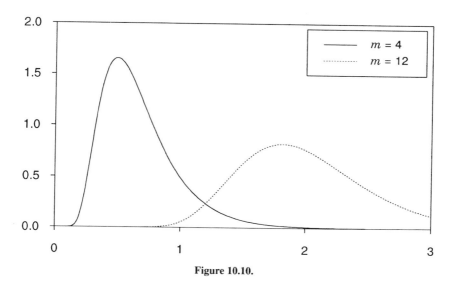

Figure 10.10.

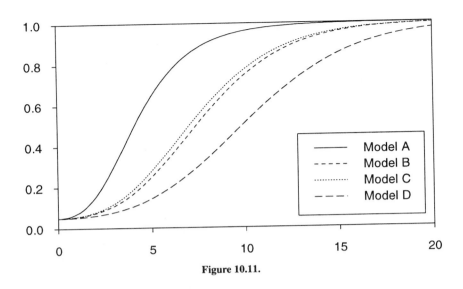

Figure 10.11.

power performance as m increases is of similar nature to that in the LBI test. The relationship $\beta_{A2}^{(m)}(\gamma) > \beta_{B2}^{(m)}(\gamma) \cong \beta_{C2}^{(m)}(\gamma) > \beta_{D2}^{(m)}(\gamma)$ holds for $m = 4$ and 12, as in $m = 1$. Table 10.7 tabulates Figures 10.11 and 10.12 for selected values of c, together with the 5% significance points under the heading x. Figures 10.13 and 10.14 draw, for $m = 4$ and 12, respectively, the power envelopes as well as the power functions of the LBIU tests at the 5% level applied to Model A. A similar remark to that given for the LBI test applies equally here.

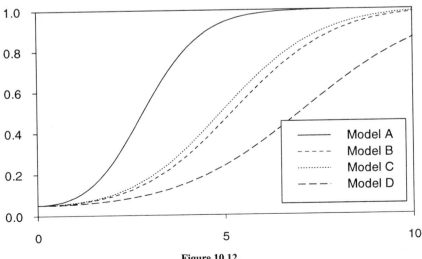

Figure 10.12.

Table 10.7. Limiting Percent Powers of the LBIU Tests at the 5% Level for $m = 4, 12$

Model	x	$c = 1$	$m = 4$ 2	5	10	15	20
A	1.2373	7.20	15.34	64.71	96.28	99.62	99.96
B	0.4227	5.56	7.40	24.85	74.96	95.70	99.44
C	0.4647	5.65	7.80	27.79	77.71	96.13	99.47
D	0.2538	5.28	6.19	14.43	51.24	85.11	97.06

Model	x	$c = 0.5$	$m = 12$ 1	2	5	8	10
A	2.9422	5.83	8.76	25.38	94.33	99.91	100.00
B	1.0592	5.22	5.91	9.19	46.16	90.04	98.29
C	1.1541	5.25	6.05	9.92	51.75	92.52	98.80
D	0.6517	5.09	5.44	7.00	24.01	63.69	85.91

PROBLEMS

5.1 Show that (10.42) is the LBI statistic for the testing problem (10.40).

5.2 Show that the limiting distribution of $V_{N1}(\theta)$ in (10.45) as $N \to \infty$ under $\alpha_m = 1 - (\theta/N)$ and $\alpha_{m0} = 1$ is given by (10.47).

5.3 Establish (10.48).

5.4 Show that (10.50) is the LBIU statistic for the testing problem (10.40).

5.5 Show that the limiting distribution of $V_{N2}(\theta)$ in (10.52) as $N \to \infty$ under $\alpha_m = 1 - (\theta/N)$ and $\alpha_{m0} = 1$ is given by (10.54).

5.6 Establish (10.55).

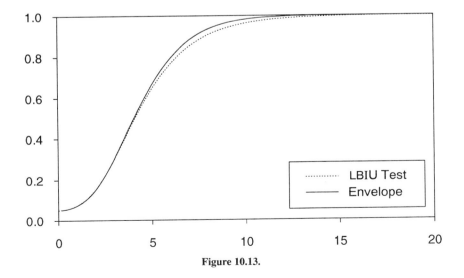

Figure 10.13.

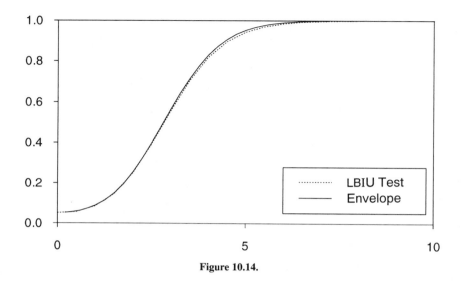

Figure 10.14.

10.6 UNIT ROOT TESTS IN THE DEPENDENT CASE

As in the AR unit root tests, we extend the time series regression model dealt with in previous sections to

$$(10.58) \quad y_j = x_j'\beta + u_j - \alpha u_{j-1}, \quad (j = 1, \ldots, T),$$

where $\{u_j\}$ is a dependent process generated by

$$(10.59) \quad u_j = \sum_{l=0}^{\infty} \phi_l \varepsilon_{j-l}, \quad \sum_{l=0}^{\infty} l|\phi_l| < \infty, \quad \sum_{l=0}^{\infty} \phi_l \neq 0,$$

with $\{\varepsilon_j\} \sim \text{NID}(0, \sigma^2)$. As for the initial value u_0 in (10.58) we consider two cases. One is the conditional case where $u_0 = 0$, and the other is the stationary case where u_0 has the same marginal distribution as $u_1, u_2, \ldots$. For both cases, the parameter space is restricted to be $-1 \leq \alpha \leq 1$. It follows that

$$(10.60) \quad y \sim \text{N}\left(X\beta, \sigma^2 C^{-1}(\alpha)\Lambda(\theta)\left(C^{-1}(\alpha)\right)'\right): \quad \text{conditional case},$$

$$(10.61) \quad y \sim \text{N}\left(X\beta, \sigma^2 D(\alpha)\Lambda^*(\theta)D'(\alpha)\right): \quad \text{stationary case},$$

where $C(\alpha)$ is the $T \times T$ matrix defined in (8.36), $\Lambda(\theta) = V(u)/\sigma^2$ with $u = (u_1, \ldots, u_T)'$, $D(\alpha) = (-\alpha e_1, C^{-1}(\alpha)) : T \times (T+1)$ with $e_1 = (1, 0, \ldots, 0)' : T \times 1$, and $\Lambda^*(\theta) = V(u^*)/\sigma^2$ with $u^* = (u_0, u')' : (T+1) \times 1$.

In the following subsections, we consider testing $H_0 : \alpha = 1$ versus $H_1 : \alpha < 1$, dealing first with the conditional case followed by the stationary case.

10.6.1 The Conditional Case

Let us first act as if $\{u_j\} = \{\varepsilon_j\}$, where $u_0 = \varepsilon_0 = 0$. Then we can devise the LBI tests suggested in Section 10.2, whose statistics depend on the regressor sequence $\{x_j\}$. We consider Models A through D in the same way as before by specifying $\{x_j\}$. Then we have the following theorem.

Theorem 10.7. *Let $S_{T1}^{(M)}$ be the LBI statistic when $\{u_j\} = \{\varepsilon_j\}$ with $u_0 = \varepsilon_0 = 0$ for Model $M = A, B, C, D$, where each $S_{T1}^{(M)}$ is defined in Theorem 10.1. Then it holds in the present case that, as $T \to \infty$ under $\alpha = 1 - (c/T)$,*

$$\mathcal{L}\left(S_{T1}^{(M)}\right) \to \frac{1}{r} a(M) \chi^2(1),$$

where $a(M)$ is defined in (10.37), while

$$r = \frac{\sigma_S^2}{\sigma_L^2}, \qquad \sigma_S^2 = \sigma^2 \sum_{l=0}^{\infty} \phi_l^2, \qquad \sigma_L^2 = \sigma^2 \left(\sum_{l=0}^{\infty} \phi_l\right)^2.$$

The nuisance parameters σ_S^2 and σ_L^2 may be consistently estimated as

$$\tilde{\sigma}_S^2 = \frac{1}{T} \tilde{u}' \tilde{u},$$

$$\tilde{\sigma}_L^2 = \tilde{\sigma}_S^2 + \frac{2}{T} \sum_{i=1}^{l} \left(1 - \frac{i}{l+1}\right) \sum_{j=i+1}^{T} \tilde{u}_j \tilde{u}_{j-i},$$

where $\tilde{u} = (\tilde{u}_1, \ldots, \tilde{u}_T)' = C(I_T - X(X'C'CX)^{-1}X'C'C)y$ and l is a lag truncation number such that $l \to \infty$ and $l/T \to 0$ as $T \to \infty$. Then we can conduct a test based on $\tilde{r} S_{T1}^{(M)}$ with $\tilde{r} = \tilde{\sigma}_S^2 / \tilde{\sigma}_L^2$. The statistic $\tilde{r} S_{T1}^{(M)}$ will have the same limiting distribution as described in Theorem 10.1 as $T \to \infty$ under $\alpha = 1 - (c/T)$.

Under finite samples, however, the statistic suffers size distortions, as in the AR unit root tests. Figure 10.15 shows the sampling null distributions of $\tilde{r} S_{T1}^{(M)}$ for Model A: $y_j = u_j - \alpha u_{j-1}$ with $u_0 = 0$ and $u_j = \varepsilon_j - \phi \varepsilon_{j-1}$ ($j \geq 1$), where three cases of $\phi = -0.8, 0,$ and 0.8 are examined with $T = 100$ for all cases. The truncation number l used is 4. The limiting null distribution is also shown. The distribution with $\phi = 0.8$ is seriously distorted, yielding nominal significance levels biased downward. Thus the null hypothesis is rarely rejected at the nominal level.

The usual remedy to overcome size distortions is to assume $\{u_j\}$ to follow an ARMA process of finite order, as in the AR unit root tests. Some attempts have been made in the stationary case discussed next.

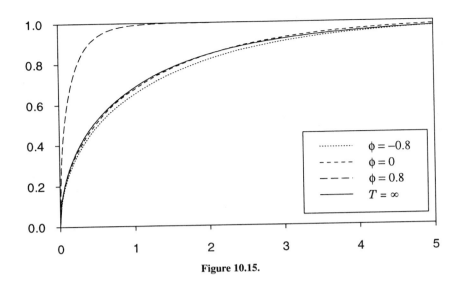

Figure 10.15.

10.6.2 The Stationary Case

Acting as if $\{u_j\} = \{\varepsilon_j\}$ in (10.58), we can devise the LBIU tests suggested in Section 10.2 for Models A through D. Then we have the following theorem.

Theorem 10.8. *Let $S_{T2}^{(M)}$ be the LBIU statistic when $\{u_j\} = \{\varepsilon_j\}$ for Model $M = A$, B, C, D, where each $S_{T2}^{(M)}$ is constructed from (10.17). Then it holds in the present case that, as $T \to \infty$ under $\alpha = 1 - (c/T)$,*

$$\mathcal{L}\left(S_{T2}^{(M)}\right) \to \mathcal{L}\left(\frac{1}{r}\int_0^1\int_0^1 \left(K_M(s,t) + c^2 K_{M(2)}(s,t)\right) dw(s)\, dw(t)\right),$$

where r is defined in Theorem 10.7, while the kernels K_M ($M = A, B, C, D$) are defined in Theorem 10.2.

A consistent estimator of r may be obtained as $\hat{r} = \hat{\sigma}_S^2/\hat{\sigma}_L^2$, where

$$\hat{\sigma}_S^2 = \frac{1}{T}\hat{u}'\hat{u},$$

$$\hat{\sigma}_L^2 = \hat{\sigma}_S^2 + \frac{2}{T}\sum_{i=1}^{l}\left(1 - \frac{i}{l+1}\right)\sum_{j=i+1}^{T}\hat{u}_j\hat{u}_{j-i},$$

$$\hat{u} = (\hat{u}_1, \ldots, \hat{u}_T)' = H\left(I_T - X(X'\Omega^{-1}X)^{-1}X'\Omega^{-1}\right)y,$$

with H defined in (10.18) and l being a lag truncation number. Note that the vector $\hat{u}$ is constructed differently from $\tilde{u}$ used in the conditional case. The Cholesky

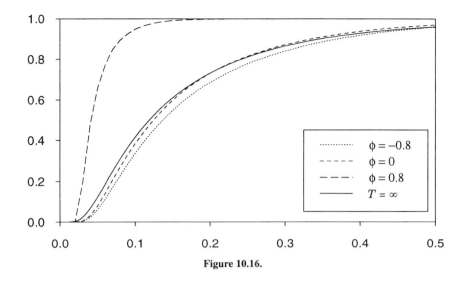

Figure 10.16.

decomposition matrix H plays an important role here. If we use

$$\tilde{u} = C(I_T - X(X'C'CX)^{-1}X'C'C)y,$$

instead of $\hat{u}$, then the resulting estimators of σ_S^2 and σ_L^2 are inconsistent. In fact, for Model A, $y_j = u_j - \alpha u_{j-1}$ with $\alpha = 1$, we have $\tilde{u}_j = y_1 + \cdots + y_j = u_j - u_0$, so that $\sum_{j=1}^{T} \tilde{u}_j^2/T \to \sigma_S^2 + u_0^2$ in probability. We now conduct a test based on $\hat{r} S_{T2}^{(M)}$ with $\hat{r} = \hat{\sigma}_S^2 / \hat{\sigma}_L^2$. The statistic $\hat{r} S_{T2}^{(M)}$ will have the same limiting distribution as described in Theorem 10.2 as $T \to \infty$ under $\alpha = 1 - (c/T)$.

The statistic, however, suffers size distortions under finite samples. Figure 10.16 shows the sampling null distributions of $\hat{r} S_{T2}^{(M)}$ for Model A with $u_j = \varepsilon_j - \phi \varepsilon_{j-1}$, where three cases of $\phi = -0.8, 0,$ and 0.8 are examined with $T = 100$ for all cases. The value of l used is 4. The limiting null distribution is also shown. The same phenomena as in the conditional case described in the last subsection are observed here.

The MA unit root tests constructed from parameterizing $\{u_j\}$ as an ARMA process of finite order may be found in Saikkonen and Luukkonen (1993a,b), Tsay (1993), and Breitung (1994).

10.7 THE RELATIONSHIP WITH TESTING IN THE STATE SPACE MODEL

As was indicated in Chapter 8, the estimation for noninvertible MA models is closely related to that for the state space models. Here we show that the relationship is carried over to testing problems.

Let us consider the state space model

(10.62)
$$y_j = x_{1j}\beta_j + x_{2j}'\gamma + \varepsilon_j,$$
$$\beta_j = \beta_{j-1} + \xi_j, \qquad (j = 1, \ldots, T),$$

where

(i) $\{x_{1j}\}$ and $\{x_{2j}\}$ are scalar and $p \times 1$ nonstochastic, fixed sequences, respectively;
(ii) $\{\varepsilon_j\} \sim \text{NID}(0, \sigma_\varepsilon^2)$ with $\sigma_\varepsilon^2 > 0$ and $\{\xi_j\} \sim \text{NID}(0, \sigma_\xi^2)$ with $\sigma_\xi^2 \geq 0$, and these are independent of each other;
(iii) γ is a $p \times 1$ unknown constant vector, whereas $\{\beta_j\}$ follows a random walk starting from an unknown constant $\beta_0 = \mu$.

The testing problem dealt with here is to test

(10.63) $\qquad H_0 : \rho = \dfrac{\sigma_\xi^2}{\sigma_\varepsilon^2} = 0 \quad \text{versus} \quad H_1 : \rho > 0.$

Note that each β_j reduces to a constant μ under H_0, while it follows a random walk under H_1. Thus this is a test for the constancy of a regression coefficient against the random walk alternative. The present problem was initially discussed in Nyblom and Mäkeläinen (1983) and Tanaka (1983b), which was generalized by Nabeya and Tanaka (1988) and Nabeya (1989). A simplified version was earlier presented in Section 1.1 of Chapter 1.

Let us first derive the LBI test for (10.63). Note that

(10.64) $\qquad y = x_1\mu + X_2\gamma + DC\xi + \varepsilon,$

where

$x_1 = (x_{11}, \ldots, x_{1T})' : T \times 1, X_2 = (x_{21}, \ldots, x_{2T})' : T \times p, D = \text{diag}(x_{11}, \ldots, x_{1T}),$

and C is the random walk generating matrix. Then the LM principle yields the LBI test, which rejects H_0 when

(10.65) $\qquad R_T = \dfrac{y'MDCC'DMy}{y'My} > c,$

where

(10.66) $\qquad M = I_T - (x_1, X_2)\big((x_1, X_2)'(x_1, X_2)\big)^{-1}(x_1, X_2)'$

with $\text{rank}(x_1, X_2) = p + 1$.

Nabeya and Tanaka (1988) discussed the asymptotic distribution of R_T under H_0 for the following simple cases:

(I) $\quad y_j = j^m \beta_j + \varepsilon_j, \qquad (m > -\tfrac{1}{2}),$
(II) $\quad y_j = \beta_j + j^m \gamma + \varepsilon_j, \qquad (m = 1, 2, 3, 4),$
(III) $\quad y_j = \beta_j + j\gamma + j^2\delta + \varepsilon_j,$

where $\beta_j = \beta_{j-1} + \xi_j$ with $\beta_0 = \mu$ unknown for all cases.

THE RELATIONSHIP WITH TESTING IN THE STATE SPACE MODEL   407

To examine the relationship with MA(1) models the above models may be expressed in differenced form as follows:

(I) $y_1 = \mu + \xi_1 + \varepsilon_1$,
  $\Delta y_j = j^m \beta_j - (j-1)^m \beta_{j-1} + \Delta \varepsilon_j$, $\left(m > -\tfrac{1}{2}\right)$,

(II) $y_1 = \mu + \gamma + \xi_1 + \varepsilon_1$,
  $\Delta y_j = (j^m - (j-1)^m)\gamma + \xi_j + \Delta \varepsilon_j$, $(m = 1, 2, 3, 4)$,

(III) $y_1 = \mu + \gamma + \delta + \xi_1 + \varepsilon_1$,
  $\Delta y_j = \gamma + (2j-1)\delta + \xi_j + \Delta \varepsilon_j$,

where $j = 2, \ldots, T$ for all cases. It is noticeable that Case (III) looks like the model (10.30) reproduced here as

(10.67) $y_1 = \mu + \gamma + \delta + \varepsilon_1$,
  $\Delta y_j = \gamma + (2j-1)\delta + \varepsilon_j - \alpha \varepsilon_{j-1}$, $(j = 2, \ldots, T)$,

apart from the error term. Case (I) with $m = 0$ and Case (II) with $m = 1$ and 2 are restricted cases of (III) in the sense that γ and/or δ are assumed to be 0; hence these cases also look like restricted versions of (10.67). Note, however, that μ is assumed to be unknown.

In the following subsections we discuss the asymptotic distribution of the LBI statistic R_T in (10.65) applied to Cases (I), (II), and (III) under a sequence of local alternatives, from which we deduce the asymptotic equivalence of the LBI tests applied to Case (I) with $m = 0$, Case (II) with $m = 1, 2$, and Case (III) with the LBIU unit root tests applied to Models A through D in the stationary case.

A question naturally arises whether the LBI tests in the case where μ is known are asymptotically equivalent to the LBI unit root tests in the conditional case. The answer is in the negative, which we shall show in Section 10.7.4.

10.7.1 Case (I)

We define the LBI statistic in the present case by

(10.68) $R_{Tm}^{(\mathrm{I})} = \dfrac{1}{T^{2m+1}} \dfrac{y'MDCC'DMy}{y'My}$,

where $M = I_T - x_1 x_1'/x_1'x_1$ with $x_1 = (1^m, \ldots, T^m)'$ and $D = \mathrm{diag}(1^m, \ldots, T^m)$.
We can first show (Problem 7.1) that, as $T \to \infty$ under $\rho = c^2/T^{2m+2}$,

(10.69) $\dfrac{1}{T} y'My \to \sigma_\varepsilon^2$ in probability.

Noting that $My = M(DC\xi + \varepsilon)$ and $DC\xi + \varepsilon \sim N\left(0, \sigma_\varepsilon^2(\rho DCC'D + I_T)\right)$ let us consider, under $\rho = c^2/T^{2m+2}$,

$$\mathcal{L}(y'MDCC'DMy/\sigma_\varepsilon^2) = \mathcal{L}\left(Z'(\rho DCC'D + I_T)^{\frac{1}{2}} MDCC'DM(\rho DCC'D + I_T)^{\frac{1}{2}} Z\right)$$

$$= \mathcal{L}\left(Z'\left(C'DMDC + \frac{c^2}{T^{2m+2}}(C'DMDC)^2\right) Z\right),$$

where $Z \sim N(0, I_T)$.

The Fredholm approach developed in Chapter 5 yields (Problem 7.2) the following theorem.

Theorem 10.9. *Consider the model*

$$y_j = j^m \beta_j + \varepsilon_j, \quad \beta_j = \beta_{j-1} + \xi_j, \quad \left(m > -\frac{1}{2}\right),$$

where μ ($= \beta_0$) is an unknown constant. Then it holds that, as $T \to \infty$ under $\rho = c^2/T^{2m+2}$,

(10.70) $\quad \mathcal{L}\left(R_{Tm}^{(I)}\right) \to \mathcal{L}\left(\int_0^1 \int_0^1 \left(K(s,t;m) + c^2 K_{(2)}(s,t;m)\right) dw(s)\,dw(t)\right),$

where $R_{Tm}^{(I)}$ is the LBI statistic in (10.68) and

$$K(s,t;m) = \frac{1}{2m+1}\left[(\min(s,t))^{2m+1} - (st)^{2m+1}\right],$$

$$K_{(2)}(s,t;m) = \int_0^1 K(s,u;m)K(u,t;m)\,du.$$

Moreover, the limiting c.f. of $R_{Tm}^{(I)}$ is given by

$$\phi(\theta; m) = \left[D(i\theta + \sqrt{-\theta^2 + 2ic^2\theta}\,; m) D(i\theta - \sqrt{-\theta^2 + 2ic^2\theta}\,; m)\right]^{-1/2},$$

(10.71)

where $D(\lambda; m)$ is the FD of $K(s,t;m)$ defined by

$$D(\lambda; m) = \Gamma\left(\frac{4m+3}{2(m+1)}\right) J_{(2m+1)/2(m+1)}\left(\frac{\sqrt{\lambda}}{m+1}\right) \Big/ \left(\frac{\sqrt{\lambda}}{2(m+1)}\right)^{(2m+1)/2(m+1)}$$

with $J_\nu(z)$ being the Bessel function of the first kind defined in (5.50).

Note that we have already encountered in (5.64) the FD appearing in the above theorem. The local alternative we have chosen here is $\rho = c^2/T^{2m+2}$. Nabeya (1989)

THE RELATIONSHIP WITH TESTING IN THE STATE SPACE MODEL

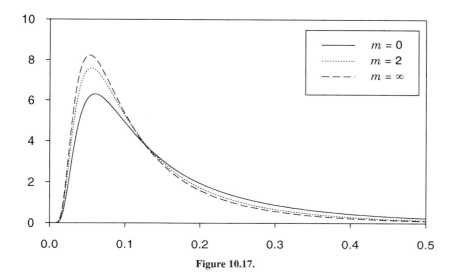

Figure 10.17.

examines some other sequences of ρ, although only the present choice of ρ makes the limiting distribution reduce to the limiting null distribution when $c = 0$. We now recognize from this theorem and Theorem 10.2 that $R_{Tm}^{(I)}$ with $m = 0$ has the same limiting distribution under $\rho = c^2/T^2$ as the LBIU unit root statistic under $\alpha = 1 - (c/T)$ applied to Model A. The statistic $R_{Tm}^{(I)}$ has no counterpart for the other values of m.

Figure 10.17 shows the limiting probability densities of $(m + 1)^2 R_{Tm}^{(I)}$ under H_0 for $m = 0$ and 2. Also shown is the limiting density of these as $m \to \infty$, whose c.f. is given by $(2J_1(\sqrt{2i\theta})/\sqrt{2i\theta})^{-1/2}$. Table 10.8 reports the limiting percent powers of the LBI tests at the 5% level for $m = 0, 1, 2, 3, 4$, and ∞. The upper 5% points of the limiting null distributions of $(m + 1)^2 R_{Tm}^{(I)}$ are also reported under the heading x. Note that the powers are given in terms of the values of c, not c^2; hence the powers for $m = 0$ coincide with those of the LBIU unit root test applied to Model A. It is seen that the powers decrease with m for each c and converge quickly to the values corresponding to $m = \infty$ computed from the limiting c.f. of $\phi\left((m + 1)^2 \theta; m\right)$ as $m \to \infty$, where ϕ is defined in (10.71).

Table 10.8. Limiting Percent Powers of the LBI Tests Based on (10.68) at the 5% Level

Case	x	$c = 1$	5	10	20	50	60
$m = 0$	0.4614	6.20	31.10	61.06	86.65	99.09	99.59
$m = 1$	0.3831	5.97	27.57	57.57	85.28	99.05	99.58
$m = 2$	0.3621	5.91	26.54	56.45	84.82	99.04	99.58
$m = 3$	0.3523	5.89	26.05	55.90	84.59	99.03	99.58
$m = 4$	0.3467	5.87	25.76	55.58	84.45	99.03	99.57
$m = \infty$	0.3258	5.81	24.66	54.29	83.89	99.01	99.57

10.7.2 Case (II)

The LBI statistic in the present case takes the form

$$(10.72) \qquad R_{Tm}^{(II)} = \frac{1}{T} \frac{y'MCC'My}{y'My},$$

where $M = I_T - (e, x_2) \left((e, x_2)'(e, x_2) \right)^{-1} (e, x_2)'$ with $e = (1, \ldots, 1)' : T \times 1$ and $x_2 = (1^m, \ldots, T^m)'$.

We can first show that $y'My/T \to \sigma_\varepsilon^2$ in probability as $T \to \infty$ under $\rho = c^2/T^2$. Since $My = M(C\xi + \varepsilon)$ and $C\xi + \varepsilon \sim N\left(0, \sigma_\varepsilon^2(\rho CC' + I_T)\right)$, we have

$$\mathcal{L}\left(y'MCC'My/\sigma_\varepsilon^2\right) = \mathcal{L}\left(Z'\left(C'MC + \frac{c^2}{T^2}(C'MC)^2\right)Z\right),$$

where $Z \sim N(0, I_T)$. Then the Fredholm approach yields (Problem 7.3) the following theorem.

Theorem 10.10. *Consider the model*

$$y_j = \beta_j + j^m \gamma + \varepsilon_j, \quad \beta_j = \beta_{j-1} + \xi_j, \quad (m = 1, 2, 3, 4),$$

where $\mu\ (= \beta_0)$ is an unknown constant. Then, as $T \to \infty$ under $\rho = c^2/T^2$, the limiting distribution of the LBI statistic $R_{Tm}^{(II)}$ in (10.72) has the same expression as the right side of (10.70), where

$$K(s, t; m) = \min(s, t) - st - \frac{2m+1}{m^2} st(1 - s^m)(1 - t^m).$$

Moreover the limiting c.f. of $R_{Tm}^{(II)}$ can be expressed as in (10.71), where the FD $D(\lambda; m)$ of $K(s, t; m)$ is given by

$$D(\lambda; 1) = \frac{12}{\lambda^2} \left(2 - \sqrt{\lambda} \sin \sqrt{\lambda} - 2 \cos \sqrt{\lambda} \right),$$

$$D(\lambda; 2) = \frac{45}{\lambda^3} \left(\sqrt{\lambda} \left(1 - \frac{\lambda}{3}\right) \sin \sqrt{\lambda} - \lambda \cos \sqrt{\lambda} \right),$$

$$D(\lambda; 3) = \frac{224}{\lambda^4} \left(4 - 2\lambda - \sqrt{\lambda} \left(2 - \frac{4\lambda}{3} + \frac{\lambda^2}{10} \right) \sin \sqrt{\lambda} \right.$$
$$\left. - \left(4 - 2\lambda + \frac{\lambda^2}{2} \right) \cos \sqrt{\lambda} \right),$$

$$D(\lambda; 4) = \frac{1350}{\lambda^5} \left(\sqrt{\lambda} \left(6 - 4\lambda + \frac{7\lambda^2}{10} - \frac{\lambda^3}{42} \right) \sin \sqrt{\lambda} \right.$$
$$\left. - \lambda \left(6 - 2\lambda + \frac{\lambda^2}{6} \right) \cos \sqrt{\lambda} \right).$$

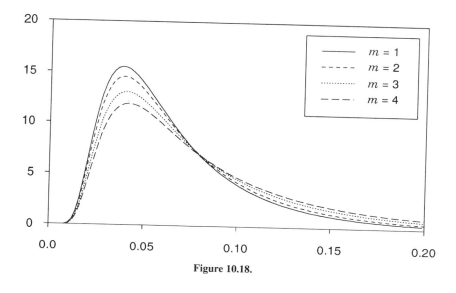

Figure 10.18.

It follows from this theorem and Theorem 10.2 that $R_{Tm}^{(II)}$ under $\rho = c^2/T^2$ with $m = 1$ has the same limiting distribution as the LBIU unit root statistic under $\alpha = 1 - (c/T)$ applied to Model B. The same relationship holds between $R_{Tm}^{(II)}$ with $m = 2$ and the LBIU unit root statistic applied to Model C. We could devise the unit root statistics which have the same limiting distributions as $R_{Tm}^{(II)}$ with $m = 3$ and 4, but those may be of little practical interest. We note in passing that the FD $D(\lambda; 4)$ in the above theorem was earlier obtained in (5.79).

Figure 10.18 shows the limiting probability densities of $R_{Tm}^{(II)}$ under H_0 for $m = 1$, 2, 3, and 4, while Table 10.9 reports the limiting percent powers of the LBI tests at the 5% level for these values of m, together with the significance points under the heading x. It is seen that the powers increase with m for a wide range of the values of c, unlike Case (I). Note that the powers corresponding to $m = 1$ and 2 are completely the same as those of the LBIU unit root tests applied to Models B and C, respectively (see Table 10.3).

Table 10.9. Limiting Percent Powers of the LBI Tests Based on (10.72) at the 5% Level

Case	x	$c = 1$	5	10	20	50	60
$m = 1$	0.1479	5.31	13.69	36.71	74.77	98.73	99.51
$m = 2$	0.1642	5.36	14.86	38.62	75.08	98.59	99.44
$m = 3$	0.1939	5.45	16.91	42.00	76.18	98.42	99.33
$m = 4$	0.2225	5.54	18.78	44.89	77.35	98.33	99.27

10.7.3 Case (III)

The LBI statistic in the present case takes the form

$$R_T^{(\mathrm{III})} = \frac{1}{T} \frac{y'MCC'My}{y'My}, \qquad (10.73)$$

where $M = I_T - (e, d, f)((e, d, f)'(e, d, f))^{-1}(e, d, f)'$ with $e = (1, \ldots, 1)' : T \times 1$, $d = (1, \ldots, T)'$ and $f = (1^2, \ldots, T^2)'$.

Proceeding in the same way as before the Fredholm approach yields (Problem 7.4) the following theorem.

Theorem 10.11. *Consider the model*

$$y_j = \beta_j + j\gamma + j^2\delta + \varepsilon_j, \quad \beta_j = \beta_{j-1} + \xi_j,$$

where $\mu \; (= \beta_0)$ is an unknown constant. Then, as $T \to \infty$ under $\rho = c^2/T^2$, the limiting distribution of the LBI statistic $R_T^{(\mathrm{III})}$ in (10.73) has the same expression as the right side of (10.70), where

$$K(s, t) = \min(s, t) - st - 2st(1 - s)(1 - t)(4 - 5s - 5t + 10st).$$

Moreover the limiting c.f. of $R_T^{(\mathrm{III})}$ can be expressed as in (10.71), where the FD of $K(s, t)$ is given by

$$D(\lambda) = \frac{8640}{\lambda^4} \left(2 + \frac{\lambda}{3} - \sqrt{\lambda}\left(2 - \frac{\lambda}{12}\right) \sin\sqrt{\lambda} - \left(2 - \frac{2\lambda}{3}\right) \cos\sqrt{\lambda} \right).$$

It follows from this theorem and Theorem 10.2 that $R_T^{(\mathrm{III})}$ under $\rho = c^2/T^2$ has the same limiting distribution as the LBIU unit root statistic under $\alpha = 1 - (c/T)$ applied to Model D. The powers in the present case are relatively low in comparison with those in Cases (I) and (II), as is seen from Table 10.3.

10.7.4 The Case of the Initial Value Known

So far we have assumed that the initial value $\mu \; (= \beta_0)$ to generate $\beta_j = \beta_{j-1} + \xi_j$ is unknown. Suppose that μ is known and is assumed to be 0. Then we have the following LBI statistics for Cases (I), (II), and (III):

$$R_{Tm}^{(\mathrm{I})} = \frac{1}{T^{2m+1}} \frac{y'DCC'Dy}{y'y}, \quad \left(m > -\frac{1}{2}\right), \qquad (10.74)$$

$$R_{Tm}^{(\mathrm{II})} = \frac{1}{T} \frac{y'M_1CC'M_1y}{y'M_1y}, \quad (m = 1, 2, 3, 4), \qquad (10.75)$$

$$R_T^{(\mathrm{III})} = \frac{1}{T} \frac{y'M_2CC'M_2y}{y'M_2y}, \qquad (10.76)$$

where $M_1 = I_T - x_1 x_1'/x_1' x_1$ and $M_2 = I_T - (d, f)\left((d, f)'(d, f)\right)^{-1} (d, f)'$ with $x_1 = (1^m, \ldots, T^m)'$, $d = (1, \ldots, T)'$ and $f = (1^2, \ldots, T^2)'$.

It might be argued that some of the above statistics have the same limiting distributions as $T \to \infty$, under a suitable sequence of ρ, as the LBI unit root statistics discussed in previous sections. This is, however, not the case. To show this let us take up the simplest model in Case (I)

(10.77)
$$y_j = \beta_j + \varepsilon_j,$$
$$\beta_j = \beta_{j-1} + \xi_j, \quad \beta_0 = 0, \quad (j = 1, \ldots, T),$$

which was introduced in Section 1.1 of Chapter 1. The LBI test rejects H_0 when

(10.78)
$$S_T = \frac{1}{T} \frac{y'CC'y}{y'y} > c,$$

and it holds that, as $T \to \infty$ under $\rho = c^2/T^2$,

$$\mathcal{L}(S_T) \to \mathcal{L}\left(\int_0^1 \int_0^1 \left(K(s,t) + c^2 K(s,t)\right) dw(s)\, dw(t)\right),$$

where $K(s, t) = 1 - \max(s, t)$. Moreover, the limiting c.f. of S_T is given by

$$\phi(\theta) = \left(\cos\sqrt{i\theta + \sqrt{-\theta^2 + 2ic^2\theta}} \cos\sqrt{i\theta - \sqrt{-\theta^2 + 2ic^2\theta}}\right)^{-1/2}.$$

It is clear that the LBI statistic S_T in (10.78) has a different distribution from any distribution of the LBI unit root statistics considered in the conditional case. The MPI test for $H_0 : \rho = 0$ versus $H_1 : \rho = \theta^2/T^2$ applied to the present model rejects H_0 when

(10.79)
$$V_T(\theta) = T \frac{y' \left[I_T - (\rho CC' + I_T)^{-1}\right] y}{y'y} > c,$$

where $\theta = T\sqrt{\rho}$. It follows from (8.100) that, as $T \to \infty$ under $\rho = \theta^2/T^2$ and the true value $\rho_0 = c^2/T^2$,

(10.80) $\mathcal{L}(V_T(\theta)) \to \mathcal{L}\left(\theta^2 \sum_{n=1}^{\infty} \frac{\left(n - \frac{1}{2}\right)^2 \pi^2 + c^2}{\left(n - \frac{1}{2}\right)^2 \pi^2 \left(\left(n - \frac{1}{2}\right)^2 \pi^2 + \theta^2\right)} Z_n^2\right),$

where $\{Z_n\} \sim \text{NID}(0, 1)$. In particular, it holds (Problem 7.5) that, as $T \to \infty$ under $\rho = \theta^2/T^2$ and $\rho_0 = 0$,

$$\lim_{T\to\infty} P\left(V_T(\theta) \le x\right) = \frac{1}{\pi} \int_0^\infty \mathrm{Re}\left[\frac{1 - \exp\left(-\frac{iux}{\theta^2}\right)}{iu} \left(\frac{\cos\sqrt{2iu - \theta^2}}{\cosh\theta}\right)^{-1/2}\right] du.$$

(10.81)

The limiting power envelope at the $100\gamma\%$ level is given (Problem 7.6) by

$$\lim_{T\to\infty} P(V_T(c) \ge x) = 1 - \frac{1}{\pi} \int_0^\infty \mathrm{Re}\left[\frac{1 - \exp\left(-\frac{iux}{c^2}\right)}{iu} \left(\cos\sqrt{2iu}\right)^{-1/2}\right] du,$$

(10.82)

where the limit is taken under $\rho = \rho_0 = c^2/T^2$, while x is the upper $100\gamma\%$ point of the limiting distribution in (10.81) with $\theta = c$.

Figure 10.19 shows the limiting power function of the LBI test in (10.78) at the 5% level, together with the limiting power envelope computed from (10.82). Also shown is the limiting power function of the POI test conducted at $\theta = 5$, which is computed (Problem 7.7) on the basis of (10.80). Table 10.10 tabulates Figure 10.19 for selected values of c. The 5% significance points for the computation of the limiting power envelope are also reported under the heading x. The POI test conducted at $\theta = 5$ performs better, as a whole, than the LBI test and attains nearly the highest possible power.

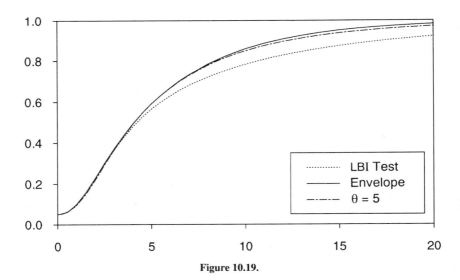

Figure 10.19.

Table 10.10. Limiting Percent Power Envelope at the 5% Level for the Model (10.77)

Test	$c = 1$	5	10	20	50	60
x	1.2059	5.5720	9.2639	15.8711	33.9668	39.7623
Envelope	9.85	59.05	85.80	98.29	100.00	100.00
LBI	9.84	56.48	78.24	92.07	99.12	99.52
$\theta = 5$	9.40	59.05	84.79	97.14	99.94	99.987

The limiting distributions of the statistics $R_{Tm}^{(I)}$, $R_{Tm}^{(II)}$, and $R_T^{(III)}$ as $T \to \infty$ under a sequence of ρ can also be derived from the results on the limiting null distributions obtained in Nabeya and Tanaka (1988). In fact, we have the following three theorems.

Theorem 10.12. *Consider the model*

$$y_j = j^m \beta_j + \varepsilon_j, \quad \beta_j = \beta_{j-1} + \xi_j, \quad \left(m > -\frac{1}{2}\right),$$

where $\beta_0 = 0$. Then, as $T \to \infty$ under $\rho = c^2/T^{2m+2}$, the limiting distribution of $R_{Tm}^{(I)}$ in (10.74) follows as in (10.70), where

$$K(s, t; m) = \frac{1}{2m + 1} \left[1 - (\max(s, t))^{2m+1}\right].$$

Moreover, the limiting c.f. of $R_{Tm}^{(I)}$ can be expressed as in (10.71), where the FD of $K(s, t; m)$ is given by

$$D(\lambda; m) = \Gamma(-\nu + 1) J_{-\nu}\left(\frac{\sqrt{\lambda}}{m+1}\right) \Big/ \left(\frac{\sqrt{\lambda}}{2(m+1)}\right)^{-\nu}$$

with $\nu = (2m + 1)/(2(m + 1))$.

Note that the FD appearing in this theorem was earlier obtained in (5.53).

Theorem 10.13. *Consider the model*

$$y_j = \beta_j + j^m \gamma + \varepsilon_j, \quad \beta_j = \beta_{j-1} + \xi_j, \quad (m = 1, 2, 3, 4),$$

where $\beta_0 = 0$. Then, as $T \to \infty$ under $\rho = c^2/T^2$, the limiting distribution of $R_{Tm}^{(II)}$ in (10.75) follows as in (10.70), where

$$K(s, t; m) = 1 - \max(s, t) - \frac{2m+1}{(m+1)^2} \left(1 - s^{m+1}\right) \left(1 - t^{m+1}\right).$$

Moreover, the limiting c.f. of $R_{Tm}^{(II)}$ can be expressed as in (10.71), where the FD $D(\lambda; m)$ of $K(s, t; m)$ is given by

$$D(\lambda; 1) = \frac{3}{\lambda^{3/2}} \left(\sin \sqrt{\lambda} - \sqrt{\lambda} \cos \sqrt{\lambda} \right),$$

$$D(\lambda; 2) = \frac{20}{\lambda^{5/2}} \left(-2\sqrt{\lambda} + (1+\lambda) \sin \sqrt{\lambda} + \sqrt{\lambda} \left(1 - \frac{\lambda}{3} \right) \cos \sqrt{\lambda} \right),$$

$$D(\lambda; 3) = \frac{126}{\lambda^{7/2}} \left(\left(2 - 2\lambda + \frac{\lambda^2}{2} \right) \sin \sqrt{\lambda} - \sqrt{\lambda} \left(2 - \frac{4\lambda}{3} + \frac{\lambda^2}{10} \right) \cos \sqrt{\lambda} \right),$$

$$D(\lambda; 4) = \frac{864}{\lambda^{9/2}} \left(\sqrt{\lambda}(-12 + 2\lambda) + \left(6 + 6\lambda - 2\lambda^2 + \frac{\lambda^3}{6} \right) \sin \sqrt{\lambda} \right.$$

$$\left. + \sqrt{\lambda} \left(6 - 4\lambda + \frac{7\lambda^2}{10} - \frac{\lambda^3}{42} \right) \cos \sqrt{\lambda} \right).$$

Theorem 10.14. *Consider the model*

$$y_j = \beta_j + j\gamma + j^2 \delta + \varepsilon_j, \quad \beta_j = \beta_{j-1} + \xi_j,$$

where $\beta_0 = 0$. Then, as $T \to \infty$ under $\rho = c^2/T^2$, the limiting distribution of $R_T^{(\mathrm{III})}$ in (10.76) follows as in (10.70), where

$$K(s,t) = 1 - \max(s,t) + \frac{2}{9}(1-s)(1-t)$$
$$\times \left(5 \left(s^2 + t^2 - 8s^2 t^2 + st(s+t) \right) - 4(1+s)(1+t) \right).$$

Moreover, the limiting c.f. of $R_T^{(\mathrm{III})}$ can be expressed as in (10.71), where the FD of $K(s,t)$ is given by

$$D(\lambda) = \frac{960}{\lambda^4} \left(2 + \lambda - \sqrt{\lambda} \left(2 + \frac{\lambda}{3} \right) \sin \sqrt{\lambda} - \left(2 - \frac{\lambda^2}{12} \right) \cos \sqrt{\lambda} \right).$$

PROBLEMS

7.1 Show that (10.69) holds.

7.2 Prove Theorem 10.9.

7.3 Prove Theorem 10.10.

7.4 Prove Theorem 10.11.

7.5 Show that the limiting distribution of $V_T(\theta)$ in (10.79) as $T \to \infty$ under $\rho = \theta^2/T^2$ and $\rho_0 = 0$ is given by (10.81).

7.6 Establish (10.82).

7.7 Describe how to compute the limiting powers of the POI test based on $V_T(\theta)$ in (10.79).

CHAPTER 11

Statistical Analysis of Cointegration

This chapter deals with cointegration, which is a phenomenon arising from linear combinations of components of multiple integrated time series. Concentrating first on the I(1) processes, we discuss estimation and testing problems associated with cointegration. Estimation problems are discussed for each of the two cases, no cointegration and cointegration, while testing for cointegration is discussed by considering two types of tests, one taking no cointegration as the null and the other taking cointegration as the null. A general procedure for determining the algebraic structure of cointegration is also described. Some of the above arguments are further extended to deal with higher order integrated series.

11.1 INTRODUCTION

Cointegration has been an important topic in multiple integrated time series since Granger (1981) first pointed out that such series may have linear combinations that are stationary. Since then a large number of works have been done and are still being undertaken. Here we discuss only a small portion of the problems concentrating on the asymptotic behavior of statistics associated with cointegration.

As the basic model, we consider the q-dimensional I(1) process $\{y_j\}$ generated by

$$(11.1) \qquad (1-L)y_j = \Delta y_j = u_j, \qquad y_0 = 0, \qquad (j = 1, \ldots, T),$$

where $\{u_j\}$ is a q-dimensional stationary process defined by

$$(11.2) \qquad u_j = \sum_{l=0}^{\infty} \Phi_l \varepsilon_{j-l} = \Phi(L)\varepsilon_j, \qquad \sum_{l=0}^{\infty} l \, \|\Phi_l\| < \infty$$

with $\{\varepsilon_j\} \sim$ i.i.d.$(0, I_q)$ and $\|\Phi_l\| = [\text{tr}(\Phi_l'\Phi_l)]^{1/2}$. Note that $V(\varepsilon_j) = I_q$ and we do not assume $\Phi_0 = I_q$ but do assume Φ_0 to be nonsingular and block lower triangular.

Let us introduce the matrix defined by

$$(11.3) \qquad A \equiv \Phi(1) = \sum_{l=0}^{\infty} \Phi_l \qquad \text{with} \qquad \text{rank}(A) = q - r.$$

It follows that A has no zero row vector. The number r in (11.3) is the dimension of the null space of A and is referred to as the *cointegration rank*, where $0 \leq r \leq q - 1$.

In Section 3.10 of Chapter 3 we have shown that

$$(11.4) \qquad \mathcal{L}(S_T) = \mathcal{L}\left(\frac{1}{T^2} \sum_{j=1}^{T} y_j y_j'\right) \to \mathcal{L}\left(A \int_0^1 w(t) w'(t)\, dt\, A'\right),$$

where $\{w(t)\}$ is the q-dimensional standard Brownian motion. Suppose that A is singular so that the limiting random matrix in (11.4) is also singular. We then recognize that the situation is much like multicollinearity in the usual regression context, which implies the existence of a linearly dependent relation among the regressor variables. In the present context, however, the linearly dependent relation is stochastic.

To put it another way, let us rewrite (11.1) as

$$(11.5) \qquad y_j = \frac{1}{1 - L} \left[A + (1 - L)\tilde{\Phi}(L) \right] \varepsilon_j$$

$$= A \frac{\varepsilon_j}{1 - L} + \tilde{\Phi}(L) \varepsilon_j,$$

where we have applied the BN decomposition introduced in Section 3.5 of Chapter 3 to $u_j = \Phi(L)\varepsilon_j$ so that

$$(11.6) \qquad \tilde{\Phi}(L) = -\sum_{l=0}^{\infty} \left(\sum_{k=l+1}^{\infty} \Phi_k \right) L^l.$$

It is seen from (11.5) that y_j has been decomposed into the nonstationary part $A\varepsilon_j/(1 - L)$ and the stationary part $\tilde{\Phi}(L)\varepsilon_j$, where the stationarity of the latter is ensured by (11.2).

Let us further define the following partitions:

$$(11.7) \qquad y_j = \begin{pmatrix} y_{1j} \\ y_{2j} \end{pmatrix}, \qquad \Phi(L) = \begin{pmatrix} \Phi_1'(L) \\ \Phi_2'(L) \end{pmatrix}, \qquad A = \begin{pmatrix} A_1' \\ A_2' \end{pmatrix},$$

where $y_{1j} : (q - 1) \times 1$ and $y_{2j} : 1 \times 1$. The matrices $\Phi(L)$ and A are also partitioned conformably with y_j. Putting $\alpha = (-\beta', 1)' : q \times 1$ with β a $(q - 1) \times 1$ vector, we obtain, by premultiplication of α on both sides of (11.5),

$$(11.8) \qquad y_{2j} = \beta' y_{1j} + \alpha' A \frac{\varepsilon_j}{1 - L} + g'(L)\varepsilon_j,$$

where $g'(L) = \alpha'\tilde{\Phi}(L)$ and $\{g'(L)\varepsilon_j\}$ is a stationary process. Then we say that $\{y_{2j}\}$ is cointegrated with $\{y_{1j}\}$ if $\alpha'A = A_2' - \beta'A_1' = 0'$ for some $\beta \neq 0$. If this is the case, A must be singular and the vector α or β is called the *cointegrating vector*. Since α belongs to the row null space of A, that is, $\alpha \in \{x : x'A = 0, x \in R^q, x \neq 0\}$, it follows from (11.8) that the cointegration rank r is the maximum number of linearly independent vectors α such that $\{\alpha'y_j\}$ becomes stationary.

Cointegration of $\{y_{1j}\}$ with $\{y_{2j}\}$ leads us from (11.8) to the following *cointegrated system*:

$$(11.9) \quad y_{2j} = \beta'y_{1j} + g'(L)\varepsilon_j, \quad \Delta y_{1j} = \Phi_1'(L)\varepsilon_j, \quad y_{10} = 0.$$

In Sections 11.3 through 11.5 we discuss the estimation problem for β assuming the cointegration rank r to be unity. Because of the correlated nature between the regressor $\{y_{1j}\}$ and the error term $g'(L)\varepsilon_j$, various estimators of β are possible. Among those are the ordinary LSE, the two-stage LSE (2SLSE), and the fully modified estimator suggested by Phillips and Hansen (1990). In particular, for the case of $q = 2$, the limiting distributions of these estimators are computed by numerical integration and percent points tabulated. These are further compared with simulation results reported in Stock (1987) and Phillips and Hansen (1990).

If A is nonsingular, there exists no cointegration and any attempt to regress $\{y_{2j}\}$ on $\{y_{1j}\}$ is spurious because of the existence of the I(1) error term on the right side of (11.8). Granger and Newbold (1974) observed this phenomenon by simulations, while Phillips (1986) developed asymptotic arguments on various statistics arising from such regressions. The case of no cointegration is dealt with in the next section prior to the case of cointegration.

In practice we need to test for the existence of cointegration. Engle and Granger (1987) proposed some testing procedures to test the null of no cointegration against the alternative of cointegration, when $q = 2$. Phillips and Ouliaris (1990) suggested some other tests and explored the asymptotic properties of those tests when q is arbitrary. We take up a few of the above test statistics in Section 11.6. We also suggest reversed tests which test the null of cointegration against the alternative of no cointegration.

Section 11.7 briefly describes a statistical procedure for determining the cointegration rank developed by Johansen (1988), while Section 11.8 discusses higher order cointegration.

11.2 CASE OF NO COINTEGRATION

In this section we assume that the matrix A in (11.3) is nonsingular; hence there exists no cointegration among the components of $\{y_j\}$. Under this assumption, let us consider the model (11.8)

$$y_{2j} = \beta'y_{1j} + \alpha'A \frac{\varepsilon_j}{1-L} + g'(L)\varepsilon_j,$$

and suppose that we estimate β by

$$\hat{\beta}_1 = \left(\sum_{j=1}^{T} y_{1j} y'_{1j}\right)^{-1} \sum_{j=1}^{T} y_{1j} y_{2j} \qquad (11.10)$$

or

$$(11.11) \quad \hat{\beta}_2 = \left(\sum_{j=1}^{T} (y_{1j} - \bar{y}_1)(y_{1j} - \bar{y}_1)'\right)^{-1} \sum_{j=1}^{T} (y_{1j} - \bar{y}_1)(y_{2j} - \bar{y}_2).$$

It is an easy matter to derive the asymptotic distributions of $\hat{\beta}_1$ and $\hat{\beta}_2$. In fact, we can establish the following theorem [Phillips (1986), Tanaka (1993) and Problem 2.1 in this chapter].

Theorem 11.1. *Under no cointegration it holds that*

$$(11.12) \qquad \mathcal{L}\left(\hat{\beta}_k\right) \to \mathcal{L}\left(\left(A'_1 W_k A_1\right)^{-1} A'_1 W_k A_2\right), \quad (k = 1, 2),$$

where A_1 and A_2 are defined in (11.7), while

$$(11.13)\, W_1 = \int_0^1 w(t) w'(t)\, dt = \int_0^1 \int_0^1 (1 - \max(s, t))\, dw(s)\, dw'(t),$$

$$(11.14)\, W_2 = W_1 - \int_0^1 w(t)\, dt \int_0^1 w'(t)\, dt = \int_0^1 \int_0^1 (\min(s, t) - st)\, dw(s)\, dw'(t).$$

This theorem implies that the LSEs $\hat{\beta}_1$ and $\hat{\beta}_2$ do not converge to constants but fluctuate around $(A'_1 A_1)^{-1} A'_1 A_2$ even asymptotically. Note that $(A'_1 A_1)^{-1} A'_1 A_2$ is the mean of the limiting distributions of $\hat{\beta}_1$ and $\hat{\beta}_2$. Thus any attempt to regress $\{y_{2j}\}$ on $\{y_{1j}\}$ is spurious in the present case. Of course, if there exists a $(q-1) \times 1$ nonzero vector β such that $A_2 = A_1 \beta$ with $\text{rank}(A_1) = q - 1$, then it follows from (11.12) that $\hat{\beta}_k$ converges in probability to β. That case corresponds to cointegration and is excluded here.

Let us restrict ourselves to the case of $q = 2$ so that $\hat{\beta}_k$s are scalar. Then we would like to compute, for $k = 1$ and 2,

$$(11.15) \qquad F_k(x) = \lim_{T \to \infty} P\left(\hat{\beta}_k \leq x\right)$$

$$= P(X_k \geq 0)$$

$$= \frac{1}{2} + \frac{1}{\pi} \int_0^\infty \frac{1}{\theta} \text{Im}(\phi_k(\theta))\, d\theta,$$

where $\phi_k(\theta)$ is the c.f. of

$$(11.16) \qquad X_k = x A'_1 W_k A_1 - \frac{1}{2}\left(A'_1 W_k A_2 + A'_2 W_k A_1\right).$$

CASE OF NO COINTEGRATION   421

The following theorem gives explicit expressions for $\phi_k(\theta)$, which can be obtained by the stochastic process or Fredholm approach [Tanaka (1993) and Problem 2.2].

Theorem 11.2. *The c.f.s $\phi_k(\theta)$ of X_k ($k = 1, 2$) in (11.16) are given by*

(11.17) $$\phi_1(\theta) = \left[\cos\sqrt{2i\delta_1\theta}\,\cos\sqrt{2i\delta_2\theta}\right]^{-1/2},$$

(11.18) $$\phi_2(\theta) = \left[\frac{\sin\sqrt{2i\delta_1\theta}}{\sqrt{2i\delta_1\theta}}\,\frac{\sin\sqrt{2i\delta_2\theta}}{\sqrt{2i\delta_2\theta}}\right]^{-1/2},$$

where δ_1 and δ_2 are the eigenvalues of the 2×2 matrix H defined by

(11.19) $$H = xA_1A_1' - \frac{1}{2}\left(A_1A_2' + A_2A_1'\right).$$

Especially if $\{u_j\}$ in (11.1) is a sequence of i.i.d.$(0, I_2)$ random variables so that $\{y_{1j}\}$ and $\{y_{2j}\}$ are independent of each other, (11.12) reduces to

(11.20) $$\mathcal{L}\left(\hat{\beta}_1\right) \to \mathcal{L}\left(\frac{\int_0^1 w_1(t)w_2(t)\,dt}{\int_0^1 w_1^2(t)\,dt}\right),$$

(11.21) $$\mathcal{L}\left(\hat{\beta}_2\right) \to \mathcal{L}\left(\frac{\int_0^1 \tilde{w}_1(t)\tilde{w}_2(t)\,dt}{\int_0^1 \tilde{w}_1^2(t)\,dt}\right),$$

where $w(t) = (w_1(t), w_2(t))'$ and $\tilde{w}(t) = w(t) - \int_0^1 w(t)\,dt$. The associated c.f.s $\phi_k(\theta)$ are given by (11.17) and (11.18) with δ_1 and δ_2 replaced by $\left(x \pm \sqrt{x^2 + 1}\right)/2$. These simple cases were earlier discussed in Section 1.4 of Chapter 1 by the eigenvalue approach, together with the limiting densities of $\hat{\beta}_k$ shown in Figure 1.5.

Moments of the limiting distributions of $\hat{\beta}_k$ ($k = 1, 2$) in (11.12) for $q = 2$ can be computed (Problem 2.3) as follows by using the formula given in (1.39).

Corollary 11.1. *Let $\mu_k(n)$ be the nth order central moment of $F_k(x)$ in (11.15). Then we have*

$$\mu_1(2) = a_1|A|^2/(A_1'A_1)^2, \quad \mu_2(2) = a_2|A|^2/(A_1'A_1)^2,$$

$$\mu_1(3) = \mu_2(3) = 0, \quad \mu_1(4) = b_1\mu_1^2(2), \quad \mu_2(4) = b_2\mu_2^2(2),$$

where

$$a_1 = \frac{1}{4}\int_0^\infty \frac{u}{\sqrt{\cosh u}}\,du - \frac{1}{2} = 0.8907,$$

$$a_2 = \frac{1}{12}\int_0^\infty \frac{u^{3/2}}{\sqrt{\sinh u}}\,du - \frac{1}{2} = 0.3965,$$

$$b_1 = \left[\frac{7}{192}\int_0^\infty \frac{u^3}{\sqrt{\cosh u}}\,du - a_1 - \frac{1}{8}\right]\Big/ a_1^2 = 4.9539,$$

$$b_2 = \left[\frac{1}{320}\int_0^\infty \frac{u^{7/2}}{\sqrt{\sinh u}}\,du - a_2 - \frac{1}{8}\right]\Big/ a_2^2 = 4.0838.$$

It is seen that the variance of F_2 is smaller, as is expected. The kurtosis $\mu_1(4)/\mu_1^2(2) - 3$ of F_1 is equal to $b_1 - 3 = 1.9539$, while that of F_2 is 1.0838. Since $\mu_1(3) = \mu_2(3) = 0$, the skewness of F_1 and F_2 is 0. In fact we can show (Problem 2.4)

Corollary 11.2. *The limiting distributions $F_1(x)$ and $F_2(x)$ in (11.15) are symmetric about the same mean $\mu = A_1'A_2/A_1'A_1$ in the sense that $F_k(x+\mu) + F_k(-x+\mu) = 1$ for any x.*

It is also found that the limiting distributions $G_k(x)$ of $(\hat{\beta}_k - \mu)/\sqrt{\mu_k(2)}$ ($k = 1, 2$) do not depend on any underlying parameters. More specifically, we have the following corollary (Problem 2.5).

Corollary 11.3. *It holds that, for $k = 1$ and 2,*

$$(11.22) \qquad G_k(x) = \lim_{T\to\infty} P\left(\frac{\hat{\beta}_k - \mu}{\sqrt{\mu_k(2)}} \leq x\right)$$

$$= \frac{1}{2} + \frac{1}{\pi}\int_0^\infty \frac{1}{\theta}\operatorname{Im}\left(\tilde{\phi}_k(\theta)\right) d\theta,$$

where $\tilde{\phi}_k(\theta)$ is $\phi_k(\theta)$ given in Theorem 11.2 with δ_1 and δ_2 replaced by $(\sqrt{a_k}\,x \pm \sqrt{a_k x^2 + 1})/2$ ($k = 1, 2$) with constants a_1 and a_2 defined in Corollary 11.1.

Figure 11.1 shows graphs of the probability densities of G_1 and G_2 together with the density of $N(0,1)$. All distributions have the mean 0, variance 1, and skewness 0. The kurtosis of G_1 is 1.9539, while that of G_2 is 1.0838, as was mentioned before.

Phillips (1986) explored asymptotic properties of some other statistics arising from the spurious regression, among which are the customary t-ratio, the coefficient of determination, and the Durbin–Watson statistic. Let us consider the regression

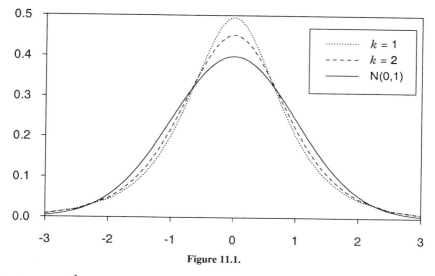

Figure 11.1.

relation $y_{2j} = \hat{\beta}_1 y_{1j} + \hat{v}_j$ for the case of $q = 2$ and put

$$t_{\hat{\beta}} = \hat{\beta}_1 \bigg/ \sqrt{\frac{1}{T}\sum_{j=1}^{T}\hat{v}_j^2 \bigg/ \sum_{j=1}^{T} y_{1j}^2},$$

$$R^2 = \hat{\beta}_1^2 \sum_{j=1}^{T} y_{1j}^2 \bigg/ \sum_{j=1}^{T} y_{2j}^2,$$

$$DW = \sum_{j=2}^{T} (\hat{v}_j - \hat{v}_{j-1})^2 \bigg/ \sum_{j=1}^{T} \hat{v}_j^2.$$

The asymptotic distributions of the above statistics may be derived by the FCLT and the continuous mapping theorem. We can show [Phillips (1986) and Problem 2.6] that

(11.23) $\quad \mathcal{L}\left(\frac{1}{\sqrt{T}} t_{\hat{\beta}}\right) \to \mathcal{L}\left(\frac{A_1' W A_2}{\sqrt{A_1' W_1 A_1 \cdot A_2' W_1 A_2 - (A_1' W_1 A_2)^2}}\right),$

(11.24) $\quad \mathcal{L}(R^2) \to \mathcal{L}\left(\frac{(A_1' W_1 A_2)^2}{A_1' W_1 A_1 \cdot A_2' W_1 A_2}\right),$

(11.25) $\quad \mathcal{L}(T \times DW) \to \mathcal{L}\left(\frac{(\zeta, -1) E(u_j u_j') \begin{pmatrix} \zeta \\ -1 \end{pmatrix}}{A_2' W_1 A_2 - \zeta^2 A_1' W_1 A_1}\right),$

where $\zeta = A_1' W_1 A_2 / A_1' W_1 A_1$.

It is deduced from the above results that

i) the t-ratio is of the stochastic order $\sqrt{T}$;
ii) the coefficient of determination, R^2, has a nondegenerate limiting distribution even if $\{y_{1j}\}$ and $\{y_{2j}\}$ are independent of each other;
iii) the Durbin–Watson statistic converges in probability to 0 at the rate of T^{-1}.

The above features remain unchanged if we consider the fitted mean regression relation $y_{2j} = \bar{y}_2 + \hat{\beta}_2 (y_{1j} - \bar{y}_1) + \hat{w}_j$ and more general multiple regressions, as was discussed in Phillips (1986).

PROBLEMS

2.1 Prove Theorem 11.1.
2.2 Prove Theorem 11.2.
2.3 Prove Corollary 11.1.
2.4 Prove Corollary 11.2.
2.5 Prove Corollary 11.3.
2.6 Establish the weak convergence results in (11.23), (11.24) and (11.25).

11.3 COINTEGRATION DISTRIBUTIONS: THE INDEPENDENT CASE

When the matrix A defined in (11.3) is singular, the cointegrated system described in (11.9) naturally arises from the original model (11.1). In this section, however, we deal with the simplest cointegrated system for which asymptotic properties of various estimators of β can be best studied.

Let us consider

$$(11.26) \qquad y_{2j} = \beta' y_{1j} + \xi_{2j}, \qquad \Delta y_{1j} = \xi_{1j}, \qquad y_{10} = 0,$$

where $\xi_{1j} : (q-1) \times 1$ and $\xi_{2j} : 1 \times 1$. We assume that $\xi_j = (\xi'_{1j}, \xi_{2j})' \sim$ i.i.d.$(0, \Sigma)$ with $\Sigma > 0$ and partition Σ conformably with ξ_{1j} and ξ_{2j} to obtain

$$(11.27) \qquad \Sigma = \begin{pmatrix} \Sigma_{11} & \Sigma_{12} \\ \Sigma_{21} & \Sigma_{22} \end{pmatrix}.$$

It follows that the regressor y_{1j} and the error term ξ_{2k} are correlated only when $j \geq k$, if $\Sigma_{12} \neq 0$. Given $y_{10} = 0$, the model (11.26) may be rewritten as

$$(11.28) \qquad Y_2 = Y_1 \beta + \Xi_2, \qquad \Delta Y_1 = \Xi_1,$$

COINTEGRATION DISTRIBUTIONS: THE INDEPENDENT CASE

where $Y_1 = (y_{11}, \ldots, y_{1T})' : T \times (q-1)$, $Y_2 = (y_{21}, \ldots, y_{2T})' : T \times 1$, $\Xi_1 = (\xi_{11}, \ldots, \xi_{1T})' : T \times (q-1)$, and $\Xi_2 = (\xi_{21}, \ldots, \xi_{2T})' : T \times 1$.

We now discuss the estimation problem for β in (11.28). For this purpose we consider the following estimators:

(11.29) $\quad \hat{\beta}_{\text{OLS}} = (Y_1'Y_1)^{-1} Y_1'Y_2 = \beta + (Y_1'Y_1)^{-1} Y_1'\Xi_2$,

(11.30) $\quad \hat{\beta}_{\text{2SLS}} = (Y_1'P_{-1}Y_1)^{-1} Y_1'P_{-1}Y_2 = \beta + (Y_1'P_{-1}Y_1)^{-1} Y_1'P_{-1}\Xi_2$,

(11.31) $\quad \hat{\beta}_{\text{ML}} = (Y_1'M_1Y_1)^{-1} Y_1'M_1Y_2 = \beta + (Y_1'M_1Y_1)^{-1} Y_1'M_1\Xi_2$,

where $P_{-1} = Y_{-1}(Y_{-1}'Y_{-1})^{-1}Y_{-1}'$ with $Y_{-1} = (0, y_{11}, \ldots, y_{1,T-1})' = LY_1$ and $M_1 = I_T - \Delta Y_1 (\Delta Y_1' \Delta Y_1)^{-1} \Delta Y_1'$. A brief description of the three estimators follows.

i) $\hat{\beta}_{\text{OLS}}$ is the ordinary LSE;
ii) $\hat{\beta}_{\text{2SLS}}$ is the 2SLSE obtained from replacing first Y_1 by $\hat{Y}_1 = P_{-1}Y_1$ and then regressing Y_2 on $\hat{Y}_1$;
iii) $\hat{\beta}_{\text{ML}}$ is the MLE under the assumption that $\{\xi_j\} \sim \text{NID}(0, \Sigma)$.

To see that $\hat{\beta}_{\text{ML}}$ is really the MLE, consider the joint density of $\text{vec}(Y_1') = (y_{11}', \ldots, y_{1T}')'$ and Y_2:

(11.32) $\quad f(\text{vec}(Y_1'), Y_2) = f_1(\text{vec}(Y_1')) f_2(Y_2|\text{vec}(Y_1'))$.

Here it can be shown (Problem 3.1) that

(11.33) $\quad \text{vec}(Y_1') \sim N(0, (CC') \otimes \Sigma_{11})$,

(11.34) $\quad Y_2|\text{vec}(Y_1') \sim N(\mu_{2\cdot 1}, \Sigma_{22\cdot 1}I_T)$,

where C is the $T \times T$ random walk generating matrix, while

$$\mu_{2\cdot 1} = Y_1\beta + \Delta Y_1 \Sigma_{11}^{-1}\Sigma_{12},$$

$$\Sigma_{22\cdot 1} = \Sigma_{22} - \Sigma_{21}\Sigma_{11}^{-1}\Sigma_{12}.$$

It turns out that the MLE of β is the maximizer of f_2 in (11.32), which is equivalent to the minimizer of $(Y_2 - Y_1\beta - \Delta Y_1\gamma)'(Y_2 - Y_1\beta - \Delta Y_1\gamma)$, where $\gamma = \Sigma_{11}^{-1}\Sigma_{12}$. This leads us to $\hat{\beta}_{\text{MLE}}$ in (11.31).

It is apparent that $\hat{\beta}_{\text{MLE}}$ is the ordinary LSE applied to the augmented model:

$$y_{2j} = \beta'y_{1j} + \gamma'\Delta y_{1j} + v_{2j} \iff Y_2 = Y_1\beta + \Delta Y_1\gamma + v_2,$$

where $\gamma = \Sigma_{11}^{-1}\Sigma_{12}$ and $v_{2j} = \xi_{2j} - \gamma'\xi_{1j}$. The introduction of the additional regressor ΔY_1 is to adjust the conditional mean of Y_2, given Y_1. Phillips (1991) discussed the ML estimation in a different way.

The asymptotic distributions of the above estimators can be derived in the following way. We first factorize Σ as $\Sigma = BB'$, where

(11.35) $$B = \begin{pmatrix} \Sigma_{11}^{1/2} & 0 \\ \Sigma_{21}\Sigma_{11}^{-1/2} & \Sigma_{22\cdot 1}^{1/2} \end{pmatrix}.$$

We then define an auxiliary process [Phillips and Durlauf (1986)]:

$$z_j = \begin{pmatrix} y_{1j} \\ x_j \end{pmatrix} = z_{j-1} + \xi_j, \qquad z_0 = 0.$$

It follows from (11.4) that

$$\mathcal{L}\left(\frac{1}{T^2}\sum_{j=1}^T z_j z_j'\right) \to \mathcal{L}\left(B \int_0^1 w(t)w'(t)\,dt B'\right).$$

In particular, we have

$$\mathcal{L}\left(\frac{1}{T^2}\sum_{j=1}^T y_{1j} y_{1j}'\right) \to \mathcal{L}\left(\Sigma_{11}^{1/2} \int_0^1 w_1(t)w_1'(t)\,dt\, \Sigma_{11}^{1/2}\right),$$

where $w(t) = (w_1'(t), w_2(t))'$ with $w_1(t) : (q-1) \times 1$ and $w_2(t) : 1 \times 1$. It can also be shown (Problem 3.2) that $Y_1' P_{-1} Y_1 / T^2$ and $Y_1' M_1 Y_1 / T^2$ appearing in (11.30) and (11.31), respectively, have the same limiting distribution as $Y_1' Y_1 / T^2$.

We have, from Theorem 3.16,

$$\mathcal{L}\left(\frac{1}{T}\sum_{j=1}^T z_{j-1}\xi_j'\right) \to \mathcal{L}\left(B \int_0^1 w(t)\,dw'(t) B'\right).$$

In particular, it holds that

$$\mathcal{L}\left(\frac{1}{T}\sum_{j=1}^T y_{1j}\xi_{2j}\right) = \mathcal{L}\left(\frac{1}{T}\sum_{j=1}^T y_{1,j-1}\xi_{2j} + \frac{1}{T}\sum_{j=1}^T \xi_{1j}\xi_{2j}\right)$$
$$\to \mathcal{L}(U_1 + U_2 + \Sigma_{12}),$$

COINTEGRATION DISTRIBUTIONS: THE INDEPENDENT CASE

where

(11.36) $$U_1 = \Sigma_{11}^{1/2} \int_0^1 w_1(t)\, dw_1'(t) \Sigma_{11}^{-1/2} \Sigma_{12},$$

(11.37) $$U_2 = \Sigma_{11}^{1/2} \int_0^1 w_1(t)\, dw_2(t) \Sigma_{22\cdot 1}^{1/2}.$$

We can also show (Problem 3.3) that

(11.38) $$\mathcal{L}\left(\frac{1}{T} Y_1' P_{-1} \Xi_2\right) \to \mathcal{L}(U_1 + U_2),$$

(11.39) $$\mathcal{L}\left(\frac{1}{T} Y_1' M_1 \Xi_2\right) \to \mathcal{L}(U_2).$$

Then, because of the joint weak convergence and the continuous mapping theorem, we can establish the following theorem.

Theorem 11.3. *For the three estimators defined in (11.29), (11.30) and (11.31), it holds that*

$$\mathcal{L}\left(T\left(\hat{\beta}_{\text{OLS}} - \beta\right)\right) \to \mathcal{L}\left(V^{-1}(U_1 + U_2 + \Sigma_{12})\right),$$
$$\mathcal{L}\left(T\left(\hat{\beta}_{\text{2SLS}} - \beta\right)\right) \to \mathcal{L}\left(V^{-1}(U_1 + U_2)\right),$$
$$\mathcal{L}\left(T\left(\hat{\beta}_{\text{ML}} - \beta\right)\right) \to \mathcal{L}\left(V^{-1} U_2\right),$$

where U_1 and U_2 are defined in (11.36) and (11.37), respectively, while

$$V = \Sigma_{11}^{1/2} \int_0^1 w_1(t) w_1'(t)\, dt\, \Sigma_{11}^{1/2}.$$

It is noticed that both $\hat{\beta}_{\text{OLS}}$ and $\hat{\beta}_{\text{2SLS}}$ depend asymptotically on the unit root distributional component $V^{-1} U_1$. This leaves the limiting distributions asymmetric. The estimator $\hat{\beta}_{\text{2SLS}}$ is expected to improve $\hat{\beta}_{\text{OLS}}$ only marginally, which will be discussed shortly. The estimator $\hat{\beta}_{\text{ML}}$ is asymptotically optimal in the sense that $T\left(\hat{\beta}_{\text{ML}} - \beta\right)$ attains the highest concentration probability about the origin as $T \to \infty$ [Prakasa Rao (1986) and Phillips (1991)].

When $\Sigma_{12} = 0$, that is, $\{y_{1j}\}$ is *strictly exogenous* in the cointegrated system (11.26), the three estimators are asymptotically equivalent. Thus there is no asymptotic advantage of employing the MLE when $\Sigma_{12} = 0$.

Let us concentrate on the case of $q = 2$ and consider

$$P\left(T\left(\hat{\beta}_{\text{OLS}} - \beta\right) \leq x\right) \to P(X_1 \geq 0),$$

where

(11.40) $X_1 = xV - U_1 - U_2 - \Sigma_{12}$

$$= a^2 x \int_0^1 w_1^2(t) \, dt - ab \int_0^1 w_1(t) \, dw_1(t) - ac \int_0^1 w_1(t) \, dw_2(t) - d$$

with

$$a = \Sigma_{11}^{1/2}, \qquad b = \Sigma_{11}^{-1/2} \Sigma_{12}, \qquad c = \Sigma_{22 \cdot 1}^{1/2}, \qquad d = \Sigma_{12}.$$

It is seen that, given $w_1 = \{w_1(t)\}$, X_1 is conditionally normal with

$$E(X_1 \mid w_1) = a^2 x \int_0^1 w_1^2(t) \, dt - \frac{ab}{2} w_1^2(1) + \frac{ab}{2} - d,$$

$$V(X_1 \mid w_1) = a^2 c^2 \int_0^1 w_1^2(t) \, dt.$$

Thus the m.g.f. $m_1(\theta)$ of X_1 is given by

$$m_1(\theta) = E\left[\exp\left\{\theta E(X_1 \mid w_1) + \theta^2 V(X_1 \mid w_1)/2\right\}\right]$$

$$= \exp\left\{\frac{\theta}{2}(ab - 2d)\right\}$$

$$\times E\left[\exp\left\{a^2\theta \left(x + \frac{c^2\theta}{2}\right) \int_0^1 w_1^2(t) \, dt - \frac{ab\theta}{2} w_1^2(1)\right\}\right].$$

Then the stochastic process approach yields the following theorem (Problem 3.4).

Theorem 11.4. *When $q = 2$, the limiting distribution function of $T(\hat{\beta}_{\text{OLS}} - \beta)$ with $\hat{\beta}_{\text{OLS}}$ defined in (11.29) can be computed as*

(11.41)
$$H_1(x) = \lim_{T \to \infty} P\left(T(\hat{\beta}_{\text{OLS}} - \beta) \leq x\right)$$

$$= \frac{1}{2} + \frac{1}{\pi} \int_0^\infty \frac{1}{\theta} \operatorname{Im}(\phi_1(\theta)) \, d\theta,$$

where $\phi_1(\theta)$ is the c.f. of X_1 in (11.40) and is given by

$$\phi_1(\theta) = \exp\left\{\frac{i\theta}{2}(ab - 2d)\right\} \left[\cos \nu + abi\theta \frac{\sin \nu}{\nu}\right]^{-1/2}$$

with $\nu = \sqrt{a^2 i\theta(2x + c^2 i\theta)}$.

The limiting distributions of $T(\hat{\beta}_{\text{2SLS}} - \beta)$ and $T(\hat{\beta}_{\text{ML}} - \beta)$ can be computed from (11.41) by putting $d = 0$ and $b = d = 0$, respectively. The first two moments of these limiting distributions are given by the following corollary (Problem 3.5).

COINTEGRATION DISTRIBUTIONS: THE INDEPENDENT CASE   429

Corollary 11.4. *Let X_{OLS}, X_{2SLS}, and X_{ML} be the limiting random variables of $T(\hat{\beta}_{\text{OLS}} - \beta)$, $T(\hat{\beta}_{\text{2SLS}} - \beta)$, and $T(\hat{\beta}_{\text{ML}} - \beta)$, respectively, when $q = 2$. Then it holds that*

$$E(X_{\text{OLS}}) = \frac{\Sigma_{12}}{2\Sigma_{11}}(c_1 + 2), \quad E(X_{\text{OLS}}^2) = \sigma_1^2 + \frac{\Sigma_{12}^2}{8\Sigma_{11}^2}(10c_1 + c_2 + 4),$$

$$E(X_{\text{2SLS}}) = \frac{\Sigma_{12}}{2\Sigma_{11}}(-c_1 + 2), \quad E(X_{\text{2SLS}}^2) = \sigma_1^2 + \frac{\Sigma_{12}^2}{8\Sigma_{11}^2}(-6c_1 + c_2 + 4),$$

$$E(X_{\text{ML}}) = 0, \quad E(X_{\text{ML}}^2) = \sigma_1^2,$$

where $\sigma_1^2 = c_1 \Sigma_{22 \cdot 1}/\Sigma_{11}$ and

$$c_1 = \int_0^\infty \frac{u}{\sqrt{\cosh u}} \, du = 5.5629,$$

$$c_2 = \int_0^\infty \frac{u^3}{\sqrt{\cosh u}} \, du = 135.6625.$$

The following conclusions may be drawn from this corollary.

i) When Σ_{12} is positive, X_{OLS} has upward bias, while X_{2SLS} downward bias. The situation is reversed when Σ_{12} is negative. The degree of the bias is determined by Σ_{12}/Σ_{11}.

ii) It holds that

$$V(X_{\text{OLS}}) = V(X_{\text{2SLS}}) = \sigma_1^2 + \frac{\Sigma_{12}^2}{8\Sigma_{11}^2}\left(c_2 - 2c_1^2 + 2c_1 - 4\right),$$

but the 2SLSE improves the ordinary LSE in terms of the MSE criterion since $E(X_{\text{OLS}}^2) > E(X_{\text{2SLS}}^2)$.

iii) The MLE $\hat{\beta}_{\text{ML}}$ is asymptotically superior to $\hat{\beta}_{\text{OLS}}$ and $\hat{\beta}_{\text{2SLS}}$ if $\Sigma_{12} \neq 0$. When $\Sigma_{12} = 0$, the three estimators are asymptotically equivalent with the variance equal to $c_1 \Sigma_{22}/\Sigma_{11}$.

The above analysis can be implemented to deal with the fitted mean cointegrated system

(11.42) $\quad y_{2j} = \delta + \beta' y_{1j} + \xi_{2j}, \quad \Delta y_{1j} = \xi_{1j}, \quad y_{10} = 0.$

Let $\tilde{\beta}_{\text{OLS}}$, $\tilde{\beta}_{\text{2SLS}}$, and $\tilde{\beta}_{\text{ML}}$ be the ordinary LSE, 2SLE, and MLE, respectively, of β for the above system. Then we obtain (Problem 3.6)

(11.43) $\quad \tilde{\beta}_{\text{OLS}} = (Y_1'MY_1)^{-1} Y_1'MY_2,$

(11.44) $\quad \tilde{\beta}_{\text{2SLS}} = (\tilde{Y}_1'M\tilde{Y}_1)^{-1} \tilde{Y}_1'MY_2,$

(11.45) $\quad \tilde{\beta}_{\text{ML}} = (Y_1'M_2Y_1)^{-1} Y_1'M_2Y_2,$

where $M = I_T - ee'/T$ with $e = (1, \ldots, 1)' : T \times 1$ and

$$\tilde{Y}_1 = (e, Y_{-1}) \begin{pmatrix} e'e & e'Y_{-1} \\ Y'_{-1}e & Y'_{-1}Y_{-1} \end{pmatrix}^{-1} \begin{pmatrix} e' \\ Y'_{-1} \end{pmatrix} Y_1,$$

$$M_2 = I_T - (e, \Delta Y_1) \begin{pmatrix} e'e & e'\Delta Y_1 \\ \Delta Y'_1 e & \Delta Y'_1 \Delta Y_1 \end{pmatrix}^{-1} \begin{pmatrix} e' \\ \Delta Y'_1 \end{pmatrix}.$$

Proceeding in the same way as before we can establish the following theorem (Problem 3.7).

Theorem 11.5. *For the estimators $\tilde{\beta}_{OLS}$, $\tilde{\beta}_{2SLS}$, and $\tilde{\beta}_{ML}$ applied to the fitted mean cointegrated system (11.42) it holds that*

$$\mathcal{L}\left(T\left(\tilde{\beta}_{OLS} - \beta\right)\right) \to \mathcal{L}\left(\tilde{V}^{-1}\left(\tilde{U}_1 + \tilde{U}_2 + \Sigma_{12}\right)\right),$$
$$\mathcal{L}\left(T\left(\tilde{\beta}_{2SLS} - \beta\right)\right) \to \mathcal{L}\left(\tilde{V}^{-1}\left(\tilde{U}_1 + \tilde{U}_2\right)\right),$$
$$\mathcal{L}\left(T\left(\tilde{\beta}_{ML} - \beta\right)\right) \to \mathcal{L}\left(\tilde{V}^{-1}\tilde{U}_2\right),$$

where

$$\tilde{U}_1 = \Sigma_{11}^{1/2} \int_0^1 \tilde{w}_1(t)\, dw'_1(t) \, \Sigma_{11}^{-1/2} \Sigma_{12},$$

$$\tilde{U}_2 = \Sigma_{11}^{1/2} \int_0^1 \tilde{w}_1(t)\, dw_2(t) \, \Sigma_{22\cdot 1}^{1/2},$$

$$\tilde{V}_1 = \Sigma_{11}^{1/2} \int_0^1 \tilde{w}_1(t) \tilde{w}'_1(t)\, dt \, \Sigma_{11}^{1/2},$$

$$\tilde{w}_1(t) = w_1(t) - \int_0^1 w_1(s)\, ds.$$

It is seen that the expressions for the limiting distributions are quite parallel to those given in Theorem 11.3. The ordinary LSE and 2SLSE still do not eliminate the higher order bias, which can be eliminated by the MLE.

When $q = 2$, the limiting distribution of $T(\tilde{\beta}_{OLS} - \beta)$ can be computed as follows (Problem 3.8).

Theorem 11.6. *When $q = 2$, it holds that*

$$H_2(x) = \lim_{T \to \infty} P\left(T\left(\tilde{\beta}_{OLS} - \beta\right) \leq x\right)$$
$$= \frac{1}{2} + \frac{1}{\pi} \int_0^\infty \frac{1}{\theta} \, Im(\phi_2(\theta))\, d\theta,$$

COINTEGRATION DISTRIBUTIONS: THE INDEPENDENT CASE 431

where $\phi_2(\theta)$ is the c.f. of

$$X_2 = a^2 x \int_0^1 \tilde{w}_1^2(t)\,dt - ab \int_0^1 \tilde{w}_1(t)\,dw_1(t) - ac \int_0^1 \tilde{w}_1(t)\,dw_2(t) - d,$$

and is given by

$$\phi_2(\theta) = \exp\left\{\frac{i\theta}{2}(ab - 2d)\right\} \left[\frac{2a^2 b^2 \theta^2}{\nu^4}(1 - \cos\nu) + \left(1 - \frac{a^2 b^2 \theta^2}{\nu^2}\right)\frac{\sin\nu}{\nu}\right]^{-1/2}$$

with $\nu = \sqrt{a^2 i\theta(2x + c^2 i\theta)}$ and $a, b, c,$ and d defined below (11.40).

The limiting distributions of $T\left(\tilde{\beta}_{2\text{SLS}} - \beta\right)$ and $T\left(\tilde{\beta}_{\text{ML}} - \beta\right)$ may be computed by putting $d = 0$ and $b = d = 0$, respectively. The first two moments may also be obtained as follows (Problem 3.9).

Corollary 11.5. *Let* Y_{OLS}, $Y_{2\text{SLS}}$, *and* Y_{ML} *be the limiting random variables of* $T\left(\tilde{\beta}_{\text{OLS}} - \beta\right)$, $T\left(\tilde{\beta}_{2\text{SLS}} - \beta\right)$, *and* $T\left(\tilde{\beta}_{\text{ML}} - \beta\right)$, *respectively, when* $q = 2$. *Then we have*

$$E(Y_{\text{OLS}}) = \frac{\Sigma_{12}}{2\Sigma_{11}} d_1, \qquad E\left(Y_{\text{OLS}}^2\right) = \sigma_2^2 + \frac{\Sigma_{12}^2}{8\Sigma_{11}^2}(4d_1 - 8d_2 + d_3),$$

$$E(Y_{2\text{SLS}}) = -E(Y_{\text{OLS}}), \qquad E(Y_{2\text{SLS}}^2) = E(Y_{\text{OLS}}^2),$$

$$E(Y_{\text{ML}}) = 0, \qquad\qquad E(Y_{\text{ML}}^2) = \sigma_2^2,$$

where $\sigma_2^2 = d_1 \Sigma_{22 \cdot 1}/\Sigma_{11}$ and

$$d_1 = \int_0^\infty u^{3/2}(\sinh u)^{-1/2}\,du = 10.7583,$$

$$d_2 = \int_0^\infty \sqrt{u}(\sinh u)^{-3/2}(\cosh u - 1)\,du = 2.6415,$$

$$d_3 = \int_0^\infty u^{7/2}(\sinh u)^{-1/2}\,du = 372.3572.$$

It is seen, unlike the zero mean case, that $\tilde{\beta}_{2\text{SLS}}$ does not improve $\tilde{\beta}_{\text{OLS}}$ in terms of the MSE criterion. Corollaries 11.4 and 11.5 yield

$$E(X_{\text{OLS}}) = 3.7814\,\Sigma_{12}/\Sigma_{11}, \qquad E(Y_{\text{OLS}}) = 5.3791\,\Sigma_{12}/\Sigma_{11},$$

$$V(X_{\text{OLS}}) = 5.5629\,\Sigma_{22}/\Sigma_{11} + 4.5493\,\Sigma_{12}^2/\Sigma_{11}^2,$$

$$V(Y_{\text{OLS}}) = 10.7583\,\Sigma_{22}/\Sigma_{11} + 9.5890\,\Sigma_{12}^2/\Sigma_{11}^2.$$

Thus the mean corrected LSE has larger bias and variance than the LSE without mean correction.

Some of the results presented in this section were earlier obtained in Section 1.5 of Chapter 1 by the eigenvalue approach. Figure 1.6 gave densities of the limiting distributions of $T\left(\hat{\beta}_{\text{OLS}} - \beta\right)$ and $T\left(\tilde{\beta}_{\text{OLS}} - \beta\right)$ when $q = 2$ and $\Sigma = I_2$. The corresponding percent points were tabulated in Table 1.6. Figure 11.2 shows limiting densities of $T\left(\hat{\beta}_{\text{OLS}} - \beta\right)/\sigma_1$ and $T\left(\tilde{\beta}_{\text{OLS}} - \beta\right)/\sigma_2$ when $q = 2$ and $\Sigma = I_2$, together with the density of N(0,1). The two limiting densities are densities of $H_k(\sigma_k x)$ in Theorems 11.4 and 11.6 with $a = c = 1$ and $b = d = 0$. Note that the three distributions in Figure 11.2 have the mean 0, variance 1, and skewness 0.

PROBLEMS

3.1 Prove (11.33) and (11.34).

3.2 Show that $Y_1' P_{-1} Y_1 / T^2$ and $Y_1' M_1 Y_1 / T^2$ have the same limiting distribution as $Y_1' Y_1 / T^2$, where these quantities are defined in (11.29) through (11.31).

3.3 Establish (11.38) and (11.39).

3.4 Prove Theorem 11.4.

3.5 Prove Corollary 11.4.

3.6 Show that $\tilde{\beta}_{\text{ML}}$ in (11.45) is the MLE of β in the model (11.42) under the assumption that $\{\xi_j\} \sim \text{NID}(0, \Sigma)$.

3.7 Prove Theorem 11.5.

3.8 Prove Theorem 11.6.

3.9 Prove Corollary 11.5.

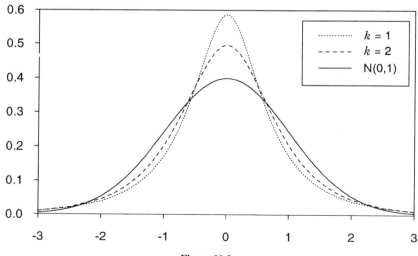

Figure 11.2.

11.4 COINTEGRATION DISTRIBUTIONS: THE DEPENDENT CASE

In this section we deal with the general cointegrated system (11.9), which is reproduced here as

(11.46) $\quad y_{2j} = \beta' y_{1j} + g'(L)\varepsilon_j, \quad \Delta y_{1j} = \Phi_1'(L)\varepsilon_j, \quad y_{10} = 0.$

The present system arises from the original model (11.1) under the assumption that there exists a $(q-1) \times 1$ nonzero vector β such that $A_2 = A_1\beta$, where $A_1 : q \times (q-1)$ and $A_2 : q \times 1$ with $A = (A_1, A_2)'$ defined in (11.3).

In the subsequent discussions, we consider the case where $\text{rank}(A) = q - 1$ so that the cointegration rank $r \; (= q - \text{rank}(A))$ is unity. We then assume, without loss of generality, that $\text{rank}(A_1) = q - 1$. Thus the parameter β in (11.46) can be uniquely determined by $\beta = (A_1' A_1)^{-1} A_1' A_2$. We also naturally require that $\text{rank}(A_1, g(1)) = q$.

Under the above assumption, one might argue that β may be estimated from the spectrum $f(\omega)$ of $\{\Delta y_j\}$ evaluated at $\omega = 0$. In fact, it holds that

$$f(0) = \frac{1}{2\pi} AA',$$

and, if we partition $f(0)$ conformably with Δy_j, we have

$$\beta = f_{11}^{-1}(0) f_{12}(0).$$

The spectral estimator of $f(\omega)$, however, follows the standard theory [Hannan (1970)]; so does the spectral estimator of β like $\hat{\beta} = \hat{f}_{11}^{-1}(0)\hat{f}_{12}(0)$. As was seen in the last section, we have estimators of β that are super consistent under the independence assumption on the error terms. The spectral estimator of β is not our concern, but the spectral method proves useful for eliminating the nuisance parameters, as is discussed later.

Let us first consider

(11.47) $\quad \hat{\beta}_{\text{OLS}} = \left(\sum_{j=1}^{T} y_{1j} y_{1j}' \right)^{-1} \sum_{j=1}^{T} y_{1j} y_{2j}$

$\qquad\qquad = \beta + \left(\sum_{j=1}^{T} y_{1j} y_{1j}' \right)^{-1} \sum_{j=1}^{T} y_{1j} g'(L)\varepsilon_j.$

To derive the asymptotic distribution of $\hat{\beta}_{\text{OLS}}$, we construct the auxiliary process

$$\Delta z_j = \begin{pmatrix} \Delta y_{1j} \\ \Delta x_j \end{pmatrix} = \begin{pmatrix} \Phi_1'(L) \\ g'(L) \end{pmatrix} \varepsilon_j, \quad z_0 = 0,$$

and define the long-run covariance matrix of $\{\Delta z_j\}$

$$\Omega = \begin{pmatrix} \Omega_{11} & \Omega_{12} \\ \Omega_{21} & \Omega_{22} \end{pmatrix} = \begin{pmatrix} A_1'A_1 & A_1'g \\ g'A_1 & g'g \end{pmatrix},$$

where $g = g(1)$. Note that Ω is positive definite by assumption. We further factorize Ω as $\Omega = DD'$, where

(11.48) $\quad D = \begin{pmatrix} \Omega_{11}^{1/2} & 0 \\ \Omega_{21}\Omega_{11}^{-1/2} & \Omega_{22\cdot1}^{1/2} \end{pmatrix}, \quad \Omega_{22\cdot1} = \Omega_{22} - \Omega_{21}\Omega_{11}^{-1}\Omega_{12}.$

Then it follows from (11.4) that

$$\mathcal{L}\left(\frac{1}{T^2}\sum_{j=1}^{T} z_j z_j'\right) \to \mathcal{L}\left(D \int_0^1 w(t)w'(t)\,dt D'\right).$$

We also have, from Theorem 3.16,

$$\mathcal{L}\left(\frac{1}{T}\sum_{j=1}^{T} z_j \left(\begin{pmatrix} \Phi_1'(L) \\ g'(L) \end{pmatrix} \varepsilon_j\right)'\right) \to \mathcal{L}\left(D \int_0^1 w(t)\,dw'(t)D' + \Lambda\right),$$

where

$$\Lambda = \sum_{k=0}^{\infty} E\left(\begin{pmatrix} \Phi_1'(L) \\ g'(L) \end{pmatrix} \varepsilon_j \left(\begin{pmatrix} \Phi_1'(L) \\ g'(L) \end{pmatrix} \varepsilon_{j+k}\right)'\right)$$

$$= \begin{pmatrix} \Lambda_{11} & \Lambda_{12} \\ \Lambda_{21} & \Lambda_{22} \end{pmatrix}.$$

Using the above results and the joint weak convergence, we can establish the following theorem (Problem 4.1).

Theorem 11.7. *Let $\hat{\beta}_{OLS}$ be the ordinary LSE of β in the cointegrated system (11.46). Then it holds that*

$$\mathcal{L}\left(T\left(\hat{\beta}_{OLS} - \beta\right)\right) \to \mathcal{L}\left(R^{-1}\left(Q_1 + Q_2 + \Lambda_{12}\right)\right),$$

where

$$Q_1 = \Omega_{11}^{1/2} \int_0^1 w_1(t)\,dw_1'(t)\, \Omega_{11}^{-1/2}\Omega_{12},$$

$$Q_2 = \Omega_{11}^{1/2} \int_0^1 w_1(t)\,dw_2(t)\, \Omega_{22\cdot 1}^{1/2},$$

$$R = \Omega_{11}^{1/2} \int_0^1 w_1(t)w_1'(t)\,dt\, \Omega_{11}^{1/2}.$$

It is seen that the limiting distribution of $T\,(\hat{\beta}_{\text{OLS}} - \beta)$ contains the unit root component $R^{-1}Q_1$, as in the independent error case. It can also be checked that, when $\Phi_1'(L)\varepsilon_j = A_1'\varepsilon_j$ and $g'(L)\varepsilon_j = g'\varepsilon_j$, the above theorem reduces to Theorem 11.3.

When $q = 2$, the limiting distribution of $T\,(\hat{\beta}_{\text{OLS}} - \beta)$ can be computed from Theorem 11.4 by putting

(11.49) $\qquad a = \Omega_{11}^{1/2},\quad b = \Omega_{11}^{-1/2}\Omega_{12},\quad c = \Omega_{22\cdot 1}^{1/2},\quad d = \Lambda_{12}.$

The corresponding first two moments are given by the following corollary (Problem 4.2).

Corollary 11.6. *Let X_{OLS} be the limiting random variable of $T\,(\hat{\beta}_{\text{OLS}} - \beta)$ given in Theorem 11.7, when $q = 2$. Then it holds that*

$$E(X_{\text{OLS}}) = \frac{2d - ab}{2a^2} c_1 + \frac{b}{a},$$

$$E\left(X_{\text{OLS}}^2\right) = \frac{b^2}{2a^2} + \left(\frac{4c^2 - 3b^2}{4a^2} + \frac{2bd}{a^3}\right) c_1 + \frac{(2d - ab)^2}{8a^4} c_2,$$

where a, b, c, and d are defined in (11.49), while c_1 and c_2 are defined in Corollary 11.4.

The ML estimation of β for the system (11.46) seems to be a formidable task. Here we attempt to modify $\hat{\beta}_{\text{OLS}}$ so that the resulting estimator contains neither the unit root component $R^{-1}Q_1$ nor the additional term $R^{-1}\Lambda_{12}$. For this purpose we first note that

$$\mathcal{L}\left(\frac{1}{T}\sum_{j=1}^T y_{1j}\Delta y_{1j}'\Omega_{11}^{-1}\Omega_{12}\right) \to \mathcal{L}\left(Q_1 + \Lambda_{11}\Omega_{11}^{-1}\Omega_{12}\right),$$

which leads us to construct the estimator

$$(11.50)\, \hat{\beta}_{FM} = \left(\sum_{j=1}^{T} y_{1j} y_{1j}'\right)^{-1}$$

$$\times \left[\sum_{j=1}^{T} y_{1j} \left(y_{2j} - \Delta y_{1j}' \hat{\Omega}_{11}^{-1} \hat{\Omega}_{12}\right) - T\left(\hat{\Lambda}_{12} - \hat{\Lambda}_{11} \hat{\Omega}_{11}^{-1} \hat{\Omega}_{12}\right)\right],$$

where $\hat{\Lambda}_{11}$, $\hat{\Lambda}_{12}$, and $\hat{\Omega}_{11}^{-1}\hat{\Omega}_{12}$ are any consistent estimators of the true parameters. The estimator $\hat{\beta}_{FM}$, called the *fully modified estimator*, was first suggested by Phillips and Hansen (1990).

We now establish the following theorem (Problem 4.3).

Theorem 11.8. *Let $\hat{\beta}_{FM}$ be the estimator defined in (11.50) for the cointegrated system (11.46). Then it holds that*

$$\mathcal{L}\left(T\left(\hat{\beta}_{FM} - \beta\right)\right) \to \mathcal{L}\left(R^{-1} Q_2\right),$$

where R and Q_2 are given in Theorem 11.7.

We can construct a consistent estimator of $\Omega_{11}^{-1}\Omega_{12} = (A_1' A_1)^{-1} A_1' g$ by the spectral method once β is estimated by OLS, noting that the joint spectrum of $\{\Delta y_{1j}\}$ and $\{y_{2j} - \beta' y_{1j}\}$ evaluated at the origin is given by $\Omega/(2\pi)$.

The nonparametric construction of $\hat{\Lambda}_{11}$ and $\hat{\Lambda}_{12}$ may be done following Park and Phillips (1988). Since

$$\Lambda_{11} = \sum_{k=0}^{\infty} E\left(\Delta y_{1j} \Delta y_{1, j+k}'\right), \quad \Lambda_{12} = \sum_{k=0}^{\infty} E\left(\Delta y_{1j} \left(y_{2, j+k} - \beta' y_{1, j+k}\right)\right),$$

Λ_{11} may be estimated consistently by

$$\hat{\Lambda}_{11} = \frac{1}{T} \sum_{i=0}^{l} \left(1 - \frac{i}{l+1}\right) \sum_{j=i+1}^{T} \Delta y_{1, j-i} \Delta y_{1j}',$$

provided $l \to \infty$ as $T \to \infty$ with $l = o(T)$. The parameter Λ_{12} can also be estimated similarly by substituting $\hat{\beta}_{OLS}$ into β.

Another feasible estimator that has the same asymptotic distribution as $\hat{\beta}_{FM}$ was suggested by Saikkonen (1991). His method is parametric and is to regress y_{2j} on y_{1j} and a suitable number of leads and lags of Δy_{1j} and Δy_{2j}.

When $q = 2$, the limiting distribution of $T(\hat{\beta}_{FM} - \beta)$ and the first two moments may be computed from Theorem 11.4 and Corollary 11.4, respectively, by putting $a = \Omega_{11}^{1/2}$, $b = 0$, $c = \Omega_{22 \cdot 1}^{1/2}$, $d = 0$, and $\Sigma = \Omega$.

The above discussions can be easily adapted to deal with the fitted mean cointegrated system

(11.51) $\quad y_{2j} = \delta + \beta' y_{1j} + g'(L)\varepsilon_j, \quad \Delta y_{1j} = \Phi_1'(L)\varepsilon_j, \quad y_{10} = 0.$

Let us consider

(11.52) $\quad \tilde{\beta}_{\text{OLS}} = \left(\sum_{j=1}^{T} (y_{1j} - \bar{y}_1)(y_{1j} - \bar{y}_1)' \right)^{-1} \sum_{j=1}^{T} (y_{1j} - \bar{y}_1)(y_{2j} - \bar{y}_2),$

$$\tilde{\beta}_{\text{FM}} = \left(\sum_{j=1}^{T} (y_{1j} - \bar{y}_1)(y_{1j} - \bar{y}_1)' \right)^{-1} \left[\sum_{j=1}^{T} (y_{1j} - \bar{y}_1) \left(y_{2j} - \Delta y_{1j}' \tilde{\Omega}_{11}^{-1} \tilde{\Omega}_{12} \right) \right.$$

(11.53) $\qquad\qquad\qquad\qquad \left. - T \left(\tilde{\Lambda}_{12} - \tilde{\Lambda}_{11} \tilde{\Omega}_{11}^{-1} \tilde{\Omega}_{12} \right) \right],$

where $\tilde{\Lambda}_{11}$, $\tilde{\Lambda}_{12}$, and $\tilde{\Omega}_{11}^{-1} \tilde{\Omega}_{12}$ are consistent estimators of the corresponding parameters, which can be constructed in a nonparametric way by using Δy_{1j} and $y_{2j} - \bar{y}_2 - \tilde{\beta}'_{\text{OLS}}(y_{1j} - \bar{y}_1)$.

We can now establish the following theorem (Problem 4.4).

Theorem 11.9. *Let the estimators $\tilde{\beta}_{\text{OLS}}$ and $\tilde{\beta}_{\text{FM}}$ of β in the fitted mean cointegrated system (11.51) be given by (11.52) and (11.53), respectively. Then it holds that*

$$\mathcal{L}\left(T\left(\tilde{\beta}_{\text{OLS}} - \beta\right)\right) \to \mathcal{L}\left(\tilde{R}^{-1}\left(\tilde{Q}_1 + \tilde{Q}_2 + \Lambda_{12}\right)\right),$$
$$\mathcal{L}\left(T\left(\tilde{\beta}_{\text{FM}} - \beta\right)\right) \to \mathcal{L}\left(\tilde{R}^{-1}\tilde{Q}_2\right),$$

where

$$\tilde{Q}_1 = \Omega_{11}^{1/2} \int_0^1 \tilde{w}_1(t)\, dw_1'(t)\, \Omega_{11}^{-1/2} \Omega_{12},$$

$$\tilde{Q}_2 = \Omega_{11}^{1/2} \int_0^1 \tilde{w}_1(t)\, dw_2(t)\, \Omega_{22\cdot 1}^{1/2},$$

$$\tilde{R} = \Omega_{11}^{1/2} \int_0^1 \tilde{w}_1(t)\tilde{w}_1'(t)\, dt\, \Omega_{11}^{1/2},$$

$$\tilde{w}_1(t) = w_1(t) - \int_0^1 w_1(s)\, ds.$$

When $q = 2$, the limiting distribuiton of $T(\tilde{\beta}_{\text{OLS}} - \beta)$ can be computed from Theorem 11.6 by putting a, b, c, and d as given in (11.49). The corresponding first two moments are given by the following corollary (Problem 4.5).

Corollary 11.7. Let Y_{OLS} be the limiting random variable of $T(\tilde{\beta}_{OLS} - \beta)$ given in Theorem 11.9, when $q = 2$. Then it holds that

$$E(Y_{OLS}) = \frac{2d - ab}{2a^2} d_1,$$

$$E(Y_{OLS}^2) = \frac{b^2 + 2c^2}{2a^2} d_1 - \frac{b^2}{a^2} d_2 + \frac{(2d - ab)^2}{8a^4} d_3,$$

where a, b, c, and d are defined in (11.49), while d_1, d_2, and d_3 are defined in Corollary 11.5.

The limiting distribution of $T(\tilde{\beta}_{FM} - \beta)$ and the corresponding first two moments for $q = 2$ can be computed from Theorem 11.6 and Corollary 11.5, respectively, by putting $a = \Omega_{11}^{1/2}$, $b = 0$; $c = \Omega_{22 \cdot 1}^{1/2}$, $d = 0$; and $\Sigma = \Omega$.

PROBLEMS

4.1 Prove Theorem 11.7.
4.2 Prove Corollary 11.6.
4.3 Prove Theorem 11.8.
4.4 Prove Theorem 11.9.
4.5 Prove Corollary 11.7.

11.5 THE SAMPLING BEHAVIOR OF COINTEGRATION DISTRIBUTIONS

In this section the finite sample properties of various cointegration distributions are examined and compared with the limiting distributions. For this purpose we concentrate on the case of $q = 2$, so that simulation studies on the finite sample distributions available in the literature and analytical results on the limiting distributions obtained in the last two sections can be used for comparison. For the finite sample distributions, we employ two sets of simulation results obtained by Stock (1987) and Phillips and Hansen (1990).

Let us first take up the model dealt with in Stock (1987):

$$(11.54) \quad \Delta y_j = \Phi(L)\varepsilon_j$$

$$= \frac{1}{\delta(L)} \begin{pmatrix} \Delta(1 - \rho L) + \gamma_2 L & -\gamma_1 L \\ \beta\gamma_2 L & \Delta(1 - \rho L) - \beta\gamma_1 L \end{pmatrix} \varepsilon_j,$$

where $y_j = (y_{1j}, y_{2j})'$, $\{\varepsilon_j\} \sim NID(0, I_2)$, $|\rho| < 1$, $\beta \neq 0$, $\gamma_2 \neq \beta\gamma_1$, and

$$\delta(L) = (1 - \rho L)\left(1 - (1 + \beta\gamma_1 - \gamma_2 + \rho)L + \rho L^2\right).$$

It is assumed that $\delta(x) \neq 0$ for $|x| \leq 1$. Since $\delta(1) = (1-\rho)(\gamma_2 - \beta\gamma_1)$ and

$$A = \Phi(1) = \frac{1}{\delta(1)} \begin{pmatrix} \gamma_2 & -\gamma_1 \\ \beta\gamma_2 & -\beta\gamma_1 \end{pmatrix},$$

it follows that $(-\beta, 1)A = 0'$. Thus there exists cointegration between $\{y_{1j}\}$ and $\{y_{2j}\}$, which yields

(11.55) $\qquad y_{2j} = \beta y_{1j} + g'(L)\varepsilon_j, \quad \Delta y_{1j} = \Phi'_1(L)\varepsilon_j, \quad y_{10} = 0,$

where

$$\Phi'_1(L) = \frac{1}{\delta(L)}\left(\Delta(1-\rho L) + \gamma_2 L, -\gamma_1 L\right),$$

$$g'(L) = \frac{(-\beta, 1)(\Phi(L) - A)}{1 - L} = \frac{(-\beta, 1)}{1-(1+\beta\gamma_1-\gamma_2+\rho)L+\rho L^2},$$

$$A'_1 = \Phi'_1(1) = \frac{1}{\delta(1)}(\gamma_2, -\gamma_1),$$

$$g' = g'(1) = \frac{1}{\gamma_2 - \beta\gamma_1}(-\beta, 1).$$

Although Stock (1987) did extensive simulation studies, we only deal here with the ordinary LSE $\hat{\beta}_{OLS}$ applied to the cointegrated system (11.55). To compare with the limiting distribution of $T(\hat{\beta}_{OLS} - \beta)$, we need the following parameters:

$$a = \sqrt{A'_1 A_1} = \frac{1}{|\delta(1)|}\sqrt{\gamma_1^2 + \gamma_2^2},$$

$$b = \frac{A'_1 g}{a} = \frac{-\gamma_1 - \beta\gamma_2}{a\delta(1)(\gamma_2 - \beta\gamma_1)},$$

$$c = \sqrt{g'g - b^2} = \sqrt{\frac{\beta^2+1}{(\gamma_2 - \beta\gamma_1)^2} - b^2},$$

$$d = \sum_{k=0}^{\infty} E\left(\Phi'_1(L)\varepsilon_j \cdot g'(L)\varepsilon_{j+k}\right).$$

The basic parameters associated with the cointegrated system (11.55) are β, γ_1, γ_2, and ρ, from which the above parameters a, b, c, and d can be determined. We put $\beta = 1$ throughout, while the other parameter values used are presented in Table 11.1. The parameter value of d was computed by using the fact that

$$E\left(\Phi'_1(L)\varepsilon_j \cdot g'(L)\varepsilon_{j+k}\right) = \frac{1}{2\pi}\int_{-\pi}^{\pi} \Phi'_1\left(e^{i\omega}\right) g\left(e^{-i\omega}\right) e^{-ik\omega} d\omega.$$

Table 11.1. Parameter Values Used to Compute Distributions in Table 11.2

Case	γ_1	γ_2	ρ	a	b	c	d
1	0	1/2	0	1	−2	2	−2
2	0	1/2	1/4	4/3	−2	2	−120/53
3	0	1/2	1/2	2	−2	2	−12/5
4	0	1/2	3/4	4	−2	2	−56/31
5	0	1/5	0	1	−5	5	−5
6	0	1/5	1/4	4/3	−5	5	−1500/241
7	0	1/5	1/2	2	−5	5	−150/19
8	0	1/5	3/4	4	−5	5	−700/83
9	−1/2	1/2	0	$1/\sqrt{2}$	0	$\sqrt{2}$	−1
10	−1/2	1/2	1/4	$2\sqrt{2}/3$	0	$\sqrt{2}$	−10/9
11	−1/2	1/2	1/2	$\sqrt{2}$	0	$\sqrt{2}$	−3/2
12	−1/2	1/2	3/4	$2\sqrt{2}$	0	$\sqrt{2}$	−14/5
13	−1/5	1/5	0	$1/\sqrt{2}$	0	$5/\sqrt{2}$	−25/16
14	−1/5	1/5	1/4	$2\sqrt{2}/3$	0	$5/\sqrt{2}$	−125/63
15	−1/5	1/5	1/2	$\sqrt{2}$	0	$5/\sqrt{2}$	−75/26
16	−1/5	1/5	3/4	$2\sqrt{2}$	0	$5/\sqrt{2}$	−175/31

It is seen that $d \neq 0$ in all cases, while $b = 0$ in some cases, so that the limiting distribution of $T(\hat{\beta}_{\text{OLS}} - \beta)$ has higher order bias. The corresponding mean μ and variance σ^2 can be easily computed from Corollary 11.6. In fact, we have

$$\mu = \frac{2d - ab}{2a^2} c_1 + \frac{b}{a},$$

$$\sigma^2 = \frac{b^2 + 4c^2}{4a^2} c_1 + \frac{(2d - ab)^2}{8a^4}(c_2 - 2c_1^2) - \frac{b^2}{2a^2},$$

where $c_1 = 5.5629$ and $c_2 = 135.6625$.

Table 11.2 gives percent points of $\mathcal{L}(T(\hat{\beta}_{\text{OLS}} - \beta))$ for $T = 200$ and ∞, together with the corresponding means and standard deviations. Results on the finite sample distributions are extracted from Stock (1987), while the limiting results are based on Tanaka (1993).

It is seen from Table 11.2 that no case produces a symmetric distribution, as was anticipated. Most cases yield downward bias because of the parameter values examined. The finite sample and limiting distributions are close to each other, although the former is more dispersed and is located to the left of the latter in the first half of the cases, while the situation is reversed in the latter half.

We note in passing that the model (11.54) can also be converted into

(11.56) $\quad (1 - \rho L)\Delta y_j = -\gamma \alpha' y_{j-1} + \varepsilon_j, \qquad \gamma = (\gamma_1, \gamma_2)', \qquad \alpha = (-\beta, 1)'.$

This is a version of the so-called *error correction model*, which describes Δy_j in terms of lags of Δy_j and a linear combination of y_{j-1}. The Granger representation

Table 11.2. Distributions of $T(\hat{\beta}_{OLS} - \beta)$ for Model (11.55)

Case	T	\multicolumn{7}{c}{Probability of a Smaller Value}	Mean	SD
		0.05	0.1	0.25	0.5	0.75	0.9	0.95		
1	200	−24.24	−18.07	−10.72	−5.64	−2.60	−0.54	0.33	−7.93	8.47
	∞	−23.01	−17.36	−10.29	−5.43	−2.56	−0.56	0.65	−7.56	7.92
2	200	−14.31	−10.93	−6.52	−3.46	−1.36	0.07	1.10	−4.66	5.24
	∞	−13.91	−10.52	−6.27	−3.29	−1.34	0.22	1.25	−4.41	4.96
3	200	−6.18	−4.77	−2.86	−1.37	−0.18	1.04	1.85	−1.69	2.64
	∞	−6.13	−4.65	−2.77	−1.32	−0.09	1.22	2.20	−1.56	2.61
4	200	−2.63	−1.91	−1.13	−0.35	0.52	1.55	2.47	−0.25	1.64
	∞	−1.63	−1.22	−0.64	−0.01	0.88	2.11	3.09	0.26	1.52
5	200	−59.96	−44.70	−26.52	−13.96	−6.44	−1.34	0.81	−19.60	20.94
	∞	−57.54	−43.39	−25.74	−13.58	−6.39	−1.39	1.63	−18.91	19.80
6	200	−41.35	−30.95	−18.43	−9.83	−4.32	−0.64	1.57	−13.44	14.61
	∞	−39.44	−29.75	−17.69	−9.32	−4.19	−0.39	1.98	−12.80	13.72
7	200	−21.69	−16.81	−10.07	−5.23	−1.86	0.52	2.12	−6.92	8.10
	∞	−21.20	−16.00	−9.55	−4.96	−1.81	0.89	2.72	−6.53	7.72
8	200	−3.73	−2.63	−1.50	−0.29	1.16	3.06	4.88	0.03	2.70
	∞	−5.61	−4.24	−2.45	−0.87	0.86	3.05	4.85	−0.71	3.23
9	200	−35.38	−27.79	−16.01	−7.09	−2.79	−0.78	−0.03	−11.22	12.40
	∞	−37.06	−27.22	−15.00	−6.84	−2.70	−0.65	0.30	−11.13	13.03
10	200	−22.16	−17.55	−9.82	−4.36	−1.67	−0.24	0.34	−7.02	7.96
	∞	−23.58	−17.29	−9.49	−4.26	−1.57	−0.14	0.58	−6.95	8.38
11	200	−13.65	−10.50	−6.01	−2.61	−0.93	−0.05	0.35	−4.21	4.87
	∞	−14.34	−10.50	−5.74	−2.56	−0.89	0.04	0.52	−4.17	5.13
12	200	−5.97	−4.79	−2.75	−1.23	−0.51	−0.16	−0.05	−1.95	2.13
	∞	−6.76	−4.95	−2.70	−1.19	−0.40	0.06	0.31	−1.95	2.43
13	200	−59.92	−44.27	−25.88	−10.78	−3.41	0.89	2.95	−17.59	21.18
	∞	−61.48	−44.93	−24.40	−10.63	−3.20	1.30	3.81	−17.38	22.34
14	200	−42.76	−31.98	−18.51	−7.67	−2.25	0.99	2.60	−12.57	15.34
	∞	−44.32	−32.37	−17.54	−7.59	−2.17	1.19	3.10	−12.42	16.19
15	200	−27.86	−20.62	−12.12	−4.98	−1.39	0.70	1.89	−8.12	10.00
	∞	−28.83	−21.03	−11.38	−4.90	−1.35	0.88	2.17	−8.02	10.56
16	200	−12.50	−9.83	−5.50	−2.43	−0.91	−0.10	0.27	−3.91	4.48
	∞	−14.54	−10.38	−5.59	−2.40	−0.64	0.47	1.12	−3.93	5.20

theorem (Engle and Granger (1987)) tells us that, if there exists cointegration, the original model (11.54) may be put into the cointegrated system (11.55) and the error correction model (11.56). Johansen (1988) employed the latter to develop inference procedures for determining the cointegration rank, which will be described in Section 11.7.

We next take up the model analyzed in Phillips and Hansen (1990):

$$\Delta y_j = \Phi_0 \varepsilon_j + \Phi_1 \varepsilon_{j-1} + \Phi_2 \varepsilon_{j-2}, \qquad (11.57)$$

where $\{\varepsilon_j\} \sim \text{NID}(0, I_2)$ and

$$\Phi_0 = \begin{pmatrix} 1 & 0 \\ \beta + \delta & \sqrt{1-\delta^2} \end{pmatrix},$$

$$\Phi_1 = \begin{pmatrix} \delta\theta + 0.6 & \theta\sqrt{1-\delta^2} \\ \beta(\delta\theta + 0.6) - 0.7\delta - 0.4 & (\beta\theta - 0.7)\sqrt{1-\delta^2} \end{pmatrix},$$

$$\Phi_2 = \begin{pmatrix} 0 & 0 \\ 0.4 - 0.3\delta & -0.3\sqrt{1-\delta^2} \end{pmatrix},$$

with $\beta \neq 0$ and $|\delta| < 1$. The role of the parameters β, δ, and θ is somewhat vague in the above formulation since it is a reformulation of Phillips and Hansen (1990), who first dealt with the cointegrated system

(11.58) $\quad y_{2j} = \beta y_{1j} + u_{2j}, \qquad \Delta y_{1j} = u_{1j},$

$$u_j = \begin{pmatrix} u_{1j} \\ u_{2j} \end{pmatrix} = \xi_j + \Theta \xi_{j-1}, \qquad \{\xi_j\} \sim \text{NID}(0, \Sigma),$$

where

$$\Theta = \begin{pmatrix} 0.6 & \theta \\ -0.4 & 0.3 \end{pmatrix}, \qquad \Sigma = \begin{pmatrix} 1 & \delta \\ \delta & 1 \end{pmatrix}.$$

Since it holds that

$$A = \Phi_0 + \Phi_1 + \Phi_2 = \begin{pmatrix} \delta\theta + 1.6 & \theta\sqrt{1-\delta^2} \\ \beta(\delta\theta + 1.6) & \beta\theta\sqrt{1-\delta^2} \end{pmatrix},$$

we have $(-\beta, 1)A = 0'$. Then we obtain the cointegrated system (11.55) or (11.58) with

$$\Phi_1'(L) = (1 + (\delta\theta + 0.6)L, \; \theta\sqrt{1-\delta^2}L),$$
$$g'(L) = \left(\delta + (0.3\delta - 0.4)L, \; \sqrt{1-\delta^2} + 0.3\sqrt{1-\delta^2}L\right),$$
$$A_1' = \left(\delta\theta + 1.6, \; \theta\sqrt{1-\delta^2}\right),$$
$$g' = \left(1.3\delta - 0.4, \; 1.3\sqrt{1-\delta^2}\right).$$

THE SAMPLING BEHAVIOR OF COINTEGRATION DISTRIBUTIONS

The parameters which determine the limiting distributions of estimators of β can be computed as

$$a = \sqrt{(\delta\theta + 1.6)^2 + \theta^2(1 - \delta^2)},$$

$$b = \frac{1}{a}\{(\delta\theta + 1.6)(1.3\delta - 0.4) + 1.3\theta(1 - \delta^2)\},$$

$$c = \sqrt{(1.3\delta - 0.4)^2 + 1.69(1 - \delta^2) - b^2},$$

$$d = (1.48 - 0.4\theta)\delta + 0.3\theta - 0.64.$$

Here we consider the ordinary LSE $\hat{\beta}_{OLS}$ and the fully modified estimator $\hat{\beta}_{FM}$ discussed in the last section. Table 11.3 records the means and standard deviations (in parentheses) for the distributions of $T(\hat{\beta}_{OLS} - \beta)$ and $T(\hat{\beta}_{FM} - \beta)$ with $T = 50$ and ∞. The true value of β is set at $\beta = 2$. The finite sample results are extracted from Phillips and Hansen (1990). The estimated densities of $\hat{\beta}_{OLS} - \beta$ and $\hat{\beta}_{FM} - \beta$ for $T = 50$ are displayed in Phillips and Hansen (1990) for $(\delta, \theta) = (-0.8, 0.8)$ and

Table 11.3. Mean and Standard Deviation of Distributions of $T(\hat{\beta}_{OLS} - \beta)$ and $T(\hat{\beta}_{FM} - \beta)$

Case	T	$\theta = 0.8$	$\theta = 0.4$	$\theta = 0.0$
		$\delta = -0.8$		
OLS	50	−6.85(6.25)	−4.50(4.45)	−2.75(3.05)
	∞	−4.85(5.31)	−3.43(3.39)	−2.36(2.14)
FM	50	−1.25(6.35)	−1.40(3.95)	−1.25(2.60)
	∞	0 (2.95)	0 (1.87)	0 (1.15)
		$\delta = -0.4$		
OLS	50	−3.35(4.05)	−2.85(3.95)	−2.00(3.05)
	∞	−1.96(3.11)	−1.92(2.78)	−1.65(2.18)
FM	50	−2.10(4.70)	−1.35(4.05)	−0.75(3.15)
	∞	0 (2.38)	0 (2.19)	0 (1.76)
		$\delta = 0.4$		
OLS	50	−1.20(2.00)	−1.00(2.30)	−0.55(2.50)
	∞	−0.38(1.44)	−0.34(1.62)	−0.24(1.79)
FM	50	−1.15(2.40)	−0.60(2.60)	0.20(3.00)
	∞	0 (1.23)	0 (1.50)	0 (1.76)
		$\delta = 0.8$		
OLS	50	−0.75(1.25)	−0.50(1.40)	−0.20(1.80)
	∞	−0.05(0.84)	0.12(0.96)	0.47(1.21)
FM	50	−0.80(1.40)	−0.25(1.50)	0.75(2.15)
	∞	0 (0.65)	0 (0.85)	0 (1.15)

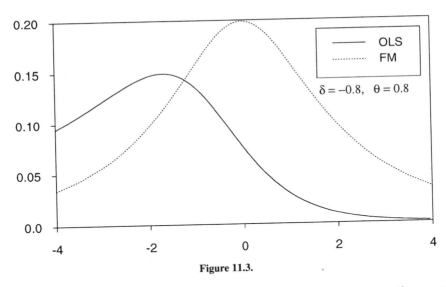

Figure 11.3.

$(-0.8, 0)$. Figure 11.3 shows limiting densities of $T\left(\hat{\beta}_{OLS} - \beta\right)$ and $T\left(\hat{\beta}_{FM} - \beta\right)$ with $(\delta, \theta) = (-0.8, 0.8)$, Figure 11.4 with $(\delta, \theta) = (-0.8, 0)$.

It is seen from Table 11.3 that $\hat{\beta}_{OLS}$ for $T = 50$ generally gives larger bias than $\hat{\beta}_{FM}$, while the standard deviations are larger in most cases for $\hat{\beta}_{FM}$. In comparison to limiting distributions, the sample size $T = 50$ is too small for good approximations. This is especially true for $\hat{\beta}_{FM}$. In fact, $\hat{\beta}_{FM}$ is much dispersed. The standard deviations of $T(\hat{\beta}_{FM} - \beta)$ with $T = 50$ are nearly twice as large as those of the limiting distributions, while those of $T(\hat{\beta}_{OLS} - \beta)$ are only 50% larger. Care should be taken when $\hat{\beta}_{FM}$ is applied under samples of the present size.

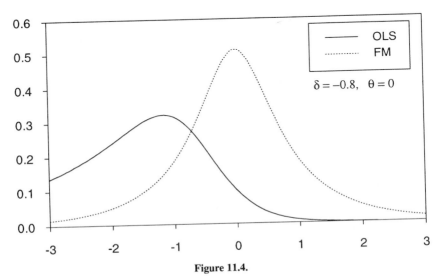

Figure 11.4.

11.6 TESTING FOR COINTEGRATION

The present section is divided into two subsections. The first is concerned with conventional tests that test the null of no cointegration against cointegration, among which are those tests suggested in Engle and Granger (1987), and residual-based tests suggested in Phillips and Ouliaris (1990). The second is concerned with reversed tests that take cointegration as the null, whose tests are discussed in Hansen (1992), Quintos and Phillips (1993), Tanaka (1993), and Shin (1994).

The above tests all assume that the cointegration rank is at most unity. In general it is unknown, and that case will be discussed in the next section.

11.6.1 Tests for the Null of No Cointegration

The model considered here is a transformed version of the original model (11.1), which is

$$(11.59) \quad y_{2j} = \beta' y_{1j} + \alpha' A \frac{\varepsilon_j}{1-L} + g'(L)\varepsilon_j, \quad \Delta y_{1j} = \Phi_1'(L)\varepsilon_j,$$

where $\alpha = (-\beta', 1)'$. We assume that $\text{rank}(A_1) = q - 1$ so that $\text{rank}(A)$ is either $q - 1$ or q, where $A = (A_1, A_2)'$.

Our purpose here is to test

$$(11.60) \quad H_0 : \text{rank}(A) = q \quad \text{versus} \quad H_1 : \text{rank}(A) = q - 1.$$

Note that there exists no cointegration under H_0, while H_1 implies cointegration. Thus the present test is a test for the null of no cointegration. The null hypothesis is equivalent, given $\text{rank}(A_1) = q - 1$, to $A_{22 \cdot 1} = A_2'A_2 - A_2'A_1(A_1'A_1)^{-1}A_1'A_2 \neq 0$, while the alternative to $A_{22 \cdot 1} = 0$.

One of the testing procedures initially suggested by Engle and Granger (1987) and extended by Phillips and Ouliaris (1990) is as follows:

i) Regress y_{2j} on y_{1j} to get the ordinary LSE $\hat{\beta}$ and compute the OLS residuals $\hat{\eta}_j = y_{2j} - \hat{\beta} y_{1j}$ $(j = 1, \ldots, T)$.

ii) Regress further $\hat{\eta}_j$ on $\hat{\eta}_{j-1}$ to obtain

$$\hat{\rho} = \sum_{j=2}^{T} \hat{\eta}_j \hat{\eta}_{j-1} \Bigg/ \sum_{j=2}^{T} \hat{\eta}_{j-1}^2$$

and compute

$$(11.61) \quad \hat{Z}_\rho = T(\hat{\rho} - 1) - \frac{\hat{\sigma}_L^2 - \hat{\sigma}_S^2}{2 \sum_{j=2}^{T} \hat{\eta}_{j-1}^2 \Bigg/ T^2},$$

where

$$\hat{\sigma}_S^2 = \frac{1}{T} \sum_{j=2}^{T} (\hat{\eta}_j - \hat{\rho}\hat{\eta}_{j-1})^2,$$

$$\hat{\sigma}_L^2 = \hat{\sigma}_S^2 + \frac{2}{T} \sum_{i=1}^{l} \left(1 - \frac{i}{l+1}\right) \sum_{j=i+1}^{T} (\hat{\eta}_j - \hat{\rho}\hat{\eta}_{j-1})(\hat{\eta}_{j-i} - \hat{\rho}\hat{\eta}_{j-i-1})$$

for some choice of the truncation number l.

iii) Reject H_0 when $\hat{Z}_\rho$ becomes smaller than a prescribed percent point tabulated in Phillips and Ouliaris (1990); otherwise accept H_0.

The statistic $\hat{Z}_\rho$ in (11.61) is much like a modified version of the AR unit root test statistic presented in Section 9.9 of Chapter 9. In fact, it can be recognized that the above test examines if there exists a unit root in the series $\{\hat{\eta}_j\} = \{y_{2j} - \hat{\beta}y_{1j}\}$. The limiting distribution of $\hat{Z}_\rho$, however, does not follow the usual unit root distribution. To see this, let us first derive the limiting null distribution of

$$T(\hat{\rho} - 1) = \frac{1}{T} \sum_{j=2}^{T} \hat{\eta}_{j-1}(\hat{\eta}_j - \hat{\eta}_{j-1}) \bigg/ \left[\frac{1}{T^2} \sum_{j=2}^{T} \hat{\eta}_{j-1}^2\right]$$

$$= \hat{\alpha}' \frac{1}{T} \sum_{j=2}^{T} y_{j-1} \Delta y_j' \hat{\alpha} \bigg/ \left[\hat{\alpha}' \frac{1}{T^2} \sum_{j=2}^{T} y_{j-1} y_{j-1}' \hat{\alpha}\right],$$

where $\hat{\alpha} = (-\hat{\beta}', 1)'$. Defining

$$B = \begin{pmatrix} B_1 & 0 \\ B_2' & B_3 \end{pmatrix} = \begin{pmatrix} (A_1'A_1)^{1/2} & 0 \\ A_2'A_1 B_1^{-1} & (A_2'A_2 - B_2'B_2)^{1/2} \end{pmatrix},$$

we obtain

$$\mathcal{L}\left(\hat{\alpha}, \frac{1}{T} \sum_{j=2}^{T} y_{j-1} \Delta y_j', \frac{1}{T^2} \sum_{j=2}^{T} y_{j-1} y_{j-1}'\right) \to \mathcal{L}\left((-X_1', 1)', X_2, X_3\right),$$

where

$$X_1 = \left(B_1 \int_0^1 w_1(t) w_1'(t) \, dt B_1\right)^{-1} \left(B_1 \int_0^1 w_1(t) \left(w_1'(t) \, dB_2 + w_2(t) \, dB_3\right)\right),$$

$$X_2 = B \int_0^1 w(t)\,dw'(t)B' + \sum_{k=1}^\infty E\left\{\Phi(L)\varepsilon_j \left(\Phi(L)\varepsilon_{j+k}\right)'\right\},$$

$$X_3 = B \int_0^1 w(t)w'(t)\,dt B',$$

with $w_1(t): (q-1) \times 1$, $w_2(t): 1 \times 1$ and $w(t) = (w_1'(t), w_2(t))'$ the q-dimensional standard Brownian motion. Thus we have (Problem 6.1) under H_0,

(11.62) $$\mathcal{L}(T(\hat{\rho}-1)) \to \mathcal{L}\left(\frac{\int_0^1 Q(t)\,dQ(t)}{\int_0^1 Q^2(t)\,dt} + R\right),$$

where

$$Q(t) = w_2(t) - \int_0^1 w_2(t)w_1'(t)\,dt \left(\int_0^1 w_1(t)w_1'(t)\,dt\right)^{-1} w_1(t),$$

$$R = \frac{(-X_1', 1)\sum_{k=1}^\infty E\left\{\Phi(L)\varepsilon_j \left(\Phi(L)\varepsilon_{j+k}\right)'\right\}\begin{pmatrix}-X_1\\1\end{pmatrix}}{(-X_1', 1)X_3\begin{pmatrix}-X_1\\1\end{pmatrix}}.$$

It is seen from (11.62) that the limiting distribution of $T(\hat{\rho}-1)$ is quite different from the unit root distribution. Even if $R = 0$, whose case occurs when $\{\Delta y_j\}$ is independent and was discussed in Engle and Granger (1987), the first term on the right side of (11.62) is different from the unit root distribution unless $Q(t) = w_2(t)$. In fact, $Q(t)$ is the residual process of the Hilbert space projection of $w_2(t)$ on the space spanned by $w_1(t)$.

We cannot use $T(\hat{\rho}-1)$ as a test statistic since its limiting distribution contains the term R which depends on nuisance parameters underlying the process $\{y_j\}$. This term can be eliminated under H_0 if we construct $\hat{Z}_\rho$ as in (11.61) since

$$\frac{\hat{\sigma}_L^2 - \hat{\sigma}_S^2}{\frac{2}{T^2}\sum_{j=2}^T \hat{\eta}_{j-1}^2} = \frac{\hat{\alpha}'\frac{1}{T}\sum_{i=1}^l \left(1 - \frac{i}{l+1}\right)\sum_{j=i+1}^T \Delta y_{j-i}\Delta y_j'\hat{\alpha}}{\frac{1}{T^2}\sum_{j=2}^T \hat{\eta}_{j-1}^2} + o_p(1),$$

which converges in distribution to R for a suitable choice of l. This last result is due to Phillips (1988). Therefore it is established that, under H_0,

(11.63) $$\mathcal{L}\left(\hat{Z}_\rho\right) \to \mathcal{L}\left(\frac{\int_0^1 Q(t)\,dQ(t)}{\int_0^1 Q^2(t)\,dt}\right).$$

Under $H_1 : A_{22 \cdot 1} = 0$, we have the cointegrated system $y_{2j} = \beta' y_{1j} + g'(L)\varepsilon_j$, and it can be shown [Phillips and Ouliaris (1990) and Problem 6.2] that

(11.64) $$\frac{1}{T}\hat{Z}_\rho \to -\frac{(\gamma(0) - \gamma(1))^2}{2\gamma^2(0)}\left(1 + \frac{g'g}{\gamma(0)}\right)$$

in probability, where $\gamma(k) = E(g'(L)\varepsilon_j \cdot g'(L)\varepsilon_{j+k})$. Thus the test statistic $\hat{Z}_\rho$ diverges to $-\infty$ at the rate of T. It is conceptually difficult to discuss the limiting distribution under the local alternative since no cointegration just means $A_{22 \cdot 1} \neq 0$. In the next subsection we consider the situation under near cointegration, which is more natural since cointegration means $A_{22 \cdot 1} = 0$.

Another test suggested by Phillips and Ouliaris (1990) is based on the t-ratio-like statistic defined by

(11.65) $$\hat{Z}_t = \left(\frac{1}{T^2 \hat{\sigma}_L^2}\sum_{j=2}^T \hat{\eta}_{j-1}^2\right)^{1/2} \hat{Z}_\rho.$$

It can be shown (Problem 6.3) that, under H_0,

(11.66) $$\mathcal{L}\left(\hat{Z}_t\right) \to \mathcal{L}\left(\frac{\int_0^1 Q(t)\,dQ(t)}{\sqrt{S'S \int_0^1 Q^2(t)\,dt}}\right),$$

where

$$S' = \left(-\int_0^1 w_2(t) w_1'(t)\,dt \left(\int_0^1 w_1(t) w_1'(t)\,dt\right)^{-1}, 1\right),$$

while, under H_1,

(11.67) $$\frac{1}{\sqrt{T}}\hat{Z}_t \to -\frac{\gamma(0) - \gamma(1)}{2\sqrt{g'g\gamma(0)}}\left(1 + \frac{g'g}{\gamma(0)}\right)$$

in probability. Thus the $\hat{Z}_\rho$ test is likely to be more powerful than the $\hat{Z}_t$ test.

11.6.2 Tests for the Null of Cointegration

To conduct tests that take cointegration as the null, we consider another variant of the original model (11.1), which is

(11.68) $\quad y_{2j} = \delta + \beta' y_{1j} + \mu_j + \gamma(L)\xi_{2j}, \qquad \Delta y_{1j} = G(L)\xi_{1j}, \qquad \Delta \mu_j = \kappa \xi_{2j},$

where $y_{10} = 0$, $\mu_0 = 0$, κ is a nonnegative constant and $\xi_j = (\xi'_{1j}, \xi_{2j})' \sim$ i.i.d.$(0, \Sigma)$ with $\Sigma > 0$, while $\gamma(L)$ and $G(L)$ are scalar and $(q-1) \times (q-1)$ lag polynomials, respectively. We assume that both $\gamma(L)\xi_{2j}$ and $G(L)\xi_{1j}$ are invertible. Then $G(1)$ is nonsingular so that there exists no cointegration among the components of $\{y_{1j}\}$.

Our purpose here is to test

(11.69) $\qquad\qquad H_0 : \kappa = 0 \qquad \text{versus} \qquad H_1 : \kappa > 0.$

It follows that $\{y_{2j}\}$ is cointegrated with $\{y_{1j}\}$ under H_0 so that $\{y_{2j} - \beta' y_{1j}\}$ is stationary, while $\{y_{2j} - \beta' y_{1j}\}$ is I(1) under H_1. A local alternative to H_0 in (11.69) may be formulated as

(11.70) $\qquad\qquad \kappa = c/T$

with c a positive constant. This gives a sequence of local alternatives of near cointegration.

To suggest test statistics, let us first consider the simplest situation where $\delta = 0$, β is specified and $\{\gamma(L)\xi_{2j}\} \sim$ NID$(0, \sigma^2)$. Then it readily follows that the test which rejects H_0 when

(11.71) $\qquad\qquad S_{T1} = \left(\sum_{j=1}^{T} v_j \right)^2 \bigg/ \sum_{j=1}^{T} v_j^2$

becomes large is LBI, where $v_j = y_{2j} - \beta' y_{1j}$. It also follows that $\mathcal{L}(S_{T1}) \to \mathcal{L}((c^2 + 3c + 3)Z^2/3)$ as $T \to \infty$ under near cointegration $\kappa = c/T$, where $Z \sim$ N$(0, 1)$.

In general, β is unknown, so v_j is unobservable. Moreover, $\{G(L)\xi_{1j}\}$ and $\{\gamma(L)\xi_{2j}\}$ are dependent; hence we need to modify the above statistic S_{T1}. Assuming still that $\delta = 0$, let us define

$$\Delta z_j = \begin{pmatrix} \Delta y_{1j} \\ \Delta x_j \end{pmatrix} = \begin{pmatrix} G(L) & 0 \\ 0' & \gamma(L) \end{pmatrix} \xi_j$$

$$= \begin{pmatrix} J_1(L) & 0 \\ J'_2(L) & J_3(L) \end{pmatrix} \varepsilon_j,$$

$$Z_T(t) = \frac{1}{\sqrt{T}} \sum_{i=1}^{j} \Delta z_i + T\left(t - \frac{j}{T}\right) \frac{\Delta z_j}{\sqrt{T}}, \quad \left(\frac{j-1}{T} \le t \le \frac{j}{T}\right),$$

where $J_1(L) = G(L)\Sigma_{11}^{1/2} : (q-1) \times (q-1)$, $J_2'(L) = \gamma(L)\Sigma_{21}\Sigma_{11}^{-1/2} : 1 \times (q-1)$, and $J_3(L) = \gamma(L)\Sigma_{22\cdot1}^{1/2} : 1 \times 1$, while $\{\varepsilon_j\} \sim$ i.i.d.$(0, I_q)$. Then the FCLT yields

$$\mathcal{L}(Z_T) \to \mathcal{L}(Jw),$$

where

$$J = \begin{pmatrix} J_1(1) & 0 \\ J_2'(1) & J_3(1) \end{pmatrix} = \begin{pmatrix} J_1 & 0 \\ J_2' & J_3 \end{pmatrix}$$

and $w = \{w(t); 0 \le t \le 1\} = (w_1', w_2)'$ is the q-dimensional standard Brownian motion with $w_1 : (q-1) \times 1$ and $w_2 : 1 \times 1$. We also define the long-run covariance matrix of $\{\Delta z_j\}$ by $\Omega = JJ'$, and partition it conformably with w.

We then consider

(11.72) $$\hat{v}_j = y_{2j} - \hat{\beta}_{FM}' y_{1j} - \hat{\Omega}_{21}\hat{\Omega}_{11}^{-1} \Delta y_{1j},$$

where $\hat{\beta}_{FM}$ is the fully modified estimator of β constructed as in (11.50), while $\hat{\Omega}_{21}\hat{\Omega}_{11}^{-1}$ is a consistent estimator of $\Omega_{21}\Omega_{11}^{-1}$ under H_0, which can be obtained from OLS residuals $y_{2j} - \hat{\beta}_{OLS}' y_{1j}$. It is seen that $\hat{v}_j$s in (11.72) are the residuals adjusted for the conditional mean, as was discussed in the ML estimation of β. We can now suggest a test which rejects H_0 when

(11.73) $$\hat{S}_{T1} = \frac{1}{T} \left(\sum_{j=1}^{T} \hat{v}_j\right)^2 \bigg/ \hat{\Omega}_{22\cdot1}$$

takes large values, where $\hat{\Omega}_{22\cdot1}$ is a consistent estimator of $\Omega_{22\cdot1} = \Omega_{22} - \Omega_{21}\Omega_{11}^{-1}\Omega_{12}$ under H_0.

The following theorem gives an expression for $\mathcal{L}(\hat{S}_{T1})$ as $T \to \infty$ under $\kappa = c/T$ (Problem 6.4).

Theorem 11.10. *Suppose that $\delta = 0$ in the model (11.68). Let the test statistic $\hat{S}_{T1}$ for the testing problem (11.69) be defined by (11.73). Then it holds that, as $T \to \infty$ under $\kappa = c/T$,*

$$\mathcal{L}(\hat{S}_{T1}) \to \mathcal{L}((Y_1 + cY_2)^2),$$

TESTING FOR COINTEGRATION

where

$$Y_1 = w_2(1) - \int_0^1 w_1'(t)\,dt \left(\int_0^1 w_1(t)w_1'(t)\,dt \right)^{-1} \int_0^1 w_1(t)dw_2(t),$$

$$Y_2 = \frac{(J_2', J_3)}{\gamma(1)J_3} \left\{ w(1) - \int_0^1 w(t)w_1'(t)\,dt \left(\int_0^1 w_1(t)w_1'(t)\,dt \right)^{-1} \int_0^1 w_1(t)\,dt \right\}.$$

An asymptotic test at the $100\alpha\%$ level may be conducted by rejecting H_0 when $\hat{S}_{T1} \geq x_\alpha$, where x_α is the upper $100\alpha\%$ point of $\mathcal{L}(Y_1^2)$. The test is consistent in the sense that $\hat{S}_{T1}$ diverges at the rate of T under a fixed alternative (Problem 6.5). Note also that $P\left((Y_1 + cY_2)^2 \geq x\right)$ converges to unity for any x as $c \to \infty$.

When the constant term δ in (11.68) is unknown, we need to devise a different test. Suppose first that $v_j = y_{2j} - \beta' y_{1j}$ is observable and $\{\gamma(L)\xi_{2j}\} \sim \text{NID}(0, \sigma^2)$. Then it can be shown (Problem 6.6) that the test which rejects H_0 when

(11.74) $$S_{T2} = \frac{1}{T} \sum_{k=1}^{T} \left\{ \sum_{j=1}^{k} (v_j - \bar{v}) \right\}^2 \Big/ \sum_{j=1}^{T} (v_j - \bar{v})^2$$

becomes large is LBIU. It then follows (Problem 6.7) that, as $T \to \infty$ under $\kappa = c/T$,

(11.75) $$\mathcal{L}(S_{T2}) \to \mathcal{L}\left(\sum_{n=1}^{\infty} \left(\frac{1}{n^2 \pi^2} + \frac{c^2}{n^4 \pi^4} \right) Z_n^2 \right),$$

where $\{Z_n\} \sim \text{NID}(0, 1)$.

Returning to the general model (11.68), where δ and β are unknown and $\gamma(L)$ and $G(L)$ are arbitrary, we suggest a test which rejects H_0 when

(11.76) $$\tilde{S}_{T2} = \frac{1}{T^2} \sum_{k=1}^{T} \left\{ \sum_{j=1}^{k} (\tilde{v}_j - \bar{\tilde{v}}) \right\}^2 \Big/ \tilde{\Omega}_{22\cdot 1}$$

becomes large, where $\tilde{\Omega}_{22\cdot 1}$ is a consistent estimator of $\Omega_{22\cdot 1}$ under H_0, while

$$\tilde{v}_j = y_{2j} - \bar{y}_2 - \tilde{\beta}_{\text{FM}}'(y_{1j} - \bar{y}_1) - \tilde{\Omega}_{21}\tilde{\Omega}_{11}^{-1} \Delta y_{1j},$$

$$\bar{\tilde{v}} = \frac{1}{T} \sum_{j=1}^{T} \tilde{v}_j = -\frac{1}{T} \tilde{\Omega}_{21}\tilde{\Omega}_{11}^{-1} y_{1T}.$$

Here $\tilde{\beta}_{\text{FM}}$ is the fitted mean fully modified estimator of β constructed as in (11.53), while $\tilde{\Omega}_{21}\tilde{\Omega}_{11}^{-1}$ is a consistent estimator of $\Omega_{21}\Omega_{11}^{-1}$ under H_0, which can be obtained from Δy_{1j} and fitted mean OLS residuals $y_{2j} - \bar{y}_2 - \tilde{\beta}_{\text{OLS}}(y_{1j} - \bar{y}_1)$. A consistent estimator of $\tilde{\Omega}_{22\cdot 1}$ in (11.76) can also be obtained similarly.

The following theorem describes the limiting distribution of $\tilde{S}_{T2}$ as $T \to \infty$ under $\kappa = c/T$ (Problem 6.8).

Theorem 11.11. *Let the test statistic $\tilde{S}_{T2}$ for the testing problem (11.69) be defined by (11.76). Then it holds that, as $T \to \infty$ under $\kappa = c/T$,*

$$\mathcal{L}\left(\tilde{S}_{T2}\right) \to \mathcal{L}\left(\int_0^1 \left(Z_1(t) + c Z_2(t)\right)^2 dt\right),$$

where

$$Z_1(t) = w_2(t) - t w_2(1) - \int_0^t \tilde{w}_1'(s)\, ds \left(\int_0^1 \tilde{w}_1(s)\tilde{w}_1'(s)\, ds\right)^{-1} \int_0^1 \tilde{w}_1(s)\, dw_2(s),$$

$$Z_2(t) = \frac{(J_2', J_3)}{\gamma(1) J_3}\left\{\int_0^t \tilde{w}(s)\, ds \right.$$

$$\left. - \int_0^1 \tilde{w}(s)\tilde{w}_1'(s)\, ds \left(\int_0^1 \tilde{w}_1(s)\tilde{w}_1'(s)\, ds\right)^{-1} \int_0^t \tilde{w}_1(s)\, ds\right\},$$

$$\tilde{w}(t) = w(t) - \int_0^1 w(s)\, ds.$$

As in the case of $\delta = 0$, an asymptotic test at the $100\alpha\%$ level may be conducted by rejecting H_0 when $\tilde{S}_{T2} \geq y_\alpha$, where y_α is the upper $100\alpha\%$ point of $\mathcal{L}(\int_0^1 Z_1^2(t)\, dt)$. The test based on $\tilde{S}_{T2}$ is consistent since $\tilde{S}_{T2}$ diverges at the rate of T under a fixed alternative.

Table 11.4 reports some estimated percent points of the limiting null distributions of $\hat{S}_{T1}$ in (11.73) and $\tilde{S}_{T2}$ in (11.76). These were obtained by simulations on the basis of Theorems 11.10 and 11.11.

Some other models for testing the null of cointegration can be found in Hansen (1992), Quintos and Phillips (1993), and Shin (1994). Because of different formulations and assumptions, different tests emerge. Test statistics are also found to have different distributions, which we do not pursue here.

PROBLEMS

6.1 Establish the weak convergence result in (11.62).
6.2 Prove (11.64).
6.3 Prove (11.66) and (11.67).
6.4 Establish Theorem 11.10.
6.5 Show that $\hat{S}_{T1} = O_p(T)$ under $H_1 : \kappa > 0$, where $\hat{S}_{T1}$ is defined in (11.73).

Table 11.4. Estimated Percent Points of the Limiting Null Distributions of $\hat{S}_{T1}$ and $\tilde{S}_{T2}$

Case	Probability of a Smaller Value				
	0.1	0.5	0.9	0.95	0.99
			$\hat{S}_{T1}$		
$q-1=1$	0.006	0.18	1.34	2.08	4.25
$q-1=2$	0.004	0.10	0.76	1.21	2.54
$q-1=3$	0.002	0.07	0.50	0.76	1.55
$q-1=4$	0.002	0.05	0.36	0.54	1.08
$q-1=5$	0.001	0.04	0.29	0.43	0.83
			$\tilde{S}_{T2}$		
$q-1=1$	0.035	0.083	0.24	0.31	0.53
$q-1=2$	0.030	0.063	0.16	0.22	0.37
$q-1=3$	0.025	0.050	0.12	0.16	0.27
$q-1=4$	0.022	0.042	0.09	0.12	0.21
$q-1=5$	0.020	0.036	0.08	0.10	0.16

6.6 Show that S_{T2} defined in (11.74) is the LBIU statistic for the testing problem (11.69) in the model (11.64), when $y_{2j} - \beta' y_{1j}$ is observable and $\{\gamma(L)\xi_{2j}\} \sim$ NID$(0, \sigma^2)$.

6.7 Prove (11.75).

6.8 Establish Theorem 11.11.

11.7 DETERMINATION OF THE COINTEGRATION RANK

In the last section we discussed various tests for cointegration assuming the cointegration rank r to be unity at most. In general, r is unknown, and we need a procedure for determining r. A general procedure for doing this was developed by Johansen (1988). The Johansen procedure is closely related to canonical correlation analysis [Anderson (1984) and Box and Tiao (1977)], reduced rank regression [Tso (1981) and Ahn and Reinsel (1990)], and the overidentification problem in simultaneous equations models [Anderson and Kunitomo (1992)]. Lütkepohl (1993), Hamilton (1994), and Hatanaka (1996) gave detailed discussions on the Johansen procedure, relating to the above references. Here we briefly describe his procedure, leaving details to the works mentioned above.

The model employed by Johansen (1988) is the pth-order vector autoregressive [VAR(p)] model

(11.77) $$y_j = B_1 y_{j-1} + \cdots + B_p y_{j-p} + \varepsilon_j,$$

where $y_j : q \times 1$ and $\{\varepsilon_j\} \sim$ i.i.d.$(0, \Sigma)$ with $\Sigma > 0$. It is assumed that the roots of $|B(x)| = 0$, where

$$B(x) = I_q - B_1 x - \cdots - B_p x^p$$

are all on or outside the unit circle. It is also assumed that $\{\Delta y_j\}$ is stationary. Thus any root on the unit circle is restricted to be unity.

We can now transform (11.77) into

$$\text{(11.78)} \qquad \Delta y_j = \sum_{l=0}^{\infty} \Phi_l \varepsilon_{j-l} = \Phi(L)\varepsilon_j .$$

It follows from (11.77) and (11.78) that

$$B(L)\Delta y_j = \Delta \varepsilon_j = B(L)\Phi(L)\varepsilon_j$$

so that

$$\text{(11.79)} \qquad B(1)\Phi(1) = 0 .$$

This means that each row of $B(1)$ belongs to the row null space of $\Phi(1)$; hence it is a cointegrating vector. It then holds that rank$(B(1))$ is the cointegration rank. Of course, if $\Phi(1)$ is of full rank, $B(1) = 0$ and there exists no cointegration. Conversely, if $B(1)$ is of full rank, then $\Phi(1) = 0$ so that $\{y_j\}$ itself becomes stationary. It follows that the number of unit roots of $|B(x)| = 0$ is exactly equal to $q - \text{rank}(B(1))$.

Our purpose here is to determine rank$(B(1))$. For this purpose, Johansen (1988) transforms (11.77) into the error correction model

$$\text{(11.80)} \qquad \Delta y_j = \gamma \alpha' y_{j-1} + \Gamma_1 \Delta y_{j-1} + \cdots + \Gamma_{p-1} \Delta y_{j-p+1} + \varepsilon_j ,$$

where γ and α are $q \times r$ matrices such that rank$(\gamma) = $ rank$(\alpha) = r$ $(\leq q)$, and $\gamma \alpha' = -B(1)$, while

$$\Gamma_k = - \sum_{i=k+1}^{p} B_i , \qquad (k = 1, \ldots, p-1) .$$

It is recognized that the r components of $\{\alpha' y_j\}$ are linearly independent stationary processes with $r = $ rank(α) being the maximum number of such relations. We call α the *cointegrating matrix* and γ the *loading matrix*.

The testing problem formulated by Johansen (1988) is

$$\text{(11.81)} \qquad H_0 : \text{rank}(\alpha) \leq r < q \qquad \text{versus} \qquad H_1 : \text{rank}(\alpha) \leq q .$$

Note that, if we assume rank(α) to be unity at most, this becomes a test for the null of no cointegration discussed in the last section.

DETERMINATION OF THE COINTEGRATION RANK 455

Johansen (1988) conducts the likelihood ratio (LR) test for (11.81) assuming $\{\varepsilon_j\} \sim \text{NID}(0, \Sigma)$. The log-likelihood for $\{\Delta y_j\}$ in (11.80) is given, except for constants, by

$$L(\theta) = -\frac{T}{2} \log |\Sigma| - \frac{1}{2} \sum_{j=1}^{T} (z_{0j} - \gamma \alpha' z_{1j} - \Gamma z_{2j})' \Sigma^{-1} (z_{0j} - \gamma \alpha' z_{1j} - \Gamma z_{2j}),$$

(11.82)

where

$$z_{0j} = \Delta y_j, \quad z_{1j} = y_{j-1}, \quad z_{2j} = (\Delta y'_{j-1}, \ldots, \Delta y'_{j-p+1})',$$
$$\Gamma = (\Gamma_1, \ldots, \Gamma_{p-1}) : q \times q(p-1).$$

Here we have assumed, for simplicity of presentation, that z_{0j}, z_{1j}, and z_{2j} are available for $j = 1, \ldots, T$. The parameters θ in (11.82) are composed of Γ, Σ, γ, and α. When $\text{rank}(\alpha) = 0$, we put $\gamma = \alpha = 0$ so that θ contains Γ and Σ only.

Let $\hat{\theta}_0$ and $\hat{\theta}_1$ be the MLEs that maximize $L(\theta)$ under $\text{rank}(\alpha) = r$ and $\text{rank}(\alpha) = q$, respectively. Then the LR test rejects H_0 when

(11.83) $$J_T = -2(L(\hat{\theta}_0) - L(\hat{\theta}_1))$$

becomes large.

The test statistic J_T can be obtained by concentrating the parameters θ out of $L(\theta)$ in the order of Σ, (Γ, γ) and α. The maximized value of L for any Γ, γ, and α is

(11.84) $$L\left(\hat{\Sigma} \mid \Gamma, \gamma, \alpha\right) = -\frac{T}{2} \log |\hat{\Sigma}| - \frac{qT}{2},$$

where

$$\hat{\Sigma} = \frac{1}{T} \sum_{j=1}^{T} (z_{0j} - \gamma \alpha' z_{1j} - \Gamma z_{2j})(z_{0j} - \gamma \alpha' z_{1j} - \Gamma z_{2j})'.$$

Suppose that $\text{rank}(\alpha) = n$ so that α and γ are $q \times n$ matrices and that α is given. Let $\hat{\gamma}(n)$ and $\hat{\Gamma}$ be the maximizers of L in (11.84) under this condition. Then it is known [Lütkepohl (1993, p.357)] that the maximized value of L is given by

(11.85) $$L\left(\hat{\Sigma}, \hat{\Gamma}, \hat{\gamma}(n) \mid \alpha\right) = -\frac{T}{2} \log |\hat{\Sigma}(n)| - \frac{qT}{2},$$

where

(11.86) $$\hat{\Sigma}(n) = \frac{1}{T} \sum_{j=1}^{T} \left(z_{0j} - \hat{\gamma}(n) \alpha' z_{1j} - \hat{\Gamma} z_{2j}\right)\left(z_{0j} - \hat{\gamma}(n) \alpha' z_{1j} - \hat{\Gamma} z_{2j}\right)'$$

$$= S_{00} - S_{01}\alpha(\alpha'S_{11}\alpha)^{-1}\alpha'S_{10},$$

$$\hat{\gamma}(n) = S_{01}\alpha(\alpha'S_{11}\alpha)^{-1},$$

$$\hat{\Gamma} = \sum_{j=1}^{T}\left(z_{0j} - \hat{\gamma}(n)\alpha'z_{1j}\right)z'_{2j}\left(\sum_{j=1}^{T}z_{2j}z'_{2j}\right)^{-1},$$

$$S_{ab} = \frac{1}{T}\sum_{j=1}^{T}\zeta_{aj}\zeta'_{bj}, \qquad (a,b = 0,1),$$

$$\zeta_{aj} = z_{aj} - \sum_{j=1}^{T}z_{aj}z'_{2j}\left(\sum_{j=1}^{T}z_{2j}z'_{2j}\right)^{-1}, \qquad (a = 0,1).$$

It may be noted that ζ_{aj} is the OLS residual obtained from the regression of z_{aj} on z_{2j}. Then it naturally holds that

$$\hat{\Sigma}(n) = \frac{1}{T}\sum_{j=1}^{T}\left(\zeta_{0j} - \hat{\gamma}(n)\alpha'\zeta_{1j}\right)\left(\zeta_{0j} - \hat{\gamma}(n)\alpha'\zeta_{1j}\right)',$$

which yields the last expression for $\hat{\Sigma}(n)$ in (11.86).

Finally we are led to maximize L in (11.85) with respect to α, which is equivalent to minimizing

(11.87)
$$\left|\hat{\Sigma}(n)\right| = \left|S_{00} - S_{01}\alpha(\alpha'S_{11}\alpha)^{-1}\alpha'S_{10}\right|$$

$$= |S_{00}|\frac{|\alpha'(S_{11} - S_{10}S_{00}^{-1}S_{01})\alpha|}{|\alpha'S_{11}\alpha|}$$

with respect to α under the condition that rank$(\alpha) = n$. Let us consider the eigenvalue problem

$$S_{10}S_{00}^{-1}S_{01}\hat{V} = S_{11}\hat{V}\hat{\Lambda}, \qquad \hat{\Lambda} = \text{diag}\left(\hat{\lambda}_1,\ldots,\hat{\lambda}_q\right),$$

where the ordered eigenvalues $\hat{\lambda}_1 \geq \cdots \geq \hat{\lambda}_q$ are the solutions to $|\lambda S_{11} - S_{10}S_{00}^{-1}S_{01}| = 0$, and $\hat{V} = (\hat{v}_1,\ldots,\hat{v}_q)$ is the matrix of the corresponding eigenvectors normalized in such a way that

$$\hat{V}'S_{11}\hat{V} = I_q.$$

It turns out that the eigenvalues $\hat{\lambda}_1 \geq \cdots \geq \hat{\lambda}_q$ give the squares of the canonical correlations between $\{\zeta_{0j}\}$ and $\{\zeta_{1j}\}$. Then it is ensured that $P(0 \leq \hat{\lambda}_i < 1) = 1$ for $i = 1,\ldots,q$.

DETERMINATION OF THE COINTEGRATION RANK

Putting $\alpha = \hat{V}\xi$, where ξ is a $q \times n$ matrix with rank$(\xi) = n$, we can see that

$$\min_\alpha |\hat{\Sigma}(n)| = |S_{00}| \min_\xi \frac{|\xi'\xi - \xi'\hat{\Lambda}\xi|}{|\xi'\xi|}$$

$$= |S_{00}| \min_{\xi'\xi=I_n} |\xi'\xi - \xi'\hat{\Lambda}\xi|$$

$$= |S_{00}| \prod_{i=1}^{n} (1 - \hat{\lambda}_i)$$

and the minimizer is given by $\hat{\xi} = (e_1, \ldots, e_n)$, where e_i is the $q \times 1$ vector with unity in the ith place and zeros elsewhere. Accordingly $|\hat{\Sigma}(n)|$ is minimized when $\hat{\alpha} = \hat{V}\hat{\xi}$.

We now deduce that

$$L(\hat{\theta}_a) = -\frac{T}{2} \sum_{i=1}^{r} \log(1 - \hat{\lambda}_i) - \frac{T}{2} \log|S_{00}| - \frac{qT}{2}, \quad (a = 0),$$

$$= -\frac{T}{2} \sum_{i=1}^{q} \log(1 - \hat{\lambda}_i) - \frac{T}{2} \log|S_{00}| - \frac{qT}{2}, \quad (a = 1),$$

so that

$$J_T = -2\left(L(\hat{\theta}_0) - L(\hat{\theta}_1)\right)$$

$$= -T \sum_{i=r+1}^{q} \log(1 - \hat{\lambda}_i).$$

Johansen (1988) proved under H_0 that

$$J_T = \sum_{i=r+1}^{q} T\hat{\lambda}_i + o_p(1)$$

and

$$(11.88) \quad \mathcal{L}(J_T) \to \mathcal{L}\left(\text{tr}\left(\int_0^1 dw(t)w'(t) \left(\int_0^1 w(t)w'(t)\,dt\right)^{-1} \int_0^1 w(t)\,dw'(t)\right)\right),$$

where $\{w(t)\}$ is the $(q - r)$-dimensional standard Brownian motion. This last limiting distribution is tabulated in Johansen (1988) on the basis of simulations. It is seen that, when $q = 1$ so that $r = 0$, the LR statistic is asymptotically the square of the unit root t-ratio statistic given below (3.16).

The present test was further extended by Johansen and Juselius (1990) and Johansen (1991) to allow for the fitted mean and trend terms in the error correction model (11.80). The power property of the Johansen tests is discussed in Hatanaka (1996).

11.8 HIGHER ORDER COINTEGRATION

In this section we extend the original model (11.1) in two directions; one considers $I(d)$ processes with d a positive integer greater than unity, and the other deals with seasonal time series. Inevitably the cointegration analysis for these models becomes much involved.

11.8.1 Cointegration in the I(d) Case

Let us consider the q-dimensional $I(d)$ process $\{y_j\}$ defined by

$$(11.89) \qquad (1-L)^d y_j = \Delta^d y_j = u_j, \qquad (j=1,\ldots,T),$$

where $\{u_j\}$ is a q-dimensional stationary process given by

$$(11.90) \qquad u_j = \sum_{l=0}^{\infty} \Phi_l \varepsilon_{j-l} = \Phi(L)\varepsilon_j, \quad \sum_{l=0}^{\infty} l^d \|\Phi_l\| < \infty$$

with $\{\varepsilon_j\} \sim$ i.i.d.$(0, I_q)$. The initial values y_j for $j = -(d-1), \ldots, 0$ are assumed to be 0.

To discuss cointegration for $I(d)$ processes, we decompose $\Phi(L)$ into

$$(11.91) \qquad \Phi(L) = A + \Psi_1 \Delta + \cdots + \Psi_{b-1}\Delta^{b-1} + \Psi_b(L)\Delta^b,$$

where b is some positive integer such that $b \leq d$, while $A = \Phi(1)$ and

$$\Psi_k = (-1)^k k! \left.\frac{d^k \Phi(x)}{dx^k}\right|_{x=1} \qquad \text{for } k = 1, \ldots, b-1.$$

The lag polynomial $\Psi_b(L)$ is determined a posteriori to attain the equality in (11.91).

Suppose that there exists a nonzero vector α_1 such that

$$(11.92) \qquad \alpha_1' \left(A, \Psi_1, \ldots, \Psi_{b_1-1}\right) = 0'.$$

Then it follows from $\Delta^d y_j = \Phi(L)\varepsilon_j$ and (11.91) that

$$(1-L)^{d-b_1} \alpha_1' y_j = \alpha_1' \Psi_{b_1}(L)\varepsilon_j,$$

that is, $\{\alpha_1' y_j\} \sim I(d - b_1)$. This may be described as cointegration with the cointegrating vector α_1 at integration order $d - b_1$. It is possible, however, that there exists another cointegrating vector α_2 such that

$$(1 - L)^{d-b_2} \alpha_2' y_j = \alpha_2' \Psi_{b_2}(L) \varepsilon_j.$$

Nonuniqueness of the resulting integration order may be overcome by choosing the smallest order. Then we simply write $\{y_j\} \sim CI(d, b)$ to mean that the $I(d)$ process $\{y_j\}$ has a cointegrating vector α such that $\{\alpha' y_j\} \sim I(d - b)$ with $d - b$ being the smallest possible integration order. Accordingly the cointegration rank is defined as the maximum number of such linearly independent αs.

Because of the complicated nature of higher order cointegration, the resulting cointegrated system also takes a complex form. Kitamura (1995) and Johansen (1995) discussed the estimation problem under such circumstances with $d = 2$. Here we consider only a simple system resulting from cointegration at integration order 0 or $\{y_j\} \sim CI(d, d)$:

$$(11.93) \qquad y_{2j} = \beta' y_{1j} + \xi_{2j}, \qquad \Delta^d y_{1j} = \xi_{1j},$$

where $y_j = (y_{1j}', y_{2j})' : q \times 1$ with $y_{1j} : (q - 1) \times 1$ and $y_{2j} : 1 \times 1$, while $\xi_j = (\xi_{1j}', \xi_{2j})' \sim$ i.i.d.$(0, \Sigma)$ with $\Sigma > 0$.

The estimation of β in (11.93) proceeds in much the same way as in the case of $d = 1$. Let us consider

$$(11.94) \quad \hat{\beta}_{\text{OLS}} = (Y_1' Y_1)^{-1} Y_1' Y_2 = \beta + (Y_1' Y_1)^{-1} Y_1' \Xi_2,$$

$$(11.95) \quad \hat{\beta}_{\text{2SLS}} = (Y_1' P_{-d} Y_1)^{-1} Y_1' P_{-d} Y_2 = \beta + (Y_1' P_{-d} Y_1)^{-1} Y_1' P_{-d} \Xi_2,$$

$$(11.96) \quad \hat{\beta}_{\text{ML}} = (Y_1' M_d Y_1)^{-1} Y_1' M_d Y_2 = \beta + (Y_1' M_d Y_1)^{-1} Y_1' M_d \Xi_2,$$

where $P_{-d} = Y_{-d}(Y_{-d}' Y_{-d})^{-1} Y_{-d}'$ with

$$Y_{-d} = L^d Y_1, M_d = I_T - \Delta^d Y_1 (\Delta^d Y_1' \Delta^d Y_1)^{-1} \Delta^d Y_1',$$

and the other matrices and vectors are defined as in (11.28). Note that $\hat{\beta}_{\text{2SLS}}$ is the 2SLSE obtained from replacing first Y_1 by $\hat{Y}_1 = P_{-d} Y_1$ and then regressing Y_2 on $\hat{Y}_1$, while $\hat{\beta}_{\text{ML}}$ is the ordinary LSE of β obtained from the augmented model

$$y_{2j} = \beta' y_{1j} + \gamma' \Delta^d y_{1j} + v_{2j},$$

where $\gamma = \Sigma_{11}^{-1} \Sigma_{12}$ and $v_{2j} = \xi_{2j} - \gamma' \xi_{1j}$. The estimator $\hat{\beta}_{\text{ML}}$ is the MLE of β for the system (11.93) if $\{\xi_j\} \sim NID(0, \Sigma)$.

The derivation of the asymptotic distributions of the above estimators is also similar to that for the case of $d = 1$. One difference is worth pointing out. Consider the auxiliary process

$$\Delta^d z_j = \Delta^d \begin{pmatrix} y_{1j} \\ x_j \end{pmatrix} = \xi_j,$$

where $\{\xi_j\} \sim$ i.i.d.$(0, \Sigma)$. By factorizing Σ as $\Sigma = BB'$ with B defined in (11.35), it follows from (3.103) that

(11.97) $$\mathcal{L}\left(\frac{1}{T^{2d}} \sum_{j=1}^{T} z_j z_j'\right) \to \mathcal{L}\left(B \int_0^1 F_{d-1}(t) F_{d-1}'(t)\, dt B'\right),$$

where $\{F_g(t)\}$ is the q-dimensional g-fold integrated Brownian motion introduced in Section 2.4 of Chapter 2. The weak convergence result in (11.97) applies to cases for $d \geq 1$, but we have

(11.98) $$\mathcal{L}\left(\frac{1}{T^d} \sum_{j=1}^{T} z_{j+h} \xi_j'\right) = \mathcal{L}\left(\frac{1}{T^d} \sum_{j=1}^{T} z_{j-1} \xi_j' + \frac{1}{T^d} \sum_{j=1}^{T} (1 - L^{h+1}) z_{j+h} \xi_j'\right)$$

$$\to \mathcal{L}\left(B \int_0^1 F_{d-1}(t)\, dw'(t) B'\right)$$

only for $d \geq 2$. Note that this result holds for any time shift h. This is because $\{(1 - L^{h+1}) z_{j+h}\}$ reduces to an I$(d - 1)$ process so that the second term on the right side of (11.98) is negligible whenever $d \geq 2$.

We can now establish the following theorem (Problem 8.1).

Theorem 11.12. *For the three estimators defined in (11.94), (11.95), and (11.96) for the system (11.93) with $d \geq 2$, it holds that*

$$\mathcal{L}\left(T^d \left(\hat{\beta}_{\text{OLS}} - \beta\right)\right) \to \mathcal{L}\left(V^{-1}(U_1 + U_2)\right),$$
$$\mathcal{L}\left(T^d \left(\hat{\beta}_{\text{2SLS}} - \beta\right)\right) \to \mathcal{L}\left(V^{-1}(U_1 + U_2)\right),$$
$$\mathcal{L}\left(T^d \left(\hat{\beta}_{\text{ML}} - \beta\right)\right) \to \mathcal{L}\left(V^{-1} U_2\right),$$

where

$$U_1 = \Sigma_{11}^{1/2} \int_0^1 F_{1,d-1}(t)\, dw_1'(t)\, \Sigma_{11}^{-1/2} \Sigma_{12},$$

$$U_2 = \Sigma_{11}^{1/2} \int_0^1 F_{1,d-1}(t)\, dw_2(t)\, \Sigma_{22 \cdot 1}^{1/2},$$

$$V = \Sigma_{11}^{1/2} \int_0^1 F_{1,d-1}(t) F_{1,d-1}'(t)\, dt\, \Sigma_{11}^{1/2},$$

HIGHER ORDER COINTEGRATION

with $F_{d-1}(t) = (F'_{1,d-1}(t), F_{2,d-1}(t))'$ and $w(t) = (w'_1(t), w_2(t))'$ being partitioned conformably with $y_j = (y'_{1j}, y_{2j})'$.

It is seen that $\hat{\beta}_{OLS}$ and $\hat{\beta}_{2SLS}$ have the same asymptotic distribution, unlike the case of $d = 1$. Since the limiting distribution contains the higher order unit root component $V^{-1}U_1$, it is biased. The estimator $\hat{\beta}_{ML}$ eliminates the higher order bias and the three estimators are asymptotically the same when $\Sigma_{12} = 0$, as in the case of $d = 1$.

The limiting distributions in Theorem 11.12 may be computed when $q = 2$. Let us consider $\hat{\beta}_{ML}$ only, for which $P(T(\hat{\beta}_{ML} - \beta) \leq x) \to P(X(d) \geq 0)$, where

$$X(d) = xV - U_2$$
$$= a^2 x \int_0^1 F^2_{1,d-1}(t)\,dt - ac \int_0^1 F_{1,d-1}(t)\,dw_2(t)$$

with $a = \Sigma_{11}^{1/2}$ and $c = \Sigma_{22\cdot1}^{1/2}$. It is easy to deduce by conditional arguments that

$$E\left[\exp\{i\theta X(d)/a^2\}\right] = E\left[\exp\left\{i\theta\left(x + \frac{c^2 i\theta}{2a^2}\right)\int_0^1 F^2_{1,d-1}(t)\,dt\right\}\right].$$

Then we have the following theorem using the results obtained in Section 5.4 of Chapter 5.

Theorem 11.13. *The limiting distributions of $T^d(\hat{\beta}_{ML} - \beta)$ for $d = 2$ and 3, when $q = 2$, can be computed as*

$$(11.99) \quad P(X(d) \geq 0) = \frac{1}{2} + \frac{1}{\pi}\int_0^\infty \frac{1}{\theta}\,\mathrm{Im}\left[\phi\left(\theta\left(x + \frac{c^2 i\theta}{2a^2}\right); d\right)\right]d\theta,$$

where

$$\phi(\theta; d) = E\left[\exp\left\{i\theta\int_0^1 F^2_{1,d-1}(t)\,dt\right\}\right]$$
$$= [D(2i\theta; d)]^{-1/2}$$

with $D(\lambda; 2)$ and $D(\lambda; 3)$ being given on the right sides of (5.89) and (5.91), respectively.

It is clear from (11.99) that the limiting distribution depends only on $(c/a)^2$ with the variance increasing with $(c/a)^2$. Figure 11.5 displays the limiting densities of $T(\hat{\beta}_{ML} - \beta)$ for $d = 2$ and 3 with $c/a = 1$.

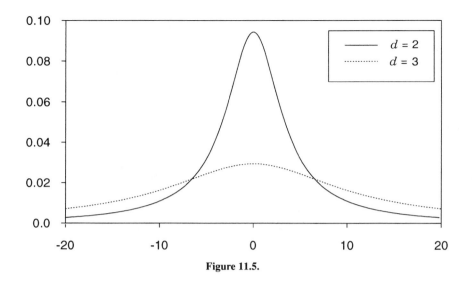

Figure 11.5.

11.8.2 Seasonal Cointegration

Another complexity arises if we consider the q-dimensional seasonal model

$$(11.100) \qquad (1 - L^m)y_j = \sum_{l=0}^{\infty} \Phi_l \varepsilon_{j-l} = \Phi(L)\varepsilon_j, \qquad (j = 1, \ldots, T),$$

where m is a positive integer greater than unity and $\{\varepsilon_j\} \sim$ i.i.d.$(0, I_q)$, while the coefficient matrices Φ_l are assumed to satisfy the summability condition described in (11.90) with d replaced by m.

Let $\theta_1, \ldots, \theta_m$ be m different roots of $x^m = 1$. Then $\Phi(L)$ may be expressed as

$$(11.101) \qquad \Phi(L) = \sum_{l=1}^{m} A_l \prod_{k \neq l}^{m} \left(1 - \frac{1}{\theta_k} L\right) + (1 - L^m)\tilde{\Phi}(L),$$

where $\tilde{\Phi}(L)$ is a lag polynomial of possibly infinite order, while

$$A_l = \Phi(\theta_l) \bigg/ \prod_{k \neq l}^{m} \left(1 - \frac{\theta_l}{\theta_k}\right).$$

The above representation for $\Phi(L)$ is originally due to J. L. Lagrange and is discussed in Hylleberg, Engle, Granger, and Yoo (1990), referred to as HEGY hereafter.

Suppose that $m = 4$, whose case corresponds to quarterly data. Then it follows [HEGY and Problem 8.2] from (11.101) that

$$(11.102) \qquad \Phi(L) = \Psi_1 \left(1 + L + L^2 + L^3\right) + \Psi_2 \left(1 - L + L^2 - L^3\right)$$
$$+ (\Psi_3 + \Psi_4 L)\left(1 - L^2\right) + \left(1 - L^4\right)\tilde{\Phi}(L),$$

where

$$\Psi_1 = \frac{1}{4}\Phi(1), \quad \Psi_2 = \frac{1}{4}\Phi(-1), \quad \Psi_3 = \frac{1}{2}\text{Re}[\Phi(i)], \quad \Psi_4 = \frac{1}{2}\text{Im}[\Phi(i)].$$

There are various possibilities of cointegration in the present case, among which are

(11.103) $\quad \alpha_1'\Phi(1) = 0' \rightarrow (1 + L + L^2 + L^3)\alpha_1'y_j \sim I(0),$

(11.104) $\quad \alpha_2'\Phi(-1) = 0' \rightarrow (1 - L + L^2 - L^3)\alpha_2'y_j \sim I(0),$

(11.105) $\quad \alpha_3'\Phi(i) = 0' \rightarrow (1 - L^2)\alpha_3'y_j \sim I(0),$

where $\alpha_1, \alpha_2,$ and α_3 are real vectors. Following HEGY, we call (11.103) cointegration at the frequency $\omega = 0$, (11.104) at $\omega = \pi$, and (11.105) at $\omega = \pi/2$. For (11.105) a weaker version of cointegration is possible by allowing for complex vectors. Thus we can relax the condition $\alpha_3'\Phi(i) = 0$ so that

(11.106) $\quad \alpha_3'(i)\Phi(i) = 0' \rightarrow (1 - L^2)\alpha_3'(L)y_j \sim I(0),$

where $\alpha_3(L)$ is a lag polynomial. Cointegration of this type is referred to as *polynomial cointegration* [Engle and Yoo (1991)]. It can be shown [HEGY and Problem 8.3] that, if $\alpha_3'(i)\Phi(i) = 0'$, then $\alpha_3(L)$ can be taken as the form of $\alpha_3(L) = \alpha_{30} + \alpha_{31}L$.

Returning to general m, let us consider the full cointegration situation where $\{\alpha'y_j\} \sim I(0)$ for some $\alpha \neq 0$. Then the following cointegrated system arises as a special case:

(11.107) $\quad y_{2j} = \beta'y_{1j} + \xi_{2j}, \quad (1 - L^m)y_{1j} = \xi_{1j},$

where $y_j = (y_{1j}', y_{2j})' : q \times 1$ with $y_{1j} : (q - 1) \times 1$ and $y_{2j} : 1 \times 1$, while $\xi_j = (\xi_{1j}', \xi_{2j})' \sim$ i.i.d.$(0, \Sigma)$ with $\Sigma > 0$.

Proceeding in the same way as before, we consider

(11.108) $\hat{\beta}_{\text{OLS}} = (Y_1'Y_1)^{-1}Y_1'Y_2 = \beta + (Y_1'Y_1)^{-1}Y_1'\Xi_2,$

(11.109) $\hat{\beta}_{\text{2SLS}} = (Y_1'P_{-m}Y_1)^{-1}Y_1'P_{-m}Y_2 = \beta + (Y_1'P_{-m}Y_1)^{-1}Y_1'P_{-m}\Xi_2,$

(11.110) $\hat{\beta}_{\text{ML}} = (Y_1'M_mY_1)^{-1}Y_1'M_mY_2 = \beta + (Y_1'M_mY_1)^{-1}Y_1'M_m\Xi_2,$

where $P_{-m} = Y_{-m}(Y_{-m}'Y_{-m})^{-1}Y_{-m}'$ with $Y_{-m} = L^mY_1, M_m = I_T - \tilde{Y}_1(\tilde{Y}_1'\tilde{Y}_1)^{-1}\tilde{Y}_1'$ with $\tilde{Y}_1 = (1 - L^m)Y_1 = \Xi_1$. Note that $\hat{\beta}_{\text{2SLS}}$ is the 2SLSE obtained from replacing first Y_1 by $\hat{Y}_1 = P_{-m}Y_1$ and then regressing Y_2 on $\hat{Y}_1$, while $\hat{\beta}_{\text{ML}}$ is the MLE of β for the system (11.107) if $\{\xi_j\} \sim \text{NID}(0, \Sigma)$.

The derivation of the asymptotic distributions of the above estimators may be done as follows: We first define

$$X_j = \begin{pmatrix} y'_{1,(j-1)m+1} \\ \vdots \\ y'_{1,jm} \end{pmatrix} : m \times (q-1), \quad z_j = \begin{pmatrix} y_{2,(j-1)m+1} \\ \vdots \\ y_{2,jm} \end{pmatrix} : m \times 1,$$

$$U_j = \begin{pmatrix} \xi'_{1,(j-1)m+1} \\ \vdots \\ \xi'_{1,jm} \end{pmatrix} : m \times (q-1), \quad \xi_{2j} = \begin{pmatrix} \xi_{2,(j-1)m+1} \\ \vdots \\ \xi_{2,jm} \end{pmatrix} : m \times 1.$$

Then the system (11.107) is equivalent, under the assumption that $T = mN$, to

(11.111) $\quad z_j = X_j \beta + \xi_{2j}, \quad X_j = X_{j-1} + U_j, \quad (j = 1, \ldots, N).$

Moreover we have

(11.112) $\quad N(\hat{\beta}_{OLS} - \beta) = \left(\frac{1}{N^2} \sum_{j=1}^{N} X'_j X_j \right)^{-1} \frac{1}{N} \sum_{j=1}^{N} X'_j \xi_{2j}.$

Let us consider the auxiliary process

(11.113) $\quad \begin{pmatrix} x_j \\ w_j \end{pmatrix} = \begin{pmatrix} x_{j-1} \\ w_{j-1} \end{pmatrix} + \begin{pmatrix} \xi_{1j} \\ \xi_{2j} \end{pmatrix}, \quad \begin{pmatrix} x_0 \\ w_0 \end{pmatrix} = 0,$

where

(11.114) $\quad x_j = (y'_{1,(j-1)m+1}, \ldots, y'_{1,jm})' = \text{vec}(X'_j),$

(11.115) $\quad \xi_{1j} = (\xi'_{1,(j-1)m+1}, \ldots, \xi'_{1,jm})' = \text{vec}(U'_j).$

Partitioning Σ conformably with ξ_j we have

$$\left\{ \begin{pmatrix} \xi_{1j} \\ \xi_{2j} \end{pmatrix} \right\} \sim \text{i.i.d.} \left(0, \begin{pmatrix} I_m \otimes \Sigma_{11} & I_m \otimes \Sigma_{12} \\ I_m \otimes \Sigma_{21} & I_m \otimes \Sigma_{22} \end{pmatrix} \right)$$

$$= \text{i.i.d.}(0, BB'),$$

where

(11.116) $\quad B = \begin{pmatrix} I_m \otimes \Sigma_{11}^{1/2} & 0 \\ I_m \otimes \Sigma_{21}\Sigma_{11}^{-1/2} & I_m \otimes \Sigma_{22\cdot 1}^{1/2} \end{pmatrix}.$

HIGHER ORDER COINTEGRATION

Then it follows from (3.103) and Theorem 3.16 that

$$\mathcal{L}\left(\frac{1}{N^2}\sum_{j=1}^{N}\binom{x_j}{w_j}(x_j', w_j')\right) \to \mathcal{L}\left(B\int_0^1 w(t)w'(t)\,dt B'\right),$$

$$\mathcal{L}\left(\frac{1}{N}\sum_{j=1}^{N}\binom{x_j}{w_j}(\xi_{1j}', \xi_{2j}')\right) \to \mathcal{L}\left(B\int_0^1 w(t)\,dw'(t) B' + BB'\right),$$

where $\{w(t)\}$ is the mq-dimensional standard Brownian motion.
In particular, we have

$$\mathcal{L}\left(\frac{1}{N^2}\sum_{j=1}^{N}x_j x_j'\right) \to \mathcal{L}\left(\left(I_m \otimes \Sigma_{11}^{1/2}\right)\int_0^1 w_1(t)w_1'(t)\,dt \left(I_m \otimes \Sigma_{11}^{1/2}\right)\right),$$

$$\mathcal{L}\left(\frac{1}{N}\sum_{j=1}^{N}x_j \xi_{2j}'\right) \to \mathcal{L}\Bigg(\left(I_m \otimes \Sigma_{11}^{1/2}\right)\int_0^1 \Big(w_1(t)\,dw_1'(t)\left(I_m \otimes \Sigma_{11}^{-1/2}\Sigma_{12}\right)$$

$$+ w_1(t)\,dw_2'(t)\left(I_m \otimes \Sigma_{22\cdot 1}^{1/2}\right)\Big) + I_m \otimes \Sigma_{12}\Bigg),$$

where $w(t) = (w_1'(t), w_2'(t))'$ with $w_1(t) : m(q-1) \times 1$ and $w_2(t) : m \times 1$.
To derive the limiting distribution of $N(\hat{\beta}_{\text{OLS}} - \beta)$ in (11.112) we note that

$$X_j' X_j = \text{sum of main diagonals of } x_j x_j',$$

$$X_j' \xi_{2j} = \text{sum of main diagonals of } x_j \xi_{2j}'.$$

This remark and the joint weak convergence of the associated quantities lead us to establish the following theorem (Problem 8.4).

Theorem 11.14. *For the three estimators defined in (11.108), (11.109), and (11.110) for the system (11.107) it holds that*

$$\mathcal{L}\left(N(\hat{\beta}_{\text{OLS}} - \beta)\right) \to \mathcal{L}\left(V^{-1}(U_1 + U_2 + m\Sigma_{12})\right),$$

$$\mathcal{L}\left(N(\hat{\beta}_{\text{2SLS}} - \beta)\right) \to \mathcal{L}\left(V^{-1}(U_1 + U_2)\right),$$

$$\mathcal{L}\left(N(\hat{\beta}_{\text{ML}} - \beta)\right) \to \mathcal{L}\left(V^{-1}U_2\right),$$

where $N = T/m$ is an integer and

$$U_1 = \Sigma_{11}^{1/2}\int_0^1 \{w_{11}(t)\,dw_{11}'(t) + \cdots + w_{1m}(t)\,dw_{1m}'(t)\}\Sigma_{11}^{-1/2}\Sigma_{12},$$

$$U_2 = \Sigma_{11}^{1/2} \int_0^1 \{w_{11}(t)\,dw_{21}(t) + \cdots + w_{1m}(t)\,dw_{2m}(t)\} \Sigma_{22\cdot1}^{1/2},$$

$$V = \Sigma_{11}^{1/2} \int_0^1 \{w_{11}(t)w'_{11}(t) + \cdots + w_{1m}(t)w'_{1m}(t)\}\,dt\,\Sigma_{11}^{1/2},$$

$$w_1(t) = \left(w'_{11}(t), \ldots, w'_{1m}(t)\right)' \; : \; m(q-1) \times 1$$

$$w_2(t) = (w_{21}(t), \ldots, w_{2m}(t))' \; : \; m \times 1.$$

The general feature of the three limiting distributions for $m \geq 2$ remains unchanged when compared with the case of $m = 1$. It is seen that U_1, U_2, and V are m-fold convolutions of the corresponding random quantities for $m = 1$.

The limiting distributions in Theorem 11.14 may be computed when $q = 2$. Let us consider $P(N(\hat{\beta}_{\text{OLS}} - \beta) \leq x) \to P(X_{\text{OLS}}(m) \geq 0)$, where

$$X_{\text{OLS}}(m) = a^2 x \int_0^1 w'_1(t)w_1(t)\,dt - ab \int_0^1 w'_1(t)\,dw_1(t)$$

$$- ac \int_0^1 w'_1(t)\,dw_2(t) - md$$

with

$$a = \Sigma_{11}^{1/2}, \qquad b = \Sigma_{12}^{-1/2}\Sigma_{12}, \qquad c = \Sigma_{22\cdot1}^{1/2}, \qquad d = \Sigma_{12}.$$

Then we can establish the following theorem (Problem 8.5).

Theorem 11.15. *When $q = 2$, the limiting distribution function of $N(\hat{\beta}_{\text{OLS}} - \beta)$ can be computed as*

$$(11.117) \quad \lim_{N \to \infty} P\left(N\left(\hat{\beta}_{\text{OLS}} - \beta\right) \leq x\right) = P\left(X_{\text{OLS}}(m) \geq 0\right)$$

$$= \frac{1}{2} + \frac{1}{\pi}\int_0^\infty \frac{1}{\theta} Im\left[\{\phi_1(\theta)\}^m\right] d\theta,$$

where $\phi_1(\theta)$ is given in Theorem 11.4.

The limiting distributions of $N\left(\hat{\beta}_{\text{2SLS}} - \beta\right)$ and $N\left(\hat{\beta}_{\text{ML}} - \beta\right)$ can be computed from (11.117) by putting $d = 0$ and $b = d = 0$, respectively. Figure 11.6 shows limiting densities of $N\left(\hat{\beta}_{\text{OLS}} - \beta\right)$ for $m = 4$ and 12 with $a = c = 1$ and $b = d = 0$.

Testing problems for seasonal cointegration are discussed in HEGY by taking no cointegration as the null hypothesis at each seasonal frequency. Engle, Granger, Hylleberg, and Lee (1993) applied the test to real data with $m = 4$, Beaulieu and

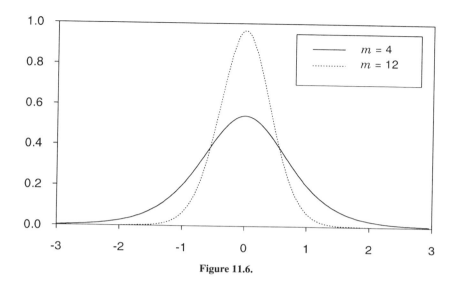

Figure 11.6.

Miron (1993) to real data with $m = 12$. The determination of the cointegration rank at each seasonal frequency is discussed in Lee (1992) by extending the Johansen procedure.

PROBLEMS

8.1 Prove Theorem 11.12.

8.2 Show that the lag polynomial $\Phi(L)$ can be expanded as in (11.102).

8.3 Show that, if $\alpha_3'(i)\Phi(i) = 0'$ holds in (11.106) for some lag polynomial $\alpha_3(L)$, we may take $\alpha_3(L) = \alpha_{30} + \alpha_{31}L$.

8.4 Prove Theorem 11.14.

8.5 Prove Theorem 11.15.

CHAPTER 12

Solutions to Problems

This chapter presents a complete set of solutions to problems given in the previous chapters. Most of the problems are concerned with corroborating the results described in the text. Thus this chapter will help make clear the details of discussions in the text.

CHAPTER 1

1.1 It follows from (1.3) that

$$D_T = |(CC')^{-1} - \lambda I_T| = (2 - \lambda)D_{T-1} - D_{T-2}$$

with $D_1 = 1 - \lambda$ and $D_2 = (2 - \lambda)(1 - \lambda) - 1$. Then we have

$$D_T = \frac{1 - x_1}{x_2 - x_1} x_1^T + \frac{x_2 - 1}{x_2 - x_1} x_2^T = \frac{\cos(T + 1/2)\omega}{\cos(\omega/2)},$$

where x_1 and x_2 are the solutions to $x^2 - (2 - \lambda)x + 1 = 0$. We can put $x_1 = e^{i\omega}$ and $x_2 = e^{-i\omega}$ so that $\cos \omega = 1 - \lambda/2$ and $\sin \omega = \sqrt{4\lambda - \lambda^2}/2$. The solutions to $D_T = 0$ are those to $\cos(T + \frac{1}{2})\omega = 0$, which yields $\omega = (t - \frac{1}{2})\pi/(T + \frac{1}{2})$ and thus $\lambda = 2 - 2\cos \omega = 4 \sin^2 \omega/2$.

1.2 We have $\phi_T(-i\theta) = (D_T(\theta))^{-1/2}$, where

$$D_T(\theta) = \left| I_T - \frac{2\theta}{(T + \frac{1}{2})^2} CC' \right| = \left| (CC')^{-1} - \frac{2\theta}{(T + \frac{1}{2})^2} I_T \right|.$$

Then it holds that

$$D_T(\theta) = \cos T\omega + \frac{(a - 2)\sin T\omega}{2 \sin \omega},$$

where $a = 2 - 2\theta/(T + \frac{1}{2})^2$, $\cos\omega = a/2$, and $\sin\omega = \sqrt{4-a^2}/2$. We now obtain

$$T\omega = T\tan^{-1}\frac{\sqrt{4-a^2}}{a} = \sqrt{2\theta}\left(1 - \frac{1}{2T} + \frac{\theta+3}{12T^2}\right) + O\left(\frac{1}{T^3}\right),$$

$$\cos T\omega = \cos\sqrt{2\theta} + \frac{\sqrt{2\theta}}{2T}\sin\sqrt{2\theta}$$
$$- \frac{1}{4T^2}\left(\theta\cos\sqrt{2\theta} + \sqrt{2\theta}\left(1 + \frac{\theta}{3}\right)\sin\sqrt{2\theta}\right) + O\left(\frac{1}{T^3}\right),$$

$$\sin T\omega = \sin\sqrt{2\theta} - \frac{\sqrt{2\theta}}{2T}\cos\sqrt{2\theta} + O\left(\frac{1}{T^2}\right),$$

$$\frac{a-2}{2\sin\omega} = \frac{a-2}{\sqrt{4-a^2}} = -\frac{\sqrt{2\theta}}{2T}\left(1 - \frac{1}{2T}\right) + O\left(\frac{1}{T^3}\right).$$

Substitution of these into $D_T(\theta)$ yields

$$D_T(\theta) = \cos\sqrt{2\theta}\left[1 + \frac{\theta}{4T^2}\left(1 - \frac{\sqrt{2\theta}}{3}\tan\sqrt{2\theta}\right)\right] + O\left(\frac{1}{T^3}\right),$$

and thus we arrive at the desired result.

1.3 Defining $\Sigma(\rho) = I_T + \rho CC'$ and noting that $z = \Sigma^{-1/2}(\rho_0)y \sim N(0, I_T)$ under H_0, we have, from (1.5),

$$\left.\frac{dL(\rho)}{d\rho}\right|_{\rho=\rho_0} = -\frac{1}{2}\operatorname{tr}\left(\Sigma^{-1}(\rho_0)CC'\right) + \frac{1}{2}z'\Sigma^{-1/2}(\rho_0)CC'\Sigma^{-1/2}(\rho_0)z$$

$$= \frac{1}{2}\sum_{t=1}^{T}\frac{\lambda_t}{1+\rho_0\lambda_t}(\xi_t^2 - 1),$$

where $\xi = (\xi_1, \ldots, \xi_T)' \sim N(0, I_T)$. The LM test for the present problem rejects H_0 for large values of the above statistic. By the Lyapunov central limit theorem it can be shown that

$$\frac{1}{\sqrt{T}}\left.\frac{dL(\rho)}{d\rho}\right|_{\rho=\rho_0} \to N(0, \sigma^2),$$

where

$$\sigma^2 = \frac{1}{2}\lim_{T\to\infty}\frac{1}{T}\sum_{t=1}^{T}\frac{\lambda_t^2}{(1+\rho_0\lambda_t)^2} = \frac{1}{2\pi}\int_0^{\pi/2}\frac{dx}{(\rho_0 + 4\sin^2 x)^2}$$

$$= \frac{\rho_0 + 2}{4(\rho_0(\rho_0+4))^{3/2}}.$$

1.4 Putting $x = (t - \frac{1}{2})\pi/(2T + 1)$, we have

$$\frac{4\lambda_t}{(2T+1)^2} - \frac{1}{(t-\frac{1}{2})^2 \pi^2} = \frac{1}{(2T+1)^2}\left(\frac{1}{\sin^2 x} - \frac{1}{x^2}\right),$$

which is positive and increasing for $0 < x < \pi/2$. Thus, for any $\delta > 0$,

$$P\left(\sum_{t=1}^{T}\left(\frac{4\lambda_t}{(2T+1)^2} - \frac{1}{(t-\frac{1}{2})^2 \pi^2}\right)\xi_t^2 \geq \delta\right)$$

$$\leq \frac{T}{\delta}\left(\frac{4\lambda_T}{(2T+1)^2} - \frac{1}{(T-\frac{1}{2})^2 \pi^2}\right)$$

$$= \frac{T}{\delta}\left(\frac{1}{(2T+1)^2}\left(\sin\frac{T-\frac{1}{2}}{2T+1}\pi\right)^{-2} - \frac{1}{(T-\frac{1}{2})^2 \pi^2}\right) \to 0,$$

which leads us to (1.8).

1.5 Put $F(x) = P(W \leq x)$ and $f(x) = F'(x)$. Then

$$\int_0^\infty e^{-\theta x} f(x)\,dx = \left(\cos\sqrt{-2\theta}\right)^{-1/2} = \left(\cosh\sqrt{2\theta}\right)^{-1/2}$$

$$= \sqrt{2}e^{-\sqrt{\theta/2}}\left(1 + e^{-2\sqrt{2\theta}}\right)^{-1/2}$$

$$= \sqrt{2}\sum_{n=0}^{\infty}\binom{-\frac{1}{2}}{n} e^{-(2n+1/2)\sqrt{2\theta}}$$

so that, taking the inverse Laplace transform,

$$f(x) = \sum_{n=0}^{\infty}\binom{-\frac{1}{2}}{n}\frac{\sqrt{2}}{2\sqrt{\pi x^3}}b_n e^{-b_n^2/(4x)}, \quad \left(b_n = \sqrt{2}\left(2n + \frac{1}{2}\right)\right),$$

$$F(x) = \int_0^x f(a)\,da$$

$$= \sum_{n=0}^{\infty}\binom{-\frac{1}{2}}{n}\frac{2}{\sqrt{\pi}}\int_{b_n/\sqrt{2x}}^{\infty} e^{-a^2/2}\,da$$

$$= 2\sqrt{2}\sum_{n=0}^{\infty}\binom{-\frac{1}{2}}{n}\Phi\left(-\frac{2n+\frac{1}{2}}{\sqrt{x}}\right).$$

2.1 By the definition of Ω we have

$$D_T = |\Omega - \lambda I_T| = (2-\lambda)D_{T-1} - D_{T-2}$$

with $D_1 = 2 - \lambda$ and $D_2 = (2-\lambda)^2 - 1$. Proceeding in the same way as in the solution to Problem 1.1, we obtain $D_T = \sin(T+1)\omega/\sin\omega$ with $\cos\omega = 1 - \lambda/2$ and $\sin\omega = \sqrt{4\lambda - \lambda^2}/2$. Thus the solutions to $D_T = 0$ are those to $\sin(T+1)\omega = 0$, which yields $\omega = t\pi/(T+1)$ and $\lambda = 2 - 2\cos\omega = 4\sin^2\frac{\omega}{2}$.

2.2 Put $\phi_T(-i\theta) = (\tilde{D}_T(\theta))^{-1/2}$. Then

$$\tilde{D}_T(\theta) = \left| I_T - \frac{2\theta}{(T+1)^2}\Omega^{-1} \right| = |\Omega|^{-1}\left|\Omega - \frac{2\theta}{(T+1)^2}I_T\right|$$

$$= \frac{1}{T+1}\left(\cos T\omega + \frac{a\sin T\omega}{2\sin\omega}\right),$$

where $a = 2 - 2\theta/(T+1)^2$, $\cos\omega = a/2$, and $\sin\omega = \sqrt{4-a^2}/2$. We now obtain

$$T\omega = T\tan^{-1}\frac{\sqrt{4-a^2}}{a} = \sqrt{2\theta}\left(1 - \frac{1}{T} + \frac{\theta+12}{12T^2}\right) + O\left(\frac{1}{T^3}\right),$$

$$\frac{\cos T\omega}{T+1} = \frac{1}{T}\cos\sqrt{2\theta} - \frac{1}{T^2}\left(\cos\sqrt{2\theta} - \sqrt{2\theta}\sin\sqrt{2\theta}\right) + O\left(\frac{1}{T^3}\right),$$

$$\sin T\omega = \sin\sqrt{2\theta} - \frac{\sqrt{2\theta}}{T}\cos\sqrt{2\theta}$$

$$+ \frac{1}{T^2}\left(\frac{\theta+12}{12}\sqrt{2\theta}\cos\sqrt{2\theta} - \theta\sin\sqrt{2\theta}\right) + O\left(\frac{1}{T^3}\right),$$

$$\frac{a}{2(T+1)\sin\omega} = \frac{a}{(T+1)\sqrt{4-a^2}} = \frac{1}{\sqrt{2\theta}} - \frac{3\sqrt{2\theta}}{8T^2} + O\left(\frac{1}{T^3}\right).$$

Substitution of these into $\tilde{D}_T(\theta)$ yields

$$\tilde{D}_T(\theta) = \frac{\sin\sqrt{2\theta}}{\sqrt{2\theta}}\left[1 + \frac{\theta}{4T^2}\left(1 + \frac{\sqrt{2\theta}}{3}\cot\sqrt{2\theta}\right)\right] + O\left(\frac{1}{T^3}\right)$$

and thus we arrive at the conclusion.

2.3 Putting $x = t\pi/(2(T+1))$, we can proceed completely in the same way as in the solution to Problem 1.4 and establish (1.16).

2.4 We have $\varepsilon'\Omega^{-2}\varepsilon = \sum_{t=1}^{T} \delta_t^2 \xi_t$, where $\xi_t \sim \text{NID}(0, 1)$ and δ_t is defined in (1.14). It can be proved, as in the solution to Problem 1.4, that

$$\text{plim}_{T\to\infty} \left(\frac{1}{T^4} \sum_{t=1}^{T} \delta_t^2 \xi_t^2 - \sum_{t=1}^{T} \frac{\xi_t^2}{t^4 \pi^4} \right) = 0$$

so that

$$\mathcal{L}\left(\frac{1}{T^4} \varepsilon'\Omega^{-2}\varepsilon\right) = \mathcal{L}\left(\frac{1}{T^4} \sum_{t=1}^{T} \delta_t^2 \xi_t^2\right) \to \mathcal{L}\left(\sum_{n=1}^{\infty} \frac{\xi_n^2}{n^4 \pi^4}\right)$$

and thus the c.f. $\phi(\theta)$ of this last random variable is

$$\phi(\theta) = \prod_{n=1}^{\infty} \left(1 - \frac{2i\theta}{n^4 \pi^4}\right)^{-1/2}$$

$$= \prod_{n=1}^{\infty} \left(1 - \frac{\sqrt{2i\theta}}{n^2 \pi^2}\right)^{-1/2} \prod_{n=1}^{\infty} \left(1 + \frac{\sqrt{2i\theta}}{n^2 \pi^2}\right)^{-1/2}$$

$$= \left(\frac{\sin(2i\theta)^{1/4}}{(2i\theta)^{1/4}}\right)^{-1/2} \left(\frac{\sinh(2i\theta)^{1/4}}{(2i\theta)^{1/4}}\right)^{-1/2}.$$

2.5 The log-likelihood $L(\alpha, \sigma^2)$ for y is

$$L(\alpha, \sigma^2) = -\frac{T}{2} \log(2\pi\sigma^2) - \frac{1}{2\sigma^2} y'\Phi^{-1}(\alpha) y,$$

where $\Phi(\alpha)$ is the same as $\Omega(\alpha)$ except the (1,1) element, which is unity. Then it is easy to obtain

$$\left.\frac{\partial L(\alpha, \sigma^2)}{\partial \alpha}\right|_{\alpha=1, \sigma^2=\hat{\sigma}^2} = \frac{1}{2\hat{\sigma}^2} y'C'C \left[C^{-1}(C^{-1})' - e_1 e_1'\right] C'Cy$$

$$= \frac{1}{2\hat{\sigma}^2} y'[C'C - C'ee'C]y$$

$$= \frac{T}{2}\left[1 - \frac{y'C'ee'Cy}{y'C'Cy}\right],$$

where $\hat{\sigma}^2 = y'C'Cy/T$ with C defined in (1.3), $e_1 = (1, 0, \ldots, 0)'$, and $e = (1, \ldots, 1)'$. The LM test rejects H_0 for small values of $\partial L(\alpha, \sigma^2)/\partial \alpha|_{\alpha=1, \sigma^2=\hat{\sigma}^2}$, which is equivalent to rejecting H_0 for large values of

$$S_T = \frac{y'C'ee'Cy}{y'C'Cy}.$$

Noting that $y \sim N(0, \sigma^2 C^{-1}(C^{-1})')$ under H_0, we have that S_T is asymptotically distributed as $\chi^2(1)$ under H_0.

2.6 The log-likelihood $L(\alpha, \sigma^2)$ for y is

$$L(\alpha, \sigma^2) = -\frac{T}{2}\log(2\pi\sigma^2) - \frac{1}{2\sigma^2}y'(C')^{-1}\Phi^{-1}(\alpha)C^{-1}y,$$

where $\Phi(\alpha)$ is defined in the solution to Problem 2.5. Then we obtain

$$\left.\frac{\partial L(\alpha, \sigma^2)}{\partial \alpha}\right|_{\alpha=1,\sigma^2=\hat{\sigma}^2} = \frac{1}{2\hat{\sigma}^2}y'(C')^{-1}C'C\left[C^{-1}(C^{-1})' - e_1e_1'\right]C'CC^{-1}y$$

$$= \frac{1}{2\hat{\sigma}^2}y'[I_T - ee']y$$

$$= \frac{T}{2}\left[1 - \frac{y'ee'y}{y'y}\right],$$

where $\hat{\sigma}^2 = y'y/T$. The LM test rejects H_0 for small values of the above quantity, which is equivalent to rejecting H_0 for large values of $y'ee'y/y'y = (\sum_{t=1}^T y_t)^2 / \sum_{t=1}^T y_t^2$.

2.7 The log-likelihood $L(\alpha, \mu, \sigma^2)$ for y is

$$L(\alpha, \mu, \sigma^2) = -\frac{T}{2}\log(2\pi\sigma^2) - \frac{1}{2\sigma^2}(y - \mu e)'(C')^{-1}\Phi^{-1}(\alpha)C^{-1}(y - \mu e).$$

The LM test considered here rejects H_0 for large values of

$$\frac{\partial^2 L(1, \hat{\mu}, \hat{\sigma}^2)}{\partial \alpha^2} = \frac{-1}{2\hat{\sigma}^2}y'M(C')^{-1}\frac{\partial^2 \Phi^{-1}(1)}{\partial \alpha^2}C^{-1}My$$

$$= \frac{1}{\hat{\sigma}^2}y'M(C')^{-1}\left[C'(CC' - ee')C - C'(I_T - ee')^2 C\right]C^{-1}My$$

$$= T\left[\frac{y'MCC'My}{y'My} - 1\right],$$

where $\hat{\mu} = \bar{y}$, $\hat{\sigma}^2 = y'My/T$, and $M = I_T - ee'/T$. Thus the LM test is equivalent to rejecting H_0 for large values of SL_T in (1.20).

2.8 Consider first

$$y'My = (C^{-1}y)'C'MC(C^{-1}y)$$

$$= \begin{pmatrix} y_1 \\ \Delta y \end{pmatrix}' C'MC \begin{pmatrix} y_1 \\ \Delta y \end{pmatrix},$$

where it can be checked that

$$C'MC = \begin{pmatrix} 0 & \cdots & \cdots & 0 \\ \vdots & & & \\ \vdots & & \Omega_*^{-1} & \\ 0 & & & \end{pmatrix}$$

CHAPTER 1

so that $y'My = (\Delta y)'\Omega_*^{-1}(\Delta y)$. Similarly we have

$$y'MCC'My = (C^{-1}y)'C'MCC'MC(C^{-1}y)$$
$$= (\Delta y)'\Omega_*^{-2}(\Delta y),$$

which establishes (1.21).

3.1 The first component of $C'M\varepsilon$ is 0 so that

$$C'M\varepsilon = \begin{pmatrix} 0 \\ C'_*(\varepsilon_* - \bar{e}e_*) \end{pmatrix},$$

where C'_*, ε_*, and e_* are the last $(T-1) \times (T-1)$, $(T-1) \times 1$, and $(T-1) \times 1$ submatrices of C', ε, and e, respectively. Noting that

$$\mathcal{L}(\varepsilon_* - \bar{e}e_*) = \mathcal{L}\left(\left(I_{T-1} - \frac{1}{T}e_*e'_*\right)^{1/2} \xi_*\right),$$

where $\xi_* \sim N(0, I_{T-1})$, we have

$$\mathcal{L}(\varepsilon'MCC'M\varepsilon) = \mathcal{L}((\varepsilon_* - \bar{e}e_*)'C_*C'_*(\varepsilon_* - \bar{e}e_*))$$
$$= \mathcal{L}\left(\xi'_*C'_*\left(I_{T-1} - \frac{1}{T}e_*e'_*\right)C_*\xi_*\right)$$
$$= \mathcal{L}(\xi'_*\Omega_*^{-1}\xi_*)$$

with Ω_* being the first $(T-1) \times (T-1)$ submatrix of Ω. Therefore (1.23) has been established.

3.2 Put $\phi_T(\theta) = a(\theta) + ib(\theta)$, where $a(\theta)$ and $b(\theta)$ are real and $a(\theta) = a(-\theta)$, $b(\theta) = -b(-\theta)$ because of the property of $\phi_T(\theta)$. Then we have

$$f_T(x) = \frac{1}{2\pi}\int_{-\infty}^{\infty}(a(\theta)\cos\theta x + b(\theta)\sin\theta x)\,d\theta$$
$$= \frac{1}{\pi}\int_0^{\infty}(a(\theta)\cos\theta x + b(\theta)\sin\theta x)\,d\theta = \frac{1}{\pi}\int_0^{\infty}\operatorname{Re}\left(e^{-i\theta x}\phi_T(\theta)\right)d\theta.$$

3.3 Consider first (1.7) and put

$$\psi_{1T}(\theta) = \log\phi_T(-i\theta)$$
$$\sim -\frac{1}{2}\log\cos\sqrt{2\theta} + \log\left[1 - \frac{\theta}{8T^2}\left(1 - \frac{\sqrt{2\theta}}{3}\tan\sqrt{2\theta}\right)\right]$$
$$\sim -\frac{1}{2}\log(1 - g(\theta)) - \frac{\theta}{8T^2}\left(1 - \frac{\sqrt{2\theta}}{3}\tan\sqrt{2\theta}\right)$$

$$\sim \frac{1}{2}\left(g(\theta) + \frac{g^2(\theta)}{2} + \frac{g^3(\theta)}{3} + \frac{g^4(\theta)}{4} + \cdots\right) - \frac{\theta}{8T^2}$$
$$+ \frac{\theta}{24T^2}\left(2\theta + \frac{4\theta^2}{3} + \frac{16\theta^3}{15} + \frac{272\theta^4}{315} + \cdots\right),$$

where

$$g(\theta) = \frac{2\theta}{2!} - \frac{4\theta^2}{4!} + \frac{8\theta^3}{6!} - \frac{16\theta^4}{8!} + \cdots.$$

Consider next (1.15) and put

$$\psi_{2T}(\theta) = \log \phi_T(-i\theta)$$

$$\sim -\frac{1}{2}\log \frac{\sin\sqrt{2\theta}}{\sqrt{2\theta}} + \log\left[1 - \frac{\theta}{8T^2}\left(1 + \frac{\sqrt{2\theta}}{3}\cot\sqrt{2\theta}\right)\right]$$

$$\sim -\frac{1}{2}\log(1 - h(\theta)) - \frac{\theta}{8T^2}\left(1 + \frac{\sqrt{2\theta}}{3}\cot\sqrt{2\theta}\right)$$

$$\sim \frac{1}{2}\left(h(\theta) + \frac{h^2(\theta)}{2} + \frac{h^3(\theta)}{3} + \frac{h^4(\theta)}{4} + \cdots\right) - \frac{\theta}{8T^2}$$

$$- \frac{\theta}{24T^2}\left(1 - \frac{2\theta}{3} - \frac{4\theta^2}{45} - \frac{16\theta^3}{945} - \cdots\right),$$

where

$$h(\theta) = \frac{2\theta}{3!} - \frac{4\theta^2}{5!} + \frac{8\theta^3}{7!} - \frac{16\theta^4}{9!} + \cdots.$$

Evaluating $\kappa_{iT}^{(j)} = d^j \psi_{iT}(0)/d\theta^j$ ($i = 1, 2$; $j = 1, 2, 3, 4$), we arrive at (1.28).

3.4 As for $\kappa_1^{(j)}$ consider

$$\psi_1(\theta) = -\frac{1}{2}\log \cos\sqrt{2\theta} = -\frac{1}{2}\sum_{n=1}^{\infty}\log\left(1 - \frac{2\theta}{(n-\frac{1}{2})^2\pi^2}\right)$$

$$= \sum_{n=1}^{\infty}\sum_{j=1}^{\infty}\frac{2^{j-1}\theta^j}{j\left((n-\frac{1}{2})\pi\right)^{2j}} = \sum_{j=1}^{\infty}\frac{\theta^j}{j!}(j-1)!\, 2^{3j-1}\sum_{n=1}^{\infty}\frac{1}{((2n-1)\pi)^{2j}}.$$

By the definition of the Bernoulli numbers, we have

$$B_j = \frac{(2j)!}{2^{2j-1}}\sum_{n=1}^{\infty}\frac{1}{(n\pi)^{2j}},$$

CHAPTER 1

so that

$$\sum_{n=1}^{\infty} \frac{1}{((2n-1)\pi)^{2j}} = \frac{1}{2(2j)!}(2^{2j}-1)B_j.$$

Therefore we obtain

$$\psi_1(\theta) = \sum_{j=1}^{\infty} \frac{\theta^j}{j!} \frac{(j-1)!\, 2^{3j-2}(2^{2j}-1)}{(2j)!} B_j,$$

which yields $\kappa_1^{(j)}$ in (1.29). The expression for $\kappa_2^{(j)}$ can be obtained similarly.

3.5 It is clear that

$$V_T = \frac{1}{T^2} y' \begin{pmatrix} 1 & & 0 \\ & \ddots & \\ & & 1 \\ 0 & & \delta \end{pmatrix} y = \frac{1}{T^2} \varepsilon' C' \begin{pmatrix} 1 & & 0 \\ & \ddots & \\ & & 1 \\ 0 & & \delta \end{pmatrix} C\varepsilon,$$

while

$$U_T = \frac{1}{T} \sum_{t=2}^{T} y_{t-1}(y_t - y_{t-1}) - \frac{\delta}{T} y_T^2$$

$$= -\frac{1}{2T} \left(\sum_{t=2}^{T} (y_t - y_{t-1})^2 - \sum_{t=2}^{T} y_t^2 + \sum_{t=2}^{T} y_{t-1}^2 \right) - \frac{\delta}{T} y_T^2$$

$$= -\frac{1}{2T} \left(\sum_{t=2}^{T} \varepsilon_t^2 - y_T^2 + y_1^2 \right) - \frac{\delta}{T} y_T^2$$

$$= \frac{1-2\delta}{2T} y_T^2 - \frac{1}{2T} \sum_{t=1}^{T} \varepsilon_t^2,$$

which leads us to the conclusion.

3.6 For any α, β, and γ let us consider

$$m_T(\theta) = \left| I_T - 2\theta \left\{ \alpha C' \begin{pmatrix} 1 & & 0 \\ & \ddots & \\ & & 1 \\ 0 & & \delta \end{pmatrix} C + \beta ee' + \gamma I_T \right\} \right|^{-1/2}$$

$$= D_T^{-1/2},$$

where

$$D_T = \left| a(CC')^{-1} + b \begin{pmatrix} 1 & & 0 \\ & \ddots & \\ & & 1 \\ 0 & & \delta \end{pmatrix} + c e_T e_T' \right|,$$

$$a = 1 - 2\gamma\theta, \quad b = -2\alpha\theta, \quad c = -2\beta\theta.$$

Since $D_T = (2a + b)D_{T-1} - a^2 D_{T-2}$ with $D_1 = a + b\delta + c$ and $D_2 = (2a + b)(a + b\delta + c) - a^2$, it can be shown that

$$D_T = \frac{1}{x_2 - x_1} \left[(x_2 - (a + b\delta + c)) x_1^T + (a + b\delta + c - x_1) x_2^T \right],$$

where x_1 and x_2 are solutions to $x^2 - (2a + b)x + a^2 = 0$. Putting $\alpha = x/T^2$, $\beta = (2\delta - 1)/(2T)$, and $\gamma = 1/(2T)$, and expressing x_1 and x_2 in polar form, we arrive at (1.33).

3.7 In the solution to Problem 3.6, we have $x_1 = re^{i\omega}$ and $x_2 = re^{-i\omega}$, where

$$r = 1 - \frac{\theta}{T}, \quad r\cos\omega = 1 - \frac{\theta}{T} - \frac{\theta x}{T^2},$$

$$r\sin\omega = \frac{1}{T}\sqrt{2\theta x \left(1 - \frac{\theta}{T} - \frac{\theta x}{2T^2}\right)}.$$

Therefore we have

$$r^T = \left(1 - \frac{\theta}{T}\right)^T = \exp\left\{T \log\left(1 - \frac{\theta}{T}\right)\right\}$$

$$= \left(1 - \frac{\theta^2}{2T}\right) e^{-\theta} + O\left(\frac{1}{T^2}\right),$$

$$T\omega = T\tan^{-1}\left[\frac{1}{T}\left(1 - \frac{\theta}{T} - \frac{\theta x}{T^2}\right)^{-1} \left(2\theta x \left(1 - \frac{\theta}{T} - \frac{\theta x}{2T^2}\right)\right)^{1/2}\right]$$

$$= \sqrt{2\theta x} + \frac{\theta\sqrt{2\theta x}}{2T} + O\left(\frac{1}{T^2}\right),$$

$$\cos T\omega = \cos\sqrt{2\theta x} - \frac{\theta\sqrt{2\theta x}}{2T}\sin\sqrt{2\theta x} + O\left(\frac{1}{T^2}\right),$$

$$\sin T\omega = \sin\sqrt{2\theta x} + \frac{\theta\sqrt{2\theta x}}{2T}\cos\sqrt{2\theta x} + O\left(\frac{1}{T^2}\right),$$

$$\frac{r\cos\omega - d}{r\sin\omega} = T\left(1 - \frac{\theta}{T} - \frac{\theta x}{T^2} - \left(1 - \frac{2\delta\theta}{T} - \frac{2\delta\theta x}{T^2}\right)\right)$$

$$\times \left(2\theta x\left(1 - \frac{\theta}{T} - \frac{\theta x}{2T^2}\right)\right)^{-1/2}$$

$$= -\frac{\theta(1-2\delta)}{\sqrt{2\theta x}} - \frac{\theta(\theta(1-2\delta) + 2x - 4\delta x)}{2T\sqrt{2\theta x}} + O\left(\frac{1}{T^2}\right).$$

Substituting these into $m_T(\theta)$ in (1.33) we arrive at (1.34).

3.8 Let the random variable on the right side of (1.42) be $Y_T = A_T/B_T$. Consider then $P(Y_T \le x) = P(xB_T - A_T \ge 0)$, where

$$xB_T - A_T = \frac{x}{T^2}\sum_{t=2}^{T} y_{t-1}^2 - \frac{1}{T}\sum_{t=2}^{T} y_{t-1}\varepsilon_t - \frac{1}{2}.$$

Comparing with (1.32) and (1.35), it is easily seen that the limiting c.f. $\phi(\theta; x)$ of this last random variable is given by

$$\phi(\theta; x) = \left[\cos\sqrt{2i\theta x} + i\theta\frac{\sin\sqrt{2i\theta x}}{\sqrt{2i\theta x}}\right]^{-1/2}$$

so that

$$\psi(\theta_1, -\theta_2) = \phi(i\theta_1; \theta_2/\theta_1)$$

$$= \left[\cosh\sqrt{2\theta_2} - \theta_1\frac{\sinh\sqrt{2\theta_2}}{\sqrt{2\theta_2}}\right]^{-1/2}.$$

Thus $E(Y)$, where $\mathcal{L}(Y_T) \to \mathcal{L}(Y)$, is given by

$$E(Y) = \int_0^\infty \frac{\partial \psi(\theta_1, -\theta_2)}{\partial \theta_1}\bigg|_{\theta_1=0} d\theta_2$$

$$= \frac{1}{2}\int_0^\infty \frac{\sinh\sqrt{2\theta}}{\sqrt{2\theta}}(\cosh\sqrt{2\theta})^{-3/2} d\theta$$

$$= 1,$$

where the last equality results from the second by putting $(\cosh\sqrt{2\theta})^{-1/2} = u$.

3.9 From (1.40) we have $\mu_{1-\delta}(1) = \mu_\delta(1) - 2(1-2\delta)$ so that (1.43) is equivalent to

(S1) $\quad \mu_{1-\delta}(2) - \mu_\delta(2) = 4(1-2\delta)(1 - 2\delta - \mu_\delta(1)).$

Putting

$$\psi_\delta(\theta_1, -\theta_2) = \phi_\delta(i\theta_1; \theta_2/\theta_1)$$

$$= e^{-\theta_1/2} \left[\cosh\sqrt{2\theta_2} - \theta_1(1-2\delta)\frac{\sinh\sqrt{2\theta_2}}{\sqrt{2\theta_2}}\right]^{-1/2},$$

it can be shown from (1.39) that the left side of (S1) is equal to

$$\int_0^\infty \theta_2 \frac{\partial^2}{\partial \theta_1^2}(\psi_{1-\delta}(\theta_1, -\theta_2) - \psi_\delta(\theta_1, -\theta_2))|_{\theta_1=0}\, d\theta_2$$

$$= (1-2\delta)\int_0^\infty \theta\frac{\sinh\sqrt{2\theta}}{\sqrt{2\theta}}(\cosh\sqrt{2\theta})^{-3/2}\, d\theta,$$

while, from the solution to Problem 3.8,

$$\mu_\delta(1) = -\frac{1}{2}\int_0^\infty \frac{1}{\sqrt{\cosh\sqrt{2\theta_1}}}\left(1 - \frac{1-2\delta}{\cosh\sqrt{2\theta_1}}\frac{\sinh\sqrt{2\theta_1}}{\sqrt{2\theta_1}}\right)d\theta_1$$

$$= -\frac{1}{2}\int_0^\infty \frac{d\theta}{\sqrt{\cosh\sqrt{2\theta}}} + (1-2\delta)$$

$$= -\frac{1}{4}\int_0^\infty \theta\frac{\sinh\sqrt{2\theta}}{\sqrt{2\theta}}(\cosh\sqrt{2\theta})^{-3/2}\, d\theta + (1-2\delta).$$

Substituting this last expression into the right side of (S1), we arrive at the conclusion.

4.1 We have

$$\phi_T(\theta) = \left|I_{2T} - \frac{i\theta}{T^2}\begin{pmatrix} 0 & C'C \\ C'C & 0 \end{pmatrix}\right|^{-1/2}$$

$$= \left|\begin{pmatrix} I_T & -\frac{i\theta}{T^2}C'C \\ -\frac{i\theta}{T^2}C'C & I_T \end{pmatrix}\right|^{-1/2}$$

$$= \left|I_T + \frac{\theta^2}{T^4}(C'C)^2\right|^{-1/2},$$

which yields the result.

CHAPTER 1

4.2 We rewrite V_T as

$$V_T = \frac{1}{T}\sum_{t=1}^{T} y_{1t}(y_{1t} - y_{1,t-1})$$

$$= \frac{1}{2T}\left(\sum_{t=1}^{T}(y_{1t} - y_{1,t-1})^2 + \sum_{t=1}^{T} y_{1t}^2 - \sum_{t=1}^{T} y_{1,t-1}^2\right)$$

$$= \frac{1}{2T}\sum_{t=1}^{T} \varepsilon_{1t}^2 + \frac{1}{2}\left(\frac{1}{\sqrt{T}}\sum_{t=1}^{T} \varepsilon_{1t}\right)^2.$$

Here the first term converges in probability to $\frac{1}{2}$, while the second converges in distribution to $\xi^2/2$, which establishes the result.

4.3 For $\phi_{1T}(\theta)$ we can proceed in the same way as in the solution to Problem 4.1. Put $e = (1, \ldots, 1)' : T \times 1$ and we have

$$\phi_{1T}(\theta) = \left|\left(I_T + \frac{i\theta}{2T}(C' + C)\right)\left(I_T - \frac{i\theta}{2T}(C' + C)\right)\right|^{-1/2}$$

$$= \left|\left(\left(1 + \frac{i\theta}{2T}\right)I_T + \frac{i\theta}{2T}ee'\right)\left(\left(1 - \frac{i\theta}{2T}\right)I_T - \frac{i\theta}{2T}ee'\right)\right|^{-1/2}$$

$$= \left|\left(1 + \frac{i\theta}{2T}\right)^{T-1}\left(1 + \frac{i\theta}{2} + \frac{i\theta}{2T}\right)\left(1 - \frac{i\theta}{2T}\right)^{T-1}\left(1 - \frac{i\theta}{2} - \frac{i\theta}{2T}\right)\right|^{-1/2},$$

which yields (1.54). For $\phi_{2T}(\theta)$ put $a = \theta/(2T)$ and consider

$$\phi_{2T}(\theta) = \left(D_T(a)D_T(-a)\right)^{-1/2},$$

where

$$D_T(a) = |I_T + aC' - aC|$$
$$= |(C')^{-1} + aI_T - a(C')^{-1}C|$$
$$= (1 + a)D_{T-1}(a) - a(1 - a)^{T-1}.$$

Noting that $D_1(a) = 1$, we can solve the above difference equation as

$$D_T(a) = \frac{1}{2}\left((1 + a)^T + (1 - a)^T\right),$$

which leads us to (1.55).

4.4 It is easy to see that

$$\psi_1(\theta) = \log \phi_1(-i\theta) = -\frac{1}{2}\log\left(1 - \frac{\theta^2}{4}\right)$$

$$= \sum_{j=1}^{\infty} \frac{\theta^{2j}}{(2j)!} \frac{(2j)!}{j \, 2^{2j+1}},$$

$$\psi_2(\theta) = \log \phi_2(-i\theta) = -\sum_{n=1}^{\infty} \log\left(1 - \frac{\theta^2}{((2n-1)\pi)^2}\right)$$

$$= \sum_{j=1}^{\infty} \frac{\theta^{2j}}{(2j)!} \frac{(2j)!}{j} \sum_{n=1}^{\infty} \frac{1}{((2n-1)\pi)^{2j}}$$

$$= \sum_{j=1}^{\infty} \frac{\theta^{2j}}{(2j)!} \frac{2^{2j} - 1}{2j} B_j,$$

where the last equality comes from the solution to Problem 3.4. Then the expressions for cumulants are easily obtained.

4.5 We have

$$\phi_{1T}(\theta; x) = \left| \left(I_T - \frac{2i\theta x}{T^2}C'C \quad \frac{i\theta}{T^2}C'C \atop \frac{i\theta}{T^2}C'C \quad I_T \right) \right|^{-1/2}$$

$$= \left| I_T - \frac{2i\theta x}{T^2}C'C + \frac{\theta^2}{T^4}(C'C)^2 \right|^{-1/2}$$

$$= \prod_{t=1}^{T} \left(1 - \frac{2i\theta x}{T^2}\lambda_t - \frac{(i\theta)^2}{T^4}\lambda_t^2\right)^{-1/2},$$

which yields (1.67).

4.6 We first note that

$$X_{2T} = \varepsilon' \begin{pmatrix} \frac{x}{T^2}C'MC & -\frac{1}{2T^2}C'MC \\ -\frac{1}{2T^2}C'MC & 0 \end{pmatrix} \varepsilon,$$

where $M = I_T - ee'/T$. Then we can proceed in the same way as in the solution to Problem 4.5 to arrive at (1.70).

4.7 From (1.39) we have

(S2) $$\mu_j(k) = \frac{1}{(k-1)!} \int_0^\infty \theta_2^{k-1} \frac{\partial^k \psi_j(\theta_1, -\theta_2)}{\partial \theta_1^k}\bigg|_{\theta_1=0} d\theta_2,$$

where

$$\psi_1(\theta_1, -\theta_2) = \phi_1(i\theta_1; \theta_2/\theta_1)$$
$$= \left[\cosh\sqrt{\theta_2 + \sqrt{\theta_1^2 + \theta_2^2}} \ \cosh\sqrt{\theta_2 - \sqrt{\theta_1^2 + \theta_2^2}} \right]^{-1/2},$$

$$\psi_2(\theta_1, -\theta_2) = \phi_2(i\theta_1; \theta_2/\theta_1)$$
$$= \left[\frac{\sinh\sqrt{\theta_2 + \sqrt{\theta_1^2 + \theta_2^2}}}{\sqrt{\theta_2 + \sqrt{\theta_1^2 + \theta_2^2}}} \frac{\sinh\sqrt{\theta_2 - \sqrt{\theta_1^2 + \theta_2^2}}}{\sqrt{\theta_2 - \sqrt{\theta_1^2 + \theta_2^2}}} \right]^{-1/2}.$$

Then we have, for instance,

$$\mu_1(2) = \frac{1}{4} \int_0^\infty \left\{ (\cosh\sqrt{2\theta})^{-1/2} - (\cosh\sqrt{2\theta})^{-3/2} \frac{\sinh\sqrt{2\theta}}{\sqrt{2\theta}} \right\} d\theta$$
$$= \frac{1}{4} \int_0^\infty \frac{u}{\sqrt{\cosh u}} du - \frac{1}{2}.$$

The other moments can be derived similarly. We used computerized algebra REDUCE to differentiate $\psi_j(\theta_1, -\theta_2)$.

4.8 We have only to show that $F_j(-x) = 1 - F_j(x)$. Because of the definition of $\phi_j(\theta; x)$, it is easy to see that $\phi_j(\theta; -x) = \phi_j(-\theta; x)$ so that

$$F_j(-x) = \frac{1}{2} + \frac{1}{\pi} \int_0^\infty \frac{1}{\theta} \operatorname{Im}(\phi_j(-\theta; x)) d\theta$$
$$= \frac{1}{2} - \frac{1}{\pi} \int_0^\infty \frac{1}{\theta} \operatorname{Im}(\phi_j(\theta; x)) d\theta$$
$$= 1 - F_j(x).$$

5.1 It is easy to see that

$$\phi_{1T}(\theta; x) = \left| \begin{pmatrix} I_T - \frac{2i\theta x}{T^2} C'C & \frac{i\theta}{T} C' \\ \frac{i\theta}{T} C & I_T \end{pmatrix} \right|^{-1/2}$$
$$= \left| I_T - \frac{2i\theta x}{T^2} C'C + \frac{\theta^2}{T^2} C'C \right|^{-1/2}$$
$$= \prod_{t=1}^T \left[1 - (2i\theta x - \theta^2) \frac{\lambda_t}{T^2} \right]^{-1/2},$$

and this last expression can be further factored as in (1.82).

5.2 We have

$$X_{2T} = \varepsilon' \begin{pmatrix} \dfrac{x}{T^2} C'MC & -\dfrac{1}{2T} C'M \\ -\dfrac{1}{2T} MC & 0 \end{pmatrix} \varepsilon,$$

where $M = I_T - ee'/T$. Then we can proceed in the same way as in the solution to Problem 5.1 to arrive at (1.85).

5.3 In (S2) we have

$$\psi_1(\theta_1, -\theta_2) = \left(\cosh\sqrt{2\theta_2 - \theta_1^2}\right)^{-1/2},$$

$$\psi_2(\theta_1, -\theta_2) = \left(\dfrac{\sinh\sqrt{2\theta_2 - \theta_1^2}}{\sqrt{2\theta_2 - \theta_1^2}}\right)^{-1/2}.$$

Then we can arrive at (1.87) after some manipulations. We used computerized algebra REDUCE to differentiate $\psi_j(\theta_1, -\theta_2)$.

5.4 We have only to show that $F_j(-x) = 1 - F_j(x)$, which can be easily checked as in the solution to Problem 4.8.

CHAPTER 2

1.1 Suppose that l.i.m. $X_n = X$ and l.i.m. $X_n = Y$. Since

$$E\left[(X - Y)^2\right] \le E\left[(X_n - X)^2\right] + 2\sqrt{E\left[(X_n - X)^2\right] E\left[(X_n - Y)^2\right]} + E\left[(X_n - Y)^2\right]$$

and the right side converges to 0, $E\left[(X - Y)^2\right] = 0$. Thus we have $P(X = Y) = 1$.

1.2 Since $E\left[(X_m - X_n)^2\right] = E\left(X_m^2 - 2X_mX_n + X_n^2\right) = 2(1 - 1/\sqrt{mn})$ for $m \ne n$, which does not tend to 0, $\{X_n\}$ does not converge in the m.s. sense. For any $\varepsilon > 0$, we have $P(|X_n| > \varepsilon) = P(X_n = \sqrt{n}) = 1/n \to 0$; hence $\{X_n\}$ converges in probability to 0.

1.3 Put $Z_n(t) = aX_n(t) + bY_n(t)$ and $Z(t) = aX(t) + bY(t)$. Note that l.i.m. $Z_n(t) = Z(t)$. For any $q \times 1$ vector c we have

$$\left|E\left[c'(Z_n(t) - Z(t))\right]\right| \le \sqrt{c'cE\left[(Z_n(t) - Z(t))'(Z_n(t) - Z(t))\right]}$$

so that $E(Z_n(t)) \to E(Z(t))$ as $n \to \infty$. Consider next

$$X_n'(t)Y_n(t) - X'(t)Y(t) = (X_n(t) - X(t))'(Y_n(t) - Y(t))$$
$$+ X'(t)(Y_n(t) - Y(t)) + (X_n(t) - X(t))'Y(t).$$

Taking expectations leads from the Cauchy–Schwarz inequality to $E(X'_n(t)Y_n(t)) \to E(X'(t)Y(t))$.

1.4 $E(Y_m(t)Y_n(t)) \to 2$ as $m = n \to \infty$, while it converges to 1 as $m, n(\neq m) \to \infty$. Thus $\{Y_n(t)\}$ does not converge in the m.s. sense.

1.5 From Theorem 2.2, it holds that $Y(t)$ is m.s. continuous if and only if $E(Y'(t + h_1)Y(t + h_2))$ converges to $E(Y'(t)Y(t))$ as $h_1, h_2 \to 0$ in any manner, which is equivalent to the condition that $E(Y'(s)Y(t))$ is continuous at (t,t).

1.6 Since $\{Y(t)\}$ is m.s. continuous at every $t \in [a, b]$,

$$\underset{h_1 \to 0}{\text{l.i.m.}} Y(s + h_1) = Y(s), \quad \underset{h_2 \to 0}{\text{l.i.m.}} Y(t + h_2) = Y(t).$$

Therefore it follows from Theorem 2.1 that $E(Y'(s + h_1)Y(t + h_2)) \to E(Y'(s)Y(t))$ as $h_1, h_2 \to 0$.

1.7 Noting that $E(X(t)) = V(X(t)) = \lambda t$, we have, for $s < t$,

$$E(X(s)X(t)) = E(X(s))E(X(t) - X(s)) + E(X^2(s))$$
$$= \lambda s + \lambda^2 st$$

so that $E(X(s)X(t)) = \lambda \min(s,t) + \lambda^2 st$, which is continuous at every (t,t). Thus $\{X(t)\}$ is m.s. continuous at all t. On the other hand, it holds that

$$E[(X(t+h) - X(t))(X(t+k) - X(t))]/(hk)$$
$$= \lambda \frac{\min(h,k) - \min(h,0) - \min(0,k)}{hk} + \lambda^2,$$

which does not converge as $h, k \to 0$. Thus $\{X(t)\}$ is nowhere m.s. differentiable.

1.8 It follows from Theorem 2.1 that

$$E(\dot{Y}(t)) = E\left(\underset{h \to 0}{\text{l.i.m.}} \frac{Y(t+h) - Y(t)}{h}\right)$$
$$= \lim_{h \to 0} E\left(\frac{Y(t+h) - Y(t)}{h}\right) = \frac{d}{dt}E(Y(t)).$$

The relation (2.3) can be proved similarly.

1.9 Note first that $E(Y(t)Y(t+h)) = \frac{1}{2}\cos \omega h$ so that

$$E\left[(Y(t+h_1) - Y(t))(Y(t+h_2) - Y(t))\right]/(h_1 h_2)$$
$$= \left[\cos \omega(h_2 - h_1) - \cos \omega h_1 - \cos \omega h_2 + 1\right]/(2h_1 h_2) \to \frac{\omega^2}{2}.$$

2.1 (a) $E[(w(t+h) - w(t))'(w(t+h) - w(t))] = q|h| \to 0$ as $h \to 0$.
(b) $E[(w(t+h) - w(t))'(w(t+h) - w(t))]/h^2 = q/|h|$ does not converge as $h \to 0$.

(c) Put $\Delta w_i = w(t_i) - w(t_{i-1})$ and $\Delta t_i = t_i - t_{i-1}$. Then

$$E\left[\left(\sum_{i=1}^n \Delta w_i' \Delta w_i - (b-a)q\right)^2\right] = E\left[\left\{\sum_{i=1}^n (\Delta w_i' \Delta w_i - q\Delta t_i)\right\}^2\right]$$

$$= \sum_{i=1}^n E\left[(\Delta w_i' \Delta w_i - q\Delta t_i)^2\right] = 2q\sum_{i=1}^n (\Delta t_i)^2 \leq 2q\Delta_n(b-a) \to 0.$$

2.2 It is clear that $w(0) \equiv 0$ and $E(w(t)) = E(\Delta w_i) = 0$, where $\Delta w_i = w(t_i) - w(t_{i-1})$. For $t_{i-1} < t_i \leq t_{k-1} < t_k$, we have

$$E(\Delta w_i \Delta w_k') = \sum_{n=1}^\infty \frac{2}{\left(\left(n-\frac{1}{2}\right)\pi\right)^2}[\sin a_{ni} - \sin a_{n,i-1}][\sin a_{nk} - \sin a_{n,k-1}]I_q$$

$$= \sum_{n=1}^\infty \frac{1}{\left(\left(n-\frac{1}{2}\right)\pi\right)^2}[\cos(a_{ni} - a_{nk}) - \cos(a_{ni} + a_{nk}) - \cos(a_{ni} - a_{n,k-1})$$

$$+ \cos(a_{ni} + a_{n,k-1}) - \cos(a_{n,i-1} - a_{n,k}) + \cos(a_{n,i-1} + a_{n,k})$$

$$+ \cos(a_{n,i-1} - a_{n,k-1}) - \cos(a_{n,i-1} + a_{n,k-1})]I_q,$$

where $a_{ni} = \left(n-\frac{1}{2}\right)\pi t_i$. Using the formula given in the problem, it can be shown that $E(\Delta w_i \Delta w_k') = 0$. Similarly we have $E(\Delta w_i \Delta w_i') = (t_i - t_{i-1})I_q$ so that Δw_i is independent $N(0, (t_i - t_{i-1})I_q)$. We also have $w(t) \sim N(0, tI_q)$. Thus $\{w(t)\}$ is the q-dimensional standard Brownian motion.

2.3 For $s \leq t$ we have

$$\begin{pmatrix} w(s) \\ w(t) \\ w(1) \end{pmatrix} \sim N(0, \Sigma), \quad \Sigma = \begin{pmatrix} s & s & s \\ s & t & t \\ s & t & 1 \end{pmatrix} \otimes I_q$$

with $\otimes$ the Kronecker product. Then it holds that $E(w(t)|w(1) = 0) = 0$ and

$$V(w(s), w(t) \mid w(1) = 0) = \begin{pmatrix} s - s^2 & s - st \\ s - st & t - t^2 \end{pmatrix} \otimes I_q$$

so that $\mathrm{Cov}(w(s), w(t) \mid w(1) = 0) = (\min(s,t) - st)I_q$.

2.4 It is clear that $\{\bar{w}(t)\}$ is a Gaussian process with $\bar{w}(0) = \bar{w}(1) \equiv 0$ and $E(\bar{w}(t)) = 0$. Moreover, for $s < t$, we have

$$E(\bar{w}(s)\bar{w}'(t)) = \sum_{n=1}^\infty \frac{1}{n^2\pi^2}(\cos n\pi(s-t) - \cos n\pi(s+t))I_q$$

$$= \left[\frac{1}{4}(t-s-1)^2 - \frac{1}{12} - \left(\frac{1}{4}(s+t-1)^2 - \frac{1}{12}\right)\right]I_q$$

$$= (s - st)I_q.$$

3.1 From Theorem 2.2, the integral in (2.7) exists if and only if $E(V'_m(t)V_n(t))$ converges to a finite function on $[a,b]$ as $m, n \to \infty$ in any manner, which is equivalent to the condition that the integral in (2.8) exists and is finite.

3.2 It follows from the solution to Problem 1.5 that $E(Y'(s)Y(t))$ is continuous on $[a,b] \times [a,b]$. Then it is clear that the double Riemann integral $\iint E(Y'(s)Y(t))\,ds\,dt$ exists and is finite. Thus $\{Y(t)\}$ is m.s. integrable by Theorem 2.3.

3.3 Since $E(w(s)w(t)) = \min(s,t)$, it is easy to obtain $E(V) = \frac{1}{2}$ and $E(W) = \frac{1}{6}$. Noting also that $E(w^2(s)w^2(t)) = 2\min^2(s,t) + st$ we obtain $E(V^2) = \frac{7}{12}$ and $E(W^2) = \frac{1}{20}$.

3.4 From Theorem 2.3 we have only to check that the integral (2.8) with $f(r,t)$ and $Y(r)$ replaced by $I_{[0,t]}(r)$ and $w(r)$, respectively, exists and is finite. The integral is

$$\int_0^1 \int_0^1 I_{[0,t]}(r) I_{[0,t]}(s) E(w'(r)w(s))\,dr\,ds = \frac{qt^3}{3}$$

so that $V(t)$ in (2.10) is well defined. From Theorem 2.4 and the above result we have that $V(t) \sim N(0, t^3 I_q/3)$.

3.5 Let us define

$$V_{1m} = \sum_{i=1}^m (1 - s'_i)\bigl(w(s_i) - w(s_{i-1})\bigr)$$

and consider

$$E(V'_{1m}V_{1n}) = \sum_{i=1}^m \sum_{j=1}^n (1-s'_i)(1-t'_j) E\left[\bigl(w(s_i) - w(s_{i-1})\bigr)' \bigl(w(t_j) - w(t_{j-1})\bigr)\right].$$

It can be checked that this last quantity converges to $q \int_0^1 (1-t)^2\,dt = q/3$ as $m, n \to \infty$. Thus V_1 is well defined.

3.6 We may assume that H is diagonal and thus we have only to show that

$$A = \int_0^1 \int_0^1 \int_0^1 \int_0^1 K(s,t)K(u,v) E\bigl(dw(s)\,dw(t)\,dw(u)\,dw(v)\bigr)$$

exists and is finite when $\{w(t)\}$ is scalar. We have

$$E\bigl(dw(s)\,dw(t)\,dw(u)\,dw(v)\bigr) = \begin{cases} 3(dt)^2 & s = t = u = v \\ ds\,du & s = t,\ u = v,\ s \neq u \\ ds\,dt & s = u,\ t = v,\ s \neq t \\ ds\,dt & s = v,\ t = u,\ s \neq t \\ 0 & \text{otherwise} \end{cases}$$

so that

$$A = 3\int_0^1 K^2(t,t)(dt)^2 + \int_0^1\int_0^1_{s\neq t} K(s,s)K(t,t)\,ds\,dt + 2\int_0^1\int_0^1_{s\neq t} K^2(s,t)\,ds\,dt$$

$$= \left(\int_0^1 K(s,s)\,ds\right)^2 + 2\int_0^1\int_0^1 K^2(s,t)\,ds\,dt < \infty.$$

3.7 For (2.22) we evaluate $\lim_{m\to\infty} E(X_{m,m})$, where $X_{m,m}$ is defined in (2.21) with $s_i = t_i$ and $s_i' = t_i'$. Putting $\Delta w_i = w(s_i) - w(s_{i-1})$ and $\Delta s_i = s_i - s_{i-1}$, we have

$$E(X_{m,m}) = \sum_{i=1}^m K(s_i', s_i')\Delta s_i\,\mathrm{tr}(H) \to E(X).$$

For (2.23) we consider

$$E\left(X_{m,m}^2\right) = \sum_{i=1}^m\sum_{j=1}^m\sum_{k=1}^m\sum_{l=1}^m K(s_i', s_j')K(s_k', s_l')E\left[\Delta w_i'H\Delta w_j\Delta w_k'H\Delta w_l\right],$$

where $E[\ ]$ is equal to

$$\sum_{a=1}^q H_{aa}^2\,E(\Delta w_{ai}\Delta w_{aj}\Delta w_{ak}\Delta w_{al}) + \sum_{a\neq b} H_{aa}H_{bb}\,E(\Delta w_{ai}\Delta w_{aj}\Delta w_{bk}\Delta w_{bl})$$
$$+ \sum_{a\neq b} H_{ab}^2\,E(\Delta w_{ai}\Delta w_{bj}\Delta w_{ak}\Delta w_{bl}) + \sum_{a\neq b} H_{ab}^2\,E(\Delta w_{ai}\Delta w_{bj}\Delta w_{bk}\Delta w_{al}).$$

Therefore we have

$$E(X_{m,m}^2) = \sum_{a=1}^q H_{aa}^2\left[3\sum_{i=1}^m K^2(s_i', s_i')(\Delta s_i)^2 + \sum_{i\neq j} K(s_i', s_i')K(s_j', s_j')\Delta s_i\Delta s_j\right.$$
$$\left.+ 2\sum_{i\neq j} K^2(s_i', s_j')\Delta s_i\Delta s_j\right] + \sum_{a\neq b} H_{aa}H_{bb}\left(\sum_{i=1}^m K(s_i', s_i')\Delta s_i\right)^2$$
$$+ 2\sum_{a\neq b} H_{ab}^2 K^2(s_i', s_j')\Delta s_i\Delta s_j$$

$$= \left(\sum_{a=1}^q H_{aa}\right)^2\left(\sum_{i=1}^m K(s_i', s_i')\Delta s_i\right)^2$$
$$+ 2\sum_{a=1}^q\sum_{b=1}^q H_{ab}^2\sum_{i=1}^m\sum_{j=1}^m K^2(s_i', s_j')\Delta s_i\Delta s_j,$$

which converges to $E\left(X^2\right)$ given in (2.23).

CHAPTER 2

3.8 It is easy to see that the left side is equal to

$$\mathcal{L}\left(\sum_{i=1}^{q} \lambda_i \int_0^1 \int_0^1 g(s)g(t)\, dw_i(s)\, dw_i(t)\right) = \mathcal{L}\left(\sum_{i=1}^{q} \lambda_i \left(\int_0^1 g(t)\, dw_i(t)\right)^2\right).$$

Since $\int_0^1 g(t)\, dw_i(t)$ $(i = 1, \ldots, q) \sim \text{NID}(0, \int_0^1 g^2(t)\, dt)$, we have the conclusion.

3.9 For (2.27) we have

$$\int_0^1 \bar{w}'(t) H \bar{w}(t)\, dt = \int_0^1 [w(t) - tw(1)]' H [w(t) - tw(1)]\, dt$$

$$= \int_0^1 \left[w'(t)Hw(t) - tw'(t)Hw(1) - w'(1)Htw(t)\right.$$

$$\left. + w'(1)Hw(1)t^2\right] dt,$$

where

$$\int_0^1 w'(t)Hw(t)\, dt = \int_0^1 \left(\int_0^t \int_0^t dw'(u) H\, dw(v)\right) dt$$

$$= \int_0^1 \int_0^1 [1 - \max(s, t)]\, dw'(s) H\, dw(t),$$

$$\int_0^1 tw'(t)Hw(1)\, dt = \int_0^1 \int_0^1 \frac{1 - s^2}{2}\, dw'(s) H\, dw(t),$$

$$\int_0^1 w'(1)Hw(1)t^2\, dt = \frac{1}{3} \int_0^1 \int_0^1 dw'(s) H\, dw(t).$$

Substituting these into the right side above, we obtain the left side of (2.27). The relation (2.26) can be proved similarly.

4.1 We prove (2.31) by induction on g. When $g = 0$, it clearly holds that $F_0(t) = w(t)$. Suppose that (2.31) holds for $g = k - 1$. Then we have

$$F_k(t) = \int_0^t F_{k-1}(s)\, ds = \int_0^t \left(\int_0^s \frac{(s-u)^{k-1}}{(k-1)!}\, dw(u)\right) ds$$

$$= \int_0^t \left(\int_u^t \frac{(s-u)^{k-1}}{(k-1)!}\, ds\right) dw(u) = \int_0^t \frac{(t-u)^k}{k!}\, dw(u),$$

which establishes (2.31).

4.2 From (2.29) and (2.31) we have

$$\int_0^1 \tilde{F}_g(t)\,dt = \int_0^1 \left[F_g(t) - tF_g(1)\right] dt$$

$$= \int_0^1 \left(\int_0^t \frac{(t-s)^g}{g!}\,dw(s)\right) dt - \frac{1}{2}\int_0^1 \frac{(1-s)^g}{g!}\,dw(s)$$

$$= \int_0^1 \left[\int_s^1 \frac{(t-s)^g}{g!}\,dt - \frac{(1-s)^g}{2(g!)}\right] dw(s)$$

$$= \int_0^1 \left[\frac{(1-s)^{g+1}}{(g+1)!} - \frac{(1-s)^g}{2(g!)}\right] dw(s).$$

We also have, from (2.30) and (2.31),

$$\int_0^1 \tilde{F}'_g(t)\tilde{F}_g(t)\,dt = \int_0^1 F'_g(t)F_g(t)\,dt - \int_0^1\int_0^1 F'_g(s)F_g(t)\,ds\,dt$$

$$= \int_0^1 \left(\int_0^t\int_0^t \frac{((t-u)(t-v))^g}{(g!)^2}\,dw'(u)\,dw(v)\right) dt$$

$$- \int_0^1\int_0^1 \frac{((1-s)(1-t))^{g+1}}{((g+1)!)^2}\,dw'(s)\,dw(t)$$

$$= \int_0^1\int_0^1 \left[K_g(s,t) - \frac{((1-s)(1-t))^{g+1}}{((g+1)!)^2}\right] dw'(s)\,dw(t).$$

4.3 Since $F_g(t)$ is continuously differentiable, we have

$$\int_0^t F_g(s)\,dF'_g(s) = F_g(s)F'_g(s)\Big|_0^t - \left(\int_0^t F_g(s)\,dF'_g(s)\right)'$$

$$= F_g(t)F'_g(t) - \left(\int_0^t F_g(s)\,dF'_g(s)\right)',$$

which proves (2.33).

5.1 Put $\tau_{i-1} = (1-a)s_{i-1} + as_i$ and $\Delta w_i = w(s_i) - w(s_{i-1})$. Then we have

$$\sum_{i=1}^m w(\tau_{i-1})\Delta w_i = \sum_{i=1}^m w(s_{i-1})\Delta w_i + \sum_{i=1}^m \left(w(\tau_{i-1}) - w(s_{i-1})\right)\Delta w_i.$$

Here the first term on the right side converges in m.s. to $\frac{1}{2}(w^2(t) - t)$, while the second term can be rewritten as

(S3) $$\sum_{i=1}^m \left(w(\tau_{i-1}) - w(s_{i-1})\right)^2 + \sum_{i=1}^m \left(w(\tau_{i-1}) - w(s_{i-1})\right)\left(w(s_i) - w(\tau_{i-1})\right).$$

Since

$$E\left[\sum_{i=1}^{m}(w(\tau_{i-1}) - w(s_{i-1}))^2\right] = \sum_{i=1}^{m}(\tau_{i-1} - s_{i-1}) = a\sum_{i=1}^{m}(s_i - s_{i-1}) = at,$$

$$V\left[\sum_{i=1}^{m}(w(\tau_{i-1}) - w(s_{i-1}))^2\right] = 2\sum_{i=1}^{m}(\tau_{i-1} - s_{i-1})^2 \leq 2a^2 t \max_i(s_i - s_{i-1}),$$

the first term in (S3) converges in m.s. to at, while the second term in (S3) can be shown to converge in m.s. to 0, which establishes (2.40).

5.2 We have already shown that $U(t)$ is m.s. continuous. For the martingale property, we first have, for $s \leq t$, $E(U(t)|U(s)) = U(s) + E(U(t) - U(s)|U(s))$. Noting that

$$U(t) - U(s) = \int_s^t X(u)\,dw(u)$$

$$= \underset{\substack{m\to\infty \\ \Delta_m \to 0}}{\text{l.i.m.}} \sum_{i=1}^{m} X(u_{i-1})(w(u_i) - w(u_{i-1})),$$

where $s = u_0 < u_1 < \cdots < u_m = t$, we can deduce that

$$E[X(u_{i-1})(w(u_i) - w(u_{i-1}))|U(s)]$$
$$= E[E[X(u_{i-1})(w(u_i) - w(u_{i-1}))|U(u_{i-1})]|U(s)]$$
$$= E[X(u_{i-1})E[w(u_i) - w(u_{i-1})|U(u_{i-1})]|U(s)]$$
$$= 0.$$

Then it is seen that $E(U(t) - U(s)|U(s)) = 0$ so that $U(t)$ is a martingale.

5.3 Putting $\Delta w_i = w(s_i) - w(s_{i-1})$ and $\Delta s_i = s_i - s_{i-1}$ we have

$$E\left[\left\{\sum_{i=1}^{m} X(s_{i-1})\left((\Delta w_i)^2 - \Delta s_i\right)\right\}^2\right]$$

$$= E\left[\sum_{i=1}^{m}\sum_{j=1}^{m} X(s_{i-1})X(s_{j-1})\left((\Delta w_i)^2 - \Delta s_i\right)\left((\Delta w_j)^2 - \Delta s_j\right)\right]$$

$$= 2\sum_{i=1}^{m} E\left(X^2(s_{i-1})\right)(\Delta s_i)^2 \leq 2\max_i E\left(X^2(s_{i-1})\right)\max_i \Delta s_i \sum_{i=1}^{m}\Delta s_i$$

$$\to 0$$

as $m \to \infty$ and $\max_i \Delta s_i \to 0$ so that (2.41) is established. The relation (2.42) can be proved similarly.

5.4 Because of the definition of y_t we have

$$\frac{1}{T}\sum_{t=1}^{T} y_{t-1}\varepsilon_t = \frac{1}{T}\sum_{t=1}^{T} y_{t-1}(y_t - y_{t-1})$$

$$= -\frac{1}{2T}\left[\sum_{t=1}^{T}(y_t - y_{t-1})^2 - \sum_{t=1}^{T} y_t^2 + \sum_{t=1}^{T} y_{t-1}^2\right]$$

$$= \frac{1}{2T}y_T^2 - \frac{1}{2T}\sum_{t=1}^{T}\varepsilon_t^2$$

$$= \frac{1}{2}\left(\frac{1}{\sqrt{T}}\sum_{t=1}^{T}\varepsilon_t\right)^2 - \frac{1}{2T}\sum_{t=1}^{T}\varepsilon_t^2,$$

which converges in distribution to $\frac{1}{2}(w^2(1) - 1)$ by the central limit theorem and the law of large numbers.

6.1 Note first that the c.f. $\phi(\theta)$ of the right-hand side in (2.46) is given by $\phi(\theta) = (\cosh\theta)^{-1/2}$. Thus it suffices to show that the c.f. $\phi_m(\theta)$ of

$$V_m(1) = \sum_{j=1}^{m} w_a(t_{j-1})(w_b(t_j) - w_b(t_{j-1})), \qquad (a \neq b)$$

converges to $\phi(\theta)$ as $m \to \infty$, where $t_j = j/m$. It is easy to check that $V_m(1)$ has the same limiting distribution as U_m, where U_m is given in (1.48). Thus we have the conclusion from the arguments there.

6.2 It is easy to obtain

$$\sum_{i=1}^{m} w(t_{i-1})\big(w(t_i) - w(t_{i-1})\big)' + \sum_{i=1}^{m}\big(w(t_i) - w(t_{i-1})\big)w'(t_{i-1})$$

$$= w(t)w'(t) - \sum_{i=1}^{m}\big(w(t_i) - w(t_{i-1})\big)\big(w(t_i) - w(t_{i-1})\big)',$$

where $0 = t_0 < t_1 < \cdots < t_m = t$. Letting $m \to \infty$ and $\Delta_m = \max_i(t_i - t_{i-1}) \to 0$, (2.47) is established by the law of large numbers.

6.3 Premultiplying A on both sides of (2.47) with $t = 1$ and taking the trace lead from $\operatorname{tr}(A) = 0$ to

$$\int_0^1 w'(s)A\,dw(s) = \frac{1}{2}w'(1)Aw(1).$$

We have $\mathcal{L}(w'(1)Aw(1)/2) = \mathcal{L}(\sum_{i=1}^{q}\lambda_i Z_i^2)$, where λ_is are the eigenvalues of $A/2$ and $\{Z_i\} \sim \text{NID}(0, 1)$.

CHAPTER 2

7.1 To establish (2.55) we use Ito's theorem, putting $f(x,t) = x^n$ with $dX(t)$ defined in (2.52). For (2.56) we put $f(x,t) = x^n$ with $dX(t) = dw(t)$ so that $\mu = 0$ and $\sigma = 1$. To prove (2.57) we put $f(x,t) = e^x$ with $dX(t) = dw(t)$, $\mu = 0$, and $\sigma = 1$.

7.2 Since the existence and uniqueness of the solution to (2.60) is ensured, we have only to show that $d\left(X(0)e^{w(t)-t/2}\right) = X(t)\,dw(t)$, which is almost trivial.

7.3 We have, by Ito's theorem,

$$dX(t) = \left(\alpha e^{\alpha t} X(0) + \beta e^{\alpha t} + \gamma \alpha e^{\alpha t} \int_0^t e^{-\alpha s}\,dw(s)\right) dt + \gamma\,dw(t)$$
$$= (\alpha X(t) + \beta)\,dt + \gamma\,dw(t),$$

which gives (2.61).

7.4 For $s \leq t$ we have, from (2.63),

$$\text{Cov}(X(s), X(t)) = e^{\alpha(s+t)} \text{Cov}\left(X(0) + \int_0^s e^{-\alpha u}\,dw(u), X(0) + \int_0^t e^{-\alpha v}\,dw(v)\right)$$
$$= e^{\alpha(s+t)} \left[V(X(0)) + \int_0^s e^{-2\alpha u}\,du\right]$$
$$= e^{\alpha(s+t)} \left[V(X(0)) + \frac{1 - e^{-2\alpha s}}{2\alpha}\right],$$

which establishes (2.64).

7.5 From the definition of $X(t)$ in (2.63), we have

$$\int_0^1 e^{2\alpha t} \left[X^2(0) + 2X(0) \int_0^t e^{-\alpha s}\,dw(s) + \int_0^t \int_0^t e^{-\alpha(u+v)}\,dw(u)\,dw(v)\right] dt$$
$$= \frac{e^{2\alpha} - 1}{2\alpha} X^2(0) + 2X(0) \int_0^1 \left(\int_s^1 e^{2\alpha t}\,dt\right) e^{-\alpha s}\,dw(s)$$
$$+ \int_0^1 \int_0^1 \left(\int_{\max(u,v)}^1 e^{2\alpha t}\,dt\right) e^{-\alpha(u+v)}\,dw(u)\,dw(v)$$
$$= \frac{e^{2\alpha} - 1}{2\alpha} X^2(0) + X(0) \int_0^1 \frac{e^{\alpha(2-s)} - e^{\alpha s}}{\alpha}\,dw(s)$$
$$+ \int_0^1 \int_0^1 \frac{e^{\alpha(2-u-v)} - e^{\alpha|u-v|}}{2\alpha}\,dw(u)\,dw(v).$$

7.6 For (2.72) we put $Y_1(t) = e^{\alpha t} w(t)$ and

$$Y_2(t) = \int_0^t e^{-\alpha s}\,dw(s)$$

so that $dY_1(t) = e^{\alpha t}(\alpha w(t)dt + dw(t))$ and $dY_2(t) = e^{-\alpha t}dw(t)$. Then (2.71) yields (2.72). For (2.73) we put $Y_1(t) = w(t)$, while $Y_2(t)$ is the same as above. Thus $dY_1(t) = dw(t)$ and $dY_2(t) = e^{-\alpha t}dw(t)$. Define $g(y,t) = \exp(y_1 y_2)$ so that $g_t = 0$ and

$$g_y = \begin{pmatrix} y_2 \\ y_1 \end{pmatrix} g, \qquad g_{yy} = \begin{pmatrix} y_2^2 & 1+y_1 y_2 \\ 1+y_1 y_2 & y_1^2 \end{pmatrix} g.$$

Then (2.70) yields

$$dg = \left\{ Y_2 dY_1 + Y_1 dY_2 + \frac{1}{2}(Y_2^2 + 2(1+Y_1 Y_2)e^{-\alpha t} + Y_1^2 e^{-2\alpha t}) dt \right\} g,$$

which leads us to (2.73).

CHAPTER 3

1.1 Since $\rho(x,y)$ is a metric, we have

$$|\rho(x,y) - \rho(\tilde{x},\tilde{y})| \le \rho(x,\tilde{x}) + \rho(y,\tilde{y})$$

which can be proved by the triangle inequalities

$$\rho(x,y) \le \rho(x,\tilde{x}) + \rho(\tilde{x},\tilde{y}) + \rho(\tilde{y},y), \quad \rho(\tilde{x},\tilde{y}) \le \rho(\tilde{x},x) + \rho(x,y) + \rho(y,\tilde{y}).$$

Then it is clear that $\rho(x,y)$ is a continuous function of x and y.

1.2 Let $\{x_n\}$ be a fundamental sequence in C, that is, $\rho(x_m, x_n) \to 0$ as $m, n \to \infty$. Because of the definition of ρ and completeness of the real line, $\{x_n(t)\}$ converges uniformly in t so that the limit $x(t)$ lies in C and $\rho(x_n, x) \to 0$. Thus C is complete. Separability follows from the Weierstrass approximation theorem which ensures that any x in C can be uniformly approximated by a polynomial with real coefficients that, in turn, can be approximated by a polynomial with coefficients of rational numbers.

4.1 We have only to show that $E[\exp\{i\theta h(X_n)\}] \to E[\exp\{i\theta h(X)\}]$ where

$$E[\exp\{i\theta h(X_n)\}] = E\{\cos \theta h(X_n)\} + iE\{\sin \theta h(X_n)\}.$$

Since $f_1(X_n) = \cos \theta h(X_n)$ and $f_2(X_n) = \sin \theta h(X_n)$ are both bounded and continuous, it must hold that $E(f_1(X_n)) \to E(f_1(X))$ and $E(f_2(X_n)) \to E(f_2(X))$, from which the conclusion follows.

4.2 Suppose that $\rho(x,y) < \varepsilon$ so that $y(t) - \varepsilon < x(t) < y(t) + \varepsilon$ for all $t \in [0,1]$. Then it follows that $|\sup x(t) - \sup y(t)| < \varepsilon$. To show that $h_2(x)$ is continuous we first have, by the triangle inequality,

$$\rho(x,0) \le \rho(x,y) + \rho(y,0), \quad \rho(y,0) \le \rho(y,x) + \rho(x,0)$$

so that $|\rho(x,0) - \rho(y,0)| \leq \rho(x,y)$, which means that

$$\left| \sup_{0 \leq t \leq 1} |x(t)| - \sup_{0 \leq t \leq 1} |y(t)| \right| \leq \sup_{0 \leq t \leq 1} |x(t) - y(t)|.$$

Thus $h_2(x)$ is shown to be continuous. The function $h_3(x)$ is the mapping which carries x of C to the point $(h_1(x), h_2(x))$ of R^2, so it is certainly continuous since $h_1(x)$ and $h_2(x)$ are both continuous.

4.3 Note that $P(\rho(X_n, c) < \varepsilon) = P(X_n \in N(c, \varepsilon))$, where $N(c, \varepsilon)$ is the open sphere with center c and radius ε. By the portmanteau theorem [Billingsley (1968, p. 24)], $\mathcal{L}(X_n) \to \mathcal{L}(c)$ implies that $P(X_n \in N(c, \varepsilon)) \to P(c \in N(c, \varepsilon))$, which is certainly unity.

4.4 Given any $\varepsilon > 0$, there exists some $\delta > 0$ such that $|X_n - c| < \delta$ implies $|h(X_n) - h(c)| < \varepsilon$. Therefore we have

$$P(|h(X_n) - h(c)| \geq \varepsilon) \leq P(|X_n - c| \geq \delta)$$

so that, by assumption, $h(X_n) \to h(c)$ in probability.

4.5 Let x be a continuity point of $P(X < x)$. Then it holds that

$$P(Y_n < x) = P(Y_n < x, X_n - Y_n < \varepsilon) + P(Y_n < x, X_n - Y_n \geq \varepsilon)$$
$$\leq P(X_n < x + \varepsilon) + P(X_n - Y_n \geq \varepsilon)$$

so that $\limsup_{n \to \infty} P(Y_n < x) \leq P(X < x + \varepsilon)$. We also have

$$P(X_n < x - \varepsilon) = P(X_n < x - \varepsilon, Y_n < x) + P(X_n < x - \varepsilon, Y_n \geq x)$$
$$\leq P(Y_n < x) + P(X_n - Y_n \leq -\varepsilon)$$

so that $\liminf_{n \to \infty} P(Y_n < x) \geq P(X < x - \varepsilon)$. Since ε is arbitrary, we have $\mathcal{L}(Y_n) \to \mathcal{L}(X)$.

4.6 Noting that

$$P\left(\max_{1 \leq j \leq T} \frac{|\varepsilon_j|}{\sqrt{T}} \leq \delta\right) = \prod_{j=1}^{T} P\left(\frac{|\varepsilon_j|}{\sqrt{T}} \leq \delta\right)$$
$$= \left\{1 - P\left(\frac{|\varepsilon_1|}{\sqrt{T}} > \delta\right)\right\}^T$$
$$\geq \left[1 - \frac{1}{\delta^2 T} E\left\{\varepsilon_1^2 I\left(\frac{|\varepsilon_1|}{\sqrt{T}} > \delta\right)\right\}\right]^T,$$

(3.11) is seen to hold.

4.7 For any fixed $c > 0$ we have, for any $\varepsilon > 0$,

$$P(|X_n Y_n| > \varepsilon) = P\left(|X_n Y_n| > \varepsilon, |Y_n| \le \frac{\varepsilon}{c}\right) + P\left(|X_n Y_n| > \varepsilon, |Y_n| > \frac{\varepsilon}{c}\right)$$

$$\le P(|X_n| > c) + P\left(|Y_n| > \frac{\varepsilon}{c}\right).$$

Thus $\lim \sup_{n \to \infty} P(|X_n Y_n| > \varepsilon) \le P(|X| > c)$. Thus it is seen that $X_n Y_n \to 0$ in probability by choosing c large [Rao (1973, p.122)].

4.8 Put $\tilde{X}_T(t) = X_T(t) - \sum_{j=1}^{T} X_T(j/T)/T$. Then $\mathcal{L}(\tilde{X}_T) \to \mathcal{L}(\tilde{w})$ by Corollary 3.2, where $\tilde{w}$ is the demeaned Brownian motion. We have

$$\left| \frac{1}{T^2} \sum_{j=1}^{T} (y_j - \bar{y})^2 - \int_0^1 \tilde{X}_T^2(t)\, dt \right|$$

$$= \left| \sum_{j=1}^{T} \int_{(j-1)/T}^{j/T} \left(\tilde{X}_T^2\left(\frac{j}{T}\right) - \tilde{X}_T^2(t) \right) dt \right|$$

$$\le 2 \sup_{0 \le t \le 1} |\tilde{X}_T(t)| \max_{1 \le j \le T} \frac{|\varepsilon_j|}{\sqrt{T}},$$

which converges in probability to 0. Thus the second relation in (3.13) is established by the continuous mapping theorem. Assuming that $\{\varepsilon_j\} \sim \text{NID}(0,1)$, we have

$$\mathcal{L}\left(\frac{1}{T^2} \sum_{j=1}^{T} (y_j - \bar{y})^2 \right) = \mathcal{L}\left(\frac{1}{T^2} y'My \right) = \mathcal{L}\left(\frac{1}{T^2} \varepsilon'C'MC\varepsilon \right)$$

$$= \mathcal{L}\left(\frac{1}{T^2} \varepsilon'MCC'M\varepsilon \right) = \mathcal{L}\left(\frac{1}{T^2} \sum_{j=1}^{T} \left(\sum_{i=j}^{T} (\varepsilon_i - \bar{\varepsilon}) \right)^2 \right)$$

$$= \mathcal{L}\left(\frac{1}{T^2} \sum_{j=1}^{T} \left(\sum_{i=1}^{j} (\varepsilon_i - \bar{\varepsilon}) \right)^2 \right),$$

where $M = I_T - ee'/T$ with $e = (1, \ldots, 1)'$ and C is defined in (1.3). Put $\bar{X}_T(t) = X_T(t) - tX_T(1)$. Then $\mathcal{L}(\bar{X}_T) \to \mathcal{L}(\bar{w})$ by Corollary 3.1, where $\bar{w} =$

$\{w(t) - tw(1)\}$. We now obtain

$$\left| \frac{1}{T^2} \sum_{j=1}^{T} \left(\sum_{i=1}^{j} (\varepsilon_i - \bar{\varepsilon}) \right)^2 - \int_0^1 \bar{X}_T^2(t)\, dt \right|$$

$$= \left| \sum_{j=1}^{T} \int_{(j-1)/T}^{j/T} \left(\bar{X}_T^2 \left(\frac{j}{T} \right) - \bar{X}_T^2(t) \right) dt \right|$$

$$\leq 2 \sup_{0 \leq t \leq 1} |\bar{X}_T(t)| \left(\max_{1 \leq j \leq T} \frac{|\varepsilon_j|}{\sqrt{T}} + \frac{1}{T} |X_T(1)| \right),$$

which converges in probability to 0. Thus the last relation in (3.13) is established by the continuous mapping theorem.

4.9 It follows from (3.8) that

$$\hat{\sigma}^2 = \frac{1}{T-1} \sum_{j=2}^{T} (y_j - y_{j-1} - (\hat{\rho} - 1) y_{j-1})^2$$

$$= \frac{1}{T-1} \left[\sum_{j=2}^{T} \varepsilon_j^2 - 2(\hat{\rho} - 1) \sum_{j=2}^{T} y_{j-1} \varepsilon_j + (\hat{\rho} - 1)^2 \sum_{j=2}^{T} y_{j-1}^2 \right].$$

Since $\hat{\rho} - 1 = O_p(T^{-1})$ and

$$\sum_{j=2}^{T} y_{j-1} \varepsilon_j = O_p(T), \quad \sum_{j=2}^{T} y_{j-1}^2 = O_p(T^2),$$

it is seen that $\hat{\sigma}^2 \to \sigma^2$ in probability.

5.1 There exists m such that $l|\alpha_l| < 1$ for all $l > m$. Thus we have

$$\sum_{l=0}^{\infty} l^2 \alpha_l^2 = \sum_{l=0}^{m} l^2 \alpha_l^2 + \sum_{l=m+1}^{\infty} l^2 \alpha_l^2 \leq \sum_{l=0}^{m} l^2 \alpha_l^2 + \sum_{l=m+1}^{\infty} l|\alpha_l| < \infty.$$

5.2 Put $\alpha(L) = \sum_{l=0}^{\infty} \alpha_l L^l$, where L is the lag operator. Then $\alpha(L) = \alpha - (\alpha - \alpha(L))$, where

$$\alpha - \alpha(L) = \sum_{k=1}^{\infty} \alpha_k \left(1 - L^k \right) = (1 - L) \sum_{k=1}^{\infty} \alpha_k \sum_{l=0}^{k-1} L^l$$

$$= (1 - L) \sum_{l=0}^{\infty} \sum_{k=l+1}^{\infty} \alpha_k L^l = (1 - L) \sum_{l=0}^{\infty} \tilde{\alpha}_l L^l,$$

which yields (3.20). Here the interchange of the order of summation is justified because of the assumption (3.19).

5.3 We first note that

$$\sum_{l=0}^{\infty} |\tilde{\alpha}_l| \leq \sum_{l=0}^{\infty} \sum_{k=l+1}^{\infty} |\alpha_k| = \sum_{k=1}^{\infty} \sum_{l=0}^{k-1} |\alpha_k|$$

$$= \sum_{k=1}^{\infty} k|\alpha_k|.$$

This last quantity is finite because of (3.19). Thus $\sum_{l=0}^{\infty} \tilde{\alpha}_l^2 < \infty$ and $\{\tilde{\varepsilon}_j\}$ is well defined in the m.s. sense so that $E(\tilde{\varepsilon}_j) = 0$ and $E(\tilde{\varepsilon}_j \tilde{\varepsilon}_{j+k}) = \sigma^2 \sum_{l=0}^{\infty} \tilde{\alpha}_l \tilde{\alpha}_{l+|k|}$.

5.4 Suppose that $\alpha(L) = \sum_{l=0}^{p} \alpha_l L^l$. Then $\alpha(L) = \alpha(1) - (\alpha(1) - \alpha(L))$, where

$$\alpha(1) - \alpha(L) = \sum_{k=1}^{p} \alpha_k \left(1 - L^k\right) = (1 - L) \sum_{k=1}^{p} \alpha_k \sum_{l=0}^{k-1} L^l$$

$$= (1 - L) \sum_{l=0}^{p-1} \sum_{k=l+1}^{p} \alpha_k L^l.$$

Thus we obtain

$$\sum_{l=0}^{p} \alpha_l \varepsilon_{j-l} = \alpha(1) \varepsilon_j - (1 - L) \sum_{l=0}^{p-1} \beta_l \varepsilon_{j-l},$$

where $\beta_l = \sum_{k=l+1}^{p} \alpha_k$.

5.5 If $\max_{1 \leq j \leq n} |Z_j| > \delta$, then there exists j such that $|Z_j| > \delta$. Thus $\sum_{j=1}^{n} Z_j^2 I(|Z_j| > \delta) \geq Z_j^2 > \delta^2$. On the other hand, if $\sum_{j=1}^{n} Z_j^2 I(|Z_j| > \delta) > \delta^2$, there must exist j such that $|Z_j| > \delta$. Thus $\max_{1 \leq j \leq n} |Z_j| > \delta$.

5.6 Using $Y_T(t)$ defined in (3.17), we have

$$\frac{1}{T^2} \sum_{j=1}^{T} y_j^2 - \int_0^1 Y_T^2(t) \, dt = \sum_{j=1}^{T} \int_{(j-1)/T}^{j/T} \left[Y_T^2 \left(\frac{j}{T} \right) - Y_T^2(t) \right] dt$$

and

$$\left| Y_T^2 \left(\frac{j}{T} \right) - Y_T^2(t) \right| \leq 2 \sup_{0 \leq t \leq 1} |Y_T(t)| \max_{1 \leq j \leq T} \frac{|u_j|}{\sqrt{T}}.$$

This last quantity converges in probability to 0 because of (3.27) and (3.28) together with $\sup |Y_T(t)| = O_p(1)$. Thus (3.30) follows from Theorem 3.7 with $\sigma = 1$ and the continuous mapping theorem. The weak convergence in (3.31)

CHAPTER 3

6.1 Putting $a_j = s_j^2/s_n^2$, we can express $\xi_n(t)$ as $\xi_n(t) = \alpha\xi_n(a_{j-1}) + \beta\xi_n(a_j)$, where $\alpha = (a_j - t)/(a_j - a_{j-1}) \geq 0$ and $\beta = 1 - \alpha$. This means that $\xi_n(t)$ is on the line joining $(a_{j-1}, \xi_n(a_{j-1}))$ and $(a_j, \xi_n(a_j))$.

6.2 When $\{\varepsilon_j\}$ is i.i.d.$(0, \sigma^2)$, we have $s_n^2 = n\sigma^2$ and (3.34) follows from the weak law of large numbers. On the other hand, (3.35) reduces to $E[\varepsilon_1^2 I(|\varepsilon_1| > \sqrt{n}\sigma\delta)] \to 0$ for every δ, which clearly holds because $E(\varepsilon_1^2) < \infty$.

6.3 We first note that

$$\varepsilon_j^2 = \varepsilon_j^2 I(|\varepsilon_j| \leq \delta s_n) + \varepsilon_j^2 I(|\varepsilon_j| > \delta s_n) \leq \delta^2 s_n^2 + \varepsilon_j^2 I(|\varepsilon_j| > \delta s_n)$$

so that

$$\max_{1 \leq j \leq n} \frac{E(\varepsilon_j^2)}{s_n^2} \leq \delta^2 + \frac{1}{s_n^2} \max_{1 \leq j \leq n} E\left[\varepsilon_j^2 I(|\varepsilon_j| > \delta s_n)\right]$$

$$\leq \delta^2 + \frac{1}{s_n^2} \sum_{j=1}^{n} E\left[\varepsilon_j^2 I(|\varepsilon_j| > \delta s_n)\right],$$

which implies (3.36) since δ is arbitrary and the Lindeberg condition is imposed.

6.4 The problem becomes completely the same as in Problem 5.5 by putting $\varepsilon_j = Z_j s_n$.

6.5 The relation (3.42) holds since

$$E\left[|X| I(|X| > \delta)\right] = \int_0^\infty P\left(|X| I(|X| > \delta) > x\right) dx$$

$$= \int_0^\delta P\left(|X| I(|X| > \delta) > x\right) dx$$

$$+ \int_\delta^\infty P\left(|X| I(|X| > \delta) > x\right) dx$$

$$= \delta P\left(|X| > \delta\right) + \int_\delta^\infty P\left(|X| > x\right) dx.$$

Then the right side of (3.42) is dominated by $cE[\eta^2 I(|\eta| > \delta)]$ so that (3.41) holds.

6.6 For a given $\gamma > 0$ and sufficiently large $\delta > 0$ we have

$$\sup_j E\left(\varepsilon_j^2\right) = \sup_j \left[E\left[\varepsilon_j^2 I\left(\varepsilon_j^2 > \delta\right)\right] + E\left[\varepsilon_j^2 I\left(\varepsilon_j^2 \leq \delta\right)\right]\right]$$

$$\leq \gamma + \delta,$$

which yields the conclusion.

6.7 To show that (3.34) holds with ε_j replaced by $\varepsilon_{jn} = j\eta_j/n$, we use the method of truncation [see, for example, Roussas (1973, p. 146)]. Define, for any $\delta > 0$,

$$Y_j = \begin{cases} \varepsilon_{jn}^2 & (\varepsilon_{jn}^2 \leq \delta n), \\ 0 & (\varepsilon_{jn}^2 > \delta n), \end{cases}$$

$$Z_j = \begin{cases} 0 & (\varepsilon_{jn}^2 \leq \delta n), \\ \varepsilon_{jn}^2 & (\varepsilon_{jn}^2 > \delta n), \end{cases}$$

so that $\varepsilon_{jn}^2 = Y_j + Z_j$. Then it holds that

$$P\left[\left|\frac{1}{s_n^2}\sum_{j=1}^n (\varepsilon_{jn}^2 - E(\varepsilon_{jn}^2))\right| > 3\gamma\right]$$

$$\leq P\left[\frac{1}{s_n^2}\left\{\left|\sum_{j=1}^n (Y_j - E(Y_j))\right| + \left|\sum_{j=1}^n Z_j\right| + \left|\sum_{j=1}^n (E(Y_j) - E(\varepsilon_{jn}^2))\right|\right\} > 3\gamma\right]$$

$$\leq P\left(\frac{1}{s_n^2}\left|\sum_{j=1}^n (Y_j - E(Y_j))\right| > \gamma\right) + P\left(\sum_{j=1}^n Z_j \neq 0\right)$$

$$+ P\left(\frac{1}{s_n^2}\left|\sum_{j=1}^n (E(Y_j) - E(\varepsilon_{jn}^2))\right| > \gamma\right).$$

Here the first term is bounded by

$$\frac{1}{\gamma^2 s_n^4}\sum_{j=1}^n V(Y_j) \leq \frac{1}{\gamma^2 s_n^4}\sum_{j=1}^n E\left[\varepsilon_{jn}^4 I(\varepsilon_{jn}^2 \leq \delta n)\right]$$

$$\leq \frac{\delta n}{\gamma^2 s_n^4}\sum_{j=1}^n E\left[\varepsilon_{jn}^2 I(\varepsilon_{jn}^2 \leq \delta n)\right]$$

$$\leq \frac{\delta n^2 \sigma^2}{\gamma^2 s_n^4} \leq \frac{c_1 \delta}{\gamma^2} \quad \text{for some } c_1 > 0.$$

The second term is bounded by

$$nP(Z_n \neq 0) = nP\left(\varepsilon_{nn}^2 > \delta n\right)$$

$$\leq \frac{1}{\delta} E\left[\varepsilon_{nn}^2 I(\varepsilon_{nn}^2 > \delta n)\right]$$

$$\leq \delta \quad \text{for } n \text{ sufficiently large.}$$

For the third term we have

$$\frac{1}{s_n^2}\left|\sum_{j=1}^n \left(E(Y_j) - E(\varepsilon_{jn}^2)\right)\right| = \frac{1}{s_n^2}\sum_{j=1}^n E\left[\varepsilon_{jn}^2 I(\varepsilon_{jn}^2 > \delta n)\right]$$

$$\leq \frac{\delta^2 n}{s_n^2} \leq c_2 \delta^2$$

for sufficiently large n and some $c_2 > 0$. Putting $\delta = \gamma^3$, we now establish (3.34) with ε_j replaced by ε_{jn}. We next consider

$$\frac{1}{s_n^2}\sum_{j=1}^n E\left[\varepsilon_{jn}^2 I(|\varepsilon_{jn}| > \delta s_n)\right] \leq \frac{1}{s_n^2}\sum_{j=1}^n E\left[\eta_j^2 I(|\eta_j| > \delta s_n)\right]$$

$$= \frac{n}{s_n^2} E\left[\eta_1^2 I(|\eta_1| > \delta s_n)\right],$$

which clearly converges to 0 for every $\delta > 0$ so that (3.35) holds with ε_j replaced by ε_{jn}.

7.1 Using the BN decomposition $u_i = \alpha\varepsilon_i + \tilde{\varepsilon}_{i-1} - \tilde{\varepsilon}_i$ in (3.20) and substituting this into (3.45), it is easy to establish (3.49).

7.2 The inequality easily follows from the definition of $R_n(t)$ in (3.51) and $0 \leq (ts_n^2 - s_{j-1}^2)/(s_j^2 - s_{j-1}^2) \leq 1$.

7.3 From the definition of $\{\tilde{\varepsilon}_j\}$ in (3.53) with $\{\varepsilon_j\}$ being a martingale difference, it is easy to derive

$$E(\tilde{\varepsilon}_j^2) = \sum_{l=0}^{\infty} \tilde{\alpha}_l^2 E(\varepsilon_{j-l}^2).$$

Moreover, we have $\sup_j E(\varepsilon_j^2) \leq cE(\eta^2)$ so that (3.55) is established.

7.4 We have, for any $\delta > 0$,

$$E\left(\tilde{\varepsilon}_j^2 I(|\tilde{\varepsilon}_j| > \delta)\right) \leq E\left(\tilde{\varepsilon}_j^2 \left(\frac{|\tilde{\varepsilon}_j|}{\delta}\right)^\gamma I(|\tilde{\varepsilon}_j| > \delta)\right) \leq \frac{1}{\delta^\gamma} E\left(|\tilde{\varepsilon}_j|^{2+\gamma}\right)$$

so that the result follows.

7.5 It can be proved easily that strong uniform integrability of $\{\varepsilon_j\}$ with a bounding variable η $(E(|\eta|^{2+\gamma}) < \infty)$ implies $\sup_j E(|\varepsilon_j|^{2+\gamma}) < \infty$. Then it follows from Hölder's inequality that

(S4) $$|\tilde{\varepsilon}_j| \leq \sum_{l=0}^{\infty} |\tilde{\alpha}_l|^{1/p} \left(|\tilde{\alpha}_l|^{1/q} |\varepsilon_{j-l}|\right) \quad \left(\frac{1}{p} + \frac{1}{q} = 1, \ q > 1\right)$$

$$\leq \left(\sum_{l=0}^{\infty} |\tilde{\alpha}_l|\right)^{1/p} \left(\sum_{l=0}^{\infty} |\tilde{\alpha}_l| |\varepsilon_{j-l}|^q\right)^{1/q}$$

so that

$$\sup_j E(|\tilde{\varepsilon}_j|^q) \le \sup_j E(|\varepsilon_j|^q) \left(\sum_{l=0}^{\infty} |\tilde{\alpha}_l|\right)^q.$$

Putting $q = 2 + \gamma$ we obtain the conclusion.

7.6 Putting $p = q = 2$ in (S4) we obtain

$$\tilde{\varepsilon}_j^2 \le \sup_j \varepsilon_j^2 \left(\sum_{l=0}^{\infty} |\tilde{\alpha}_l|\right)^2 \equiv |X|.$$

Thus we have

$$\frac{1}{s_n^2} \sum_{j=1}^{n} E\left[\tilde{\varepsilon}_j^2 I\left(\tilde{\varepsilon}_j^2 > s_n^2 \delta\right)\right] \le E\left[|X| I\left(|X| > s_n^2 \delta\right)\right] \left(\frac{s_n^2}{n}\right)^{-1},$$

which converges to 0 because $E(|X|) < \infty$ by assumption and s_n^2 is the same order as n.

8.1 Note that $(1-L)y_j^{(d)} = \varepsilon_j/(1-L)^{d-1} = y_j^{(d-1)}$, which yields $y_j^{(d)} = y_{j-1}^{(d)} + y_j^{(d-1)}$. By back substitution, this produces $y_j^{(d)} = y_1^{(d-1)} + \cdots + y_j^{(d-1)}$.

8.2 It is easy to establish the first inequality, which is bounded by

$$\sum_{i=1}^{j} \int_{(i-1)/n}^{i/n} \left|Y_n^{(1)}\left(\frac{i}{n}\right) - Y_n^{(1)}(s)\right| ds + \int_t^{j/n} \left|Y_n^{(1)}(s)\right| ds + \frac{1}{\sqrt{n}} \max_{1 \le j \le n} |\varepsilon_j|$$

$$\le \frac{2}{\sqrt{n}} \max_{1 \le j \le n} |\varepsilon_j| + \frac{1}{n} \sup_{0 \le t \le 1} \left|Y_n^{(1)}(t)\right|.$$

8.3 It is sufficient to show that the quantities on the right side of (3.63) converge in probability to 0. The first term does because of (3.11) (see Problem 4.6), while $\sup_{0 \le t \le 1} |Y_n^{(1)}(t)| = O_p(1)$ because $\mathcal{L}(\sup_{0 \le t \le 1} |Y_n^{(1)}(t)|) \to \mathcal{L}(\sigma \sup_{0 \le t \le 1} |w(t)|)$. Thus the second term also converges in probability to 0.

8.4 It is obvious that the first inequality holds. Since

$$y_j^{(k)} = \frac{\varepsilon_j}{(1-L)^k} = \Sigma \cdots \Sigma \varepsilon_i, \quad (k \ \Sigma s),$$

it holds that $|y_j^{(k)}| \le n^k \max_{1 \le j \le n} |\varepsilon_j|$, which establishes the second inequality.

8.5 Using the BN decomposition, we obtain

$$\text{(S5)} \quad y_j^{(d)} = \frac{u_j}{(1-L)^d} = \frac{1}{(1-L)^d} [\alpha \varepsilon_j - (1-L)\tilde{\varepsilon}_j]$$

$$= \alpha \frac{\varepsilon_j}{(1-L)^d} - \frac{\tilde{\varepsilon}_j}{(1-L)^{d-1}} = \alpha x_j^{(d)} - z_j^{(d-1)},$$

where $(1-L)^d x_j^{(d)} = \varepsilon_j$, $(1-L)^{d-1} z_j^{(d-1)} = \tilde{\varepsilon}_j$. Therefore we have

$$Y_n^{(d)}(t) = \frac{1}{n^{d-1/2}} y_{[nt]}^{(d)} + (nt - [nt]) \frac{1}{n^{d-1/2}} y_{[nt]+1}^{(d-1)}$$

$$= \alpha \left[\frac{1}{n^{d-1/2}} x_{[nt]}^{(d)} + (nt - [nt]) \frac{1}{n^{d-1/2}} x_{[nt]+1}^{(d-1)} \right] + R_n(t),$$

where

$$|R_n(t)| = \frac{1}{n^{d-1/2}} \left| z_{[nt]}^{(d-1)} + (nt - [nt]) z_{[nt]+1}^{(d-2)} \right|$$

$$\leq \frac{1}{\sqrt{n}} \max_{0 \leq j \leq n} |\tilde{\varepsilon}_j| + \frac{1}{n\sqrt{n}} \max_{0 \leq j \leq n} |\tilde{\varepsilon}_j|.$$

Since $\sup_{0 \leq t \leq 1} |R_n(t)|$ converges in probability to 0 [see (3.27)], Theorem 3.10 follows from the result for the case $u_j = \varepsilon_j$ and the continuous mapping theorem.

8.6 We have only to show that the right side of (3.72) converges in probability to 0. Since $\mathcal{L}(\sup_{0 \leq t \leq 1} |Y_T^{(d)}(t)|) \to \mathcal{L}(|\alpha|\sigma \sup_{0 \leq t \leq 1} |F_{d-1}(t)|)$ so that $\sup_{0 \leq t \leq 1} |Y_T^{(d)}(t)| = O_p(1)$ and

$$\frac{1}{T^{d-1/2}} \max_{1 \leq j \leq T} \left| y_j^{(d-1)} \right| \leq \frac{1}{\sqrt{T}} \max_{1 \leq j \leq T} |u_j|,$$

we obtain the conclusion, noting that $\max_{1 \leq j \leq T} |u_j|/\sqrt{T}$ converges in probability to 0 because of strict and second-order stationarity of $\{u_j\}$.

9.1 Abel's transformation corresponds to the partial integration formula and can be proved easily. Putting $a_i = \rho_n^{j-i}$ and $b_i = S_i$, we obtain

$$a_{j+1} b_j - a_1 b_0 = \rho_n^{-1} S_j,$$

$$(a_{i+1} - a_i) b_i = (1 - \rho_n) \rho_n^{j-i-1} S_i,$$

which establishes (3.78).

9.2 Consider $|h_t(x; \gamma) - h_t(y; \gamma)|$ for $x, y \in C$, which is bounded by $(2 + e^{|\beta|}) \rho(x, y)$ so that h is a continuous mapping defined on C.

9.3 The partial integration formula yields

$$\int_0^t e^{\beta s} dw(s) = e^{\beta t} w(t) - \beta \int_0^t e^{\beta s} w(s) \, ds,$$

which leads us to the conclusion.

9.4 We need to prove that the right side of (3.84) converges in probability to 0 uniformly in j as $n \to \infty$. Consider

$$A_{jn} \leq \sup_{0 \leq t \leq 1} \left| \exp\left\{ [nt] \log\left(1 - \frac{\beta}{n}\right) \right\} - e^{-\beta t} \right|,$$

where $\log(1 - \beta/n) = -\beta/n + O(n^{-2})$. Then it holds that

$$A_{jn} \leq c |\exp(O(n^{-1})) - 1| \to 0$$

with c a positive constant. We can show similarly that $C_{jn} \to 0$ in probability, while it is almost obvious that $B_{jn} \to 0$ and $D_{jn} \to 0$ in probability.

9.5 We have

$$|V_T - h(X_T)| = \left| \frac{1}{T} \sum_{j=1}^{T} X_T^2 \left(\frac{j}{T}\right) - \int_0^1 X_T^2(t)\, dt \right|$$

$$\leq 2 \sup_{0 \leq t \leq 1} |X_T(t)| \frac{1}{\sqrt{T}} \max_{1 \leq j \leq T} |y_j - y_{j-1}|$$

$$\leq 2 \sup_{0 \leq t \leq 1} |X_T(t)| \left[\frac{|\beta|}{T} \sup_{0 \leq t \leq 1} |X_T(t)| + \frac{1}{\sqrt{T}} \max_{1 \leq j \leq T} |\varepsilon_j| \right],$$

which converges in probability to 0. Then $\mathcal{L}(V_T/\sigma^2) \to \mathcal{L}(h(X))$ since $\mathcal{L}(X_T/\sigma) \to \mathcal{L}(X)$ and $\mathcal{L}(h(X_T/\sigma)) \to \mathcal{L}(h(X))$.

9.6 The stochastic process $\{X_n(t)\}$ in the theorem can be rewritten as in (3.86) with $X(0)$ replaced by $\alpha X(0)$, where $Y_n(t)$ is now defined, using the BN decomposition, as

$$Y_n(t) = \frac{1}{\sqrt{n}} \sum_{j=1}^{[nt]} u_j + (nt - [nt]) \frac{1}{\sqrt{n}} u_{[nt]+1}$$

$$= \alpha Z_n(t) + R_n(t).$$

Here $Z_n(t)$ is given by the right side of (3.82), while $R_n(t)$ is the remainder term defined by (3.25). We also have

$$y_j = \alpha x_j + \sum_{i=1}^{j} \rho_n^{j-i} (\tilde{\varepsilon}_{i-1} - \tilde{\varepsilon}_i),$$

$$x_j = \left(1 - \frac{\beta}{n}\right) x_{j-1} + \varepsilon_j, \qquad x_0 = \sqrt{n}\, \sigma X(0).$$

Then $X_n(t) = \alpha U_n(t) + M_n(t)$, where

$$U_n(t) = \rho_n^{j-1}\sigma X(0) + \rho_n^{-1} Z_n\left(\frac{j-1}{n}\right) - \frac{\beta}{n}\sum_{i=1}^{j-1}\rho_n^{j-i-2}Z_n\left(\frac{i}{n}\right)$$

$$+ n\left(t - \frac{j-1}{n}\right)\frac{x_j - x_{j-1}}{\sqrt{n}},$$

$$M_n(t) = \rho_n^{-1}R_n\left(\frac{j-1}{n}\right) - \frac{\beta}{n}\sum_{i=1}^{j-1}\rho_n^{j-i-2}R_n\left(\frac{i}{n}\right)$$

$$+ n\left(t - \frac{j-1}{n}\right)\frac{1}{\sqrt{n}}$$

$$\times \left(\sum_{i=1}^{j}\rho_n^{j-i}(\tilde{\varepsilon}_{i-1} - \tilde{\varepsilon}_i) - \sum_{i=1}^{j-1}\rho_n^{j-i-1}(\tilde{\varepsilon}_{i-1} - \tilde{\varepsilon}_i)\right).$$

Using the fact that $\sup_{0 \le t \le 1} |R_n(t)| \to 0$ in probability, we can show that $\sup_{0 \le t \le 1} |M_n(t)| \to 0$ in probability. Thus $\mathcal{L}(X_n/\sigma) \to \mathcal{L}(\alpha X)$ by the continuous mapping theorem since $\mathcal{L}(U_n/\sigma) \to \mathcal{L}(X)$.

10.1 Define

$$h(x) = \text{tr}\left(\int_0^1 x(t)x'(t)\,dt\right), \qquad x \in C^q.$$

Then we note that

$$\frac{1}{T^2}\sum_{j=1}^{T} y_j'H'Hy_j = \text{tr}\left(H'H\frac{1}{T^2}\sum_{j=1}^{T} y_j y_j'\right),$$

which converges in distribution to $h(H\Sigma^{1/2}w)$. This establishes (3.94).

10.2 We first note that

$$\sum_{l=0}^{\infty}\|\tilde{A}_l\| \le \sum_{l=0}^{\infty}\sum_{k=l+1}^{\infty}\|A_k\| = \sum_{k=1}^{\infty}\sum_{l=0}^{k-1}\|A_k\|$$

$$= \sum_{k=1}^{\infty} k\|A_k\| < \infty.$$

Thus $\sum_{l=0}^{\infty} A_l A_l'$ converges and $\{\tilde{\varepsilon}_j\}$ is well defined in the m.s. sense so that $E(\tilde{\varepsilon}_j) = 0$ and

$$E(\tilde{\varepsilon}_j \tilde{\varepsilon}_{j+k}') = \begin{cases} \sum_{l=0}^{\infty} A_l A_{l+k}' & (k \geq 0), \\ \sum_{l=0}^{\infty} A_{l-k} A_l' & (k < 0). \end{cases}$$

10.3 The inequality follows from the triangle inequality and the Cauchy–Schwarz inequality.

10.4 Using the relation

$$aa' - bb' = (a-b)(a-b)' + b(a-b)' + (a-b)b',$$

it is easy to obtain that, for x fixed,

$$\|h(x) - h(y)\| \leq \left[\rho_q^2(x, y) + 2\rho_q(x, y) \sup_{0 \leq t \leq 1} \|x(t)\|\right] \times q$$

so that h is a continuous mapping defined on C^q.

10.5 The first inequality is obvious, while the second comes from the fact that

$$\left|Y_{kT}\left(\frac{j}{T}\right) - Y_{kT}(t)\right| \leq \|A^{-1}\| \frac{1}{\sqrt{T}} \max_{1 \leq j \leq T} \|u_j\|.$$

The right side above converges in probability to 0 if $\sum_{j=1}^{T} u_j' u_j I(u_j' u_j > T\delta)/T$ converges in probability to 0 for any $\delta > 0$, which follows from second-order stationarity of $\{u_j\}$ and the Markov inequality.

10.6 Define $x_j^{(d)} = \varepsilon_j/(1-L)^d$ with $x_{-(d-1)}^{(d)} = x_{-(d-2)}^{(d)} = \cdots = x_0^{(d)} = 0$ and put, for $d \geq 2$,

$$X_n^{(d)}(t) = \frac{1}{n^{d-1/2}} x_{[nt]}^{(d)} + (nt - [nt]) \frac{1}{n^{d-1/2}} x_{[nt]+1}^{(d-1)}$$

$$= \frac{1}{n} \sum_{j=1}^{[nt]} X_n^{(d-1)}\left(\frac{j}{n}\right) + (nt - [nt]) \frac{1}{n^{d-1/2}} x_{[nt]+1}^{(d-1)},$$

where

$$X_n^{(1)}(t) = \frac{1}{\sqrt{n}} \sum_{j=1}^{[nt]} \varepsilon_j + (nt - [nt]) \frac{1}{\sqrt{n}} \varepsilon_{[nt]+1}.$$

Using the BN decomposition, we have $y_j^{(d)} = u_j/(1-L)^d = Ax_j^{(d)} - z_j^{(d-1)}$ with $z_j^{(d-1)} = \tilde{\varepsilon}_j/(1-L)^{d-1}$ so that $Y_n^{(d)}(t) = AX_n^{(d)}(t) + R_n(t)$, where

$$|R_{in}(t)| \leq \frac{1}{n^{d-1/2}} \left\| z_{[nt]}^{(d-1)} + (nt - [nt]) z_{[nt]+1}^{(d-2)} \right\|$$

$$\leq \frac{1}{\sqrt{n}} \max_{0 \leq j \leq n} \|\tilde{\varepsilon}_j\| + \frac{1}{n\sqrt{n}} \max_{0 \leq j \leq n} \|\tilde{\varepsilon}_j\|.$$

It is seen that $\rho_q(Y_n^{(d)}, AX_n^{(d)}) \to 0$ in probability. Define now

$$G_{dn}(t) = \int_0^t X_n^{(d)}(s)\, ds,$$

where $\mathcal{L}(G_{1n}) \to \mathcal{L}(F_1)$. Since it can be shown that $\rho(X_n^{(2)}, G_{1n}) \to 0$ in probability, it holds that $\mathcal{L}(Y_n^{(2)}) \to \mathcal{L}(AF_1)$.

10.7 Suppose that the theorem holds for $d = k - 1 \,(\geq 3)$. Using the notations in the solution to Problem 10.6, we have $Y_n^{(k)}(t) = AX_n^{(k)}(t) + R_n(t)$ with $\rho_q(Y_n^{(k)}, AX_n^{(k)}) \to 0$ in probability and

$$\|X_n^{(k)}(t) - G_{k-1,n}(t)\| \leq \frac{2}{\sqrt{n}} \max_{1 \leq j \leq n} \|\varepsilon_j\| + \frac{1}{n} \sup_{0 \leq t \leq 1} \|X_n^{(k-1)}(t)\|,$$

which converges in probability to 0. Since $\mathcal{L}(X_n^{(k-1)}) \to \mathcal{L}(F_{k-2})$ and $\mathcal{L}(G_{k-1,n}) \to \mathcal{L}(F_{k-1})$ by assumption, we can conclude that $\mathcal{L}(X_n^{(k)}) \to \mathcal{L}(F_{k-1})$ and thus $\mathcal{L}(Y_n^{(k)}) \to \mathcal{L}(AF_{k-1})$.

10.8 The first inequality is obvious. The second inequality can be obtained by using the relation $aa' - bb' = (a-b)(a-b)' + b(a-b)' + (a-b)b'$.

10.9 Note that, for $(j-1)/T \leq t \leq j/T$,

$$Y_T^{(d)}(t) = \frac{1}{T^{d-1/2}} y_{j-1}^{(d)} + T\left(t - \frac{j-1}{T}\right) \frac{1}{T^{d-1/2}} y_j^{(d-1)},$$

so that $dY_T^{(d)}(t)/dt = y_j^{(d-1)}/T^{d-3/2} = (y_j^{(d)} - y_{j-1}^{(d)})/T^{d-3/2}$. Thus the left side of (3.110) is equal to

$$\sum_{j=1}^T \int_{(j-1)/T}^{j/T} \left[\frac{y_{j-1}^{(d)}}{T^{d-1/2}} + T\left(t - \frac{j-1}{T}\right) \frac{y_j^{(d-1)}}{T^{d-1/2}} \right] \frac{(y_j^{(d)} - y_{j-1}^{(d)})'}{T^{d-3/2}} dt$$

$$= \frac{1}{T^{2d-1}} \sum_{j=1}^T y_{j-1}^{(d)} \left(y_j^{(d)} - y_{j-1}^{(d)}\right)' + \frac{1}{2T^{2d-1}} \sum_{j=1}^T y_j^{(d-1)} \left(y_j^{(d)} - y_{j-1}^{(d)}\right)'$$

$$= U_T^{(d)} + \frac{1}{2T^{2d-1}} \sum_{j=1}^T y_j^{(d-1)} \left(y_j^{(d-1)}\right)'.$$

10.10 It holds that

$$\sum_{j=1}^{T}\left[y_{j-1}^{(d)}\left(y_{j}^{(d)}-y_{j-1}^{(d)}\right)'+\left(y_{j}^{(d)}-y_{j-1}^{(d)}\right)\left(y_{j-1}^{(d)}\right)'\right]$$

$$=-\left[\sum_{j=1}^{T}\left(y_{j}^{(d)}-y_{j-1}^{(d)}\right)\left(y_{j}^{(d)}-y_{j-1}^{(d)}\right)'\right.$$

$$\left.-\sum_{j=1}^{T}y_{j}^{(d)}\left(y_{j}^{(d)}\right)'+\sum_{j=1}^{T}y_{j-1}^{(d)}\left(y_{j-1}^{(d)}\right)'\right]$$

so that

$$U_{T}^{(d)}+\left(U_{T}^{(d)}\right)'=\frac{1}{T^{2d-1}}y_{T}^{(d)}\left(y_{T}^{(d)}\right)'-\frac{1}{T^{2d-1}}\sum_{j=1}^{T}y_{j}^{(d-1)}\left(y_{j}^{(d-1)}\right)'$$

$$=Y_{T}^{(d)}(1)\left(Y_{T}^{(d)}(1)\right)'+O_{p}\left(\frac{1}{T}\right).$$

We now have (3.111) because of (3.108).

11.1 Noting that $dX_T(t)/dt = \sqrt{T}(x_j - x_{j-1})$ for $(j-1)/T \le t \le j/T$, we have

$$\int_{0}^{1}X_{T}(t)\,dX_{T}(t)$$

$$=\sum_{j=1}^{T}\int_{(j-1)/T}^{j/T}\left[\frac{x_{j-1}}{\sqrt{T}}+T\left(t-\frac{j-1}{T}\right)\frac{x_j-x_{j-1}}{\sqrt{T}}\right]dt\,\sqrt{T}(x_j-x_{j-1})$$

$$=\frac{1}{T}\sum_{j=1}^{T}x_{j-1}(x_j-x_{j-1})+\frac{1}{2T}\sum_{j=1}^{T}(x_j-x_{j-1})^2,$$

which yields the right side of (3.113).

11.2 Noting that $\varepsilon_j = x_j - x_{j-1} + \beta x_{j-1}/T$, we have

(S6) $$\frac{1}{T}\sum_{j=1}^{T}x_{j-1}\varepsilon_{j}=\frac{1}{T}\sum_{j=1}^{T}x_{j-1}(x_{j}-x_{j-1})+\frac{\beta}{T^{2}}\sum_{j=1}^{T}x_{j-1}^{2},$$

where

$$\mathcal{L}\left(\frac{1}{T}\sum_{j=1}^{T}x_{j-1}(x_j-x_{j-1}),\,\frac{\beta}{T^2}\sum_{j=1}^{T}x_{j-1}^2\right)$$

$$\rightarrow \mathcal{L}\left(\int_{0}^{1}X(t)\,dX(t),\,\beta\int_{0}^{1}X^{2}(t)\,dt\right).$$

It follows from the continuous mapping theorem that

$$\mathcal{L}\left(\frac{1}{T}\sum_{j=1}^{T} x_{j-1}\varepsilon_j\right) \to \mathcal{L}\left(\int_0^1 X(t)\,dX(t) + \beta \int_0^1 X^2(t)\,dt\right)$$

$$= \mathcal{L}\left(\int_0^1 X(t)\,dw(t)\right).$$

11.3 Note that

$$\frac{1}{T}\sum_{j=1}^{T} x_{j-1}(x_j - x_{j-1}) = \frac{1}{2T}(x_T^2 - x_0^2) - \frac{1}{2T}\sum_{j=1}^{T}\left(-\frac{\beta}{T}x_{j-1} + u_j\right)^2$$

$$= \frac{1}{2T}(x_T^2 - x_0^2) - \frac{1}{2T}\sum_{j=1}^{T} u_j^2 + R_T,$$

where

$$R_T = -\frac{\beta^2}{2T^3}\sum_{j=1}^{T} x_{j-1}^2 + \frac{\beta}{T^2}\sum_{j=1}^{T} x_{j-1}u_j.$$

It follows from the results in Section 3.9 that $R_T \to 0$ in probability and

$$\mathcal{L}\left(\frac{1}{T}x_T^2, \frac{1}{T}x_0^2\right) \to \mathcal{L}\left(\alpha^2 X^2(1), \alpha^2 X^2(0)\right),$$

$$\frac{1}{T}\sum_{j=1}^{T} u_j^2 \to \sum_{l=0}^{\infty} \alpha_l^2 \quad \text{in probability.}$$

Thus Theorem 3.15 follows.

11.4 Since (S6) holds with ε_j replaced by u_j and

$$\mathcal{L}\left(\frac{1}{T}\sum_{j=1}^{T} x_{j-1}(x_j - x_{j-1}),\, \frac{\beta}{T^2}\sum_{j=1}^{T} x_{j-1}^2\right)$$

$$\to \mathcal{L}\left(\alpha^2 \int_0^1 X(t)\,dX(t) + \frac{1}{2}\left(\alpha^2 - \sum_{l=0}^{\infty}\alpha_l^2\right),\, \alpha^2\beta\int_0^1 X^2(t)\,dt\right),$$

(3.117) follows from the continuous mapping theorem.

11.5 Since it holds that

$$\frac{1}{T}\sum_{j=1}^{T} y_j' H\varepsilon_j = \text{tr}\left(H\frac{1}{T}\sum_{j=1}^{T}\varepsilon_j y_{j-1}'\right) + \text{tr}\left(H\frac{1}{T}\sum_{j=1}^{T}\varepsilon_j \varepsilon_j'\right),$$

(3.120) follows from (3.119), the weak law of large numbers and the continuous mapping theorem.

11.6 The first equality is obvious and the remainder term in the second equality is

$$\frac{1}{T}\left[A\sum_{j=1}^{T} z_{j-1}(\tilde{\varepsilon}_{j-1} - \tilde{\varepsilon}_j)' - A\sum_{j=1}^{T}\varepsilon_j\tilde{\varepsilon}_j' + \tilde{\varepsilon}_0\sum_{j=1}^{T} u_j'\right]$$

$$= \frac{1}{T}\left[A\sum_{j=1}^{T} z_{j-1}\tilde{\varepsilon}_{j-1}' - A\sum_{j=1}^{T}(z_j - \varepsilon_j)\tilde{\varepsilon}_j' - A\sum_{j=1}^{T}\varepsilon_j\tilde{\varepsilon}_j' + \tilde{\varepsilon}_0\sum_{j=1}^{T} u_j'\right]$$

$$= \frac{1}{T}\left[A(z_0\tilde{\varepsilon}_0' - z_T\tilde{\varepsilon}_T') + \tilde{\varepsilon}_0\sum_{j=1}^{T}\tilde{u}_j'\right],$$

which is evidently $o_p(1)$.

11.7 Define $v_j = (\varepsilon_j', \tilde{\varepsilon}_j')'$ and $w_j = (\tilde{\varepsilon}_{j-1}', u_j')'$. Then $\{v_j\}$ and $\{w_j\}$ are strictly stationary with zero means and finite second moments. It now follows from Theorem 2 of Hannan (1970, p. 203) that

$$\frac{1}{T}\sum_{j=1}^{T}\varepsilon_j\tilde{\varepsilon}_j' \to E(\varepsilon_j\tilde{\varepsilon}_j') = E\left[\varepsilon_j\left(\sum_{k=0}^{\infty}\tilde{A}_k\varepsilon_{j-k}\right)'\right] \quad \text{(a.s.)}$$

$$= \tilde{A}_0 = A - A_0,$$

$$\frac{1}{T}\sum_{j=1}^{T}\tilde{\varepsilon}_{j-1}u_j' \to E(\tilde{\varepsilon}_{j-1}u_j') = E\left[\sum_{l=0}^{\infty}\tilde{A}_l\varepsilon_{j-1-l}\left(\sum_{m=0}^{\infty}A_m\varepsilon_{j-m}\right)'\right] \quad \text{(a.s.)}$$

$$= \sum_{l=0}^{\infty}\tilde{A}_l A_{l+1}' = \sum_{l=0}^{\infty}\left(\sum_{k=l+1}^{\infty}A_k\right)A_{l+1}'.$$

11.8 We have only to show that

$$\sum_{k=1}^{\infty}\sum_{l=0}^{\infty}\alpha_l\alpha_{k+l} = \frac{1}{2}\left(\alpha^2 - \sum_{l=0}^{\infty}\alpha_l^2\right).$$

Since the right side is equal to

$$\sum_{l<k}\alpha_l\alpha_k = \sum_{k=1}^{\infty}\sum_{l=0}^{\infty}\alpha_l\alpha_{k+l},$$

CHAPTER 4

1.1 It is easy to obtain

$$\frac{l_n(\alpha)}{l_n(\beta)} = \exp\left[\frac{\beta-\alpha}{n}\sum_{j=1}^{n} y_{j-1}(y_j - y_{j-1}) - \frac{\alpha^2-\beta^2}{2n^2}\sum_{j=1}^{n} y_{j-1}^2\right],$$

where it holds by the result of Section 3.11 and the continuous mapping theorem that

$$\mathcal{L}\left(\frac{1}{n}\sum_{j=1}^{n} y_{j-1}(y_j - y_{j-1}), \frac{1}{n^2}\sum_{j=1}^{n} y_{j-1}^2\right) \to \mathcal{L}\left(\int_0^1 y(t)\,dy(t), \int_0^1 y^2(t)\,dt\right).$$

Thus we establish (4.7) using again the continuous mapping theorem.

1.2 The expressions up to the next to last line are a consequence of Theorem 4.1 and the Ito calculus $d(Y^2(t)) = 2Y(t)dY(t) + dt$. Since

$$Y(1) = \kappa e^{-\beta} + e^{-\beta}\int_0^1 e^{\beta s}\,dw(s) \sim N(\mu, \sigma^2),$$

where $\mu = \kappa e^{-\beta}$ and $\sigma^2 = (1 - e^{-2\beta})/(2\beta)$, we have

$$E\left(e^{\theta S_1}\right) = \left[1 - (\beta-\alpha)\sigma^2\right]^{-1/2}\exp\left[\frac{\alpha-\beta}{2}\left(\kappa^2 + 1 - \frac{\mu^2}{1-(\beta-\alpha)\sigma^2}\right)\right],$$

which yields the last expression in (4.8).

1.3 Putting $X = Y(1)$ and $Y = \int_0^1 Y(t)\,dt$, we have that $(X, Y)' \sim N(\mu, \Sigma)$, where

$$\mu = \begin{pmatrix} \kappa e^{-\beta} \\ (1-e^{-\beta})\frac{\kappa}{\beta} \end{pmatrix},$$

$$\Sigma = \begin{pmatrix} \frac{1-e^{-2\beta}}{2\beta} & \frac{1}{\beta^2}\left(\frac{1}{2} - e^{-\beta} + \frac{e^{-2\beta}}{2}\right) \\ \frac{1}{\beta^2}\left(\frac{1}{2} - e^{-\beta} + \frac{e^{-2\beta}}{2}\right) & \frac{1}{\beta^3}\left(\beta - \frac{3}{2} + 2e^{-\beta} - \frac{e^{-2\beta}}{2}\right) \end{pmatrix}.$$

Therefore we obtain

$$E\left[\exp\left\{\frac{\beta-\alpha}{2}X^2 - \theta Y^2\right\}\right]$$

$$= |I_2 - \Sigma\Lambda|^{-1/2}\exp\left[\frac{1}{2}\mu'\Sigma^{-1}\{(I_2 - \Sigma\Lambda)^{-1} - I_2\}\mu\right],$$

where $\Lambda = \text{diag}(\beta - \alpha, -2\theta)$. We can arrive at the last expression in (4.11) after some manipulations.

1.4 Using Theorem 4.1 we have

$$E\left[\exp\left\{\theta \int_0^1 (w(t) - tw(1))^2 \, dt\right\}\right]$$

$$= E\left[\exp\left\{\left(\frac{\beta}{2} + \frac{\theta}{3}\right) Y^2(1) - 2\theta Y(1) \int_0^1 tY(t) \, dt - \frac{\beta}{2}\right\}\right],$$

where $\beta = \sqrt{-2\theta}$. Putting

$$X = Y(1) = e^{-\beta} \int_0^1 e^{\beta s} \, dw(s),$$

$$Y = \int_0^1 tY(t) \, dt = \int_0^1 \left\{\frac{s}{\beta} + \frac{1}{\beta^2} - \left(\frac{1}{\beta} + \frac{1}{\beta^2}\right) e^{-\beta(1-s)}\right\} dw(s),$$

we obtain

$$V(X) = \frac{1 - e^{-2\beta}}{2\beta},$$

$$\text{Cov}(X, Y) = \frac{1}{2\beta^2} - \frac{1}{2\beta^3} + \left(\frac{1}{2\beta^2} + \frac{1}{2\beta^3}\right) e^{-2\beta},$$

$$V(Y) = \frac{1}{3\beta^2} - \frac{1}{2\beta^3} + \frac{1}{2\beta^5} - \left(\frac{1}{2\beta^3} + \frac{1}{\beta^4} + \frac{1}{2\beta^5}\right) e^{-2\beta},$$

from which we can arrive at the result after some algebra.

1.5 We obtain

$$V(Y(0)) = \frac{1}{2\alpha}, \qquad V(Y(1)) = \frac{e^{-2\beta}}{2\alpha} + \frac{1 - e^{-2\beta}}{2\beta},$$

$$V\left(\int_0^1 Y(t) \, dt\right)$$

$$= \frac{1}{\beta^2}\left[1 + \frac{1}{2\alpha} - \frac{3}{2\beta} + \left(\frac{2}{\beta} - \frac{1}{\alpha}\right) e^{-\beta} + \left(\frac{1}{2\alpha} - \frac{1}{2\beta}\right) e^{-2\beta}\right],$$

$$\text{Cov}(Y(0), Y(1)) = \frac{e^{-\beta}}{2\alpha}, \qquad \text{Cov}\left(Y(0), \int_0^1 Y(t) \, dt\right) = \frac{1}{2\alpha\beta}(1 - e^{-\beta}),$$

$$\text{Cov}\left(Y(1), \int_0^1 Y(t) \, dt\right) = \frac{1}{\beta^2}\left[\frac{1}{2} + \left(\frac{\beta}{2\alpha} - 1\right) e^{-\beta} + \left(\frac{1}{2} - \frac{\beta}{2\alpha}\right) e^{-2\beta}\right].$$

Noting that the above three random variables are normally distributed with means 0, some manipulations yield (4.14) and (4.15).

1.6 We can proceed in the same way as in the solution to Problem 1.2. A different definition of β gives a different final expression.

1.7 Since $E[\exp(\theta_1 U + \theta_2 S_1)] = E[\exp\{\theta(xS_1 - U)\}]$ with $\theta = -\theta_1$ and $x = -\theta_2/\theta_1$, the joint m.g.f. of U and S_1 is derived from (4.17), replacing θ, x, and β by $-\theta_1$, $-\theta_2/\theta_1$, and $\sqrt{\alpha^2 - 2\theta_2}$, respectively.

1.8 Replacing κ in (4.17) by $X(0)$ and noting that $E\left[\exp\{\theta X^2(0)\}\right] = (1 - \theta/\alpha)^{-1/2}$, we obtain, from (4.17),

$$E\left[\exp\{\theta(xS_1 - U)\}\right]$$
$$= E\left[E\left[\exp\{\theta(xS_1 - U)\} \mid X(0)\right]\right]$$
$$= \exp\left(\frac{\alpha + \theta}{2}\right)\left[\left(1 - \frac{\theta\left(\alpha + \frac{\theta}{2} + x\right)}{\alpha g(\theta)}\frac{\sinh\beta}{\beta}\right)g(\theta)\right]^{-1/2}$$
$$= \exp\left(\frac{\alpha + \theta}{2}\right)\left[g(\theta) - \frac{\theta\left(\alpha + \frac{\theta}{2} + x\right)}{\alpha}\frac{\sinh\beta}{\beta}\right]^{-1/2},$$

where $g(\theta) = \cosh\beta + (\alpha + \theta)\sinh\beta/\beta$. This gives us (4.19).

1.9 Noting that

$$T(\hat{\rho} - 1) = \frac{\frac{1}{T}\sum_{j=1}^{T} y_{j-1}(y_j - y_{j-1})}{\frac{1}{T^2}\sum_{j=1}^{T} y_{j-1}^2},$$

$$\mathcal{L}\left(\frac{1}{T}\sum_{j=1}^{T} y_{j-1}(y_j - y_{j-1}), \frac{1}{T^2}\sum_{j=1}^{T} y_{j-1}^2\right)$$
$$\to \mathcal{L}\left(\int_0^1 X(t)\,dX(t), \int_0^1 X^2(t)\,dt\right),$$

we can establish (4.20) by the continuous mapping theorem.

1.10 By the conditional argument leading to (4.24) we have

$$E\left[\exp\{i\theta(xS_1 - V)\}\right] = E\left[\exp\left\{i\left(\theta x + \frac{i\theta^2}{2}\right)S_1\right\}\right]$$

and thus, replacing θ by $i(\theta x + (i\theta^2)/2)$ in (4.14), we obtain (4.25). Note that $\cosh\sqrt{-\theta} = \cos\sqrt{\theta}$ and $\sinh\sqrt{-\theta}/\sqrt{-\theta} = \sin\sqrt{\theta}/\sqrt{\theta}$.

1.11 Noting that

$$T(\hat{\beta} - \beta) = \frac{\frac{1}{T}\sum_{j=1}^{T} y_{1j}\varepsilon_{2j}}{\frac{1}{T^2}\sum_{j=1}^{T} y_{1j}^2},$$

$$\mathcal{L}\left(\frac{1}{T}\sum_{j=1}^{T} y_{1j}\varepsilon_{2j}, \frac{1}{T^2}\sum_{j=1}^{T} y_{1j}^2\right) \to \mathcal{L}\left(\int_0^1 X_1(t)\,dw_2(t), \int_0^1 X_1^2(t)\,dt\right),$$

we can establish (4.26) by the continuous mapping theorem.

2.1 Since $dY_g(t)/dt = \beta Y_g(t) + F_{g-1}(t)$ and $d^{g-1}F_{g-1}(t)/dt^{g-1} = w(t)$, (4.33) follows by differentiation.

2.2 Noting that $(1-L)^g(1-(1+(\beta/T))L)y_j = (1-(1+(\beta/T))L)z_j = \varepsilon_j$ we obtain

$$\left.\frac{l_T(0)}{l_T(\alpha)}\right|_\beta = \exp\left[\frac{-\alpha}{T}\sum_{j=1}^{T} z_{j-1}(z_j - z_{j-1}) + \frac{\alpha^2}{2T^2}\sum_{j=1}^{T} z_{j-1}^2\right],$$

where $z_j = (1+(\beta/T))z_{j-1} + \varepsilon_j$. Therefore (4.39) holds by the continuous mapping theorem. Since $z_j = (1-L)^g y_j$, $\{Z(t)\}$ must satisfy $Z(t) = d^g Y_g(t)/dt^g$.

2.3 The expressions up to the third equality are a consequence of Theorem 4.2. Since $dY_1(t) = (\beta Y_1(t) + w(t))\,dt$, we have

$$\int_0^1 \frac{dY_1(t)}{dt}\,d\left(\frac{dY_1(t)}{dt}\right) = \frac{1}{2}\left[(\beta Y_1(1) + w(1))^2 - 1\right],$$

$$\int_0^1 \left(\frac{dY_1(t)}{dt}\right)^2 dt = \int_0^1 (\beta Y_1(t) + w(t))^2\,dt$$

$$= \beta^2 \int_0^1 Y_1^2(t)\,dt + 2\beta \int_0^1 Y_1(t)(dY_1(t) - \beta Y_1(t)\,dt)$$

$$+ \int_0^1 w^2(t)\,dt$$

$$= -\beta^2 \int_0^1 Y_1^2(t)\,dt + \beta Y_1^2(1) + \int_0^1 w^2(t)\,dt,$$

which yields the last expression in (4.41).

2.4 The first equality is a consequence of Theorem 4.1. Since

$$X = X(1) = e^\gamma \int_0^1 e^{-\gamma s} dw(s),$$

$$Y = \int_0^1 e^{-\beta t} X(t) dt = \frac{1}{\beta - \gamma} \int_0^1 (e^{-\beta t} - e^{\gamma - \beta} e^{-\gamma t}) dw(t),$$

we obtain $(X, Y)' \sim N(0, \Sigma)$, where

$$\Sigma_{11} = \frac{e^{2\gamma} - 1}{2\gamma}, \quad \Sigma_{12} = \frac{1}{\beta - \gamma} \left[\frac{e^{-\beta}(1 - e^{2\gamma})}{2\gamma} - \frac{e^{-\beta} - e^\gamma}{\beta + \gamma} \right],$$

$$\Sigma_{22} = \frac{1}{(\beta - \gamma)^2} \left[\frac{1 - e^{-2\beta}}{2\beta} + \frac{2e^{-\beta}(e^{-\beta} - e^\gamma)}{\beta + \gamma} + \frac{e^{-2\beta}(e^{2\gamma} - 1)}{2\gamma} \right].$$

Then

$$m_1(\theta) = |I_2 - 2A\Sigma|^{-1/2} \exp\left(\frac{\beta + \gamma}{2}\right),$$

where $A_{11} = (-\beta - \gamma)/2$, $A_{12} = A_{21} = -\beta^2 e^\beta/2$, and $A_{22} = 0$. Some manipulations yield the last expression in (4.42).

2.5 The derivation is almost the same as in the solution to Problem 2.4. We arrive at

$$E\left[\exp\{\theta(xS(F_1) - U)\}\right] = |I_2 - 2B\Sigma|^{-1/2} \exp\left(\frac{\beta + \gamma}{2}\right),$$

where $B_{11} = (-\beta - \gamma)/2$, $B_{22} = -\theta e^{2\beta}/2$, $B_{12} = B_{21} = -\beta^2 e^\beta/2$, while Σ is defined in the solution to Problem 2.4. The last expression in (4.44) is obtained after some calculations.

2.6 Noting that

$$T(\hat\rho - 1) = \frac{\frac{1}{T^3} \sum_{j=1}^T y_{j-1}(y_j - y_{j-1})}{\frac{1}{T^4} \sum_{j=1}^T y_{j-1}^2},$$

$$\mathcal{L}\left(\frac{1}{T^3} \sum_{j=1}^T y_{j-1}(y_j - y_{j-1}), \frac{1}{T^4} \sum_{j=1}^T y_{j-1}^2\right)$$

$$\to \mathcal{L}\left(\int_0^1 F_1(t) dF_1(t), \int_0^1 F_1^2(t) dt\right),$$

we can establish (4.46) by the continuous mapping theorem.

2.7 Noting that

$$T^2(\hat{\beta} - \beta) = \frac{\frac{1}{T^2}\sum_{j=1}^{T} y_{1j}\varepsilon_{2j}}{\frac{1}{T^4}\sum_{j=1}^{T} y_{1j}^2},$$

$$\mathcal{L}\left(\frac{1}{T^2}\sum_{j=1}^{T} y_{1j}\varepsilon_{2j}, \frac{1}{T^4}\sum_{j=1}^{T} y_{1j}^2\right) \to \mathcal{L}\left(\int_0^1 F_1(t)\,dw_2(t), \int_0^1 F_1^2(t)\,dt\right),$$

we can establish (4.50) by the continuous mapping theorem.

3.1 Since it holds that

$$\mathcal{L}\left(\int_0^1 w'(t)Hw(t)\,dt\right) = \mathcal{L}\left(\int_0^1 w'(t)\Lambda w(t)\,dt\right),$$

where Λ is the diagonal matrix with the eigenvalues of H on diagonals, we have (4.58) because of (4.10) and the independence property of components of $\{w(t)\}$.

3.2 We have, from the matrix version of Ito's theorem,

$$\int_0^1 X(t)\,dX'(t) + \left(\int_0^1 X(t)\,dX'(t)\right)' = X(1)X'(1) - I_q.$$

Premultiplying A and taking the trace yield the conclusion.

3.3 It follows from (4.54) that

$$X(1) \sim N\left(0, \frac{1}{2}A^{-1}(e^{2A} - I_q)\right)$$

so that

$$E(e^{\theta S_1}) = \left|I_q + \frac{1}{2}(e^{2A} - I_q)\right|^{-1/2} \exp\left\{\frac{1}{2}\operatorname{tr}(A)\right\}$$

$$= \prod_{j=1}^{q}\left[\frac{1}{2}\left\{\exp(2\sqrt{-2\theta\lambda_j}) + 1\right\}\exp(-\sqrt{-2\theta\lambda_j})\right]^{-1/2}$$

$$= \prod_{j=1}^{q}(\cosh\sqrt{-2\theta\lambda_j})^{-1/2},$$

which gives (4.58).

3.4 The eigenvalues of H are $\frac{1}{2}$ and $-\frac{1}{2}$, which yields (4.61) because of (4.58).

3.5 When G is symmetric, it follows from (4.59) that $S_3 = [w'(1)Gw(1) - \text{tr}(G)]/2$. Since $w(1) \sim N(0, I_q)$, we have

$$E(e^{\theta S_3}) = \exp\left\{-\frac{\theta}{2}\text{tr}(G)\right\} \prod_{j=1}^{q}(1 - \theta\lambda_j)^{-1/2},$$

where λ_js are the eigenvalues of G.

3.6 Noting that

$$S_4 \mid \{w_1(t)\} \sim N\left(0, \int_0^1 w_1^2(t)\,dt\right),$$

we obtain, from (4.58),

$$E(e^{\theta S_4}) = E\left[\exp\left\{\frac{\theta^2}{2}\int_0^1 w_1^2(t)\,dt\right\}\right]$$
$$= (\cos\theta)^{-1/2}.$$

3.7 Using the relation that

$$E\bigl(w_2(s)\,dw_2(t)\bigr) = \begin{cases} 0 & s < t \\ dt & s \geq t, \end{cases}$$

we obtain

$$\int_0^1\int_0^1 w_1(t)\,dw_1(s)E\bigl(w_2(s)\,dw_2(t)\bigr) = \int_0^1\left\{\int_t^1 dw_1(s)\right\}w_1(t)\,dt$$
$$= \int_0^1 \bigl(w_1(1) - w_1(t)\bigr)w_1(t)\,dt,$$

from which (4.65) follows.

3.8 Define $dX(t) = -\beta X(t)\,dt + dw_1(t)$ with $X(0) = 0$. Then we need to compute

$$E\bigl(e^{\theta S_5}\bigr) = E\left[E\left[e^{\theta S_5}|\{w_1(t)\}\right]\right]$$
$$= E\left[\exp\left\{\frac{\theta^2}{2}\int_0^1\left(w_1(t) - \frac{1}{2}w_1(1)\right)^2 dt\right\}\right]$$
$$= E\left[\exp\left\{\frac{\theta^2 + 4\beta}{8}X^2(1) - \frac{\theta^2}{2}X(1)\int_0^1 X(t)\,dt - \frac{\beta}{2}\right\}\right],$$

where $\beta = i\theta$ and $(X,Y)' \sim N(0,\Sigma)$ with

$$X = X(1) = e^{-\beta}\int_0^1 e^{\beta s}\,dw_1(s),$$

$$Y = \int_0^1 X(t)\,dt = \frac{e^{-\beta}}{\beta}\int_0^1 \left(e^\beta - e^{\beta s}\right) dw_1(s),$$

$$V(X) = \frac{1-e^{-2\beta}}{2\beta}, \quad \mathrm{Cov}(X,Y) = \frac{e^{-\beta}}{2\beta^2}(e^{-\beta}-2) + \frac{1}{2\beta^2},$$

$$V(Y) = \frac{e^{-\beta}}{2\beta^3}(-e^{-\beta}+4) + \frac{1}{\beta^2} - \frac{3}{2\beta^3}.$$

Since $E(e^{\theta S_5}) = |I_2 - 2A\Sigma|^{-1/2}e^{-\beta/2}$, where

$$A = \begin{pmatrix} \dfrac{\theta^2 + 4\beta}{8} & -\dfrac{\theta^2}{4} \\ -\dfrac{\theta^2}{4} & 0 \end{pmatrix},$$

we obtain $E(e^{\theta S_5}) = (\cos(\theta/2))^{-1}$ after some algebra.

CHAPTER 5

1.1 Since

$$X(t) = e^{-\beta t}\int_0^t e^{\beta s}\,dw(s),$$

we have

$$\int_0^1 X^2(t)\,dt = \int_0^1 e^{-2\beta t}\left\{\int_0^t\int_0^t e^{\beta(u+v)}\,dw(u)\,dw(v)\right\}dt$$

$$= \int_0^1\int_0^1\left\{\int_{\max(u,v)}^1 e^{-2\beta t}\,dt\right\}e^{\beta(u+v)}\,dw(u)\,dw(v)$$

$$= \int_0^1\int_0^1 \frac{e^{-\beta|s-t|} - e^{-\beta(2-s-t)}}{2\beta}\,dw(s)\,dw(t).$$

1.2 Putting $w_1(t) = w(t)$, we have, from (4.65)

$$V\left[S_2\,|\,\{w(t)\}\right]$$

$$= \int_0^1 w^2(t)\,dt - w(1)\int_0^1 w(t)\,dt + \frac{1}{4}w^2(1)$$

$$= \frac{1}{4}\int_0^1\int_0^1 \left[4(1-\max(s,t)) - 2(1-s+1-t) + 1\right] dw(s)\,dw(t)$$

$$= \int_0^1\int_0^1 \frac{1}{4}[1 - 2|s-t|]\,dw(s)\,dw(t).$$

1.3 From the expression given above (5.7), we have

$$S_3 = \frac{1}{2}g(1)w^2(1) - \frac{1}{2}\int_0^1 \left(g'(t)w^2(t) + g(t)\right) dt,$$

where

$$\int_0^1 g'(t)w^2(t)\,dt = \int_0^1\int_0^1 \left[g(1) - g(\max(s,t))\right] dw(s)\,dw(t)$$

$$= g(1)w^2(1) - \int_0^1\int_0^1 g(\max(s,t))\,dw(s)\,dw(t).$$

Thus we can arrive at the right side of (5.7).

1.4 Note first that

$$S_{T5} = \frac{1}{T^2}\varepsilon'CMC'\varepsilon,$$

where $M = I_T - ee'/T$ with $e = (1,\ldots,1)'$ and C is defined in (1.3). Defining $y_j = y_{j-1} + \varepsilon_j$ with $y_0 = 0$, the FCLT and the continuous mapping theorem yield

$$\mathcal{L}\left(\frac{1}{T^2}\varepsilon'C'MC\varepsilon\right) = \mathcal{L}\left(\frac{1}{T^2}\sum_{j=1}^T (y_j - \bar{y})^2\right)$$

$$\to \mathcal{L}\left(\int_0^1 w^2(t)\,dt - \left(\int_0^1 w(t)\,dt\right)^2\right).$$

Thus $\mathcal{L}(S_{T5})$ also has the same limiting distribution as above.

2.1 Since $a_1(T)$ is the coefficient of λ in the expansion of $D_T(\lambda) = |I_T - \lambda K_T/T|$, (5.14) clearly holds. For (5.15) we use the formula

$$\frac{d^2 D_T(\lambda)}{d\lambda^2} = \frac{d}{d\lambda}\sum_{j=1}^T D_{jT}(\lambda),$$

where $D_{jT}(\lambda)$ is the determinant of $I_T - \lambda K_T/T$ with the jth column replaced by its derivative with respect to λ. Evaluating at $\lambda = 0$ leads to (5.15).

2.2 Defining

(S7) $$G(t) = \int_0^t g(s)\,ds, \qquad (G(0) = 0),$$

we have

$$\int_0^1 \int_0^1 \big[1 - \max(s,t)\big] g(s)g(t)\,ds\,dt$$

$$= \int_0^1 \left[\int_0^1 g(s)\,ds - t\int_0^t g(s)\,ds - \int_t^1 sg(s)\,ds \right] g(t)\,dt$$

$$= \int_0^1 \left(\int_t^1 G(s)\,ds \right) g(t)\,dt = \int_0^1 G^2(t)\,dt \geq 0.$$

The other cases can be proved similarly.

2.3 It follows from (5.21) that

$$\int_0^1 K(t,t)\,dt = \sum_{n=1}^\infty \frac{1}{\lambda_n} \int_0^1 f_n^2(t)\,dt = \sum_{n=1}^\infty \frac{1}{\lambda_n}.$$

We also have

$$\int_0^1 \int_0^1 K^2(s,t)\,ds\,dt = \sum_{m=1}^\infty \sum_{n=1}^\infty \frac{1}{\lambda_m \lambda_n} \left(\int_0^1 f_m(s) f_n(s)\,ds \right)^2$$

$$= \sum_{n=1}^\infty \frac{1}{\lambda_n^2}.$$

3.1 Let us consider

$$S_N = \int_0^1 \int_0^1 K_N(s,t)\,dw(s)\,dw(t), \quad K_N(s,t) = \sum_{n=1}^N \frac{1}{\lambda_n} f_n(s) f_n(t).$$

Then Mercer's theorem ensures that $\text{l.i.m.}_{N\to\infty} S_N = S$, where S is defined in (5.24). Moreover, we have

$$S_N = \int_0^1 \int_0^1 \sum_{n=1}^N \frac{1}{\lambda_n} f_n(s) f_n(t)\,dw(s)\,dw(t) = \sum_{n=1}^N \frac{1}{\lambda_n} \left\{ \int_0^1 f_n(t)\,dw(t) \right\}^2,$$

which leads us to the conclusion.

3.2 Following the second definition of $D(\lambda)$ in (5.26), we have

$$D(\lambda) = \sum_{n=0}^{\infty} \frac{(-1)^n \lambda^n}{n!} \int_0^1 \cdots \int_0^1 \left| \begin{pmatrix} g(t_1) \\ \vdots \\ g(t_n) \end{pmatrix} (g(t_1), \ldots, g(t_n)) \right| dt_1 \cdots dt_n$$

$$= 1 - \lambda \int_0^1 g^2(t)\, dt \,.$$

3.3 We have only to show that (5.30) implies (5.10). We have

$$\int_0^1 [1 - \max(s,t)] f(s)\, ds = -\frac{1}{\lambda} \int_0^1 [1 - \max(s,t)] f''(s)\, ds$$

$$= -\frac{1}{\lambda} \left[f'(1) - tf'(t) - \int_t^1 s f''(s)\, ds \right]$$

$$= \frac{1}{\lambda} f(t),$$

which implies (5.10).

3.4 The integral equation (5.10) with $K(s,t) = \min(s,t) - st$ is equivalent to $f(t) = c_1 \cos\sqrt{\lambda}\, t + c_2 \sin\sqrt{\lambda}\, t$ with $f(0) = f(1) = 0$. Thus $\lambda\ (\neq 0)$ is an eigenvalue if and only if $\sin\sqrt{\lambda} = 0$, from which we obtain $D(\lambda) = \sin\sqrt{\lambda}/\sqrt{\lambda}$ as the FD of K so that (5.34) results.

3.5 Defining $G(t)$ as in (S7), we obtain

$$\int_0^1 \int_0^1 K(s,t) g(s) g(t)\, ds\, dt = \frac{1}{4} \int_0^1 (G(1) - 2G(t))^2\, dt \geq 0.$$

3.6 We show that (5.36) implies (5.10). We obtain

$$\int_0^1 K(s,t) f(s)\, ds = \frac{-1}{4\lambda} \int_0^1 [1 - 2|s - t|] f''(s)\, ds$$

$$= \frac{-1}{4\lambda} \left[f'(1) - f'(0) - 2 \left\{ \int_0^t (t-s) f''(s)\, ds \right. \right.$$

$$\left. \left. + \int_t^1 (s-t) f''(s)\, ds \right\} \right]$$

$$= \frac{1}{\lambda} f(t).$$

3.7 The integral equation (5.10) with $K(s,t)$ given in (5.39) yields $f(t) = c_1 \cos\sqrt{\lambda}\, t + c_2 \sin\sqrt{\lambda}\, t + c_3$, where $f(0) = f(1)$, $f'(0) = f'(1)$, and

$$f(0) = \lambda \left[\frac{c_3}{12} + \frac{1}{2} \int_0^1 (t^2 - t) f(t)\, dt \right].$$

Then we have $M(\lambda)c = 0$, where $c = (c_1, c_2, c_3)'$ and

$$|M(\lambda)| = \begin{vmatrix} 1 - \cos\sqrt{\lambda} & -\sin\sqrt{\lambda} & 0 \\ \sin\sqrt{\lambda} & 1 - \cos\sqrt{\lambda} & 0 \\ \cos\sqrt{\lambda} - \dfrac{2\sin\sqrt{\lambda}}{\sqrt{\lambda}} - 1 & \sin\sqrt{\lambda} + \dfrac{2\cos\sqrt{\lambda}}{\sqrt{\lambda}} - \dfrac{2}{\sqrt{\lambda}} & -2 \end{vmatrix}$$

$$= -8\left(\sin\dfrac{\sqrt{\lambda}}{2}\right)^2.$$

Thus the eigenvalues are given by $\lambda_n = 4n^2\pi^2$ $(n = 1, 2, \ldots)$. Since $\operatorname{rank}(M(\lambda_n)) = 1$ for each λ_n, the multiplicity of each eigenvalue is two. In fact, $\int_0^1 K(t,t)\,dt = \frac{1}{12}$, while $\sum_{n=1}^\infty 1/(4n^2\pi^2) = \frac{1}{24}$. Then we obtain the FD of K as in (5.40).

3.8 We show that (5.42) implies (5.10). We obtain

$$\int_0^1 K(s,t) f(s)\,ds = -\dfrac{1}{\lambda}\int_0^1 [1 - \max(s,t) + b]\,f''(s)\,ds$$

$$= -\dfrac{1}{\lambda}\left[bf'(1) + f(1) - f(t)\right],$$

where $f(1) = -bf'(1)$ since

$$f(1) = \lambda b \int_0^1 f(s)\,ds = -b\int_0^1 f''(s)\,ds = -bf'(1).$$

3.9 When $b = 0$, it is clear that the zeros of $h(z)$ are all simple. Suppose that $b \neq 0$. If the zeros of $h(z)$ are not simple, it holds that $h(z) = h'(z) = 0$, that is, $\cos z - bz\sin z = 0$ and $bz\cos z + (b+1)\sin z = 0$. Then it follows that $b + 1 + b^2 z^2 = 0$ so that $z = \pm\sqrt{-b - 1/b}$ and $\sin^2(\sqrt{-b-1/b}) = -1/b$, from which we have contradiction. In fact, $\sin^2(\sqrt{-b-1/b}) < -1/b$ for any real b.

3.10 The zeros of $h(z)$ satisfy $\tan z = 1/(bz)$, where z is real or purely imaginary. Suppose first that b is positive. Then no purely imaginary number $z = ix$ satisfies $\tan ix = 1/(bix) \Leftrightarrow \tanh x = -1/(bx)$, as can be seen from the graphs of $\tanh x$ and $-1/(bx)$ with $b > 0$. When b is negative, the graphs of $\tanh x$ and $-1/(bx)$ cross at two points, $\pm a$, say, which yields the conclusion.

4.1 Putting $l(t) = t^m$ and $L(t) = \int_0^t l^2(s)\,ds$, we consider

$$\int_0^1 \int_0^1 \left\{\int_{\max(s,t)}^1 l^2(u)\,du\right\} g(s)g(t)\,ds\,dt$$

$$= \int_0^1 \left[L(1)G(1) - L(t)G(t) - \int_t^1 L(s)g(s)\,ds\right]g(t)\,dt$$

$$= L(1)G^2(1) - 2\int_0^1 L(t)G(t)g(t)\,dt,$$

where $G(t)$ is defined in (S7). Here we have

$$\int_0^1 L(t)G(t)g(t)\,dt = \frac{1}{2}\left[L(1)G^2(1) - \int_0^1 G^2(t)l^2(t)\,dt\right],$$

which implies that the kernel appearing in (5.45) is positive definite.

4.2 It is easy to derive (5.47) from (5.10). Suppose that (5.47) holds. Then, using the two boundary conditions and noting that

$$\left(\frac{f'(t)}{t^{2m}}\right)' + \lambda f(t) = 0,$$

$$t^{2m}f(t) = -\frac{1}{\lambda}\left(f''(t) - \frac{2m}{t}f'(t)\right),$$

we have

$$\lambda\int_0^1 [1 - (\max(s,t))^{2m+1}]f(s)\,ds$$

$$= -\int_0^1 \left(\frac{f'(s)}{s^{2m}}\right)'ds + t^{2m+1}\int_0^t \left(\frac{f'(s)}{s^{2m}}\right)'ds + \int_t^1 (sf''(s) - 2mf'(s))\,ds$$

$$= -f'(1) + t^{2m+1}\frac{f'(t)}{t^{2m}} + f'(1) - tf'(t) - (2m+1)(f(1) - f(t))$$

$$= (2m+1)f(t).$$

4.3 It is easy to see that $f'(t)$ takes the form

$$f'(t) = c_1\left(\frac{\sqrt{\lambda}}{2(m+1)}\right)^\nu \frac{2m+1}{\Gamma(\nu+1)} t^{2m}\left[1 + t \times \{\text{polynomials in } t\}\right]$$

so that $f'(t)/t^{2m} \to 0$ as $t \to 0$ implies the first row of $M(\lambda)$. Since

$$f(1) = c_1 J_\nu\left(\frac{\sqrt{\lambda}}{m+1}\right) + c_2 J_{-\nu}\left(\frac{\sqrt{\lambda}}{m+1}\right) = 0,$$

we also have the second row of $M(\lambda)$.

4.4 Using (5.48) and (5.49), we obtain the general solution to (5.56) as

$$f(t) = t^{m/2} \left\{ c_1 J_\nu \left(\frac{\sqrt{-2\lambda m}}{m+1} t^{(m+1)/2} \right) + c_2 J_{-\nu} \left(\frac{\sqrt{-2\lambda m}}{m+1} t^{(m+1)/2} \right) \right\},$$

where $\nu = -m/(m+1)$. Then $f'(t)$ takes the form:

$$f'(t) = c_2 \left(\frac{\sqrt{-2\lambda m}}{2(m+1)} \right)^{m/(m+1)} \frac{m t^{m-1}}{\Gamma\left(\frac{2m+1}{m+1}\right)} \left[1 + t \times \{\text{polynomials in } t\} \right]$$

so that $f'(t)/t^{m-1} \to 0$ as $t \to 0$ implies

$$\left(\frac{\sqrt{-2\lambda m}}{2(m+1)} \right)^{m/(m+1)} \frac{m}{\Gamma\left(\frac{2m+1}{m+1}\right)} c_2 = 0.$$

Since it holds that, when $c_2 = 0$,

$$f'(1) = -c_1 \frac{\sqrt{-2\lambda m}}{2} J_{\nu+1}\left(\frac{\sqrt{-2\lambda m}}{m+1} \right),$$

the other condition $f'(1) = mf(1)$ yields, after some algebra, the FD $D_2(\lambda)$ as in (5.57), where we have used the relation described in the problem.

4.5 By the definition of the Bessel function, we have

$$D_2(\lambda) = \Gamma(\nu) \left[\frac{1}{\Gamma(\nu)} + \frac{\frac{\lambda m}{2(m+1)^2}}{\Gamma(\nu+1)} + \frac{\left(\frac{\lambda m}{2(m+1)^2}\right)^2}{2!\Gamma(\nu+2)} + \cdots \right]$$

$$= 1 + \frac{\lambda m}{2\nu(m+1)^2} + \sum_{k=2}^{\infty} \frac{\left(\frac{\lambda m}{2(m+1)^2}\right)^k}{k!(\nu+k-1)(\nu+k-2)\cdots\nu},$$

where $\nu = -m/(m+1)$. Then it is seen that $D_2(\lambda)$ reduces to $1 - \lambda/2$ when $m = 0$.

4.6 It can be shown that the integral equation (5.10) with $K = K_3$ is equivalent to

$$h''(t) + \lambda t^{2m} h(t) = 0, \qquad h(0) = h(1) = 0,$$

where $h(t) = t^{-m} f(t)$. The general solution is given by

$$h(t) = \sqrt{t} \left\{ c_1 J_\nu \left(\frac{\sqrt{\lambda}}{m+1} t^{m+1} \right) + c_2 J_{-\nu} \left(\frac{\sqrt{\lambda}}{m+1} t^{m+1} \right) \right\},$$

where $\nu = 1/(2(m+1))$. The boundary condition $h(0) = 0$ implies $c_2 = 0$ and the other condition $h(1) = 0$ yields, when $c_2 = 0$, $J_\nu(\sqrt{\lambda}/(m+1))c_1 = 0$. Then we can obtain the FD $D_3(\lambda)$ of K_3 as in (5.60) so that the c.f. of U_3 is given by $(D_3(2i\theta))^{-1/2}$.

4.7 The integral equation (5.10) with the kernel appearing on the right side of (5.62) is equivalent to

$$h''(t) + \frac{2m+2}{2m+1}\frac{h'(t)}{t} + \frac{\lambda}{(2m+1)^2} t^{-2m/(2m+1)} h(t) = 0,$$

with the boundary conditions $h(1) = 0$ and $t^{1/(2m+1)}h(t) \to 0$ as $t \to 0$, where $h(t) = f(t)/t^{1/(2m+1)}$. The general solution is

$$h(t) = t^{-1/2(2m+1)}\left\{c_1 J_\nu\left(\frac{\sqrt{\lambda}}{m+1}t^{(m+1)/(2m+1)}\right)\right.$$

$$\left. + c_2 J_{-\nu}\left(\frac{\sqrt{\lambda}}{m+1}t^{(m+1)/(2m+1)}\right)\right\},$$

where $\nu = 1/(2(m+1))$. Then the boundary condition $t^{1/(2m+1)}h(t) \to 0$ as $t \to 0$ implies $c_2 = 0$ so that we obtain the same FD $D_3(\lambda)$ from $h(1) = 0$.

4.8 Consider the integral equation (5.10) with $K = K_4$. We obtain

$$f'(t)t^{-2m} = \lambda\left[\int_t^1 f(s)\,ds - \int_0^1 s^{2m+1} f(s)\,ds\right],$$

from which it follows that

$$f''(t) - \frac{2m}{t} f'(t) + \lambda t^{2m} f(t) = 0, \qquad f(0) = f(1) = 0.$$

The general solution is given by

$$f(t) = t^{(2m+1)/2}\left\{c_1 J_\nu\left(\frac{\sqrt{\lambda}}{m+1}t^{m+1}\right) + c_2 J_{-\nu}\left(\frac{\sqrt{\lambda}}{m+1}t^{m+1}\right)\right\},$$

where $\nu = (2m+1)/(2(m+1))$. The boundary condition $f(0) = 0$ implies $c_2 = 0$ and thus we obtain the FD $D_4(\lambda)$ as in (5.64) from $f(1) = 0$. Thus the c.f. of U_4 is given by $(D_4(2i\theta))^{-1/2}$.

4.9 From the boundary conditions $f(0) = f(1) = 0$ and the first condition in (5.76), we obtain $M(\lambda)c = 0$, where $c = (a_1, c_1, c_2)'$ and

$$M(\lambda) = \begin{pmatrix} 0 & 1 & 0 \\ \frac{45}{4}\left(1 - \frac{6}{\lambda}\right) & \cos\sqrt{\lambda} & \sin\sqrt{\lambda} \\ -\frac{90}{7\lambda} & M_{33}(\lambda) & M_{34}(\lambda) \end{pmatrix},$$

with $M_{33}(\lambda)$ and $M_{34}(\lambda)$ defined in (5.78). Making use of REDUCE, we obtain

$$|M(\lambda)| = \frac{45}{7\lambda^4}\left[35\sqrt{\lambda}(\lambda^2 - 12\lambda + 36)\cos\sqrt{\lambda}\right.$$
$$\left. + (5\lambda^3 - 147\lambda^2 + 840\lambda - 1260)\sin\sqrt{\lambda}\right],$$

which yields the FD $\tilde{D}_5(\lambda)$ in (5.79).

4.10 We are led to consider $f''(t) + \lambda f(t) = -4\lambda a_1 \cos 2\pi t$ with the boundary conditions $f(0) = f(1) = 0$, where

$$a_1 = \int_0^1 \sin^2 \pi s f(s)\, ds.$$

When $\lambda \neq 4\pi^2$, the general solution is given by

$$f(t) = c_1 \cos\sqrt{\lambda}\, t + c_2 \sin\sqrt{\lambda}\, t + \frac{4\lambda a_1}{4\pi^2 - \lambda}\cos 2\pi t$$

and we have $M(\lambda)c = 0$, where $c = (a_1, c_1, c_2)'$ and

$$M(\lambda) = \begin{pmatrix} \dfrac{4\lambda}{4\pi^2 - \lambda} & 1 & 0 \\ \dfrac{4\lambda}{4\pi^2 - \lambda} & \cos\sqrt{\lambda} & \sin\sqrt{\lambda} \\ \dfrac{-4\pi^2}{4\pi^2 - \lambda} & h_1(\lambda) & h_2(\lambda) \end{pmatrix},$$

$$h_1(\lambda) = \frac{2\pi^2 \sin\sqrt{\lambda}}{\sqrt{\lambda}(4\pi^2 - \lambda)}, \qquad h_2(\lambda) = \frac{2\pi^2(1 - \cos\sqrt{\lambda})}{\sqrt{\lambda}(4\pi^2 - \lambda)}.$$

Therefore we obtain

$$|M(\lambda)| = \frac{-4\pi^2}{4\pi^2 - \lambda}\left[\sin\sqrt{\lambda} + \frac{4\sqrt{\lambda}}{4\pi^2 - \lambda}(1 - \cos\sqrt{\lambda})\right].$$

When $\lambda = 4\pi^2$, the general solution is given by

$$f(t) = c_1 \cos 2\pi t + c_2 \sin 2\pi t - 4\pi a_1 t \sin 2\pi t,$$

and the three conditions yield $Nc = 0$ with $|N| = 0$ so that $\lambda = 4\pi^2$ is found to be an eigenvalue of multiplicity 1. Then we can obtain the FD $D_7(\lambda)$ as in (5.85). Note that $D_7(4\pi^2) = 0$.

CHAPTER 5

4.11 We are led to consider

$$f''(t) + \lambda f(t) = -4\lambda a_1 \cos 2\pi t + 2\lambda a_2 \sin 2\pi t$$

with the boundary conditions $f(0) = f(1) = 0$ and

$$a_1 = \int_0^1 \sin^2 \pi s \, f(s) \, ds, \qquad a_2 = \int_0^1 \sin 2\pi s \, f(s) \, ds.$$

When $\lambda \neq 4\pi^2$, the general solution is given by

$$f(t) = c_1 \cos \sqrt{\lambda} \, t + c_2 \sin \sqrt{\lambda} \, t + \frac{4\lambda a_1}{4\pi^2 - \lambda} \cos 2\pi t - \frac{2\lambda a_2}{4\pi^2 - \lambda} \sin 2\pi t$$

and we have $M(\lambda)c = 0$, where $c = (a_1, a_2, c_1, c_2)'$ and

$$M(\lambda) = \begin{pmatrix} \dfrac{4\lambda}{4\pi^2 - \lambda} & 0 & 1 & 0 \\ \dfrac{4\lambda}{4\pi^2 - \lambda} & 0 & \cos\sqrt{\lambda} & \sin\sqrt{\lambda} \\ \dfrac{-4\pi^2}{4\pi^2 - \lambda} & 0 & h_1(\lambda) & h_2(\lambda) \\ 0 & \dfrac{-4\pi^2}{4\pi^2 - \lambda} & \dfrac{\sqrt{\lambda}}{\pi} h_2(\lambda) & -\dfrac{\sqrt{\lambda}}{\pi} h_1(\lambda) \end{pmatrix}$$

with $h_1(\lambda)$ and $h_2(\lambda)$ defined in the solution to Problem 4.10. Therefore we obtain

$$|M(\lambda)| = \left(\frac{4\pi^2}{4\pi^2 - \lambda}\right)^2 \left[\sin\sqrt{\lambda} + \frac{4\sqrt{\lambda}}{4\pi^2 - \lambda}(1 - \cos\sqrt{\lambda})\right].$$

When $\lambda = 4\pi^2$, the general solution is given by

$$f(t) = c_1 \cos 2\pi t + c_2 \sin 2\pi t - 4\pi a_1 t \sin 2\pi t - 2\pi a_2 t \cos 2\pi t.$$

The four conditions above yield $Nc = 0$, where $c = (a_1, a_2, c_1, c_2)'$ and

$$N = \begin{pmatrix} 0 & 0 & 1 & 0 \\ 0 & -2\pi & 1 & 0 \\ -\dfrac{1}{4} & \dfrac{\pi}{4} & -\dfrac{1}{4} & 0 \\ -\pi & -\dfrac{3}{4} & 0 & \dfrac{1}{2} \end{pmatrix}, \qquad |N| = -\frac{\pi}{4}.$$

Thus $\lambda = 4\pi^2$ cannot be an eigenvalue. Then we obtain the FD $D_8(\lambda)$ as in (5.87). Note that $D_8(4\pi^2) \neq 0$.

5.1 Noting that $Z_n \sim N(0,1)$, we have

$$E\left[\exp\left\{\frac{i\theta}{\lambda_n}\left(Z_n^2 + af_n(0)Z_n\right)\right\}\right]$$

$$= \frac{1}{\sqrt{2\pi}}\int_{-\infty}^{\infty}\exp\left[-\frac{1}{2}\left\{x^2 - \frac{2i\theta}{\lambda_n}\left(x^2 + af_n(0)x\right)\right\}^2\right]dx$$

$$= \left(1 - \frac{2i\theta}{\lambda_n}\right)^{-1/2}\exp\left\{\frac{(ia\theta)^2}{2}\frac{f_n^2(0)}{\lambda_n(\lambda_n - 2i\theta)}\right\},$$

which yields (5.102).

5.2 Using the second relation in (5.98) and Mercer's theorem, we have

$$\sum_{n=1}^{\infty}\frac{f_n^2(0)}{\lambda_n(\lambda_n - 2i\theta)} = \frac{1}{2i\theta}\left[\sum_{n=1}^{\infty}\frac{f_n^2(0)}{\lambda_n - 2i\theta} - \sum_{n=1}^{\infty}\frac{f_n^2(0)}{\lambda_n}\right]$$

$$= \frac{1}{2i\theta}\left\{\Gamma(0,0;2i\theta) - K(0,0)\right\},$$

which proves the theorem.

5.3 That (5.104) implies (5.105) is easily established. Suppose that (5.105) holds. Using the two boundary conditions in (5.105) and noting that

$$\left(\frac{h'(t)}{t^{2m}}\right)' + \lambda h(t) = 0, \quad t^{2m}h(t) = -\frac{1}{\lambda}\left(h''(t) - \frac{2m}{t}h'(t)\right),$$

we can show that

$$\lambda\int_0^1\left[1 - (\max(s,t))^{2m+1}\right]h(s)\,ds = t^{2m+1} - 1 + (2m+1)h(t).$$

5.4 We have only to derive the resolvent $\Gamma(s,t;\lambda)$ of $K(s,t) = (\max(s,t))^m/2$ evaluated at the origin. Putting $h(t) = \Gamma(0,t;\lambda)$, we consider

$$h(t) = K(0,t) + \lambda\int_0^t h(s)K(s,t)\,ds$$

$$= \frac{1}{2}t^m + \lambda\left[\frac{1}{2}t^m\int_0^t h(s)\,ds + \frac{1}{2}\int_t^1 s^m h(s)\,ds\right],$$

which is equivalent to

$$h''(t) - \frac{m-1}{t}h'(t) - \frac{\lambda m}{2}t^{m-1}h(t) = 0, \quad \lim_{t\to 0}\frac{h'(t)}{t^{m-1}} = \frac{m}{2}, \quad h'(1) = mh(1).$$

The general solution is given by

$$h(t) = t^{m/2} \left\{ c_1 J_\nu \left(\frac{\sqrt{-2\lambda m}}{m+1} t^{(m+1)/2} \right) + c_2 J_{-\nu} \left(\frac{\sqrt{-2\lambda m}}{m+1} t^{(m+1)/2} \right) \right\},$$

where $\nu = -m/(m+1)$. From the two boundary conditions, we can determine c_1 and c_2 uniquely. Then

$$\Gamma(0,0;\lambda) = h(0) = \frac{c_1}{\Gamma(\nu+1)} \left(\frac{\sqrt{-2\lambda m}}{2(m+1)} \right)^\nu$$

$$= \frac{1}{2} \frac{\Gamma(-\nu+1)}{\Gamma(\nu+1)} \frac{J_{-\nu+1}\left(\frac{\sqrt{-2\lambda m}}{m+1}\right)}{J_{\nu-1}\left(\frac{\sqrt{-2\lambda m}}{m+1}\right)} \left(\frac{\sqrt{-2\lambda m}}{2(m+1)} \right)^{2\nu}$$

so that (5.111) is established by Theorem 5.8 and (5.57).

5.5 It is easy to deduce that

$$\int_0^1 q(t) r(t)\, dt = \int_0^1 \sum_{n=1}^\infty \frac{c_n}{\sqrt{\lambda_n}} f_n(t) \left(m(t) - \sum_{m=1}^\infty \frac{c_m}{\sqrt{\lambda_m}} f_m(t) \right) dt$$

$$= \sum_{n=1}^\infty \frac{c_n^2}{\lambda_n} - \sum_{m,n=1}^\infty \frac{c_m}{\sqrt{\lambda_m}} \frac{c_n}{\sqrt{\lambda_n}} \int_0^1 f_m(t) f_n(t)\, dt$$

$$= \sum_{n=1}^\infty \frac{c_n^2}{\lambda_n} - \sum_{n=1}^\infty \frac{c_n^2}{\lambda_n} = 0.$$

5.6 We first obtain

$$E\left[\exp\left\{ \frac{i\theta}{\lambda_n}(Z_n + c_n)^2 \right\} \right] = \left(1 - \frac{2i\theta}{\lambda_n} \right)^{-1/2} \exp\left\{ \frac{i\theta c_n^2}{\lambda_n - 2i\theta} \right\}$$

so that

$$E\left(e^{i\theta S_Z} \right) = \prod_{n=1}^\infty \left(1 - \frac{2i\theta}{\lambda_n} \right)^{-1/2} \exp\left\{ \sum_{n=1}^\infty \frac{i\theta c_n^2}{\lambda_n - 2i\theta} + i\theta \int_0^1 r^2(t)\, dt \right\}$$

$$= (D(2i\theta))^{-1/2} \exp\left\{ i\theta \int_0^1 m^2(t)\, dt + i\theta \sum_{n=1}^\infty \left(\frac{c_n^2}{\lambda_n - 2i\theta} - \frac{c_n^2}{\lambda_n} \right) \right\}$$

$$= (D(2i\theta))^{-1/2} \exp\left\{ i\theta \int_0^1 m^2(t)\, dt - 2\theta^2 \sum_{n=1}^\infty \frac{c_n^2}{\lambda_n(\lambda_n - 2i\theta)} \right\}.$$

5.7 The integral equation (5.117) with $K(s,t) = \text{Cov}(w(s), w(t)) = \min(s,t) - st$ and $m(s) = a + bs$ is equivalent to $h''(t) + \lambda h(t) = -a - bt$ with $h(0) = h(1) = 0$, where the general solution is given by $h(t) = c_1 \cos \sqrt{\lambda} t + c_2 \sin \sqrt{\lambda} t - (a + bt)/\lambda$. From the boundary conditions we have

$$c_1 = \frac{a}{\lambda}, \quad c_2 = \frac{a+b}{\lambda \sin \sqrt{\lambda}} - \frac{a}{\lambda} \cot \sqrt{\lambda}.$$

Then

$$\frac{\lambda}{2} \int_0^1 \{m^2(t) + \lambda h(t) m(t)\} dt$$

$$= \frac{1}{\cos \sqrt{\lambda} + 1} \left[a(a+b) \sqrt{\lambda} \sin \sqrt{\lambda} \right.$$

$$\left. + \frac{b^2}{2} \left(\frac{\sqrt{\lambda} \cos \sqrt{\lambda} \sin \sqrt{\lambda}}{\cos \sqrt{\lambda} - 1} + \cos \sqrt{\lambda} + 1 \right) \right],$$

which yields the c.f. given in (5.123).

5.8 The integral equation (5.117) with $K = K_6$ is equivalent to

$$h''(t) + \lambda h(t) = -\frac{a}{2\pi}(1 - \cos 2\pi t) + 2\lambda c_3 \sin 2\pi t, \quad h(0) = h(1) = 0$$

and

$$c_3 = \int_0^1 h(s) \sin 2\pi s \, ds.$$

The general solution is

$$h(t) = c_1 \cos \sqrt{\lambda} t + c_2 \sin \sqrt{\lambda} t - \frac{a}{2\lambda \pi} + \frac{a}{2\pi} \frac{\cos 2\pi t}{\lambda - 4\pi^2} + 2\lambda c_3 \frac{\sin 2\pi t}{\lambda - 4\pi^2}.$$

From the above conditions we have

$$c_1 = \frac{2\pi a}{\lambda(4\pi^2 - \lambda)}, \quad c_2 = \frac{c_1(1 - \cos \sqrt{\lambda})}{\sin \sqrt{\lambda}}$$

and thus

$$\frac{\lambda}{2} \int_0^1 \left\{ \frac{a^2(1 - \cos 2\pi t)^2}{4\pi^2} + \lambda h(t) \frac{a}{2\pi}(1 - \cos 2\pi t) \right\} dt$$

$$= a^2 \left[\frac{\lambda}{4(4\pi^2 - \lambda)} + \frac{4\pi^2 \sqrt{\lambda}}{(4\pi^2 - \lambda)^2} \frac{1 - \cos \sqrt{\lambda}}{\sin \sqrt{\lambda}} \right],$$

which yields (5.125).

5.9 Let us define

$$S_N = \sum_{n=1}^{N} \frac{1}{\lambda_n} \left(Z_n^2 + a f_n(0) Z_n Z \right) + bZ^2 = W'AW,$$

where $W = (Z_1, \ldots, Z_N, Z)'$ and

$$A = \begin{pmatrix} \Lambda & h \\ h' & b \end{pmatrix}, \quad \Lambda = \text{diag}\left(\frac{1}{\lambda_1}, \ldots, \frac{1}{\lambda_N}\right),$$

$$h = \left(\frac{a f_1(0)}{2\lambda_1}, \ldots, \frac{a f_N(0)}{2\lambda_N}\right)'.$$

Since

$$E\left(e^{i\theta S_N}\right) = |I_{N+1} - 2i\theta A|^{-1/2}$$

$$= \left[|I_N - 2i\theta \Lambda| \{1 - 2ib\theta + 4\theta^2 h'(I_N - 2i\theta\Lambda)^{-1}h\}\right]^{-1/2}$$

$$= \left[\prod_{n=1}^{N}\left(1 - \frac{2i\theta}{\lambda_n}\right)\right]^{-1/2} \left[1 - 2ib\theta + a^2\theta^2 \sum_{n=1}^{N} \frac{f_n^2(0)}{\lambda_n(\lambda_n - 2i\theta)}\right]^{-1/2}$$

and S_N converges in probability to S as $N \to \infty$, (5.128) is established.

5.10 Using the definition of $m(t) = q(t) + r(t)$ with $q(t)$ defined below (5.112), we have

$$S_Z = \int_0^1 \left\{ \sum_{n=1}^{\infty} \frac{f_n(t)}{\sqrt{\lambda_n}} Z_n + (q(t) + r(t))Z \right\}^2 dt$$

$$= \int_0^1 \left\{ \sum_{n=1}^{\infty} \frac{f_n(t)}{\sqrt{\lambda_n}} (Z_n + c_n Z)^2 \right\} dt + Z^2 \int_0^1 r^2(t)\, dt$$

$$+ 2Z \int_0^1 r(t) \left(\sum_{n=1}^{\infty} \frac{f_n(t)}{\sqrt{\lambda_n}} Z_n + q(t)Z \right) dt$$

$$= \sum_{n=1}^{\infty} \frac{1}{\lambda_n} (Z_n + c_n Z)^2 + Z^2 \int_0^1 r^2(t)\, dt,$$

where use has been made of the facts that

$$\int_0^1 f_m(t) f_n(t)\, dt = \delta_{mn}, \quad \int_0^1 r(t) q(t)\, dt = 0,$$

$$\int_0^1 r(t) f_n(t)\, dt = \int_0^1 (m(t) - q(t)) f_n(t)\, dt = 0.$$

5.11 Defining

$$S_N = \sum_{n=1}^{N} \frac{1}{\lambda_n}(Z_n + c_n Z)^2 + Z^2 \int_0^1 r^2(t)\,dt = W'AW,$$

where $W = (Z_1, \ldots, Z_N, Z)'$ and

$$A = \begin{pmatrix} \Lambda & h \\ h' & \gamma \end{pmatrix}, \quad \Lambda = \text{diag}\left(\frac{1}{\lambda_1}, \ldots, \frac{1}{\lambda_N}\right),$$

$$h = \left(\frac{c_1}{\lambda_1}, \ldots, \frac{c_N}{\lambda_N}\right)', \quad \gamma = \int_0^1 r^2(t)\,dt + \sum_{n=1}^{N} \frac{c_n^2}{\lambda_n},$$

we obtain

$$E\left(e^{i\theta S_N}\right) = |I_{N+1} - 2i\theta A|^{-1/2}$$

$$= \left[\prod_{n=1}^{N}\left(1 - \frac{2i\theta}{\lambda_n}\right)\right]^{-1/2} \left[1 - 2i\theta\gamma + 4\theta^2 \sum_{n=1}^{N} \frac{c_n^2}{\lambda_n(\lambda_n - 2i\theta)}\right]^{-1/2}.$$

Noting that, as $N \to \infty$,

$$\gamma \to \int_0^1 r^2(t)\,dt + \sum_{n=1}^{\infty} \frac{c_n^2}{\lambda_n} = \int_0^1 m^2(t)\,dt,$$

we have

$$E\left(e^{i\theta S}\right) = (D(2i\theta))^{-1/2}\left[1 - 2i\theta \int_0^1 m^2(t)\,dt + 4\theta^2 \sum_{n=1}^{\infty} \frac{c_n^2}{\lambda_n(\lambda_n - 2i\theta)}\right]^{-1/2}.$$

Since

$$\frac{c_n^2}{\lambda_n} = \int_0^1 \int_0^1 m(s)m(t) f_n(s) f_n(t)\,ds\,dt,$$

we can establish Theorem 5.11 using the second relation for the resolvent in (5.98).

6.1 Since $\log(1 - (\beta/T)) = -\beta/T + O(T^{-2})$, we have

$$\left|B_T(j,k) - K\left(\frac{j}{T}, \frac{k}{T}\right)\right|$$

$$= \left|\exp\left\{|j - k|\log\left(1 - \frac{\beta}{T}\right)\right\} - \exp\left\{-\frac{\beta}{T}|j - k|\right\}\right|$$

$$= \exp\left\{-\frac{\beta}{T}|j - k|\right\}\left|\exp\left\{|j - k|O(T^{-2})\right\} - 1\right|,$$

which evidently goes to 0 uniformly for all j, k as $T \to \infty$.

6.2 Putting $d_T(j,k) = B_T(j,k) - K(j/T, k/T)$ and $\delta_T = \max |d_T(j,k)|$ we have

$$R_T = \frac{1}{T} \sum_{j,k=1}^{T} d_T(j,k)\varepsilon_j\varepsilon_k$$

$$= \frac{1}{T} \sum_{j=1}^{T} d_T(j,j)\varepsilon_j^2 + \frac{1}{T} \sum_{j \neq k} d_T(j,k)\varepsilon_j\varepsilon_k$$

$$= Q_1 + Q_2,$$

where

$$E(|Q_1|) \leq \delta_T, \qquad E(Q_2^2) \leq \frac{2T(T-1)}{T^2} \delta_T^2 \leq 2\delta_T^2.$$

Thus we have $E(|R_T|) \leq (1 + \sqrt{2})\delta_T \to 0$ so that the conclusion follows from Markov's inequality.

6.3 We consider, putting $s = j/T$ and $t = k/T$,

$$\left| \frac{\rho^{|j-k|} - \rho^{2T-j-k+2}}{T(1-\rho^2)} - \frac{e^{-\beta|s-t|} - e^{-\beta(2-s-t)}}{2\beta} \right|$$

$$\leq \left| \frac{\rho^{|j-k|}}{T(1-\rho^2)} - \frac{e^{-\beta|s-t|}}{2\beta} \right| + \left| \frac{\rho^{2T-j-k+2}}{T(1-\rho^2)} - \frac{1}{2\beta} e^{-\beta(2-s-t)} \right|.$$

The quantities on the right side can be shown to converge to 0 uniformly for all j, k as $T \to \infty$, as in the solution to Problem 6.1.

6.4 Put $B_T = \Omega^{-1}/T = [(B_T(j,k))]$ and consider the absolute difference of the (j,k)th element of Ω^{-2}/T^3 and $K_{(2)}(j/T, k/T)$, which is

$$\left| \frac{1}{T} \sum_{l=1}^{T} B_T(j,l)B_T(l,k) - \int_0^1 K\left(\frac{j}{T}, u\right) K\left(u, \frac{k}{T}\right) du \right|$$

$$= \left| \sum_{l=1}^{T} \int_{(l-1)/T}^{l/T} \left\{ B_T(j,l)B_T(l,k) - K\left(\frac{j}{T}, u\right) K\left(u, \frac{k}{T}\right) \right\} du \right|$$

$$\leq \sum_{l=1}^{T} \int_{(l-1)/T}^{l/T} \left| \left\{ B_T(j,l) - K\left(\frac{j}{T}, u\right) \right\} B_T(l,k) \right.$$

$$\left. + K\left(\frac{j}{T}, u\right) \left\{ B_T(l,k) - K\left(u, \frac{k}{T}\right) \right\} \right| du.$$

This last quantity is shown to converge to 0 uniformly for all j, k as $T \to \infty$. Then we consider, instead of S_{T4},

$$S'_{T4} = \frac{1}{T} \sum_{j,k=1}^{T} \left[K\left(\frac{j}{T}, \frac{k}{T}\right) + \gamma K_{(2)}\left(\frac{j}{T}, \frac{k}{T}\right) \right] \varepsilon_j \varepsilon_k$$

$$= \sum_{n=1}^{\infty} \left(\frac{1}{\lambda_n} + \frac{\gamma}{\lambda_n^2} \right) \left(\frac{1}{\sqrt{T}} \sum_{j=1}^{T} f_n\left(\frac{j}{T}\right) \varepsilon_j \right)^2,$$

which yields the first expression in (5.150). The distributional equivalence of the first and second expressions is obvious.

6.5 We have only to establish (5.152), which is easily proved from (5.151) and the definition of the FD given in (5.25).

6.6 We can show easily that

$$R_T = \frac{1}{T} \sum_{j,k=1}^{T} \left\{ B_T(j,k) - K\left(\frac{j}{T}, \frac{k}{T}\right) \right\} \varepsilon'_j H \varepsilon_k$$

converges in probability to 0. Thus we consider

$$\frac{1}{T} \sum_{j,k=1}^{T} K\left(\frac{j}{T}, \frac{k}{T}\right) \varepsilon'_j H \varepsilon_k$$

$$= \sum_{n=1}^{\infty} \frac{1}{\lambda_n} \frac{1}{\sqrt{T}} \sum_{j=1}^{T} f_n\left(\frac{j}{T}\right) \varepsilon'_j H \frac{1}{\sqrt{T}} \sum_{k=1}^{T} f_n\left(\frac{k}{T}\right) \varepsilon_k,$$

which converges in distribution to $\sum_{n=1}^{\infty} Z'_n H Z_n / \lambda_n$, where $\{Z_n\} \sim \text{NID}(0, I_q)$. Then Mercer's theorem establishes (5.154).

6.7 Noting that

$$\mathcal{L}\left(\int_0^1 \int_0^1 K(s,t) \, dw'(s) H \, dw(t) \right) = \mathcal{L}\left(\sum_{n=1}^{\infty} \frac{1}{\lambda_n} Z'_n H Z_n \right)$$

$$= \mathcal{L}\left(\sum_{n=1}^{\infty} \frac{1}{\lambda_n} Z'_n \Lambda Z_n \right),$$

where $\Lambda = \text{diag}(\delta_1, \ldots, \delta_q)$, we obtain, as the c.f. of this last distribution,

$$\prod_{n=1}^{\infty} \left\{ \prod_{j=1}^{q} \left(1 - \frac{2i\delta_j \theta}{\lambda_n} \right)^{-1/2} \right\} = \prod_{j=1}^{q} \left(D(2i\delta_j \theta) \right)^{-1/2}.$$

6.8 Put $B_N = C'(\rho)C(\rho)/N = [(B_N(j,k))]$. Then we have

$$\frac{1}{N^2}y'y = \frac{1}{N}\sum_{j,k=1}^{T} B_N(j,k)\boldsymbol{\varepsilon}_j'\boldsymbol{\varepsilon}_k,$$

where $\boldsymbol{\varepsilon}_j = (\varepsilon_{(j-1)m+1}, \ldots, \varepsilon_{jm})' : m \times 1$. Thus (5.157) follows from (5.148) and (5.154).

6.9 Putting $d_T(j,k) = B_T(j,k) - K(j/T, k/T)$ and $\delta_T = \max |d_T(j,k)|$, we have

$$R_T = \sum_{l,m=0}^{\infty} \bigl(Q_1(l,m) + Q_2(l,m)\bigr),$$

where

$$Q_1(l,m) = \frac{\alpha_l \alpha_m}{T} \sum_{j=\max(1-l,1-m)}^{\min(T-l,T-m)} d_T(j+l, j+m)\varepsilon_j^2,$$

$$Q_2(l,m) = \frac{\alpha_l \alpha_m}{T} \sum_{\substack{j=1-l \\ j \neq k}}^{T-l} \sum_{k=1-m}^{T-m} d_T(j+l, k+m)\varepsilon_j\varepsilon_k.$$

Then we can establish that

$$E(|Q_1|) \leq c_1 \delta_T |\alpha_l| |\alpha_m|,$$
$$E(Q_2^2) \leq c_2 (\delta_T |\alpha_l| |\alpha_m|)^2,$$

for some positive constants c_1 and c_2. Therefore, by Schwarz's and Markov's inequalities, we see that, for every $x > 0$,

$$P(|R_T| > x) \leq \frac{c_1 + \sqrt{c_2}}{x} \delta_T \left(\sum_{l=0}^{\infty} |\alpha_l|\right)^2 \to 0.$$

6.10 We consider

$$V_T' = \frac{1}{T} \sum_{j,k=1}^{T} K\left(\frac{j}{T}, \frac{k}{T}\right) u_j u_k$$

$$= \sum_{l,m=0}^{\infty} \alpha_l \alpha_m \frac{1}{T} \sum_{j,k=1}^{T} K\left(\frac{j}{T}, \frac{k}{T}\right) \varepsilon_{j-l}\varepsilon_{k-m}$$

$$= V_{T,M}' + R_{T,M},$$

where

$$V_{T,M}^I = \sum_{l,m=0}^{M} \alpha_l \alpha_m \frac{1}{T} \sum_{j,k=1}^{T} K\left(\frac{j}{T}, \frac{k}{T}\right) \varepsilon_{j-l} \varepsilon_{k-m}$$

and $R_{T,M}$ is the remainder term. There exists a sequence $\{a_M\}$ such that $E(|R_{T,M}|) \le a_M$ for all T and $a_M \to 0$ as $M \to \infty$. We further deduce that, for M fixed,

$$\mathcal{L}(V_{T,M}^I) \to \mathcal{L}\left(\left(\sum_{l=0}^{M} \alpha_l\right)^2 \int_0^1 \int_0^1 K(s,t)\, dw(s)\, dw(t)\right),$$

which yields (5.161).

CHAPTER 6

3.1 It is easy to deduce that, for $x \ge 0$,

$$P(X^2 - Y^2 \le x) = \int_{-\infty}^{\infty} \left\{ \int_{-\sqrt{x+t^2}}^{\sqrt{x+t^2}} \frac{1}{\sqrt{2\pi}} e^{-s^2/2} ds \right\} \frac{1}{\sqrt{2\pi}} e^{-t^2/2} dt$$

$$= \int_{-\infty}^{\infty} \left\{ 1 - 2\Phi(-\sqrt{x+t^2}) \right\} \frac{1}{\sqrt{2\pi}} e^{-t^2/2} dt$$

$$= 1 - \frac{4}{\sqrt{2\pi}} \int_0^{\infty} \Phi(-\sqrt{x+t^2}) e^{-t^2/2} dt.$$

3.2 Lévy's inversion formula (6.1) yields

$$\operatorname{Re}\left[\frac{1 - e^{-i\theta x}}{i\theta} \phi_3(\theta)\right] = \operatorname{Re}\left[\frac{\sin \theta x - i(1 - \cos \theta x)}{\theta} \phi_3(\theta)\right]$$

$$= \frac{1}{\theta}\left[\operatorname{Re}\{\phi_3(\theta)\} \sin \theta x + \operatorname{Im}\{\phi_3(\theta)\}(1 - \cos \theta x)\right]$$

and we obtain the second equality in (6.21) by transforming θ into $\theta = u^4$.

3.3 Consider

$$\sum_{k=0}^{\infty} (-1)^{k+N} V_{k+N} = (-1)^N (V_N - V_{N+1} + V_{N+2} - \cdots)$$

$$= (-1)^N (1 - F + F^2 - \cdots) V_N$$

$$= (-1)^N \frac{V_N}{1 + F}.$$

CHAPTER 6

Since

$$\frac{1}{1+x} = \sum_{k=0}^{\infty} \frac{1}{k!} \frac{d^k}{dx^k}\left(\frac{1}{1+x}\right)\bigg|_{x=1} (x-1)^k$$

$$= \sum_{k=0}^{\infty} \frac{(-1)^k}{2^{k+1}} (x-1)^k,$$

we can establish the last equality in (6.26).

4.1 Defining an auxiliary process $dX(t) = -\beta X(t)\,dt + dw(t)$ with $X(0) = 0$, we have

$$E\left[\exp\{\theta(xV_6 - U_6)\}\right] = E\left[\exp\left\{\theta x \int_0^1 w^2(t)\,dt - \theta \int_0^1 w(t)\,dw(t)\right\}\right]$$

$$= E\left[\exp\left\{\theta x \int_0^1 X^2(t)\,dt - \theta \int_0^1 X(t)\,dX(t)\right\} \frac{d\mu_w}{d\mu_X}(X)\right]$$

$$= E\left[\exp\left\{\frac{\beta - \theta}{2} X^2(1)\right\}\right] \exp\left(\frac{\theta - \beta}{2}\right),$$

where $\beta = \sqrt{-2\theta x}$. Since $X(1) \sim N(0, (1-e^{-2\beta})/(2\beta))$, it holds that

$$E\left[\exp\left\{\frac{\beta - \theta}{2} X^2(1)\right\}\right] e^{-\beta/2} = \left[1 - \frac{(\beta - \theta)(1 - e^{-2\beta})}{2\beta}\right]^{-1/2} e^{-\beta/2}$$

$$= \left(\cosh \beta + \frac{\theta}{\beta} \sinh \beta\right)^{-1/2},$$

which yields (6.40).

4.2 Noting that $\operatorname{Im}[\phi_6(0; x)] = 0$, we obtain

$$\lim_{u \to 0} \operatorname{Im}\left[\frac{\phi_6(u^2; x)}{u}\right] = \operatorname{Im}\left[\lim_{u \to 0}\left\{\frac{\phi_6(u^2; x)}{u^2} \times u\right\}\right]$$

$$= \operatorname{Im}\left[\frac{\partial \phi_6(\theta; x)}{\partial \theta}\bigg|_{\theta=0} \times \lim_{u \to 0} u\right]$$

$$= 0,$$

where $\partial \phi_6(\theta; x)/\partial \theta|_{\theta=0}$ is given in the solution to Problem 4.3.

4.3 Proceeding in the same way as in the solution to Problem 4.2, we have

$$\lim_{\theta \to 0} \frac{1}{\theta} \operatorname{Im}\left[\phi_6(\theta; x)\right] = \operatorname{Im}\left[\frac{\partial \phi_6(\theta; x)}{\partial \theta}\bigg|_{\theta=0}\right],$$

where

$$\frac{\partial \phi_6(\theta; x)}{\partial \theta} = e^{i\theta/2} A^{-1/2} \left[\frac{i}{2} - \frac{1}{2} A^{-1} \frac{\partial}{\partial \theta} \left\{ \left(1 - i\theta x - \frac{\theta^2 x^2}{6} + \cdots \right) \right. \right.$$
$$\left. \left. + i\theta \left(1 - \frac{i\theta x}{3} + \cdots \right) \right\} \right],$$

$$A = \cos\sqrt{2i\theta x} + i\theta \frac{\sin\sqrt{2i\theta x}}{\sqrt{2i\theta x}}.$$

Therefore $\partial\phi_6(\theta; x)/\partial\theta|_{\theta=0} = ix/2$ so that (6.43) holds.

4.4 Noting that $F_6(0) = P(S_6 \leq 0) = P(w^2(1) \leq 1)$, we obtain $F_6(0) = 1 - 2\Phi(-1) = 1 - 2 \times 0.15866 = 0.68268$.

4.5 We have only to show that

(S8) $\quad \text{Im}\left[\cos(-\theta^2)^{1/4} \cosh(-\theta^2)^{1/4}\right] = 0.$

Since $-\theta^2 = \theta^2 \exp\{i(2n+1)\pi\}$ for $n = 0, \pm 1, \pm 2, \ldots$, we have

$$\cos(-\theta^2)^{1/4} = \cos(x + iy)$$
$$= \cos x \cosh y - i \sin x \sinh y,$$
$$\cosh(-\theta^2)^{1/4} = \cosh(x + iy)$$
$$= \cosh x \cos y + i \sinh x \sin y,$$

where $x = \sqrt{\theta} \cos\{(2n+1)\pi/4\}$ and $y = \sqrt{\theta} \sin\{(2n+1)\pi/4\}$. Thus $y = x$ or $y = -x$ and it can be easily checked that (S8) holds.

4.6 We show that $F_7(-x) = 1 - F_7(x)$ for any x. Since $g_7(\theta; -x) = g_7(-\theta; x)$, we have

$$F_7(-x) = \frac{1}{2} + \frac{1}{\pi} \int_0^\infty \frac{1}{\theta} \text{Im}\left[\phi_7(-\theta; x)\right] d\theta$$
$$= \frac{1}{2} - \frac{1}{\pi} \int_0^\infty \frac{1}{\theta} \text{Im}\left[\phi_7(\theta; x)\right] d\theta$$
$$= 1 - F_7(x).$$

CHAPTER 7

1.1 Putting $x_j = 1$ in (7.2) with $u_j = \varepsilon_j$, we have $y_j = \rho y_{j-1} + (1-\rho)\beta + \varepsilon_j$. Thus $\alpha = (1-\rho)\beta$ must be 0 when $\rho = 1$. Putting $x_j = (1, j)'$ and $\beta = (\beta_1, \beta_2)'$ in (7.2) with $u_j = \varepsilon_j$ leads us to $y_j = \rho y_{j-1} + \beta_1(1-\rho) + \beta_2\rho + j\beta_2(1-\rho) + \varepsilon_j$ so that $\gamma = (1-\rho)\beta_2$ must be 0 when $\rho = 1$.

1.2 The LSE $\hat{\rho}$ is given by

$$\hat{\rho} = \frac{\sum_{j=2}^{T}(y_{j-1}-\bar{y}_{-1})(y_j-\bar{y}_0)}{\sum_{j=2}^{T}(y_{j-1}-\bar{y}_{-1})^2} = \frac{\sum_{j=2}^{T}(y_{j-1}-\bar{y}_{-1})\varepsilon_j}{\sum_{j=2}^{T}y_{j-1}^2-(T-1)\bar{y}_{-1}^2}+1,$$

where $\bar{y}_{-1} = \sum_{j=2}^{T} y_{j-1}/(T-1)$ and $\bar{y}_0 = \sum_{j=2}^{T} y_j/(T-1)$. Since $y_j = \alpha j + \varepsilon_1 + \cdots + \varepsilon_j$, we have

$$\plim_{T\to\infty} \frac{1}{T^3}\sum_{j=2}^{T} y_{j-1}^2 = \frac{\alpha^2}{3}, \quad \plim_{T\to\infty} \frac{1}{T^2}\bar{y}_{-1}^2 = \frac{\alpha^2}{4},$$

$$\frac{1}{T\sqrt{T}}\sum_{j=2}^{T}(y_{j-1}-\bar{y}_{-1})\varepsilon_j = \frac{\alpha}{T\sqrt{T}}\sum_{j=2}^{T}\left(j-1-\frac{T}{2}\right)\varepsilon_j + o_p(1)$$

$$\to N\left(0, \frac{\alpha^2\sigma^2}{12}\right).$$

Thus we obtain $T\sqrt{T}(\hat{\rho}-1) \to N(0, 12\sigma^2/\alpha^2)$.

1.3 The first equality is obvious. Consider

$$2\sum_{j=2}^{T}\hat{\eta}_{j-1}(\hat{\eta}_j-\hat{\eta}_{j-1}) = -\sum_{j=2}^{T}(\hat{\eta}_j-\hat{\eta}_{j-1})^2 + \sum_{j=2}^{T}\hat{\eta}_j^2 - \sum_{j=2}^{T}\hat{\eta}_{j-1}^2$$

$$= -\sum_{j=2}^{T}(\hat{\eta}_j-\hat{\eta}_{j-1})^2 + \hat{\eta}_T^2 - \hat{\eta}_1^2,$$

which yields U_T in (7.12).

1.4 Consider first

$$\frac{1}{T}E\left(\sum_{j=1}^{T}u_j\right)^2 = \frac{1}{T}\sum_{j,k=1}^{T}\gamma_{j-k} = \sum_{j=-(T-1)}^{T-1}\left(1-\frac{|j|}{T}\right)\gamma_j,$$

which converges to $\sum_{j=-\infty}^{\infty}\gamma_j = 2\pi f(0) = \sigma^2\left(\sum_{l=0}^{\infty}\alpha_l\right)^2 = \sigma_L^2$. Since $E(u_j^2) = \gamma_0 = \sigma^2\sum_{l=0}^{\infty}\alpha_l^2$, we have $\sigma_S^2 = \gamma_0$.

2.1 It is easy to see that

$$\frac{1}{T}\sum_{j=2}^{T}(\eta_j-\eta_{j-1})^2 = \frac{1}{T}\sum_{j=2}^{T}\left(-\frac{c}{T}\eta_{j-1}+u_j\right)^2$$

$$= \frac{1}{T}\sum_{j=2}^{T} u_j^2 + o_p(1) \to \sigma_S^2 \quad \text{in probability.}$$

We also have that

$$\frac{\eta_1^2}{T} - \sigma_L^2 Z^2 = \frac{1}{T}\left(\sqrt{T}\rho\sigma_L Z + u_1\right)^2 - \sigma_L^2 Z^2$$

$$= (\rho^2 - 1)\sigma_L^2 Z^2 + \frac{2}{\sqrt{T}}\rho\sigma_L Z u_1 + \frac{1}{T}u_1^2,$$

which evidently converges in probability to 0.

2.2 Defining a continuous function on C:

$$h(x) = \left(\frac{1}{2}(x^2(1) - x^2(0) - \sigma_S^2), \int_0^1 x^2(t)\,dt\right),$$

we can deduce that $h(Y_T) - (U_{1T}, V_{1T}) \to 0$ in probability. Since $\mathcal{L}(Y_T) \to \mathcal{L}(\sigma_L Y)$, we can establish (7.29) by the continuous mapping theorem.

2.3 It is easy to see that

$$\mathcal{L}\left(\gamma\left(\frac{U_1}{V_1} + c\right)\right) \to \mathcal{L}\left(\frac{\int_0^1 e^{-ct}\,dw(t)}{\int_0^1 e^{-2ct}\,dt}\right),$$

where

$$\int_0^1 e^{-2ct}\,dt = \frac{\sinh c}{ce^c}, \quad \int_0^1 e^{-ct}\,dw(t) \sim N\left(0, \frac{\sinh c}{ce^c}\right),$$

which establishes (7.34).

2.4 Since $\hat{\eta}_j - \hat{\eta}_{j-1} = \eta_j - \eta_{j-1}$, (7.39) holds because of (7.27). Noting that $\hat{\eta}_1 = \eta_1 - \bar{\eta} = \sqrt{T}\rho\sigma_L Y(0) + u_1 - \bar{\eta}$ and $\bar{\eta} = \sum_{j=1}^{T} Y_T(j/T)/\sqrt{T}$, we can also establish (7.40).

2.5 It follows from (7.23) and (7.24) that

$$\gamma\left(\frac{U_2}{V_2} + c\right) = \frac{\gamma\left\{\int_0^1 Y(t)\,dw(t) - w(1)\int_0^1 Y(t)\,dt + \frac{1}{2}(1-r)\right\}}{\int_0^1 Y^2(t)\,dt - \left(\int_0^1 Y(t)\,dt\right)^2}$$

$$= \frac{\int_0^1 e^{-ct}\,dw(t) - w(1)\int_0^1 e^{-ct}\,dt}{\int_0^1 e^{-2ct}\,dt - \left(\int_0^1 e^{-ct}\,dt\right)^2} + o_p(1),$$

where

$$\int_0^1 e^{-ct}\,dw(t) - w(1)\int_0^1 e^{-ct}\,dt \sim N\left(0, \int_0^1 e^{-2ct}\,dt - \left(\int_0^1 e^{-ct}\,dt\right)^2\right),$$

$$\int_0^1 e^{-2ct}\,dt - \left(\int_0^1 e^{-ct}\,dt\right)^2 = \frac{c\sinh c - 2\cosh c + 2}{c^2 e^c}$$

This leads us to the conclusion.

2.6 Fuller's estimator $\tilde{\rho}$ gives

$$T(\tilde{\rho} - 1) = \frac{\frac{1}{T}\sum_{j=2}^{T}(y_{j-1} - \bar{y}_{-1})y_j}{\frac{1}{T^2}\sum_{j=2}^{T}(y_{j-1} - \bar{y}_{-1})^2} = \frac{\frac{1}{T}\sum_{j=2}^{T}(\eta_{j-1} - \bar{\eta}_{-1})\eta_j}{\frac{1}{T^2}\sum_{j=2}^{T}(\eta_{j-1} - \bar{\eta}_{-1})^2},$$

where $\bar{y}_{-1} = \sum_{j=2}^{T} y_{j-1}/(T-1)$ and $\bar{\eta}_{-1} = \sum_{j=2}^{T} \eta_{j-1}/(T-1)$. Since

$$T(\hat{\rho} - 1) = \frac{\frac{1}{T}\sum_{j=2}^{T}(\eta_{j-1} - \bar{\eta})(\eta_j - \bar{\eta})}{\frac{1}{T^2}\sum_{j=2}^{T}(\eta_{j-1} - \bar{\eta})^2},$$

it can be checked easily that the limiting distribution of $T(\tilde{\rho} - 1)$ is the same as that of $T(\hat{\rho} - 1)$ given in (7.41).

2.7 It holds that

$$\hat{\eta}_j - \hat{\eta}_{j-1} = \eta_j - \eta_{j-1} - \left(\frac{3}{T^3} + O\left(\frac{1}{T^4}\right)\right)\sum_{k=1}^{T} k\eta_k,$$

where $\sum_{k=1}^{T} k\eta_k = O_p\left(T^2\sqrt{T}\right)$. Then it is seen that (7.49) holds because of (7.27). Since

$$\frac{\hat{\eta}_1^2}{T} = \frac{1}{T}\left(\sqrt{T}\rho\sigma_L Y(0) + u_1 - \sum_{k=1}^{T} k\eta_k \bigg/ \sum_{k=1}^{T} k^2\right)^2$$

$$= \sigma_L^2 Y^2(0) + o_p(1),$$

(7.50) can also be established.

2.8 Since $c = 0$ and $Y(0) = \gamma$, we have $Y(t) = \gamma + w(t)$ so that, as $|\gamma| \to \infty$,

$$\frac{U_3}{\gamma^2} = \frac{1}{2}\left(\left(1 - 3\int_0^1 t\,dt\right)^2 - 1\right) + o_p(1) = -\frac{3}{8} + o_p(1),$$

$$\frac{V_3}{\gamma^2} = 1 - 3\left(\int_0^1 t\,dt\right)^2 + o_p(1) = \frac{1}{4} + o_p(1).$$

Thus U_3/V_3 converges in probability to $-\frac{3}{2}$ as $|\gamma| \to \infty$.

2.9 The LSE $\tilde{\rho}$ in the present case gives

$$T(\tilde{\rho} - 1) = \frac{1}{T}\sum_{j=2}^{T}\left(y_{j-1} - \frac{j\sum_{k=2}^{T} ky_{k-1}}{\sum_{k=2}^{T} k^2}\right)$$

$$\times (y_j - y_{j-1}) \bigg/ \left[\frac{1}{T^2}\sum_{j=2}^{T}\left(y_{j-1} - \frac{j\sum_{k=2}^{T} ky_{k-1}}{\sum_{k=2}^{T} k^2}\right)^2\right],$$

where $y_j = \rho y_{j-1} + \varepsilon_j$. Then the FCLT and the continuous mapping theorem yield

$$\mathcal{L}(T(\tilde{\rho} - 1)) \to \mathcal{L}\left(\frac{\int_0^1 Y(t)\,dY(t) - 3\int_0^1 tY(t)\,dt \int_0^1 t\,dY(t)}{\int_0^1 \left(Y(t) - 3t\int_0^1 sY(s)\,ds\right)^2 dt}\right),$$

where $dY(t) = -cY(t)\,dt + dw(t)$.

2.10 It can be checked that

$$\hat{\eta}_j - \hat{\eta}_{j-1} = \eta_j - \eta_{j-1} + \left(\frac{12}{T^4} + o\left(\frac{1}{T^5}\right)\right)\left(\frac{T^2}{2}\sum_{k=1}^{T}\eta_k - T\sum_{k=1}^{T}k\eta_k\right),$$

where $\sum_{k=1}^{T}\eta_k = O_p(T\sqrt{T})$ and $\sum_{k=1}^{T}k\eta_k = O_p(T^2\sqrt{T})$. Thus we obtain (7.55) because of (7.27). We also have

$$\hat{\eta}_1 = \eta_1 - \frac{12}{T^4}\left[\left(\frac{T^3}{3} - \frac{T^2}{2}\right)\sum_{k=1}^{T}\eta_k + \left(T - \frac{T^2}{2}\right)\sum_{k=1}^{T}k\eta_k\right] + o_p(\sqrt{T})$$

$$= \sqrt{T}\sigma_L Y(0) - \frac{4}{T}\sum_{k=1}^{T}\eta_k + \frac{6}{T^2}\sum_{k=1}^{T}k\eta_k + o_p(\sqrt{T})$$

CHAPTER 7

so that (7.56) holds.

2.11 The normal equations for the LSEs $\tilde{a}, \tilde{b},$ and $-\tilde{c}$ in the model (7.59) are

$$\begin{pmatrix} 1 & \frac{1}{2} & \int_0^1 \tilde{Y}(t)\,dt \\ \frac{1}{2} & \frac{1}{3} & \int_0^1 t\tilde{Y}(t)\,dt \\ \int_0^1 \tilde{Y}(t)\,dt & \int_0^1 t\tilde{Y}(t)\,dt & \int_0^1 \tilde{Y}^2(t)\,dt \end{pmatrix} \begin{pmatrix} \tilde{a} \\ \tilde{b} \\ -\tilde{c} \end{pmatrix} = \begin{pmatrix} \int_0^1 d\tilde{Y}(t) \\ \int_0^1 t\,d\tilde{Y}(t) \\ \int_0^1 \tilde{Y}(t)\,d\tilde{Y}(t) \end{pmatrix}.$$

Solving for $-\tilde{c}$, we obtain $-\tilde{c} = U_4/V_4$ with Y and r replaced by $\tilde{Y}$ and 1, respectively, in the definitions of U_4 and V_4.

3.1 We have only to show that W_T can be expressed as in (7.62). Since

$$W_T = -\frac{1}{T} \sum_{j=2}^T \hat{\eta}_{j-1}(\hat{\eta}_j + \hat{\eta}_{j-1})$$

$$= -\frac{1}{2T}\left[\sum_{j=2}^T (\hat{\eta}_j + \hat{\eta}_{j-1})^2 - \sum_{j=2}^T \hat{\eta}_j^2 + \sum_{j=2}^T \hat{\eta}_{j-1}^2\right],$$

this last expression yields (7.62).

3.2 Suppose that T is even so that $N = T/2$ is an integer. Then we have, using (7.65),

$$\bar{\eta} = \frac{1}{T}\sum_{j=1}^N (\xi_{2j} - \xi_{2j-1}) = \frac{1}{T}\sum_{j=1}^N \left(-\frac{c}{T}\xi_{2j-1} + v_{2j}\right),$$

which is clearly $O_p(1/\sqrt{T})$. The case of T odd can also be proved similarly.

3.3 In the present case $-T(\hat{\rho} + 1) = W_T/X_T$, where

$$W_T = \frac{1}{2T}\left[(\eta_T - \bar{\eta})^2 - (\eta_1 - \bar{\eta})^2 - \sum_{j=2}^T (\eta_j + \eta_{j-1} - 2\bar{\eta})^2\right]$$

$$= \frac{1}{2T}\left[\eta_T^2 - \eta_1^2 - \sum_{j=2}^T (\eta_j + \eta_{j-1})^2\right] + o_p(1),$$

$$X_T = \frac{1}{T^2}\left[\sum_{j=1}^T (\eta_j - \bar{\eta})^2 - (\eta_T - \bar{\eta})^2\right]$$

$$= \frac{1}{T^2}\sum_{j=1}^T \eta_j^2 + o_p(1).$$

Here we have used the fact that $\bar{\eta} = O_p(1/\sqrt{T})$. Then we can deduce (7.67) as in Model A.

3.4 Suppose that T is even so that $N = T/2$ is an integer. Then (7.65) leads us to

$$\sum_{k=1}^{T} k\eta_k = \sum_{j=1}^{N} (2j\xi_{2j} - (2j-1)\xi_{2j-1})$$

$$= -\frac{2c}{T} \sum_{j=1}^{N} j\xi_{2j-1} + 2\sum_{j=1}^{N} jv_{2j} + \sum_{j=1}^{N} \xi_{2j-1},$$

which is $O_p(T\sqrt{T})$. The case of T odd can be dealt with similarly.

3.5 We have $-T(\hat{\rho} + 1) = W_T/X_T$, where

$$W_T = \frac{1}{2T} \left[\left(\eta_T - \frac{T\sum_{k=1}^{T} k\eta_k}{\sum_{k=1}^{T} k^2} \right)^2 - \left(\eta_1 - \frac{\sum_{k=1}^{T} k\eta_k}{\sum_{k=1}^{T} k^2} \right)^2 \right.$$

$$\left. - \sum_{j=2}^{T} \left(\eta_j + \eta_{j-1} - \frac{(2j-1)\sum_{k=1}^{T} k\eta_k}{\sum_{k=1}^{T} k^2} \right)^2 \right]$$

$$= \frac{1}{2T} \left[\eta_T^2 - \eta_1^2 - \sum_{j=2}^{T} (\eta_j + \eta_{j-1})^2 \right] + o_p(1),$$

$$X_T = \frac{1}{T^2} \sum_{j=1}^{T} \left(\eta_j - \frac{j\sum_{k=1}^{T} k\eta_k}{\sum_{k=1}^{T} k^2} \right)^2 + o_p(1)$$

$$= \frac{1}{T^2} \sum_{j=1}^{T} \eta_j^2 + o_p(1).$$

Here use has been made of the facts that $\sum_{k=1}^{T} k^2 = T^3/3 + O(T^2)$ and $\sum_{k=1}^{T} k\eta_k = O_p(T\sqrt{T})$. We can now deduce (7.67) as in Model A.

3.6 The present case is a mixture of Model B and Model C. We can prove (7.67) using the facts that $\bar{\eta} = O_p(1/\sqrt{T})$, $\sum_{k=1}^{T} k^2 = T^3/3 + O(T^2)$, and $\sum_{k=1}^{T} k\eta_k = O_p(T\sqrt{T})$.

4.1 Recalling the definitions of U_2 and V_2 in Theorem 7.2 and using (7.73), we have

$$m_{21}(\theta_1, \theta_2) = E\left[\exp\{\theta_1 U(Z) + \theta_2 V(Z)\}\frac{d\mu_Y}{d\mu_Z}(Z)\right],$$

where

$$U(Z) = \frac{1}{2}(Z^2(1) - \gamma^2 - 1) - (Z(1) - \gamma)\int_0^1 Z(t)\,dt + \frac{1}{2}(1-r),$$

$$V(Z) = \int_0^1 Z^2(t)\,dt - \left(\int_0^1 Z(t)\,dt\right)^2,$$

$$\frac{d\mu_Y}{d\mu_Z}(Z) = \exp\left\{\frac{\beta-c}{2}(Z^2(1) - \gamma^2 - 1) + \frac{\beta^2-c^2}{2}\int_0^1 Z^2(t)\,dt\right\}.$$

Putting $\beta = \sqrt{c^2 - 2\theta_2}$ yields the last expression in (7.75).

4.2 Noting that $W_l \sim N(\gamma\kappa_l, \Omega_l)$, we consider

$$(w - \gamma\kappa_l)'\Omega_l^{-1}(w - \gamma\kappa_l) - w'A_l w - 2\gamma h_l' w$$
$$= (w - \gamma g)'(\Omega_l^{-1} - A_l)(w - \gamma g) - \gamma^2 g'(\Omega_l^{-1} - A_l)g + \gamma^2 \kappa_l'\Omega_l^{-1}\kappa_l,$$

where $g = (\Omega_l^{-1} - A_l)^{-1}(\Omega_l^{-1}\kappa_l + h_l)$. Then (7.75) leads us to the first equality in (7.76). The second equality can be obtained by substituting $a = \beta + \theta_1 - c$ and

$$g'(\Omega_l^{-1} - A_l)g = (\Omega_l^{-1}\kappa_l + h_l)'(\Omega_l^{-1} - A_l)^{-1}(\Omega_l^{-1}\kappa_l + h_l)$$
$$= (\kappa_l + \Omega_l h_l)'(B_l - A_l\Omega_l)^{-1}\Omega_l^{-1}(\kappa_l + \Omega_l h_l)$$
$$= (\kappa_l + \Omega_l h_l)'(\Omega_l^{-1} + A_l(B_l - \Omega_l A_l)^{-1})(\kappa_l + \Omega_l h_l)$$
$$= \kappa_l'\Omega_l^{-1}\kappa_l + 2h_l'\kappa_l + h_l'\Omega_l h_l$$
$$+ (\kappa_l + \Omega_l h_l)'A_l(B_l - \Omega_l A_l)^{-1}(\kappa_l + \Omega_l h_l),$$

where we have used the matrix identity $(B_l - A_l\Omega_l)^{-1} = B_l + A_l(B_l - \Omega_l A_l)^{-1}\Omega_l$.

4.3 Substituting $h_1 = 0$ and $A_1 = a$ into the last expression in (7.76), we have

$$m_{11}(\theta_1, \theta_2)$$
$$= \exp\left[\frac{1}{2}\left\{c - r\theta_1 + \gamma^2\left(-a + \frac{a\kappa_1^2}{1 - a\Omega_1}\right)\right\}\right][e^\beta(1 - a\Omega_1)]^{-1/2},$$

where $a = \beta + \theta_1 - c$, $\kappa_1 = e^{-\beta}$, and

$$1 - a\Omega_1 = 1 - (\beta + \theta_1 - c)\frac{1 - e^{-2\beta}}{2\beta}$$

$$= e^{-\beta}\left[\cosh\beta + (c - \theta_1)\frac{\sinh\beta}{\beta}\right].$$

Then we can arrive at (7.77) easily.

4.4 The first equality is obvious, while the second equality comes from (7.75). This leads us to the last equality in (7.76) with γ^2 replaced by $Z^2(0)$. Then, noting that $E[\exp\{bZ^2(0)\}] = (1 - b/c)^{-1/2}$, we have the next to last equality, which has the interpretation as given on the right side of the last equality.

5.1 For the case of $l = 1$ we consider the limiting distribution of

$$\frac{1}{\sqrt{2c}}\left(\frac{U_1}{V_1} + cr\right) = \frac{\sqrt{\frac{c}{2}}(U_1 + crV_1)}{cV_1}.$$

For this purpose we have, from (7.82),

$$m_{12}\left(\sqrt{\frac{c}{2}}\theta_1, \sqrt{\frac{c}{2}}cr\theta_1 + c\theta_2\right)$$

$$= \exp\left\{\frac{1}{2}\left(c - \sqrt{\frac{c}{2}}r\theta_1\right)\right\}$$

$$\times \left[\cosh\nu + \frac{2c^2 - \frac{c}{2}\theta_1^2 - 2\sqrt{\frac{c}{2}}cr\theta_1 - 2c\theta_2}{2c}\frac{\sinh\nu}{\nu}\right]^{-1/2},$$

where

$$\nu = \left(c^2 - 2\sqrt{\frac{c}{2}}cr\theta_1 - 2c\theta_2\right)^{1/2}$$

$$= c - \sqrt{\frac{c}{2}}r\theta_1 - \theta_2 - \frac{r^2\theta_1^2}{4} + O\left(\frac{1}{\sqrt{c}}\right).$$

Then we can deduce that

$$m_{12}\left(\sqrt{\frac{c}{2}}\theta_1, \sqrt{\frac{c}{2}}cr\theta_1 + c\theta_2\right) \to \exp\left[\frac{\theta_1^2}{2}\frac{r^2}{4} + \frac{\theta_2}{2}\right],$$

which implies that the limiting distribution is $N(0, r^2)$.

5.2 Consider

$$m_{11}(-4ce^{2c}\theta_1, 4c^2e^{2c}\theta_2 \mid \gamma = 0) = \exp\left\{\frac{1}{2}(c + 4cre^{2c}\theta_1)\right\} H_1^{-1/2},$$

where

$$H_1 = \cosh \nu + (c + 4ce^{2c}\theta_1) \frac{\sinh \nu}{\nu}, \quad \nu = \sqrt{c^2 - 8c^2 e^{2c}\theta_2}.$$

Since

$$\frac{1}{\nu} = (c^2 - 8c^2 e^{2c}\theta_2)^{-1/2} = -\frac{1}{c}\left(1 + 4e^{2c}\theta_2 + O(e^{4c})\right),$$

we obtain $H_1 = (1 - 2\theta_1 - 2\theta_2)e^c + O(e^{3c})$. Thus it holds that $m_{11} \to (1 - 2\theta_1 - 2\theta_2)^{-1/2}$, which leads us to the conclusion.

5.3 Let us consider

$$m_{41}(ce^c\theta_1, c^2 e^c\theta_1 + 2c^2 e^{2c}\theta_2 \mid \gamma = 0) = \exp\left\{\frac{c - rce^c\theta_1}{2}\right\} A^{-1/2},$$

where

$$A = \left(\frac{A_1}{\nu^5} - \frac{A_2}{\nu^7}\right) \sinh \nu + \left(\frac{c^4}{\nu^4} + \frac{A_3}{\nu^6} + \frac{A_2}{\nu^8}\right) \cosh \nu + \frac{A_4}{\nu^6} - \frac{A_2}{\nu^8},$$

$$\nu = (c^2 - 2c^2 e^c \theta_1 - 4c^2 e^{2c}\theta_2)^{1/2} = -c + O(ce^c),$$

$$\frac{c^4}{\nu^4} = 1 + 4e^c\theta_1 + 8e^{2c}\theta_2 + 12e^{2c}\theta_1^2 + O(e^{3c}),$$

$$\frac{c^5}{\nu^5} = -\left(1 + 5e^c\theta_1 + 10e^{2c}\theta_2 + \frac{35}{2}e^{2c}\theta_1^2 + O(e^{3c})\right),$$

$$\frac{c^6}{\nu^6} = 1 + O(e^c), \quad \frac{c^7}{\nu^7} = -1 + O(e^c), \quad \frac{c^8}{\nu^8} = 1 + O(e^c),$$

$$A_1 = c^5 - c^5 e^c\theta_1 - 8c^2(c^2 - 3c - 3)e^c\theta_1 + O(c^4 e^{2c}),$$

$$A_2 = 24c^5 e^c\theta_1 + O(c^5 e^{2c}),$$

$$A_3 = -8c^5 e^c\theta_1 + O(c^5 e^{2c}), \quad A_4 = O(c^5 e^c).$$

We then have

$$A = \frac{1}{2}\left[\left\{\frac{-1}{c^5}\left(1 + 5e^c\theta_1 + 10e^{2c}\theta_2 + \frac{35}{2}e^{2c}\theta_1^2 + O(e^{3c})\right)\right.\right.$$

$$\times \left(c^5 - c^5 e^c\theta_1 - 8c^2(c^2 - 3c - 3)e^c\theta_1 + O(c^4 e^{2c})\right)$$

$$+ \frac{1}{c^7}(1 + O(e^c)) \left(24c^5 e^c \theta_1 + O(c^5 e^{2c})\right)\bigg\} (e^\nu - e^{-\nu})$$

$$+ \bigg\{ 1 + 4e^c \theta_1 + 8e^{2c} \theta_2 + 12e^{2c} \theta_1^2 + O(e^{3c})$$

$$+ \frac{1}{c^6}(1 + O(e^c)) \left(-8c^5 e^c \theta_1 + O(c^5 e^{2c})\right)$$

$$+ \frac{1}{c^8}(1 + O(e^c)) \left(24c^5 e^c \theta_1 + O(c^5 e^{2c})\right)\bigg\}(e^\nu + e^{-\nu}) \bigg] + O\left(\frac{e^c}{c}\right)$$

$$= \left(1 - \theta_2 - \frac{1}{4}\theta_1^2\right) e^c + O\left(\frac{e^c}{c}\right)$$

so that $m_{41} \to (1 - \theta_2 - \theta_1^2/4)^{-1/2}$. This last m.g.f. is that of $(XY/2, X^2/2)$, where $(X, Y)' \sim N(0, I_2)$. Thus we obtain the required result.

5.4 Consider

$$m_{31}\left(c^3 e^{2c}\theta, c^4 e^{2c}\theta \mid \gamma = 0\right) = \exp\left\{\frac{1}{2}\left(c - rc^3 e^{2c}\theta\right)\right\} H_3^{-1/2},$$

where

$$H_3 = \frac{c^3}{\nu^3} \left\{1 - e^{2c}\theta\left(c^2 + 3c + 3\right)\right\} \sinh \nu + \frac{c^2}{\nu^2} \cosh \nu$$

$$- 3c^3 e^{2c}\theta \left\{c^2 + 3c + 3 - 2c(c + 1)\right\} \left(\frac{\sinh \nu}{\nu^5} - \frac{\cosh \nu}{\nu^4}\right),$$

$$\nu = \sqrt{c^2 \left(1 - 2c^2 e^{2c}\theta\right)}.$$

Since we have

$$\frac{1}{\nu^k} = \frac{(-1)^k}{c^k}\left(1 + kc^2 e^{2c}\theta + O\left(c^4 e^{4c}\right)\right), \quad (k = 1, 2, \ldots),$$

we obtain

$$H_3 = e^c \left\{1 + \frac{3}{2}\theta + O\left(\frac{1}{c}\right)\right\}.$$

Thus $m_{31} \to (1 + 3\theta/2)^{-1/2}$, which is the required result.

5.5 Consider

$$m_{31}\left(c^2 e^{2c}\theta, c^3 e^{2c}\theta\right) = \exp\left\{\frac{1}{2}\left(c - rc^2 e^{2c}\theta + \frac{G\gamma^2}{H_3}\right)\right\} H_3^{-1/2},$$

where

$$G = e^{2c}\theta \left[\{2c^5 - (c^2 + 3c + 3)(c^4 e^{2c}\theta - 3c^2 + 2c^3) + 3c^4\} \left(-\frac{\sinh \nu}{\nu^3}\right) \right.$$
$$- 3\{2c^5 - c^6 e^{2c}\theta + 3c^4 - 2c^5 - 3c^5 e^{2c}\theta - 3c^4 e^{2c}\theta\} \frac{\sinh \nu}{\nu^5}$$
$$+ 3\{2c^5 - c^6 e^{2c}\theta + 5c^4 - 2c^5\}$$
$$- (c+1)(3c^4 e^{2c}\theta - 6c^2 + 4c^3)\} \frac{\cosh \nu}{\nu^5}$$
$$\left. - \frac{6}{\nu^4} \{c^4 + (c+1)(3c^2 - 2c^3)\} \right],$$

$$H_3 = \{c^2 e^{2c}\theta (c^2 + 3c + 3) - c^3\} \left(-\frac{\sinh \nu}{\nu^3}\right) + \frac{c^2}{\nu^2} \cosh \nu$$
$$- 3c^2 e^{2c}\theta \{c^2 + 3c + 3 - 2c(c+1)\} \left(\frac{\sinh \nu}{\nu^5} - \frac{\cosh \nu}{\nu^4}\right),$$

$$\frac{1}{\nu^k} = (c^2 - 2c^3 e^{2c}\theta)^{-k/2}$$
$$= \frac{(-1)^k}{c^k} \left(1 + kc e^{2c}\theta + O(c^2 e^{4c})\right), \quad (k = 1, 2, \ldots).$$

Then we obtain

$$G = e^c \left\{3\theta + O\left(\frac{1}{c}\right)\right\}, \quad H_3 = e^c \left\{1 + O\left(\frac{1}{c}\right)\right\},$$

so that $m_{31} \to \exp(3\gamma^2\theta/2)$.

5.6 Let us consider

$$m_{21}\left(\sqrt{-c}\, e^c \theta_1,\, c\sqrt{-c}\, e^c \theta_1 + c e^{2c} \theta_2\right)$$
$$= \exp\left[\frac{c - r\sqrt{-c}\, e^c \theta_1}{2} - \frac{\gamma^2 c^3 e^{2c}(\theta_1^2 - 2\theta_2)}{2H_2} A\right] H_2^{-1/2},$$

where

$$A = \frac{\sinh \nu}{\nu^3} - \frac{2}{\nu^4}(\cosh \nu - 1),$$
$$\nu = (c^2 - 2c\sqrt{-c}\, e^c \theta_1 - 2c e^{2c} \theta_2)^{1/2} = -c + O(\sqrt{-c}\, e^c),$$
$$\frac{c}{\nu} = -1 + \frac{e^c}{\sqrt{-c}} \theta_1 - \frac{e^{2c}}{c} \theta_2 + \frac{3e^{2c}}{2c} \theta_1^2 + O\left(\frac{e^{3c}}{c\sqrt{-c}}\right),$$

$$H_2 = \frac{-ce^{2c}\theta_1^2 + c^2\sqrt{-c}\,e^c\theta_1 - c^3 + 2c\sqrt{-c}\,e^c\theta_1 + 2ce^{2c}\theta_2}{-\nu^2} \frac{\sinh \nu}{\nu}$$

$$+ \frac{c^2}{\nu^2}\cosh \nu$$

$$+ \frac{2(-ce^{2c}\theta_1^2 + c^2\sqrt{-c}\,e^c\theta_1 - 2c^2\sqrt{-c}\,e^c\theta_1 - 2c^2e^{2c}\theta_2)}{\nu^4}$$

$$\times (\cosh \nu - 1).$$

Since

$$\frac{c^2}{\nu^2} = 1 - \frac{2e^c}{\sqrt{-c}}\theta_1 + O\left(\frac{e^{2c}}{c}\right),$$

$$\frac{c^3}{\nu^3} = -1 + \frac{3e^c}{\sqrt{-c}}\theta_1 + O\left(\frac{e^{2c}}{c}\right),$$

$$\frac{c^4}{\nu^4} = 1 - \frac{4e^c}{\sqrt{-c}}\theta_1 + O\left(\frac{e^{2c}}{c}\right),$$

we obtain $H_2 = e^c + O(e^c/c)$ and

$$c^3e^{2c}A = -\frac{1}{2}e^c + O\left(\frac{e^c}{c}\right).$$

Then we have $m_{21} \to \exp\{\gamma^2(\theta_1^2 - 2\theta_2)/4\}$, as in the case of $l = 1$.

6.1 We have only to identify the matrices A_l and B_l given below (7.92). Consider the case of $l = 1$. Putting $y = (y_0, y_1, \ldots, y_T)'$ with $y_0 = \varepsilon_0/\sqrt{1-\rho^2}$ and $\varepsilon = (\varepsilon_0, \varepsilon_1, \ldots, \varepsilon_T)'$, we have

$$T(\tilde{\rho}_1 - \rho) = \frac{1}{T\sigma^2}\sum_{j=1}^{T} y_{j-1}\varepsilon_j \bigg/ \left[\frac{1}{T^2\sigma^2}\sum_{j=1}^{T} y_{j-1}^2\right],$$

where

$$\sum_{j=1}^{T} y_{j-1}\varepsilon_j = \frac{1}{2}\left[y'\begin{pmatrix} 0 & I_T \\ 0 & 0' \end{pmatrix}\varepsilon + \varepsilon'\begin{pmatrix} 0' & 0 \\ I_T & 0 \end{pmatrix}y\right],$$

$$\sum_{j=1}^{T} y_{j-1}^2 = y'\begin{pmatrix} I_T & 0 \\ 0' & 0 \end{pmatrix}y.$$

Noting that

$$y = \begin{pmatrix} \rho^T/\sqrt{1-\rho^2} & C_{11} & & 0 \\ & \rho^{T-1} & \cdots & \rho & 1 \end{pmatrix}\varepsilon,$$

we obtain $P(T(\tilde{\rho}_1 - \rho) \leq x) = P(\varepsilon'(xB_1 - A_1)\varepsilon/\sigma^2 \geq 0)$, which yields (7.92). The case of $l = 2$ can be proved similarly.

6.2 Noting that $1/\sqrt{1-\rho^2} = \sqrt{T}/\sqrt{2c} + O(1/\sqrt{T})$, we may put $y_0 = \varepsilon_0/\sqrt{1-\rho^2} = \sqrt{T}\sigma Z + R_T$, where $Z \sim N(0, 1/(2c))$ and $R_T = O_p(1/\sqrt{T})$. Then it is evident that $\mathcal{L}(T(\tilde{\rho}_l - 1)) \to \mathcal{L}(U_l/V_l)$, where U_l and V_l are defined in Section 7.2 with $Y(0) \sim N(0, 1/(2c))$ and $r = 1$. Since $T(\tilde{\rho}_l - \rho) = T(\tilde{\rho}_l - 1) + c$, we have that $P(T(\tilde{\rho}_l - \rho) \leq x) \to P(zV_l - U_l \geq 0)$, which yields (7.95).

7.1 For Model D we have

$$y_j = \beta_0 + j\beta_1 + \eta_j = ((1, j) \otimes I_m)\beta + \eta_j$$

so that $y = ((e, d) \otimes I_m)\beta + \eta$. Since

$$\eta_j = \rho_m \eta_{j-1} + u_j = \rho_m^{j-1} u_1 + \rho_m^{j-2} u_2 + \cdots + \rho_m u_{j-1} + u_j,$$

it is easy to see that $\eta = (C(\rho_m) \otimes I_m) u$.

7.2 Because of the definition of $\hat{\rho}_m$, we have

$$N(\hat{\rho}_m - 1) = \frac{1}{N} \sum_{j=2}^{N} \hat{\eta}'_{j-1} (\hat{\eta}_j - \hat{\eta}_{j-1}) \bigg/ \left[\frac{1}{N^2} \sum_{j=2}^{N} \hat{\eta}'_{j-1} \hat{\eta}_{j-1}\right],$$

where

$$2 \sum_{j=2}^{N} \hat{\eta}'_{j-1} (\hat{\eta}_j - \hat{\eta}_{j-1})$$

$$= -\sum_{j=2}^{N} (\hat{\eta}_j - \hat{\eta}_{j-1})' (\hat{\eta}_j - \hat{\eta}_{j-1}) + \sum_{j=2}^{N} \hat{\eta}'_j \hat{\eta}_j - \sum_{j=2}^{N} \hat{\eta}'_{j-1} \hat{\eta}_{j-1}$$

$$= \hat{\eta}'_N \hat{\eta}_N - \sum_{j=1}^{N} (\hat{\eta}_j - \hat{\eta}_{j-1})' (\hat{\eta}_j - \hat{\eta}_{j-1}) .$$

Since $\hat{\eta}_N = (e'_N \otimes I_m)\hat{\eta} = (e'_N \otimes I_m)(\bar{M}C(\rho_m) \otimes I_m)u$, we arrive at the last expression for U_N. The expressions for V_N can be verified easily.

7.3 Since $\hat{\beta} = \beta + (d'd)^{-1}(d' \otimes I_m)\eta$ and $y_j = \beta j + \eta_j$, we first have

$$\hat{\eta}_j = y_j - j\hat{\beta} = \eta_j - j(d'd)^{-1}(d' \otimes I_m)\eta = \eta_j - j(d'd)^{-1} \sum_{k=1}^{N} k\eta_k$$

so that

$$\hat{\eta}_j - \hat{\eta}_{j-1} = \eta_j - \eta_{j-1} - \left(\frac{3}{N^3} + O\left(\frac{1}{N^4}\right)\right) \sum_{k=1}^{N} k\eta_k ,$$

where $\sum_{k=1}^{N} k\eta_k = O_p(T^2\sqrt{T})$. Then the weak law of large numbers ensures that

$$\frac{1}{N}\sum_{j=1}^{N}(\hat{\boldsymbol{\eta}}_j - \hat{\boldsymbol{\eta}}_{j-1})'(\hat{\boldsymbol{\eta}}_j - \hat{\boldsymbol{\eta}}_{j-1}) = \frac{1}{N}\sum_{j=1}^{N}(\boldsymbol{\eta}_j - \boldsymbol{\eta}_{j-1})'(\boldsymbol{\eta}_j - \boldsymbol{\eta}_{j-1}) + o_p(1)$$

$$= \frac{1}{N}\sum_{j=1}^{N} u'_j u_j + o_p(1)$$

$$\to m\sigma^2 \sum_{l=0}^{\infty} \alpha_l^2$$

in probability.

7.4 In the present case we have $\bar{M} = I_N$ and

$$K_N = xK_{2N} - K_{1N} = \frac{x}{N}C'(\rho_m)C(\rho_m) - \frac{1}{2}C'(\rho_m)e_N e'_N C(\rho_m).$$

It can be shown after some algebra [Nabeya and Tanaka (1990a)] that

$$\lim_{N\to\infty} \max_{j,k} \left| K_N(j,k) - K\left(\frac{j}{N}, \frac{k}{N}\right) \right| = 0,$$

where $K(s,t) = \{x(e^{-c|s-t|} - e^{-c(2-s-t)})/c - e^{-c(2-s-t)}\}/2$. Since

$$xV_N - U_N = \frac{1}{N}\sum_{j,k=1}^{N} K_N(j,k)u'_j u_k + \frac{m\sigma^2}{2}\sum_{l=0}^{\infty} \alpha_l^2 + o_p(1),$$

(7.107) follows from the arguments in Section 5.6 of Chapter 5.

7.5 Put $F(x) = P(\int_0^1 w'(t)w(t)\,dt \le x)$ and $f(x) = dF(x)/dx$. Then

$$\int_0^{\infty} e^{-\theta x} f(x)\,dx = \left(\cosh\sqrt{2\theta}\right)^{-m/2}$$

$$= 2^{m/2} e^{-m\sqrt{\theta/2}} \left(1 + e^{-2\sqrt{2\theta}}\right)^{-m/2}$$

$$= 2^{m/2} \sum_{k=0}^{\infty} \binom{-\frac{m}{2}}{k} e^{-(2k+m/2)\sqrt{2\theta}}$$

so that, taking the inverse Laplace transform,

$$f(x) = \sum_{k=0}^{\infty} \binom{-\frac{m}{2}}{k} \frac{2^{m/2}}{2\sqrt{\pi x^3}} b_k e^{-b_k^2/(4x)}, \quad \left(b_k = \sqrt{2}\left(2k + \frac{m}{2}\right)\right),$$

$$F(x) = \int_0^x f(a)\,da$$

$$= \sum_{k=0}^{\infty} \binom{-\frac{m}{2}}{k} \frac{2^{(m+1)/2}}{\sqrt{\pi}} \int_{b_k/\sqrt{2x}}^{\infty} e^{-a^2/2}\,da$$

$$= 2^{(m+2)/2} \sum_{k=0}^{\infty} \binom{-\frac{m}{2}}{k} \Phi\left(-\frac{2k+\frac{m}{2}}{\sqrt{x}}\right).$$

7.6 Noting that $y_0 = 0$ and $y_j = y_{j-1} + \varepsilon_j = \varepsilon_1 + \cdots + \varepsilon_j$, we obtain

$$\sum_{j=2}^{T} y_{j-1} y_j = \sum_{j=1}^{N} \sum_{k=0}^{m-1} y_{(j-1)m+k}\, y_{(j-1)m+k+1}$$

$$= \sum_{j=1}^{N} \sum_{k=1}^{m-1} y_{(j-1)m+k}\, y_{(j-1)m+k+1} + \sum_{j=1}^{N} y_{(j-1)m}\, y_{(j-1)m+1}$$

$$= \sum_{j=1}^{N} \sum_{k=1}^{m-1} \mathbf{y}_j' e_k e_{k+1}' \mathbf{y}_j + \sum_{j=1}^{N} \mathbf{y}_{j-1}' e_m e_1' \mathbf{y}_j$$

$$= \sum_{j=1}^{N} \mathbf{y}_j' \left(\sum_{k=1}^{m-1} e_k e_{k+1}' + e_m e_1' \right) \mathbf{y}_j - \sum_{j=1}^{N} \boldsymbol{\varepsilon}_j' e_m e_1' \mathbf{y}_j,$$

which yields (7.114).

7.7 Note first that

$$\sum_{j=1}^{N} \boldsymbol{\varepsilon}_j' e_m e_1' \mathbf{y}_j = \sum_{j=1}^{N} \boldsymbol{\varepsilon}_j' e_m e_1' (\boldsymbol{\varepsilon}_1 + \cdots + \boldsymbol{\varepsilon}_j)$$

$$= \sum_{j=1}^{N} \varepsilon_{jm}(\varepsilon_1 + \varepsilon_{m+1} + \cdots + \varepsilon_{(j-1)m+1})$$

$$= \sum_{j=1}^{N} \xi_j(\eta_1 + \eta_2 + \cdots + \eta_j),$$

where $\xi_j = \varepsilon_{jm}$ and $\eta_j = \varepsilon_{(j-1)m+1}$. Since $\{\xi_j\}$ and $\{\eta_j\}$ are i.i.d.$(0, \sigma^2)$ sequences and are independent of each other, it follows from the weak convergence result in Chapter 3 that $\sum_{j=1}^{N} \boldsymbol{\varepsilon}_j' e_m e_1' \mathbf{y}_j / N$ converges to a nondegenerate distribution. Therefore (7.115) holds.

7.8 Let us put $y_0 = 0$ and suppose first that $0 \le l < m$. Then

$$\sum_{j=l+1}^{T} y_{j-l} y_j = \sum_{j=1}^{N-1} \sum_{k=0}^{m-1} y_{(j-1)m+k} y_{(j-1)m+k+l} + \sum_{k=0}^{m-l} y_{(N-1)m+k} y_{(N-1)m+k+l}$$

$$= \sum_{j=1}^{N-1} y_j' H_l y_j + R_N$$

$$= \sum_{j,k=1}^{N-1} \min(N-j, N-k) \varepsilon_j' H_l \varepsilon_k + R_N,$$

where R_N/T^2 converges in probability to 0. Thus (7.117) holds because of the same reasoning as in (7.116). When $l \ge m$, we may put $l = im + n$ ($i = 1, 2, \ldots$; $n = 0, 1, \ldots, m-1$). Since

$$y_j = y_{j-im} + \sum_{k=0}^{i-1} \varepsilon_{j-km},$$

we have

$$\frac{1}{T^2} \sum_{j=l+1}^{T} y_{j-l} y_j = \frac{1}{T^2} \sum_{j=l+1}^{T} y_{j-im-n} y_{j-im} + \frac{1}{T^2} \sum_{j=l+1}^{T} \sum_{k=0}^{i-1} y_{j-l} \varepsilon_{j-km}$$

$$= \frac{1}{T^2} \sum_{j=n+1}^{T} y_{j-n} y_j + o_p(1)$$

so that (7.117) holds for general l.

7.9 Defining $dX_M(u) = -\beta X_M(u) du + dw(u)$ with $X_M(0) = 0$, Girsanov's theorem gives

$$E\left[\exp\left\{\frac{\theta_1}{M} \int_0^M Z_M(u) dZ_M(u) + \frac{\theta_2}{M^2} \int_0^M Z_M^2(u) du\right\}\right]$$

$$= \exp\left\{M\left(-\frac{\beta - c}{2} - \frac{\theta_1}{2M}\right)\right\} E\left[\exp\left\{\left(\frac{\beta - c}{2} + \frac{\theta_1}{2M}\right) X_M^2(M)\right\}\right],$$

where $\beta = \sqrt{c^2 - 2\theta_2/M^2}$ and $X_M(M) \sim N(0, (1 - e^{-2\beta M})/(2\beta))$. Then this m.g.f. is shown to be identical with that given in (7.121). Thus (7.123) is established.

7.10 It follows from Problem 7.9 and (7.121) that

$$E\left[\exp\left\{\frac{\theta_1}{\sqrt{M}} \left(\int_0^M Z_M(u) dZ_M(u) + c \int_0^M Z_M^2(u) du\right) + \frac{\theta_2}{M} \int_0^M Z_M^2(u) du\right\}\right]$$

$$= \exp\left\{\frac{cM - \sqrt{M}\theta_1}{2}\right\}\left[\cosh\nu + (cM - \sqrt{M}\theta_1)\frac{\sinh\nu}{\nu}\right]^{-1/2},$$

where

$$\nu = (c^2M^2 - 2cM\sqrt{M}\theta_1 - 2M\theta_2)^{1/2}$$

$$= cM - \sqrt{M}\theta_1 - \frac{\theta_2}{c} - \frac{\theta_1^2}{2c} + O\left(\frac{1}{\sqrt{M}}\right).$$

Then the above m.g.f. converges to $\exp\{(\theta_1^2/2 + \theta_2)/(2c)\}$, which implies (7.126). Similarly we can show that

$$E\left[\exp\left\{\frac{\theta_1}{M}\int_0^M Z_M(u)\,dZ_M(u) + \frac{\theta_2}{M}\int_0^M Z_M^2(u)\,du\right\}\right]$$

$$\to \exp\left\{-\frac{1}{2}\left(\theta_1 - \frac{\theta_2}{c}\right)\right\},$$

which implies (7.125).

7.11 It follows from (7.121) and (7.124) that

$$E\left[\exp\left\{ce^{cM}\theta_1\left(\int_0^M Z_M(u)\,dZ_M(u) + c\int_0^M Z_M^2(u)\,du\right)\right.\right.$$

$$\left.\left.+ 2c^2e^{2cM}\theta_2\int_0^M Z_M^2(u)\,du\right\}\right]$$

$$= \exp\left\{\frac{cM}{2}(1 - e^{cM}\theta_1)\right\}\left[\cosh\nu + cM(1 - e^{cM}\theta_1)\frac{\sinh\nu}{\nu}\right]^{-1/2},$$

where

$$\frac{1}{\nu} = (c^2M^2 - 2c^2M^2e^{cM}\theta_1 - 4c^2M^2e^{2cM}\theta_2)^{-1/2}$$

$$= \frac{1}{cM}\left(1 + e^{cM}\theta_1 + 2e^{2cM}\theta_2 + \frac{3}{2}e^{2cM}\theta_1^2 + O(e^{3cM})\right).$$

Then we have

$$\cosh\nu + cM(1 - e^{cM}\theta_1)\frac{\sinh\nu}{\nu} = e^{cM}\left(1 - \theta_2 - \frac{1}{4}\theta_1^2\right) + O(e^{2cM}),$$

so that the above m.g.f. converges to $(1 - \theta_2 - \theta_1^2/4)^{-1/2}$, which is the joint m.g.f. of $(XY/2, X^2/2)$, where $(X,Y)' \sim N(0, I_2)$. Thus (7.127) is established.

8.1 Putting $\varepsilon_j = 0$ for $j \leq 0$, we have

$$y_j = \frac{\varepsilon_j}{(1-e^{i\theta}L)(1-e^{-i\theta}L)} = \frac{1}{2i\sin\theta}\left[\frac{e^{i\theta}}{1-e^{i\theta}L} - \frac{e^{-i\theta}}{1-e^{-i\theta}L}\right]\varepsilon_j$$

$$= \frac{e^{i\theta}}{2i\sin\theta}\left[e^{i(j-1)\theta}\varepsilon_1 + e^{i(j-2)\theta}\varepsilon_2 + \cdots + e^{i\theta}\varepsilon_{j-1} + \varepsilon_j\right]$$

$$- \frac{e^{-i\theta}}{2i\sin\theta}\cdot\left[e^{-i(j-1)\theta}\varepsilon_1 + e^{-i(j-2)\theta}\varepsilon_2 + \cdots + e^{-i\theta}\varepsilon_{j-1} + \varepsilon_j\right]$$

$$= \frac{1}{\sin\theta}\sum_{k=1}^{j}\varepsilon_k\sin(j-k+1)\theta = \frac{1}{\sin\theta}\left[X_j\sin(j+1)\theta - Y_j\cos(j+1)\theta\right].$$

8.2 It follows from (7.129) and (7.132) that

$$\sum_{j=2}^{T}y_{j-1}y_j = \frac{1}{\sin^2\theta}\sum_{j=2}^{T}(X_{j-1}\sin j\theta - Y_{j-1}\cos j\theta)$$

$$\times (X_j\sin(j+1)\theta - Y_j\cos(j+1)\theta)$$

$$= \frac{1}{2\sin^2\theta}\sum_{j=2}^{T}\left[(X_{j-1}X_j + Y_{j-1}Y_j)\cos\theta + (X_{j-1}Y_j - X_jY_{j-1})\sin\theta\right.$$

$$- (X_{j-1}X_j - Y_{j-1}Y_j)\cos(2j+1)\theta$$

$$\left.- (X_{j-1}Y_j + X_jY_{j-1})\sin(2j+1)\theta\right]$$

$$= \frac{\sigma^2 T\cos\theta}{4\sin^2\theta}\sum_{j=1}^{T}Z'_T\left(\frac{j}{T}\right)Z_T\left(\frac{j}{T}\right) + o_p(T),$$

which implies (7.133), where we have used $X_j = X_{j-1} + \varepsilon_j\cos j\theta$ and $Y_j = Y_{j-1} + \varepsilon_j\sin j\theta$. Consider next

$$\sum_{j=2}^{T}y_{j-1}\varepsilon_j = \frac{1}{\sin\theta}\sum_{j=2}^{T}(X_{j-1}\sin j\theta - Y_{j-1}\cos j\theta)\varepsilon_j$$

$$= \frac{1}{\sin\theta}\sum_{j=2}^{T}(X_{j-1}(Y_j - Y_{j-1}) - Y_{j-1}(X_j - X_{j-1})),$$

which leads us to (7.134). Finally we have

$$\sum_{j=3}^{T}y_{j-2}\varepsilon_j = \frac{1}{\sin\theta}\sum_{j=3}^{T}[X_{j-2}(\sin j\theta\cos\theta - \cos j\theta\sin\theta)$$

$$- Y_{j-2}(\cos j\theta\cos\theta + \sin j\theta\sin\theta)]\varepsilon_j$$

$$= \frac{1}{\sin\theta} \sum_{j=3}^{T} [\cos\theta\{X_{j-2}\Delta Y_j - Y_{j-2}\Delta X_j\}$$
$$- \sin\theta\{X_{j-2}\Delta X_j + Y_{j-2}\Delta Y_j\}],$$

which yields (7.135).

9.1 Putting $u_j = \varepsilon_j/\beta(L)$, it follows that, for any $k > 0$,

$$y_{j-k} = \frac{u_{j-k}}{(1-L)^d} = \frac{u_{j-1} - (1-L)(u_{j-1} + \cdots + u_{j-k+1})}{(1-L)^d}$$
$$= \frac{u_{j-1}}{(1-L)^d} - \frac{u_{j-1} + \cdots + u_{j-k+1}}{(1-L)^{d-1}}.$$

Then it can be verified that

$$\frac{1}{T^{2d}\sigma^2} \sum_{j=p+1}^{T} y_{j-1} y'_{j-1} = \frac{1}{T^{2d}\sigma^2} \sum_{j=p+1}^{T} \left(\frac{u_{j-1}}{(1-L)^d}\right)^2 ee' + o_p(1),$$

which establishes (7.140).

9.2 When $d = 1$, it holds that

$$\delta = M'\phi = \begin{pmatrix} 1 & 1 & & & \\ & -1 & \cdot & & 0 \\ & & \cdot & \cdot & \\ & & & \cdot & 1 \\ & 0 & & & -1 \end{pmatrix} \phi,$$

from which we have $\delta_1 = \phi_1 + \phi_2$, $\delta_{q+1} = -\phi_{q+1}$, and $\delta_k = -\phi_k + \phi_{k+1}$ ($k = 2,\cdots,q$). Solving for ϕ_ks we obtain (7.142). When $d = 2$, it holds that

$$\delta = M'\phi = \begin{pmatrix} 1 & 1 & 1 & & & \\ & -1 & -2 & \cdot & & 0 \\ & & 1 & \cdot & \cdot & \\ & & & \cdot & \cdot & \\ & & & & \cdot & 1 \\ & 0 & & & & -2 \\ & & & & & 1 \end{pmatrix} \phi,$$

from which we have $\delta_1 = \phi_1 + \phi_2 + \phi_3$, $\delta_2 = -\phi_2 - 2\phi_3 + \phi_4$, $\delta_{q+1} = \phi_{q+1} - 2\phi_{q+2}$, $\delta_{q+2} = \phi_{q+2}$, and $\delta_k = \phi_k - 2\phi_{k+1} + \phi_{k+2}$ ($k = 3,\ldots,q$), which yields (7.143).

9.3 Since $\{\Delta^k y_{i-1}\} \sim I(d-k)$ and it holds that

$$\frac{1}{T^{d-k}\sqrt{T}} \sum_{j=p+1}^{T} \Delta^k y_{j-1} \Delta^d y_{j-m} \to 0 \quad \text{in probability}$$

for $k = 0, 1, \ldots, d-1$ and $m = 1, \ldots, q$, the off-block diagonal elements in the limiting distribution reduce to 0. Since $\{u_{j-1}\}$ is second-order stationary with i.i.d. innovations, we have, by the weak law of large numbers,

$$\frac{1}{T} \sum_{j=p+1}^{T} u_{j-1} u'_{j-1} \to \Gamma \quad \text{in probability.}$$

Finally the FCLT and the continuous mapping theorem gives the joint weak convergence

$$\mathcal{L}\left(G_T^{-1} \sum_{j=p+1}^{T} x_{j-1} x'_{j-1} G_T^{-1}\right) \to \mathcal{L}(\alpha^2 \sigma^2 F),$$

where $G_T = \text{diag}(T^d, \ldots, T): d \times d$. Thus we obtain the conclusion.

9.4 In view of (7.145), we have only to show that $\bar{M}_1 Q_1 y_{j-1} = z_{j-1}$. It is easy to see that

$$\bar{M}_1 Q_1 \begin{pmatrix} y_{j-1} \\ \vdots \\ y_{j-p} \end{pmatrix} = \bar{M}_1 \frac{1}{(1-L)^d} \begin{pmatrix} \varepsilon_{j-1} \\ \vdots \\ \varepsilon_{j-d} \end{pmatrix}$$

$$= \left(\frac{\varepsilon_{j-1}}{(1-L)^d}, \ldots, \frac{\varepsilon_{j-1}}{1-L}\right)' = z_{j-1}.$$

9.5 We first have

$$(\bar{M}'_1 G_T^{-1})^{-1} (\hat{\delta} - \delta) = \left(G_T^{-1} \bar{M}_1 \sum_j y_{j-1} y'_{j-1} \bar{M}'_1 G_T^{-1}\right)^{-1} G_T^{-1} \bar{M}_1 \sum_j y_{j-1} u_j$$

$$= \left(G_T^{-1} \sum_j x_{j-1} x'_{j-1} G_T^{-1}\right)^{-1} G_T^{-1} \sum_j x_{j-1} u_j,$$

where

$$x_{j-1} = (y_{j-1}, \Delta y_{j-1}, \ldots, \Delta^{d-1} y_{j-1})'$$

$$= \left(\frac{u_{j-1}}{(1-L)^d}, \frac{u_{j-1}}{(1-L)^{d-1}}, \ldots, \frac{u_{j-1}}{1-L}\right)'$$

with $u_j = \varepsilon_j / \beta(L)$. It is evident that

$$\mathcal{L}\left(\frac{1}{T}\sum_j \frac{u_{j-1}}{1-L}u_j\right) \to \mathcal{L}\left(\alpha^2\sigma^2\left(\int_0^1 w(t)\,dw(t) + \frac{1-r}{2}\right)\right).$$

We show that, for $k = 2, \ldots, d$,

(S9) $\quad \mathcal{L}\left(\dfrac{1}{T^k}\sum_j \dfrac{u_{j-1}}{(1-L)^k}u_j\right) \to \mathcal{L}\left(\alpha^2\sigma^2 \int_0^1 F_{k-1}(t)\,dw(t)\right).$

Let us consider the case of $k = 2$. Using the BN decomposition we have

$$\sum_j \frac{u_{j-1}}{(1-L)^2}u_j = \sum_j \left(\sum_{l=1}^{j-1}\sum_{m=1}^{l} u_m\right)u_j$$

$$= \sum_j \sum_{l=1}^{j-1}\sum_{m=1}^{l}(\alpha\varepsilon_m + \tilde{\varepsilon}_{m-1} - \tilde{\varepsilon}_m)u_j$$

$$= \alpha^2\sum_j \frac{\varepsilon_{j-1}}{(1-L)^2}\varepsilon_j + \alpha\sum_j \frac{\varepsilon_{j-1}}{(1-L)^2}(\tilde{\varepsilon}_{j-1} - \tilde{\varepsilon}_j)$$

$$+ \tilde{\varepsilon}_0\sum_j(j-1)u_j - \sum_j \frac{\tilde{\varepsilon}_j}{1-L}u_j$$

$$= \alpha^2\sum_j \frac{\varepsilon_{j-1}}{(1-L)^2}\varepsilon_j + O_p(T\sqrt{T}),$$

where we have used the fact that

$$\sum_j \frac{\varepsilon_{j-1}}{(1-L)^2}(\tilde{\varepsilon}_{j-1} - \tilde{\varepsilon}_j) = \sum_j \frac{\varepsilon_{j-1}}{(1-L)^2}\tilde{\varepsilon}_{j-1} - \sum_j \left(\frac{\varepsilon_j}{(1-L)^2} - \frac{\varepsilon_j}{1-L}\right)\tilde{\varepsilon}_j$$

$$= \frac{\varepsilon_0}{(1-L)^2}\tilde{\varepsilon}_0 - \frac{\varepsilon_T}{(1-L)^2}\tilde{\varepsilon}_T + \sum_j \frac{\varepsilon_j}{1-L}\tilde{\varepsilon}_j$$

$$= O_p(T\sqrt{T}).$$

Thus (S9) is established for $k = 2$. The case of $k \geq 3$ can also be proved by induction. Then we obtain (7.150) by the FCLT and the continuous mapping theorem.

9.6 Since

$$T(\hat{\delta} - 1) = \frac{1}{T^{2d-1}} \sum_{j=2}^{T} y_{j-1}(y_j - y_{j-1}) \bigg/ \left[\frac{1}{T^{2d}} \sum_{j=2}^{T} y_{j-1}^2 \right]$$

$$= \frac{1}{2T^{2d-1}} \left[y_T^2 - \sum_{j=1}^{T} (y_j - y_{j-1})^2 \right] \bigg/ \left[\frac{1}{T^{2d}} \sum_{j=2}^{T} y_{j-1}^2 \right]$$

and $\sum_{j=2}^{T}(y_j - y_{j-1})^2 = O_p(T^{2(d-1)})$, (7.151) follows from the FCLT and the continuous mapping theorem.

9.7 It follows from Theorem 4.2 that

$$E\left(e^{\theta Y_2}\right) = E\left[\exp\left\{\theta x \int_0^1 F_1^2(t)\,dt - \frac{\theta}{2} F_1^2(1)\right\}\right]$$

$$= E\left[\exp\left\{\theta x \int_0^1 X^2(t)\,dt - \frac{\theta}{2} X^2(1) - \beta \int_0^1 \frac{dX(t)}{dt}\,d\left(\frac{dX(t)}{dt}\right)\right.\right.$$

$$\left.\left. + \frac{\beta^2}{2} \int_0^1 \left(\frac{dX(t)}{dt}\right)^2 dt \right\}\right],$$

where $dX(t)/dt = \beta X(t) + w(t)$ and

$$X(t) = e^{\beta t} \int_0^t e^{-\beta s} w(s)\,ds.$$

Noting that

$$\int_0^1 \frac{dX(t)}{dt}\,d\left(\frac{dX(t)}{dt}\right) = \frac{1}{2}\left\{(\beta X(1) + w(1))^2 - 1\right\},$$

$$\int_0^1 \left(\frac{dX(t)}{dt}\right)^2 dt = \int_0^1 (\beta X(t) + w(t))^2\,dt$$

$$= \beta^2 \int_0^1 X^2(t)\,dt + 2\beta \int_0^1 X(t)(dX(t) - \beta X(t)\,dt) + \int_0^1 w^2(t)\,dt$$

$$= -\beta^2 \int_0^1 X^2(t)\,dt + \beta X^2(1) + \int_0^1 w^2(t)\,dt,$$

we obtain the first equality in (7.152), where $\beta = (2\theta x)^{1/4}$. Applying Girsanov's theorem again we arrive at the second equality, where $\gamma = i\beta$. The last equality is obvious.

CHAPTER 8

1.1 Since we have $E(y_1^2) = \sigma^2$, $E(y_j^2) = (1 + \alpha^2)\sigma^2$, and $E(y_j y_{j-1}) = -\alpha\sigma^2$ for $j \geq 2$, α can be uniquely determined as $\alpha = -E(y_j y_{j-1})/E(y_1^2)$, which may take any value. Note that, in the stationary case, we have $E(y_j^2) = (1 + \alpha^2)\sigma^2$ for $j \geq 1$ and $E(y_k y_{k-1}) = -\alpha\sigma^2$ for $k \geq 2$ so that the parameter vectors (α, σ^2) and $(1/\alpha, \alpha^2\sigma^2)$ give the same model.

1.2 It is easy to see that $y \sim N(0, \sigma^2 \Phi(\alpha))$, where $\Phi(\alpha) = \Omega(\alpha) - \alpha^2 e_1 e_1' = C^{-1}(\alpha)(C^{-1}(\alpha))'$ with $C(\alpha)$ defined in (8.36). Since $|C(\alpha)| = 1$ so that $\log |\Phi(\alpha)| = 0$, we arrive at (8.4).

1.3 Let us put $\Omega(\alpha) = (1 + \alpha^2) I_T - 2\alpha B$. Then we have

$$D_T = |B - \lambda I_T| = -\lambda D_{T-1} - \frac{1}{4} D_{T-2},$$

with $D_1 = -\lambda$ and $D_2 = \lambda^2 - 1/4$, from which we obtain

$$D_T = \frac{\sin(T+1)\theta}{2^T \sin \theta}, \quad \cos\theta = -\lambda, \quad \sin\theta = \sqrt{1 - \lambda^2}, \qquad 0 < \theta < \pi.$$

Thus $D_T = 0$ yields $\theta = j\pi/(T+1)$ $(j = 1, \ldots, T)$ so that the eigenvalues of B are given by $\cos(j\pi/(T+1))$. Therefore the eigenvalues of $\Omega(\alpha)$ are given by $1 + \alpha^2 - 2\alpha \cos(j\pi/(T+1))$, $(j = 1, \ldots, T)$.

1.4 It is easy to obtain $D_T = |\Omega(\alpha)| = (1 + \alpha^2) D_{T-1} - \alpha^2 D_{T-2}$ with $D_1 = 1 + \alpha^2$ and $D_2 = 1 + \alpha^2 + \alpha^4$. Then, if $|\alpha| \neq 1$, we derive

$$D_T = (1 - \alpha^{2(T+1)})/(1 - \alpha^2).$$

When $|\alpha| = 1$, we have $D_T = T + 1$.

2.1 Putting $\delta_{jT} = \cos(j\pi/(T+1))$ and noting that $y \sim N(0, \sigma^2 \Omega(\alpha_0))$, we deduce

$$\mathcal{L}\left(\frac{1}{T} y' \Omega^{-1}(1) y\right) = \mathcal{L}\left(\frac{\sigma^2}{T} Z' \Omega^{1/2}(\alpha_0) \Omega^{-1}(1) \Omega^{1/2}(\alpha_0) Z\right)$$

$$= \mathcal{L}\left(\frac{\sigma^2}{T} \sum_{j=1}^{T} \frac{(1 - \alpha_0)^2 + 2\alpha_0(1 - \delta_{jT})}{2(1 - \delta_{jT})} Z_j^2\right)$$

$$= \mathcal{L}\left(\frac{\alpha_0 \sigma^2}{T} \sum_{j=1}^{T} Z_j^2 + o_p(1)\right),$$

which gives (8.12), where $\{Z_j\} \sim \text{NID}(0, 1)$. We also deduce

$$\frac{1}{T+1} \sum_{j=0}^{T} \alpha^{2j} = \frac{1}{T+1} \frac{1 - \alpha^{2(T+1)}}{1 - \alpha^2} = \frac{1 - \left(1 - \frac{\theta}{T}\right)^{2(T+1)}}{2\theta + O\left(\frac{1}{T}\right)}$$

$$\rightarrow \frac{1-e^{-2\theta}}{2\theta} = \frac{\sinh\theta}{\theta e^{\theta}}.$$

2.2 Let us consider

$$Y_T = \sum_{j=1}^{K_T} A_{jT} Z_j^2 + \sum_{j=K_T+1}^{T} A_{jT} Z_j^2,$$

where K_T is a sequence of integers such that $K_T \to \infty$, $K_T/T \to 0$, and $K_T^2/T \to \infty$ as $T \to \infty$, while

$$A_{jT} = \frac{c^2 + 4(T^2 - cT)s_{jT}^2}{4T^2 s_{jT}^2 \left(\theta^2 + 4(T^2 - \theta T)s_{jT}^2\right)}.$$

For $j = 1, \ldots, K_T$ it holds that

$$4(T+1)^2 s_{jT}^2 = j^2\pi^2 + j^4 O(T^{-2}) = j^2\pi^2 \left(1 + O\left(\left(\frac{K_T}{T}\right)^2\right)\right).$$

We also have

$$P\left(\sum_{j=K_T+1}^{T} A_{jT} Z_j^2 > \varepsilon\right) < \frac{1}{\varepsilon}\sum_{j=K_T+1}^{T} A_{jT}$$

$$< \frac{1}{\varepsilon}(T - K_T)A_{K_T T}$$

$$= \frac{1}{\varepsilon}O\left(\frac{T}{K_T^2}\right) \to 0.$$

Then we can deduce that

$$\mathcal{L}(Y_T) \to \mathcal{L}\left(\sum_{n=1}^{\infty} \frac{n^2\pi^2 + c^2}{n^2\pi^2 (n^2\pi^2 + \theta^2)} Z_n^2\right).$$

Since the second and third terms of S_{T1} in (8.14) converge in probability to θ and 0, respectively, we establish (8.15).

2.3 We have only to show that

$$\mathcal{L}\left(\frac{1}{\sigma^2} y'\left(\Omega^{-1}(\alpha) - \Omega^{-1}(1)\right) y\right) \to \mathcal{L}\left(\int_0^1 \int_0^1 \bar{K}_1(s,t;\theta) dw(s) dw(t) + \theta\right),$$

(S10)

where $\bar{K}_1 = -2K_1$. Noting that $y = -\alpha_0\varepsilon_0 e_1 + C^{-1}(\alpha_0)\varepsilon = D(\alpha_0)\varepsilon^*$, where $e_1 = (1, 0, \ldots, 0)' : T \times 1$, $D(\alpha_0) = \left(-\alpha_0 e_1, C^{-1}(\alpha_0)\right)$, $\varepsilon^* = (\varepsilon_0, \varepsilon')'$,

CHAPTER 8

and $C(\alpha)$ is defined in (8.36), consider

$$\frac{1}{\sigma^2} y' \left(\Omega^{-1}(\alpha) - \Omega^{-1}(1) \right) y = \frac{1}{\sigma^2} \varepsilon^{*\,\prime} D'(\alpha_0) \left(\Omega^{-1}(\alpha) - \Omega^{-1}(1) \right) D(\alpha_0) \varepsilon^*$$

$$= \frac{1}{T} Z' B_T Z + \frac{\theta}{T} Z'Z + R_T,$$

where $Z = \varepsilon/\sigma \sim N(0, I_T)$ and

$$B_T = T \left[(C^{-1}(\alpha_0))' \left(\Omega^{-1}(\alpha) - \Omega^{-1}(1) \right) C^{-1}(\alpha_0) \right] - \theta I_T,$$

$$R_T = \frac{1}{T\sigma^2} \alpha_0^2 \varepsilon_0^2 e_1' \left(\Omega^{-1}(\alpha) - \Omega^{-1}(1) \right) e_1$$

$$- \frac{2}{T\sigma^2} \alpha_0 \varepsilon_0 e_1' \left(\Omega^{-1}(\alpha) - \Omega^{-1}(1) \right) C^{-1}(\alpha_0) \varepsilon.$$

Using the fact that

$$\Omega^{-1}(\alpha) = C'(\alpha)C(\alpha) - \frac{C'(\alpha) d_\alpha d'_\alpha C(\alpha)}{1 + d'_\alpha d_\alpha},$$

where $d_\alpha = (\alpha, \alpha^2, \ldots, \alpha^T)'$, it can be checked that $R_T \to 0$ in probability. Then Theorem 5.12 establishes (S10) after some algebra.

2.4 Let us consider

$$E \left[\exp \left\{ iu \left(X_1(\theta) + \frac{1}{2} \log \frac{\sinh \theta}{\theta} \right) \right\} \right]$$

$$= \prod_{n=1}^{\infty} \left[1 - \frac{iu\theta^2(n^2\pi^2 + c^2)}{n^2\pi^2(n^2\pi^2 + \theta^2)} \right]^{-1/2} = \prod_{n=1}^{\infty} \left[\frac{1 + \frac{(1-iu)\theta^2}{n^2\pi^2} - \frac{ic^2\theta^2 u}{n^4\pi^4}}{1 + \frac{\theta^2}{n^2\pi^2}} \right]^{-1/2}$$

$$= \prod_{n=1}^{\infty} \left[\frac{\left(1 - \frac{a(u) + b(u)}{n^2\pi^2} \right) \left(1 - \frac{a(u) - b(u)}{n^2\pi^2} \right)}{1 + \frac{\theta^2}{n^2\pi^2}} \right]^{-1/2},$$

which leads us to (8.18).

2.5 We first have

$$\frac{dg_{T1}(\alpha)}{d\alpha} = \frac{T}{2} \frac{y' \Omega^{-1}(\alpha) \frac{d\Omega(\alpha)}{d\alpha} \Omega^{-1}(\alpha) y}{y' \Omega^{-1}(\alpha) y} - \frac{1}{2} \mathrm{tr} \left(\Omega^{-1}(\alpha) \frac{d\Omega(\alpha)}{d\alpha} \right).$$

Noting that $d\Omega(\alpha)/d\alpha \,|_{\alpha=1} = \Omega(1)$ and $d\Omega(\alpha)/d\alpha \,|_{\alpha=-1} = -\Omega(-1)$, we obtain the conclusion.

2.6 When $l = 1$, we have

$$\frac{dh_{T1}(\theta)}{d\theta} = \frac{1}{2}\frac{y'\dfrac{d\Omega^{-1}(\alpha)}{d\alpha}y}{y'\Omega^{-1}(\alpha)y} + \frac{1}{T}\frac{\displaystyle\sum_{i=1}^{T} i\alpha^{2i-1}}{\displaystyle\sum_{i=0}^{T}\alpha^{2i}}.$$

We first have

$$\frac{1}{T^2}\sum_{i=1}^{T} i\alpha^{2i-1} = \frac{1}{2T^2}\frac{d}{d\alpha}\sum_{i=1}^{T}\alpha^{2i}$$

$$= \frac{1}{2T^2}\frac{-2(T+1)(1-\alpha^2)\alpha^{2T+1} + 2\alpha(1-\alpha^{2(T+1)})}{(1-\alpha^2)^2}$$

$$\to \frac{1 - (2\theta+1)e^{-2\theta}}{4\theta^2},$$

which yields, using (8.13),

$$\frac{1}{T}\frac{\displaystyle\sum_{i=1}^{T} i\alpha^{2i-1}}{\displaystyle\sum_{i=0}^{T}\alpha^{2i}} \to \frac{1-(2\theta+1)e^{-2\theta}}{4\theta^2} \times \frac{\theta e^{\theta}}{\sinh\theta}$$

$$= \frac{1}{2}\left(1 + \frac{1}{\theta} - \coth\theta\right).$$

Putting $\delta_j = \cos(j\pi/(T+1))$ and $s_j = \sin(j\pi/2(T+1))$, we consider

$$\frac{1}{T\sigma^2}y'\frac{d\Omega^{-1}(\alpha)}{d\alpha}y = -\frac{1}{T}\sum_{j=1}^{T}\frac{(1+\alpha_0^2-2\alpha_0\delta_j)(2\alpha-2\delta_j)}{(1+\alpha^2-2\alpha\delta_j)^2}Z_j^2$$

$$= -\frac{1}{T}\sum_{j=1}^{T}\frac{\left(c^2+4(T^2-cT)s_j^2\right)\left(4T^2s_j^2-2\theta T\right)}{\left(\theta^2+4(T^2-\theta T)s_j^2\right)^2}Z_j^2$$

$$= \sum_{j=1}^{T}\frac{2\theta\left(c^2+4(T^2-cT)s_j^2\right)}{\left(\theta^2+4(T^2-\theta T)s_j^2\right)^2} - \frac{1}{T}\sum_{j=1}^{T}Z_j^2$$

$$-\frac{1}{T}\sum_{j=1}^{T}\left[\frac{4T^2s_j^2\left(c^2+4(T^2-cT)s_j^2\right)}{\left(\theta^2+4(T^2-\theta T)s_j^2\right)^2}-1\right]Z_j^2,$$

which is shown to converge in distribution to

$$\sum_{n=1}^{\infty} \frac{2\theta(n^2\pi^2 + c^2)}{(n^2\pi^2 + \theta^2)^2} Z_n^2 - 1.$$

Since $y'\Omega^{-1}(\alpha)y/T \to \sigma^2$ in probability, we have proved (8.21) for $l = 1$. The case of $l = 2$ can be proved similarly.

2.7 Since the c.f. of $\sum_{n=1}^{\infty} Z_n^2/(n^2\pi^2)$ is given by $((\sin\sqrt{2i\theta})/\sqrt{2i\theta})^{-1/2}$, we immediately obtain (8.26) due to Theorem 5.13.

2.8 The first and second derivatives of

$$\log \sinh \theta = \log \theta + \sum_{n=1}^{\infty} \log\left(\frac{n^2\pi^2 + \theta^2}{n^2\pi^2}\right)$$

with respect to θ are

$$\coth \theta = \frac{1}{\theta} + \sum_{n=1}^{\infty} \frac{2\theta}{n^2\pi^2 + \theta^2},$$

$$-\operatorname{cosech}^2 \theta = -\frac{1}{\theta^2} + \sum_{n=1}^{\infty} \frac{2}{n^2\pi^2 + \theta^2} - \sum_{n=1}^{\infty} \frac{4\theta^2}{(n^2\pi^2 + \theta^2)^2}.$$

Thus we obtain

$$\sum_{n=1}^{\infty} \frac{\theta^3}{(n^2\pi^2 + \theta^2)^2} = \frac{\theta}{4}\operatorname{cosech}^2\theta + \frac{1}{4}\coth\theta - \frac{1}{2\theta},$$

which yields (8.30).

2.9 It is easy to deduce

$$\phi(\theta; x) = \prod_{n=1}^{\infty} \left[1 - \frac{2i\theta(n^2\pi^2 + c^2)}{(n^2\pi^2 + x^2)^2}\right]^{-1/2}$$

$$= \prod_{n=1}^{\infty} \left[\frac{\left(1 - \frac{c(\theta) + d(\theta)}{n^2\pi^2}\right)\left(1 - \frac{c(\theta) - d(\theta)}{n^2\pi^2}\right)}{\left(1 + \frac{x^2}{n^2\pi^2}\right)^2}\right]^{-1/2}$$

$$= \frac{\sinh x}{x}\left[\frac{\sin\sqrt{c(\theta) + d(\theta)}}{\sqrt{c(\theta) + d(\theta)}} \frac{\sin\sqrt{c(\theta) - d(\theta)}}{\sqrt{c(\theta) - d(\theta)}}\right]^{-1/2}$$

3.1 Let G_{jk} be the (j,k)th element of $(C(1)C^{-1}(\alpha_0))'C(1)C^{-1}(\alpha_0)$. Then we have

$$G_{jj} = 1 + (1-\alpha_0)^2(T-j), \quad G_{jk} = 1 - \alpha_0 + (1-\alpha_0)^2(T-k), \quad (j<k).$$
(S11)

Thus it holds that

$$\frac{1}{T}\sum_{j,k=1}^{T} G_{jk}\varepsilon_j\varepsilon_k$$

$$= \frac{1}{T}\left[\sum_{j=1}^{T}\left\{1 + \frac{c^2}{T^2}(T-j)\right\}\varepsilon_j^2 + 2\sum_{j<k}\left\{\frac{c}{T} + \frac{c^2}{T^2}(T-k)\right\}\varepsilon_j\varepsilon_k\right],$$

which clearly converges in probability to σ^2.

3.2 Let H_{jk} be the (j,k)th element of $(C(\alpha)C^{-1}(\alpha_0))'C(\alpha)C^{-1}(\alpha_0)$. Then we have

(S12) $$H_{jj} = 1 + (\alpha - \alpha_0)^2 \frac{1 - \alpha^{2(T-j)}}{1 - \alpha^2},$$

(S13) $$H_{jk} = (\alpha - \alpha_0)\alpha^{k-j-1} + (\alpha - \alpha_0)^2 \frac{\alpha^{k-j} - \alpha^{2T-j-k}}{1 - \alpha^2}, \quad (j<k).$$

Thus, using (S11), we obtain

$$B_T(j,j) = T(H_{jj} - G_{jj}) = -\frac{c^2}{T}(T-j) + \frac{(c-\theta)^2}{T}\frac{1 - \alpha^{2(T-j)}}{1 - \alpha^2},$$

$$B_T(j,k) = T(H_{jk} - G_{jk})$$

$$= -c - \frac{c^2}{T}(T-k) + (c-\theta)\alpha^{k-j-1}$$

$$+ \frac{(c-\theta)^2}{T}\frac{\alpha^{k-j} - \alpha^{2T-j-k}}{1-\alpha^2}, \quad (j<k).$$

If we replace $B_T(j,j)$ by $B_T(j,j) - \theta$, then we can find the uniform limit of the modified $B_T(j,k)$ as $-2K_2(s,t;\theta)$ with K_2 defined in (8.40). It follows from Theorem 5.12 that

$$\mathcal{L}(S_{T2}) \to \mathcal{L}\left(-2\int_0^1\int_0^1 K_2(s,t;\theta)\,dw(s)\,dw(t) + \theta\right),$$

which yields (8.38).

3.3 We consider the integral equation

$$f(t) = \lambda \int_0^1 K_2(s,t;\theta)f(s)\,ds$$

$$= \frac{\lambda\theta}{2}\left[\cosh\theta(1-t)\int_0^t e^{-\theta(1-s)}f(s)\,ds + e^{-\theta(1-t)}\int_t^1 \cosh\theta(1-s)f(s)\,ds\right],$$

which is shown to be equivalent to

$$f''(t) - \theta^2\left(1 - \frac{\lambda}{2}\right)f(t) = 0, \qquad f'(0) = \theta f(0), \qquad f'(1) = 0.$$

Suppose first that $\lambda \neq 2$. Then the general solution is

$$f(t) = c_1 e^{At} + c_2 e^{-At}, \qquad A = \theta\sqrt{1 - \frac{\lambda}{2}}.$$

The two boundary conditions yield $M(\lambda)c = 0$, where $c = (c_1, c_2)'$ and

$$M(\lambda) = \begin{pmatrix} A - \theta & -A - \theta \\ Ae^A & -Ae^{-A} \end{pmatrix}.$$

Since $|M(\lambda)| = 2A(\theta\cosh A + A\sinh A)$, the candidate for the FD is given as in (8.42). When $\lambda = 2$, we have $f''(t) = 0$ with $f'(0) = \theta f(0)$ and $f'(1) = 0$. This implies that $\theta = 0$, which is a contradiction. Checking the second condition in Theorem 5.5, we can conclude that (8.42) is the FD of $K_2(s,t;\theta)$ in (8.41).

3.4 Let us consider

$$\frac{dh_{T2}(\theta)}{d\theta} = \frac{1}{2}\frac{d}{d\alpha}\log y'\Phi^{-1}(\alpha)y = \frac{1}{2}\frac{\varepsilon'\dfrac{dH(\alpha)}{d\alpha}\varepsilon}{y'\Phi^{-1}(\alpha)y},$$

where $H(\alpha) = (C(\alpha)C^{-1}(\alpha_0))'C(\alpha)C^{-1}(\alpha_0)$. It can be shown as in (8.37) that $y'\Phi^{-1}(\alpha)y/T \to \sigma^2$ in probability. Let the (j,k)th element of $dH(\alpha)/d\alpha$ be F_{jk}. Using (S12) and (S13), we obtain

$$F_{jj} = \left(\frac{2(\alpha - \alpha_0)}{1 - \alpha^2} + \frac{2\alpha(\alpha - \alpha_0)^2}{(1 - \alpha^2)^2}\right)\left(1 - \alpha^{2(T-j)}\right)$$

$$+ \frac{(\alpha - \alpha_0)^2}{1 - \alpha^2}(-2(T-j))\alpha^{2T-2j-1},$$

$$F_{jk} = \left(\alpha + (\alpha - \alpha_0)(k - j - 1)\right)\alpha^{k-j-2}$$

$$+ \left(\frac{2(\alpha - \alpha_0)}{1 - \alpha^2} + \frac{2\alpha(\alpha - \alpha_0)^2}{(1 - \alpha^2)^2}\right)\left(\alpha^{k-j} - \alpha^{2T-j-k}\right)$$

$$+ \frac{(\alpha - \alpha_0)^2}{1 - \alpha^2}\left((k-j)\alpha^{k-j-1} - (2T - j - k)\alpha^{2T-j-k-1}\right), \quad (j < k).$$

Then it holds that

$$\mathcal{L}\left(\frac{1}{T\sigma^2}\varepsilon'\frac{dH(\alpha)}{d\alpha}\varepsilon\right) = \mathcal{L}\left(\frac{1}{T\sigma^2}\varepsilon'\left(\frac{dH(\alpha)}{d\alpha}+I_T\right)\varepsilon - \frac{1}{T\sigma^2}\varepsilon'\varepsilon\right)$$

$$\to \mathcal{L}\left(\int_0^1\int_0^1 J(s,t;\theta)\,dw(s)\,dw(t) - 1\right),$$

where $J(s,t;\theta) = 2\partial K_2(s,t;\theta)/\partial\theta$. Thus (8.44) is established for $l = 1$.

3.5 Consider the integral equation

$$f(t) = \lambda\int_0^1\left\{\frac{1}{2}+c-\frac{c}{2}(s+t)+\frac{c^2}{2}(1-s)(1-t)\right\}f(s)\,ds,$$

which is equivalent to $f(t) = a + bt$ with

$$f(0) = \frac{\lambda}{2}\int_0^1(1+2c+c^2-cs-c^2s)f(s)\,ds = a,$$

$$f(1) = \frac{\lambda}{2}\int_0^1(1+c-cs)f(s)\,ds = a+b.$$

Then $\lambda(\neq 0)$ is an eigenvalue if and only if

$$|M(\lambda)| = \left|\begin{pmatrix}\lambda\left(\frac{1}{2}+\frac{3c}{4}+\frac{c^2}{4}\right)-1 & \lambda\left(\frac{1}{4}+\frac{c}{3}+\frac{c^2}{12}\right)\\ \lambda\left(\frac{1}{2}+\frac{c}{4}\right)-1 & \lambda\left(\frac{1}{4}+\frac{c}{12}\right)-1\end{pmatrix}\right|$$

$$= 1 - \frac{\lambda}{6}(c^2+3c+3),$$

which is the FD of $dX_2(\theta)/d\theta|_{\theta=0} + \frac{1}{2}$. Thus (8.48) is established.

3.6 Let us consider the integral equation

$$f(t) = \lambda\int_0^1\left.\frac{\partial K_2(s,t;\theta)}{\partial\theta}\right|_{\theta=-x}f(s)\,ds,$$

which is equivalent to $f''''(t) - (2x^2+\lambda x)f''(t) + x^4 f(t) = 0$ with $f'(1) = f'''(1) = 0$ and

$$f(1) = \frac{\lambda}{2}\int_0^1(1+x-xs)e^{x(1-s)}f(s)\,ds,$$

$$f'''(0) = -xf''(0) + (x^2+\lambda x)f'(0) + (x^3+\lambda x^2)f(0).$$

The general solution is given by
$$f(t) = c_1 e^{At} + c_2 e^{-At} + c_3 e^{Bt} + c_4 e^{-Bt},$$
where
$$A^2 = \frac{x}{2}\left(\lambda + 2x + \sqrt{\lambda^2 + 4\lambda x}\right), \qquad B^2 = \frac{x}{2}\left(\lambda + 2x - \sqrt{\lambda^2 + 4\lambda x}\right).$$

We then have $M(\lambda)c = 0$, where $c = (c_1, c_2, c_3, c_4)'$ and $M(\lambda)$ is a 4×4 matrix constructed from the four boundary conditions given above. Evaluating $|M(\lambda)|$ by REDUCE, we obtain (8.50).

4.1 Noting that $y \sim N(0, \sigma^2(\Omega(\alpha_{m0}) \otimes I_m))$ and $|\Omega(\alpha_m) \otimes I_m| = |\Omega(\alpha_m)|^m$, we easily obtain the log-likelihood for α_m and σ^2, from which (8.57) results by concentrating σ^2 out.

4.2 Noting that
$$E\left[\exp\left\{i\theta \sum_{n=1}^{\infty} \frac{n^2\pi^2 + c^2}{(n^2\pi^2 + x^2)^2} Z'_n Z_n\right\}\right] = \prod_{n=1}^{\infty}\left[1 - \frac{2i\theta(n^2\pi^2 + c^2)}{(n^2\pi^2 + x^2)^2}\right]^{-m/2},$$

(8.63) can be easily established because of (8.33).

4.3 Let us consider
$$\frac{1}{N} Z'\{(B - \theta I_N) \otimes I_m\} Z = \frac{1}{N} \sum_{j,k=1}^{N} A_{jk} Z'_j Z_k,$$

where $\{Z_j\} \sim \text{NID}(0, I_m)$. The weak convergence of this quantity for $m = 1$ has already been established in (8.38). Noting that $Z'Z/N \to m$ in probability, (8.65) follows from (5.154).

4.4 It follows from (5.155) that
$$E\left[\exp\left\{i\theta \int_0^1 \int_0^1 \left.\frac{\partial K_2(s,t;\theta)}{\partial \theta}\right|_{\theta=0} dw'(s)\, dw(t)\right\}\right] = (D(2i\theta))^{-m/2},$$

where $D(\lambda)$ is the FD of $\partial K_2(s,t;\theta)/\partial\theta|_{\theta=0}$. We have already obtained $D(\lambda) = 1 - \lambda(c^2 + 3c + 3)/6$, which establishes (8.68).

4.5 Suppose that $c = 0$. Then it holds that
$$\frac{dX_{m2}(\theta)}{d\theta} = \int_0^1 \int_0^1 \frac{\partial K_2(s,t;\theta)}{\partial \theta} dw'(s)\, dw(t),$$

where $\partial K_2(s,t;\theta)/\partial\theta$ is defined by (8.46) with $c = 0$. Then we obtain
$$E\left(\frac{dX_{m2}(\theta)}{d\theta}\right) = -\frac{m}{4}(1 - e^{-2\theta}) < 0,$$
$$V\left(\frac{dX_{m2}(\theta)}{d\theta}\right) = \frac{m}{16\theta} + O\left(\frac{1}{\theta^2}\right).$$

Thus the consistency proof for $m = 1$ can be used in the present case.

4.6 Note first that

$$E\left[\exp\left\{i\theta \int_0^1 \int_0^1 \frac{\partial K_2(s,t;\theta)}{\partial \theta}\bigg|_{\theta=-x} dw'(s)\,dw(t)\right\}\right] = (D(2i\theta))^{-m/2},$$

where $D(\lambda)$ is the FD of $\partial K_2(s,t;\theta)/\partial \theta|_{\theta=-x}$. Then (8.71) can be established because of (8.50).

4.7 It is easy to obtain

$$\frac{d^2 X_{M1}(\theta)}{d\theta^2}\bigg|_{\theta=0} = M^2 \left[\sum_{n=1}^{\infty}\left(\frac{1}{n^2\pi^2} + \frac{c^2 M^2}{n^4\pi^4}\right) Z_n^2 - \frac{1}{6}\right].$$

Thus (8.77) follows from (8.25) and (8.26). We can prove (8.78) similarly.

5.1 Let us put

$$\frac{1}{T} y'\Omega^{-1}(1)y = \frac{1}{T} u' A_T u + R_T,$$

where $A_T = (C^{-1}(\alpha_0))' \left[C'(1)C(1) - C'(1)ee'C(1)/(T+1)\right] C^{-1}(\alpha_0)$ and

$$R_T = \frac{1}{T}\alpha_0^2 u_0^2 e_1' \left[C'(1)C(1) - \frac{1}{T+1}C'(1)ee'C(1)\right] e_1$$

$$- \frac{2}{T}\alpha_0 u_0 e_1' \left[C'(1)C(1) - \frac{1}{T+1}C'(1)ee'C(1)\right] C^{-1}(\alpha_0) u$$

$$= \frac{\alpha_0^2 u_0^2}{T+1} - \frac{2\alpha_0 u_0}{T(T+1)} \sum_{j=1}^{T} ((T-j)(1-\alpha_0)+1) u_j$$

$$= o_p(1).$$

By the weak law of large numbers it holds that

$$\frac{1}{T(T+1)} u' (C^{-1}(\alpha_0))' C'(1)ee'C(1)C^{-1}(\alpha_0) u$$

$$= \frac{1}{T(T+1)} \left\{\sum_{j=1}^{T}\left(c\left(1-\frac{j}{T}\right)+1\right) u_j\right\}^2$$

$$\to 0 \quad \text{in probability}.$$

Thus, using (S11), we have

$$\frac{1}{T} y'\Omega^{-1}(1)y = \frac{1}{T}\sum_{j,k=1}^{T} G_{jk} u_j u_k + o_p(1)$$

$$= \frac{1}{T} \sum_{j=1}^{T} \left(1 + \frac{c^2}{T^2}(T-j)\right) u_j^2$$

$$+ 2 \sum_{j<k} \left(\frac{c}{T} + \frac{c^2}{T^2}(T-k)\right) u_j u_k \bigg] + o_p(1)$$

$$= \frac{1}{T} \sum_{j=1}^{T} u_j^2 + o_p(1),$$

which yields (8.81).

5.2 The first equality can be proved by noting that $y = -\alpha_0 u_0 e_1 + C^{-1}(\alpha_0)u$. Since $u'u/T \to \sigma^2 \sum_{l=0}^{\infty} \phi_l^2$ in probability, (8.82) follows from (8.17) and (5.161).

5.3 It is easy to establish (8.87) by using (S11) and the weak law of large numbers. We can also prove (8.88) using (8.38) and (5.161).

6.1 It follows from (8.57) that

$$\frac{d^2 l_{T1}(\alpha)}{d\alpha^2}\bigg|_{\alpha=1} = T \left(\frac{y' \left(\Omega^{-2}(1) \otimes I_m\right) y}{y' \left(\Omega^{-1}(1) \otimes I_m\right) y} - \frac{1}{2}\right) - \frac{m(N^2 - N)}{6}.$$

Then it holds that

$$P \left(\frac{d^2 l_{T1}(\alpha)}{d\alpha^2}\bigg|_{\alpha=1} < 0\right)$$

$$= P \left(\frac{N+2}{6} y' \left(\Omega^{-1}(1) \otimes I_m\right) y - y' \left(\Omega^{-2}(1) \otimes I_m\right) y > 0\right),$$

which leads us to (8.91).

6.2 Noting that $\Phi^{-1}(1) = C'C$ and $d\Phi(\alpha)/d\alpha |_{\alpha=1} = (C'C)^{-1} - e_1 e_1'$ with $C = C(1)$ and $e_1 = (1, 0, \ldots, 0)' : N \times 1$, we obtain, from (8.64),

$$P \left(\frac{dl_{T2}(\alpha)}{d\alpha}\bigg|_{\alpha=1} > 0\right) = P \left(\varepsilon'(A_N \otimes I_m)\varepsilon > 0\right),$$

where

$$A_N = \left(C^{-1}(\alpha_0)\right)' \left[C'C \left((C'C)^{-1} - e_1 e_1'\right) C'C\right] C^{-1}(\alpha_0)$$

$$= \left(CC^{-1}(\alpha_0)\right)' CC^{-1}(\alpha_0) - \left(CC^{-1}(\alpha_0)\right)' ee' CC^{-1}(\alpha_0),$$

with $e = (1, \ldots, 1)' : N \times 1$. Then (8.92) follows from (S11) and the fact that the jth component of $(CC^{-1}(\alpha_0))'e$ is given by $(N - j)(1 - \alpha_0) + 1$.

7.1 The first two equalities are obvious. Let us consider

$$\sum_{j=1}^{T}(1+\rho_0\lambda_{jT})\left(\frac{1}{1+\rho\lambda_{jT}}-1\right)Z_j^2$$

$$= -\theta^2 \sum_{j=1}^{T} \frac{c^2+4T^2 s_{jT}^2}{4T^2 s_{jT}^2 \left(\theta^2+4T^2 s_{jT}^2\right)} Z_j^2,$$

where $s_{jT} = \sin((j-\frac{1}{2})\pi/(2T+1))$. Using the same arguments as in the solution to Problem 2.2, we can establish (8.100).

7.2 We first obtain

$$\frac{dY_1(\theta)}{d\theta} = \sum_{n=1}^{\infty} \frac{\theta\left((n-\frac{1}{2})^2 \pi^2 + c^2\right)}{\left((n-\frac{1}{2})^2 \pi^2 + \theta^2\right)^2} Z_n^2 - \sum_{n=1}^{\infty} \frac{\theta}{(n-\frac{1}{2})^2 \pi^2 + \theta^2},$$

which yields

$$E\left(\frac{dY_1(\theta)}{d\theta}\right) = -\sum_{n=1}^{\infty} \frac{\theta(\theta^2 - c^2)}{\left((n-\frac{1}{2})^2 \pi^2 + \theta^2\right)^2} < 0 \quad \text{for} \quad \theta > c,$$

$$V\left(\frac{dY_1(\theta)}{d\theta}\right) = \sum_{n=1}^{\infty} \frac{2\theta^2\left((n-\frac{1}{2})^2 \pi^2 + c^2\right)^2}{\left((n-\frac{1}{2})^2 \pi^2 + \theta^2\right)^4}$$

$$\leq \sum_{n=1}^{\infty} \frac{2\theta^2}{\left((n-\frac{1}{2})^2 \pi^2 + \theta^2\right)^2}$$

$$= \frac{1}{2\theta}(\tanh\theta - \theta\,\text{sech}^2\theta) = O\left(\frac{1}{\theta}\right).$$

Then, using the same reasoning as before, we can ensure the existence of the local maximum $\hat{\kappa}$ such that $\hat{\kappa} = O_p(T^{-1})$.

7.3 We first obtain

$$E\left[\exp\left\{i\theta \sum_{n=1}^{\infty} \frac{Z_n^2}{(n-\frac{1}{2})^2 \pi^2}\right\}\right] = \left(\cos\sqrt{2i\theta}\right)^{-1/2}.$$

Then (8.104) follows from Theorem 5.13.

CHAPTER 8

7.4 It is easy to deduce that

$$\psi_1(\theta) = \prod_{n=1}^{\infty} \left[1 - \frac{2i\theta \left(\left(n - \frac{1}{2}\right)^2 \pi^2 + c^2 \right)}{\left(\left(n - \frac{1}{2}\right)^2 \pi^2 + x^2 \right)^2} \right]^{-1/2}$$

$$= \prod_{n=1}^{\infty} \left[\frac{\left(1 - \frac{a(\theta) + b(\theta)}{\left(n - \frac{1}{2}\right)^2 \pi^2}\right) \left(1 - \frac{a(\theta) - b(\theta)}{\left(n - \frac{1}{2}\right)^2 \pi^2}\right)}{\left(1 + \frac{x^2}{\left(n - \frac{1}{2}\right)^2 \pi^2}\right)^2} \right]^{-1/2}$$

$$= \cosh x \left[\cos \sqrt{a(\theta) + b(\theta)} \, \cos \sqrt{a(\theta) - b(\theta)} \right]^{-1/2}.$$

7.5 Noting that $\Delta y \sim N(0, \sigma_\varepsilon^2(\Omega + \rho I_{T-1}))$, we have

$$\mathcal{L}\left(\frac{1}{\sigma_\varepsilon^2} \Delta y' \left((\Omega + \rho I_{T-1})^{-1} - \Omega^{-1} \right) \Delta y \right)$$

$$= \mathcal{L}\left(\sum_{j=1}^{T-1} (\rho_0 + 4s_{jT}^2) \left(\frac{1}{\rho + 4s_{jT}^2} - \frac{1}{4s_{jT}^2} \right) Z_j^2 \right)$$

$$= \mathcal{L}\left(-\theta^2 \sum_{j=1}^{T-1} \frac{c^2 + 4T^2 s_{jT}^2}{4T^2 s_{jT}^2 \left(\theta^2 + 4T^2 s_{jT}^2 \right)} Z_j^2 \right),$$

where $s_{jT} = \sin(j\pi/(2T))$. Using the same arguments as in the solution to Problem 2.2, we can establish (8.109).

7.6 Let us consider $|\Omega + \rho I_{T-1}| = (\rho + 2)^{T-1} D_T$, where $D_T = D_{T-1} - a^2 D_{T-2}$ with $a = -1/(\rho + 2)$. When $\rho \neq 0$, we have $D_T = c_1 x_1^{T-1} + c_2 x_2^{T-1}$, where $D_1 = 1$, $D_2 = 1 - a^2$, and

$$x_1 = \frac{\rho + 2 + \sqrt{\rho^2 + 4\rho}}{2(\rho + 2)} = \frac{1}{\rho + 2}\left(1 + \frac{\theta}{T} + O\left(\frac{1}{T^2}\right) \right),$$

$$x_2 = \frac{\rho + 2 - \sqrt{\rho^2 + 4\rho}}{2(\rho + 2)} = \frac{1}{\rho + 2}\left(1 - \frac{\theta}{T} + O\left(\frac{1}{T^2}\right) \right),$$

$$c_1 = \frac{\rho^2 + 4\rho + 2 + (\rho + 2)\sqrt{\rho^2 + 4\rho}}{2(\rho + 2)\sqrt{\rho^2 + 4\rho}} = \frac{T}{2\theta}\left(1 + O\left(\frac{1}{T}\right) \right),$$

$$c_2 = \frac{-\rho^2 - 4\rho - 2 + (\rho + 2)\sqrt{\rho^2 + 4\rho}}{2(\rho + 2)\sqrt{\rho^2 + 4\rho}} = -\frac{T}{2\theta}\left(1 + O\left(\frac{1}{T}\right) \right).$$

Then (8.110) is established by noting that $|\Omega| = T$.

CHAPTER 9

2.1 The (j,k)th element $\Phi_{jk}(\rho)$ of $\Phi(\rho)$ is given for $j \leq k$ by

$$\Phi_{jk}(\rho) = \sum_{i=0}^{j-1} \rho^{k-j+2i}.$$

Therefore we obtain, for $j \leq k$,

(S14) $$\frac{d\Phi_{jk}(\rho)}{d\rho} = \sum_{i=0}^{j-1} (k-j+2i)\rho^{k-j+2i-1}$$

so that $d\Phi_{jk}(\rho)/d\rho\big|_{\rho=1} = jk - j = jk - \min(j,k)$. This yields (9.12).

2.2 Under H_0 it holds that $S_{T1}/T = X/(X+Y)$, where

$$X = \frac{1}{T\sigma^2} y_T^2 = \left(\frac{1}{\sqrt{T}\sigma} \sum_{j=1}^{T} \varepsilon_j\right)^2 \sim \chi^2(1),$$

$$Y = \frac{1}{\sigma^2} \sum_{j=1}^{T} \left(\varepsilon_j - \frac{1}{T}\sum_{i=1}^{T} \varepsilon_i\right)^2 \sim \chi^2(T-1).$$

The conclusion follows from X and Y being independent.

2.3 It is easy to obtain $Q_2'\Phi(1)Q_2 = I_{T-1}$ and $Q_2Q_2' = (C^{-1}(1))'C^{-1}(1) - e_1 e_1'$, where $e_1 = (1, 0, \ldots, 0)' : T \times 1$. Thus $y'Q_2Q_2'y = \sum_{j=2}^{T}(y_j - y_{j-1})^2$ and $y'Q_2Q_2'dd'Q_2Q_2'y = (y_T - y_1)^2$.

2.4 Under H_0 it holds that $S_{T2}/T = X/(X+Y)$, where

$$X = \frac{1}{(T-1)\sigma^2}(y_T - y_1)^2 = \left(\frac{1}{\sqrt{T-1}\sigma}\sum_{j=2}^{T}\varepsilon_j\right)^2 \sim \chi^2(1),$$

$$Y = \frac{1}{\sigma^2}\sum_{j=2}^{T}\left(\varepsilon_j - \frac{1}{T-1}\sum_{i=2}^{T}\varepsilon_i\right)^2 \sim \chi^2(T-2).$$

The conclusion follows from the independence of X and Y.

2.5 It follows from (S14) that, for $j \leq k$,

$$\frac{d^2\Phi_{jk}(\rho)}{d\rho^2}\bigg|_{\rho=1} = \sum_{i=0}^{j-1}(k-j+2i)(k-j+2i-1)$$

$$= jk(k-3) + \frac{j(j^2+5)}{3},$$

which yields (9.18).

2.6 It can be shown that

$$(Q_3'\Phi(1)Q_3)^{-1}Q_3' = (-I_{T-1}, 0) + \frac{1}{T}(0,\ldots,0,\tilde{d}),$$

where $\tilde{d} = (1,\ldots,T-1)'$. This yields the last expression for S_{T3}.

2.7 Let us put $Z = (Q_3'\Phi(1)Q_3)^{-1/2}Q_3'y/\sigma$ so that

$$S_{T3} = \frac{1}{T}\frac{Z'(Q_3'\Phi(1)Q_3)^{-1}Z}{Z'Z}.$$

Since $Z \sim N(0, I_{T-1})$ under H_0, $Z'Z/T$ converges in probability to unity. Noting that the (j,k)th element of $(Q_3'\Phi(1)Q_3)^{-1}$ is $\min(j,k) - jk/T$, we can establish (9.23) from Theorem 5.12.

2.8 It can be shown after some algebra that

$$(Q_4'\Phi(1)Q_4)^{-1}Q_4' = \begin{pmatrix} 0 & -1 & & & 0 \\ \vdots & & \ddots & & \vdots \\ 0 & & & -1 & 0 \end{pmatrix}$$

$$+ \frac{1}{T-1}\begin{pmatrix} T-2 \\ \vdots & & 0 \\ 1 \end{pmatrix} + \frac{1}{T-1}\begin{pmatrix} & & 1 \\ 0 & & \vdots \\ & & T-2 \end{pmatrix},$$

which yields the last expression for S_{T4}.

2.9 Let us put $Z = (Q_4'\Phi(1)Q_4)^{-1/2}Q_4'y/\sigma$ so that

$$S_{T4} = \frac{1}{T}\frac{Z'(Q_4'\Phi(1)Q_4)^{-1}Z}{Z'Z}.$$

Noting that the (j,k)th element of $(Q_4'\Phi(1)Q_4)^{-1}$ is $\min(j,k) - jk/(T-1)$, we can deduce that the limiting null distribution of S_{T4} is the same as that of S_{T3} given in (9.23).

3.1 It can be easily checked that Lemma 9.1 holds if Q is replaced by XG, where G is a $p \times p$ nonsingular matrix and $P'X = 0$. Then (9.34) follows from (9.36) by putting $P = H$, $A = \Sigma(\theta_0)$, and $Q = XG$. Noting that $H(H'\Sigma(\theta_0)H)^{-1}H' = \tilde{M}'\Sigma^{-1}(\theta_0)\tilde{M} = \tilde{M}'\Sigma^{-1}(\theta_0) = \Sigma^{-1}(\theta_0)\tilde{M}$, we can also prove (9.35).

3.2 The first equality comes from the fact that

$$d\Phi^{-1}(\rho)/d\rho = -\Phi^{-1}(\rho)d\Phi(\rho)/d\rho\Phi^{-1}(\rho).$$

Since $d\Phi^{-1}(\rho)/d\rho\big|_{\rho=1} = \Phi^{-1}(1) - e_T e_T'$, we have

$$LM_1' = -\frac{\tilde{\eta}'\left(\Phi^{-1}(1) - e_T e_T'\right)\tilde{\eta}}{\tilde{\eta}'\Phi^{-1}(1)\tilde{\eta}},$$

which yields the second equality.

3.3 We first have $\tilde{\eta}_T = e_T' \tilde{M} y$, where it holds that $X' \Phi^{-1}(1) \tilde{M} = 0$. Thus $\tilde{\eta}_T$ reduces to 0 if e_T belongs to the column space of $\Phi^{-1}(1)X$ or if $\Phi(1)e_T = d$ belongs to the column space of X.

3.4 For Model C we have

$$\tilde{\eta} = \left[I_T - d \left(d' \Phi^{-1}(1) d \right)^{-1} d' \Phi^{-1}(1) \right] y$$

$$= \left(I_T - \frac{1}{T} d e_T' \right) y = y - \frac{1}{T} y_T d,$$

which gives S_{T3} in (9.22). For Model D we have

$$\tilde{\eta} = \left[I_T - (e, d) \left(\begin{pmatrix} e' \\ d' \end{pmatrix} \Phi^{-1}(1)(e, d) \right)^{-1} \begin{pmatrix} e' \\ d' \end{pmatrix} \Phi^{-1}(1) \right] y$$

$$= \left[I_T - \frac{1}{T-1} (T e e_1' - d e_1' - e e_T' + d e_T') \right] y$$

$$= y - y_1 e - \frac{1}{T-1}(y_T - y_1)(d - e),$$

which gives S_{T4} in (9.26).

5.1 We have already proved (9.46) in Section 7.2 of Chapter 7. For (9.47) it can be shown that $\tilde{\eta}_1 = O_p(1)$ for all models. For $j \geq 2$ we have

$$\tilde{\eta}_j - \tilde{\eta}_{j-1} = \begin{cases} \eta_j - \eta_{j-1}, & \text{Models A, B,} \\ \eta_j - \eta_{j-1} - \frac{1}{T} \eta_T, & \text{Model C,} \\ \eta_j - \eta_{j-1} - \frac{1}{T-1}(\eta_T - \eta_1), & \text{Model D.} \end{cases}$$

Noting that $\eta_j - \eta_{j-1} = -c \eta_{j-1}/T + \varepsilon_j$ and $\eta_T = O_p(\sqrt{T})$, we can establish (9.47).

5.2 Let us first consider

$$\frac{1}{T\sigma^2} \tilde{\eta}_T^2 = \frac{1}{T\sigma^2} \sum_{j,k=1}^T \left(1 - \frac{c}{T} \right)^{2T-j-k} \varepsilon_j \varepsilon_k,$$

which converges in distribution to

$$\int_0^1 \int_0^1 e^{-c(2-s-t)} dw(s) \, dw(t) \sim \frac{1 - e^{-2c}}{2c} \chi^2(1).$$

Consider next

$$\frac{1}{T^2\sigma^2}\sum_{j=1}^{T}(\tilde{\eta}_j - \bar{\tilde{\eta}})^2 = \frac{1}{T^2\sigma^2}\sum_{j=1}^{T}(\eta_j - \bar{\eta})^2$$

$$= \frac{1}{T}\sum_{j=1}^{T}\left(Y_T\left(\frac{j}{T}\right) - \frac{1}{T}\sum_{k=1}^{T}Y_T\left(\frac{k}{T}\right)\right)^2,$$

where

$$Y_T(t) = \frac{1}{\sqrt{T}\sigma}\eta_{j-1} + T\left(t - \frac{j-1}{T}\right)\frac{\eta_j - \eta_{j-1}}{\sqrt{T}\sigma}, \quad \left(\frac{j-1}{T} \le t \le \frac{j}{T}\right).$$

(S15)

Since $\mathcal{L}(Y_T) \to \mathcal{L}(Y)$ by the FCLT, the weak convergence result on R_3 follows from the continuous mapping theorem.

5.3 Let us define $dZ(t) = -\gamma Z(t)\,dt + dw(t)$ with $Z(0) = 0$. Then Girsanov's theorem yields

$$E\left[\exp\left\{\theta\int_0^1 Y^2(t)\,dt\right\}\right] = E\left[\exp\left\{\frac{\gamma - c}{2}(Z^2(1) - 1)\right\}\right]$$

$$= e^{c/2}\left[\cosh\gamma + c\frac{\sinh\gamma}{\gamma}\right]^{-1/2},$$

where $\gamma = \sqrt{c^2 - 2\theta}$. This gives us the expression for $\beta_2(\alpha)$. The expression for $\beta_3(\alpha)$ can be proved similarly.

5.4 It is easy to see that $2c/(1 - e^{-2c}) \to \infty$ as $c \to \infty$ so that $\beta_1(\alpha) \to 1$. In Section 7.5 of Chapter 7 we have proved that, as $c \to \infty$,

$$c\int_0^1 Y^2(t)\,dt \to \frac{1}{2}, \quad c\int_0^1\left(Y(t) - \int_0^1 Y(s)\,ds\right)^2 dt \to \frac{1}{2},$$

$$\int_0^1 Y(t)\,dY(t)\bigg/\left(c\int_0^1 Y^2(t)\,dt\right) \to -1,$$

in probability. The above facts imply that $\beta_k(\alpha) \to 1$ as $c \to \infty$ for $k = 2, \ldots, 6$.

5.5 Noting that $\hat{\eta}_j = \eta_j - \bar{\eta}$, we have

$$\frac{1}{T^2\sigma^2}\sum_{j=1}^{T}\hat{\eta}_j^2 = \frac{1}{T}\sum_{j=1}^{T}\left(Y_T\left(\frac{j}{T}\right) - \frac{1}{T}\sum_{k=1}^{T}Y_T\left(\frac{k}{T}\right)\right)^2,$$

$$\frac{1}{T\sigma^2}\sum_{j=1}^{T}(\hat{\eta}_j - \hat{\eta}_{j-1})^2 = \left(\frac{1}{T}\sum_{j=1}^{T}Y_T\left(\frac{j}{T}\right)\right)^2 + \frac{1}{T\sigma^2}\sum_{j=2}^{T}(\eta_j - \eta_{j-1})^2 + o_p(1),$$

where $Y_T(t)$ is defined in (S15). Then the weak convergence result on R_6 follows from the FCLT and the continuous mapping theorem.

5.6 Defining $dZ(t) = -\gamma Z(t)\,dt + dw(t)$ with $Z(0) = 0$, we consider

$$E\left[\exp\left\{\theta x + \theta x \left(\int_0^1 Y(t)\,dt\right)^2 - \theta \int_0^1 \left(Y(t) - \int_0^1 Y(s)\,ds\right)^2 dt\right\}\right]$$

$$= e^{\theta x} E\left[\exp\left\{\frac{\gamma - c}{2}(Z^2(1) - 1) + \theta(x+1)\left(\int_0^1 Z(t)\,dt\right)^2\right\}\right],$$

where $\gamma = \sqrt{c^2 + 2\theta}$. This leads us to the expressions for $\psi_1(\theta; x)$ and $\beta_6(\alpha)$.

5.7 Noting that $\tilde{\eta}_j = \eta_j - j\eta_T/T$, we have

$$\frac{1}{T^2\sigma^2}\sum_{j=1}^T (\tilde{\eta}_j - \bar{\tilde{\eta}})^2 = \frac{1}{T^2\sigma^2}\sum_{j=1}^T \left\{\eta_j - \frac{j}{T}\eta_T - \left(\bar{\eta} - \frac{T+1}{2T}\eta_T\right)\right\}^2$$

$$= \frac{1}{T}\sum_{j=1}^T \left\{Y_T\left(\frac{j}{T}\right) - \frac{1}{T}\sum_{k=1}^T Y_T\left(\frac{k}{T}\right)\right.$$

$$\left. - \left(\frac{j}{T} - \frac{1}{2}\right)Y_T(1)\right\}^2 + o_p(1),$$

where $Y_T(t)$ is defined in (S15). Then the weak convergence result on R_3 is easily established.

5.8 Defining $dZ(t) = -\gamma Z(t)\,dt + dw(t)$ with $Z(0) = 0$, we consider

$$E\left[\exp\left\{\theta \int_0^1 \left(Y(t) - \int_0^1 Y(s)\,ds - \left(t - \frac{1}{2}\right)Y(1)\right)^2 dt\right\}\right]$$

$$= E\left[\exp\left\{\frac{\gamma - c}{2}(Z^2(1) - 1) - \theta \left(\int_0^1 Z(t)\,dt\right)^2\right.\right.$$

$$\left.\left. - 2\theta Z(1)\int_0^1 \left(t - \frac{1}{2}\right)Z(t)\,dt + \frac{\theta}{12}Z^2(1)\right\}\right],$$

where $\gamma = \sqrt{c^2 - 2\theta}$. This yields $\phi_3(-i\theta)$ after some algebra.

5.9 Noting that

$$\hat{\eta}_j = \eta_j + \left(\frac{6j}{T} - 4\right)\frac{1}{T}\sum_{j=1}^T \eta_j - \left(\frac{12j}{T} - 6\right)\frac{1}{T^2}\sum_{j=1}^T j\eta_j + o_p(1),$$

we have

$$\hat{\eta}_1 = -\frac{4}{T}\sum_{j=1}^{T}\eta_j + \frac{6}{T^2}\sum_{j=1}^{T}j\eta_j + O_p(1),$$

$$\hat{\eta}_j - \hat{\eta}_{j-1} = \eta_j - \eta_{j-1} + o_p(1), \qquad (j \geq 2).$$

The weak convergence result on R_6 follows from the above relations.

5.10 We consider

$$E\left[\exp\left\{\theta x + \theta x \left(4\int_0^1 Y(t)\,dt - 6\int_0^1 tY(t)\,dt\right)^2 - \theta V_4\right\}\right]$$

$$= e^{\theta x}E\left[\exp\left\{\frac{\gamma - c}{2}(Z^2(1) - 1) + 4\theta(4x + 1)\left(\int_0^1 Z(t)\,dt\right)^2\right.\right.$$

$$\left.\left. + 12\theta(3x + 1)\left(\int_0^1 tZ(t)\,dt\right)^2 - 12\theta(4x + 1)\int_0^1 Z(t)\,dt \int_0^1 tZ(t)\,dt\right\}\right],$$

where $dZ(t) = -\gamma Z(t)\,dt + dw(t)$ with $Z(0) = 0$ and $\gamma = \sqrt{c^2 + 2\theta}$. This yields $e^{\theta x}\psi_2(-i\theta; x)$ after some algebra; hence we obtain the expression for $\beta_6(\alpha)$.

6.1 Let $f(v|\rho)$ be the density of the maximal invariant $v = H'y/\sqrt{y'HH'y}$, where $f(v|\rho)$ is defined as in (9.10). Then the Neyman–Pearson lemma ensures that the test which rejects H_0 for large values of $f(v|1 - (\theta/T))/f(v|1)$ is MPI. By using Lemma 9.1, this is seen to be equivalent to rejecting H_0 when $V_T^{(M)}(\theta)$ in (9.51) takes large values.

6.2 The weak convergence result on $V_T^{(A)}(\theta)$ is proved in the text. Consider $V_T^{(B)}(\theta)$ in (9.51), where $\tilde{\eta}_j^{(0)} = \eta_j - \eta_1$ and

$$\tilde{\eta}_j^{(1)} = \eta_j - \frac{1}{1 + (T-1)(1-\rho)^2}$$

$$\times \left[(1 - \rho + \rho^2)\eta_1 + (1 - \rho)^2(\eta_2 + \cdots + \eta_{T-1}) + (1 - \rho)\eta_T\right]$$

with $\rho = 1 - (\theta/T)$. Then it is easy to deduce that $\mathcal{L}(V_T^{(B)}(\theta)) \to \mathcal{L}(V^{(B)}(c, \theta)) = \mathcal{L}(V^{(A)}(c, \theta))$. For Model C we have $\tilde{\eta}_j^{(0)} = \eta_j - j\eta_T/T$ so that the denominator of $V_T^{(C)}(\theta)$ divided by T converges in probability to σ^2. Since

$$\tilde{\eta}_j^{(1)} = y_j - j\tilde{\beta}^{(1)} = \eta_j - \frac{j}{\delta T + O(1)}\sum_{j=1}^{T}(1 + (1-\rho)j)(\eta_j - \rho\eta_{j-1})$$

$$= \eta_j - \frac{j}{\sqrt{T}}A_T,$$

the numerator of $V_T^{(C)}(\theta)$ is

$$\sum_{j=1}^{T}\left(\eta_j - \eta_{j-1} - \frac{1}{T}\eta_j\right)^2 - \sum_{j=1}^{T}\left(\tilde{\eta}_j^{(1)} - \tilde{\eta}_{j-1}^{(1)} + (1-\rho)\tilde{\eta}_{j-1}^{(1)}\right)^2$$

$$= -\frac{1}{T}\eta_T^2 + \frac{2}{\sqrt{T}}A_T\,\eta_T - A_T^2 - \theta\left(\frac{1}{\sqrt{T}}\eta_T - A_T\right)^2$$

$$-\frac{\theta^2}{T}\sum_{j=1}^{T}\left(\frac{1}{\sqrt{T}}\eta_j - \frac{j}{T}A_T\right)^2 + \theta + o_p(1).$$

The joint weak convergence and the fact that

$$\mathcal{L}(A_T) \to \mathcal{L}\left(\frac{\sigma}{\delta}\left((\theta+1)Y(1) + \theta^2\int_0^1 tY(t)\,dt\right)\right)$$

lead us to deduce that $\mathcal{L}(V_T^{(C)}(\theta)) \to \mathcal{L}(V^{(C)}(c,\theta))$. For Model D it can be checked that

$$\tilde{\eta}_j^{(0)} - \tilde{\eta}_{j-1}^{(0)} = \eta_j - \eta_{j-1} - \frac{1}{T-1}(\eta_T - \eta_1),$$

$$\tilde{\eta}_j^{(1)} - \tilde{\eta}_{j-1}^{(1)} = \eta_j - \eta_{j-1} - \frac{1}{\sqrt{T}}A_T + O_p\left(\frac{1}{T}\right).$$

Thus it holds that $\mathcal{L}(V_T^{(D)}(\theta)) \to \mathcal{L}(V^{(D)}(c,\theta)) = \mathcal{L}\left(V^{(C)}(c,\theta)\right)$.

6.3 For Models A and B we easily obtain, by Girsanov's theorem,

$$E\left[\exp\left\{u\left(c - V^{(A)}(c,c)\right)/c^2\right\}\right] = E\left[\exp\left\{u\int_0^1 Y^2(t)\,dt + \frac{u}{c}Y^2(1)\right\}\right]$$

$$= \left[\left(\cos\mu - \frac{\mu}{c}\sin\mu\right)/e^c\right]^{-1/2},$$

$$E\left[\exp\left\{u\left(c - V^{(A)}(0,c)\right)/c^2\right\}\right] = E\left[\exp\left\{u\int_0^1 w^2(t)\,dt + \frac{u}{c}w^2(1)\right\}\right]$$

$$= \left[\cos\nu - \frac{\nu}{c}\sin\nu\right]^{-1/2}.$$

We can compute $E\left[\exp\{u(c - V^{(C)}(c,c))/c^2\}\right]$ and $E[\exp\{u(c - V^{(C)}(0,c))/c^2\}]$ similarly, which establishes the theorem.

6.4 For Models C and D we obtain, by Girsanov's theorem,

$$E\left[\exp\left\{u\left(\theta - V^{(C)}(c,\theta)\right)/\theta^2\right\}\right]$$

$$= E\left[\exp\left\{u\left(\int_0^1 Y^2(t)\,dt + \frac{\theta+1}{3\delta}Y^2(1)\right.\right.\right.$$

$$\left.\left.\left.-\frac{2(\theta+1)}{\delta}Y(1)\int_0^1 tY(t)\,dt - \frac{\theta^2}{\delta}\left(\int_0^1 tY(t)\,dt\right)^2\right)\right\}\right]$$

$$= \exp\left(\frac{c-\beta}{2}\right)E\left[\exp\left\{\left(\frac{\beta-c}{2} + \frac{u(\theta+1)}{3\delta}\right)Z^2(1) - \frac{u\theta^2}{\delta}\left(\int_0^1 tZ(t)\,dt\right)^2\right.\right.$$

$$\left.\left.-\frac{2u(\theta+1)}{\delta}Z(1)\int_0^1 tZ(t)\,dt\right\}\right],$$

where $dZ(t) = -\beta Z(t)\,dt + dw(t)$ with $Z(0) = 0$ and $\beta = \sqrt{c^2 - 2u}$. We can arrive, after some algebra, at $\phi^{(C)}(-iu; c, \theta)$. We can compute $E[\exp\{u(\theta - V^{(A)}(c,\theta))/\theta^2\}]$ similarly, which establishes the theorem.

7.1 Let us consider $T(\hat{\rho}(\delta) - 1) = U_T/V_T$, where

$$U_T = \frac{1}{T\sigma^2}\left[\sum_{j=2}^T y_{j-1}(y_j - y_{j-1}) - \delta y_T^2\right]$$

$$= \frac{1}{2}\left(Y_T^2(1) - 1\right) - \delta Y_T^2(1) + o_p(1),$$

$$V_T = \frac{1}{T^2\sigma^2}\left[\sum_{j=2}^T y_{j-1}^2 + \delta y_T^2\right]$$

$$= \frac{1}{T}\sum_{j=1}^T Y_T^2\left(\frac{j}{T}\right) + o_p(1),$$

with $Y_T(t)$ defined in (S15). Then we can establish the first equality in (9.59). The second equality can also be proved by using Girsanov's theorem.

7.2 Let us put $x_j = (y_{j-1}, y_{j-2})'$ and $G = (G_1, G_2)$, where $G_1 = (1, -\rho)'$ and $G_2 = (1, 0)'$. Then $G'x_j = (\varepsilon_{j-1}, y_{j-1})'$ and

$$\begin{pmatrix}\hat{\rho}_1 - \rho \\ \hat{\rho}_2\end{pmatrix} = \left(\Sigma x_j x_j'\right)^{-1}\Sigma x_j \varepsilon_j$$

$$= G\begin{pmatrix}\Sigma \varepsilon_{j-1}^2 & \Sigma \varepsilon_{j-1} y_{j-1} \\ \Sigma \varepsilon_{j-1} y_{j-1} & \Sigma y_{j-1}^2\end{pmatrix}^{-1}\begin{pmatrix}\Sigma \varepsilon_{j-1} \varepsilon_j \\ \Sigma y_{j-1} \varepsilon_j\end{pmatrix}.$$

Thus we obtain $\sqrt{T}(\hat{\rho}_1 - \rho) = A/B$, where

$$A = \left(\frac{1}{T}\sum y_{j-1}^2 - \frac{1}{T}\sum \varepsilon_{j-1}y_{j-1}\right)\frac{1}{\sqrt{T}}\sum \varepsilon_{j-1}\varepsilon_j$$

$$+ \left(\frac{1}{T}\sum \varepsilon_{j-1}^2 - \frac{1}{T}\sum \varepsilon_{j-1}y_{j-1}\right)\frac{1}{\sqrt{T}}\sum y_{j-1}\varepsilon_j$$

$$= \frac{\rho^2\sigma^2}{1-\rho^2}\frac{1}{\sqrt{T}}\sum \varepsilon_{j-1}\varepsilon_j + o_p(1),$$

$$B = \frac{1}{T}\sum \varepsilon_{j-1}^2 \frac{1}{T}\sum y_{j-1}^2 - \left(\frac{1}{T}\sum \varepsilon_{j-1}y_{j-1}\right)^2$$

$$= \frac{\sigma^4\rho^2}{1-\rho^2} + o_p(1).$$

Then we can deduce that $\sqrt{T}(\hat{\rho}_1 - \rho) \to N(0,1)$ so that

$$P\left(\sqrt{T}(\hat{\rho}_1 - 1) \le x\right) = P\left(\sqrt{T}(\hat{\rho}_1 - \rho) \le x + \sqrt{T}(1-\rho)\right)$$
$$\cong \Phi\left(x + \sqrt{T}(1-\rho)\right).$$

8.1 The LBI test rejects H_0 when

$$S_T = -\frac{\tilde{\eta}'\left(\left.\frac{d\Phi^{-1}(\rho_m)}{d\rho_m}\right|_{\rho_m=1} \otimes I_m\right)\tilde{\eta}}{\tilde{\eta}'\left(\Phi^{-1}(1) \otimes I_m\right)\tilde{\eta}} < c.$$

Since $d\Phi^{-1}(\rho)/d\rho|_{\rho=1} = \Phi^{-1}(1) - e_N e_N'$ with $e_N = (0,\ldots,0,1)' : N \times 1$, the above test is seen to be equivalent to the one given in Theorem 9.17.

8.2 Let us put

$$\sum_{j=1}^{m} \tilde{\eta}_{T-m+j}^2 = a_N' a_N,$$

where $a_N = (e_N' \otimes I_m)(\tilde{M} \otimes I_m) y = (e_N' \tilde{M} \otimes I_m) y$ with $e_N = (0,\cdots,0,1)' : N \times 1$. Since $\bar{X}'\Phi^{-1}(1)\tilde{M} = 0$, a_N reduces to 0 if e_N belongs to the column space of $\Phi^{-1}(1)\bar{X}$ or if $\Phi(1)e_N = d$ belongs to the column space of $\bar{X}$.

8.3 The LBIU test rejects H_0 when

$$S_T = -\frac{\tilde{\eta}'\left(\left.\frac{d^2\Phi^{-1}(\rho_m)}{d\rho_m^2}\right|_{\rho_m=1} \otimes I_m\right)\tilde{\eta}}{\tilde{\eta}'\left(\Phi^{-1}(1) \otimes I_m\right)\tilde{\eta}} > c.$$

Since $d^2\Phi^{-1}(\rho)/d\rho^2\big|_{\rho=1} = 2(I_N - e_N e'_N)$ and $\tilde{\eta}'(e_N e'_N \otimes I_m)\tilde{\eta} = 0$, the above test is seen to be equivalent to the one given in Theorem 9.18.

8.4 Since it can be shown that

$$\tilde{\eta} = \left[I_T - (d \otimes I_m)\left(d'\Phi^{-1}(1)d \otimes I_m\right)^{-1}\left(d'\Phi^{-1}(1) \otimes I_m\right)\right]y$$

$$= \left(I_T - \frac{1}{N}(d \otimes I_m)(e'_N \otimes I_m)\right)y$$

$$= y - \frac{1}{N}(d \otimes I_m)y_N ,$$

we can obtain the rejection region (9.71) from Theorem 9.18.

8.5 Putting $\bar{X} = (e, d)$ we can show that

$$\tilde{\eta} = \left[I_T - (\bar{X} \otimes I_m)(\bar{X}'\Phi^{-1}(1)\bar{X} \otimes I_m)^{-1}(\bar{X}'\Phi^{-1}(1) \otimes I_m)\right]y$$

$$= \left[I_T - ee'_1 \otimes I_m - \frac{1}{N-1}((d-e)(e'_N - e'_1)) \otimes I_m\right]y$$

$$= y - (e \otimes I_m)y_1 - \frac{1}{N-1}((d-e) \otimes I_m)(y_N - y_1) ,$$

which yields the rejection region (9.72).

8.6 We first note that R_{C2} may be rewritten as

$$R_{C2} = \frac{m}{N} \frac{\tilde{\eta}'\tilde{\eta}}{\tilde{\eta}'\left(\Phi^{-1}(1) \otimes I_m\right)\tilde{\eta}}$$

$$= \frac{m}{N} \frac{\varepsilon'(B_N \otimes I_m)\varepsilon}{\varepsilon'(A_N \otimes I_m)\varepsilon} ,$$

where $A_N = C'(\rho_m)\tilde{M}'\Phi^{-1}(1)\tilde{M}C(\rho_m)$ and $B_N = C'(\rho_m)\tilde{M}'\tilde{M}C(\rho_m)$ with $\tilde{M} = I_N - d(d'\Phi^{-1}(1)d)^{-1}d'\Phi^{-1}(1)$. We have already shown that, when $m = 1$,

$$\frac{1}{N\sigma^2}\varepsilon'A_N\varepsilon \to 1 \quad \text{in probability}$$

so that, for general m,

$$\frac{1}{N\sigma^2}\varepsilon'(A_N \otimes I_m)\varepsilon \to m \quad \text{in probability.}$$

We also have

$$\mathcal{L}\left(\frac{1}{N^2\sigma^2}\varepsilon'(B_N \otimes I_m)\varepsilon\right) \to \mathcal{L}\left(\int_0^1 \int_0^1 K(s,t)\, d\mathbf{w}'(s)\, d\mathbf{w}(t)\right) ,$$

where $K(s,t)$ is a positive definite kernel and $\{\mathbf{w}(t)\}$ is the m-dimensional standard Brownian motion. Since the c.f. of this last limiting distribution is given by $(\phi_2(\theta))^m$, we obtain the conclusion.

9.1 The statistic R_2 may be rewritten as

$$R_2 = \frac{1}{T^2} u' B_T u \Big/ \frac{1}{T} u' A_T u,$$

where $A_T = C'(\rho)\tilde{M}'\Phi^{-1}(1)\tilde{M}C(\rho)$ and $B_T = C'(\rho)\tilde{M}'\tilde{M}C(\rho)$. We have shown that

$$\frac{1}{T\sigma^2} \varepsilon' A_T \varepsilon \to 1 \quad \text{in probability}, \quad \mathcal{L}\left(\frac{1}{T^2\sigma^2} \varepsilon' B_T \varepsilon\right) \to \mathcal{L}(W_2).$$

Then $u' A_T u/T \to \sigma_S^2$ in probability, and it follows from (5.161) that

$$\mathcal{L}\left(\frac{1}{T^2} u' B_T u\right) \to \mathcal{L}(\sigma_L^2 W_2)$$

so that $\mathcal{L}(R_2) \to \mathcal{L}(W_2/r)$. The weak convergence result on R_6 can be proved similarly.

9.2 The weak convergence results for Models A and C are obvious. As for Models B and D, let us consider

$$\frac{1}{R_6} + \frac{\tilde{\sigma}_L^2 - \tilde{\sigma}_S^2}{\sum_{j=1}^{T} \hat{\eta}_j^2 \Big/ T^2} = \left(\frac{1}{T}\sum_{j=1}^{T}(\hat{\eta}_j - \hat{\eta}_{j-1})^2 + \tilde{\sigma}_L^2 - \tilde{\sigma}_S^2\right) \Big/ \frac{1}{T^2}\sum_{j=1}^{T} \hat{\eta}_j^2.$$

This converges in distribution to

$$(\sigma_L^2 X_6 + \sigma_S^2 + \sigma_L^2 - \sigma_S^2)/(\sigma_L^2 W_6) = (X_6 + 1)/W_6,$$

which yields the conclusion.

9.3 It is easy to deduce from (9.51) that $V_T^{(A)}(\theta) = U_T/V_T$, where

$$V_T = \frac{1}{T}\sum_{j=1}^{T}(y_j - y_{j-1})^2 \to \sigma_S^2 \quad \text{in probability},$$

$$\mathcal{L}(U_T) = \mathcal{L}\left(-\frac{\theta^2}{T^2}\sum_{j=2}^{T} y_{j-1}^2 + \frac{\theta}{T}\sum_{j=2}^{T}(y_j - y_{j-1})^2 - \frac{\theta}{T}y_T^2\right)$$

$$\to \mathcal{L}\left(-\theta^2 \sigma_L^2 \int_0^1 Y^2(t)\,dt - \theta\sigma_L^2 Y^2(1) + \theta\sigma_S^2\right).$$

Thus we can establish (9.81) for Model A.

10.1 The LBI test rejects H_0 when

$$S_T = -\frac{(y - X\hat{\beta})' \frac{d\Omega^{-1}(\rho)}{d\rho}\big|_{\rho=0} (y - X\hat{\beta})}{\hat{\sigma}^2} > c,$$

where $\hat{\beta} = (X'X)^{-1}X'y$, $\hat{\sigma}^2 = (y - X\hat{\beta})'(y - X\hat{\beta})/T$, and $\Omega(\rho) = I_T + \rho CC'$. It is easily seen that the above test is equivalent to the test based on U_T.

10.2 Since $y'My = (\varepsilon + C\xi)'M(\varepsilon + C\xi)$ and $\varepsilon + C\xi \sim N(0, \sigma_\varepsilon^2(I_T + \rho CC'))$, we have

$$\mathcal{L}\left(\frac{1}{T} y'My\right) = \mathcal{L}\left(\frac{\sigma_\varepsilon^2}{T} Z'(I_T + \rho CC')^{1/2} M(I_T + \rho CC')^{1/2} Z\right)$$

$$= \mathcal{L}\left(\frac{\sigma_\varepsilon^2}{T} Z'MZ + \frac{c^2 \sigma_\varepsilon^2}{T^3} Z'MCC'MZ\right),$$

where $Z \sim N(0, I_T)$. It holds that

$$\frac{1}{T} Z'MZ \to 1 \quad \text{in probability}, \qquad \frac{1}{T^3} Z'MCC'MZ = O_p\left(\frac{1}{T}\right).$$

Thus (9.88) is established.

10.3 Let us consider

$$B_T = \frac{1}{T} C'MC = \frac{1}{T}\left[C'C - C'X(X'X)^{-1}X'C\right],$$

where $X = (e, d)$. It is seen that $K(s, t)$ in (9.91) satisfies

$$\lim_{T \to \infty} \max_{j,k} |B_T(j, k) - K(j/T, k/T)| = 0.$$

Moreover the symmetric and continuous kernel $K(s, t)$ is shown to be positive definite. Thus (9.90) follows from Theorem 5.13.

10.4 Because of Theorem 5.13, we have only to prove that the FD of $K(s, t)$ is given by (9.94). The integral equation (5.10) is shown to be equivalent to

$$f(t) = c_1 \cos \sqrt{\lambda} t + c_2 \sin \sqrt{\lambda} t + 6a,$$

$$f(0) = f(1) = 0, \qquad a = \int_0^1 (s - s^2) f(s) \, ds.$$

Then the approach taken in Section 5.4 of Chapter 5 leads us to obtain the FD of $K(s, t)$ as in (9.94).

10.5 The LBI test, if it exists, rejects H_0 when

$$R_T = -\frac{y'M(C')^{-1} \left.\dfrac{d\Sigma^{-1}(\alpha)}{d\alpha}\right|_{\alpha=1} C^{-1}My}{y'My} < c,$$

where $\Sigma^{-1}(\alpha) = C'(\alpha)C(\alpha)$. Since

$$\left.\frac{d\Sigma^{-1}(\alpha)}{d\alpha}\right|_{\alpha=1} = -\Sigma^{-1}(1)\left.\frac{d\Sigma(\alpha)}{d\alpha}\right|_{\alpha=1}\Sigma^{-1}(1)$$

$$= -C'(I_T - ee')C,$$

it is seen that $R_T = y'M(I_T - ee')My/y'My = 1$. Thus we consider the LBIU test that rejects H_0 when

$$S_T = -\frac{y'M(C')^{-1} \left.\dfrac{d^2\Sigma^{-1}(\alpha)}{d\alpha^2}\right|_{\alpha=1} C^{-1}My}{y'My} > c.$$

Since it can be shown that

$$\left.\frac{d^2\Sigma^{-1}(\alpha)}{d\alpha^2}\right|_{\alpha=1} = 2C'(I_T - ee')^2C - 2C'(CC' - ee')C,$$

it is seen that the above test is equivalent to the test based on V_T.

10.6 Note first that $\mathcal{L}(y'My) = \mathcal{L}(\varepsilon'MC(C'(\alpha)C(\alpha))^{-1}C'M\varepsilon)$, where

$$\text{(S16)} \quad (C'(\alpha)C(\alpha))^{-1} = \alpha\,(C'C)^{-1} + (1-\alpha)^2 I_T + \alpha(1-\alpha)e_1e_1'$$

$$= \left(1 - \frac{c}{T}\right)(C'C)^{-1} + \frac{c^2}{T^2}I_T + \frac{c}{T}\left(1 - \frac{c}{T}\right)e_1e_1'.$$

Then we obtain

$$\mathcal{L}\left(\frac{1}{T}y'My\right) = \mathcal{L}\left(\frac{1}{T}\varepsilon'M\varepsilon + o_p(1)\right),$$

which establishes (9.99).

10.7 Using (S16), we can deduce that

$$\mathcal{L}\left(\frac{1}{T^2\sigma_\varepsilon^2}y'MCC'My\right) = \mathcal{L}\left(\frac{1}{T^2}Z'C'M\left(I_T + \frac{c^2}{T^2}CC'\right)MCZ + o_p(1)\right).$$

Thus we can establish the weak convergence result (9.100) from (9.89) and (9.90).

CHAPTER 10

2.1 The LM principle yields the LBI test which rejects H_0 when

$$-\frac{y'\bar{M}'\frac{d}{d\alpha}C'(\alpha)C(\alpha)\big|_{\alpha=1}\bar{M}y}{y'\bar{M}'C'C\bar{M}y} < c,$$

where $\bar{M} = I_T - X(X'C'CX)^{-1}X'C'C$. Since $dC'(\alpha)C(\alpha)/d\alpha|_{\alpha=1} = C'ee'C - C'C$ and $C\bar{M} = \tilde{M}C$ with $\tilde{M}^2 = \tilde{M}$, the above test implies (10.6).

2.2 Noting that $\tilde{M}Cy = \tilde{M}CC^{-1}(\alpha_0)\varepsilon$ and $CC^{-1}(\alpha_0) = I_T + (1-\alpha_0)(C - I_T)$, we have

$$\frac{1}{T}y'C'\tilde{M}Cy = \frac{1}{T}\varepsilon'\left[I_T + \frac{c}{T}(C'-I_T)\right]\tilde{M}\left[I_T + \frac{c}{T}(C-I_T)\right]\varepsilon$$

$$= \frac{1}{T}\varepsilon'\left[\tilde{M} + \frac{c}{T}\{\tilde{M}(C-I_T) + (C'-I_T)\tilde{M}\}\right.$$

$$\left. + \frac{c^2}{T^2}(C'-I_T)\tilde{M}(C-I_T)\right]\varepsilon.$$

Here it holds that $\text{plim}\,(\varepsilon'\tilde{M}\varepsilon/T) = \sigma^2$, while the other terms converge in probability to 0. Thus we establish (10.7).

2.3 Let us consider

$$\mathcal{L}\left(\frac{1}{T\sigma^2}y'C'\tilde{M}ee'\tilde{M}Cy\right) = \mathcal{L}\left(\frac{1}{T\sigma^2}(e'\tilde{M}CC^{-1}(\alpha_0)\varepsilon)^2\right)$$

$$= \mathcal{L}\left(\left(\frac{1}{\sqrt{T}}\sum_{j=1}^T a_j Z_j\right)^2\right) \to A\chi^2(1),$$

where $\{Z_j\} \sim \text{NID}(0,1)$ and

$$A = \lim_{T\to\infty}\frac{1}{T}\sum_{j=1}^T a_j^2 = \lim_{T\to\infty}\frac{1}{T}e'\tilde{M}C\left(C'(\alpha_0)C(\alpha_0)\right)^{-1}C'\tilde{M}e$$

$$= \lim_{T\to\infty}\frac{1}{T}e'\tilde{M}C\left[(C'C)^{-1} + \frac{c}{T}e_1 e_1' + \frac{c^2}{T^2}I_T\right]C'\tilde{M}e.$$

The computation of the value A for each model establishes Theorem 10.1.

2.4 It follows from (8.42) that the limiting c.f. of $X_T = V_{T1}(\theta) + \theta$ for $c = 0$ is given by

$$\phi(u) = \left[\left(\cos\theta\sqrt{2iu-1} - \sqrt{2iu-1}\sin\theta\sqrt{2iu-1}\right)\big/e^\theta\right]^{-1/2}.$$

Thus $P(V_{T1}(\theta) \leq x) = P(X_T/\theta^2 \leq (\theta+x)/\theta^2)$ yields (10.11).

2.5 When $\theta = c$, the kernel $K(s,t;\theta)$ in (10.10) takes the form

$$K(s,t) = c + c^2 - c^2 \max(s,t),$$

whose FD is found to be

$$D(\lambda) = \cos c\sqrt{\lambda} - \sqrt{\lambda}\sin c\sqrt{\lambda}.$$

Thus the limiting c.f. of $Y_T = V_{T1}(c) + c$ is given by $(D(2iu))^{-1/2}$. Since $P(V_{T1}(c) \geq x) = P(Y_T/c^2 \geq (c+x)/c^2)$, (10.12) is established.

2.6 Let $P_{lm}(j|k)$ be the (l,m)th element of $P(j|k)$. Then the Kalman filter algorithm yields

$$P(j|j-1) = \begin{pmatrix} P_{22}(j-1|j-1) + \sigma^2 & \sigma^2 \\ \sigma^2 & \sigma^2 \end{pmatrix},$$

$$P(j|j) = \begin{pmatrix} 0 & 0 \\ 0 & P_{22}(j|j-1) - \dfrac{P_{12}^2(j|j-1)}{P_{11}(j|j-1)} \end{pmatrix}.$$

Thus we obtain

$$P_{11}(j|j-1) = 2\sigma^2 - \frac{\sigma^4}{P_{11}(j-1|j-2)},$$

where $P_{11}(1|0) = V(y_1) = 2\sigma^2$. We now have $P_{11}(j|j-1) = (j+1)\sigma^2/j$. Putting $\beta(j|k) = (\beta_1(j|k), \beta_2(j|k))'$, we can also derive

$$\beta(j|j-1) = \begin{pmatrix} -\beta_2(j-1|j-1) \\ 0 \end{pmatrix},$$

$$\beta(j|j) = \beta(j|j-1) + \begin{pmatrix} 1 \\ P_{21}(j|j-1)/P_{11}(j|j-1) \end{pmatrix}(y_j - \beta_1(j|j-1))$$

$$= \begin{pmatrix} y_j \\ \dfrac{j}{j+1}(y_j + \beta_2(j-1|j-1)) \end{pmatrix},$$

so that

$$\beta_2(j|j) = \frac{j}{j+1}\beta_2(j-1|j-1) + \frac{j}{j+1}y_j$$

$$= \frac{1}{j+1}(y_1 + 2y_2 + \cdots + jy_j),$$

CHAPTER 10   589

$$y_j - a'\beta(j\mid j-1) = y_j + \beta_2(j-1\mid j-1)$$
$$= \frac{1}{j}(y_1 + 2y_2 + \cdots + jy_j).$$

Therefore we obtain

$$y'\Omega^{-1}y = \sigma^2 \sum_{j=1}^{T} \frac{(y_j - a'\beta(j\mid j-1))^2}{a'P(j\mid j-1)a}$$
$$= \sum_{j=1}^{T} \frac{1}{j(j+1)}(y_1 + 2y_2 + \cdots + jy_j)^2.$$

2.7 Noting that $B(\alpha)B'(\alpha) = \Omega(\alpha) = \alpha\Omega + (1-\alpha)^2 I_T$, we have

$$\mathcal{L}\left(\frac{1}{T}\tilde{\eta}'\Omega^{-1}\tilde{\eta}\right) = \mathcal{L}\left(\frac{1}{T}\varepsilon'M\Omega^{-1/2}\left(\alpha\Omega + (1-\alpha)^2 I_T\right)\Omega^{-1/2}M\varepsilon\right)$$
$$= \mathcal{L}\left(\frac{\alpha}{T}\varepsilon'M\varepsilon + \frac{c^2}{T^3}\varepsilon'M\Omega^{-1}M\varepsilon\right),$$

where $M = I_T - \Omega^{-1/2}X(X'\Omega^{-1}X)^{-1}X'\Omega^{-1/2} = M^2$. It clearly holds that plim $(\alpha\varepsilon'M\varepsilon/T) = \sigma^2$, while $\varepsilon'M\Omega^{-1}M\varepsilon = O_p(T^2)$, which establishes (10.21).

2.8 The case of Model A can be easily proved. For Model B we obtain

$$A_T = \Omega^{-1} - \Omega^{-1}ee'\Omega^{-1}/e'\Omega^{-1}e,$$

where

$$\Omega^{-1}e = \left(C'C - \frac{1}{T+1}C'ee'C\right)e = C'd - \frac{T}{2}C'e,$$
$$e'\Omega^{-1}e = e'C'Ce - \frac{1}{T+1}(e'Ce)^2 = \frac{T(T+1)(T+2)}{12}.$$

Thus we have

$$A_T(j,k) = \min(j,k) - \frac{jk}{T+1}$$
$$-\frac{3\{(T+j)(T-j+1) - T(T-j+1)\}\{(T+k)(T-k+1) - T(T-k+1)\}}{T(T+1)(T+2)}.$$

Then it can be checked that $A_T(j,k)/T$ has the uniform limit $K_B(s,t)$ in the sense of (10.23). The integral equation (5.10) with $K = K_B$ is shown to be equivalent to

$$f(t) = c_1 \cos \sqrt{\lambda}\, t + c_2 \sin \sqrt{\lambda}\, t + 6a,$$

$$f(0) = f(1) = 0, \qquad a = \int_0^1 (t - t^2) f(t)\, dt.$$

Then the Fredholm approach yields the FD $D_B(\lambda)$ of K_B. The case of Model C can be similarly proved. For Model D it is not hard to find the kernel $K_D(s, t)$. To obtain the FD of K_D we need to evaluate the determinant of a 4×4 matrix, which we have done by REDUCE, and arrive at $D_D(\lambda)$ in the theorem.

2.9 It follows from (8.15) and (8.18) that the limiting c.f. of $X_T = V_{T2}(\theta) + \theta$ for $c = 0$ is given by

$$\phi(u) = \left[\frac{\sin \theta \sqrt{2iu - 1}}{\theta \sqrt{2iu - 1}} \bigg/ \frac{\sinh \theta}{\theta} \right]^{-1/2}.$$

Thus $P(V_{T2}(\theta) \le x) = P(X_T/\theta^2 \le (\theta + x)/\theta^2)$ yields (10.28).

2.10 When $\theta = c$, it follows from (10.27) that

$$\mathcal{L}(V_{T2}(c)) \to \mathcal{L}\left(c^2 \sum_{n=1}^{\infty} \frac{1}{n^2 \pi^2} Z_n^2 - c \right),$$

which evidently yields (10.29).

4.1 It follows from (10.26) that, under $\alpha = 1 - (\theta/T)$ and $\alpha_0 = 1$,

$$P(V_{T2}(\theta) \ge x) = P\left(\frac{x}{T} y' \Omega^{-1} y - y' \left(\Omega^{-1} - \Omega^{-1}(\alpha) \right) y \le 0 \right)$$

$$= P\left(\sum_{j=1}^{T} \left(\left(\frac{x}{T} - 1 \right) + \frac{2 - 2\delta_j}{1 + \alpha^2 - 2\alpha \delta_j} \right) Z_j^2 \le 0 \right),$$

where $\delta_j = \cos(j\pi/(T+1))$. Then the upper 5% point x can be computed for each α and T by Imhof's formula described in (7.92).

4.2 Using (10.28) we first obtain the upper 5% point x for each $\theta = c$. Then the limiting power envelope can be computed following (10.29) for each combination of c and x.

4.3 Using (10.28) we first obtain the upper 5% point x for a fixed θ at which the point optimal test is conducted. It follows from (10.27) and (8.18) that the limiting c.f. of $X_T = V_{2T}(\theta) + \theta$ as $T \to \infty$ under $\alpha = 1 - (\theta/T)$ and $\alpha_0 = 1 - (c/T)$ is given by

$$\phi(u) = \left[\frac{\sin \sqrt{a+b}}{\sqrt{a+b}} \frac{\sin \sqrt{a-b}}{\sqrt{a-b}} \bigg/ \frac{\sinh \theta}{\theta} \right]^{-1/2},$$

where

$$a = \frac{\theta^2(2iu - 1)}{2}, \quad b = \frac{\theta\sqrt{\theta^2(2iu-1)^2 + 8ic^2u}}{2}.$$

Thus the limiting powers can be computed as

$$\lim_{T\to\infty} P(V_{T2}(\theta) \geq x) = \lim_{T\to\infty} P\left(\frac{X_T}{\theta^2} \geq \frac{\theta + x}{\theta^2}\right)$$

$$= 1 - \frac{1}{\pi}\int_0^\infty \operatorname{Re}\left[\frac{1 - \exp\left\{-\frac{iu(\theta+x)}{\theta^2}\right\}}{iu} \phi\left(\frac{u}{\theta^2}\right)\right] du.$$

5.1 Since $y \sim N((\bar{X} \otimes I_m)\beta, \sigma^2(C'(\alpha_m)C(\alpha_m))^{-1} \otimes I_m)$, the LBI test rejects H_0 when

$$-\frac{y'\left[\left(\bar{M}'\frac{d}{d\alpha}C'(\alpha)C(\alpha)\bigg|_{\alpha=1}\bar{M}\right) \otimes I_m\right] y}{y'\left[(\bar{M}'C'C\bar{M}) \otimes I_m\right] y} < c,$$

where $\bar{M} = I_N - \bar{X}(\bar{X}'C'C\bar{X})^{-1}\bar{X}'C'C$. Substituting $dC'(\alpha)C(\alpha)/d\alpha|_{\alpha=1} = C'ee'C - C'C$, we obtain the LBI statistic S_{N1}.

5.2 Since the limiting distribution of $X_N = V_{N1}(\theta) + m\theta$ is the m-fold convolution of the limiting distribution of $V_{T1}(\theta) + \theta$ in (10.11), (10.47) follows immediately from (10.11) by noting that $P(V_{N1}(\theta) \leq x) = P(X_N/\theta^2 \leq (m\theta + x)/\theta^2)$.

5.3 The limiting distribution of $Y_N = V_{N1}(c) + cm$ is the m-fold convolution of the limiting distribution of $V_{T1}(c) + c$ in (10.12). Thus (10.48) follows from (10.12) by noting that $P(V_{N1}(c) \geq x) = P(Y_N/c^2 \geq (cm + x)/c^2)$.

5.4 Since $y \sim N((\bar{X} \otimes I_m)\beta, \sigma^2\Omega(\alpha_m) \otimes I_m)$, the LBIU test rejects H_0 when

$$-\frac{y'\left[\left(\tilde{N}'\frac{d^2}{d\alpha^2}\Omega^{-1}(\alpha)\bigg|_{\alpha=1}\tilde{N}\right) \otimes I_m\right] y}{y'\left[(\tilde{N}'\Omega^{-1}\tilde{N}) \otimes I_m\right] y} > c.$$

Since $d^2\Omega^{-1}(\alpha)/d\alpha^2\big|_{\alpha=1} = 2\Omega^{-1} - 2\Omega^{-2}$, S_{N2} is shown to be the LBIU statistic.

5.5 The limiting distribution of $X_N = V_{N2}(\theta) + m\theta$ is the m-fold convolution of the limiting distribution of $V_{T2}(\theta) + \theta$ in (10.28). Thus (10.54) follows from (10.28).

5.6 The limiting distribution of $Y_N = V_{N2}(c) + cm$ is the m-fold convolution of the limiting distribution of $V_{T2}(c) + c$ in (10.29). Thus (10.55) follows from (10.29).

7.1 Since it holds that

$$\mathcal{L}\left(\frac{y'My}{T\sigma_\varepsilon^2}\right) = \mathcal{L}\left(\frac{1}{T}Z'M(I_T + \rho DCC'D)MZ\right)$$

$$= \mathcal{L}\left(\frac{1}{T}Z'MZ + \frac{c^2}{T^{2m+3}}Z'MDCC'DMZ\right),$$

where $Z \sim N(0, I_T)$, we establish (10.69) noting that $\text{plim}(Z'MZ/T) = 1$ and $Z'MDCC'DMZ = O_p(T^{2m+2})$.

7.2 It is easy to deduce that the (j,k)th element of $C'DMDC$ is given by

$$B_T(j,k) = \sum_{l=\max(j,k)}^{T} l^{2m} - \sum_{l=j}^{T} l^{2m} \sum_{l=k}^{T} l^{2m} \bigg/ \sum_{l=1}^{T} l^{2m}.$$

Thus $B_T(j,k)/T^{2m+1}$ converges uniformly to $K(s,t;m)$, which proves (10.70) because of Theorem 5.13. The associated FD was earlier obtained in (5.64).

7.3 The (j,k)th element of $C'MC$ is given by

$$B_T(j,k) = T + 1 - \max(j,k) - \frac{1}{T\sum_{l=1}^{T} l^{2m} - \left(\sum_{l=1}^{T} l^m\right)^2}$$

$$\times \left[(T-k+1)\left\{(T-j+1)\sum_{l=1}^{T} l^{2m} - \sum_{l=j}^{T} l^m \sum_{l=1}^{T} l^m\right\}\right.$$

$$\left. + \sum_{l=k}^{T} l^m \left\{T\sum_{l=j}^{T} l^m - (T-j+1)\sum_{l=1}^{T} l^m\right\}\right].$$

Then it can be checked that $B_T(j,k)/T$ converges uniformly to $K(s,t;m)$. The associated FDs for $m = 1$ and 2 are available in Theorem 10.2. Consider the case of $m = 3$. The integral equation (5.10) with $K(s,t)$ replaced by $K(s,t;3)$ is shown to be equivalent to

$$f(t) = c_1 \cos \sqrt{\lambda}\, t + c_2 \sin \sqrt{\lambda}\, t + at^2 - 2a/\lambda,$$

$$f(0) = f(1) = 0, \qquad a = \frac{28}{3}\int_0^1 t(1-t^3)f(t)\,dt.$$

Then we obtain $D(\lambda; 3)$ after some algebra. The case of $m = 4$ can be dealt with similarly.

7.4 Since it can be shown that $\text{plim}(y'My/T) = \sigma_\varepsilon^2$ as $T \to \infty$ under $\rho = c^2/T^2$, we concentrate on

$$\mathcal{L}\left(\frac{1}{T^2\sigma_\varepsilon^2}y'MCC'My\right) = \mathcal{L}\left(\frac{1}{T^2}Z'\left(C'MC + \frac{c^2}{T^2}(C'MC)^2\right)Z\right),$$

where $Z \sim N(0, I_T)$. Evaluating the (j,k)th element of

$$C'MC = C'C - C'(e,d,f)\left((e,d,f)'(e,d,f)\right)^{-1}(e,d,f)'C,$$

we obtain the kernel $K(s,t)$. Then the Fredholm approach yields the FD given in the theorem.

7.5 It follows from (10.80) that the limiting c.f. of $V_T(\theta)/\theta^2$ for $c = 0$ is given by

$$\phi(u) = \prod_{n=1}^{\infty}\left[1 - \frac{2iu}{\left(n - \frac{1}{2}\right)^2\pi^2 + \theta^2}\right]^{-1/2}$$

$$= \prod_{n=1}^{\infty}\left[\left\{1 - \frac{2iu - \theta^2}{\left(n - \frac{1}{2}\right)^2\pi^2}\right\}\bigg/\left\{1 + \frac{\theta^2}{\left(n - \frac{1}{2}\right)^2\pi^2}\right\}\right]^{-1/2}$$

$$= \left[\cos\sqrt{2iu - \theta^2}\bigg/\cosh\theta\right]^{-1/2}.$$

Thus (10.81) follows from $P(V_T(\theta) \leq x) = P\left(V_T(\theta)/\theta^2 \leq x/\theta^2\right)$.

7.6 When $\theta = c$, it follows from (10.80) that

$$\mathcal{L}(V_T(c)) \to \mathcal{L}\left(c^2\sum_{n=1}^{\infty}\frac{1}{\left(n - \frac{1}{2}\right)^2\pi^2}Z_n^2\right).$$

Then $P(V_T(c) \geq x) = P(V_T(c)/c^2 \geq x/c^2)$ implies (10.82).

7.7 It follows from (10.80) that the limiting c.f. of $V_T(\theta)$ is given by

$$\phi(u) = \prod_{n=1}^{\infty}\left[1 - \frac{2iu\theta^2\left(\left(n - \frac{1}{2}\right)^2\pi^2 + c^2\right)}{\left(n - \frac{1}{2}\right)^2\pi^2\left(\left(n - \frac{1}{2}\right)^2\pi^2 + \theta^2\right)}\right]^{-1/2}$$

$$= \prod_{n=1}^{\infty}\left[\frac{1 + \frac{(1 - 2iu)\theta^2}{\left(n - \frac{1}{2}\right)^2\pi^2} - \frac{2ic^2u\theta^2}{\left(n - \frac{1}{2}\right)^4\pi^4}}{1 + \frac{\theta^2}{\left(n - \frac{1}{2}\right)^2\pi^2}}\right]^{-1/2}$$

$$= \left(\cos\sqrt{a+b}\cos\sqrt{a-b}\bigg/\cosh\theta\right)^{-1/2},$$

where

$$a = \frac{\theta^2}{2}(2iu - 1), \quad b = \frac{\theta}{2}\sqrt{\theta^2(2iu - 1)^2 + 8ic^2 u}.$$

Then the limiting powers of the POI test conducted at θ under the $100\gamma\%$ significance level can be computed as

$$\lim_{T \to \infty} P\left(\frac{V_T(\theta)}{\theta^2} \geq \frac{x}{\theta^2}\right) = 1 - \frac{1}{\pi}\int_0^\infty \mathrm{Re}\left[\frac{1 - \exp\left(-\frac{iux}{\theta^2}\right)}{iu}\phi\left(\frac{u}{\theta^2}\right)\right] du,$$

where x is the upper $100\gamma\%$ point of the limiting distribution in (10.81).

CHAPTER 11

2.1 Let us construct the partial sum process

$$Y_T(t) = \frac{1}{\sqrt{T}} y_j + T\left(t - \frac{j}{T}\right)\frac{1}{\sqrt{T}} u_j, \quad \left(\frac{j-1}{T} \leq t \leq \frac{j}{T}\right).$$

Then it follows that $\mathcal{L}(Y_T) \to \mathcal{L}(Aw)$ and

$$\mathcal{L}\left(\frac{1}{T^2}\sum_{j=1}^T y_j y_j'\right) = \mathcal{L}\left(\frac{1}{T}\sum_{j=1}^T Y_T\left(\frac{j}{T}\right)Y_T'\left(\frac{j}{T}\right)\right)$$

$$\to \mathcal{L}\left(A\int_0^1 w(t)w'(t)\,dt A'\right).$$

The continuous mapping theorem now establishes (11.12) for $k = 1$. If we construct, for $(j-1)/T \leq t \leq j/T$,

$$\tilde{Y}_T(t) = \frac{1}{\sqrt{T}}(y_j - \bar{y}) + T\left(t - \frac{j}{T}\right)\frac{1}{\sqrt{T}} u_j,$$

we have that $\mathcal{L}(\tilde{Y}_T) \to \mathcal{L}(A\tilde{w})$ and

$$\mathcal{L}\left(\frac{1}{T^2}\sum_{j=1}^T (y_j - \bar{y})(y_j - \bar{y})'\right) = \mathcal{L}\left(\frac{1}{T}\sum_{j=1}^T \tilde{Y}_T\left(\frac{j}{T}\right)\tilde{Y}_T'\left(\frac{j}{T}\right)\right)$$

$$\to \mathcal{L}\left(A\int_0^1 \tilde{w}(t)\tilde{w}'(t)\,dt A'\right),$$

where $\tilde{w}(t) = w(t) - \int_0^1 w(s)\,ds$. Then (11.12) also holds for $k = 2$ because of the continuous mapping theorem.

2.2 For $k = 1$ it follows from Theorem 4.4 that

$$E\left(e^{\theta X_k}\right) = E\left[\exp\left\{\theta \int_0^1 w'(t)Hw(t)\,dt\right\}\right]$$

$$= \prod_{a=1}^{2} E\left[\exp\left\{\theta\delta_a \int_0^1 w_a^2(t)\,dt\right\}\right],$$

where $w(t) = (w_1(t), w_2(t))'$ is the two-dimensional standard Brownian motion. Then we obtain (11.17) from (4.10). The case of $k = 2$ can be proved similarly.

2.3 Let $\nu_k(n)$ be the nth order raw moment of $F_k(x)$ in (11.15). Then, we have, from (1.39),

(S17) $\quad \nu_k(n) = \dfrac{1}{(n-1)!} \int_0^\infty \theta_2^{n-1} \left.\dfrac{\partial^n \psi_k(\theta_1, -\theta_2)}{\partial \theta_1^n}\right|_{\theta_1=0} d\theta_2,$

where

$$\psi_1(\theta_1, -\theta_2) = \left[\cos\sqrt{a+b}\,\cos\sqrt{a-b}\right]^{-1/2},$$

$$\psi_2(\theta_1, -\theta_2) = \left[\dfrac{\sin\sqrt{a+b}}{\sqrt{a+b}} \dfrac{\sin\sqrt{a-b}}{\sqrt{a-b}}\right]^{-1/2},$$

$$a = -\theta_2 A_1' A_1 + \theta_1 A_1' A_2, \qquad b = \sqrt{a^2 + \theta_1^2 |A|^2}.$$

Using any computerized algebra, we can easily obtain partial derivatives of ψ_k. Then we compute, for instance,

$$\nu_1(1) = \dfrac{1}{2} A_1' A_2 \int_0^\infty \dfrac{\sinh\sqrt{2\theta A_1' A_1}}{\sqrt{2\theta A_1' A_1}} \left(\cosh\sqrt{2\theta A_1' A_1}\right)^{-3/2} d\theta = \dfrac{A_1' A_2}{A_1' A_1}.$$

Finally we can obtain $\mu_k(n)$ from $\nu_k(n)$.

2.4 Mercer's theorem (Theorem 5.2) gives us

(S18) $\quad F_k(x) = P\left(\displaystyle\sum_{n=1}^\infty \dfrac{1}{\lambda_{kn}}\left(\delta_1(x)X_n^2 + \delta_2(x)Y_n^2\right) \geq 0\right),$

where $(X_n, Y_n)' \sim \text{NID}(0, I_2)$, $\lambda_{1n} = (n - \tfrac{1}{2})^2 \pi^2$ and $\lambda_{2n} = n^2 \pi^2$, while $\delta_1(x)$ and $\delta_2(x)$ are the eigenvalues of the matrix H given in (11.19). Then it can be checked easily that $F_k(x + \mu) = 1 - F_k(-x + \mu)$.

2.5 It follows from (S18) that

$$G_k(x) = F_k(\sigma_k x + \mu)$$
$$= P\left(\sum_{n=1}^{\infty} \frac{1}{\lambda_{kn}} \left(\delta_1 (\sigma_k x + \mu) X_n^2 + \delta_2 (\sigma_k x + \mu) Y_n^2\right) \geq 0\right),$$

where $\sigma_k = \sqrt{\mu_k(2)}$. Since

$$\delta_1 (\sigma_k x + \mu), \delta_2 (\sigma_k x + \mu) = \frac{1}{2}\left[\sigma_k x A_1' A_1 \pm \sqrt{(\sigma_k x A_1' A_1)^2 + |A|^2}\right]$$
$$= \frac{|\det(A)|}{2}\left(\sqrt{a_k}x \pm \sqrt{a_k x^2 + 1}\right),$$

where a_k is defined in Corollary 11.1, (11.22) is seen to hold.

2.6 We first note that $\hat{v}_j = y_{2j} - \hat{\beta}_1 y_{1j}$; hence

$$\frac{1}{T^2}\sum_{j=1}^{T} \hat{v}_j^2 = \frac{1}{T^2}\sum_{j=1}^{T} y_{2j}^2 - \hat{\beta}_1^2 \frac{1}{T^2}\sum_{j=1}^{T} y_{1j}^2.$$

Since it holds that

$$\mathcal{L}\left(\hat{\beta}_1, \frac{1}{T^2}\sum_{j=1}^{T} y_{1j}^2, \frac{1}{T^2}\sum_{j=1}^{T} y_{2j}^2, \frac{1}{T^2}\sum_{j=1}^{T} \hat{v}_j^2\right)$$
$$\to \mathcal{L}\left(\zeta, A_1' W_1 A_1, A_2' W_1 A_2, A_2' W_1 A_2 - \zeta^2 A_1' W_1 A_1\right),$$

this proves (11.23) and (11.24). Moreover

$$\frac{1}{T}\sum_{j=2}^{T} (\hat{v}_j - \hat{v}_{j-1})^2 = (\hat{\beta}_1, -1) \frac{1}{T}\sum_{j=1}^{T} u_j u_j' \begin{pmatrix} \hat{\beta}_1 \\ -1 \end{pmatrix},$$

which yields (11.25).

3.1 Since Y_1' can be expressed as $Y_1' = \Xi_1' C'$, we have

$$\text{vec}\,(Y_1') = \text{vec}\,(\Xi_1' C')$$
$$= (C \otimes I_q) \text{vec}\,(\Xi_1').$$

Noting that $\text{vec}\,(\Xi_1') \sim N(0, I_T \otimes \Sigma_{11})$, we obtain (11.33). Moreover, it can be shown that

$$\begin{pmatrix} \text{vec}\,(Y_1') \\ \Xi_2 \end{pmatrix} \sim N\left(0, \begin{pmatrix} (CC') \otimes \Sigma_{11} & C \otimes \Sigma_{12} \\ C' \otimes \Sigma_{21} & I_T \otimes \Sigma_{22} \end{pmatrix}\right),$$

which proves (11.34).

3.2 It is easy to deduce that

$$\frac{1}{T^2}Y_1'P_{-1}Y_1 = \frac{1}{T^2}Y_1'Y_{-1}\left(Y_{-1}'Y_{-1}\right)^{-1}Y_{-1}'Y_1$$

$$= \frac{1}{T^2}Y_1'(Y_1 - \Xi_1)\left\{\frac{1}{T^2}(Y_1 - \Xi_1)'(Y_1 - \Xi_1)\right\}^{-1}\frac{1}{T^2}(Y_1 - \Xi_1)'Y_1$$

$$= \frac{1}{T^2}Y_1'Y_1 + o_p(1),$$

$$\frac{1}{T^2}Y_1'M_1Y_1 = \frac{1}{T^2}Y_1'Y_1 - \frac{1}{T}Y_1'\Xi_1\left(\frac{1}{T}\Xi_1'\Xi_1\right)^{-1}\frac{1}{T}\Xi_1'Y_1 \times \frac{1}{T}$$

$$= \frac{1}{T^2}Y_1'Y_1 + o_p(1).$$

3.3 We can show easily that

$$\mathcal{L}\left(\frac{1}{T}Y_1'P_{-1}\Xi_2\right) = \mathcal{L}\left(\frac{1}{T}Y_1'Y_{-1}\left(Y_{-1}'Y_{-1}\right)^{-1}Y_{-1}'\Xi_2\right)$$

$$= \mathcal{L}\left(\frac{1}{T}Y_{-1}'\Xi_2 + o_p(1)\right)$$

$$\to \mathcal{L}(U_1 + U_2),$$

$$\mathcal{L}\left(\frac{1}{T}Y_1'M_1\Xi_2\right)$$

$$= \mathcal{L}\left(\frac{1}{T}Y_1'\Xi_2 - \frac{1}{T}Y_1'\Xi_1\left(\frac{1}{T}\Xi_1'\Xi_1\right)^{-1}\frac{1}{T}\Xi_1'\Xi_2\right)$$

$$\to \mathcal{L}\left(U_1 + U_2 + \Sigma_{12} - \left(\Sigma_{11}^{1/2}\int_0^1 w_1(t)\,dw_1'(t)\Sigma_{11}^{1/2} + \Sigma_{11}\right)\Sigma_{11}^{-1}\Sigma_{12}\right)$$

$$= \mathcal{L}(U_2).$$

3.4 We have only to show that $m_1(\theta) = \phi_1(-i\theta)$, where $m_1(\theta)$ is given below (11.40). Girsanov's theorem yields

$$m_1(\theta) = \exp\left\{\frac{\theta}{2}(ab - 2d) - \frac{\gamma}{2}\right\}E\left[\exp\left\{\frac{1}{2}(\gamma - ab\theta)Z^2(1)\right\}\right],$$

where $\gamma = \sqrt{-a^2\theta(2x + c^2\theta)}$ and $dZ(t) = -\gamma Z(t)dt + dw_1(t)$ with $Z(0) = 0$. Since $Z(1) \sim N(0, (1 - e^{-2\gamma})/(2\gamma))$, we can easily obtain the conclusion.

3.5 We compute the right side of (S17), where X_{OLS} has

(S19) $\psi_1(\theta_1, -\theta_2) = \exp\left\{\dfrac{\theta_1}{2}(2d - ab)\right\}\left[\cos\mu - ab\theta_1\dfrac{\sin\mu}{\mu}\right]^{-1/2},$

$$\mu = \sqrt{a^2(c^2\theta_1^2 - 2\theta_2)}.$$

Then we obtain, for instance,

$$E(X_{\text{OLS}}) = \int_0^\infty \left.\dfrac{\partial \psi_1(\theta_1, -\theta_2)}{\partial \theta_1}\right|_{\theta_1=0} d\theta_2$$

$$= \dfrac{\Sigma_{12}}{2}\int_0^\infty \left(\cosh a\sqrt{2\theta}\right)^{-1/2}\left(1 + \dfrac{1}{\cosh a\sqrt{2\theta}}\dfrac{\sinh a\sqrt{2\theta}}{a\sqrt{2\theta}}\right)d\theta$$

$$= \dfrac{\Sigma_{12}}{2\Sigma_{11}}\left[\int_0^\infty \dfrac{u}{\sqrt{\cosh u}}du + \int_0^\infty (\cosh u)^{-3/2}\sinh u\, du\right]$$

$$= \dfrac{\Sigma_{12}}{2\Sigma_{11}}(c_1 + 2).$$

We can compute $E(X_{\text{OLS}}^2)$ and moments of X_{2SLS} and X_{ML} similarly.

3.6 It follows from (11.34) and (11.42) that

$$Y_2 \mid \text{vec}(Y_1') \sim N\left(\delta e + Y_1\beta + \Delta Y_1 \Sigma_{11}^{-1}\Sigma_{12}, \Sigma_{22\cdot 1}I_T\right).$$

Since $f(\text{vec}(Y_1'), Y_2) = f_1(\text{vec}(Y_1'))f_2(Y_2 \mid \text{vec}(Y_1'))$, the MLE of β is the ordinary LSE of β obtained from $Y_2 = \delta e + Y_1\beta + \Delta Y_1\gamma + v_2$, where $\gamma = \Sigma_{11}^{-1}\Sigma_{12}$ and $v_2 = \Xi_2 - \Xi_1\gamma$. This gives us $\tilde{\beta}_{\text{ML}}$ in (11.45).

3.7 Let us consider first

$$T(\tilde{\beta}_{\text{OLS}} - \beta) = \left(\dfrac{1}{T^2}Y_1'MY_1\right)^{-1}\dfrac{1}{T}Y_1'M\Xi_2,$$

where it holds that

$$\mathcal{L}\left(\dfrac{1}{T^2}Y_1'MY_1, \dfrac{1}{T}Y_1'M\Xi_2\right)$$

$$= \mathcal{L}\left(\dfrac{1}{T^2}\sum_{j=1}^T (y_{1j} - \bar{y}_1)(y_{1j} - \bar{y}_1)', \dfrac{1}{T}\sum_{j=1}^T (y_{1,j-1} - \bar{y}_1 + \xi_{1j})\xi_{2j}\right)$$

$$\to \mathcal{L}(\tilde{V}, \tilde{U}_1 + \tilde{U}_2 + \Sigma_{12}).$$

Thus it follows that $\mathcal{L}(T(\tilde{\beta}_{\text{OLS}} - \beta)) \to \mathcal{L}(\tilde{V}^{-1}(\tilde{U}_1 + \tilde{U}_2 + \Sigma_{12}))$. We consider next

$$T(\tilde{\beta}_{\text{2SLS}} - \beta) = \left(\frac{1}{T^2}Y_1'\tilde{P}M\tilde{P}Y_1\right)^{-1}\frac{1}{T}Y_1'\tilde{P}M\Xi_2,$$

where $\tilde{Y}_1 = \tilde{P}Y_1$. Since $(e, Y_{-1})'\tilde{P} = (e, Y_{-1})'$, we find

$$\frac{1}{T^2}Y_1'\tilde{P}M\tilde{P}Y_1 = \frac{1}{T^2}Y_1'MY_1 + o_p(1),$$

$$\frac{1}{T}Y_1'\tilde{P}M\Xi_2 = \frac{1}{T}Y_{-1}'\Xi_2 - \frac{1}{T}Y_1'ee'\Xi_2 + o_p(1)$$

so that $\mathcal{L}(T(\tilde{\beta}_{\text{2SLS}} - \beta)) \to \mathcal{L}(\tilde{V}^{-1}(\tilde{U}_1 + \tilde{U}_2))$. Finally we consider

$$T(\tilde{\beta}_{\text{ML}} - \beta) = \left(\frac{1}{T^2}Y_1'M_2Y_1\right)^{-1}\frac{1}{T}Y_1'M_2\Xi\kappa,$$

where $\Xi = (\Xi_1, \Xi_2)$, $\kappa = (-\gamma', 1)'$ with $\gamma = \Sigma_{11}^{-1}\Sigma_{12}$ and

$$\frac{1}{T^2}Y_1'M_2Y_1 = \frac{1}{T^2}Y_1'MY_1 + o_p(1),$$

$$\frac{1}{T}Y_1'M_2\Xi = \frac{1}{T}Y_1'\Xi - \frac{1}{T}(Y_1'e, Y_1'\Xi_1)\begin{pmatrix} e'e & e'\Xi_1 \\ \Xi_1'e & \Xi_1'\Xi_1 \end{pmatrix}^{-1}\begin{pmatrix} e'\Xi \\ \Xi_1'\Xi \end{pmatrix}$$

$$= \frac{1}{T}Y_1'\Xi - \frac{1}{T^2}Y_1'ee'\Xi - \frac{1}{T}Y_1'\Xi_1(\Xi_1'\Xi_1)^{-1}\Xi_1'\Xi + o_p(1).$$

Since $Y_1'M_2\Xi\kappa/T$ converges in distribution to

$$\left[\Sigma_{11}^{1/2}\int_0^1 w_1(t)dw'(t)B' + (\Sigma_{11}, \Sigma_{12}) - \Sigma_{11}^{1/2}\int_0^1 w_1(t)dtw'(1)B'\right.$$
$$\left. - \left(\Sigma_{11}^{1/2}\int_0^1 w_1(t)dw_1'(t)\Sigma_{11}^{1/2} + \Sigma_{11}\right)\Sigma_{11}^{-1}(\Sigma_{11}, \Sigma_{12})\right]\begin{pmatrix} -\gamma \\ 1 \end{pmatrix}$$

$$= \tilde{U}_2,$$

where B is defined in (11.35), we can deduce that $\mathcal{L}(T(\tilde{\beta}_{\text{ML}} - \beta)) \to \mathcal{L}(\tilde{V}^{-1}\tilde{U}_2)$.

3.8 Let us compute

$$m_2(\theta) = E\left[E\left\{e^{\theta X_2} | w_1\right\}\right]$$

$$= E\left[\exp\left\{\theta E(X_2|w_1) + \frac{\theta^2}{2}V(X_2|w_1)\right\}\right]$$

$$= \exp\left\{\frac{\theta}{2}(ab - 2d)\right\} E\left[\exp\left\{c_1 \int_0^1 \tilde{w}_1^2(t)\,dt\right.\right.$$
$$\left.\left. + c_2 w_1^2(1) + c_3 w_1(1) \int_0^1 w_1(t)\,dt\right\}\right],$$

where

$$c_1 = a^2\theta\left(x + \frac{c^2\theta}{2}\right), \qquad c_2 = -\frac{ab\theta}{2}, \qquad c_3 = ab\theta.$$

Girsanov's theorem leads us to

$$m_2(\theta) = \exp\left\{\frac{\theta}{2}(ab - 2d) - \frac{\gamma}{2}\right\}$$
$$\times E\left[\exp\left\{-c_1\left(\int_0^1 Z(t)\,dt\right)^2 + \left(c_2 + \frac{\gamma}{2}\right) Z^2(1) + c_3 Z(1) \int_0^1 Z(t)\,dt\right\}\right],$$

where $dZ(t) = -\gamma Z(t)\,dt + dw_1(t)$ with $\gamma = \sqrt{-2c_1}$ and $Z(0) = 0$. We obtain $m_2(\theta) = \phi_2(-i\theta)$ after some algebra.

3.9 We compute the right side of (S17), where Y_{OLS} has

$$\psi_1(\theta_1, -\theta_2)$$
$$= \exp\left\{\frac{\theta_1}{2}(2d - ab)\right\}\left[\frac{2a^2 b^2 \theta_1^2}{\mu^4}(\cos\mu - 1) + \left(1 + \frac{a^2 b^2 \theta_1^2}{\mu^2}\right)\frac{\sin\mu}{\mu}\right]^{-1/2},$$
$$\mu = \sqrt{a^2(c^2\theta_1^2 - 2\theta_2)}.$$

(S20)

Proceeding in the same way as in the solution to Problem 3.5, we obtain moments of Y_{OLS}, Y_{2SLS} and Y_{ML}.

4.1 It follows from (11.47) that

$$T(\hat{\beta}_{\text{OLS}} - \beta) = \left(\frac{1}{T^2}\sum_{j=1}^T y_{1j} y'_{1j}\right)^{-1} \frac{1}{T}\sum_{j=1}^T y_{1j} g'(L)\varepsilon_j.$$

Using the weak convergence results on the auxiliary process $\{z_j\}$ introduced below (11.47), we can deduce that

$$\mathcal{L}\left(\frac{1}{T^2}\sum_{j=1}^T y_{1j} y'_{1j}, \frac{1}{T}\sum_{j=1}^T y_{1j} g'(L)\varepsilon_j\right) \to \mathcal{L}(R, Q_1 + Q_2 + \Lambda_{12}),$$

which establishes the theorem.

4.2 We compute the right side of (S17) when $\psi_1(\theta_1, -\theta_2)$ is given by (S19), where $a, b, c,$ and d are defined in (11.49). We obtain, for instance,

$$\left.\frac{\partial \psi_1(\theta_1, -\theta_2)}{\partial \theta_1}\right|_{\theta_1=0} = \frac{\left(\cosh a\sqrt{2\theta_2}\right)^{-1/2}}{2}\left(2d - ab + \frac{ab}{\cosh a\sqrt{2\theta_2}} \cdot \frac{\sinh a\sqrt{2\theta_2}}{a\sqrt{2\theta_2}}\right),$$

which yields $E(X_{\text{OLS}})$. We can compute $E(X_{\text{OLS}}^2)$ similarly.

4.3 It follows from (11.50) that $T(\hat{\beta}_{\text{FM}} - \beta) = V_T^{-1} U_T$, where

$$U_T = \frac{1}{T}\sum_{j=1}^{T} y_{1j}\left(g'(L)\varepsilon_j - \hat{\Omega}_{21}\hat{\Omega}_{11}^{-1}\Phi_1'(L)\varepsilon_j\right) - \hat{\Lambda}_{12} + \hat{\Lambda}_{11}\hat{\Omega}_{11}^{-1}\hat{\Omega}_{12},$$

$$V_T = \frac{1}{T^2}\sum_{j=1}^{T} y_{1j}y_{1j}'.$$

Because of the weak convergence results on the auxiliary process $\{z_j\}$ and Theorem 11.7, we can deduce that

$$\mathcal{L}(U_T, V_T) \to \mathcal{L}\left(Q_1 + Q_2 + \Lambda_{12} - \left(\Omega_{11}^{1/2}\int_0^1 w_1(t)\,dw_1'(t)\Omega_{11}^{1/2} + \Lambda_{11}\right)\Omega_{11}^{-1}\Omega_{12}\right.$$

$$\left.- \Lambda_{12} + \Lambda_{11}\Omega_{11}^{-1}\Omega_{12}, R\right)$$

$$= \mathcal{L}(Q_2, R),$$

which establishes the theorem.

4.4 Let us consider $T(\tilde{\beta}_{\text{OLS}} - \beta) = \tilde{V}_T^{-1}\tilde{U}_{1T}$ and $T(\tilde{\beta}_{\text{FM}} - \beta) = \tilde{V}_T^{-1}\tilde{U}_{2T}$, where

$$\tilde{U}_{1T} = \frac{1}{T}\sum_{j=1}^{T}(y_{1j} - \bar{y}_1)g'(L)\varepsilon_j,$$

$$\tilde{U}_{2T} = \frac{1}{T}\sum_{j=1}^{T}(y_{1j} - \bar{y}_1)\left(g'(L)\varepsilon_j - \tilde{\Omega}_{21}\tilde{\Omega}_{11}^{-1}\Phi_1'(L)\varepsilon_j\right) - \tilde{\Lambda}_{12} + \tilde{\Lambda}_{11}\tilde{\Omega}_{11}^{-1}\tilde{\Omega}_{12},$$

$$\tilde{V}_T = \frac{1}{T^2}\sum_{j=1}^{T}(y_{1j} - \bar{y}_1)(y_{1j} - \bar{y}_1)'.$$

If we construct the auxiliary process

$$\Delta z_j = \begin{pmatrix}\Delta y_{1j} \\ \Delta x_j\end{pmatrix} = \begin{pmatrix}\Phi_1'(L) \\ g'(L)\end{pmatrix}\varepsilon_j, \qquad z_0 = 0,$$

we have that

$$\mathcal{L}\left(\frac{1}{T^2}\sum_{j=1}^{T}(z_j-\bar{z})(z_j-\bar{z})'\right) \to \mathcal{L}\left(D\int_0^1 \tilde{w}(t)\tilde{w}'(t)\,dt D'\right),$$

$$\mathcal{L}\left(\frac{1}{T}\sum_{j=1}^{T}(z_j-\bar{z})\Delta z_j'\right) \to \mathcal{L}\left(D\int_0^1 \tilde{w}(t)\,dw'(t)D' + \Lambda\right).$$

Then it is easy to deduce that

$$\mathcal{L}\left(\tilde{U}_{1T},\tilde{V}_T\right) \to \mathcal{L}\left(\tilde{Q}_1 + \tilde{Q}_2 + \Lambda_{12},\tilde{R}\right),$$

$$\mathcal{L}\left(\tilde{U}_{2T},\tilde{V}_T\right) \to \mathcal{L}\Big(\tilde{Q}_1 + \tilde{Q}_2 + \Lambda_{12} - \left(\Omega_{11}^{1/2}\int_0^1 \tilde{w}_1(t)\,dw_1'(t)\Omega_{11}^{1/2} + \Lambda_{11}\right)$$

$$\times \Omega_{11}^{-1}\Omega_{12} - \Lambda_{12} + \Lambda_{11}\Omega_{11}^{-1}\Omega_{12},\tilde{R}\Big)$$

$$= \mathcal{L}\left(\tilde{Q}_2,\tilde{R}\right).$$

Thus the theorem is established.

4.5 We compute the right side of (S17) when $\psi_1(\theta_1,-\theta_2)$ is given by (S20), where $a,b,c,$ and d are defined in (11.49). We obtain, for instance,

$$\left.\frac{\partial \psi_1(\theta_1,-\theta_2)}{\partial \theta_1}\right|_{\theta_1=0} = \frac{2d-ab}{2}\frac{\sqrt{a\sqrt{2\theta_2}}}{\sqrt{\sinh a\sqrt{2\theta_2}}}$$

so that

$$E(Y_{\text{OLS}}) = \frac{2d-ab}{2a^2}\int_0^\infty \frac{u^{3/2}}{\sqrt{\sinh u}}\,du = \frac{2d-ab}{2a^2}d_1.$$

We can compute $E(Y_{\text{OLS}}^2)$ similarly.

6.1 It follows from the text that

$$\mathcal{L}(T(\hat{\rho}-1)) \to \mathcal{L}\left(\frac{(-X_1',1)B\int_0^1 w(t)\,dw'(t)B'\begin{pmatrix}-X_1\\1\end{pmatrix}}{(-X_1',1)B\int_0^1 w(t)w'(t)\,dt\,B'\begin{pmatrix}-X_1\\1\end{pmatrix}} + R\right).$$

Since we have

$$(-X_1', 1) Bw(t) = \left(-\int_0^1 (B_2'w_1(t) + B_3w_2(t)) w_1'(t) dt B_1\right.$$

$$\times \left(B_1 \int_0^1 w_1(t) w_1'(t) dt B_1\right)^{-1}, 1\right) \begin{pmatrix} B_1 w_1(t) \\ B_2'w_1(t) + B_3w_2(t) \end{pmatrix}$$

$$= B_3 Q(t),$$

we can prove (11.62).

6.2 Consider

$$\frac{1}{T}\hat{Z}_\rho = \hat{\rho} - 1 - (\hat{\sigma}_L^2 - \hat{\sigma}_S^2) \Big/ \left(\frac{2}{T} \sum_{j=2}^T \hat{\eta}_{j-1}^2\right).$$

Noting that

$$\hat{\eta}_j = y_{2j} - \hat{\beta}' y_{1j} = g'(L)\varepsilon_j - (\hat{\beta} - \beta)' y_{1j},$$

we obtain

$$\frac{1}{T} \sum_{j=2}^T \hat{\eta}_{j-1}^2 \to \gamma(0), \qquad \frac{1}{T} \sum_{j=2}^T \hat{\eta}_{j-1}\hat{\eta}_j \to \gamma(1)$$

in probability so that plim $\hat{\rho} = \gamma(1)/\gamma(0) = \rho$. Since

$$\hat{\eta}_j - \hat{\rho}\hat{\eta}_{j-1} = g'(L)\varepsilon_j - \rho g'(L)\varepsilon_{j-1} - (\hat{\beta} - \beta)'(y_j - \rho y_{1, j-1}),$$

we have

$$\hat{\sigma}_S^2 = \frac{1}{T} \sum_{j=2}^T (\hat{\eta}_j - \hat{\rho}\hat{\eta}_{j-1})^2 \to E\left\{(g'(L)\varepsilon_j - \rho g'(L)\varepsilon_{j-1})^2\right\}$$

$$= (\gamma^2(0) - \gamma^2(1))/\gamma(0).$$

The estimator $\hat{\sigma}_L^2$ converges in probability to 2π times the spectrum of $\{g'(L)\varepsilon_j - \rho g'(L)\varepsilon_{j-1}\}$ evaluated at the origin, that is

$$\text{plim } \hat{\sigma}_L^2 = (g' - \rho g')(g - \rho g) = (\gamma(0) - \gamma(1))^2 g'g \Big/ \gamma^2(0).$$

Therefore it follows that

$$\text{plim}\left(\frac{1}{T}\hat{Z}_\rho\right) = \frac{\gamma(1)}{\gamma(0)} - 1 - \frac{1}{2\gamma(0)}\left[\frac{(\gamma(0)-\gamma(1))^2}{\gamma^2(0)}g'g - \frac{\gamma^2(0)-\gamma^2(1)}{\gamma(0)}\right]$$

$$= -\frac{(\gamma(0)-\gamma(1))^2}{2\gamma^2(0)}\left(1+\frac{g'g}{\gamma(0)}\right).$$

6.3 We can deduce from the arguments leading to (11.63) that

$$\mathcal{L}\left(\frac{1}{T^2}\sum_{j=2}^{T}\hat{\eta}_{j-1}^2, \hat{\sigma}_L^2\right) \to \mathcal{L}\left(B_3^2\int_0^1 Q^2(t)\,dt, B_3^2 S'S\right),$$

which proves (11.66) by the continuous mapping theorem. Under H_1 we have

$$\frac{1}{\sqrt{T}}\hat{Z}_t = \left(\frac{1}{\hat{\sigma}_L^2}\frac{1}{T}\sum_{j=2}^{T}\hat{\eta}_{j-1}^2\right)^{1/2}\frac{1}{T}\hat{Z}_\rho$$

$$\to \sqrt{\frac{\gamma^3(0)}{(\gamma(0)-\gamma(1))^2 g'g}}\left(-\frac{(\gamma(0)-\gamma(1))^2}{2\gamma^2(0)}\right)\left(1+\frac{g'g}{\gamma(0)}\right)$$

$$= -\frac{\gamma(0)-\gamma(1)}{2\sqrt{\gamma(0)g'g}}\left(1+\frac{g'g}{\gamma(0)}\right) \qquad \text{in probability}.$$

6.4 Let us consider

$$\hat{v}_j = y_{2j} - \hat{\beta}'_{\text{FM}}y_{1j} - \hat{\Omega}_{21}\hat{\Omega}_{11}^{-1}\Delta y_{1j}$$

$$= \frac{c}{T}\frac{\xi_{2j}}{1-L} + \gamma(L)\xi_{2j} - (\hat{\beta}_{\text{FM}} - \beta)'y_{1j} - \hat{\Omega}_{21}\hat{\Omega}_{11}^{-1}G(L)\xi_{1j},$$

where

$$T(\hat{\beta}_{\text{FM}} - \beta) = \left(\frac{1}{T^2}\sum_{j=1}^{T}y_{1j}y'_{1j}\right)^{-1}\left[\frac{1}{T}\sum_{j=1}^{T}y_{1j}\right.$$

$$\times \left(\frac{c}{T}\frac{\xi_{2j}}{1-L} + \gamma(L)\xi_{2j} - \hat{\Omega}_{21}\hat{\Omega}_{11}^{-1}G(L)\xi_{1j}\right)$$

$$\left. - \hat{\Lambda}_{12} + \hat{\Lambda}_{11}\hat{\Omega}_{11}^{-1}\hat{\Omega}_{12}\right].$$

Defining the auxiliary process $\{z_j\}$ as in the text, we deduce that

$$\mathcal{L}\left(T(\hat{\beta}_{\text{FM}} - \beta)\right) \to \mathcal{L}\left((J_1')^{-1}Y\right),$$

where

$$Y = J_3 \left(\int_0^1 w_1(t) w_1'(t) \, dt \right)^{-1} \int_0^1 w_1(t) \, dw_2(t)$$
$$+ \frac{c}{\gamma(1)} \left(\int_0^1 w_1(t) w_1'(t) \, dt \right)^{-1} \int_0^1 w_1(t) w'(t) \, dt \begin{pmatrix} J_2 \\ J_3 \end{pmatrix}.$$

Then it follows that

$$\mathcal{L}\left(\frac{1}{T} \left(\sum_{j=1}^T \hat{v}_j \right)^2 \right) \to \mathcal{L}\left((J_2', J_3) \, w(1) - Y' \int_0^1 w_1(t) \, dt - J_2' w_1(1) \right.$$
$$\left. + \frac{c}{\gamma(1)} (J_2', J_3) \, w(1) \right)^2$$
$$= \mathcal{L}\left(J_3^2 (Y_1 + cY_2)^2 \right).$$

Since $\text{plim } \hat{\Omega}_{22 \cdot 1} = \Omega_{22 \cdot 1} = J_3^2$, the theorem is established.

6.5 Under the fixed alternative, we have that $\{\hat{v}_j\}$ is I(1) so that $\sum_{j=1}^T \hat{v}_j = O_p(T\sqrt{T})$. Since $\hat{\Omega}_{22 \cdot 1}$ is constructed from the long-run variance of $\{y_{2j} - \hat{\beta}_{FM}' y_{1j}\}$, it holds that $\hat{\Omega}_{22 \cdot 1} = O_p(T)$. Thus $\hat{S}_{T1} = O_p(T)$.

6.6 The present model may be expressed as

(S21) $\qquad v = \delta e + (\kappa C + I_T) \Xi_2 \sim N\left(\delta e, \sigma^2 \Omega(\kappa)\right),$

where $e = (1, \ldots, 1)' : T \times 1$, $\Xi_2 = (\xi_{21}, \ldots, \xi_{2T})'$, and $\Omega(\kappa) = (\kappa C + I_T)(\kappa C' + I_T)$ with C being the random walk generating matrix. Let $L(\kappa, \delta, \sigma^2)$ be the log-likelihood for v. Then we have

$$L(\kappa, \delta, \sigma^2) = -\frac{T}{2} \log(2\pi\sigma^2) - \frac{1}{2} \log |\Omega(\kappa)|$$
$$- \frac{1}{2\sigma^2} (v - \delta e)' \Omega^{-1}(\kappa) (v - \delta e).$$

It is easy to obtain

$$\left. \frac{\partial^2 L}{\partial \kappa^2} \right|_{H_0} = \text{constant} + T \frac{v' MCC'Mv}{v' Mv},$$

where $M = I_T - ee'/T$. This gives us the LBIU statistic S_{T2}.

6.7 Consider

$$S_{T2} = \frac{1}{T^2} v'MCC'Mv \Big/ \frac{1}{T} v'Mv,$$

where v is defined in (S21) with κ replaced by c/T. We have

$$\frac{1}{T} v'Mv = \frac{1}{T} \Xi_2' \left(\frac{c}{T} C' + I_T\right) M \left(\frac{c}{T} C + I_T\right) \Xi_2$$

$$\to \sigma^2 \quad \text{in probability}.$$

Moreover, it follows from Theorem 5.13 that

$$\mathcal{L}\left(\frac{1}{T^2 \sigma^2} v'MCC'Mv\right) = \mathcal{L}\left(\frac{1}{T^2 \sigma^2} \Xi_2' \left(C'MC + \frac{c^2}{T^2}(C'MC)^2\right) \Xi_2 + o_p(1)\right)$$

$$\to \mathcal{L}\left(\sum_{n=1}^{\infty} \left(\frac{1}{\lambda_n} + \frac{c^2}{\lambda_n^2}\right) Z_n^2\right),$$

where $\{\lambda_n\}$ is a sequence of eigenvalues of the kernel $K(s,t) = \min(s,t) - st$. Since $\lambda_n = n^2 \pi^2$ in the present case, (11.75) is established.

6.8 We first note that

$$\tilde{v}_j - \bar{\tilde{v}} = \frac{c}{T} \frac{\tilde{\xi}_{2j} - \bar{\tilde{\xi}}_2}{1-L} + \gamma(L)(\tilde{\xi}_{2j} - \bar{\tilde{\xi}}_2) - (\tilde{\beta}_{FM} - \beta)'(y_{1j} - \bar{y}_1)$$
$$- \tilde{\Omega}_{21} \tilde{\Omega}_{11}^{-1} (\Delta y_{1j} - \Delta \bar{y}_1),$$

where

$$T(\tilde{\beta}_{FM} - \beta) = \left(\frac{1}{T^2} \sum_{j=1}^{T} (y_{1j} - \bar{y}_1)(y_{1j} - \bar{y}_1)'\right)^{-1} \left[\frac{1}{T} \sum_{j=1}^{T} (y_{1j} - \bar{y}_1)\right.$$

$$\times \left(\frac{c}{T} \frac{\tilde{\xi}_{2j}}{1-L} + \gamma(L)\tilde{\xi}_{2j} - \tilde{\Omega}_{21} \tilde{\Omega}_{11}^{-1} \Delta y_{1j}\right)$$

$$\left. - \tilde{\Lambda}_{12} + \tilde{\Lambda}_{11} \tilde{\Omega}_{11}^{-1} \tilde{\Omega}_{12}\right].$$

We can deduce that $\mathcal{L}(T(\tilde{\beta}_{FM} - \beta)) \to \mathcal{L}((J_1')^{-1} Z)$, where

$$Z = J_3 \left(\int_0^1 \tilde{w}_1(t) \tilde{w}_1'(t) \, dt\right)^{-1} \int_0^1 \tilde{w}_1(t) \, dw_2(t)$$

$$+ \frac{c}{\gamma(1)} \left(\int_0^1 \tilde{w}_1(t) \tilde{w}_1'(t) \, dt\right)^{-1} \int_0^1 \tilde{w}_1(t) \tilde{w}'(t) \, dt \begin{pmatrix} J_2 \\ J_3 \end{pmatrix}.$$

Defining the partial sum process

$$X_T(t) = \frac{1}{\sqrt{T}} \sum_{j=1}^{[Tt]} (\tilde{v}_j - \bar{\tilde{v}}) + (Tt - [Tt])\frac{1}{\sqrt{T}} (\tilde{v}_{[Tt]+1} - \bar{\tilde{v}}),$$

we obtain $\mathcal{L}(X_T) \to \mathcal{L}(X)$, where

$$X(t) = (J_2', J_3)(w(t) - tw(1)) - Z' \int_0^t \tilde{w}_1(s)\,ds - J_2'(w_1(t) - tw_1(1))$$

$$+ \frac{c}{\gamma(1)} (J_2', J_3) \int_0^t \tilde{w}(s)\,ds$$

$$= J_3(Z_1(t) + cZ_2(t)).$$

Then it follows that

$$\mathcal{L}(\tilde{S}_{T2}) = \mathcal{L}\left(\frac{1}{T} \sum_{j=1}^T X_T^2\left(\frac{j}{T}\right) \Big/ \tilde{\Omega}_{22\cdot 1}\right)$$

$$\to \mathcal{L}\left(\int_0^1 X^2(t)\,dt \Big/ J_3^2\right) = \mathcal{L}\left(\int_0^1 (Z_1(t) + cZ_2(t))^2\,dt\right).$$

8.1 It follows from (11.97) and (11.98) that

$$\mathcal{L}\left(\frac{1}{T^{2d}} Y_1'Y_1, \frac{1}{T^d} Y_1'\Xi_2\right) \to \mathcal{L}(V, U_1 + U_2).$$

Thus $\mathcal{L}(T^d(\hat{\beta}_{OLS} - \beta)) \to \mathcal{L}(V^{-1}(U_1 + U_2))$. Since

$$\frac{1}{T^{2d}} Y_1'P_{-d}Y_1 = \frac{1}{T^{2d}} Y_1'Y_1 + o_p(1), \qquad \frac{1}{T^d} Y_1'P_{-d}\Xi_2 = \frac{1}{T^d} Y_1'\Xi_2 + o_p(1),$$

we also have $\mathcal{L}(T^d(\hat{\beta}_{2SLS} - \beta)) \to \mathcal{L}(V^{-1}(U_1 + U_2))$. Noting that

$$\frac{1}{T^{2d}} Y_1'M_dY_1 = \frac{1}{T^{2d}} Y_1'Y_1 + o_p(1),$$

$$\frac{1}{T^d} Y_1'M_d\Xi_2 = \frac{1}{T^d} Y_1'\Xi_2 - \frac{1}{T^d} Y_1'\Xi_1 \left(\frac{1}{T} \Xi_1'\Xi_1\right)^{-1} \frac{1}{T} \Xi_1'\Xi_2,$$

we deduce that $\mathcal{L}(T^d(\hat{\beta}_{ML} - \beta)) \to \mathcal{L}(V^{-1}U_2)$.

8.2 Let us put

$$1 - L^4 = \prod_{k=1}^4 \left(1 - \frac{1}{\theta_k} L\right),$$

where $\theta_1 = 1$, $\theta_2 = -1$, $\theta_3 = i$, and $\theta_4 = -i$. Then it follows from (11.101) that

$$\Phi(L) = A_1 \left(1 + L + L^2 + L^3\right) + A_2 \left(1 - L + L^2 - L^3\right)$$
$$+ A_3 \left(1 - L^2\right)(1 - iL) + A_4 \left(1 - L^2\right)(1 + iL) + \left(1 - L^4\right) \tilde{\Phi}(L),$$

where A_4 must be the complex conjugate of A_3 since the coefficients of $\Phi(L)$ are real. Thus we may put $A_3 = G + iH$ and $A_4 = G - iH$ with G and H being real, which leads us to the expansion in (11.102).

8.3 Using (11.101) we can expand $\alpha_3(L)$ as

$$\alpha_3(L) = \gamma_1(1 - iL) + \gamma_2(1 + iL) + \left(1 + L^2\right) \tilde{\alpha}_3(L)$$
$$= \alpha_{30} + \alpha_{31}L + \left(1 + L^2\right) \tilde{\alpha}_3(L).$$

Then it is seen that $\alpha_3'(i)\Phi(i) = 0'$ is equivalent to $(\alpha_{30} + \alpha_{31}i)'\Phi(i) = 0'$.

8.4 The fact that $\mathcal{L}(N(\hat{\beta}_{OLS} - \beta)) \to \mathcal{L}(V^{-1}(U_1 + U_2 + m\Sigma_{12}))$ comes from the continuous mapping theorem and the remark above this theorem. Since

$$\frac{1}{T^2}Y_1'P_{-m}Y_1 = \frac{1}{T^2}Y_1'Y_1 + o_p(1), \qquad \frac{1}{T}Y_1'P_{-m}\Xi_2 = \frac{1}{T}Y_{-m}'\Xi_2 + o_p(1),$$

we obtain $\mathcal{L}\left(N\left(\hat{\beta}_{2SLS} - \beta\right)\right) \to \mathcal{L}\left(V^{-1}(U_1 + U_2)\right)$. Noting that

$$\frac{1}{T^2}Y_1'M_mY_1 = \frac{1}{T^2}Y_1'Y_1 + o_p(1),$$

$$\frac{1}{T}Y_1'M_m\Xi_2 = \frac{1}{T}Y_1'\Xi_2 - \frac{1}{T}Y_1'\Xi_1 \left(\frac{1}{T}\Xi_1'\Xi_1\right)^{-1} \frac{1}{T}\Xi_1'\Xi_2,$$

we deduce that

$$\mathcal{L}\left(\frac{1}{T^2}Y_1'M_mY_1, \frac{1}{T}Y_1'M_m\Xi_2\right) \to \mathcal{L}(V, U_2).$$

Thus we have $\mathcal{L}(N(\hat{\beta}_{ML} - \beta)) \to \mathcal{L}(V^{-1}U_2)$.

8.5 Given $w_1 = \{w_1(t)\}$, the quantity $X_{OLS}(m)$ is normal with

$$E(X_{OLS}(m) \mid w_1) = a^2 x \int_0^1 w_1'(t) w_1(t)\, dt - \frac{ab}{2} w_1'(1) w_1(1) + \frac{m}{2}(ab - 2d),$$

$$V(X_{OLS}(m) \mid w_1) = a^2 c^2 \int_0^1 w_1'(t) w_1(t)\, dt.$$

Then we obtain $E\{\exp(i\theta X_{OLS}(m))\} = \{\phi_1(\theta)\}^m$, which proves the theorem.

References

Abramowitz, M. and Stegun, I. A. (1972). *Handbook of Mathematical Functions*, Dover, New York.

Ahn, S. K. and Reinsel, G. C. (1990). "Estimation of partially nonstationary multivariate autoregressive model," *Journal of the American Statistical Association*, **85**, 813–823.

Ahtola, J. and Tiao, G. C. (1987). "Distributions of least squares estimators of autoregressive parameters for a process with complex roots on the unit circle," *Journal of Time Series Analysis*, **8**, 1–14.

Anderson, T. W. (1959). "On asymptotic distributions of estimates of parameters of stochastic difference equations," *Annals of Mathematical Statistics*, **30**, 676–687.

Anderson, T. W. (1971). *The Statistical Analysis of Time Series*, Wiley, New York.

Anderson, T. W. (1984). *An Introduction to Multivariate Statistical Analysis*, 2d ed., Wiley, New York.

Anderson, T. W. and Darling, D. A. (1952). "Asymptotic theory of certain 'goodness of fit' criteria based on stochastic processes," *Annals of Mathematical Statistics*, **23**, 193–212.

Anderson, T. W. and Kunitomo, N. (1992). "Tests of overidentification and predeterminedness in simultaneous equation models," *Journal of Econometrics*, **54**, 49–78.

Anderson, T. W. and Takemura, A. (1986). "Why do noninvertible estimated moving averages occur?," *Journal of Time Series Analysis*, **7**, 235–254.

Arnold, L. (1974). *Stochastic Differential Equations: Theory and Applications*, Wiley, New York.

Athreya, K. B. and Pantula, S. G. (1986). "A note on strong mixing of ARMA processes," *Statistics and Probability Letters*, **4**, 187–190.

Beaulieu, J. J. and Miron, J. A. (1993). "Seasonal unit roots in aggregate U.S. data," *Journal of Econometrics*, **55**, 305–328.

Bellman, R. (1970). *Introduction to Matrix Analysis*, 2d ed., McGraw-Hill, New York.

Beveridge, S. and Nelson, C. R. (1981). "A new approach to decomposition of economic time series into permanent and transitory components with particular attention to measurement of the 'business cycle'," *Journal of Monetary Economics*, **7**, 151–174.

Bhargava, A. (1986). "On the theory of testing for unit roots in observed time series," *Review of Economic Studies*, **53**, 369-384.

Billingsley, P. (1968). *Convergence of Probability Measures*, Wiley, New York.

Billingsley, P. (1986). *Probability and Measure*, 2d ed., Wiley, New York.

Bobkoski, M. J. (1983). "Hypothesis testing in nonstationary time series, " Ph.D. Thesis, University of Wisconsin.

Box, G. E. P. and Tiao, G. C. (1977). "A canonical analysis of multiple time series," *Biometrika*, **64**, 355–365.

Breitung, J. (1994). "Some simple tests of the MA unit root hypothesis," *Journal of Time Series Analysis*, **15**, 351–370.

Brown, B. M. (1971). "Martingale central limit theorems," *Annals of Mathematical Statistics*, **42**, 59–66.

Chan, N. H. and Wei, C. Z. (1988). "Limiting distributions of least squares estimates of unstable autoregressive processes," *Annals of Statistics*, **16**, 367–401.

Choi, I. (1993). "Asymptotic normality of the least-squares estimates for higher order autoregressive integrated processes with some applications," *Econometric Theory*, **9**, 263–282.

Chow, Y. S. and Teicher, H. (1988). *Probability Theory*, 2d ed., Springer-Verlag, New York.

Courant, R. and Hilbert, D. (1953). *Methods of Mathematical Physics*, Vol. I, Wiley, New York.

Cryer, J. D. and Ledolter, J. (1981). "Small sample properties of the maximum-likelihood estimator in the first-order moving average model," *Biometrika*, **68**, 191–194.

Daniels, H. E. (1954). "Saddlepoint approximations in statistics," *Annals of Mathematical Statistics*, **25**, 631–650.

Darling, D.A. (1955). "The Cramér-Smirnov test in the parametric case," *Annals of Mathematical Statistics*, **26**, 1–20.

Davis, R. A. and Dunsmuir, W. T. M. (1996). "Maximum likelihood estimation for MA(1) processes with a root on or near the unit circle," *Econometric Theory*, **12**, 1–29.

Dickey, D. A. (1976). "Estimation and hypothesis testing in nonstationary time series," Ph.D. Thesis, Iowa State University.

Dickey, D. A. and Fuller, W. A. (1979). "Distribution of the estimators for autoregressive time series with a unit root," *Journal of the American Statistical Association*, **74**, 427–431.

Dickey, D. A. and Fuller, W. A. (1981). "Likelihood ratio statistics for autoregressive time series with a unit root," *Econometrica*, **49**, 1057–1072.

Dickey, D. A., Bell, W. R., and Miller, R. B. (1986). "Unit roots in time series models: Tests and implications," *The American Statistician*, **40**, 12–26.

Dickey, D. A., Hasza, D. P., and Fuller, W. A. (1984). "Testing for unit roots in seasonal time series," *Journal of the American Statistical Association*, **79**, 355–367.

Donsker, M. D. (1951). "An invariance principle for certain probability limit theorems," *Memoirs of the American Mathematical Society*, **6**, 1–12.

Donsker, M. D. (1952). "Justification and extension of Doob's heuristic approach to the Kolmogorov–Smirnov theorems," *Annals of Mathematical Statistics*, **23**, 277–281.

Elliott, G., Rothenberg, T. J., and Stock, J. H. (1992). "Efficient tests for an autoregressive unit root," *mimeo*, National Bureau of Economic Research, Cambridge, MA.

Engle, R. F. and Granger, C. W. J. (1987). "Co-integration and error correction: Representation, estimation, and testing," *Econometrica*, **55**, 251–276.

Engle, R. F., Granger, C. W. J., Hylleberg, S., and Lee, H. S. (1993). "Seasonal cointegration," *Journal of Econometrics*, **55**, 275–298.

Engle, R. F. and Yoo, B. S. (1991). "Cointegrated economic time series: An overview with new results," in *Long-Run Economic Relationships*, Engle, R. F. and Granger, C. W. J. eds., Oxford University Press, Oxford.

Erdös, P. and Kac, M. (1946). "On certain limit theorems of the theory of probability," *Bulletin of the American Mathematical Society*, **52**, 292–302.

Evans, G. B. A. and Savin, N. E. (1981a). "The calculation of the limiting distribution of the least squares estimator of the parameter in a random walk model," *Annals of Statistics*, **9**, 1114–1118.

Evans, G. B. A. and Savin, N. E. (1981b). "Testing for unit roots: 1," *Econometrica*, **49**, 753–779.

Evans, G. B. A. and Savin, N. E. (1984). "Testing for unit roots: 2," *Econometrica*, **52**, 1241–1269.

Ferguson, T. S. (1967). *Mathematical Statistics: A Decision Theoretic Approach*, Academic Press, New York.

Fuller, W. A. (1976). *Introduction to Statistical Time Series*, Wiley, New York.

Fuller, W. A. (1985). "Nonstationary autoregressive time series," in *Handbook of Statistics 5*, Hannan, E. J., Krishnaiah, P. R., and Rao, M. M., eds., North-Holland, Amsterdam.

Gardner, L. A. (1969). "On detecting changes in the mean of normal variates," *Annals of Mathematical Statistics*, **40**, 116–126.

Girsanov, I. V. (1960). "On transforming a certain class of stochastic processes by absolutely continuous substitution of measures," *Theory of Probability and Its Applications*, **5**, 285–301.

Granger, C. W. J. (1981). "Some properties of time series data and their use in econometric model specification," *Journal of Econometrics*, **16**, 121–130.

Granger, C. W. J. and Newbold, P. (1974). "Spurious regressions in econometrics," *Journal of Econometrics*, **2**, 111–120.

Hall, A. (1989). "Testing for a unit root in the presence of moving average errors," *Biometrika*, **76**, 49–56.

Hall, P. and Heyde, C. C. (1980). *Martingale Limit Theory and Its Application*, Academic Press, New York.

Hamilton, J. D. (1994). *Time Series Analysis*, Princeton University Press, Princeton.

Hannan, E. J. (1970). *Multiple Time Series*, Wiley, New York.

Hannan, E. J. and Heyde, C. C. (1972). "On limit theorems for quadratic functions of discrete time series," *Annals of Mathematical Statistics*, **43**, 2058–2066.

Hansen, B. E. (1992). "Tests for parameter instability in regressions with I(1) processes," *Journal of Business and Economic Statistics*, **10**, 321–335.

Hatanaka, M. (1996). *Time-Series-Based Econometrics: Unit Roots and Co-Integrations*, Oxford University Press, Oxford.

Helland, I. S. (1982). "Central limit theorems for martingales with discrete or continuous time," *Scandinavian Journal of Statistics*, **9**, 79–94.

Helstrom, C. W. (1978). "Approximate evaluation of detection probabilities in radar and optical communications," *IEEE Transactions on Aerospace and Electronic Systems*, **14**, 630–640.

Helstrom, C. W. (1995). *Elements of Signal Detection and Estimation*, Prentice Hall, Englewood Cliffs, NJ.

Hida, T. (1980). *Brownian Motion*, Springer-Verlag, New York.

Hochstadt, H. (1973). *Integral Equations*, Wiley, New York.

Huber, P. J. (1964). "Robust estimation of a location parameter," *Annals of Mathematical Statistics*, **35**, 73–101.

Hylleberg, S., Engle, R. F., Granger, C. W. J., and Yoo, B. S. (1990). "Seasonal integration and cointegration," *Journal of Econometrics*, **44**, 215–238.

Imhof, J. P. (1961). "Computing the distribution of quadratic forms in normal variables," *Biometrika*, **48**, 419–426.

Jazwinski, A. H. (1970). *Stochastic Processes and Filtering Theory*, Academic Press, New York.

Jeganathan, P. (1991). "On the asymptotic behavior of least-squares estimators in AR time series with roots near the unit circle," *Econometric Theory*, **7**, 269–306.

Johansen, S. (1988). "Statistical analysis of cointegrating vectors," *Journal of Economic Dynamics and Control*, **12**, 231–254.

Johansen, S. (1991). "Estimation and hypothesis testing of cointegration vectors in Gaussian vector autoregressive model," *Econometrica*, **59**, 1551–1580.

Johansen, S. (1995). "A statistical analysis of cointegration for I(2) variables," *Econometric Theory*, **11**, 25–59.

Johansen, S. and Juselius, K. (1990). "Maximum likelihood estimation and inference on cointegration with applications to the demand for money," *Oxford Bulletin of Economics and Statistics*, **52**, 109–210.

Kac, M., Kiefer, J., and Wolfowitz, J. (1955). "On tests of normality and other tests of goodness of fit based on distance methods," *Annals of Mathematical Statistics*, **26**, 189–211.

Kang, K. M. (1975). "A comparison of estimators for moving average processes," unpublished technical report, Australian Bureau of Statistics, Canberra.

Kariya, T. (1980). "Locally robust tests for serial correlation in least squares regression," *Annals of Statistics*, **8**, 1065–1070.

King, M.A. (1980). "Robust tests for spherical symmetry and their application to least squares regression," *Annals of Statistics*, **8**, 1265–1271.

King, M.A. (1988). "Towards a theory of point optimal tests," *Econometric Reviews*, **6**, 169–218.

King, M.A. and Hillier, G. H. (1985). "Locally best invariant tests of the error covariance matrix of the linear regression model," *Journal of the Royal Statistical Society*, (B), **47**, 98–102.

Kitamura, Y. (1995). "Estimation of cointegrated systems with I(2) processes," *Econometric Theory*, **11**, 1–24.

Knight, J. L. and Satchell, S. E. (1993). "Asymptotic expansions for random walks with normal errors," *Econometric Theory*, **9**, 363–376.

Kwiatkowski, D., Phillips, P. C. B., Schmidt, P., and Shin, Y. (1992). "Testing the null hypothesis of stationarity against the alternative of a unit root," *Journal of Econometrics*, **54**, 159–178.

Lee, H. S. (1992). "Maximum likelihood inference on cointegration and seasonal cointegration," *Journal of Econometrics*, **54**, 1–47.

Liptser, R. S. and Shiryayev, A. N. (1977). *Statistics of Random Processes I: General Theory*, Springer-Verlag, New York.

Liptser, R. S. and Shiryayev, A. N. (1978). *Statistics of Random Processes II: Applications*, Springer-Verlag, New York.

Loève, M. (1977). *Probability Theory I*, 4th ed., Springer-Verlag, New York.

Loève, M. (1978). *Probability Theory II*, 4th ed., Springer-Verlag, New York.

Longman, I. M. (1956). "Note on a method for computing infinite integrals of oscillatory functions," *Proceedings of the Cambridge Philosophical Society*, **52**, 764–768.

Lütkepohl, H. (1993). *Introduction to Multiple Time Series Analysis*, 2d ed., Springer-Verlag, New York.

MacNeill, I. B. (1974). "Tests for change of parameter at unknown times and distributions of some related functionals on Brownian motion," *Annals of Statistics*, **2**, 950–962.

MacNeill, I. B. (1978). "Properties of sequences of partial sums of polynomial regression residuals with applications to tests for change of regression at unknown times," *Annals of Statistics*, **6**, 422–433.

Mandelbrot, B. B. and Van Ness, J. W. (1968). "Fractional Brownian motions, fractional Brownian noises and applications," *SIAM Review*, **10**, 422–437.

REFERENCES

McLeish, D. L. (1975a). "A maximal inequality and dependent strong laws," *Annals of Probability*, **3**, 829–839.

McLeish, D. L. (1975b). "Invariance principles for dependent variables," *Zeitschrift für Wahrscheinlichkeitstheorie und Verwandte Gebiete*, **32**, 165–178.

McLeish, D. L. (1977). "On the invariance principle for nonstationary mixingales," *Annals of Probability*, **5**, 616–621.

Nabeya, S. (1989). "Asymptotic distributions of test statistics for the constancy of regression coefficients under a sequence of random walk alternatives," *Journal of the Japan Statistical Society*, **19**, 23–33.

Nabeya, S. (1992). "Limiting moment generating function of Cramér–von Mises–Smirnov goodness of fit statistics under null and local alternatives," *Journal of the Japan Statistical Society*, **22**, 113–122.

Nabeya, S. and Perron, P. (1994). "Local asymptotic distributions related to the AR(1) model with dependent errors," *Journal of Econometrics*, **62**, 229–264.

Nabeya, S. and Sørensen, B. E. (1994). "Asymptotic distributions of the least squares estimators and test statistics in the near unit root model with non-zero initial value and local drift and trend," *Econometric Theory*, **11**, 937–966.

Nabeya, S. and Tanaka, K. (1988). "Asymptotic theory of a test for the constancy of regression coefficients against the random walk alternative," *Annals of Statistics*, **16**, 218–235.

Nabeya, S. and Tanaka, K. (1990a). "A general approach to the limiting distribution for estimators in time series regression with nonstable autoregressive errors," *Econometrica*, **58**, 145–163.

Nabeya, S. and Tanaka, K. (1990b). "Limiting powers of unit-root tests in time-series regression," *Journal of Econometrics*, **46**, 247–271.

Nyblom, J. and Mäkeläinen, T. (1983). "Comparisons of tests for the presence of random walk coefficients in a simple linear model," *Journal of the American Statistical Association*, **78**, 856–864.

Park, J. Y. and Phillips, P. C. B. (1988). "Statistical inference in regressions with integrated processes: Part 1," *Econometric Theory*, **4**, 468–497.

Perron, P. (1989). "The calculation of the limiting distribution of the least-squares estimator in a near-integrated model," *Econometric Theory*, **5**, 241–255.

Perron, P. (1991a). "A continuous-time approximation to the unstable first-order autoregressive model: The case without an intercept," *Econometrica*, **59**, 211–236.

Perron, P. (1991b). "A continuous-time approximation to the stationary first-order autoregressive model," *Econometric Theory*, **7**, 236–252.

Phillips, P. C. B. (1977). "Approximations to some finite sample distributions associated with a first-order stochastic difference equation," *Econometrica*, **45**, 463–485.

Phillips, P. C. B. (1978). "Edgeworth and saddlepoint approximations in the first-order non-circular autoregression," *Biometrika*, **65**, 91–98.

Phillips, P. C. B. (1986). "Understanding spurious regressions in econometrics," *Journal of Econometrics*, **33**, 311–340.

Phillips, P. C. B. (1987a). "Time series regression with a unit root," *Econometrica*, **55**, 277–301.

Phillips, P.C. B. (1987b). "Towards a unified asymptotic theory for autoregression," *Biometrika*, **74**, 535–547.

Phillips, P. C. B. (1988). "Weak convergence of sample covariance matrices to stochastic integrals via martingale approximations," *Econometric Theory*, **4**, 528–533.

Phillips, P. C. B. (1989). "Partially identified econometric models," *Econometric Theory*, **5**, 181–240.

Phillips, P. C. B. (1991). "Optimal inference in cointegrated systems," *Econometrica*, **59**, 283–306.

Phillips, P. C. B. and Durlauf, S. N. (1986). "Multiple time series regression with integrated processes," *Review of Economic Studies*, **53**, 473–495.

Phillips, P. C. B. and Hansen, B. E. (1990). "Statistical inference in instrumental variables regression with I(1) processes," *Review of Economic Studies*, **57**, 99–125.

Phillips, P. C. B. and Ouliaris, S. (1990). "Asymptotic properties of residual based tests for cointegration," *Econometrica*, **58**, 165–193.

Phillips, P. C. B. and Perron, P. (1988). "Testing for a unit root in time series regression," *Biometrika*, **75**, 335–346.

Phillips, P. C. B. and Solo, V. (1992). "Asymptotics for linear processes," *Annals of Statistics*, **20**, 971–1001.

Pötscher, B. M. (1991). "Noninvertibility and pseudo-maximum likelihood estimation of misspecified ARMA models," *Econometric Theory*, **7**, 435–449.

Prakasa Rao, B. L. S. (1986). *Asymptotic Theory of Statistical Inference*, Wiley, New York.

Quintos, C. E. and Phillips, P. C. B. (1993). "Parameter constancy in cointegrating regressions," *Empirical Economics*, **18**, 675–706.

Rao, C. R. (1973). *Linear Statistical Inference and Its Applications*, 2d ed., Wiley, New York.

Roussas, G. G. (1973). *A First Course in Mathematical Statistics*, Addison-Wesley, Reading, MA.

Rutherford, D. E. (1946). "Some continuant determinants arising in physics and chemistry," *Proceedings of the Royal Society of Edinburgh*, **A-62**, 229–236.

Said, E. S. and Dickey, D. A. (1984). "Testing for unit roots in autoregressive-moving average models of unknown order," *Biometrika*, **71**, 599–607.

Saikkonen, P. (1991). "Asymptotically efficient estimation of cointegration regressions," *Econometric Theory*, **7**, 1–21.

Saikkonen, P. and Luukkonen, R. (1993a). "Testing for a moving average unit root in autoregressive integrated moving average models," *Journal of the American Statistical Association*, **88**, 596–601.

Saikkonen, P. and Luukkonen, R. (1993b). "Point optimal tests for testing the order of differencing in ARIMA models," *Econometric Theory*, **9**, 343–362.

Sargan, J. D. and Bhargava, A. (1983). "Maximum likelihood estimation of regression models with first order moving average errors when the root lies on the unit circle," *Econometrica*, **51**, 799–820.

Schweppe, F. C. (1965). "Evaluation of likelihood functions for Gaussian signals," *IEEE Information Theory*, **IT-11**, 61–70.

Schwert, G. W. (1989). "Tests for unit roots: A Monte Carlo investigation," *Journal of Business and Economic Statistics*, **7**, 147–159.

Sen, A. K. and Srivastava, M. S. (1973). "On multivariate tests for detecting change in mean," *Sankhya A*, **35**, 173–186.

Shephard, N. (1993). "Distribution of the ML estimator of an MA(1) and a local level model," *Econometric Theory*, **9**, 377–401.

Shephard, N. and Harvey, A. C. (1990). "On the probability of estimating a deterministic component in the local level model," *Journal of Time Series Analysis*, **11**, 339–347.

Shin, Y. (1994). "A residual-based test of the null of cointegration against the alternative of no cointegration," *Econometric Theory*, **10**, 91–115.

Shiryayev, A. N. (1984). *Probability*, Springer-Verlag, New York.

Shorack, G. R. and Wellner, J. A. (1986). *Empirical Processes with Applications to Statistics*, Wiley, New York.

Sims, C. A., Stock, J. H., and Watson, M. W. (1990). "Inference in linear time series models with some unit roots," *Econometrica*, **58**, 113–144.
Solo, V. (1984). "The order of differencing in ARIMA models," *Journal of the American Statistical Association*, **79**, 916–921.
Soong, T. T. (1973). *Random Differential Equations in Science and Engineering*, Academic Press, New York.
Stock, J. H. (1987). "Asymptotic properties of least squares estimators of cointegrating vectors," *Econometrica*, **55**, 1035–1056.
Tanaka, K. (1983a). "The one-sided Lagrange multiplier test of the AR(p) model vs the AR(p) model with measurement error," *Journal of the Royal Statistical Society*, (B), **45**, 77–80.
Tanaka, K. (1983b). "Non-normality of the Lagrange multiplier statistic for testing the constancy of regression coefficients," *Econometrica*, **51**, 1577–1582.
Tanaka, K. (1983c). "Asymptotic expansions associated with the AR(1) model with unknown mean," *Econometrica*, **51**, 1221–1231.
Tanaka, K. (1984). "An asymptotic expansion associated with the maximum likelihood estimators in ARMA models," *Journal of the Royal Statistical Society*, (B), **46**, 58–67.
Tanaka, K. (1990a). "The Fredholm approach to asymptotic inference on nonstationary and noninvertible time series models," *Econometric Theory*, **6**, 411–432.
Tanaka, K. (1990b). "Testing for a moving average unit root," *Econometric Theory*, **6**, 433–444.
Tanaka, K. (1993). "An alternative approach to the asymptotic theory of spurious regression, cointegration, and near cointegration," *Econometric Theory*, **9**, 36–61.
Tanaka, K. (1995). "The optimality of extended score tests with applications to testing for a moving average unit root," in *Advances in Econometrics and Quantitative Economics*, Maddala, G. S., Phillips, P. C. B., and Srinivasan, T. N., eds., Blackwell, Oxford.
Tanaka, K. and Satchell, S.E. (1989). "Asymptotic properties of the maximum-likelihood and nonlinear least-squares estimators for noninvertible moving average models," *Econometric Theory*, **5**, 333–353.
Tsay, R. S. (1993). "Testing for noninvertible models with applications," *Journal of Business and Economic Statistics*, **11**, 225–233.
Tso, M. K. S. (1981). "Reduced-rank regression and canonical analysis," *Journal of the Royal Statistical Society*, (B), **43**, 183–189.
Watson, G. N. (1958). *A Treatise on the Theory of Bessel Functions*, 2d ed., Cambridge University Press, London.
Watson, G. S. (1961). "Goodness-of-fit tests on a circle," *Biometrika*, **48**, 109–114.
White, J. S. (1958). "The limiting distribution of the serial correlation coefficient in the explosive case," *Annals of Mathematical Statistics*, **29**, 1188–1197.
Whittaker, E. T. and Watson, G. N. (1958). *A Course of Modern Analysis*, 4th ed., Cambridge University Press, London.
Withers, C. S. (1981). "Conditions for linear processes to be strong-mixing," *Zeitschrift für Wahrscheinlichkeitstheorie und Verwandte Gebiete*, **57**, 477–480.
Yoshihara, K. (1992). *Weakly Dependent Stochastic Sequences and Their Applications* Vol. I, Sanseido, Tokyo.
Yoshihara, K. (1993). *Weakly Dependent Stochastic Sequences and Their Applications* Vol. II, Sanseido, Tokyo.

Author Index

Abramowitz, M., 145
Ahn, S. K., 453
Ahtola, J., 264, 267
Anderson, T. W., 6, 7, 8, 46, 136, 148, 216, 248, 280, 281, 364, 453
Arnold, L., 121, 122
Athreya, K. B., 70

Beaulieu, J. J., 466
Bell, W. R., 367
Bellman, R., 124
Beveridge, S., 77
Bhargava, A., 6, 281, 287
Billingsley, P., 39, 66, 67, 68, 69, 70, 72, 85, 93, 95, 96, 97, 287, 495
Bobkoski, M. J., 17, 91, 216, 217, 237
Box, G. E. P., 453
Breitung, J., 405
Brown, B. M., 80

Chan, N. H., 39, 49, 105, 259, 264, 265, 266, 270, 271
Choi, I., 272
Chow, Y. S., 82
Courant, R., 131, 159
Cryer, J. D., 6, 281

Daniels, H. E., 207
Darling, D. A., 8, 46, 136, 148, 153
Davis, R. A., 6, 281, 284, 288, 289, 290, 293
Dickey, D. A., 216, 253, 262, 322, 335, 355, 361, 367
Donsker, M. D., 68
Dunsmuir, W. T. M., 6, 281, 284, 288, 289, 290, 293
Durlauf, S. N., 426

Elliott, G., 344, 345

Engle, R. F., 28, 419, 441, 445, 447, 462, 463, 466
Erdös, P., 3, 10
Evans, G. B. A., 16, 17, 215, 216, 322
Ferguson, T. S., 324, 325
Fuller, W. A., 17, 77, 214, 215, 216, 224, 229, 231, 253, 262, 268, 322, 335, 355, 361, 366, 367

Gardner, L. A., 4
Girsanov, I. V., 110
Granger, C. W. J., 24, 28, 417, 419, 441, 445, 447, 462, 466

Hall, A., 367
Hall, P., 67, 68, 80, 82
Hamilton, J. D., 322, 453
Hannan, E. J., 77, 82, 433, 510
Hansen, B. E., 419, 436, 438, 441, 442, 443, 445, 452
Harvey, A. C., 316
Hasza, D. P., 253, 262, 355, 361
Hatanaka, M., 322, 366, 453, 458
Helland, I. S., 265
Helstrom, C. W., 207, 211
Heyde, C. C., 67, 68, 80, 82
Hida, T., 22, 39
Hilbert, D., 131, 159
Hillier, G. H., 329
Hochstadt, H., 131, 133, 134, 135, 159
Huber, P. J., 289
Hylleberg, S., 462, 466

Imhof, J. P., 2, 188, 249

Jazwinski, A. H., 53, 57, 58, 62, 379
Jeganathan, P., 270

Johansen, S., 419, 441, 453, 454, 455, 457, 458, 459
Juselius, K., 458

Kac, M., 3, 10, 149
Kang, K. M., 5, 281
Kariya, T., 323
Kiefer, J., 149
King, M. A., 323, 329, 345
Kitamura, Y., 459
Knight, J. L., 15
Kunitomo, N., 453
Kwiatkowski, D., 5, 368

Ledolter, J., 6, 281
Lee, H. S., 466, 467
Liptser, R. S., 110, 111, 122, 124, 126, 217
Loève, M., 36, 37
Longman, I. M., 191
Lütkepohl, H., 453, 455
Luukkonen, R., 9, 370, 382, 383, 405

MacNeill, I. B., 4, 145, 186
Mäkeläinen, T., 3, 8, 406
Mandelbrot, B. B., 50
McLeish, D. L., 70
Miller, R. B., 367
Miron, J. A., 467

Nabeya, S., 3, 18, 137, 145, 148, 150, 153, 154, 163, 164, 165, 166, 167, 171, 172, 175, 183, 186, 214, 217, 237, 259, 274, 322, 333, 335, 363, 368, 381, 406, 408, 415, 552
Nelson, C. R., 77
Newbold, P., 24, 419
Nyblom, J., 8, 406

Ouliaris, S., 419, 445, 446, 448

Pantula, S. G., 70
Park, J. Y., 436
Perron, P., 17, 114, 217, 237, 239, 259, 261, 274, 364, 367
Phillips, P. C. B., 5, 70, 77, 81, 85, 105, 115, 216, 217, 237, 245, 247, 250, 259, 364, 367, 368, 419, 420, 422, 423, 424, 426, 427, 436, 438, 441, 442, 443, 445, 446, 447, 448, 452
Pötscher, B. M., 309
Prakasa Rao, B. L. S., 427

Quintos, C. E., 445, 452

Rao, C. R., 71, 72, 96, 330, 496

Reinsel, G. C., 453
Rothenberg, T. J., 344, 345
Roussas, G. G., 500
Rutherford, D. E., 2

Said, E. S., 367
Saikkonen, P., 9, 369, 382, 383, 405, 436
Sargan, J. D., 6, 281, 287
Satchell, S. E., 6, 15, 281, 284, 297, 311
Savin, N. E., 16, 17, 215, 216, 322
Schmidt, P., 5, 368
Schweppe, F. C., 379
Schwert, G. W., 367
Sen, A. K., 4, 257
Shephard, N., 289, 316, 317
Shin, Y., 5, 368, 445, 452
Shiryayev, A. N., 66, 110, 111, 122, 124, 126, 217
Shorack, G. R., 73
Sims, C. A., 272
Solo, V., 70, 77, 81, 85, 367
Soong, T. T., 41, 42, 43, 45, 53, 71
Sørensen, B. E., 237
Srivastava, M. S., 4, 257
Stegun, I. A., 145
Stock, J. H., 272, 344, 345, 419, 438, 439, 440

Takemura, A., 6, 281
Tanaka, K., 3, 6, 18, 137, 145, 148, 150, 160, 167, 171, 172, 174, 175, 177, 183, 186, 214, 217, 237, 250, 281, 284, 290, 297, 311, 322, 329, 333, 335, 353, 363, 368, 381, 406, 415, 420, 421, 440, 445, 552
Teicher, H., 82
Tiao, G. C., 264, 267, 453
Tsay, R. S., 405
Tso, M. K. S., 453

Van Ness, J. W., 50

Watson, G. N., 131, 134, 142, 146
Watson, G. S., 141
Watson, M. W., 272
Wei, C. Z., 39, 49, 105, 259, 264, 265, 266, 270, 271
Wellner, J. A., 73
White, J. S., 14, 15, 114, 216, 237
Whittaker, E. T., 131, 134, 142
Withers, C. S., 70
Wolfowitz, J., 149

Yoo, B. S., 462, 463
Yoshihara, K., 70

Subject Index

Abel's transformation, 90
Accumulation point, 135
Adapted, 79
Analytic function, *see* Function
Anderson-Darling statistic, *see* Statistic
Approximate distribution, *see* Distribution
AR model, 176
  nonstationary, 213
  seasonal, 176, 253, 355
AR unit root, *see* Unit root
ARMA process, 281, 366, 403
Asymptotic expansion, 2, 7, 12, 15
Augmented model, 354, 425
Autocorrelation, 217
Auxiliary process, 109, 126, 426

Bernoulli number, 14, 23
Bessel function:
  modified, 8
  of the first kind, 145
  of the second kind, 145
Bessel's equation, 145
Beveridge-Nelson (BN) decomposition, 77, 98
Bilinear form, 129, 157, 178
Bisection method, 203
Block lower triangular, 97, 417
Boundary condition, 138
Boundary-crossing probability, 73
Bounded variation, 44
Brownian bridge, 8, 39, 69
Brownian motion, 4, 39, 69

Cameron-Martin formula, 124
Canonical correlation, 453
Cauchy-Schwarz inequality, 485, 506
Central limit theorem (CLT), 470, 492
  Lyapunov, 470

Characteristic function (c.f.), 2
  limiting, 14
Chebyshev's inequality, 80, 288
Cholesky decomposition, 378, 404
Cointegrated system, 28, 116, 419
Cointegrating matrix, 454
Cointegrating vector, 419
Cointegration, 33, 419
  distribution, 424, 433
  full, 463
  higher order, 116, 121, 458
  no, 419
  polynomial, 463
  rank, 418
  seasonal, 462
Column space, 325, 331, 332, 334, 357
Commute, 36, 42, 281, 295
Compactness, 66
  relative, 67, 97
Completeness, 66, 95, 97
Complex plane, 142
Complex roots on the unit circle, 264
Component space, 95
Computerized algebra, 121, 151, 483, 595
Concentrated log-likelihood, *see* Likelihood
Concentration probability, 427
Conditional argument, 114, 120, 237, 461
Conditional case, 279, 294, 373
Consistency, 281, 302, 309, 311
Continuous kernel, *see* Kernel
Continuous mapping, 70, 92
Continuous mapping theorem, 70
Continuous record asymptotics, 259, 304
Convergence:
  determining class, 67
  in distribution, 67
  in mean square (m.s.), 35
  in probability, 36

619

Convergence (*Continued*)
  with probability one, 74
  of stochastic processes, 66, 86, 90, 95, 102
Convolution, 256, 359, 360, 466
Cramér-Wold device, 287
Cumulant, 13, 23

Definite kernel, *see* Kernel
Degenerate kernel, *see* Kernel
Demeaned Brownian motion, 27, 70
Dependent process, 362, 402
Determining class, 67, 144
Deterministic trend, 214
Dickey-Fuller test, 334
Difference stationarity, 367
Differenced form, 383
Distribution:
  approximate, 251, 290
  beta, 325, 375
  binomial, 214
  Cauchy, 247, 261
  χ^2, 3
  exact, 2, 12, 249
  finite-dimensional, 67, 95, 96, 144
  finite sample, 215, 249
  limiting, 3, 8, 10, 12, 22, 31
  limiting null, 333, 335, 377, 388, 399, 403, 446
  normal, 1
  Poisson, 214
  symmetric, 19, 26, 422, 440
  uniform, 69
Donsker's theorem, 69
Durbin-Watson test, 334

Edgeworth-type approximation, 250, 290
Eigenfunction, 132
  orthonormal, 144
Eigenvalue, 132
  approach, 2, 7, 281
  of a kernel, 132
  problem, 456
Entire function, *see* Function
Ergodic, 77
Error correction model, 440, 454
Euler's transformation, 191
Exact distribution, *see* Distribution

Finite-dimensional distribution, *see* Distribution
Finite sample distribution, *see* Distribution
Finite sample power, 351, 390
Fixed alternative, 451
Fixed initial value case, 233
Fortran, 186, 199, 203

Fractional integral, 50
Fredholm:
  approach, 129, 171
  determinant (FD), 132
  minor, 159
  resolvent, 159
  theory, 132, 157
Frequency, 364, 463
Full cointegration, *see* Cointegration
Fuller's estimator, 215, 224, 229
Fuller's representation, 268, 270
Fully modified estimator, 436
Function:
  analytic, 151, 159
  entire, 134
  integral, 134
  meromorphic, 159
  space, 65
Functional central limit theorem (FCLT), 68, 77, 79, 83, 96, 97, 99
Fundamental sequence, 66

Gaussian process, 36
Genus, 134, 136
Geometric Brownian motion, 59
Girsanov's theorem, 109, 117, 121
Global MLE, *see* MLE
GLS residual, *see* Residual
Granger representation theorem, 440
Group of transformations, 323, 334

Higher order:
  bias, 430, 440
  cointegration, *see* Cointegration
  integration, 121
  nonstationarity, 253, 259
Hilbert space projection, 447
Hölder's inequality, 85, 501
Homogeneous case, 79, 132

I(d) process, 28, 271, 458
Identifiable, 5, 279
Imhof's formula, 2, 188, 249
Independent increments, 38, 39, 60
Independent process, 36, 83
Indicator function, 43, 78
Induced measure, 66, 95, 110, 117, 122
Infinite product, 134, 135, 140
Integral equation, 132
  of Fredholm type, 131
Integral function, *see* Function
Integrated Brownian bridge, 50
Integrated Brownian motion, 43, 48, 86, 117, 131, 154
Integrated demeaned Brownian motion, 50

SUBJECT INDEX

Integrated process, 28, 51, 78
Integration:
  by parts, 45
  order, 9
Invariance principle (IP), 68
Invariant, 323, 334
Invariant unit root test, 342
Inversion formula, 11, 182
Ito calculus, 57
Ito integral, 53, 55
Ito stochastic differential equation (SDE), 58
Ito's theorem, 58
  extended version of, 62
  matrix version of, 122

Johansen procedure, 453

Kalman filter algorithm, 378
Kernel, 132
  continuous, 134
  definite, 135
  degenerate, 46, 134
  nearly definite, 135
  nondegenerate, 46, 134
  symmetric, 134
Kronecker:
  product, 486
  structure, 355
Kurtosis, 14

L_2:
  -completeness theorem, 36
  space, 35
Lag:
  polynomial, 449
  truncation number, 363, 403
Lagrange multiplier (LM) test, 2, 6, 324, 375
Laplace transform, 5, 264
Law of large numbers, 104, 492, 499, 510, 552, 558, 570
LBI and unbiased (LBIU), 325, 357, 370, 381, 395, 451
Least squares estimator (LSE), 14, 24, 218
  ordinary, 425
  two-stage, 425
Likelihood, 110, 118
  concentrated log-, 282, 294
  log-, 2, 6, 280, 379, 455
  ratio statistic, 2, 110, 335, 455
Limiting distribution, *see* Distribution
Limiting local power, 335, 385
Limiting power envelope, *see* Power envelope
Lindeberg condition, 80
Linearly independent, 132, 138, 140

Linear process, 77, 87, 97, 177
Linear time series model, 1, 279, 321, 373
Loading matrix, 454
Local alternative, 176
Locally best invariant (LBI), 324, 356, 368, 376, 393, 406, 449
Local MLE, *see* MLE
Local optimality, 322, 348
Local power, 175
Log-likelihood, *see* Likelihood
Long-run variance, 217

MA model, 5
  noninvertible, 5, 280
  seasonal, 300, 392
Marginal probability measure, 95, 97
Markov's inequality, 78, 506, 533, 535
Martingale, 54, 79
  difference, 79
MA unit root, *see* Unit root
Maximal invariant, 323
Maximum likelihood estimator (MLE), 6
  global, 281, 289
  local, 282, 294
  pseudolocal, 307
Mean square (m.s.):
  continuity, 37
  differentiability, 37
  Ito integral, *see* Ito integral
  Riemann integral, 42
  Riemann-Stieltjes integral, 43
Measurable, 71
Measurable mapping, 65, 72
Measurable space, 65, 95
Mercer's theorem, 135
Meromorphic function, *see* Function
Method of truncation, 500
Metric, 65, 72
Metric space, 72
Mirror-image property, 231
Mixing, 70
Mixingale, 70
Moment generating function (m.g.f.), 15
Most powerful invariant (MPI) test, 343, 376, 381
Multiple unit roots, 267
Multiplicity, 132
  of eigenvalues, 132, 139

Near cointegration, 449
Near integrated process, 61, 94, 115
Near random walk, 61, 93
Nearly definite kernel, *see* Kernel
Nearly noninvertible, 280

Negative unit root, see Unit root
Newton's method, 203
Neyman-Pearson lemma, 342, 376
No cointegration, see Cointegration
Nondegenerate kernel, see Kernel
Nonhomogeneous case, 79, 157
Noninvertible MA model, see MA model
Nonparametric construction, 369, 436
Nonstationary AR model, see AR model
Nuisance parameter, 323, 367, 403, 447
Null space, 139
Numerical derivative, 197
Numerical integration, 181

OLS residual, see Residual
Optimal test, 323
Ornstein-Uhlenbeck (O-U) process, 59
Orthonormal sequence, 135
Oscillating case, 186
Overdifferencing, 9, 279
Overidentification, 453

Partial sum process, 68
Partition, 45
Period, 176
Periodogram, 364
Point optimal invariant (POI) test, 345, 359, 391
Poisson process, 38
Polynomial cointegration, see Cointegration
Portmanteau theorem, 495
Power envelope, 342, 388, 390
Probability:
  measure, 66, 95
  space, 33, 65
Product measure, 95
Prohorov's theorem, 67
Pseudolocal MLE, see MLE

Quadratic form, 2, 131, 170, 334
Quadratic functional, 129, 136, 144
Quadrature, 125

Radon-Nikodym derivative, 110, 126, 282
Random walk generating matrix, 2
Random walk plus noise model, 1
Ratio statistic, 113, 115, 119
REDUCE, 151, 483, 484
Regression:
  model, 213
  spurious, 214, 420, 422
  time series, 333
Relative compactness, see Compactness

Residual:
  -based test, 445
  GLS, 333
  OLS, 333
Residual process, 447
Resolvent, see Fredholm resolvent
Reversed unit root test, 368
Riccati equation, 124

Saddlepoint, 208
  approximation, 207
  method, 207
Sample integral, 71
Sample path, 37, 40, 49
Score test, 2, 324
Seasonal AR model, see AR model
Seasonal cointegration, see Cointegration
Seasonal MA model, see MA model
Secant method, 210
Separability, 66, 95, 97
Short-run variance, 217
σ-field, 65, 79
Signal-to-noise ratio, 314
Simple closed curve, 142
Simpson's rule, 186
Simultaneous equation, 453
Singularity point, 207
Size distortion, 364, 403, 405
Skewness, 14
Skew symmetric, 56
Slowly convergent alternating series, 191
Spectral density, 217
Spectral estimator, 433
Spectral method, 433, 436
Spectrum, 364, 433
Spurious regression, see Regression
Square integrable, 79
Square integrable martingale difference, 80
Square root of a complex variable, 11, 182
Standard Brownian motion, see Brownian motion
State space model, 1, 314, 405
Stationarity:
  second-order, 77
  strict, 77
Stationary case, 237, 279, 282, 377
Stationary process, 60, 213, 417
Statistic:
  Anderson-Darling, 148
  Dickey-Fuller, 334
  Durbin-Watson, 334
  LBI, 333
  LBIU, 333

SUBJECT INDEX

t-ratio, 76, 335, 422, 448, 457
von Neumann ratio, 334
Stochastic area, 22, 123, 140
Stochastic differential, 58
Stochastic integral, *see* Mean square integral
Stochastic order, 424
Stochastic process, 65
Stochastic process approach, 76, 109
Stochastic trend, 214
Stratonovich integral, 54
Strictly stationary, *see* Stationarity
Strongly uniformly integrable (s.u.i.), 81
Summability condition, 78, 462
Super consistent, 433
Symmetric distribution, *see* Distribution
Symmetric kernel, *see* Kernel

Tangency, 349, 391
Test:
  for an AR unit root, 321
  for cointegration, 449
  for an MA unit root, 5, 373
  for no cointegration, 445
  for parameter constancy, 1, 406
  for a seasonal AR unit root, 355
  for a seaonal MA unit root, 392
Tightness, 67, 95, 97
Transformation of measures, 109

Trend stationarity, 214
Triangle inequality, 494
Triangular array, 83

Uniform limit, 176
Uniformly integrable, 81
Uniform metric, 65, 95
Unit root:
  AR, 4, 322
  component, 427, 435
  distribution, 239, 427
  MA, 5, 373
  near, 230
  negative, 230
  seasonal AR, 355
  seasonal MA, 392
  test, *see* Test

VAR model, 453
Von Neumann ratio, *see* Statistic

Weak convergence:
  in function space, 67
  of statistics, 76, 170
Weber's function, 145
Weierstrass approximation theorem, 494

Yule-Walker estimator, 14, 352

WILEY SERIES IN PROBABILITY AND STATISTICS

ESTABLISHED BY WALTER A. SHEWHART AND SAMUEL S. WILKS

Editors
*Vic Barnett, Ralph A. Bradley, Nicholas I. Fisher, J. Stuart Hunter,
J. B. Kadane, David G. Kendall, David W. Scott, Adrian F. M. Smith,
Jozef L. Teugels, Geoffrey S. Watson*

Probability and Statistics
   ANDERSON · An Introduction to Multivariate Statistical Analysis, *Second Edition*
 *ANDERSON · The Statistical Analysis of Time Series
   ARNOLD, BALAKRISHNAN, and NAGARAJA · A First Course in Order Statistics
   BACCELLI, COHEN, OLSDER, and QUADRAT · Synchronization and Linearity:
     An Algebra for Discrete Event Systems
   BARTOSZYNSKI and NIEWIADOMSKA-BUGAJ · Probability and Statistical Inference
   BERNARDO and SMITH · Bayesian Statistical Concepts and Theory
   BHATTACHARYYA and JOHNSON · Statistical Concepts and Methods
   BILLINGSLEY · Convergence of Probability Measures
   BILLINGSLEY · Probability and Measure, *Second Edition*
   BOROVKOV · Asymptotic Methods in Queuing Theory
   BRANDT, FRANKEN, and LISEK · Stationary Stochastic Models
   CAINES · Linear Stochastic Systems
   CAIROLI and DALANG · Sequential Stochastic Optimization
   CHEN · Recursive Estimation and Control for Stochastic Systems
   CONSTANTINE · Combinatorial Theory and Statistical Design
   COOK and WEISBERG · An Introduction to Regression Graphics
   COVER and THOMAS · Elements of Information Theory
   CSÖRGŐ and HORVÁTH · Weighted Approximations in Probability Statistics
 *DOOB · Stochastic Processes
   DUDEWICZ and MISHRA · Modern Mathematical Statistics
   ETHIER and KURTZ · Markov Processes: Characterization and Convergence
   FELLER · An Introduction to Probability Theory and Its Applications, Volume 1,
     Third Edition, Revised; Volume II, *Second Edition*
   FREEMAN and SMITH · Aspects of Uncertainty: A Tribute to D. V. Lindley
   FULLER · Introduction to Statistical Time Series, *Second Edition*
   FULLER · Measurement Error Models
   GIFI · Nonlinear Multivariate Analysis
   GUTTORP · Statistical Inference for Branching Processes
   HALD · A History of Probability and Statistics and Their Applications before 1750
   HALL · Introduction to the Theory of Coverage Processes
   HANNAN and DEISTLER · The Statistical Theory of Linear Systems
   HEDAYAT and SINHA · Design and Inference in Finite Population Sampling
   HOEL · Introduction to Mathematical Statistics, *Fifth Edition*
   HUBER · Robust Statistics
   IMAN and CONOVER · A Modern Approach to Statistics
   JUREK and MASON · Operator-Limit Distributions in Probability Theory
   KAUFMAN and ROUSSEEUW · Finding Groups in Data: An Introduction to Cluster
     Analysis
   LAMPERTI · Probability: A Survey of the Mathematical Theory
   LARSON · Introduction to Probability Theory and Statistical Inference, *Third Edition*
   LESSLER and KALSBEEK · Nonsampling Error in Surveys
   LINDVALL · Lectures on the Coupling Method
   MANTON, WOODBURY, and TOLLEY · Statistical Applications Using Fuzzy Sets

*Now available in a lower priced paperback edition in the Wiley Classics Library.

Probability and Statistics (Continued)
   MARDIA · The Art of Statistical Science: A Tribute to G. S. Watson
   MORGENTHALER and TUKEY · Configural Polysampling: A Route to Practical Robustness
   MUIRHEAD · Aspects of Multivariate Statistical Theory
   OLIVER and SMITH · Influence Diagrams, Belief Nets and Decision Analysis
   *PARZEN · Modern Probability Theory and Its Applications
   PRESS · Bayesian Statistics: Principles, Models, and Applications
   PUKELSHEIM · Optimal Experimental Design
   PURI and SEN · Nonparametric Methods in General Linear Models
   PURI, VILAPLANA, and WERTZ · New Perspectives in Theoretical and Applied Statistics
   RAO · Asymptotic Theory of Statistical Inference
   RAO · Linear Statistical Inference and Its Applications, *Second Edition*
   *RAO and SHANBHAG · Choquet-Deny Type Functional Equations with Applications to Stochastic Models
   RENCHER · Methods of Multivariate Analysis
   ROBERTSON, WRIGHT, and DYKSTRA · Order Restricted Statistical Inference
   ROGERS and WILLIAMS · Diffusions, Markov Processes, and Martingales, Volume I: Foundations, *Second Edition;* Volume II: Îto Calculus
   ROHATGI · An Introduction to Probability Theory and Mathematical Statistics
   ROSS · Stochastic Processes
   RUBINSTEIN · Simulation and the Monte Carlo Method
   RUBINSTEIN and SHAPIRO · Discrete Event Systems: Sensitivity Analysis and Stochastic Optimization by the Score Function Method
   RUZSA and SZEKELY · Algebraic Probability Theory
   SCHEFFE · The Analysis of Variance
   SEBER · Linear Regression Analysis
   SEBER · Multivariate Observations
   SEBER and WILD · Nonlinear Regression
   SERFLING · Approximation Theorems of Mathematical Statistics
   SHORACK and WELLNER · Empirical Processes with Applications to Statistics
   SMALL and McLEISH · Hilbert Space Methods in Probability and Statistical Inference
   STAPLETON · Linear Statistical Models
   STAUDTE and SHEATHER · Robust Estimation and Testing
   STOYANOV · Counterexamples in Probability
   STYAN · The Collected Papers of T. W. Anderson: 1943–1985
   TANAKA · Time Series Analysis: Nonstationary and Noninvertible Distribution Theory
   THOMPSON and SEBER · Adaptive Sampling
   WELSH · Aspects of Statistical Inference
   WHITTAKER · Graphical Models in Applied Multivariate Statistics
   YANG · The Construction Theory of Denumerable Markov Processes

Applied Probability and Statistics
   ABRAHAM and LEDOLTER · Statistical Methods for Forecasting
   AGRESTI · Analysis of Ordinal Categorical Data
   AGRESTI · Categorical Data Analysis
   AGRESTI · An Introduction to Categorical Data Analysis
   ANDERSON and LOYNES · The Teaching of Practical Statistics
   ANDERSON, AUQUIER, HAUCK, OAKES, VANDAELE, and WEISBERG · Statistical Methods for Comparative Studies
   ARMITAGE and DAVID · Advances in Biometry
   *ARTHANARI and DODGE · Mathematical Programming in Statistics
   ASMUSSEN · Applied Probability and Queues

*Now available in a lower priced paperback edition in the Wiley Classics Library.

Applied Probability and Statistics (Continued)

BAILEY · The Elements of Stochastic Processes with Applications to the Natural Sciences

BARNETT and LEWIS · Outliers in Statistical Data, *Second Edition*

BARTHOLOMEW, FORBES, and McLEAN · Statistical Techniques for Manpower Planning, *Second Edition*

BATES and WATTS · Nonlinear Regression Analysis and Its Applications

BECHHOFER, SANTNER, and GOLDSMAN · Design and Analysis of Experiments for Statistical Selection, Screening, and Multiple Comparisons

BELSLEY · Conditioning Diagnostics: Collinearity and Weak Data in Regression

BELSLEY, KUH, and WELSCH · Regression Diagnostics: Identifying Influential Data and Sources of Collinearity

BERRY · Bayesian Analysis in Statistics and Econometrics: Essays in Honor of Arnold Zellner

BERRY, CHALONER, and GEWEKE · Bayesian Analysis in Statistics and Econometrics: Essays in Honor of Arnold Zellner

BHAT · Elements of Applied Stochastic Processes, *Second Edition*

BHATTACHARYA and WAYMIRE · Stochastic Processes with Applications

BIEMER, GROVES, LYBERG, MATHIOWETZ, and SUDMAN · Measurement Errors in Surveys

BIRKES and DODGE · Alternative Methods of Regression

BLOOMFIELD · Fourier Analysis of Time Series: An Introduction

BOLLEN · Structural Equations with Latent Variables

BOULEAU · Numerical Methods for Stochastic Processes

BOX · R. A. Fisher, the Life of a Scientist

BOX and DRAPER · Empirical Model-Building and Response Surfaces

BOX and DRAPER · Evolutionary Operation: A Statistical Method for Process Improvement

BOX, HUNTER, and HUNTER · Statistics for Experimenters: An Introduction to Design, Data Analysis, and Model Building

BROWN and HOLLANDER · Statistics: A Biomedical Introduction

BUCKLEW · Large Deviation Techniques in Decision, Simulation, and Estimation

BUNKE and BUNKE · Nonlinear Regression, Functional Relations and Robust Methods: Statistical Methods of Model Building

CHATTERJEE and HADI · Sensitivity Analysis in Linear Regression

CHATTERJEE and PRICE · Regression Analysis by Example, *Second Edition*

CLARKE and DISNEY · Probability and Random Processes: A First Course with Applications, *Second Edition*

COCHRAN · Sampling Techniques, *Third Edition*

*COCHRAN and COX · Experimental Designs, *Second Edition*

CONOVER · Practical Nonparametric Statistics, *Second Edition*

CONOVER and IMAN · Introduction to Modern Business Statistics

CORNELL · Experiments with Mixtures, Designs, Models, and the Analysis of Mixture Data, *Second Edition*

COX · A Handbook of Introductory Statistical Methods

*COX · Planning of Experiments

COX, BINDER, CHINNAPPA, CHRISTIANSON, COLLEDGE, and KOTT · Business Survey Methods

CRESSIE · Statistics for Spatial Data, *Revised Edition*

DANIEL · Applications of Statistics to Industrial Experimentation

DANIEL · Biostatistics: A Foundation for Analysis in the Health Sciences, *Sixth Edition*

DAVID · Order Statistics, *Second Edition*

*DEGROOT, FIENBERG, and KADANE · Statistics and the Law

*DEMING · Sample Design in Business Research

DILLON and GOLDSTEIN · Multivariate Analysis: Methods and Applications

*Now available in a lower priced paperback edition in the Wiley Classics Library.

Applied Probability and Statistics (Continued)

DODGE and ROMIG · Sampling Inspection Tables, *Second Edition*
DOWDY and WEARDEN · Statistics for Research, *Second Edition*
DRAPER and SMITH · Applied Regression Analysis, *Second Edition*
DUNN · Basic Statistics: A Primer for the Biomedical Sciences, *Second Edition*
DUNN and CLARK · Applied Statistics: Analysis of Variance and Regression, *Second Edition*
ELANDT-JOHNSON and JOHNSON · Survival Models and Data Analysis
EVANS, PEACOCK, and HASTINGS · Statistical Distributions, *Second Edition*
FISHER and VAN BELLE · Biostatistics: A Methodology for the Health Sciences
FLEISS · The Design and Analysis of Clinical Experiments
FLEISS · Statistical Methods for Rates and Proportions, *Second Edition*
FLEMING and HARRINGTON · Counting Processes and Survival Analysis
FLURY · Common Principal Components and Related Multivariate Models
GALLANT · Nonlinear Statistical Models
GLASSERMAN and YAO · Monotone Structure in Discrete-Event Systems
GREENWOOD and NIKULIN · A Guide to Chi-Squared Testing
GROSS and HARRIS · Fundamentals of Queueing Theory, *Second Edition*
GROVES · Survey Errors and Survey Costs
GROVES, BIEMER, LYBERG, MASSEY, NICHOLLS, and WAKSBERG · Telephone Survey Methodology
HAHN and MEEKER · Statistical Intervals: A Guide for Practitioners
HAND · Discrimination and Classification
*HANSEN, HURWITZ, and MADOW · Sample Survey Methods and Theory, Volume 1: Methods and Applications
*HANSEN, HURWITZ, and MADOW · Sample Survey Methods and Theory, Volume II: Theory
HEIBERGER · Computation for the Analysis of Designed Experiments
HELLER · MACSYMA for Statisticians
HINKELMAN and KEMPTHORNE: · Design and Analysis of Experiments, Volume 1: Introduction to Experimental Design
HOAGLIN, MOSTELLER, and TUKEY · Exploratory Approach to Analysis of Variance
HOAGLIN, MOSTELLER, and TUKEY · Exploring Data Tables, Trends and Shapes
HOAGLIN, MOSTELLER, and TUKEY · Understanding Robust and Exploratory Data Analysis
HOCHBERG and TAMHANE · Multiple Comparison Procedures
HOCKING · Methods and Applications of Linear Models: Regression and the Analysis of Variables
HOEL · Elementary Statistics, *Fifth Edition*
HOGG and KLUGMAN · Loss Distributions
HOLLANDER and WOLFE · Nonparametric Statistical Methods
HOSMER and LEMESHOW · Applied Logistic Regression
HØYLAND and RAUSAND · System Reliability Theory: Models and Statistical Methods
HUBERTY · Applied Discriminant Analysis
IMAN and CONOVER · Modern Business Statistics
JACKSON · A User's Guide to Principle Components
JOHN · Statistical Methods in Engineering and Quality Assurance
JOHNSON · Multivariate Statistical Simulation
JOHNSON and KOTZ · Distributions in Statistics
          Continuous Univariate Distributions—2
          Continuous Multivariate Distributions
JOHNSON, KOTZ, and BALAKRISHNAN · Continuous Univariate Distributions, Volume 1, *Second Edition*
JOHNSON, KOTZ, and BALAKRISHNAN · Discrete Multivariate Distributions

*Now available in a lower priced paperback edition in the Wiley Classics Library.

Applied Probability and Statistics (Continued)

JOHNSON, KOTZ, and KEMP · Univariate Discrete Distributions, *Second Edition*

JUDGE, GRIFFITHS, HILL, LÜTKEPOHL, and LEE · The Theory and Practice of Econometrics, *Second Edition*

JUDGE, HILL, GRIFFITHS, LÜTKEPOHL, and LEE · Introduction to the Theory and Practice of Econometrics, *Second Edition*

JUREČKOVÁ and SEN · Robust Statistical Procedures: Aymptotics and Interrelations

KADANE · Bayesian Methods and Ethics in a Clinical Trial Design

KADANE AND SCHUM · A Probabilistic Analysis of the Sacco and Vanzetti Evidence

KALBFLEISCH and PRENTICE · The Statistical Analysis of Failure Time Data

KASPRZYK, DUNCAN, KALTON, and SINGH · Panel Surveys

KISH · Statistical Design for Research

*KISH · Survey Sampling

LAD · Operational Subjective Statistical Methods: A Mathematical, Philosophical, and Historical Introduction

LANGE, RYAN, BILLARD, BRILLINGER, CONQUEST, and GREENHOUSE · Case Studies in Biometry

LAWLESS · Statistical Models and Methods for Lifetime Data

LEBART, MORINEAU., and WARWICK · Multivariate Descriptive Statistical Analysis: Correspondence Analysis and Related Techniques for Large Matrices

LEE · Statistical Methods for Survival Data Analysis, *Second Edition*

LePAGE and BILLARD · Exploring the Limits of Bootstrap

LEVY and LEMESHOW · Sampling of Populations: Methods and Applications

LINHART and ZUCCHINI · Model Selection

LITTLE and RUBIN · Statistical Analysis with Missing Data

MAGNUS and NEUDECKER · Matrix Differential Calculus with Applications in Statistics and Econometrics

MAINDONALD · Statistical Computation

MALLOWS · Design, Data, and Analysis by Some Friends of Cuthbert Daniel

MANN, SCHAFER, and SINGPURWALLA · Methods for Statistical Analysis of Reliability and Life Data

MASON, GUNST, and HESS · Statistical Design and Analysis of Experiments with Applications to Engineering and Science

McLACHLAN and KRISHNAN · The EM Algorithm and Extensions

McLACHLAN · Discriminant Analysis and Statistical Pattern Recognition

McNEIL · Epidemiological Research Methods

MILLER · Survival Analysis

MONTGOMERY and MYERS · Response Surface Methodology: Process and Product in Optimization Using Designed Experiments

MONTGOMERY and PECK · Introduction to Linear Regression Analysis, *Second Edition*

NELSON · Accelerated Testing, Statistical Models, Test Plans, and Data Analyses

NELSON · Applied Life Data Analysis

OCHI · Applied Probability and Stochastic Processes in Engineering and Physical Sciences

OKABE, BOOTS, and SUGIHARA · Spatial Tesselations: Concepts and Applications of Voronoi Diagrams

OSBORNE · Finite Algorithms in Optimization and Data Analysis

PANKRATZ · Forecasting with Dynamic Regression Models

PANKRATZ · Forecasting with Univariate Box-Jenkins Models: Concepts and Cases

PORT · Theoretical Probability for Applications

PUTERMAN · Markov Decision Processes: Discrete Stochastic Dynamic Programming

RACHEV · Probability Metrics and the Stability of Stochastic Models

RÉNYI · A Diary on Information Theory

RIPLEY · Spatial Statistics

*Now available in a lower priced paperback edition in the Wiley Classics Library.

Applied Probability and Statistics (Continued)
   RIPLEY · Stochastic Simulation
   ROSS · Introduction to Probability and Statistics for Engineers and Scientists
   ROUSSEEUW and LEROY · Robust Regression and Outlier Detection
   RUBIN · Multiple Imputation for Nonresponse in Surveys
   RYAN · Modern Regression Methods
   RYAN · Statistical Methods for Quality Improvement
   SCHUSS - Theory and Applications of Stochastic Differential Equations
   SCOTT · Multivariate Density Estimation: Theory, Practice, and Visualization
   SEARLE · Linear Models
   SEARLE · Linear Models for Unbalanced Data
   SEARLE · Matrix Algebra Useful for Statistics
   SEARLE, CASELLA, and McCULLOCH · Variance Components
   SKINNER, HOLT, and SMITH · Analysis of Complex Surveys
   STOYAN, KENDALL, and MECKE · Stochastic Geometry and Its Applications, *Second Edition*
   STOYAN and STOYAN · Fractals, Random Shapes and Point Fields: Methods of Geometrical Statistics
   THOMPSON · Empirical Model Building
   THOMPSON · Sampling
   TIERNEY · LISP-STAT: An Object-Oriented Environment for Statistical Computing and Dynamic Graphics
   TIJMS · Stochastic Modeling and Analysis: A Computational Approach
   TITTERINGTON, SMITH, and MAKOV · Statistical Analysis of Finite Mixture Distributions
   UPTON and FINGLETON · Spatial Data Analysis by Example, Volume 1: Point Pattern and Quantitative Data
   UPTON and FINGLETON · Spatial Data Analysis by Example, Volume II: Categorical and Directional Data
   VAN RIJCKEVORSEL and DE LEEUW · Component and Correspondence Analysis
   WEISBERG · Applied Linear Regression, *Second Edition*
   WESTFALL and YOUNG · Resampling-Based Multiple Testing: Examples and Methods for *p*-Value Adjustment
   WHITTLE · Optimization Over Time: Dynamic Programming and Stochastic Control, Volume I and Volume II
   WHITTLE · Systems in Stochastic Equilibrium
   WONNACOTT and WONNACOTT · Econometrics, *Second Edition*
   WONNACOTT and WONNACOTT · Introductory Statistics, *Fifth Edition*
   WONNACOTT and WONNACOTT · Introductory Statistics for Business and Economics, *Fourth Edition*
   WOODING · Planning Pharmaceutical Clinical Trials: Basic Statistical Principles
   WOOLSON · Statistical Methods for the Analysis of Biomedical Data
  *ZELLNER · An Introduction to Bayesian Inference in Econometrics
Tracts on Probability and Statistics
   BILLINGSLEY · Convergence of Probability Measures
   KELLY · Reversability and Stochastic Networks
   TOUTENBURG · Prior Information in Linear Models